THERMOPHYSICAL PROPERTIES OF NEON, ARGON, KRYPTON, AND XENON

NATIONAL STANDARD REFERENCE DATA SERVICE OF THE USSR: A Series of Property Tables

1. Thermodynamic Properties of Helium
2. Thermodynamic Properties of Nitrogen
3. Thermodynamic Properties of Methane
4. Thermodynamic Properties of Ethane
5. Thermodynamic Properties of Oxygen
6. Thermodynamic Properties of Air
7. Thermodynamic Properties of Ethylene
8. Thermophysical Properties of Freons, Part 1
9. Thermophysical Properties of Freons, Part 2
10. Thermophysical Properties of Neon, Argon, Krypton, and Xenon

In Preparation

Thermodynamic Properties of n-Hexane
Thermodynamic Properties of Propane
Thermodynamic Properties of Hydrogen

THERMOPHYSICAL PROPERTIES OF NEON, ARGON, KRYPTON, AND XENON

V. A. Rabinovich
A. A. Vasserman
V. I. Nedostup
L. S. Veksler

Theodore B. Selover, Jr.
English-Language Edition Editor

HEMISPHERE PUBLISHING CORPORATION
A subsidiary of Harper & Row, Publishers, Inc.
Washington New York London

DISTRIBUTION OUTSIDE NORTH AMERICA
SPRINGER-VERLAG
Berlin Heidelberg New York London Paris Tokyo

THERMOPHYSICAL PROPERTIES OF NEON, ARGON, KRYPTON, AND XENON

Originally published by Standards Publishers, Moscow, 1976, as Teplofizicheskie svoistva neona, argona, kriptona i ksenona in the Monograph Series of the National Standard Reference Data Service of the USSR: State Committee on Standards of the Council of Ministers of the USSR.

1 2 3 4 5 6 7 8 9 0 B C B C 8 9 8 7

This book was set in Times Roman by The Sheridan Press.
BookCrafters, Inc. was printer and binder.

Library of Congress Cataloging-in-Publication Data

Teplofizicheskie svoistva neona, argona, kriptona i ksenona. English.
Thermophysical properties of neon, argon, krypton, and xenon.

(National standard reference data service of the USSR)
Translation of: Teplofizicheskie svoistva neona, argona, kriptona i ksenona.
Bibliography: p.
Includes index.
1. Neon. 2. Argon. 3. Krypton. 4. Xenon.
I. Rabinovich, V. A. (Viktor Abramovich) II. Selover, Theodore B. III. Title. IV. Series.
QD181.N5T4713 1987 546'.75 87-17723
ISBN 0-89116-675-0 Hemisphere Publishing Corporation

DISTRIBUTION OUTSIDE NORTH AMERICA:
ISBN 3-540-19297-7 Springer-Verlag Berlin

CONTENTS

PART II. NONEQUILIBRIUM PROPERTIES

PART III. TABLES OF THERMOPHYSICAL PROPERTIES OF NEON, ARGON, KRYPTON, AND XENON

PREFACE TO THE SERIES

This treatise is part of a continuing series on thermodynamic properties of technologically important fluids. These very important contributions by scientists and engineers working through the aegis of the Soviet National Service for Standard Reference Data have been released to Hemisphere Publishing Corporation for translation to make them available to the English reading technical community. The authors are Soviet experts in the field.

While a team of translators was involved in producing the English versions, the overall series is being published under the technical editorship of T. B. Selover, Jr.

Each volume presents a comprehensive survey of the world's literature up to its publication date. A special effort has been made to give a thorough presentation of Russian as well as other work. Many studies not previously known to Western counterparts are included. The results have been to broaden the range of applicability of data and to improve upon equations of state to provide accurate computation methods for generating smoothed tables of properties.

For some volumes there are no equivalent comprehensive surveys available in English. Thus, a valuable service has been fulfilled to workers in the fields of process design, equipment development, custody transfer, and safety.

Each volume is set up in the same way with Part I dealing with a study of all necessary aspects of experimental data interpretation and analysis. Then in Part II the fundamental constants, symbols with units, and data tables can be found. The use of SI units is consistent throughout.

The section on experimental data is particularly important because it covers, in detail, the key studies. Possible errors in measurement or data analysis that could have led to inaccurate tabular results in publications are pointed out.

Methods of constructing the equation of state and procedures for computing the data tables are covered thoroughly. There is a detailed error analysis of data generated from the equation of state relative to literature values.

Properties of each fluid in phase equilibria are treated at the freezing curve and the saturation curve with tables given for both temperature and pressure dependence. Properties cover the range of triple point to critical point in the condensed phase. Single phase properties in each volume cover a range of temperature and pressure wide enough to include most practical applications. Ideal gas property equations and calculation methods are included. The scope of properties generated and covered in the tables is more comprehensive than is typically found in any one English-language treatise. This gives the advantage of internal consistency.

An extensive bibliographic listing is included with each volume. All Russian citations have been translated. In some cases the availability in English of translated Russian sources is given.

Frequent reference to key English-language references has been made to help in providing correct translations. In some cases the original symbols have been changed to avoid confusion within the text or to avoid misuse of terms commonly found in English. Where mistakes in the original Russian text have been found, they have been corrected and so noted as editor's changes. Added descriptive clarification has been used in some places for tables, figures, and text to clarify meaning.

Although we recognize that key review papers have been published in 1986 for some of the fluids in this set, the series stands on its own merit. It represents a vast accumulation of knowledge never before available to Western countries. Through careful study of these volumes, workers in the field will develop an appreciation for both the scope of studies and where differences exist. Hopefully, this will lead to more dialogue between the authors and their Western counterparts.

Theodore B. Selover, Jr.

PREFACE

Elements of the helium group occur infrequently on the earth and therefore were discovered later than other elements which were used by Mendeleyev for composing the periodic table of the elements. The history of the discovery of these elements is interesting.

In 1893, Lord Rayleigh, who at that time served as professor in the Cavendish laboratory, investigated the density of different gases and found that nitrogen prepared by the removal of oxygen, carbon dioxide, and water vapor from the air had a standard density slightly greater than that of nitrogen prepared by the decomposition of one of its compounds, ammonia. Ramsey, who in 1894 removed nitrogen and oxygen from a known quantity of air, became interested in Rayleigh's discovery. The residual gas was subjected to spectral analysis which indicated the presence of a new element. The gas discovered did not react with other elements. For this reason, Ramsey named this element argon (the idle one in Greek). The periodic table did not have a place for this element. That suggested to Ramsey that there may be an entire vertical column of elements missing from the table.

A year later, Ramsey discovered another gas released upon heating cleveite, which is a rare mineral. Spectral analysis of the new substance showed that it was identical with the spectrum of helium, discovered in the sun as far back as 1868. Ramsey, together with Travers, continued the investigation of air and soon was able to separate helium from air.

As the fractionation technique for separation of air became more refined, Ramsey and Travers were able during a very short time to single out three new elements, which they termed neon (new), krypton (hidden) and xenon (stranger). The last and heaviest element of this group was discovered by Dorn in 1900, who termed it radon since it was released in the course of decomposition of radium and was radioactive.

This means that all six elements of the zero valent group of the periodic table were discovered during the last six years of the past century and have continued to serve as the focus of attention of scientists all over the world. However, not all these gases have been investigated to the same degree. The most fundamental studies are those of helium. In particular, its thermophysical properties are extensively presented in books (W. H. Keesom, Helium, Elsevier, 1942; N. V. Tsederberg, V. N. Popov, and N. A. Morozova, Thermodynamic and Thermophysical Properties of Helium, Halsted Press, N.Y. 1971; also NTIS TT70-50096). For this reason, the properties of helium are not considered in this book. On the other hand, there are no experimental data on the thermophysical properties of radon. This makes it impossible to perform calculations with the required accuracy. The remaining four elements are the subject of the present study. The most extensively investigated of these is argon, which was used as the basis substance when using relative methods in this book.

Under standard conditions neon, argon, krypton, and xenon exist as gases. The atoms of these substances have saturated electron shells and for this reason they almost never enter into chemical reactions. Due to the spherical symmetry the potential energy of interaction between the atoms of inert gases is a function only of the distance between them, and not of their mutual orientation. The results of the molecular-kinetic theory of gases and liquids are applicable precisely to such systems. For this reason, experimental data on equilibrium and nonequilibrium properties of inert gases serve as the basis for checking various theoretical concepts which in the early stage of the development of theory required performance of experimental studies.

Since the sixties, due to the explosive growth of new and promising technologies, inert gases have come into ever increasing practical use (see, among others, the book by V. G. Fastovskiy, A. Ye. Rovinskiy, and Yu. V. Petrovskiy, Inert Gases, Halsted Press, N.Y., 1967). Their use has been aided, to a large degree, by the relative simplicity of obtaining inert gases as side products in the separation of air in large-capacity plants. The above circumstances stimulated new experimental and analytic studies of thermophysical properties of inert gases in the Soviet Union and in the West over various ranges of parameters. This has resulted in a quite large body of data. However, for neon, krypton, and xenon they do not cover the entire range needed for practical purposes and are of highly nonuniform reliability. At the same time, no studies were performed for critically analyzing and correlating all the available data by means of analytic expressions which can be used for practical calculations.

Although publication and accumulation of experimental results continue, it appears possible at present to start the systematization of available data, select the most reliable, investigate previously unexplored ranges of parameters, and compile interpolation equations for calculating tables of thermophysical properties suitable for practical use.

Given the extreme practical importance of this information, in addition to its scientific significance, The International Union of Pure and Applied Chemistry (IUPAC) organized in 1964 a committee for the compilation of tables of thermodynamic properties of substances of importance in engineering. These substances also include inert gases. The goal of this committee which is comprised of representatives of many countries, including the USSR, is to compile uniform international tables of a number of substances for the specified substances. For this reason, the present study, which is concerned with comprehensive investigation of equilibrium and nonequilibrium properties of neon, argon, krypton, and xenon over a wide range of parameters of state, can be regarded as a contribution which brings us closer to the intended goal.

The book consists of three parts. The first part describes the investigation of thermodynamic properties of the four substances in the three phases. The second is concerned with the viscosity and thermal conductivity of these substances in the liquid and gaseous state, and the third contains detailed reference tables of thermophysical properties of neon, argon, krypton, and xenon. Particular attention was paid in the first and second parts of the book to the interrelationship and consistency of all the properties analyzed in it.

The survey of original studies containing data on thermophysical properties of inert gases has performed in chronological sequence and reflects the state of the problem to the middle of 1973. The most important recent publications which were not analyzed by anyone are considered in the book in enough detail to provide the reader with sufficient information on the details of the experiments and on the accuracy of data listed in original articles.

Original analytic methods, which provide insight into general relationships governing the behavior of the four inert gases and allow extrapolation of data given over a relatively narrow range of parameters, were used when possible. The extent to which the results of the molecular-kinetic theory can be used for calculating numerical values of various properties of elements of the helium group was also checked.

The first part considers the specifics of calculation of thermodynamic properties of solidified gases and analyzes and processes experimental data on various equilibrium properties of the substances under study in the crystal state over a wide temperature and pressure range. The analytic tools leading to final equations of state for calculating thermodynamic properties from 0 K to the melting curve at 1000 bar and above are described. The possible error of the calculated values is given.

Reliable data on the phase equilibrium curve are selected and matched for obtaining equations describing the temperature dependence of pressure along sublimation, melting and saturation curves.

A general approach to the compilation of an equation of state for the gas phase is given. It is shown that an analytic description of the liquid and dense-gas region is best treated as a problem of its own. Experimental data on the compressibility of the four inert gases are correlated, and the most reliable of these are pointed out. The need for additional experimental studies, a part of which was performed by the present authors, is justified. The virial equations of state that are obtained describe the most reliable experimental data for each gas within the experimental accuracy over a density range from zero to the second critical.

Independent equations of state are recommended for describing the thermodynamic properties of the substances under study at high densities. The form of these equations can be suggested from analysis of certain theories of the liquid state. The detailed comparison of analytic and experimental data performed by the authors leads to the conclusion that the equations of state for liquid neon, argon, krypton, and xenon obtained in the book exhibit acceptable accuracy.

In the second part, data on viscosity and thermal conductivity are correlated on the assumption that the same function of interatomic interaction applies to the entire group of inert gases and that monatomic gases at low pressures ($p \leq 1$ atm) obey the familiar relationship between the viscosity and thermal conductivity. A method is suggested which allows the selection of the most reliable experimental data and results in a single analytic expression of the temperature dependence of the above transport properties, applicable to all four gases.

The correlation of data on viscosity at moderate and high pressures was preceded by a detailed analysis that showed that the original data exhibit significant disagreement even in that range of parameters where there is a simple relationship between the excess viscosity and density. For this reason, we performed additional experimental studies of the viscosity of the four gases at temperatures between 300 and 550 K and pressure to 600 bar. The equation used for these calculations was the Enskog equation which contains a relationship between the diameter of the solid spheres and the temperature and density found from selected experimental values of viscosity.

During the past decade a large volume of experimental data was obtained on the thermal conductivity of gaseous and liquid argon, krypton, xenon, and gaseous neon over a wide range of temperature and pressure. For this reason, the compilation of the equation for calculating the thermal conductivities of these substances was preceded by analysis of experimental data and also by a survey of methods of their workup and correlation. It was established in constructing the graphs in reduced coordinates that the excess thermal conductivity is a single function for the four substances. This makes it possible to compile a simple equation for calculating the thermal conductivity of the four monatomic gases and liquids over a wide range of parameters, with the exception of the critical range.

The third part of the book presents analytic expressions for determining the calorific quantities and reference tables of thermophysical properties of neon, argon, krypton, and xenon. For each of these substances, the tabulated data pertain to thermodynamic properties for the solid state from 0 K to the melting curve at $p \leq 1000$ bar, along the melting curve from the pressure at the triple point to $p \sim 1000$ bar, along the saturation curve from the triple point to the critical, and also in the single-phase region (liquid or gas) over the temperature range from the triple point to 1300 K at pressures to 1000 bar. The analytic values of thermodynamic functions obtained from equations of state for the gas and liquid have to be matched graphically at the gas-liquid interface. For this reason, they are less accurate as compared with "purely" analytic equations. Tabulated data pertaining to the critical region are also less accurate, since the starting equations of state do not fully satisfy the critical point and the critical conditions and do not reflect the newest tenets of the scaling theory.

The tables of values of transport properties—dynamic viscosity and thermal conductivity—for the four substances in the liquid and gaseous states are represented from the triple points to $T = 1300$ K and $p = 1000$ bar. The thermal conductivity was calculated from an equation which makes no allowance for the anomalous rise in this quantity in the critical region. For this reason the tables for each substance designate this region by a dashed line and the data in them should be regarded as roughly approximate.

This book was prepared by the staff of the Thermophysical and Chemico-Physical Studies of the All-Union Research Institute of Physical Engineering and Radio Engineering Measurements, Department of Applied Thermodynamics of the Problems Laboratory of the Odessa Institute of Marine Engineer and the Odessa Department of the Engineering Thermophysics Institute of the Ukrainian Academy of Sciences. The authors wish to thank the staff members of these institutions for assistance in performing this study, and wish to express their thanks to N. M. Akulova, L. S. Kanevskaya, L. V. Kocharova, and A. A. Tolmacheva of the Institute of Physical Engineering and Radio Engineering Measurements. Particular thanks are due to the reviewing editor, Candidate of Technical Sciences, A. D. Kozlov, for useful comments and advice.

The authors make no claim to perfection for the book and would be grateful for critical comments which will be taken into account in subsequent work.

NOMENCLATURE

B_2	second virial coefficient
c_p	isobaric specific heat
c_{p0}	ideal gas isobaric specific heat
c_s	saturation specific heat
C_v	isochoric specific heat
e	density
e_0	gas density at 0 K
F	Helmholtz free energy
g	gram
Gr	Grashof number
h, H	enthalpy
kg	kilogram
lg	$\log_{10}$
ln	$\log_e$
p	pressure
p_g	pressure gauge
p_h	hydrostatic pressure
Pr	Prandtl number
p_s	sublimation vapor pressure
p_v	liquid saturation vapor pressure
S	entropy
t	temperature (°C)
T	temperature (K)
T_B	Boyle temperature
U	internal energy

ω, w	velocity of sound
z	compressibility
z_T	dynamic viscosity at T
α_{exp}	thermal expansion coefficient
α_i	integral adiabatic Joule–Thomson coefficient
β_i	differential adiabatic Joule–Thomson coefficient
β_T	isothermal compression
λ_T	thermal conductivity at T
Θ	T_{cr}/T
ω	V_{cr}/V

SUBSCRIPTS

attr	attractive
c	crystal
calc	calculated
conf	configuration
cr	critical
e	density
eff	effective
exptl	experimental
g	gas
HT	high temperature
id	ideal
in	inner
l	liquid
LT	low temperature
m	mean or melting
rep	repulsive
s	sublimation
sph	spherical
tr	triple point
υ, v	volume

PART ONE

EQUILIBRIUM PROPERTIES

CHAPTER

ONE

THERMODYNAMIC PROPERTIES OF THE SOLID PHASE AND IN THE STATE OF PHASE EQUILIBRIUM

In spite of the significant achievements attained lately in the USSR and the West in developing the theory of solid state, up to now no methods have been worked out for the exact calculation of thermodynamic functions without resorting to experimental results. At the same time, experimental values for the crystals under consideration are highly limited. Thus, properties such as the specific heat $c_s (c_s \cong c_p)$ and the thermal expansion coefficient α_{exp} at the sublimation pressure were measured from the limiting low temperature to melting temperature. However, the pressure (density) dependence of this function was not explored experimentally. Data on the isothermal compressibility (and pv curves) as a rule, are available at two–three values of temperature to a pressure of about 10 kbar for each of the four crystalline substances. Such a state of affairs makes it impossible to obtain a reliable empirical equation of state capable of representing the pressure dependence of the specific heat of the crystal. Moreover, in the empirical approach one encounters significant difficulties as early as when attempting to provide an analytic description of the temperature dependence of the c_s of crystals. The peculiar shape of this curve makes it very difficult to describe it by a single analytic expression (in the form of a power-law polynomial) over the temperature range from zero to melting.

At present it is actually possible to calculate with a quite high accuracy the thermodynamic functions of ideal crystals at low temperatures, when the contribution of anharmonic effects is negligible [1.1], or when it suffices to incorporate only the first correction of aharmonicity [1.2]. However, it was shown by

Klein et al. [1.3], that for crystals and particles which interact according to a Lennard-Jones type potential, the retention of even four terms of the expansions of the potential energy of the crystal in powers of the shifts of particles from the lattice nodes does not suffice at temperatures close to that at the melting curve. Also, the computational error may be due not only to the limitations of the theory of perturbations of the 2nd order, but also to inaccuracy of the initial (quite arbitrary) potential function. This circumstance is less important in the harmonic approximation, since the contribution to thermodynamic functions at low temperatures is made by a much narrower segment of the two-particle potential than at high temperatures, and it suffices to use a paired two-parametric potential with parameters determined from data on the heat of sublimation and the lattice constant as $T \to 0$ [1.4].

This means that the temperature range adjoining the melting curve is the most vulnerable in the modern theory of ideal crystals. It is unknown whether the theory of perturbations, in which the harmonic approximation is the zero approximation, is convergent for this region. A consistent statistical theory for such a region should yield a nonanalytic free energy which corresponds to melting. As is known, such a theory has not yet been developed. Abovskiy et al. [1.5, 1.6] have actually obtained a form of equations describing the thermodynamic functions of an ideal crystal in the analytic region near the melting curve, but the results were initially used for calculating the properties of the liquid phase near the solidification curve. The equations consist of rapidly converging asymptotic relationships, the applicability of which extends in the pT diagram to the strip, including the melting curve. The equations contain adjustment parameters, approximate values of which can be determined from data on the potential function. Let us analyze the application of these equations together with equations in the low-temperature region for describing the thermodynamic properties of the four elements of the zero valent group in the crystal state.

1.1 EQUATIONS FOR THERMODYNAMIC FUNCTIONS†

The "High-Temperature" Region

We shall now present individual prerequisite considerations and approximations used in obtaining the equations in references [1.5–1.7], dropping the details of the derivation.

1. It is assumed that the potential of interaction between two particles satisfies the following conditions: upon a small shift $|\mathbf{r}|$ of the atom from the coordinate of the minimum R_e of the potential function ($|\mathbf{r}| \approx 0.1\,R_e$)

† Secs. 1.1 and 1.2 were written by V. A. Abovskiy.

$$\Phi(R) = \Phi_0 \exp\left(-\alpha \frac{R}{R_e}\right),$$
$$\alpha \gg 1 \text{ at } R < R_e, \tag{1.1}$$
$$\frac{1}{\alpha}[\Phi(R - |r|) - \Phi(R_e)] \approx [\Phi(R + |r|) - \Phi(R_e)]$$

i.e., the repulsion branch of the two-particle potential is approximated by the Born-Mayer approximation at high values of ''steepness'' parameter α, and the attraction wave near the minimum is much steeper than the repulsion branch.

These conditions are satisfied by model potentials usually employed for describing the interaction between two atoms with filled electron shells such as the Lennard-Jones, Buckingham, Morse, etc., potentials.

Nonadditivity of multiparticle interactions which may be commensurable by the order of magnitude with the additive component of the potential function is permitted.

2. The statistical part of the problem is treated in the classical approximation. The configuration integral is transformed in accordance with the cell-cluster theory [1.8], with the configuration entropy given in the general case by the expression [1.6]:

$$S_{\text{config}} = Nk\left[\ln \frac{\Delta_f}{v} + \ln(1 + \Sigma \langle f_{ij} \rangle + \ldots)\right] \tag{1.2}$$

where N is the number of particles in the system and k is the Boltzmann constant. The first term is the contribution to the entropy of the so-called Lennard-Jones and Devonshire approximation; here v is the volume of the system devolving upon a particle, whereas Δ_f is the domain of the configuration space which makes a nonzero contribution to the configuration integral. The second term is the contribution to the entropy of correlations in the locations of particles not included in the Lennard-Jones and Devonshire approximation. This term permits only an approximate estimate. In particular, by introducing the so-called approximation of independent binary correlations [1.8] and taking into consideration, due to the short range of potential (1.1) only the interaction between the closest neighbors, we find:

$$S_{\text{config}} = Nk \ln \frac{\Delta f}{v}[1 + O(\varphi)] \tag{1.3}$$

where $\varphi = \dfrac{\frac{c}{2}\ln(1 + \langle f \rangle)}{\left|\ln\left(\frac{\Delta f}{v}\right)\right|}$; c is the coordination number and $\langle f \rangle$ is the two-particle correlation correction. For the Lennard-Jones potential [1.7], as for the potential of solid spheres, $\langle f \rangle \simeq 0.1$. Condition (1.1) makes it possible to clarify [1.6] that at values of density corresponding to the crystal state ($v \approx R_e^3$), $\frac{\Delta_f}{v} \approx 10^{-2} - 10^{-3}$ for which reason $\varphi \ll 1$.

Equation (1.3) was written on the assumption that the approximation of an independently-paired correlation provides a correct estimate of the order of magnitude of the correlation term in Eq. (1.2). This holds in the case of harmonic forces [1.8]. However, potential (1.1), even at small displacements from R_e, is by far not harmonic. Equation (1.3) acquires a more general meaning if $\langle f \rangle$ is taken to represent some "effective" correction, which makes allowance for the exact value of the second term in Eq. (1.2). Henceforth, the only item of importance is the smallness of parameter φ, which holds also at significant differences between $\langle f \rangle$ and the two-particle correlation correction.

Equation (1.3) is thus an asymptotic expression for the entropy of the system at low values of φ.

Function Δ_f can be obtained in explicit form by satisfying the condition

$$\tau = \frac{LkT}{\Phi_0 \exp\left(-\alpha \frac{l}{R_e}\right)} \gg 1 \tag{1.4}$$

where $l = (\gamma v)^{\frac{1}{3}}$ is the distance between the closest neighbors (here γ is the structure factor of the lattice); $L \gg 1$ is the cell potential "cutoff" parameter. This condition defines the domain of applicability of the final results. In the case of a crystal this domain is the region of parameters of state adjoining the melting curve and having a width of the order of $0.1T_m$ to $0.2T_m$.

The final result for the isochoric specific heat

$$C_v = Nk\left[\frac{3}{2} + \frac{3D}{v^{1/3} - C + D\ln T} + O(\varphi) + O\left(\frac{1}{\tau}\right)\right] \tag{1.5}$$

where $D = \frac{R_e}{\alpha\gamma^{1/3}}$ and $C = \frac{R_e}{\alpha\gamma^{1/3}} \ln \frac{\Phi_0}{kL}$ are fitting parameters, the indeterminacy of which is due to the indeterminacy of the paired potential parameters (1.1) and of parameter L. The small correction terms can in the first approximation be replaced by the constant $|n| \ll 1$. The nonadditivity of the interactions at $\tau \gg 1$ is represented in Eq. (1.5) by the term $O(1/\tau)$ [1.6].

For the subsequent presentation it is best to write C_v as

$$C_v = \frac{3}{2}Nk + \frac{3}{2}Nk\left(\frac{1}{\ln \tau} - n\right) \tag{1.6}$$

with τ, defined by Eq. (1.4), expressed in terms of C and D as:

$$\tau = \frac{T}{\exp\left(\frac{C - v^{1/3}}{D}\right)} \tag{1.7}$$

The above shows that the specific heat C_v of an ideal crystal at high temperatures is a function of a single variable—the reduced temperature τ. A rather general analysis [1.5, 1.9] shows that such a situation prevails for the "thermal" component of the free energy. This makes it possible to introduce the high-temperature analog of the Grüneisen parameter which defines the behavior of the temperature derivatives of the free energy with respect to density:

$$\Gamma_{HT} = \frac{\partial \ln \Theta_{HT}}{\partial \ln v} = \frac{\alpha}{3}\frac{(\gamma v)^{1/3}}{R_e} \tag{1.8}$$

where

$$\Theta_{HT} = \frac{T}{\tau} = \frac{\Phi_0}{kL}\exp\left(-\alpha\frac{l}{R_e}\right) = \exp\left(\frac{C - v^{1/3}}{D}\right) \tag{1.9}$$

The above formulas suffice for constructing interpolation equations which approximate the thermodynamic functions of the crystal. For this reason, there is no need to reproduce here other equations obtained in [1.5, 1.6].

Low-Temperature Region

As previously pointed out, the low-temperature region is described almost exactly by the harmonic approximation [1.1]. However, the computations that this involves, due to the need to calculate the spectrum of the natural frequencies of oscillations of the crystal lattice [1.10] are very cumbersome. Thermodynamic functions can be calculated without computing the spectrum of natural frequencies.

A method that is very effective at extremely low temperatures is that due to Houston [1.11], involving expanding the frequency distribution function in four powers of frequency, which is actually a generalization of the Debye theory. At higher temperatures $T > \Theta/2\pi$, where $\Theta = \frac{h\omega_L}{k}$; h is the Planck constant and ω_L is the cutoff frequency, one usually uses an expansion of thermodynamic functions with respect to the spectral moments (the Tyrring series [1.1]); however the Tyrring series was observed to exhibit rapid convergence only at temperatures at which anharmonic corrections must be made.

The above shows that available classical methods do not allow approximation of the temperature derivatives of the free energy of a crystal by single analytic expressions over the entire range of applicability of the harmonic approximation.

Allowance for the density dependence of the spectrum of natural frequencies (quasiharmonic approximation [1.1]) within the framework of the above method involves cumbersome recalculations of the integrals contained in the expansions. The approximation of experimental data comes down to the selection of parameters of the potential function of the chosen form which defines the force constants of the lattice and the frequency spectrum.

This makes it much more attractive to use a method for an approximation of thermodynamic functions corresponding to the harmonic approximation suggested by Peresada [1.12]. He used the relationship between the distribution function of the frequency of vibrations and the spectral densities of special manifold Jacobian matrices for obtaining, within the framework of the model of closest neighbors, analytic expressions for thermodynamic functions which contain only one adjustment parameter—the characteristic temperature corresponding to the maximum frequency. Unfortunately, as noted by the author, this expression is capable of a satisfactory description of experimental values of specific heat at low temperatures only for xenon. It is unclear whether the reason for this is the fact that only interactions between the closest neighbors are considered. The Houston method and the Tyrring series which incorporate the closest neighbors and all the particles of the system yield virtually identical results [1.4]. In addition, it is unclear whether the density dependence of the characteristic temperature (which is formally equivalent to the Debye temperature) can be described by the familiar expression for the Grüneisen parameter [1.13, 1.14]. Finally, all the aforementioned methods are suitable only for the temperature range where the anharmonicity is negligible.

Since the above formalism is suitable only for calculations over a relatively narrow range of extremely strong anharmonicity, it can be concluded that the methods being considered are insufficient for solving the problem at hand. We shall therefore consider a method suggested for the approximation of thermodynamic functions at low and moderate temperatures.

Since the temperature dependence of thermodynamic functions such as the specific heat and the thermal expansion coefficient is known, the calculations have as their purpose reliable determination of their density (pressure) dependence. Note that a simple, analytic, semiquantitative description of thermodynamic functions is possible in the Debye approximation [1.15] for a wide range of temperatures. Over a narrow range of rather low temperatures this approximation becomes exact.

It is well known that the reason for deviation from experimental results and exact theory in the harmonic region consists of the fact that the real frequency distribution function is different from the parabolic used in the Debye theory. The traditional measure of deviation is the quantity $\Theta_D(T)$, the form of which is determined by the ratio of the mean squares of frequencies of the real and the Debye spectra. For crystals of the zero valent group this relationship has a characteristic shape (Fig. 1), obtained both experimentally and analytically [1.4, 1.15, 1.16]. Following a narrow segment of applicability of the Debye theory

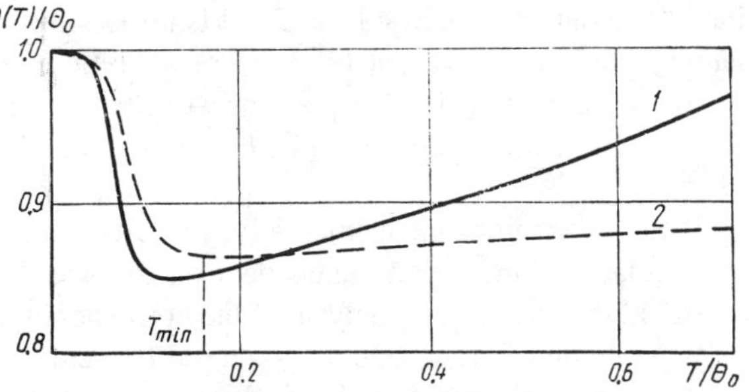

FIGURE 1. Characteristic dependence of the effective Debye temperature of crystals of the zero valent group on T, defined from: 1) data on the isochoric specific heat of crystalline krypton ($\Theta_{ef}^{C_v}$); 2) results of calculation of the free energy (Θ_{ef}^{F}).

$\left(T \lesssim \frac{\Theta_D}{20}\right)$ the value of Θ_D decreases rather steeply. This is followed by a region with only slight variation in Θ_D—the "effective Debye region" which extends over a rather wide temperature range over which anharmonicity becomes significant [1.2]. It should be emphasized that the dependence of the "effective" Debye temperature on T can be determined from any known thermodynamic function. It is significant that the corresponding relationships for two functions, one of which is a temperature derivative of the other, should in general be basically different (see Fig. 1).

Since the deviations of the "effective" Debye temperature as $T \to 0$ calculated from experimental data and analytically are small† (i.e., the Debye model turns out to be a rather good zero approximation for describing the thermodynamic functions over a rather wide temperature range), it is possible to suggest a simple method of approximation, ensuring at the same time correct transition to low temperatures. It suffices to describe the behavior of $\Theta_D(T)$. Since allowance for it is, in fact, a correction, the approximation can be performed with a small number of fitting parameters.

Obviously, it is most convenient to perform the approximation on the basis of the effective Debye temperature corresponding to the free energy, i.e., determined from the condition that

$$F_{\text{exptl.}} = F\left[\frac{\Theta_D(T)}{T}\right] \tag{1.10}$$

In this case, the other thermodynamic functions are determined by differentiating the starting equation. However, if the calculations are based on an effec-

† This holds provided that $\Theta_D(T)$ is determined from data on the entropy [1.17] or the free energy. If however, this is done from data on specific heat C_v, then the value of $\Theta_D(T)$ increases significantly over the high-temperature range [1.18]. This is due to the specific form of the Debye specific heat at temperatures of $T \cong \Theta_D$, where C_v^D is virtually independent of the temperature.

tive Debye temperature defined from data on specific heat C_v, it is impossible to obtain the other thermodynamic functions in explicit form, since analytic integration becomes impossible. Such a selection is also poor because $\Theta_D(T)$ increases at high temperature, when this relationship cannot be treated as a correction.

Although the free energy is not determined experimentally, its values at the sublimation pressure can be calculated from experimental data on the specific heat and the density of the crystal at this pressure. Analysis of the corresponding values of $\Theta_D(T)$, as previously mentioned, allows one to single out two characteristic regions of variation in this quantity† (see Fig. 1). Writing

$$\Theta_D(T) = \Theta_{min} + \Delta\Theta(T) \tag{1.11}$$

we shall, as usual, describe correction $\Delta\Theta(T)$ by a Taylor series expansion. Its absolute value over a wide temperature range is not too high, but its rapid change over the low-temperature range leads to the assumption that it would be more effective to use an expansion for the logarithm of $\Delta\Theta(T)$, restricting oneself to the linear approximation. The parameter of this expansion should be T^2. In this case, when $\Delta\Theta(T) \approx \exp(-bT^2)$ is expanded in a power series the structure of the expansion becomes identical to the analytic result obtained for $\Delta\Theta(T)$ in the Houston method [1.11, 1.16]. Linear approximation for the Debye temperature turns out to be sufficient in the "effective Debye region." With reference to the normalization for $\Delta\Theta(T)$ ($\Delta\Theta \approx \Theta_0 - \Theta_{min}$ as $T \to 0$ and $\Delta\Theta \cong 0$ at $T = T_{min}$) and to the fact that the weak linear term has virtually no effect on the exponential shape of the function over the narrow range of low temperatures, we can write the approximating expression in the form

$$\Theta_D(T) = \Theta_D(0) + a[\exp(-bT^2) - 1] + cT \tag{1.12}$$

Let us consider the density dependence of Θ_D. It is known that at $T \to 0$, this behavior is controlled by the Grüneisen parameter with different methods of computation yielding similar expressions for this coefficient in terms of the derivatives of the lattice pressure of the system with respect to volume [1.14]. Retaining the classical definition of the Grüneisen coefficient

$$\Gamma = -\frac{\partial \ln \Theta_D}{\partial \ln v} \tag{1.13}$$

for the effective Debye temperature, we also introduce the relationship governing $\gamma(T)$. The latter was investigated analytically only when the effective Debye temperature is determined from data on specific heat [1.13]. The temper-

† The density changes very little along the sublimation curve at low and moderate temperatures. For this reason, all the above pertains to the isochoric behavior of $\Theta_D(T)$.

ature dependence of it turns out to be much weaker than the corresponding behavior of $\Theta(T)$†.

Since, when correction $\Delta\Theta(T)$ is small, incorporation of $\gamma(T)$ introduces a second-order infinitesimal into $\Theta(T)$. We shall represent the Grüneisen parameter by its value at $T \to 0$. In this case

$$\Theta_D = \Theta(T) f(v) \tag{1.14}$$

where

$$f(v) = \exp \int_v^{v_0} \frac{\gamma}{v} dv \tag{1.15}$$

We shall calculate the Grüneisen parameter from the generalized equation recommended by Zharkov and Kalinin [1.14]:

$$\Gamma = -\frac{v}{2} \frac{\partial^2 (p_p x^{2m/3})/\partial v^2}{\partial (p_p x^{2m/3})/\partial v} + \frac{1}{3}(m - 2) + \delta \tag{1.16}$$

where p_p is the potential (lattice) pressure of the system; $x = v/v_0$ is the dimensionless volume and δ is a normalizing factor; and the constant m may take integral values of the order of unity. Equation (1.16) was introduced on assumptions corresponding to the harmonic approximation. For this reason it is unknown whether it can be used for calculations in highly anharmonic regions. The possibility of performing such calculations will be proven below in comparing with results obtained for the high-temperature region.

To obtain an explicit relationship for γ as a function of the volume and the fitting parameters, one must specify a certain model of a two-particle potential (neglecting for the moment multiparticle forces). When the potential is substituted into Eq. (1.16) its form is important only over the range of distances corresponding to the "potential well," i.e., interaction between close neighbors. For a Lennard-Jones type potential, such a range corresponds to a density change of 30%. That corresponds to the maximum pressure attained in known experiments with these crystals.

† Certain experimentalists [1.17, 1.18] identify the Grüneisen parameter with the ratio $\frac{v}{C_v}\left(\frac{\partial p}{\partial T}\right)_v$. If Θ_D is independent of the temperature, this ratio becomes identical to the derivative $-\frac{\partial \ln \Theta}{\partial \ln v}$. When Θ_D is temperature dependent, this inequality no longer holds, irrespective of the thermodynamic functions used for determining $\Theta_D(T)$. It is, hence, incorrect to attempt to interpret the behavior of $\frac{v}{C_v}\left(\frac{\partial p}{\partial T}\right)_v$ on the basis of analytic results [1.4, 1.13] for $\Gamma(T) = -[\partial \ln \Theta_D(T)/\partial \ln v]$, which was done by Losee and Simmons [1.17].

It is convenient, in calculations with Eq. (1.16) over the above density range, to use a special approximation of the two-particle potential. The use of two-term Lennard-Jones or Buckingham type functions within the framework of the above problem gives rise to significant difficulties due to varying the potential parameters (in order to best approximate the experimental data) and the impossibility of analytically integrating Eq. (1.15) if the parameters of the potential function are arbitrary. The integration difficulties are due to the need to retain two rapidly varying terms of the previously mentioned potential functions in the derivatives of Eq. (1.16). Usually the first of these terms ($\sim R^{-12}$) for the Lennard-Jones function is identified with repulsion and the second ($\sim R^{-6}$) with attraction. However, it should be emphasized that such an interpretation has meaning only for regions at significant distance from the potential well, where the above terms are not commensurable. Near the minimum, the Lennard-Jones (or Buckingham) potential function is formed by both terms "on equal rights." Since both terms depend highly on R, the same situation prevails with respect to the derivatives in Eq. (1.16). At the same time, the resultant potential exhibits properties expressed by Eq. (1.1), i.e., the real repulsion branch is significantly more steep than the attraction branch. It is therefore best to select such an approximation of the potential in the vicinity of the well that the main contribution to the derivatives of the potential at $R \lesssim R_e$ would be due to a single, rapidly changing repulsion term. In addition to a better conformance with the real interaction pattern in this region†, such an approximation significantly simplifies computations with Eqs. (1.16) and (1.15).

The above conditions of the approximation are sufficiently satisfied by a function, whose part, depending on the coordinate, is selected in the form of the potential of ionic crystals [1.14]:

$$\Phi(R) = \Phi_0 \exp\left[\alpha\left(1 - \frac{R}{R_e}\right)\right] - \left(\Phi_1 \frac{R_e}{R} - \Phi_2\right) \tag{1.17}$$

in which the first term can be approximately interpreted as the repulsion term and the second, as the attraction term. Constant Φ_2 must be introduced to obtain correct values of the binding energy. The parameters of function (1.17) can be easily estimated by solving the system of equations compiled, to the effect that function (1.17) in the vicinity of the potential well ($R = R_e \pm 0.1R_e$) reproduces the Lennard-Jones potential:

$$\left.\begin{array}{ll} \text{a)} & R_e = R_e^{\text{L-D}} \\ \text{b)} & \Phi(R = R_e) = -\varepsilon \\ \text{c)} & \left.\dfrac{\partial \Phi}{\partial R}\right|_{R=R_e} = 0 \\ \text{d)} & \left.\begin{array}{l} \Phi(R_e + |r|) - \Phi(R_e) \approx 0{,}3\varepsilon \\ \Phi(R_e - |r|) - \Phi(R_e) \approx \varepsilon \end{array}\right\} |r| \approx 0{,}1R_e \end{array}\right\} \tag{1.18}$$

† We have reference to a better justified interpretation of attraction and repulsion [1.9] than in the case of a Lennard-Jones function formed by a sum of terms which have no physical meaning in the region under consideration.

where ϵ is the depth of the potential well. Equations a) through c) in (1.18) ensure agreement between minima of function (1.17) and the Lennard-Jones potential, whereas equation d) provides for equality of these functions at the boundaries of the region of interatomic distances under study. To solve Eqs. (1.18) it is convenient to introduce the additional parameter n from the condition $|r|/R_e = n/\alpha$. Then equation d) is reduced to a transcendental equation in n:

$$\frac{0.3\varepsilon}{n+e^{-n}-1} = \frac{\varepsilon}{e^{n}-n-1} \tag{1.19}$$

from which we find that n $\simeq$ 1.8. Then $\alpha = 18$, $\Phi_0 \approx 0.3\epsilon$, $\Phi_1 = \alpha\Phi_0 \simeq 6\epsilon$, and $\Phi_2 = 4.7\epsilon$. (The value of (Φ_2) is of no importance in calculations using Eq. (1.16).)

Grüneisen parameter (1.16) for function (1.17) was calculated by Zharkov and Kalinin [1.14]:

$$\Gamma = \frac{1}{6}\,\frac{r(x)}{\xi(x)} + \delta \tag{1.20}$$

where $r(x) = \frac{1}{3}\alpha\Phi_0[\alpha^2 x^{\frac{1}{3}} + 2\alpha(1-m) + 2(1-m)x^{-\frac{1}{3}}]\times$

$$\times \exp[\alpha(1-x^{\frac{1}{3}})] - 2\Phi_1(2-m)x^{-1}$$

$$\xi(x) = \frac{1}{3}\alpha\Phi_0[\alpha + 2(1-m)x^{-\frac{1}{3}}]\exp[\alpha(1-x^{\frac{1}{3}})] -$$

$$-\frac{2}{3}\Phi_1(2-m)x^{-1}$$

Using the solution of Eqs. (1.18) ($\alpha \gg 1$), we can express functions $r(x)$ and $\xi(x)$ at $x \leqslant 1$, and $m \simeq 1$ can be expressed more simply:

$$r(x) = \frac{1}{3}\alpha^3\Phi_0 x^{\frac{1}{3}}\left[1+O\left(\frac{1}{\alpha}\right)\right]\exp[\alpha(1-x^{\frac{1}{3}})]$$

$$\xi(x) = \frac{1}{3}\alpha^2\Phi_0\left[1+O\left(\frac{1}{\alpha}\right)\right]\exp[\alpha(1-x^{\frac{1}{3}})]$$

We substitute these values into Eq. (1.20)

$$\Gamma = \frac{\alpha}{6}x^{\frac{1}{3}}\left[1+O\left(\frac{1}{\alpha}\right)\right] + \delta \tag{1.21}$$

Expanding function $\frac{\alpha}{6}x^{\frac{1}{3}}O\left(\frac{1}{\alpha}\right)$ in a power series in $x^{\frac{1}{3}} - 1$ and retaining the linear approximation which is permissible since this function is small as compared with the principal term in Eq. (1.21), we find

$$\Gamma = \frac{\alpha'}{6} x^{\frac{1}{3}} + \delta' \tag{1.22}$$

where the fitting parameters α' and δ' differ little from α and δ in Eq. (1.21).

We shall now compare Eqs. (1.22) and (1.8). If it is assumed that over the wide range of values of energy—from the region of the well to $\sim LkT$, for which approximation (1.1) is of significance—the repulsion branch can be described by a single exponent, then the values of α from Eq. (1.1) and of α' from Eq. (1.22) should be quite close. Hence, to within the small normalization constant δ'†

$$\Gamma = \frac{1}{2} \Gamma_{HT} \tag{1.23}$$

The appearance of factor 1/2 is due to the fact that Γ_{HT} is calculated in the classical approximation, and this coefficient defines the density dependence only of the configuration part of the thermodynamic functions. Coefficient Γ corresponds to the quantum approach and defines the density dependence of the total value of the thermodynamic functions. Since the law of equipartition of energy is approximately satisfied at high temperatures [1.19], we arrive at the conclusion that the density dependence of the thermodynamic properties of an ideal crystal both at low and high temperatures is defined by the same density function. The slight difference between α and α' can be due only to an impossibility to describe the potential repulsion branch over a wide range of values of energy by the simple Born-Mayer function. However, in this case a single potential function is not necessary, since the statistical solution of the problem is discrete. The low- and high-temperature regions are analyzed independently.

Result (1.23) validates the use of Eq. (1.22) for calculations in the region with significant anharmonicity. Since the Grüneisen parameter is identical at low temperatures also in the region of limiting anharmonicity; there is no basis for the assumption that it cannot be described by Eq. (1.22) in the intermediate region.

Up to now, the calculations were performed on the basis of the hypothesis of additivity of the potential function. The results obtained by Belogolovskiy et al. [1.20] indicate that crystals of the zero group are satisfactorily described by

† The value of Γ should be normalized to its theoretical values as $T \rightarrow 0$ or to quantity $\frac{v}{C_v}\left(\frac{\partial p}{\partial T}\right)_v$, determined experimentally at low temperatures. In both cases, for the majority of crystals, $\Gamma \cong 3$ [1.4, 1.17, 1.18]. Since the first term in the right-hand side of Eq. (1.22) is also close to 3, δ' should be small.

the Cauchy equations, deviations from which are a measure of noncentrality of intermolecular forces. As a result of this, it can be assumed that the nonadditivity of interactions makes a contribution to the Grüneisen parameter of the order of $1/\alpha$, represented by fitting parameters in Eq. (1.22).

The calculation of the Grüneisen parameter actually exhausts the problem of calculating the density dependence of thermodynamic functions. The final expression for the effective Debye temperature (1.14) used in describing the thermodynamics of the low-temperature system is

$$\Theta_D = \frac{\Theta_0 + a\,[\exp(-bT^2) - 1] + cT}{\exp\left\{\frac{\alpha'}{2}\left[\left(\frac{v}{v_0}\right)^{\frac{1}{3}} - 1\right]\right\}\left(\frac{v}{v_0}\right)^{\delta'}} \tag{1.24}$$

Due to the temperature dependence of Θ_D only the expressions for the free energy and for its derivatives with respect to volume, pressure, and the isothermal compressibility coefficient, have the same form as the Debye functions. The formulas for entropy, specific heat and the thermal pressure coefficient contain additional terms, containing the derivatives $\partial\Theta_D/\partial T$ and $\partial^2\Theta_D/\partial T^2$. We shall now list the equations for calculating thermodynamic functions, introducing the notation $t = \dfrac{\Theta_D}{T}$.

Free energy

$$F = E_0(v) + NkT\,\{3\ln(1 - e^{-t}) - D(t)\} \tag{1.25}$$

Energy

$$E = E_0(v) + NkT\left\{\frac{9}{8}\frac{\Theta(0)}{T} + 3D(t)\right\} - \\ - NkTD(t)\frac{3}{t}\frac{\partial\Theta}{\partial T} \tag{1.25a}$$

Entropy

$$S = Nk\,\{4D(t) - 3\ln(1 - e^{-t})\} - NkD(t)\frac{3}{t}\frac{\partial\Theta}{\partial T} \tag{1.26}$$

Isochoric specific heat

$$C_v = 3Nk\left[4D(t) - \frac{3t}{e^t - 1}\right] - 6Nk\left[\frac{4}{t}D(t) - \\ - \frac{3}{e^t - 1}\right]\frac{\partial\Theta}{\partial T} + 3Nk\left[\frac{4D(t)}{t^2} - \frac{3}{(e^t - 1)\,t}\right]\left(\frac{\partial\Theta}{\partial T}\right)^2 - \\ - NkTD(t)\frac{3}{t}\frac{\partial^2\Theta}{\partial T^2} \tag{1.27}$$

Pressure

$$p = p_0(v) + Nk\frac{\Gamma}{v}\left[\frac{9}{8}\Theta(0) + 3TD(t)\right] \tag{1.28}$$

Thermal pressure coefficient

$$\gamma_v = \left(\frac{\partial p}{\partial T}\right)_v = 3Nk\frac{\Gamma}{v}\left\{\left[4D(t) - \frac{3t}{e^t - 1}\right] + \left[\frac{3}{e^t - 1} - \frac{3}{t}D(t)\right]\frac{\partial\theta}{\partial T}\right\} \tag{1.29}$$

Volumetric compression modulus

$$K_T = -v\left(\frac{\partial p}{\partial v}\right)_T = K_0(v) + Nk\frac{\Gamma}{v}\left\{\left[\frac{9}{8}\Theta(0) + 3TD(t)\right]\left(1 + \Gamma - \frac{\partial \ln \Gamma}{\partial \ln x}\right) - 12TD(t)\Gamma + \frac{9\theta\Gamma}{e^t - 1}\right\} \tag{1.30}$$

The remaining thermodynamic functions can be easily calculated using the above expressions and familiar thermodynamic equations. In Eqs. (1.25)–(1.30) $D(t)$ is the Debye function, calculated according to the recommendation by De Launau [1.21]

$$D(t) = \int_0^t \frac{t^3 dt}{e^t - 1} = \frac{\pi^4}{5}t^{-3} - 3\sum_{n=1}^{\infty}\frac{1}{n}\left[1 + \frac{3}{n}\frac{1}{t} + \frac{6}{n^2}\left(\frac{1}{t}\right)^2 + \frac{6}{n^3}\left(\frac{1}{t}\right)^3\right]e^{-nt} \tag{1.31}$$

The use of a computer makes it possible to easily retain in series (1.31) a number of terms sufficient for obtaining the result with the desired accuracy.

Let us consider the calculation of lattice energy $E_0(v)$ and pressure $p_0(v)$. As previously noted, the forms of these functions can be established by specifying a model for the potential energy. In the case of paired additive interaction it suffices to select a two-particle potential function; it is not advisable to use the approximation suggested for calculating the Grüneisen parameter for the following reasons:

1. The advantage due to convenience and simplicity of calculations, which is evident in calculating the Grüneisen parameter and effective Debye temperature, is lost;

2. The lattice sums which describe the interaction between far neighbors for potential (1.17) are unknown (although these were calculated for the Lennard-Jones potential [1.22]);

It is unknown whether function (1.17) describes correctly the interaction between far neighbors, since it was used for approximating the two-particle potential only near its minimum. (This interaction is correctly described by the $\sim R^{-6}$ term in the Lennard-Jones potential.)

The last two circumstances play no role in calculating the Grüneisen parameter, but are very important in calculating functions $E_0(v)$ and $p_0(v)$. The only significant factor in interaction of far neighbors is the long-range attractive interaction. It is obvious that, even when the contribution to the energy of the attractive interactions of close and far neighbors is the same their contribution to the Grüneisen parameter still remains of the order of $1/\alpha$ (see Eqs. (1.20) and (1.21)). At the same time, the energy and pressure of the system as described by Lennard-Jones type two-particle potential functions or such functions as described by Eqs. (1.17) are, over the range of densities under study, small differences of large quantities, namely the repulsion and attraction parts of the volume potential. In this case the accuracy of calculation of the attraction branch and allowance for interaction between far neighbors play a significant role in reliable determination of the sought functions.

On the basis of the above considerations it is convenient to use the Lennard-Jones (n − 6) potential for describing the lattice energy and pressure. The sought functions become:

$$E_0(v) = A\left(\frac{v}{v_0}\right)^{-\frac{n}{3}} - B\left(\frac{v}{v_0}\right)^{-2} \tag{1.32}$$

$$p_0(v) = \frac{n}{3}A\left(\frac{v}{v_0}\right)^{-\left(\frac{n}{3}+1\right)} - 2B\left(\frac{v}{v_0}\right)^{-3} \tag{1.33}$$

where v_0 is some fixed value of the volume (it is convenient to select $v_0 = v$ at $T \to 0$ and $p \to 0$).

The zero approximation of n, A and B is best determined using the published [1.23] parameters of the 12–6 potential:

$$\begin{gathered} n = 12 \\ A = \frac{c}{2} N\varepsilon \left(\frac{R_e{}^3}{\gamma v_0}\right)^4 \sigma_{\text{rep}} \\ B = cN\varepsilon \left(\frac{R_e{}^3}{\gamma v_0}\right)^2 \sigma_{\text{attr}} \end{gathered} \tag{1.34}$$

where the coordination number is $c = 12$; the structure factor of the FCC lattice is $\gamma = \sqrt{2}$, and $\sigma_{\text{rep}} \cong 1.01$ and $\sigma_{\text{attr}} \cong 1.20$ are the lattice sums for repulsion and attraction, respectively. Since the nonadditive forces in the crystals under study are small [1.20] and since the values of n, A and B are, in the last count, determined on the basis of experimental data, it can be assumed that the nonadditive corrections to the energy and pressure of the lattice will be "effectively" represented by these parameters. Such an assumption can also be based on the rather convincing assertions concerning the "effectiveness" of the two-particle Lennard-Jones 6–12 function, in the sense of its representation of multiparticle forces [1.24].

It can be expected that the formalism suggested here will permit a description of thermodynamic functions of crystals without structural phase transitions at low and moderate temperatures, including the region of perceptible anharmonism. The computational equations contain nine fitting parameters. Six of these are contained directly in the equation for Θ_D and define the "thermal" components of the thermodynamic functions (Θ_0, a, b, c, α', δ'), whereas three parameters define the energy and pressure of the lattice.

Matching of Equations for "High" and Low Temperatures

It is natural that different calculation methods of thermodynamic functions of crystals at low and high temperatures make it necessary to splice together the results. This mating obviously must involve a correct transition to the low- and high-temperature limits. The empirical approach requires introducing some additional functions, the argument of which should most conveniently be the dimensionless temperature τ, defined by Eqs. (1.4) and (1.7). We recall that at "high" temperatures $\tau \gg 1$, whereas at low temperatures $\tau \lesssim 1$.

The first approximation of the correct description of the limiting results ($\tau \ll 1$ and $\tau \gg 1$) can apparently be obtained using the scheme

$$f = f_{\mathrm{LT}} + \exp\left(-\frac{g}{\tau}\right)(f_{\mathrm{HT}} - f_{\mathrm{LT}}) \tag{1.35}$$

where f_{LT} and f_{HT} designate some thermodynamic function at low and high temperatures, respectively; g is a fitting parameter of the order of unity. This elementary scheme, however, does not yield correct values of f at $\tau \simeq 1$, where $\exp(-g/\tau)$ also turns out to be of the order of unity, and function f_{HT}, which has no meaning in this region, may have a significant effect on the results of the calculations.

It can be easily noted that the most exact results are obtained by replacing the constant g with function $g(\tau)$, satisfying the conditions

$$\left.\begin{aligned} g(\tau) &\approx 1 \text{ at } \tau \gg 1 \\ g(\tau) &\gg 1 \text{ at } \tau \lesssim 1 \end{aligned}\right\} \tag{1.36}$$

Expanding $g(\tau)$ with respect to τ and retaining the linear term

$$g(\tau) = g_0 - g_1\tau \tag{1.37}$$

we find from conditions (1.36) the relationship between parameters g_0 and g_1:

$$g_0 \approx g_1\tau_m, \; g_0 \gg 1, \tag{1.38}$$

where τ_m is the value of τ at the melting curve.

Obviously, the introduction of function $g(\tau)$ supresses the effect of f_{HT} at $\tau \approx 1$ and ensures the transition $f \rightarrow f_{HT}$ at $\tau \gg 1$.

It is most rational to select as function f the specific heat of the crystal at constant volume, since the expression for the free energy at high temperatures is so far unknown, whereas the use of a formula for the entropy involves introducing two additional fitting parameters. In addition, in selecting C_v it is possible to significantly simplify the calculations from Eq. (1.35), substituting into the second term of its right-hand side the limiting values of the principal (Debye) term of Eq. (1.27) at $\tau \gg 1$, i.e., $3Nk$ (instead of f_{LT}). Since $\partial\Theta/\partial T$ and $\partial^2\Theta/\partial T^2$ are small at high temperatures, it can be assumed that such a replacement will result in a small error, compensated for by determining parameter n from Eq. (1.6), and also g_0 and g_1 from experimental data.

Still another advantage of selecting C_v is the fact that it points to the need to empirically modify Eq. (1.6) which is not obvious in other cases. It can be easily noticed that Eq. (1.6) exhibits a singularity at $\tau \rightarrow 1$, i.e., in the region where it has no meaning. In this region, the values of $\exp[-g/(\tau)/\tau]$ are small but finite, and the singularity in Eq. (1.35) prevails. This shortcoming can be eliminated by replacing $1/\ln\tau$ in Eq. (1.6) by $\dfrac{1}{\ln \tau + 1/\tau}$. At $\tau \simeq 10$, when using Eq. (1.6), this correction may result in an error of $\delta C_v \approx 2\%$. This is easily compensated for by adjusting the parameters, and the singularity vanishes at low temperatures.

We shall now list formulas for thermodynamic functions which follow from Eq. (1.35), when C_v is substituted into it as function f.

Specific heat at constant volume

$$C_v = \text{Eq. (1.27)} + 3Nk \exp\left(-\frac{g_0}{\tau} + g_1\right)\left[\frac{1}{\ln \tau + 1/\tau} - \frac{1}{2} - \frac{n}{3}\right] \tag{1.39}$$

Thermal pressure coefficient

$$\gamma_v = \text{Eq. (1.29)} + 3Nk \exp\left(-\frac{g_0}{\tau} + g_1\right)\left[\frac{1}{\ln \tau + 1/\tau} - \frac{1}{2} - \frac{n}{3}\right]\frac{\partial \ln \tau}{\partial v} \tag{1.40}$$

where, in accordance with Eq. (1.7) $\dfrac{\partial \ln \tau}{\partial v} = \dfrac{1}{3Dv^{2/3}}$.

The calculation of energy, entropy, pressure and its derivative with respect to volume involves evaluating the indefinite integrals

$\int \frac{\exp\left(-\frac{g_0}{\tau}\right)}{\ln \tau + \frac{1}{\tau}} \frac{d\tau}{\tau}$ and $\int \frac{\exp\left(-\frac{g_0}{\tau}\right)}{\ln \tau + \frac{1}{\tau}}$. This difficulty can be easily circumvented. If it is remembered that the range of applicability of the high-temperature corrections is very narrow, it allows one to evaluate the above integrals with sufficient accuracy by removing the mean value of the smoothly varying part of the integrand $\frac{1}{\ln \frac{\tau}{2} + \frac{2}{\tau}}$ from under the integration sign. We now list the expressions obtained as a result.

Entropy

$$S = \text{Eq.}\,(1.26) + 3Nk \exp(g_1)\, Ei\left(-\frac{g_0}{\tau}\right)\left[\frac{1}{\ln \frac{\tau}{2} + \frac{2}{\tau}} - \frac{1}{2} - \frac{n}{3}\right] \tag{1.41}$$

Energy

$$E = \text{Eq.}\,(1.25\text{a}) + 3NkT \exp(g_1)\left[\frac{1}{g_0}\exp\left(-\frac{g_0}{\tau}\right) + \frac{1}{\tau} Ei\left(-\frac{g_0}{\tau}\right)\right]\left[\frac{1}{\ln \frac{\tau}{2} + \frac{2}{\tau}} - \frac{1}{2} - \frac{n}{3}\right] \tag{1.42}$$

Pressure

$$p = \text{Eq.}\,(1.28) + \frac{NkT}{Dv^{2/3}} \exp(g_1)\left[\frac{1}{g_0}\exp\left(-\frac{g_0}{\tau}\right) + \frac{1}{\tau} Ei\left(-\frac{g_0}{\tau}\right)\right]\left[\frac{1}{\ln \frac{\tau}{2} + \frac{2}{\tau}} - \frac{1}{2} - \frac{n}{3}\right] \tag{1.43}$$

Bulk compression modulus

$$K_T = -v\left(\frac{\partial p}{\partial v}\right)_T = \text{Eq.}\,(1.30) - \frac{NkTv}{3\,(Dv^{2/3})^2} \times$$

$$\times \left[\frac{1}{\ln \frac{\tau}{2} + \frac{2}{\tau}} - \frac{1}{2} - \frac{n}{3}\right]\left\{\exp\left(g_1 - \frac{g_0}{\tau}\right) - \exp(g_1)\left(1 + \right.\right.$$

$$+2\frac{D}{v^{1.3}}\Big)\left[\frac{1}{g_0}\exp\left(-\frac{g_0}{\tau}\right)+\frac{1}{\tau}Ei\left(-\frac{g_0}{\tau}\right)\right]\Big\} \tag{1.44}$$

These formulas are intended for describing the thermodynamic functions of the ideal crystal lattice. To describe experimental data on real crystals it is necessary to make allowance for nonideality of their structure which has a significant effect on the thermodynamic properties near the melting curve [1.17]. For our purposes it is convenient to use formulas of the phenomenological theory of defects [1.14, 1.25, 1.26]. The parameters of this theory—entropy S_d and energy ϵ_d of the formation of defects—complement the set of fitting parameters contained in Eqs. (1.39)–(1.44). Presented below are computational formulas incorporating the contributions of the defects:

to the entropy

$$\Delta S = Nk\exp\left(\frac{S_d}{k} - \frac{\epsilon_d + pv_d}{kT}\right)\left(1 + \frac{\epsilon_d + pv_d}{kT}\right) \tag{1.45}$$

to the energy

$$\Delta E = N\epsilon_d\exp\left(\frac{S_d}{k} - \frac{\epsilon_d + pv_d}{kT}\right) \tag{1.46}$$

to the isochoric specific heat

$$\Delta C_v \approx \frac{N}{kT^2}\left[\epsilon_d + pv_d - Tv\left(\frac{\partial p}{\partial T}\right)_v\right]^2\exp\left(\frac{S_d}{k} - \frac{\epsilon_d + pv_d}{kT}\right) \tag{1.47}$$

to the isobaric specific heat

$$\Delta C_p = N\frac{(\epsilon_d + pv_d)^2}{kT}\exp\left(\frac{S_d}{k} - \frac{\epsilon_d + pv_d}{kT}\right) \tag{1.48}$$

to the pressure

$$\Delta p = \frac{Nv_d}{v}\exp\left(\frac{S_d}{k} - \frac{\epsilon_d + pv_d}{kT}\right)p\left[\frac{pv_d}{kT}\frac{k_T}{p} + \left(\frac{\partial \ln v_d}{\partial \ln v}\right)_T\right] \tag{1.49}$$

to the thermal expansion coefficient

$$\Delta\alpha_{\exp} = \frac{v_d}{v}\frac{1}{kT^2}\exp\left[\frac{S_d}{k} - \frac{\epsilon_d + pv_d}{kT}\right](\epsilon_d + pv_d) \tag{1.50}$$

to the isothermal compressibility coefficient

$$\Delta\beta_T = \frac{v_d}{kT}\exp\left(\frac{S_d}{k} - \frac{\epsilon_d + pv_d}{kT}\right) \qquad (1.51)$$

Since the contribution of defects to functions most sensitive to them at the melting curve do not exceed 10%, the defect volume v_d in Eqs. (1.45)–(1.51) can be replaced by the specific volume of the system [1.41], whereas p, γ_v and K_T can be represented by the corresponding functions of an ideal lattice.

1.2. APPROXIMATION OF EXPERIMENTAL DATA AND CALCULATION OF THERMODYNAMIC PROPERTIES OF CRYSTALS

The use of formulas obtained in 1.1 for workup of experimental data for actual substances requires consideration of additional conditions.

1. The results given for the "high-temperature" region were obtained in the classical approximation and can be used with satisfactory confidence for krypton and xenon crystals. The criterion for satisfaction of classical relationships in this can apparently be the value of T_m/Θ_0, where T_m is the melting temperature (the triple-point temperature). For krypton and xenon the value of this ratio is larger than unity, whereas for neon and argon it is smaller than unity. The classical behavior of specific heat C_v consists of reduction in its value with rising temperature, whereas in the quantum region, C_v rises with T [1.17, 1.18, 1.27, 1.28]. Hence, the results presented in 1.1 for the "high-temperature" region are inapplicable to neon and argon, which makes it necessary to approximate data on crystalline neon and argon only on the basis of "low-temperature" equations obtained in Sec. 1.1. Given this, the far extrapolation into the high-density region on the basis of the above equations near the melting curve is less reliable than for "classical" crystals. For example, in the quantum region C_v is a rapidly decreasing function of density, whereas in the classical region C_v increases with the density, which shifts the isochores of C_v to the region of higher temperatures with increasing density. For this region the low-temperature equations cannot in principle be used for describing data on krypton and xenon near the melting curve.

2. In the procedure for approximating experimental data it was convenient to replace the lattice energy (1.32) by its power expansion in

$$\frac{v - v_0}{v_0} = x - 1:$$

$$E_0(v) = E_0(x = 1) + \sum_{i=1}^{n} \left.\frac{\partial^i E}{\partial x^i}\right|_{x=1} \frac{1}{i!}(x - 1)^i \qquad (1.52)$$

where $$\frac{\partial^i E}{\partial x^i} = A \prod_1^i \left(-\frac{n}{3} - i\right) - B \prod_1^i (-2 - i) \tag{1.53}$$

the values of A and B are calculated from Eqs. (1.34). The coefficients in expansion (1.52) are determined in the zero approximation from known parameters of the Lennard-Jones potential [1.23]. Expansion (1.52) significantly facilitates the matching of different experimental data since its first term, less the energy of zero vibrations, is equal to the heat of sublimation of the crystals as $T \to 0$. The coefficient of the second term multiplied by $1/v_0$ is equal to the pressure of zero vibrations taken with opposite sign as $T \to 0$ and $p \to 0$. The coefficient of the third term defines the isothermal compressibility of the crystal under the same conditions. It is best to use these experimental data together with parameters of the two-particle potential for determining the zero approximations of the coefficients of expansion (1.52) when the entire body of available experimental results is approximated.

Neon. The values of thermodynamic properties of crystalline neon were approximated on the basis of the following experimental data: pv curves along the 4 and 18 K isotherms [1.29, 1.30]; the expression for $C_s(T)$ given by Batchelder et al. [1.28] as a result of smoothing data in [1.31, 1.32]; values of density, coefficients of thermal expansion α_{exp} and isothermal compressibility β_T along the sublimation curve [1.28] obtained by X-ray diffractometry and values of α_{exp} along the sublimation curve [1.33]. Approximations on the basis of the above analytic equations were obtained by using standard programs for searching for the minimum of a function of many variables. No experimental information is available on the parameters of thermal defects of the neon lattice. For this reason, the zero approximation of ϵ_d was obtained from the law of corresponding states with triple-point temperature used as the reduction parameter.

Figure 2† shows the deviations of the computed values of C_s, α_{exp} and β_T from those in [1.28] and [1.33]. The deviations for C_s lie within the limits of experimental error. For β_T at $T \simeq 20$ K, $\delta\beta_T^{\text{exptl}} \simeq \pm 10\%$. Figure 2 does not show the deviations of computational values of C_s from data in [1.34] since the latter exhibit a significant systematic deviation (to 7%) relative to subsequent results [1.31, 1.32, 1.28].

At low temperatures ($T < 8$ K) the error of experimental values of α_{exp}, according to estimates of Batchelder et al. [1.28], lies within the limits of 10–30%. We refer to the fact that all three functions in the principal approximation (without consideration of defects, which are significant only at $T \simeq T_m$) are interrelated by the expression

$$\beta_T = v\alpha_{exp}/[\Gamma(v)\,(C_s - \alpha^2_{exp} vT/\beta_T)] \tag{1.54}$$

† In Figs. 2 through 9 $\delta f = \dfrac{f_{\text{exptl}} - f_{\text{calc}}}{f_{\text{exptl}}} \cdot 100\%$.

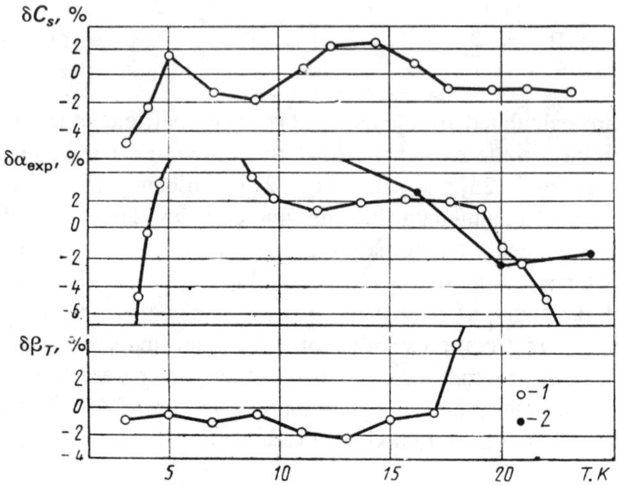

FIGURE 2. Deviations of computational values of C_s, β_T and α_{exp} of crystalline neon from data of Batchelder et al. [1.28] (1) and Bolz and Mauer [1.33] (2).

We shall also make use of the fact that: (1) the computational equations provide a satisfactory description of C_s and β_T, the experimental error of which at low temperatures ranges between 2 and 3%; (2) that $\Gamma(v)$ changes little over the range under study and the numerical values of this function which provide a satisfactory description of α_{exp} and β_T at $T > 10$ K are reliable; (3) and that $C_s \gg \alpha_{exp}{}^2 vT/\beta_T$ up to the melting curve. It can be concluded from the above that at $T < 8$ K the analytic values of α_{exp} are more exact than the experimental. This situation also applies to the remaining crystals of elements of the zero valent group.

The fact that the computed values of α_{exp} at $T > 20$ K exceed the experimental [1.28] by an amount exceeding the experimental error ($\delta\alpha_{exp}^{exptl} \cong 2\%$) is apparently due to insensitivity of the X-ray diffraction method to the thermal defects of the structure. It is known [1.17, 1.18] that the actual thermal expansion of a crystal can be measured by the dilatometer method, and the difference between the results of the dilatometer and X-ray diffraction measurements reflects the effect of thermal defects.

The specific heat C_s is as sensitive to structural effects as α_{exp}. The satisfactory agreement between the computational and experimental data at $T > 20$ K validates the computational values of C_s which include the contributions of lattice defects to thermodynamic functions. The mutual compensation of contributions of different nature (for example, of $\Theta(T)$ and defects) cannot be significant since the exponential nature of functions controlled by thermal defects is highly different from the smooth (at high temperatures) functions of the argument $t = \Theta(T)/T$ contained in the computational equations.

The deviations of the computational values of density from data of [1.28] along the sublimation curve and calculated from the measured values of α_{exp} and the lattice constant at $T = 4$ K are shown in Fig. 3. The mean deviation at

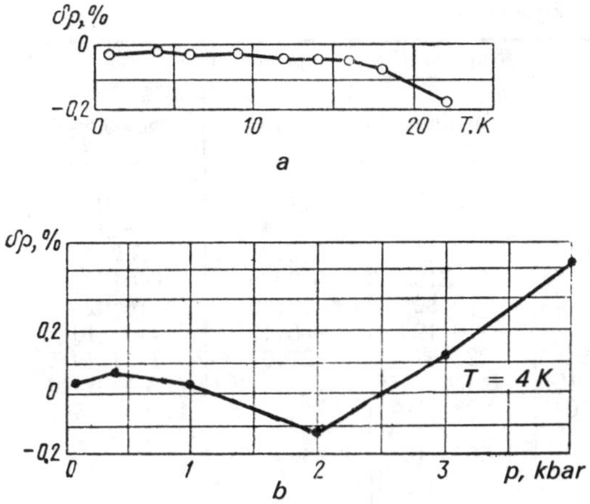

FIGURE 3. Deviations of computational values of the density of crystalline neon from data of Batchelder et al. [1.28] (a) and Stewart [1.29] (b).

$T < 20$ K is 0.03%. The increase in $|\delta v|$ at $T \simeq 20$ K is due to the fact that Batchelder et al. [1.28] make no allowance for thermal defects. The deviations from data of Stewart [1.29] along the 4 K isotherm lie within the limits of the possible error of these data. They do not agree with later data [1.30] at $T = 18$ K—the isotherms intersect. The deviations of computational data from those obtained by Stewart [1.30] exceed severalfold the deviations from data in his previous paper [1.29]. The latter are also in better agreement with the entire set of properties along the sublimation curve than the data in [1.30]. It cannot be claimed that the data in [1.29] are more exact since Stewart states otherwise [1.30]. A more convincing check apparently requires additional experimental work.

The parameters of computational equations for neon are listed in Table 1. The properties along the melting curve were calculated from Eq. (1.65) in Sec. 1.4 representing the pT behavior. The thermodynamic properties of crystalline neon at pressures to 1000 bar are given in Table I (Part III). The reference base for enthalpy was its value as $T \to 0$ and $p \to 0$. Note that the properties at pressures below 10 bar and along the sublimation curve agree within the limits of computational error. On the basis of comparison of the computational values with the experimental values analyzed here and of analysis of deviations, we recommend the following mean errors of tabulated values listed in part three of the book:

$$\delta\rho = \pm 0.05\%; \ \delta c_p = \delta c_v = \pm 2\%^\dagger$$
$$\delta s = \pm 0.3\%; \ \Delta h = \pm 0.5 \text{ J/gm (at } T \approx T_m); \ \delta\alpha_{exp} \approx \delta\beta_T = \pm 3\%$$

† The above estimate of the error for C_v is confirmed by comparison with results of recent measurements performed by Fugate and Swenson (J. Low. Temp. Phys., *10*, p. 137, 1973) published after the tables presented here were calculated.

TABLE 1. Parameters of Computational Equations

Parameter and dimensions	Numerical value of parameter			
	for neon	for argon	for krypton	for xenon
Θ_0, K	74.6	93.3	71.7	64.0
a, K	7.7	11.3	7.4	8.2
b, K^{-2}	0.0190	0.0111	0.0199	0.0275
c	0.0390	0.0310	0.0335	0.0178
α'	17.04	16.80	17.16	16.02
δ'	—0.189	—0.190	—0.197	—0.210
v_0, cm³/mol	13.389	22.557	27.09	34.717
E_0, J/mol	$-2.591 \cdot 10^3$	$-8.625 \cdot 10^3$	$-1.184 \cdot 10^4$	$-1.645 \cdot 10^4$
$\partial E/\partial x$, J/mol	$1.843 \cdot 10^3$	$2.298 \cdot 10^3$	$1.746 \cdot 10^3$	$1.595 \cdot 10^3$
$\frac{1}{2}\partial^2 E/\partial x^2$, J/mol	$4.11 \cdot 10^3$	$2.605 \cdot 10^4$	$4.290 \cdot 10^4$	$5.72 \cdot 10^4$
$\frac{1}{6}\partial^3 E/\partial x^3$, J/mol	$-1.95 \cdot 10^4$	$-7.98 \cdot 10^4$	$-1.270 \cdot 10^5$	$-1.66 \cdot 10^5$
$\frac{1}{24}\partial^4 E/\partial x^4$, J/mol	$3.79 \cdot 10^4$	$1.90 \cdot 10^5$	$2.47 \cdot 10^5$	$2.90 \cdot 10^5$
$\frac{1}{120}\partial^5 E/\partial x^5$, J/mol	$-1.07 \cdot 10^5$	$-3.59 \cdot 10^5$	$-4.94 \cdot 10^5$	$-6.94 \cdot 10^5$
S_d/k	1.70	2.00	2.025	2.02
ϵ_d/k, K	205	660	875	1280
C, cm/(mol)$^{1/3}$	—	—	3.790	4.090
D, cm/(mol)$^{1/3}$	—	—	0.211	0.206
n	—	—	0.315	0.206
g_0	—	—	10	10
g_1	—	—	1.55	1.55

Argon. The thermodynamic properties of crystalline argon have been studied experimentally more thoroughly than the properties of other crystals of the zero valent group. Specific heat C_s along the sublimation curve was measured by Flubacher et al. [1.35] with a mean error of ~2% at $T < 10$ K and of less than 1% at $T > 20$ K. Results are available for dilatometer measurements of α_{exp} at $20 \lesssim T$ 50 K [1.18] and also of X-ray diffraction measurements of α_{exp} and β_T [1.27] from 4 K to the triple point. The pv behavior along the 38, 65 and 77 K isotherms has been investigated by Stewart [1.29, 1.30]. The pvT behavior along the melting curve was measured by Van Witzenburg and Stryland [1.36], Crawford and Daniels [1.37, 1.38] and Stishov et al. [1.39] at pressures from p_{tr} to 17 kbar.

The computed values of C_s, α_{exp} and β_T are compared with the above experimental data in Fig. 4. The mean deviation for C_s lies within the limits of the error specified by Flubacher et al. [1.35]. The same applies to $\delta\beta_T$. The deviation from values of α_{exp} of Peterson et al. [1.27] near the melting point is the same as for the data for neon and is due to the same factors. The values of α_{exp} [1.18] at $T \leqslant 20$ K differ from the highly mutually compatible data in [1.27, 1.35] on properties along the sublimation curve.

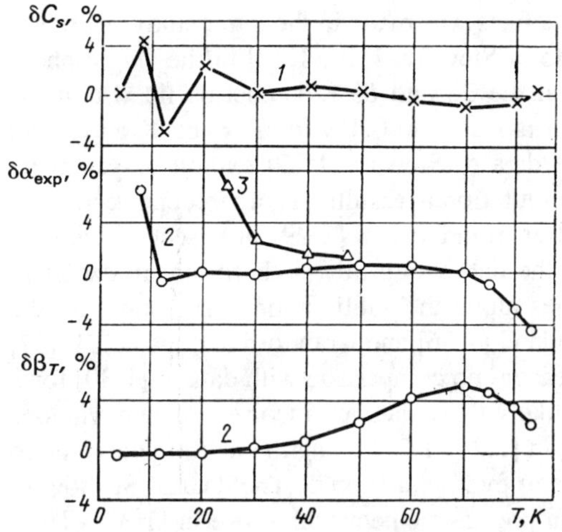

FIGURE 4. Deviations of computational values of the thermodynamic functions of crystalline argon along the sublimation curve from data of: Flubacher et al [1.35] (1) Peterson et al. [1.27] (2), and Manzheliy [1.18] (3).

Figure 5 shows the deviations of computed values of the specific volume along the sublimation curve from data in [1.18, 1.27] which basically do not exceed the error of the corresponding experiments. The small difference from the results of Peterson et al. [1.27] at $T > 70$ K, as for neon, is due to the fact

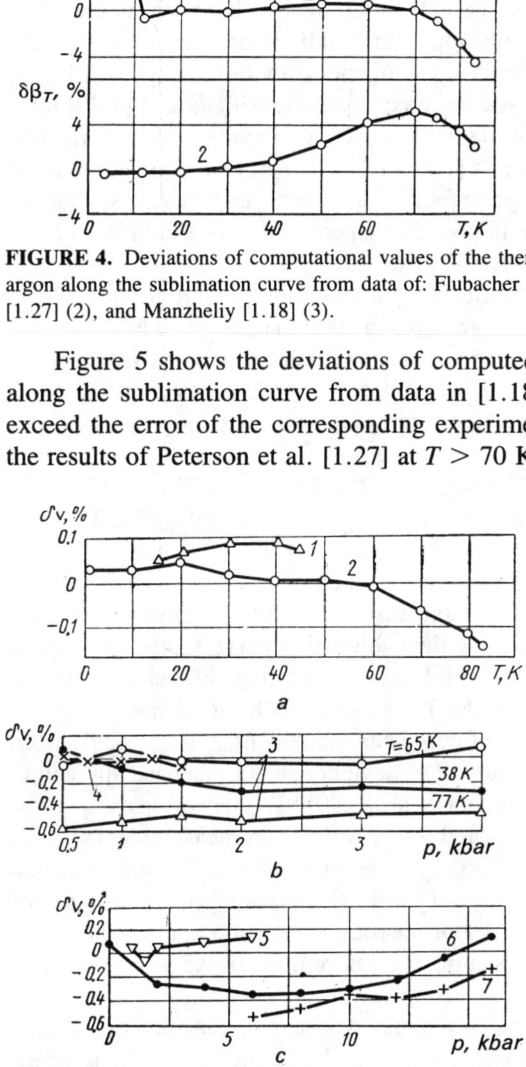

FIGURE 5. Deviation of computed values of the specific volume of crystalline argon from experimental data. a) Along the sublimation curve; b) in the single-phase region; c) along the melting curve; 1) [1.18]; 2) [1.27]; 3) and 6) [1.29, 1.30]; 4) [1.36]; 5) [1.37, 1.38]; 7) [1.39].

that no allowance for the effect of defects was made in the latter study. Comparison of the analytic values and data of Stewart [1.29, 1.30] in the single-phase region shows that the deviations at the 38 and 65 K isotherms lie within the limits of experimental error, which is less than 0.1% for $T = 65$ K and about 0.2% for 38 K. At $T = 77$ K the data of Stewart [1.29] exhibit a systematic deviation (~0.5%) from the computational results. More recent results of Stewart [1.30] for this isotherm differ from those in [1.29], intersecting with the 65 K isotherm. At the same time the data along this isotherm are in excellent agreement with the set of properties along the sublimation curve and with the results of Van Witzenburg and Stryland [1.36] and Crawford and Daniels [1.37] along the melting curve. For this reason, no comparison with data in [1.30] for $T = 77$ K is shown in Fig. 5; most likely these data are incorrect. The deviations of the computed values from those in [1.36, 1.37] along the melting curve up to $p = 6$ kbar amount on the average to $\delta v_m = \pm 0.05\%$. The data of Stishov, et al. [1.39] at $p = 6$–16 kbar are in poor agreement with those in [1.37]. Their deviation from the computed is systematic and $\delta v_m = -0.4\%$.

The parameters of the computational equations for argon are listed in Table 1. The thermodynamic functions at pressures to 1000 bar are given in Table VIII (Part III). The properties along the melting curve were calculated from the pT behavior expressed by Eq. (1.64) in Sec. 1.4.

The probable values of the mean errors of tabulated data are:

$$\delta\rho = \pm 0.05\%;\ \delta c_p = \pm 1\%;\ \delta c_v = \pm 2\%;\ \delta s = \pm 0.2\%;$$
$$\Delta h = \pm 0.3 \text{ J/gm (at } T \approx T_m);\ \delta\alpha_{exp} \approx \pm 1\%;\ \delta\beta_T = \pm 2\%$$

Krypton. The parameters of computational equations defining the thermodynamic properties of crystalline krypton were selected by working up the following experimental data: on C_s along the sublimation curve [1.40]; on α_{exp}, ρ and β_T along the sublimation curve [1.17, 1.18]; on the pvT behavior in the single-phase region [1.29, 1.30] on the 77, 81 and 115 K isotherms.

Figure 6 shows the deviations of computed values of C_s, α_{exp} and β_T from the corresponding experimental values. The deviations for C_s lie basically within the limits of error of data of Beaumont et al. [1.40]. The deviations shown for α_{exp} are from results of dilatometer and X-ray diffraction measurements [1.17] (the difference between them at $T \simeq T_m$ is as high as ~13%) and from the results of dilatometer measurements in [1.18]. The accuracy of description of these data is the same as the accuracy of measurements and validates the selection of parameters S_d and ϵ_d, representing the contribution of the thermal defects to the function under study.

Figure 7 shows the deviations of computed values of the specific volume along the sublimation curve from data in [1.17, 1.18] and in the single-phase region [1.29, 1.30]. The deviations on the 77, 81 and 115 K isotherms are systematic and become as high as $\delta v \simeq 1\%$ at p between 5 and 10 kbar. Varying the coefficients in expansion (1.52) provides a better description of data in

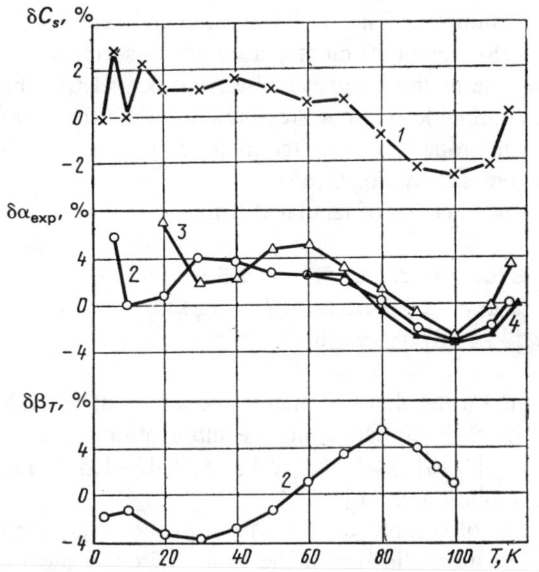

FIGURE 6. Deviation of computed values of thermodynamic functions of crystalline krypton along the sublimation curve from the data of: Beaumont et al. [1.40] (1); Losee and Simmons [1.17] (2), (4) and Manzheliy [1.18] (3).

[1.29, 1.30]. However, this produces a significant systematic deviation for α_{exp} and β_T [1.17, 1.18] along the sublimation curve which exceeds the error of these results.

The parameters of the computational equations are listed in Table 1. Note

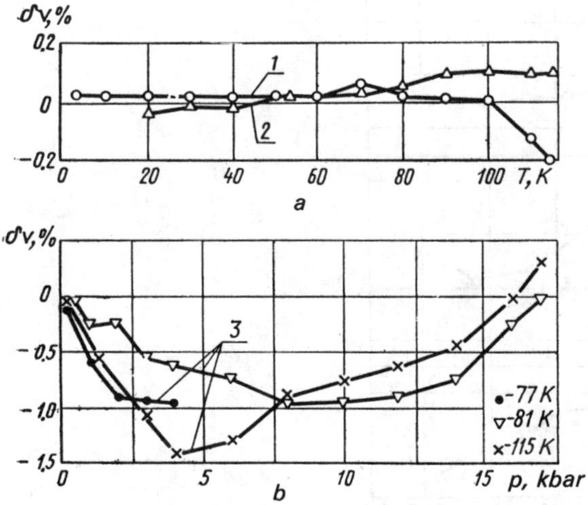

FIGURE 7. Deviation of computed values of the specific volume of crystalline krypton from experimental data. a) Along the sublimation curve; b) in the single-phase region; 1) [1.17]; 2) [1.18]; 3) [1.29, 1.30].

that the zero approximations of parameters C and D for krypton and xenon were determined by recalculating analogous coefficients, used for describing the properties of the liquid phase by the method suggested by Abovskiy [1.6]. The thermodynamic functions of crystalline krypton at pressures up to 1000 bar are listed in Table XIV (Part III). The properties along the melting curve were calculated from the pT behavior expressed by Eq. (1.65).

The probable values of the mean errors of tabulated data:

$$\delta\rho = \pm 0.05\%;\ \delta c_p = \pm 1\%;\ \delta c_v = \pm 3\%;$$
$$\delta s = \pm 0.3\%;\ \Delta h = \pm 0.3 \text{ J/gm (at } T \approx T_m\text{)};$$
$$\delta\alpha_{\text{exp}} = \delta\beta_T = \pm 3\%.$$

Xenon. Experimental results on the thermodynamic properties of crystalline xenon were worked up on the basis of: data along the sublimation curve on specific heat C_s [1.32, 1.41]; α_{exp} [1.18]; β_T [1.42]; ρ [1.18, 1.42–1.44]; and on the pvT behavior in the single-phase region [1.42].

Figure 8 shows the deviations of computed values from the corresponding experimental results. All the deviations lie within the limits of experimental error. Deviation $\delta\beta_T$ ranges from 7% at $T < 25$ K to 14% at $T = 150$ K. Such a systematic difference is due to lack of agreement between values obtained by extrapolation and data of the same study [1.42] in the single-phase region (Fig. 9) and with data in [1.42–1.44] on the density along the sublimation curve.

Figure 9 shows the deviations of values of the specific volume. The signifi-

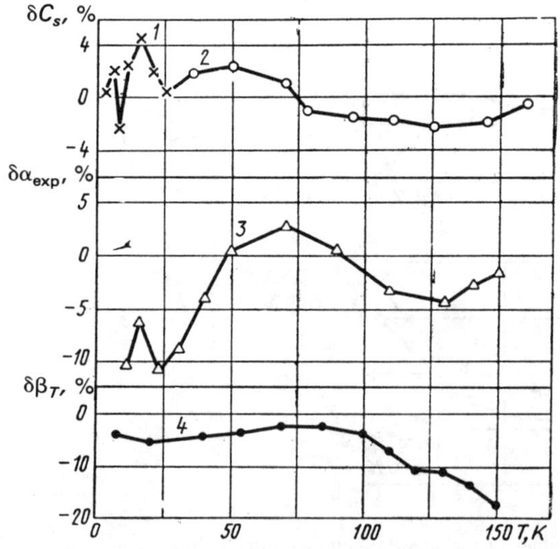

FIGURE 8. Deviations of computed values of thermodynamic functions of crystalline xenon along the sublimation curve from data of: Fenichel and Serin [1.32] (1); Beaumont et al. [1.40] (2); Manzheliy [1.18] (3); Packard and Swenson [1.42] (4).

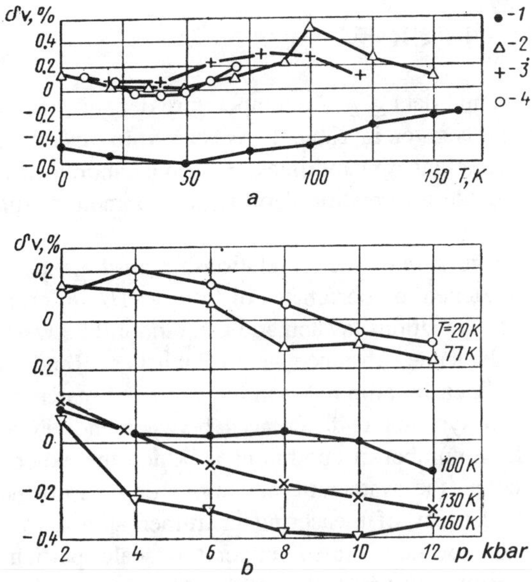

FIGURE 9. Deviations of computed values of the specific volume of crystalline xenon from experimental data. a) Along the sublimation curve; b) in the single-phase region [1.42]; 1) [1.18]; 2) [1.42]; 3) [1.43]; 4) [1.44].

cant systematic deviations of data of Manzheley [1.18] from those of Packard and Swenson [1.42], Eatwell and Smith [1.43], and Sears and Klug [1.44] along the sublimation curve are due to differences in the methods of obtaining results. Function $v_s(T)$ in [1.18] was obtained by integrating the measured behavior of $\alpha_{exp}(T)$ from the reference value of density at $T = 160$ K, in [1.43, 1.44], on the basis of X-ray diffraction measurements of the lattice constant which apparently ensures the higher accuracy in determining $v_s(T)$. The latest data are in satisfactory agreement with the results in [1.42] to zero pressure.

The data of Packard and Swenson [1.42] for the single-phase region exhibit better agreement over a rather wide pressure range (to 12 kbar) than similar results for other crystals of the zero valent group [1.29, 1.30]. Only the 40 K isotherm which intersects the isotherm at 20 K has been dropped from consideration. As seen from Fig. 9b, the deviations along four isotherms (20, 77, 100 and 130 K) are not higher than $\pm 0.2\%$.

The parameters of the computational equations for xenon are listed in Table 1. The thermodynamic functions of solid xenon at pressures to 1000 bar are listed in Table XXI (Part III). The properties along the melting curve were calculated using the pT behavior described by Eq. (1.69). The probable mean errors of tabulated values:

$$\delta\rho = \pm 0.10\%; \ \delta c_p = \pm 2\%; \ \delta c_v = 4\%;$$
$$\delta s = \pm 0.4\%; \ \Delta h = \pm 0.4 \text{ J/gm (at } T \approx T_m);$$
$$\delta\alpha_{exp} = \pm 4\%; \ \delta\beta_T = \pm 5\%$$

1.3. PHASE EQUILIBRIUM ALONG THE SUBLIMATION CURVE

A list of studies in which experimental data give the temperature dependence of the vapor pressure above the crystal is given in Table 2. As seen from this table, each of the four substances was investigated many times in various laboratories. For this reason, the currently available information permits development of the most probable expression for $p_s = f(T)$.

For the purpose of analytic expression of such a relationship all the experimental data were compared and matched in coordinates of lg p_s vs $1/T$ in terms of which this curve straightens out. Deviations (systematic and random) between data of different investigators for the same substance were established. Random measurement errors can be relatively easily eliminated in the course of matching the original data. With respect to systematic differences between individual clusters of data one must consider a number of conditions at which the experiments were performed. In particular, the systematic deviations depend on the purity of the test substance and accuracy of measuring instrumentations, to a significant measure on the difference of the assumed temperature scale, particularly at low pressures (below atmospheric), where the error of temperature measurement significantly affects the experimental accuracy. The commonly used estimation of the accuracy of equations for $p_s = f(T)$ only on the basis of ΔT is insufficient, since frequently this does not bring to light the low accuracy of equations. For this reason, we worked up data along the sublimation curve with correction of the experimental values of temperatures so as to allow for the deviations of temperatures listed by different investigators at reference points from those currently used. Such a correction results in better agreement between results of earlier and more recent work. Nevertheless, this also involved difficulties which were encountered in finding the most probable expression for lg $p_s = f(T)$ for each of the substances.

As noted above, when plotted as lg p_s vs $1/T$, the experimental data along the sublimation curve exhibit linear behavior over some temperature range. The parameters of this straight line, in their turn, depend on the original data and on the temperature range in which they are given. When no allowance is made for this, workup of the same data yields significantly different parameters of the straight line. This results in poor agreement between the computed and experimental values of the heat of phase transitions, for example, at the triple point. For this reason, reliable values of the heat of sublimation were used as an additional criterion in selecting data at the sublimation curve.

The variation in pressure along the phase equilibrium curve is obtained from the Clausius-Clapeyron relationship

$$\left(\frac{dp}{dT}\right)_s = \frac{\Delta H}{T\,(v_2 - v_1)} \qquad (1.55)$$

TABLE 2. Experimental Determination of the Vapor Pressure above the Crystal

Year	Author	Temperature, K	Reference
	Neon		
1903	Travers and Jaquerod	15—20	[1.45]
1923	Crommelin	14—20	[1.46]
1927	Crommelin and Gibson	16—25	[1.47]
1935	Keesom and Haantjes	15—24	[1.48]
1936	Henning and Otto	22—24	[1.49]
1961	Bigeleisen and Roth	16—24	[1.50]
1962	Grilly	20—25	[1.51]
	Argon		
1913,1914	Crommelin	67—84	[1.52]
1922	Born	66—84	[1.53]
1951	Clark et al.	70—84	[1.54]
1956	Freeman and Halsey	82—84	[1.55]
1960	Heastie and Lefebvre	81—84	[1.56]
1961	Flubacher et al.	66—84	[1.35]
1962	Boato et al.	72—84	[1.57]
	Krypton		
1901	Ramsey and Travers	84—113	[1.58]
1930	Petters and Weil	81—116	[1.59]
1931	Allen and Moore	114—116	[1.60]
1935	Keesom et al	79—116	[1.61]
1956	Freeman and Halsey	87—116	[1.55]
1958	Fisher and MacMillan	63—80	[1.62]
1961	Beaumont et al.	83—116	[1.40]
	Xenon		
1930	Petters and Weil	95—161	[1.59]
1931	Allen and Moore	154—161	[1.60]
1956	Freeman and Halsey	110—161	[1.55]
1961	Podgurski and Davis	70—90	[1.63]

where s = sublimation. For the crystal-vapor phase transition: $p = p_s$, $\Delta H = \Delta H_s$, $v_2 = v_g$ and $v_1 = v_c$. If it is remembered that $v_g \gg v_c$, it is assumed that the specific volume of the vapor along the sublimation curve can be calculated from the expression $v_g = RT/p$ and that ΔH_s can be taken to be a constant, then, following integration, Eq. (1.55) becomes

$$\ln p_s = a - \frac{\Delta H_s}{RT} \tag{1.56}$$

In general, the heat of phase transition ΔH_s is not constant, for which reason Eq. (1.56) can be valid only over a limited temperature range. When data on the temperature dependence of the specific heat of the crystal and of the vapor in equilibrium with it are available

$$\Delta H_s = \Delta H_s^0 + \int_0^T C_p^g dT - \int_0^T C_p^c dT - RT^2 p \frac{dB}{dT} \tag{1.57}$$

where ΔH_s^0 is the heat of sublimation at $T = 0$, and the last term represents the deviation of saturated vapor from the ideal-gas state (using B as the second virial coefficient). Combining Eqs. (1.55) and (1.57) we obtain the expression

$$\ln p_s = -\frac{\Delta H_s^0}{RT} + \int_0^T \frac{dT}{RT^2} \int_0^T C_p^g dT -$$

$$- \int_0^T \frac{dT}{RT^2} \int_0^T C_p^c dT - Bp + 2.303j \tag{1.58}$$

where j is a chemical constant whose theoretical value is calculated from the expression $j = -1.587 + 1.5 \lg M$ (here M is the atomic weight). Since the specific heat of monatomic gases at $p = 0$ is independent of the temperature and equal to $C_p^g/R = 2.5$, the second term of the above equation lends itself easily to integration. The third term can be represented as a sum of two integrals by integration by parts. Accordingly, Eq. (1.58) for monatomic gases becomes

$$\ln p_s = -\frac{\Delta H_s^0}{RT} + 2.5 \ln T - \frac{1}{R} \int_0^T \frac{C_p^c}{T} dT +$$

$$+ \frac{1}{RT} \int_0^T C_p^c dT - Bp + 2.303j \tag{1.59}$$

In principle Eq. (1.59) can be used for determining $p_s(T)$ over the entire temperature range from absolute zero to the triple point. However, the available experimental data on the temperature dependence of the saturated vapor pressure of the substances under study are represented at $T/T_{tr} > 0.44$ and are described by simpler equations.

Thus, Salter [1.64] derived the following equation for an ideal crystal undergoing quasi-harmonic lattice vibrations

$$\ln p = a + \frac{b}{T} - 0.5 \ln T \tag{1.60}$$

and showed that at $T \geqslant \Theta_D/2$ coefficients a and b are related respectively to the mean geometric frequency of vibratory spectrum of the lattice and its energy. If anharmonic vibrations need to be incorporated, an additional term, dT, appears in the right-hand side of Eq. (1.60).

In practice, the number of terms of this equation and the numerical values of the coefficients contained in it are determined from experimental data. Such equations are presented below for each of the four substances.

The most recent study of neon, by Grilly [1.51], gives, in addition to original data on the temperature dependence of the vapor pressure of neon above the crystal, described with an rms error of 0.21%, a comparison of his data with the results of others. When the Grilly equations are extrapolated to lower temperatures, the results deviate significantly from those in [1.47, 1.48, 1.50]. In addition, the value of the derivative dp_s/dT obtained from the Grilly equation is somewhat on the low side. This can be easily seen by comparing the value of ΔH_s in the triple point calculated from Eq. (1.55) with the experimental. Therefore, it appeared best to obtain an equation which describes the experimental data on the vapor pressure of neon with the required accuracy over the entire pressure range and yielding a more reliable value of dp_s/dT at the triple point.

The behavior of $p_s(T)$ was determined from the results in [1.47–1.51]. At $T > 20$ K more weight was assigned to the data of Grilly [1.51] and Henning and Otto [1.49], whereas at lower temperatures the averaged data in the remaining studies were used. The resulting equation (where p_s is in mm Hg)

$$\lg p_s = 35.4505 - \frac{210.9546}{T} - 21.3930 \lg T + 0.21962T. \quad (1.61)$$

is valid at 16–24 K.

Figure 10† shows the deviations of experimental data from those calculated from this equation. For the overwhelming majority of points the deviations do not exceed ±2%. The larger deviations are due to systematic differences between data in [1.47] and [1.48] which increase with reduction in temperature, and poor internal agreement of results in [1.50].

Information on the vapor pressure of crystalline argon is available over a narrower range of reduced temperature T/T_{tr} than for neon. In coordinates of lg p_s vs $1/T$ the results of seven studies (see Table 2) are in satisfactory agreement with one another and exhibit linear behavior. In accordance with the slope of this straight line, when using reliable values of v_g and v_c at the triple point we obtain a value of ΔH_s which is almost identical to the experimental given in [1.40].

The equation of the straight line (p_s is in mm Hg) was obtained for the temperature range of 66 to 84 K

† In Figs. 10 to 14 $\delta f = \frac{f_{calc} - f_{exptl}}{f_{exptl}} \cdot 100\%$.

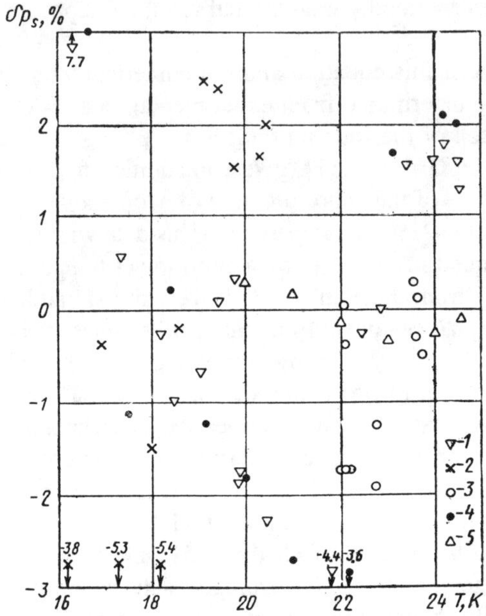

FIGURE 10. Deviations of experimental values of the vapor pressure of crystalline neon from those calculated from Eq. (1.61). 1) [1.47]; 2) [1.48]; 3) [1.49]; 4) [1.50]; 5) [1.51].

$$\lg p_s = 7.6937 - \frac{417.30}{T} \tag{1.62}$$

and describes more accurately the results of [1.54, 1.55, 1.35].

The first two studies of krypton listed in Table 2 are not too accurate and are only of historic value. The remaining results are in satisfactory mutual agreement. The minimum disagreement is that between the results of Beaumont et al. [1.40] and Keesom et al. [1.61]. The Freeman-Halsey equation (p_s is in mm Hg) [1.55]

$$\lg p_s = 7.7447 - \frac{576.6}{T} \tag{1.63}$$

provides an optimum description of both groups of data for which reason it should be used for calculating the behavior of $p_s(T)$ at 83 to 116 K. The validity of this equation is also supported by the fact that the value of the heat of sublimation at the triple point calculated from Eq. (1.63) is in satisfactory agreement with experimental data in [1.40].

The temperature dependence of the vapor pressure of crystalline xenon has been less fully investigated than for the three preceeding substances. It is seen from data listed in Table 2 that the results of Petters and Weil [1.59] are not reliable, whereas those of Allen and Moore [1.60] were obtained for a narrow temperature range. In fact, equations for describing this behavior are presented only in the last two studies. The temperature ranges do not overlap. An attempt to extrapolate the equations, for example, to 100 K yields significantly different values of lg p_s. This is not surprising since the slopes of the straight lines listed in [1.55] and [1.63] are perceptibly different from one another. Apparently the entire temperature range between 70 and 161 K cannot be described by a linear equation.

The above fact was noticed by Grutter and Shorrock [1.65] who, in order to check the accuracy of the above results, calculated the vapor pressure using Eq. (1.59) and calorimetric data of Clusius and Riccoboni [1.41]. Comparison showed that, starting with 145 K, the data of Freeman and Halsey [1.55] are systematically on the high side, with the differences increasing with decreasing temperature. On the other hand, the Podgurski-Davis equation [1.63] provides good agreement with results computed from Eq. (1.59) over the temperature range used in the experiments.

For this reason, the equations of the vapor pressure curve above the crystal for xenon were corrected as recommended by Grutter and Shorrock [1.65]. The value of the heat of sublimation at the triple point, obtained as the sum of values of the heat of melting and vaporization, was used as an additional criterion for the accuracy of the equation. The equation (p_s is in mm Hg)

$$\lg p_s = 9.5446 - \frac{851.35}{T} - 0.6702 \lg T \qquad (1.64)$$

is valid from 70 K to the triple point.

1.4. PRESSURE DEPENDENCE OF THE MELTING TEMPERATURE

The melting curve, separating the crystalline and liquid domains, rises steeply on the pT diagram with temperature. It is difficult to determine its form over a wide temperature range. It followed from Tamman's original theory that at high pressures the curve should turn to the ordinate axis and attain the maximum critical melting pressure at $T = 0$, thus delimiting the solid state region on the pT diagram. A similar point of view is held by Ebert [1.66] and Grilly and Mills [1.67], who assume that the melting curve has a critical point in the pressure range which was not yet experimentally investigated. However, Bridgman [1.68] and Simon [1.69] deny the existence of such a point on the basis of experimental data obtained by them for many substances at pressures up to 10–50 kbar. This opinion is supported by analytic calculations performed by Gilvarry [1.70]. Apparently, the melting temperature increases monotonically with rising pressure and direct transition from the liquid to the solid state is impossible.

To be able to investigate the shape of the melting curve over a wide temperature range, Simon et al. [1.71] developed a method for blocking the capillary, by means of which they obtained data on the melting pressure of a number of volatile substances (He, Ne, Ar, H_2, N_2) up to $p \simeq 10$ kbar. Their experimental data confirmed the form of equation for describing the melting curve obtained by Simon and Glatzel [1.72]

$$\lg(p_m + a) = c \lg T + b \tag{1.65}$$

Equation (1.65) is frequently used in a somewhat different form:

$$\frac{p_m}{a} = \left(\frac{T}{T_0}\right)^c - 1 \tag{1.66}$$

where c is a characteristic quantity for the given substance; T_0 and a are normalizing parameters, with a being close to the internal pressure and T_0 being equal to the triple-point temperature; P_m is the phase transition (melting) pressure.

The pressure dependence of melting temperature for neon was investigated for the first time by Simon et al. [1.73], employing the capillary blocking technique. This technique consists of the following. Compressed gas is fed to a narrow thick-walled steel capillary. Two pressure gauges are connected to the ends of the capillary; their readings are the same as the pressure is raised, until the capillary becomes blocked by the solidified substance. The maximum reading of the stopped pointer of one of the pressure gauges gives the melting (solidification) pressure at the given bath temperature. The experimental data are represented in [1.73] on the graph and in the tables at 0 to 4.9 kbar at 24 to 69 K. The results are not too accurate and are in poor mutual agreement.

More exact measurements were performed by Keesom and Lisman [1.74]. However, their data are given for a narrow pressure range ($p < 200$ bar).

Mills and Grilly [1.75] investigated the melting curves of a number of substances, including neon, for which they obtained 37 data points at pressures from 0 to 3.5 kbar. The test specimen consisted of a mixture of three isotopes of neon (90.9 Ne^{20}, 0.28 Ne^{21} and 8.82 Ne^{22}) and had an impurity content of He < 0.09%, H_2 < 0.05% and N_2 < 0.05%. Unfortunately, no experimental data are given, the results being represented by the equation

$$p = -1057.99 + 6.289415T^{1.599916} \tag{1.67}$$

(where p is in kg/cm^2) describing them with an rms error of 4.91 kg/cm^2. It follows from the graph plotted in [1.75] that individual data points deviate from the computed by as much as ±10 kg/cm^2.

Crawford and Daniels [1.76] obtained experimental data on the melting curve of neon over a much wider range than in [1.73–1.75]—from 629 to 9954 bar at T between 31.1 and 110.0 K (34 points at 17 values of T). The neon investigated by them contained 0.0018% argon. The temperature error does not exceed 0.002 K.

Comparison of experimental data of different investigators shows that the values of Simon et al. [1.73] are systematically higher than those in [1.76]. The differences are as much as 400 bar at $p \simeq 4$ kbar. Data of Mills and Grilly [1.75] at pressures to 2 kbar are in satisfactory agreement with the results in [1.76]. At higher pressures the differences rise to 25 bar. Crawford and Daniels point out that the differences due to the presence of from 0.09 to 0.19% impurities in the neon investigated in [1.75] and also due to errors in measuring the temperature (from 0.03 to 0.10 K) could be 15 bars which is significantly less than the maximum observed difference.

The melting curve of argon was investigated over a wider temperature and pressure range than the melting curves of other inert gases. Thus, as far back as 1930, Simon with his coworkers [1.73] investigated this curve from the triple-point temperature to 150 K (at p = 3.4 kbar) by the capillary blocking technique.

Bridgman [1.77] broadened the temperature range to 185 K (p = 5.5 kbar), using the traveling piston method which he developed. The liquid is compressed to one side of the piston which travels uniformly until the liquid starts freezing. Then a discontinuous change in density produces a break in the curve, expressing the piston travel as a function of pressure at $p = p_m$. In Bridgman's opinion, this method yields the most reliable results since, unlike the capillary blocking method, there are no shear stresses which usually cause the data to be on the high side. However, subsequent experiments by Swenson [1.78] showed that the effect of these stresses is negligible and that the differences between the data in [1.73] and [1.77] cannot be eliminated by correcting for these stresses. Bridgman also lists values of Δv in the course of the "crystal-liquid" phase transition.

Clusius and Weigand [1.79] investigated the melting curves of argon, krypton, and xenon over narrow temperature ranges at pressures to ~200 bar, using the results to determine the slopes of the initial segments of the melting curves and to calculate the densities of the solid substances at triple points using the Clausius-Clapeyron relationship (1.55).

Robinson [1.80] determined the melting pressure of argon at temperatures between 83 and 234 K (p = 0–8.5 kbar). The experimental arrangement was unique. A magnetic sphere was placed in a freezing chamber. The sphere was freely lifted and lowered into the test liquid by an external magnet until the liquid started freezing. The complete stop of the sphere was established from cessation of signals from a built-in microphone. The data points are given in tables and in a graph. Robinson described his results by Eq. (1.66).

A quite exact pressure dependence of the melting temperature for argon (and also krypton and zenon) was obtained by Michels and Prins [1.81] by the capillary blocking technique. The maximum pressure was 1500 bar. The experimental data can be described with high accuracy by an equation such as (1.65).

The melting curve of argon at higher pressures was investigated by Lahr and Eversole [1.82] by the traveling piston technique. The experimental data are given at temperatures from 137 to 360 K at pressures between 2.4 and 18 kbar. In addition to the *pT* curve they also give the values of Δv, the possible error of which is estimated by them at $\pm 5\%$. The significant differences (from 35 to 100%) from analogous data of Bridgman [1.77] (the latter are on the low side) are attributed by Lahr and Eversole to errors in Bridgman's measurements.

The pressure, temperature and molar volume were measured by Lahr and Eversole, both upon solidification and melting. They found that the density increase upon freezing sets in at a somewhat lower pressure than its reduction upon melting at the same temperature. Since the melting and freezing conditions should be identical, the melting curve was obtained by averaging of the "hysteresis loop." Lahr and Eversole claim that their experiments do not contain systematic errors, on the basis that they tested pure experiments by extensively verified methods. They judged random errors from deviations of experimental points, from the curve averaging them, drew in *pT* coordinates, and estimated them at $\pm 2\%$. The data points over the entire pressure range are described by an equation such as (1.66).

Crawford and Daniels [1.37, 1.38] in their investigation of the *pvT* behavior of argon over a wide range of parameters obtained experimental data on the melting pressure, and also on molar volumes of the liquid and crystal on the phase transition curve at temperatures of 94.73 to 201.32 K (at p = 0.45–6.34 kbar). The experimental technique is analyzed in Chapter 3. The melting pressure was determined by constructing experimental data along the isotherms in pressure—density coordinates. As Lahr and Eversole did, Crawford and Daniels measured the pressure and molar volume along isotherms for solidification and melting. However, the points they obtained did not form a hysteresis loop, but fitted general curves. The experimental data for the liquid-crystal transition curve are given in tabulated form.

Subsequently, Hardy, Crawford and Daniels [1.83] designed equipment for the precise measurement of the melting pressure of substances, and used it for investigating the melting curve of argon at 0.59 to 11.42 kbar (T = 97.8–273.0 K). The temperature and pressure were recorded after they were maintained constant within ± 0.001 K and ± 0.05 bar for 30 minutes. The temperature was measured by a platinum resistance thermometer, calibrated against the 1968 International Practical Temperature Scale, and the pressure by a manganin pressure gauge whose readings were reproduced with a deviation of ± 0.3 bar. Together with the calibration error (from the melting curve for mercury) they estimated the error of pressure measurement at ± 1 bar.

Hardy et al. [1.83] used high-purity (99.999%) argon. After each experiment, the argon was subjected to a mass-spectroscope analysis which confirmed the absence of impurities. This means that the measured values of the melting temperature can differ from the corresponding temperature for ideally pure argon by not more than 0.002 K. Eighty-four experimental points were obtained in five series of experiments. The data of the different series exhibit a scatter within ±0.4 bar.

Hardy et al. [1.83] performed a graphical comparison of their data with the results of others. The values in [1.73, 1.80] at temperatures to 200 K are systematically higher and at T = 140–150 K the differences are as high as 250 to 270 bar. The difference from the data of Michels and Prins [1.81] are much smaller (to 5 bar). However, they are higher than the errors specified by these investigators (0.01% for pressure and 0.001 K for temperature). By correcting for the difference in temperature scales and the presence of impurity in the argon used by Michels and Prins, these differences are reduced to 0.5–1.0 bar. The results given in [1.37, 1.38] are basically in satisfactory agreement with those in [1.83], but at temperatures above 200 K the differences are as high as 10 bar, due to the poor accuracy of calibration of the pressure gauge in [1.37, 1.38].

Hardy and his coworkers [1.83] investigated the feasibility of analytic description of their data by means of Eqs. (1.65) and (1.66) and established that, when this is done, the error of approximation significantly exceeds the error of the experimental data presented over such a wide pressure range. For this reason they decided to modify Simon's equation, by introducing a fourth parameter:

$$p_{\mathrm{m}} = A(T + D)^{c} + B \tag{1.68}$$

This equation reflects the experimental data in [1.83] to 11.42 kbar, with a mean arithmetic error of 0.65 bar and a maximum error of 1.25 bar.

The melting curve of krypton was investigated experimentally to 12 kbar. As mentioned above, this curve at low pressures was investigated by Clusius and Weigand [1.79] and at a pressure to 1.5 kbar by Michels and Prins [1.81].

Stryland and his coworkers [1.84] applied a modified Robinson technique [1.80] for investigating the pT behavior of krypton at temperatures of 120–193 K and at pressures of 0.2 to 3.0 kbar. The authors described their data by an equation in the form of Eq. (1.66) with a mean error of ±5 bar. The maximum differences are observed at low pressures, where the experimental data are higher than the computed by up to 40 bar.

The melting curve of krypton over the widest range of parameters (T = 169–362 K, p = 2–12 kbar) was investigated by Lahr and Eversole [1.82]. The experimental results are presented in graphs and tables which also contain the values of the molar volumes of liquid and solid krypton on the coexistence curves. On the basis of these data, they calculated the increment of entropy, enthalpy and internal energy in the course of the crystal-liquid phase transition.

Crawford and Daniels [1.76] investigated the melting curve of krypton with the equipment used by Hardy et al. [1.83] for determining the melting pressure of argon. The krypton specimen contained 0.0053% neon which could have resulted in a maximum error of 0.012 K in determining the melting temperature. The coefficients of Eq. (1.68), which describe the original data with an rms error of 0.31 and a maximum error of 0.75 bar were determined from 23 data points at $p = 0.79-8.01$ kbar (at $T = 138.3-291.7$ K).

A comparison with data of preceding studies showed that the differences from the results of Lahr and Eversol [1.82] at 3–4 kbar exceed 100 bar, and from those by Stryland et al. [1.84] to about 30 bar. The best agreement was observed with results of Michels and Prins [1.81]. The deviations lie within the limits of 1–8 bar and are caused, in the opinion of Crawford and Daniels [1.76], by the presence of impurities in the krypton investigated in [1.81] and a slight difference in the pressure scales.

The melting curve of xenon was investigated [1.79, 1.81, 1.84] over the same pressure ranges as for krypton, and in [1.82] to a pressure of 6.5 kbar. All these investigators describe the results of their measurements by the Simons equation with a different error. The most exact equation was presented by Michels and Prins, who presented their data over the entire temperature range with a mean error of ± 0.07 bar. Only at low pressures ($p < 70$ bar) are the test data higher than the computed, and at the triple point this difference is as high as 1.8 bar.

Comparison of pT coordinates showed that the results in [1.79, 1.81] are in satisfactory agreement. The data in [1.82, 1.84] are in satisfactory mutual agreement, but they are systematically higher than the values listed in the first two studies (13 bar, on the average).

In selecting equations for calculating the temperature dependence of the melting pressure, we were guided by the need to obtain more exact values of p_m over the range from the triple point to 1000 bar. In connection with this, an attempt was made to improve the accuracy of the Mills-Grilly equation (1.67) for neon by reducing the temperature range and including data of Crawford and Daniels [1.76] at $p \leqslant 2000$ bar. The equation obtained in form (1.65) satisfies the triple point and describes the experimental data in [1.75, 1.76] with a mean error of 1 bar at $p = 0.43-2000$ bar. The coefficients of this equation are listed in Table 3. We recommend use of the equations of Michels and Prins [1.81], describing the original data with a mean error of 0.11 and 0.08 bar, respectively, for argon and krypton at pressures to 1000 bar.† The coefficients of these equations are also listed in Table 3.

For xenon, as previously noted, the values of p_m calculated from the Michels-Prins equation at low pressures are lower than the experimental, and the differences increase continuously as the triple point is approached. To eliminate this shortcoming, we added to the Michels-Prins equation a correction function which decreases rapidly with rising temperatures; following this, the equation took the form (p_m is in bar):

$$p_m = -2610.13 + 0.808906T^{1.589165} + 1.839 \exp[-0.2706(T - T_{tr})^{1.38}] \quad (1.69)$$

† The previously mentioned differences between data in [1.81] and [1.76, 1.83] are observed basically at $p > 1000$ bar.

TABLE 3. Coefficients of Equations in the Form of (1.65) for Neon, Argon, and Krypton (p, bar; T, K)

Substance	Coefficient		
	a	b	c
Neon	995.33	0.725504	1.634940
Argon	2114.73	0.261024	1.593293
Krypton	2376.07	0.039332	1.616984

It fully satisfies the triple point, and at $p < 70$ bar the order of divergence from the experimental data is the same as at higher pressures.

We recommend that the values of p_m for neon, argon, and krypton at $p > 1000$ bar be calculated from an equation in the form of (1.68) obtained in [1.76, 1.83]. The coefficients of this equation are listed in Table 4, which also gives the maximum values of p_m up to which the equations of melting curves of neon, argon, and krypton are valid.

1.5. THE LIQUID-VAPOR PHASE EQUILIBRIUM

The vapor pressures of the four inert gases have been investigated experimentally in a large number of studies which makes it impossible to analyze each of them in detail. For argon alone, about 20 investigations were performed, most of them after 1950. In spite of the fact that the experimeters used highly pure specimens, reliable techniques and precise apparatus, the differences between the results of different investigations are as high as 0.1 bar. Given this, we matched the test values in coordinates of lg p_v vs $1/T$, in which the vapor pressure curve is close to a straight line.

The main criterion in selecting the most reliable data on the saturated vapor pressure, in addition to analysis of the experimental technique, purity of the test

TABLE 4. Coefficients of Equations in the Form of (1.68) for Neon, Argon, and Krypton (p, bar; T, K)

Substance	Coefficient				Limit of applicability of equation with respect to pressure, kbar
	A	B	C	D	
Neon	15.70774	−587.70	1.41852	−11.685	9.95
Argon	4.99313	−1484.94	1.43057	−30.179	11.42
Krypton	3.36253	−1778.71	1.44084	−38.096	8.01

specimen and the degree of reproducibility of results of measurements, was their compatibility. This criterion was basically applied for selecting data, the difference between which would not exceed ±0.2%. The method of least squares was applied for obtaining analytic expressions for the temperature dependence of the saturated vapor pressure for each of the four substances.

Analysis of various forms of equations for $p_v = f(T)$ of certain simple liquids was performed by Garside and Smith [1.85]. They showed, on the basis of data for xenon [1.86], that by using equations of three forms

$$\lg p_v = A + BT^{-1} + C \lg T + DT \quad (1.70)$$

$$\lg p_v = A_1 + B_1T^{-1} + C_1T + D_1T^2 \quad (1.71)$$

$$\lg p_v = \sum_{i=1}^{n} A_i T^{i-2} \quad (1.72)$$

($n = 7$) it is possible to describe the input information with the same accuracy as that of the experimental error.

Verbeke [1.87] assumed that for a better description of experimental data, upon approaching the critical point the equation of the vapor pressure curve should contain the term $(T_{cr} - T)^\alpha$, in which, according to the "scaling" theory, $\alpha \simeq 2$. Analyzing various forms of equations on the basis of experimental data on the saturated vapor pressure of oxygen [1.88], Verbeke suggested the equation

$$\lg p_v = A_2 + B_2T^{-1} + C_2T + D_2T^3 + E(T_{cr} - \mathrm{T})^\alpha \quad (1.73)$$

which describes with experimental accuracy the p_v values of not only oxygen, but also of equilibrium hydrogen and argon over a temperature range from the triple point to the critical temperature. This resulted in the following values of α: 1.91680 for O_2, 1.80140 for H_2 and 1.90798 for Ar. It can be easily seen that at $\alpha = 2$, Eq. (1.73) transforms to the form of (1.72) at $n = 5$, and in this case they are equivalent. However, if the accuracy of the original data is sufficiently high ($\Delta p_v < 0.002$ bar), particularly when close to critical value, the use of Eq. (1.73) may turn out to be justified.

Preliminary calculations performed by the present authors [1.89] showed that Eq. (1.70) can be most effectively used for describing the vapor pressure curves of argon, krypton, and xenon at temperatures from the triple point to the critical. The coefficients of Eq. (1.70) were determined from experimental data by computer and are listed in Table 5.

We shall now present a brief analysis of experimental data on the vapor pressure for each of the four substances. We shall point out the accuracy of description of the data and validate the selection of parameters of the triple points, normal boiling points and critical points.

Neon. The earliest measurements of vapor pressure of natural neon were performed by Kamerlingh-Onnes and Crommelin [1.90] at T between 24.43 and

TABLE 5. Coefficients of Eqs. (1.71) for Neon and (1.70) for Argon, Xenon, and Krypton (p bar, T, K)

Substance	Coefficient			
	A	B	C	D
Neon*	4.586063	−106.090	−0.0356616	0.000411092
Argon	21.57638	−548.18391	−8.651115	0.0171929
Krypton	20.41183	−723.68513	−7.539334	0.0109024
Xenon	21.17109	−1000.1252	−7.381810	0.00766133

* The equation of the vapor pressure of neon was obtained by Grilly [1.51].

27.42 K. This range was extended by Cath with Kamerlingh-Onnes [1.91] to the critical temperature. The equation suggested by Crommelin [1.92] for the pressure from the triple point to 2 atm describes these data with a large error. Crommelin and Gibson [1.47] published experimental data on the vapor pressure of neon at T = 15.47 to 27.31 K. On the basis of these results Verschaffelt [1.93] suggested an equation of the vapor pressure curve which is valid from the triple to the critical point.

Subsequently Keesom and Haantjes [1.48] published new data on the vapor pressure of neon for T between 15.09 and 27.09 K. The temperature in the experiments was measured by a platinum resistance thermometer with an error of 0.02–0.03 K. Bigeleisen and Roth [1.50] investigated the vapor pressure of natural neon, Ne^{20}, and Ne^{22} over approximately the same temperature range. Data at T = 22.04–27.49 K were obtained by Henning and Otto [1.49].

Precision measurements of the vapor pressure of liquid neon for T = 24.54–44.38 K were performed by Grilly [1.51]. They obtained 61 experimental points. The data proper are not given in his paper. However, the author compiled an equation in the form of (1.71) which describes them with a standard deviation of 0.085 and a maximum error of 0.23%. The temperature was measured by a platinum resistance thermometer which was first calibrated to the NBS-39 temperature scale. To reduce the values to the NBS-55 scale, Grilly lowered the scale of the resistance thermometer by 0.01 K.

The values of pressure in [1.90, 1.92, 1.47] are, on the average, 1% lower than those in [1.51], and in individual points 2–4% lower. The results in [1.48, 1.49, 1.51] are in satisfactory mutual agreement.

Gibbons [1.94], in investigating the vapor pressure curve of neon at T between 31.98 and 44.51 K, obtained 41 experimental points, but placed in the table 33 of them. In the opinion of the author, they agree with the results of Grilly [1.51] to within 0.1%. According to our comparison (Fig. 11) the difference at $T \leqslant 40$ K is greater. For individual points the deviations are greater than ±0.5%. It is possible that the table of Gibbons suffers from printing mistakes. Figure 11 also shows the deviations of the results of Henning and Otto [1.49] from the values obtained from the Grilly equation [1.51].

In conclusion, we note the study by Furukawa [1.95], who investigated a

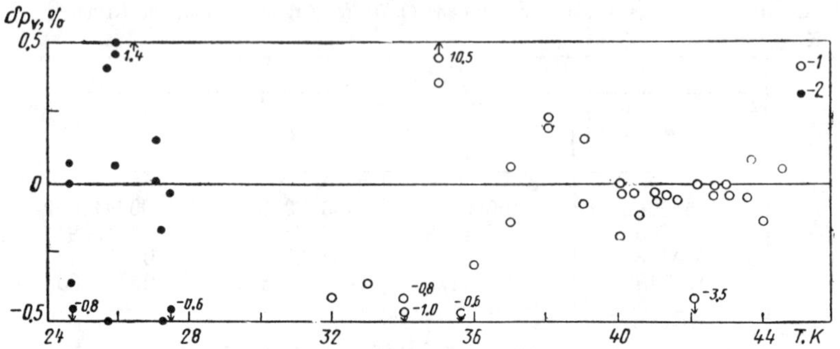

FIGURE 11. Deviations of experimental values of the saturated vapor pressure of neon calculated from the Grilly equation [1.51]: from data of Gibbons [1.94] (1), and Henning and Otto [1.49] (2).

small segment of the vapor pressure curve of neon from the triple point to the normal boiling point with the utmost precision. In two series of experiments (when a current of 1 and 2 mA was passed through the platinum resistance thermometer) he obtained 117 experimental points with a scatter not exceeding 0.05 mm Hg. The temperature error is estimated by Furukawa at ± 0.001 K on the NBS-55 scale. The resulting values of p_v were used for compiling an equation which describes them with experimental accuracy. Comparison of analytic results of Furukawa [1.95] and Grilly [1.51] shows that the deviation does not exceed 0.005 K.

On the basis of above, subsequent calculations were performed on the basis of the Grilly equation in form (1.71), which is valid from the triple to the critical temperature points and is based on rather reliable experimental results verified by Gibbons [1.94] (at $T \geqslant 40$ K) and Furukawa [1.95]. For this reason, the values of the parameters of characteristic points also based on the results of Grilly [1.51] are in excellent agreement with the data in [1.95–1.97].

Argon. The vapor pressure curve of argon at T = 134–151 K was investigated as far back as the end of the 19th century by Olszewski [1.98]. Then Ramsey and Travers [1.99] expanded the range to 78 K. In 1922, Born [1.53] measured the vapor pressure of argon above a crystal and liquid at temperatures of 65 to 90 K. All these data are not too accurate but are of historic value.

Crommelin [1.52] investigated the temperature dependence of the vapor pressure of argon above crystal and liquid over the wide range of T = 68–150 K. His results are rather reliable and are of significance up to the present. Clusius and Frank [1.100] determined very precisely the triple-point temperature (83.78 K) and the normal boiling temperature (87.29 K). Freeman and Halsey [1.55] investigated the behavior of $p_v = f(T)$ for a number of inert gases, including argon. Narinskiy [1.101] determined the saturated vapor pressure of argon at T = 90–120 K. The argon vapor pressure curve was investigated in detail by Clark et al. [1.54], who obtained data on the vapor pressure above the crystal and liquid at T = 70–150.72 K. A similar study at T = 66–86 K was performed by Flubacher et al. [1.35]. In the critical region, the vapor pressure curve was investigated by Michels and his coworkers [1.102] who found that T_{cr} = 150.86 K and p_{cr} = 48.34 atm.

A number of investigators obtained analytic expressions for the temperature behavior of the saturated vapor pressure for the ranges of variables covered by their experiments. However, no equation was available which would describe the vapor pressure curve of argon from the triple to the critical point. For this reason, the results of these studies were analyzed and correlated by the present authors [1.103, 1.104]. These publications give an equation in the form of (1.72), contain five terms, and compare experimental and computational values.

Experimental studies of the vapor pressure curve of argon were continued, but the new results not only did not suffice to significantly reduce the confidence interval of measuring error, but, in certain cases, even increased it. Thus, the experimental data of Van Itterbeek et al. [1.105] for a wide temperature range at $T < 124$ K lie systematically above the results of preceding studies. In the opinion of Verbeke [1.87], these systematic deviations are due to the impurity of the substance. Experimental data of McGain and Ziegler [1.106] are in satisfactory agreement with the preceding results. The values of p_v given by Verbeke et al. [1.107] over a narrow temperature range (nine data points at 149.23 to 150.77 K) complemented the measurements in a previous study by Verbeke and his coworkers [1.108] and performed previously in the same laboratory. Bowman, et al. [1.109] investigated the vapor pressure curve of argon at T = 84.22–150.05 K. On the basis of 36 data points they compiled an exact equation in the form of (1.70).

The temperature and pressure of argon at the triple point were determined by Clark et al. [1.54], Michels and Prins [1.81], Stryland et al. [1.84], Clusius and Frank [1.100], Pool et al. [1.110], and Thomaes and Steenwinkel [1.111], all of whom employed different techniques for approaching the triple point.

The coordinates of the point of intersection of the boiling, melting and sublimation curves, constructed on the basis of the most reliable of the aforementioned studies were used as parameters.

The normal boiling temperature of liquid argon was measured by many investigators, and all the values are in satisfactory agreement with one another and with the general shape of the vapor pressure curve.

To establish the most reliable critical parameters, we analyzed published experimental data on the critical point of argon (Table 6). The critical pressure obtained by Radovskiy by the acoustic method [1.112] is on the high side and does not fit the general behavior of the saturated pressure as a function of temperature. The values of critical parameters of Michels et al. [1.102], recommended in the majority of handbooks, are also on the high side. This follows not only from comparison with data of subsequent studies, but also from careful analysis of the experimental technique which made no allowance for the gravity effect. The critical parameters recommended by us are based on the results of

TABLE 6. Critical Parameters of Argon According to Data of Different Investigators

Year	Author	T_{cr}, K	p_{cr}, atm	Reference	Year	Author	T_{cr}, K	p_{cr}, atm	Reference
1912	Mathias et al.	150.68	48.0	[1.113]	1967	McGain and Ziegler	150.65	47.92	[1.106]
1958	Michels et al.	150.86	48.34	[1.102]	1968	Chashkin	150.64	—	[1.114]
1966	Radovskiy	150.62	48.62	[1.112]	1969	Verbeke et al.	150.653	47.936	[1.107]

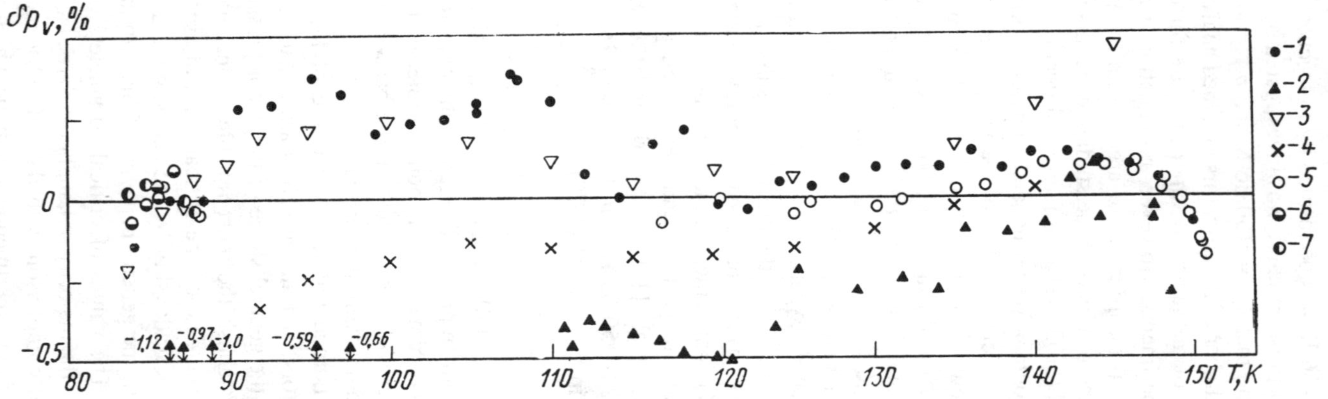

FIGURE 12. Deviations of experimental values of saturated vapor pressure of argon from those calculated from Eq. (1.70). 1) [1.109]; 2) [1.105]; 3) [1.54] (London); 4) [1.54] (Amsterdam; 5) [1.102]; 6) [1.35]; 7) [1.55].

McGain and Ziegler [1.106], Verbeke et al. [1.107], Mathias et al. [1.113], and Chashkin [1.114] and are compatible with the general form of the vapor pressure curve.

The analytic expression for $p_v = f(T)$ was based on experimental data which agreed with one another within $\pm 0.2\%$ with respect to pressure. Particular attention was paid to characteristic points. Figure 12 shows the deviations of the principal experimental data of a number of investigators from those calculated from Eq. (1.70) over a temperature range from the triple to the critical point. It is seen from the figure that the underlying experimental results are described by our equation with minimum mean deviation.

Krypton. Experimental data on the saturated vapor pressure of krypton are given in [1.40, 1.55, 1.59–1.61, 1.79, 1.109, 1.115–1.118]. Of these, the vapor pressure curve of krypton was investigated over the entire temperature range from the triple to the critical point only in [1.109, 1.116–1.118]. The results of measurements performed during the first half of the thirties pertain to the range from the triple to the normal boiling point and are not too accurate. This applies in the first place to the results of Petters and Weil [1.59], which deviate significantly from all the subsequent data. The experimental data of Allen and Moore [1.60] were found unsuitable for analytic workup due to large scatter of points and poor reproducibility of experiments. This can be judged on the basis of the significant differences between two series of experiments. Analysis showed that the values of p_v listed by Justi [1.115] are erroneous. Better agreement with results of subsequent measurements is exhibited by data of Keesom et al. [1.61] and Freeman and Halsey [1.55]. In the last study, the vapor pressure curve of krypton was investigated from 116 to 121 K. The experimental results are represented by an interpolation $\lg p_v = A + B/T$.

In 1937 Meihuizen and Crommelin [1.116] investigated for the first time the behavior of $p_v = f(T)$ from the triple to the critical point. Unfortunately, they do not list the purity of the specimen and the accuracy of their data. Michels and his coworkers [1.117] investigated the vapor pressure curve at T between 115.98 and 208.89 K and obtained 29 data points on the basis of which they compiled an equation in the form of (1.70), which describes the original data with an error of less than 0.1% with respect to p_v.

Subsequently, a similar study was performed by Bowman et al. [1.109] and by Theeuwes and Berman [1.118]. The experiments by Bowman et al. were performed with particularly pure krypton (99.998%) and involved measuring the vapor pressure at different levels of liquid in the cryostat. The temperature was measured by a platinum resistance thermometer to within ± 0.001 K and the pressure with an error of $\pm 0.04\%$. They obtained 40 experimental points at temperatures between 116.28 and 209.06 K and described them by Eq. (1.70) with an error not exceeding ± 0.08 atm.

Theeuwes and Bearman [1.118] investigated the vapor pressure curve of krypton with the same high precision. The temperature was measured to within

±0.01 K by means of a 25-ohm platinum resistance thermometer and the pressure by a piston-type pressure gauge with an error of ±0.01%. The sensing element consisted of a differential pressure indicator, the zero position of which was determined to within ±0.005 atm. The results of measurements are reproducible with the same accuracy which, at low temperatures, exceeds two to threefold the relative error of ±0.01% specified in [1.118]. The krypton they used contained only 0.0009% Ne, 0.0002% O_2 and 0.0015% Xe. The purity of the specimen was investigated by a gas chromatograph during the experiments and no changes of composition were found. The experimental data (26 points) were obtained at 126.68 to 208.15 K and are described by Eq. (1.72) with an error of ±0.005 atm for $n = 7$.

The results in [1.109, 1.117, 1.118] are in satisfactory mutual agreement over the entire temperature range from the triple to the critical point and were taken as the reference. Near the triple point, where the absolute values of p_v are small and hence the relative error is high, these results were supplemented by the data of Beaumont et al. [1.40].

It is seen from Fig. 13 that our equation describes the overwhelming majority of experimental values within ±0.2%. Somewhat greater relative deviations are observed in the vicinity of the triple point. However, this equation averages the results in [1.40, 1.109, 1.117] which deviate to 1%. To be able to estimate the accuracy of experimental data of Meihuizen and Crommelin [1.116], their deviations from the computed are also given in Fig. 13.

Rather reliable values of the parameters of the triple point are given by Beaumont et al. [1.40]. These are in good agreement with data obtained by Thomaes and Steenwinkel [1.119] in investigating the sublimation curve and by Michels and Prins [1.81] in their study of the melting curve. The values which we assumed were obtained by averaging the results of the above studies. The normal boiling point for krypton was taken on the basis of data in [1.109, 1.117] which agree in its vicinity with a deviation of ±0.02 K. The critical parameters were determined only in the paper by Meihuizen and Crommelin [1.116]. They are in satisfactory agreement with data obtained by slight extrapolation of the

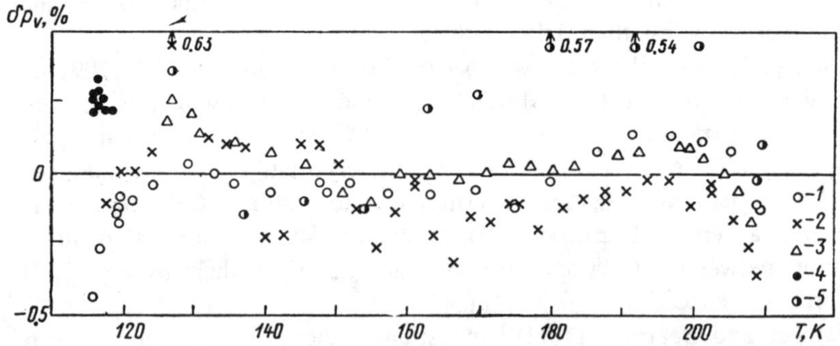

FIGURE 13. Deviations of experimental values of the saturated vapor pressure of krypton from those calculated from Eq. (1.70). 1) [1.117]; 2) [1.109]; 3) [1.118]; 4) [1.40]; 5) [1.116].

curve of $p_v = f(T)$, based on results of subsequent measurements [1.117, 1.118].

Xenon. Experimental data on the vapor pressure of liquid xenon are listed in [1.59, 1.60, 1.55, 1.79, 1.86, 1.109, 1.120, 1.121]. As previously noted, the results in the first two are not very reliable. Freeman and Halsey [1.55] and Heuse and Otto [1.120] investigated the behavior of $p_v = f(T)$ over a narrow temperature range (160.8 to 166.0 K) which was significantly broadened by Clusius and Weigand [1.79] (1939).

Precision measurements of the temperature dependence of the vapor pressure of liquid xenon were performed by Michels and Wassenaar [1.86] over the entire temperature range from the triple to the critical point. They obtained 38 data points and described them with a high accuracy by Eq. (1.70). A similar study was subsequently performed by Bowman et al. [1.109] and by Theeuwes and Bearman [1.121]. The first group of investigators used particularly pure xenon (99.998%) and obtained 35 points over the temperature range of 162.02 to 288.97 K. The experimental curve of $p_v = f(T)$ is described by Eq. (1.70) with an error of less than ± 0.04 atm. Only at one point ($T = 288.97$ K) does the difference between the experimental and computed values rise as high as 0.184 atm.

Theeuwes and Bearman [1.121] investigated the vapor pressure curve of high-purity (99.996%) xenon for $T = 175{-}289.5$ K. The accuracy of measurement of the principal parameters is the same as in experiments with krypton [1.118]. A total of 30 data points were obtained for p_v and these are represented by Eq. (1.72) using $n = 8$ with an error of less than ± 0.006 atm.

The results in [1.86, 1.109, 1.121] are in good mutual agreement and were described by us by Eq. (1.70) with an error of $\pm 0.2\%$ (Fig. 14).

The parameters of the triple point of xenon are listed in many studies. The most reliable are the results in [1.56, 1.81, 1.120]. The value of the normal boiling temperature for xenon was taken from data in [1.86]. The critical parameters of xenon have been determined rather reliably, since several studies [1.122–1.124] are devoted entirely to the investigation of the critical region

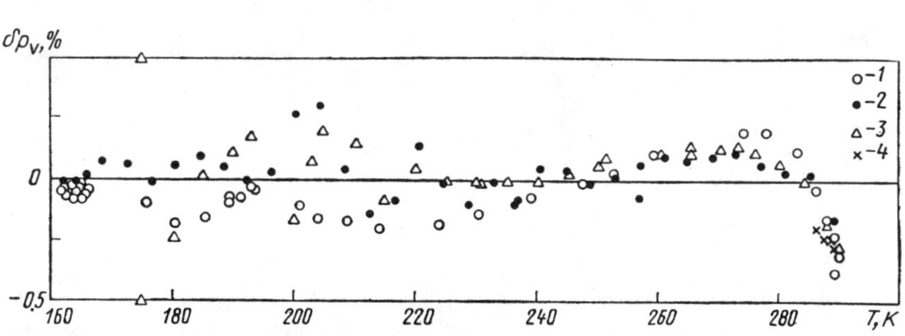

FIGURE 14. Deviation of experimental values of saturated vapor pressure of xenon from those calculated from Eq. (1.70). 1) [1.86]; 2) [1.109]; 3) [1.121]; 4) [1.124].

TABLE 7. Atomic Weight and Parameters of the Four Substances at Their Characteristic Points

Quantity	Value of parameter for			
	neon	argon	krypton	xenon
Atomic weight	20.183	39.948	83.80	131.3
Triple point temperature, K	24.55±0.01	83.81±0.01	115.76±0.02	161.35±0.02
pressure, atm	0.427±0.001	0.682±0.002	0.722±0.002	0.805±0.002
Normal boiling temperature, K(p = 1 atm)	27.09±0.01	87.29±0.02	119.78±0.02	165.03±0.02
Critical point temperature, K	44.40±0.02	150.65±0.02	209.39±0.02	289.74±0.02
pressure, atm	26.2±0.05	48.0±0.1	54.2±0.1	57.5±0.1

with consideration of the specifics of behavior of the substances in it. Figure 14 shows the deviations of computed values of p_v from the results in [1.124].

The most probable values of the parameters of characteristic points of the four substances are listed in Table 7, which also gives their atomic weight.

CHAPTER

TWO

THERMODYNAMIC PROPERTIES IN THE GASEOUS STATE

The thermal equation of state can be used for calculating the thermodynamic properties of any substance over a range of variables for which the starting equation is valid, if the temperature behavior of specific heat c_{p0} is known. Practical needs brought about the appearance of a large number of empirical equations of state, among others, by Beattie and Bridgeman [2.1], Benedict, Webb, Rubin [2.2], and Martin and Hou [2.3], etc. These are analyzed in detail by Vukalovich and Novikov [2.4]. The most significant is the Van der Waals equation, since, in spite of its simplicity, it transmits almost all the qualitative features of the thermodynamic behavior of individual substances. This includes continuous transition from the gaseous to the liquid state and vice versa. The latter circumstance is important, since it points to the feasibility of obtaining a unified equation of state which describes with experimental accuracy the properties of both the liquid and gaseous phases.

Such a macroscopic description of a thermodynamic system conceals a rather complex mechanism of interparticle interaction. In fact, at high temperature and moderate pressure, when paired and triple interactions between particles predominate, the system can be described by a simple equation of state, containing two virial coefficients. When the pressure rises and the temperature decreases, the picture changes for two reasons: (1) the interaction between particles becomes enhanced due to increasing density, and (2) the reduction in temperature decreases the role of the kinetic energy as compared with the particle interaction energy. Under certain conditions the substance becomes liquid, which is characterized by a strong correlation of particles causing cohesion in the liquid. This demonstrates the difficulty of compiling a unified equation of state.

Recently many attempts have been made to solve this problem by resorting to computers. Such work has been performed primarily in the USSR and the USA. In the USSR the work is done on the basis of the virial equation of state, verified analytically by Mayer [2.5] and Bogolyubov [2.6],

$$Z=\frac{pv}{RT}=1+B_2(T)\rho+B_3(T)\rho^2+B_4(T)\rho^3+\ldots \qquad (2.1)$$

or the equation of state expressed in terms of the so-called elementary functions, suggested by Kazavchinskiy [2.7, 2.8],

$$Z = \alpha_0(\omega) + \alpha_1(\omega)\vartheta + \beta(\omega)\psi(\vartheta) + \gamma(\omega)\varphi(\vartheta) + \ldots \qquad (2.2)$$

where $\omega = v_{cr}/v$ and $\vartheta = T_{cr}/T$. It is possible to convert from one equation to the other on the condition that all the virial coefficients comprise a linear combination of ϑ and similar temperature functions $\psi(\vartheta)$, $\varphi(\vartheta)$ . . .

Vukalovich et al. [2.9] suggested three methods for approximating *pvT* data by a virial equation of state whose coefficients are determined by computer. The first method provides for sequential singling out of several lower-order virial coefficients. For a wider range of parameters, it is more effective to use a method for compiling the equation of state in which all the coefficients of the equation are determined simultaneously by requiring that the sum of squares of deviations of the computed values of Z from the experimental be at minimum. The calculations are performed with consideration of the weights of experimental points. These are inversely proportional to the square of the error of data. The third method is essentially a combination of the first two methods and leads to an equation of state with a virial component.

For a long time, the fact that the coefficients of the virial equation were obtained by computer by the method of least squares caused the equation to describe experimental data reliably only over a relatively narrow range of parameters since it contained less than 25 coefficients. Only recently were Altunin and Gadetskiy [2.10] able to compile an equation of state containing up to 50 constants which opens new possibilities for compiling a unified equation of state.

The method of representing Eq. (2.2) in terms of the so-called conditional elementary functions suggested by Kazavchinskiy [2.11] made it possible to attain greater success. Vasserman [2.12] was the first to use this technique for obtaining a unified equation for nitrogen describing experimental data to $\omega = 2.8$ and $\vartheta = 1.47$. However, broadening this range in order to include data at higher density and at the crystallization curve necessitated the use of a more complex form of the equation of state [2.13].

Better results were obtained when using Kazavchinskiy's technique [2.11] in the case of data on compressibility of hydrogen [2.14] and helium [2.15]. The equation of state for parahydrogen contains four volume functions and two temperature functions and describes the gaseous and liquid phases to $\omega = 2.8$. The equation of state for helium between 2.2 and 20 K contains five volume and three temperature functions while at $T > 20$ K two of the temperature functions become zero.

The unified equation of state being compiled at the NBS (USA) is based on an empirical equation suggested by Strobridge [2.16]:

$$p = RT\rho + \left(n_1 T + n_2 + \frac{n_3}{T} + \frac{n_4}{T^2} + \frac{n_5}{T^4}\right)\rho^2 +$$
$$+ (n_6 T + n_7)\rho^3 + n_8 T\rho^4 + \rho^3\left(\frac{n_9}{T^2} + \frac{n_{10}}{T^3} + \frac{n_{11}}{T^4}\right)\exp(-n_{16}\rho^2) +$$
$$+ \rho^5\left(\frac{n_{12}}{T^2} + \frac{n_{13}}{T^3} + \frac{n_{14}}{T^4}\right)\exp(-n_{16}\rho^2) + n_{15}\rho^6 \qquad (2.3)$$

This equation was used for the first time to describe data on the compressibility of nitrogen for 64 to 350 K at pressures to 300 bar, and subsequently for describing the *pvT* behavior of helium, parahydrogen, and carbon monoxide to 300 bar and 350 K.

In order to describe new *pvT* data of Weber [2.17] for gaseous and liquid oxygen at 80–300 K and 6–340 bar, Stewart [2.18] used a more complex Eq. (2.3), to which he added linear and nonlinear terms and raised the number of constants to 27. However, there is no confidence that further broadening of ranges can be attained by making Strobridge's equation even more complex.

The simplest way to calculate the thermodynamic properties of the four monatomic gases from the triple point to 1300 K at maximum pressure of 1000 bar consists of describing the liquid and gas phases by independent equations of state. Then one should restrict oneself for the gas to $\rho_{max} \simeq 2\rho_{cr}$. Then obtaining an exact equation of state does not involve particular difficulties when using an equation in the form of (2.1) or (2.2). But, the artificial separation of the description of properties of a liquid and highly compressed gas, while significantly facilitating the solution of the problem as a whole, involves additional work in matching the computed data. The amount of work required is commensurate with that needed for correcting values obtained from a unified equation of state, which currently does not ensure equality of chemical potentials at the left and right branches of the liquid-vapor coexistence curve (see, among others, [2.19]).

To be able to determine the coefficients of the equation of state one must have compatible experimental thermal data. The compatibility of experimental data listed by various investigators can be checked directly upon computer compilation of the equation. However, preliminary analysis of experimental results still remains significant, since it allows clarifying which of the data should be preferred and whether they can be used for compiling an equation of state for the intended range of variables.

2.1. BRIEF ANALYSIS OF EXPERIMENTAL DATA

As of now, a large body of experimental data has been accumulated on the thermodynamic properties of monatomic gases. Since 1961, following the appearance of the most complete handbook on the properties of inert gases [2.20], the number of publications has doubled. In spite of this, the entire range of variables of practical interest was not investigated experimentally. In addition, almost all the investigators were concerned with only the thermal properties of substances. On top of this, the results are not of equal value from the point of view of accuracy. For this reason, in our analysis particular attention was paid to the reliability of the experimental technique, the purity of the test gas, and also the internal and mutual compatibility of data.

Unfortunately, such an analysis is impossible for a number of studies, due to the lack of information on the experimental equipment, measuring instrumentation and their accuracy and methods of correction, and the accuracy of determining the corrections. In addition, information on the purity of the test specimens is frequently missing. In such cases, the error has to be estimated by comparing the data with one another or with results of the most exact measurements.

All the experimental data were corrected so as to bring them into line with the IPTS-68. If the study employed a relative method and the quantity to which the results are referred is very reliable, a recalculation was performed. In certain cases this results in better matching of data.

Neon. The studies published up to now on the experimental determination of properties of gaseous neon are listed in Table 8. The first data on compressibility [1.90, 2.21] over a wide temperature range are given on nine isotherms. However, information on the measurement error and the purity of the test specimens is missing.

Holborn and Otto [2.22] initially investigated the *pvT* behavior of a binary mixture made up of 72.39% neon and 27.61% helium. Using these data and values obtained previously for pure helium the authors determined the density of neon. In a subsequent study [2.23] they measured the compressibility of pure neon, with 0.1% helium as the main impurity. The experimental data, whose maximum error is estimated by the authors at 0.2%, are represented by ten isotherms at pressures to 100 atm†. Holborn and Otto [2.24] measured the compressibility of gaseous helium at $T \simeq 65$ K. The results of [2.23, 2.24] along isotherms were described and were reduced to dependence on pressure using an equation such as

$$pv_A = A + Bp + Cp^2 + Dp^4 \tag{2.4}$$

† Otto [2.35] lists data on compressibility of neon at 273.15 and 373.15 K at pressures to 188 atm.

TABLE 8. Experimental Studies of Thermodynamic Properties of Neon

Year	Author	Property	Temperature, K	Pressure, bar	Reference
1915	Onnes and Crommelin	p, v, T	56—293	22—93	[1.90]
1920	Crommelin et al.	p, v, T	56—170	21—79	[2.21]
1924	Holborn and Otto	p, v, T	273—673	22—99	[2.22]
1925	Holborn and Otto	p, v, T	91—673	20—99	[2.23]
1926	Holborn and Otto	p, v, T	65	20—89	[2.24]
1928	Michels and Gibson	p, v, T	273—373	32—493	[2.25]
1954	Nicholson and Schneider	p, v, T	273—973	10—80	[2.26]*
1960	Michels et al.	p, v, T	273—423	24—2900	[2.27]*
1966	Gladun	α_i	30—90	1—200	[2.28]
1967	Sullivan and Sonntag	p, v, T	70—120	1—300	[2.29]
1968	Runge	α_i	27—100	1—240	[2.30]
1969	Gibbons	p, v, T	27—70	4—200	[1.94]*
1970	Onosovskiy and Moroz	p, v, T	65—273	10—250	[2.31]*
1971	Onosovskiy	p, v, T	52	11—252	[2.32]*
1971	Streett	p, v, T	80—130	73—2045	[2.33]*
1972	Dedikov et al.	c_p	30—70	20—120	[2.34]
	» »	β_i	70	2—120	[2.34]

* Here and in similar tables (9, 11, 12) for argon, krypton, and xenon the asterisk designates studies which shall be used for compiling the equations of state.

α_i is the integral adiabatic Joule-Thomson effect.

β_i is the differential adiabatic Joule-Thomson effect.

Michels and Gibson [2.25] significantly broadened the pressure range under study. However, they treated their values of compressibility of spectrally pure neon as preliminary, since the experimental procedure was adjusted and refined in the course of these experiments.

The results of the above studies are in satisfactory mutual agreement in the overlapping temperature ranges. Thus, the differences between data in [2.23] and [2.25] do not exceed $\pm 0.1\%$. At $T = 273$ K they are in agreement with the data of Kamerlingh-Onnes and Crommelin [1.90] within $\pm 0.2\%$. However, as the temperature is reduced, the deviations of these data from those listed in [2.23, 2.24] increase and amount to 0.5% at $T = 65$ K.

On the basis of experimental results in [1.90, 2.21–2.25, 2.35], Yendall [2.36] compiled an equation of state which is valid from the normal boiling point to 320 K at pressures to 200 bar. Comparison of the values of pressure calculated from this equation with the experimental shows that the mean deviation from data in [1.90, 2.21] amounts to 0.19%, 0.22% from those in [2.23, 2.24], and 0.53% in [2.25] at maximum deviations of 1, 2 and 1.5%, respectively.

In their studies of compressibility, Nicholson and Schneider [2.26] attained the highest temperatures and Michels et al. [2.27] the highest pressure, as of now. Both investigations used reliable methods of determination of the compressibility of spectrally pure neon. For these reasons, the results are highly accurate and deviate from one another by less than $\pm 0.1\%$.

In 1963, McCarty and his coworkers [2.37] compiled tables of thermodynamic properties of neon from 30 to 300 K and 0.1 to 200 bar. In addition to experimental data, they also used the law of corresponding states to calculate properties for the range not included in the experiments. They showed, using argon and nitrogen as the basis substances and comparing similar cross sections of the surfaces of state, that the deviations of thermodynamic surface of neon from the surfaces of the basis substances are systematic. This made it possible to ensure an acceptable coincidence of the surfaces being compared by means of a function of reduced parameters.

The equation of state obtained by McCarty et al. [2.37] was used for calculating the thermodynamic properties of neon above the aforementioned range of parameters. The mean deviations between experimental and computed values of Z lies within $\pm 0.16\%$, with a maximum of 1%. In their opinion, the possible computational error for ranges outside those covered in the experiment may be as high as 1.5–2%.

Recently, several authors have published on experimental study of the pvT behavior of neon at low temperatures. Sullivan and Sonntag [2.29] employed Burnett's modified method for measuring the compressibility of high-purity neon and represented experimental data for four isotherms: 70, 80, 100 and 120 K. They calculated the possible experimental error which amounted to 0.13% at $T = 80$ K. Comparison of their results with the data of preceding studies shows that, on the average, their numerical values are higher. The discrepancies increase with reduction of temperature and pressure and exceed 1%. The data on compressibility of argon obtained with the same equipment at the 163.15 K isotherm are also on the high side, as compared with the reliable data of the Amsterdam laboratory. This suggests that a systematic error has been committed in the experiments of Sullivan and Sonntag [2.29].

Gibbons [1.94] investigated the compressibility of neon at low temperatures using a constant-volume piezometer. The impurity content ranged from 0.0009 to 0.0085% by volume in the different series of experiments. The experimental data, the error of which is estimated by the author at 0.13%, pertain to the liquid and gaseous regions. The tables also list experimental values of the saturated vapor pressure and the density of saturated liquid. The latter were obtained analytically by extrapolating "quasi-isochores" to intersection at the vapor pressure curve. Comparison of these results with the data of Sullivan and Sonntag [2.29] at the 70 K isotherm showed that the latter are systematically higher with the deviations increasing with the pressure and being as high as ~2% at $p = 200$ bar.

The pvT behavior of gaseous neon below room temperature was investigated in detail by Onosovskiy and Moroz [2.31], who used a nonrelieved constant-volume piezometer. The results are presented on six isotherms. The possible error does not exceed $\pm 0.1\%$. The authors arrived at this conclusion by comparing their data with those of other investigators at the 273.15 K isotherm and also on the basis of analysis of computational formulas and the experimental technique.

Subsequently, Onosovskiy [2.32] investigated the compressibility of neon at a single isotherm (52 K) and calculated that the experimental error is 0.15%. He also lists values along the critical isotherm, obtained by graphic matching of the experimental results with data on the saturation curve.

Recently, Streett [2.33] investigated the *pvT* behavior of neon at five isotherms by the constant-volume piezometer method. A special feature of his technique consisted of discharging the gas from the piezometer to a gas meter after obtaining each test point in order to be able to calculate the weight of the gas by means of an equation of state for the low-pressure region.

The gas used in the experiments was 99.97% pure. The discrepancies of the experimental values of density amounted to hundredths of a percent, and the main source of absolute error consisted of incorrect determination of the volume of the system. To reduce these errors, the experimental equipment was calibrated by determining the density of saturated liquid argon and nitrogen at several temperatures near 100 K and by comparing with the most precise experimental data. These density values, determined with an error of less than 0.01% by extrapolation of experimental isotherms in coordinates (p, ρ) to the saturation pressure, systematically deviated by about 0.25% from the most reliable experimental values. Streett corrected the volume of the piezometer, since the inaccuracy in determining it, which amounted to 3.6 cm^3, could have been the most probable error source. This means that actually he used the relative method. The error of his experimental data on density is estimated by Streett at 0.1%.

Analysis and comparison of the results of the aforementioned studies leads to drawing the following conclusions. At temperatures exceeding 273 K, the results agree within ±0.2%. The data of Michels et al. [2.27] are the most accurate and were used for comparing the data of different investigators at the 273.15 K isotherm (Fig. 15).

At lower temperatures (to 70 K) the test data in [1.94, 2.21, 2.23, 2.31] are

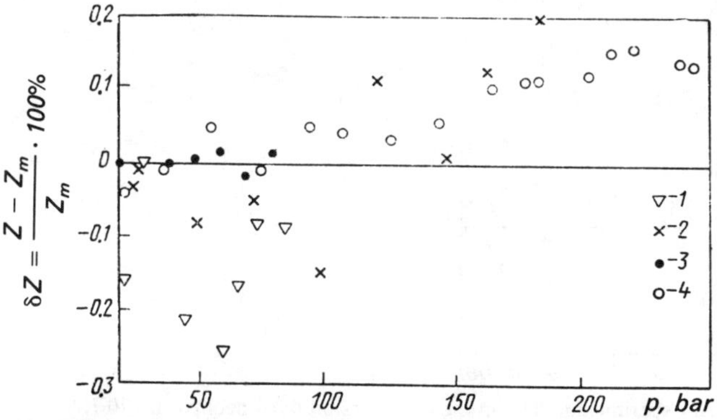

FIGURE 15. Comparison of experimental data on the compressibility of neon at the 273.15 K isotherm with those of Michels [2.27] identified as Z_m. 1) [1.90]; 2) [2.23]; 3) [2.26]; 4) [2.31].

in mutual agreement to within ±0.25%. Comparison of the 70 K isotherm (Fig. 16) was performed relative to data of Onosovskiy and Moroz [2.31], obtained at this temperature over the widest pressure range. The results of Sullivan and Sonntag [2.29] are up to 1.3% higher than other data which are in satisfactory mutual agreement. For this reason, we did not use the results in [2.29] for subsequent work. At $T < 70$ K the agreement is somewhat poor. This is noticeable already at the 60 K isotherm where the difference between the data in [1.94] and those in [2.31, 2.32] becomes as high as 0.4%. The values of Z in [1.90, 2.21] at the lowest temperature (55.6 K) are not reliable since they are in poor agreement with the data in [1.94, 2.31] at the closest isotherms and at the saturation curve.

The results of Streett [2.33] are approximately 0.3% higher than the data of Holborn and Otto [2.23] for the same temperature range and 0.4–0.6% lower than the results of Sullivan and Sonntag [2.29]. The differences between the data of Streett and the results of Onovoskiy and Moroz [2.31] are systematic and amount to 0.3–0.4%. Streett's data on Z at the 80 K isotherm are up to 2% higher.

As seen from Table 8, experimental studies of calorific properties of neon have only recently been performed and are not numerous. The experimental facility employed by Gladun [2.28] is based on a method developed by Roebuck

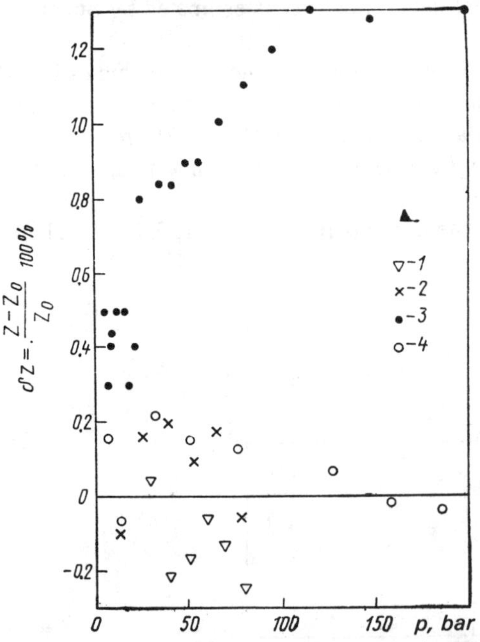

FIGURE 16. Comparison of experimental data on the compressibility of neon at the 70 K isotherm with those of Onosovskiy [2.31] identified as Z_0. 1) [2.21]; 2) [2.23, 2.24]; 3) [2.29]; 4) [1.94].

[2.39]. To bring the gas to the required initial temperature and pressure, it is passed through a heat exchanger, placed in a bath with boiling nitrogen, argon or oxygen, and is then expanded under adiabatic conditions to close to atmospheric pressure. The temperature of the gas is maintained constant (within ±0.02 K), and its pressure, prior to entry into the cryostat, is adjusted automatically with a deviation of 0.1 atm.

The pressure differences over the 2–190 bar range were measured by a Bourdon-type pressure gauge with an error of ±0.5%. Lower pressures were measured by a mercury manometer. The temperature was measured by indium thermometers calibrated against data on the vapor pressure of neon and nitrogen. The overall error of temperature measurement is estimated by Gladun at ±0.04 K.

The gas flowrate in these experiments was 2 m^3/hr. It was established experimentally that varying the flowrate within 40% has virtually no effect on the results. The experimental values of temperature and pressure at nine values of enthalpy are given in [2.28] in tabulated form.

The equipment used by Runge [2.30], unlike that of Gladun, was capable of measurements at lower specific enthalpy when the final state of the substance after expansion is two-phase.

The measuring instrument is based on a throttling valve and two cells interconnected by long thin tubes. The design of the facility and precise design of certain components made it possible to eliminate errors due to influx of heat. Adiabatic conditions were produced by radiation shields and vacuum insulation.

The compressed gas flowed into the high-pressure cell, where the initial p and T were measured. Then the gas was passed through a throttling valve and following expansion entered the low-pressure cells from whence it was redirected into the circulation loop through a coil welded to the casing of the gas thermometer. The absolute temperature of the expanded gas was measured by a gas thermometer, and the Joule-Thomson effect by an iron-constantan thermocouple. The latter was calibrated by placing it in a sleeve specially provided for this purpose in the center of the gas thermometer. The temperature was measured within 0.01 K. Experiments performed at different flowrates demonstrated that the measured parameters are independent of the latter.

Unfortunately, Runge [2.30] did not tabulate his measurements and also presented a graph of $\alpha_i(T)$ at rounded off values of pressure. He constructed an h-T diagram and a table of values of enthalpy at rounded off values of temperature and pressure. Runge estimates the possible error of measurements at ±0.1 K. His comparison with data in [2.28] showed that the mean deviation is 0.04 K and the maximum deviation is 0.2 K. Comparison with computed data of McCarty and Stewart [2.40] yields a mean deviation of 0.7 K with a maximum of 0.8 K.

Runge used the h-T diagram and applicable thermodynamic relationships for numerical calculation of the values of entropy and of the product pv at 28 to 100 K at pressures from 1 to 240 bar. The values of the specific heat C_p and the differential Joule-Thomson effect β_i are given only in graphical form. In his conclusion the author makes a statement, in our opinion controversial, to the effect that his method of calculation is preferable to that for calculating thermodynamic properties of substances using an equation of state compiled on the basis of reliable thermal data. Without going into details, we wish to point out that the facility used for measuring α_i is cumbersome and is difficult both to construct and operate (particularly in studies at low enthalpies). In addition, it is less accurate than equipment used for determining the *pvT* behavior. On the basis of data on the Joule-Thomson effect, in calculations of quantities such as density or entropy, one must have expressions for $\rho = \rho(p)$ and $s = s(p)$ at least along one isotherm. For this reason, the error in determining these functions enters entirely into the computed values of ρ and s.

Dedikov with his coworkers [2.34] determined the isobaric specific heat and the adiabatic differential Joule-Thomson effect of neon with an impurity content not exceeding 0.1%. The facility employed the flow-through calorimeter technique with closed cycle of the test substance. This ensured purity and constant composition of the gas at all the stages of the experiment. The calorimeter and the throttling device are connected in series in the direction of gas flow and are placed between two sleeves with 100-ohm platinum resistance thermometers. From these the temperature difference ΔT can be read within 0.003 K.

Particular attention in the study was given to ensuring a constant gas flowrate and for maintaining the calorimeter and the throttling device under adiabatic conditions. The temperature of the two devices was maintained equal to the temperature of the shield surrounding them. The gas flowrate was measured by a semi-automatic volumetric flowmeter with an error not exceeding 0.1–0.2%. A specially designed bellow-type pressure regulator was used for rapidly setting and maintaining the specified pressure difference across the throttling device. The flowrate of the gas through the throttling device did not exceed 5 to 7% of the total flowrate. The pressure drop across the throttling device was changed from 1.3 to 2.5 atm and was measured by a high-pressure differential mercury manometer with an error of $\Delta p = \pm 0.5$ mm Hg. The pressure in the experiments by Dedikov et al. was measured by MP-60 piston-type pressure gauges (0.02 precision class) and MP-600 (0.05 precision class), separated from the gas system by a bellows-type separator. The temperature of the constant temperature bath was maintained by means of a VRT-2 off-the-shelf automatic regulator.

The experimental data on specific heat C_p of neon are presented on six isobars. Their error, as estimated by the authors, does not exceed $\pm 1.2\%$. The values of β_i were obtained only along one isotherm (20 points) and deviate from the curve averaging it by not more than 1%. Comparison of these with computed data of Runge [2.30] showed that the deviation does not exceed 2%.

Argon. The thermodynamic properties of argon were investigated more thoroughly than those of other inert gases (Table 9). As for neon, the over-

TABLE 9. Experimental Studies of Thermodynamic Properties of Argon

Year	Author	Property	Temperature, K	Pressure, bar	Reference
1901	Ramsey and Traverse	p, v, T	284; 510	30—103	[1.58, 1.99]
1910	Onnes and Crommelin	p, v, T	123—294	11—62	[2.41]
1915	Holborn and Schultze	p, v, T	273—473	8—100	[2.42]
1922—1925	Holborn and Otto	p, v, T	90—673	10—100	[2.22, 2.23, 2.43]
1923	Masson and Dolly	p, v, T	298	20—125	[2.44]
1924—1935	Bridgman	p, v, T	123—341	3000—15000	[2.45, 2.46]
1930	Tanner and Masson	p, v, T	298—448	30—125	[2.47]
1934	Roebuck and Osterberg	α_i	123—573	1—200	[2.48]
1943	Clark and Katz	p, v, T	297	1—20	[2.49]
1949	Michels et al.	C_p/C_v	273—423	19—2600	[2.50]*
1953	Whalley et al.	p, v, T	273—873	10—80	[2.51]*
1956	Getzen	p, v, T	273—573	1—400	[2.20]
1957	Ishkin and Rogovaya	δ_T	133—299	1—50	[2.52]
1958	Michels et al.	p, v, T	133—248	4—1060	[2.53]*
1960	Lecocq	p, v, T	573—1223	21—920	[2.54]*
1961	Rogovaya and Kaganer	p, v, T	90—248	17—200	[2.55]
1962	Bagatskiy et al.	C_v	Transcritical region		[2.56]
1963	Lestz	w	273; 304	1—12	[2.57]
1963—1966	Radovskiy	p, v, T; w	84—173	0,6—60	[1.112, 2.58]
1965	Voronel' et al.	C_v	133—152	23—50	[2.59]
1966	Crain and Sonntag	p, v, T	143—273	2—515	[2.38]
1966	Van Dael et al.	w	150; 671	49—52	[2.60]
1967—1970	Polyakov and Tsiklis	p, v, T	373—673	1500—10000	[2.61]
1968	Carome et al.	w	90—140	5—110	[2.62]
1969	Crawford and Daniels	p, v, T	95—210	200—6194	[1.37]
1969	Thoen and Van Dael	w	100—150	8—509	[2.63]
1969	Robertson et al.	p, v, T	308—673	1500—10000	[2.64]

δ_T is the differential isothermal Joule-Thomson effect; w is the velocity of sound.

whelming majority of experimental studies on argon was performed for determining the pvT behavior.

The experimental data obtained at $p \leqslant 125$ atm up to the forties, with the exception of the results in [1.58, 1.99] and a part of data in [2.41] (the latter should be corrected, as recommended by Cragoe [2.41]), exhibit acceptable accuracy. For this reason, they are in mutual agreement within the overall error of independent measurements. Holborn and his coworkers [2.22, 2.23, 2.42, 2.43] described experimental data on the compressibility of argon along isotherms by a polynomial such as (2.4). Tanner and Masson [2.47], as well as Holborn and Otto, compiled equations of isotherms in the form of polynomials of pressure. All these, and earlier data, are currently only of historic value. Subsequently, very accurate results on the thermal properties of argon over a wider range of variables were obtained by Michels et al. [2.50, 2.53].

Initially, Michels and his coworkers measured the compressibility of spectrally pure argon over the temperature range from 0 to 150°C, which is standard for their laboratory, and represented the data along seven isotherms spaced 25°C apart. The equations for all the isotherms were obtained in the form of polynomials with respect to density, reflecting the input data with an error not exceeding 0.05% with respect to pressure. Subsequently, the same investigators measured the thermal properties of argon at temperatures below 273 K [2.53]. The range of pressure along isotherms narrowed down with reduction in temperature and was limited to 159 bar at $T = 133$ K. Particular attention in this study was given to the transcritical region where the temperature intervals were reduced. Michels et al. [2.53] also present the previously mentioned (Chapter 1) thermal properties along the saturation curve to 120 K and certain data on the compressibility of liquid argon. This means that the experiments described in [2.50, 2.53] cover a temperature range from 120 to 423 K, including the saturation curve. Due to their high accuracy they serve as the final stage in investigating the thermal properties of argon over the above range of parameters.

Rogovaya and Kaganer [2.55] compared experimental data on the compressibility of gaseous and liquid argon along 10 isotherms with the results of Michels et al. [2.53]. They found that up to 173 K the deviations basically do not exceed 0.1%, whereas in the transcritical region they are as high as 1% for individual points. They do not analyze the reasons for these differences, and the claim that the experimental values in [2.53] at $T = 163$ K deviate to 0.5% from those computed from the equation for this isotherm is in error.

Crain and Sonntag [2.38] investigated the *pvT* behavior of nitrogen, argon, and their mixtures in various propertions and represented the data on compressibility of gaseous argon along the four isotherms: 273.15, 203.15, 163.15 and 143.15 K. For the last isotherms their data are limited to a maximum pressure of 35.76 atm, whereas at $T = 163.15$ K detailed measurements were performed to a pressure of 152.38 atm, above which only one data point is given. They estimate the error of their measurements at $\pm 0.1\%$.

Our comparison of data in [2.38] with those in [2.53] showed that the differences basically do not exceed $\pm 0.1\%$ (Fig. 17)†. However, systematic differences, as much as 1% and more, were observed along the 163.15 K isotherm at pressures above 50 bar. As seen from the figure, the data of Crain and Sonntag [2.38] and of Rogovaya and Kaganer [2.55] deviate from the results of Michels et al. [2.53] in different directions, going past the tolerance limits which is due to difficulties in performing experiments in the transcritical region.

The most complete study of compressibility of argon at temperatures above 423 K was performed by Lecocq [2.54] along six isotherms (300, 400, 500, 650, 800 and 950°C). As did certain investigators before him, Lecocq compiled equations of isotherms which describe the experimental data with high accuracy. In addition, he also compared the virial coefficients, determined from experimental data, with those computed on the basis of various functions of intermo-

† The deviations at points $p = 308.10$ and 507.46 atm at $T = 203.15$ K are of the same order of magnitude and are not shown in the figure.

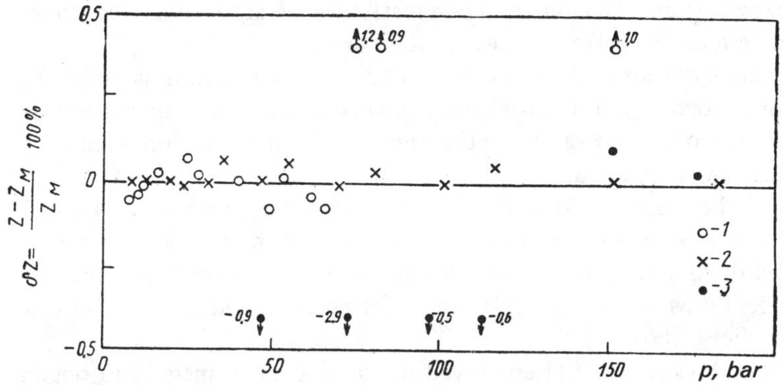

FIGURE 17. Deviations of data on the compressibility of argon obtained by Crain and Sonntag [2.38] at the 163.15 (1) and 203.15 K (2) isotherms and those of Rogovaya and Kaganer [2.55] at the 163.15 K isotherm (2) from experimental data of Michels et al. [2.53] identified as Z_m

lecular interaction. The analytic values obtained from the Lennard-Jones potential are in satisfactory agreement with the experimental. However, better agreement is obtained by replacing the $1/r^{12}$ function, representing repulsion, by an exponential function.

The data of Whalley et al. [2.51] agree, at pressures to 80 bar, with the published data [2.50, 2.54] with a deviation within $\pm 0.1\%$. The compressibility of gaseous argon at higher pressures (to 400 bar) was investigated by Getzen. These results were not published but were used by the author for compiling a virial equation of state given in Cook's handbook [2.20]. We used this equation for calculating the values of Z and for comparing them along the 0 and 150°C isotherms with data in [2.50], and along the 300°C isotherm with those in [2.54] (Table 10). The agreement is basically satisfactory with the exception of the 0°C isotherm where systematic deviations, going outside the permissible limits, start at $p > 200$ bar. Apparently, Getzen's equation is not capable of

TABLE 10. Comparison of the Values of Z Calculated from Getzen's Equation [2.20] (line 1) with Experimental Results of Michels et al. [2.50] and Lecocq [2.54]

Pressure, atm	Z at t, °C			Pressure, atm	Z at t, °C		
	0	150	300		0	150	300
50	0.9562	1.0035	1.0128	250	0.9270	1.0532	1.0744
	0.9568	1.0038	1.0113		0.9301	1.0539	1.0731
100	0.9241	1.0107	1.0266	300	0.9581	1.0739	1.0927
	0.9249	1.0112	1.0244		0.9620	1.0750	1.0919
200	0.9095	1.0358	1.0575	350	0.9912	1.0969	1.1120
	0.9113	1.0367	1.0554		1.0033	1.0970	1.1116
				400	1.0418	1.1226	1.1328
					1.0507	1.1238	1.1324

describing experimental data on the compressibility of argon over the entire 0–300°C temperature range to pressure of 400 bar.

The isochores constructed from data in [2.50, 2.53] in coordinates (Z, T) smoothly join Lecocq's [2.54] experimental points, forming a compatible thermodynamic plane over almost the entire range of parameters for which the equation of state was obtained.

A class of their own is formed by studies presenting results on the pvT behavior of argon at $p > 1500$ bar. The values of Z in the 1924 study by Bridgman [2.45] along the 341 K isotherm and the points obtained subsequently at lower temperatures contain errors which affected the accuracy of results as noted by Benedict [2.66].

Polyakov and Tsiklis [2.61] employed the displacement method to obtain data on the molar volumes of argon and presented smoothed values at the 100, 200, 300 and 400°C isotherms. According to their estimates, the error of their data up to $p < 2500$ bar is 1.5% and at higher pressures, 0.3%. Robertson et al. [2.64] performed their investigations over approximately the same range of parameters and estimated the error of their data at 0.7%. However, the differences between the results in [2.61] and [2.64] with respect to density are as high as 3% and Robertson et al. tend to attribute this to insufficient purity of the argon and the use of less refined measuring techniques by Polyakov and Tsiklis.

The data of Crawford and Daniels [1.37] at lower temperatures are represented along 13 isotherms with an error, in their opinion, of ±0.1% with respect to density. They complement the results in [2.60, 2.64] at very high pressures.

A number of investigators measured the speed of sound in argon. Radovskiy [1.112, 2.58] used data on the speed of sound in gaseous argon and on the density to construct a T-S diagram. The measured values of the speed of sound, basically in liquid argon, are listed in [2.57, 2.61–2.63]. As a rule, these data are very accurate and are used for calculating the adiabatic compressibility and the ratio of specific heats. Lately, data on the speed of sound are used for checking and correcting the thermal equations of state.

Results of experimental studies of calorific properties of argon are quite limited. Clark and Katz [2.49] measured the ratio C_p/C_v along the 24.1°C isotherm and described data by a simple analytic expression. Roebuck and Osterberg in their experiments [2.48], as in experiments with air and nitrogen, committed a systematic error in pressure measurement [2.67]. For this reason, data on α_i should be corrected by multiplying by 1.033. The differential isothermal Joule-Thomson effect of argon and of the argon-nitrogen mixture was investigated by Ishkin and Rogovaya [2.52]. Their data are approximately 5% on the high side due to error in calibrating the gas meter of the experimental facility [2.68]. The differences between the values in [2.48] and [2.52] can be reduced by correcting these data.

Bagatskiy et al. [2.56] performed a very precise study of the isochoric specific heat of argon in the vicinity of the critical point and found a steep rise in C_v

TABLE 11. Experimental Studies of the Compressibility of Krypton

Year	Author	Property	Temperature, K	Pressure, bar	Reference
1901	Ramsey and Traverse	p, v, T	284; 510	30—100	[1.58, 1.99]
1952	Beattie et al.	p, v, T	273—573	21—415	[2.69]
1954	Whalley and Schneider	p, v, T	273—873	10—80	[2.70]*
1963	Lestz	w	273; 304	1—5	[2.57]
1966	Trappeniers et al.	p, v, T	273—423	19—3072	[2.71]*
1970	Theeuwes and Bearman	p, v, T	130—240	12—294	[2.72]*
1971	Streett and Staveley	p, v, T	120—220	5—3728	[2.73]

near the saturation curve and a jump upon transition through it. When worked up in coordinates $[C_v, \lg (T - T_{cr})]$ the resulting data form two, almost parallel, straight lines. Based on this, Bagatskiy et al. hypothesized that C_v increases to infinity at the critical point†. In a subsequent study, Voronel' and his coworkers [2.59] investigated C_v more completely and showed that this singularity at the critical point is universal. The experimental values of C_v are listed in [2.59] along three isochores close to the critical: 0.530, 0.533 and 0.538 gm/cm^3.

Krypton. A list of experimental studies is given in Table 11. It was mentioned previously that the data of Ramsey and Traverse [1.58, 1.99], due to their poor accuracy, are only of historical interest.

The pvT behavior of krypton was investigated most completely by Beattie and his coworkers [2.69]. The krypton used by them contained 0.9% xenon as the main impurity. The experimental data are presented along 14 isochores over a temperature range of 273 to 573 K with a uniform step of 25 K. On the basis of these results, the authors compiled an equation of state in virial form, containing three virial coefficients. The highest pressure was, up until recently, attained by Trappeniers et al. [2.71], who investigated the compressibility of spectrally pure krypton by a variable-volume piezometer. The data points are presented along seven isotherms. Equations were compiled in the form of polynomials as a function of density describing the original data with high accuracy. The results in [2.69–2.71] are in satisfactory agreement with one another over the overlapping temperature range at pressures to 80 bar. However, at higher pressures the experimental data of Beattie et al. [2.69] start to deviate systematically from the data of Trappeniers et al. [2.71] and at individual points the differences are as high as 0.6%. These systematic deviations apparently occur because no correction was made in [2.69] for the presence of mercury vapor in the test gas. According to calculations by Jepson and Rowlinson [2.74], such a correction is greatest at 300°C and ranges from -0.6 to $\pm 0.2\%$ at densities of 1 and 10 mole/liter, respectively. It must also be remembered that the presence 0.9% molar percent of xenon in the krypton could be responsible for an addi-

† Up to present not a single equation of state which is valid over a wide range of parameters transmits this singularity at the critical point.

tional error in the results of [2.69]. This error can be easily determined by assuming that the property of the mixture can be calculated from the Amagat rule. Thus, at $p = 50$ bar and $t = 50$°C the error due to the presence of xenon in the test specimen is 0.13%.

Up until recently, the *pvT* behavior of krypton in the single-phase region at $T < 273$ K was not investigated by anyone. Recently Theeuwes and Bearman [2.72] and Streett and Staveley [2.73] measured the compressibility of liquid and gaseous krypton at pressures to 290 and 3679 atm, respectively. These studies will be analyzed in detail in Chapter 3 since the majority of points pertain to the liquid phase.

With results listed in [2.71, 2.72] and data along the saturation curve, it is possible to draw smooth isochores in coordinates (Z, T) which show that the starting data are compatible. However, if it is remembered that the isochores in the region of $\omega \simeq 1-2$ and $\tau \simeq 1.1-1.6$ change their sign, then the values of Z obtained by interpolation over the temperature range of 240 to 273 K by far do not exhibit the same accuracy as the experimental data.

Xenon. Of the seven studies pertaining to gaseous xenon (Table 12), the results obtained by Michels et al. [2.77] cover a wide range of parameters and are most reliable. Data on the compressibility of xenon are presented on ten isotherms. Two of these (0 and 13.5°C) display values pertaining to the liquid region and are described by separate polynomials. In the vicinity of the critical point, the results in the study by Chynaweth and Schneider [2.76] of the speed of sound are given as a graph. Voronel' [2.79] investigated the isochoric specific heat, also in the vicinity of the critical point, from the single- and two-phase side regions.

Habgood and Schneider [1.124] performed a special study of the behavior of xenon in the transcritical region and analyzed the influence of the hydrostatic effect due to gravity on the accuracy of determining the compressibility coefficient. The authors represent the experimental data along certain isotherms and isobars, arranged densely in the immediate vicinity of the critical point. The results in [1.124, 2.77] are in mutual agreement within 0.1–0.2%.

TABLE 12. Experimental Studies of the Compressibility of Xenon

Year	Author	Property	Temperature, K	Pressure, bar	Reference
1901	Ramsey and Travers	p, v, T	284. 510	30—100	[1.58, 1.99]
1951	Beattie et al.	p, v, T	290—573	21—406	[2.75]
1952	Chynaweth and Schneider	w	Transcritical region		[2.76]
1954	Michels et al.	p, v, T	273—423	17—2828	[2.77]*
1954	Habgood and Schneider	p, v, T	286—291	54—60	[1.124]*
1955	Walley et al.	p, v, T	273—973	4—50	[2.78]*
1961	Voronel'	C_v	Transcritical region		[2.79]

Whalley et al. [2.78] investigated the thermal properties of xenon by the Burnett method over a wider temperature range at pressures to 50 bar. Instead of results of direct measurements, they list equations of isotherms in the form of a three-membered quadratic function of pressure. The computed values of Z were compared by us with data in [2.77]. At temperatures between 0 and 100°C the deviations do not exceed 0.1%, and at $t = 150$°C and $p = 1$ bar the results in [2.78] are lower by 0.2%.

Beattie et al. [2.75] presented their experimental data along 13 isotherms at rounded off values of density. Comparison with results of Michels and his co-workers [2.77] showed that at all temperatures the deviations increase with the pressure and become as high as 0.8% at $t \leqslant 50$°C. The possible reasons for such differences have been mentioned before, but it should be mentioned that at temperatures close to the critical the presence of 0.14% of krypton in the xenon sample may have a significant effect. At temperatures of 200 and 300°C the data in [2.75] are up to 0.25% higher than those in [2.78].

The above brief analysis of studies of thermodynamic properties of the four inert gases leads to the following conclusions. Of these substances, only argon has been investigated experimentally practically over the entire range of parameters for which tables of thermodynamic properties need be calculated. In spite of this, Altunin and Gvozdkova [2.80] express doubts about the reliability of data of Lecocq [2.54]. For the three other substances there are simply no reliable experimental pvT data at temperatures and pressures above 423 K and 100 bar since the results of Beattie et al. [2.69, 2.75] exhibit a significant error. For this reason, we performed an additional experimental study of the compressibility of inert gases at elevated temperature and pressure (see Sec. 2.2).

2.2. EXPERIMENTAL STUDY OF COMPRESSIBILITY AT ELEVATED TEMPERATURE AND PRESSURE

In view of the high accuracy of experimental data obtained in the Amsterdam laboratory for the four substances, it appeared best to use the relative method with room temperature as the lower limit. The maximum temperature was 800 K, which allowed the use of standard 10-ohm resistance thermometer. The maximum experimental pressure was governed by the design of the non-relieved piezometer and did not exceed 600 bar. The test facility had to satisfy two primary requirements: (1) that it allow acquisition of highly accurate data, and (2) that the experimental technique be as simple as possible. With reference to the above, we analyzed the equipment used by Lecocq [2.54], Vukalovich and Altunin [2.81], and Tsymarnyy with Zagoruchenko [2.82]. We gave preference to the last since it permitted the simplest solution of the problem [2.83].

Figure 18 presents a schematic diagram of the test arrangement. Its main element is a stainless steel 1Kh18N9T piezometer 10 without a dead space

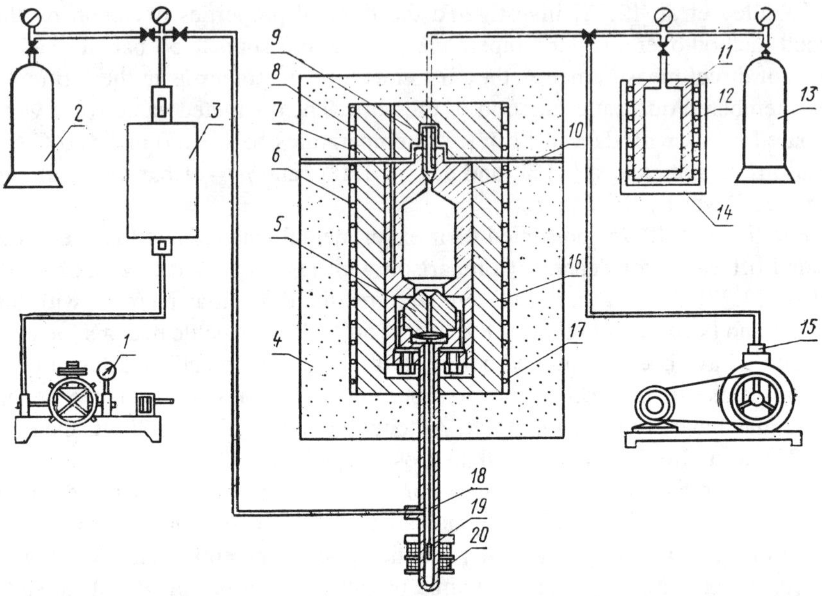

FIGURE 18. Schematic diagram of test arrangement. 1) Dead-weight pressure gauge; 2) tank; 3) oil-type booster; 4) thin-walled asbestos insulation; 5) differential manometer; 6) main heater in porcelain insulation; 7) cover; 8) top guard heater; 9) needle valve; 10) piezometer; 11) thermocompressor; 12) electric heater; 13) tank; 14) vacuum-jacketed Dewar; 15) pump; 16) copper slab; 17) bottom guard heater; 81) connecting rod; 19) core; 20) sensor.

volume. Its dimensions were: outer diameter 100 mm, inner diameter 45 mm, and volume 110 cm^3. Before the experiments were started, with cover (7) removed, thick-walled stainless steel (1Kh18N9T) thermocompressor (11) and the connecting communications were evacuated by vacuum roughing pump (15) to a pressure of 10^{-2} mm Hg. Then the test gas from the tank was supplied to the liquid nitrogen-cooled thermocompressor to obtain the required initial pressure. Then, a specific quantity of gas, which remained constant during the experiment, was fed from the thermocompressor through shutoff valve (9) to the piezometer. A 0.1 mm thick stainless steel membrane was argon-arc welded to the housing of differential manometer (5), connected to the piezometer and to a limiter which had spherical surfaces with large radius. These surfaces permitted it to move freely to both sides of the equilibrium position within the elastic limits and protect it from failure at significant one-sided overloads. In order to avoid the latter, the pressure in the piezometer and in the balancing system was raised slowly and the travel of the membrane (by way of connecting rod (18) and iron core (19)) was monitored by reading off induction transducer (20). The differential manometer and the transducer (two inductive resistance connected into an ordinary bridge system) allow reliable measurement of pressure differences starting with 0.05 atm. The pressure in the filling system was roughly determined by a reference spring-type pressure gauge and precisely by the dead-weight pressure gauge (1). The membrane of the differential manometer separates the piezometer cavity from the oil-type booster (3), the lower part of which is connected directly to pressure gauge (1), and the upper to compressed gas (nitrogen) tank (2) and to the differential manometer. The oil level in the booster was changed by using the manual plunger pump of pressure gauge (1). This action regulated the pressure of the gas beneath the membrane. The geometric dimensions of the booster allow raising the initial pressure tenfold.

After filling with gas, the piezometer cavity was cut off by valve (9) and the tee, packing gland and the communications were disconnected. The cover of the constant-temperature bath was lowered and the initial temperature and pressure were measured precisely. Then the data obtained in the Amsterdam laboratory were used to determine the density.

The piezometer and the gas contained in it were heated by supplying power separately to each section of the heater through autotransformers connected to an electronic voltage regulator. When approaching the specified temperature, the temperature field over the height of the constant temperature bath was equalized by adjusting the voltage fed to the main and to the guard heaters. Uniformity of temperature was monitored from readings of differential chromel-alumel thermocouples. The attainment of steady conditions was additionally monitored on the basis of steadiness of the zero position of the membrane and readings of a PTS-10 class 1 platinum resistance thermometer (R_0 = 9.9954 abs. ohm).

The sensing element of the thermometer averaged the temperature in the working part of the piezometer. The thermometer was connected to an ordinary potentiometer circuit and had, in series with it, a reference quality resistance coil, placed in an oil-type constant-temperature bath. The voltage drops at the thermometer and coil terminals were measured by a class 0.002 self-correcting, low-resistance potentiometer P-308, and the value of the emf was determined as the arithmetic mean of several measurements with direct and opposite polarity.

The actual pressure in the piezometer cavity consisted of: (1) gauge pressure p_g, measured by a 0.05 precision class reference quality dead-weight pressure gauge MP-600; (2) atmospheric pressure p_{at}, measured by a reference quality barometer with an error of less than 0.25 mm Hg; and (3) hydrostatic pressure p_h, due to the difference in the oil levels in the booster and in the dead-weight pressure gauge.

The value of p_g was corrected for deformation of the pressure gauge connecting rod, deviation of temperature and acceleration of gravity from values assumed in calibrating the pressure gauge. Pressure p_h was reduced to a minimum (0.03 to 0.2 atm) by proper positioning of the booster and the dead-weight pressure gauge. A transparent Plexiglas ruler was placed in the oil tank of the dead-weight gauge to permit the exact determination of the oil level which uniquely represents the oil level in the booster. This allows the determination of the difference in levels to within ± 5 mm representing a pressure of $\sim 0.5 \cdot 10^{-3}$ atm.

A family of constant-mass curves—"quasi-isochores"—was obtained by measuring the pressure corresponding to different levels of piezometer filling at selected temperatures. The measurements at each such quasi-isochore were performed several times, permitting judgment of the reproducibility of results.

The piezometer volume changes due to the test pressures and temperatures requiring a correction of the specific volume for each temperature-pressure set. The first correction ΔV_t is due to isobaric expansion of the piezometer as its temperature is raised.

$$\Delta V_t = V_t - V_0 = 3\alpha V_0(t - 20) \tag{2.5}$$

where V_0 is the initial volume of the piezometer at room temperature; V_t is the volume of the piezometer at test temperature t and α is the linear expansion coefficient of 1Kh18N9T [stainless] steel, taken from [2.84].

The second correction ΔV_p is due to isothermal change of the piezometer volume due to noncompensated internal pressure and is calculated from a formula obtained on the basis of the theory of thick-walled cylinders:

$$\Delta V_p = \pi l[(r + \Delta r)^2 - r^2] \tag{2.6}$$

where l is the length of the cylinder; r is its inner radius; Δr is the change in r due to pressure p,

$$\Delta r = \frac{pr}{E(R^2 - r^2)}[(1 - 2\mu)r^2 + (1 + \mu)R^2] \tag{2.7}$$

R is the outer radius of the cylinder; μ is Poisson's ratio and E is the elastic modulus. This correction is calculated with $\mu = 0.3$ and E as a function of temperature from data in [2.85].

Accordingly, the volume of the piezometer at the test temperature t and pressure p is given by the sum

$$V_{p,t} = V_0 + \Delta V_t + \Delta V_p \tag{2.8}$$

The neon, argon, krypton, and xenon used in the experiments were high purity and had an impurity content % (according to the supplier's certificate) of less than:

	Ne	Ar	Kr	Xe
helium	0.1	—	—	—
hydrogen	0.001	—	0.006	—
oxygen	0.001	0.001	0.001	0.001
nitrogen	0.01	0.005	0.023	0.032
krypton	—	—	—	0.046
xenon	—	—	0.67	—

The specimens also contained moisture: Ne and Ar less than 0.02 g/m^3; Kr, less than 0.01 g/m^3, and Xe, less than 0.009 g/m^3. As is seen, krypton had the lowest purity. For this reason, the pvT data for it were corrected on the basis of additivity of properties of binary mixtures.

Tables 13 and 14 list experimental data on the compressibility of neon and argon. The initial specific volumes at filling v_0 were determined from data in

TABLE 13. Experimental Data on Compressibility of Neon (t, °C; p, atm; ρ, g/cm^3)

t	p	ρ	t	p	ρ
18.46	107.9	0.0866	167.32	255.1	0.1301
73.61	128.3	0.0863	203.93	276.2	0.1298
102.70	139.0	0.0862	292.67	326.1	0.1292
156.14	158.5	0.0859	396.26	383.4	0.1285
208.18	177.4	0.0857	444.48	409.8	0.1282
291.64	207.2	0.0853	10.08	163.8	0.1312
390.60	242.1	0.0848	55.26	190.2	0.1309
452.72	263.7	0.0846	104.36	218.8	0.1304
80.36	130.8	0.0863	165.10	253.7	0.1301
168.01	162.6	0.0859	208.05	278.5	0.1298
254.01	193.7	0.0855	298.31	329.1	0.1292
344.20	225.7	0.0851	394.01	382.2	0.1285
426.60	254.6	0.0847	438.62	406.4	0.1282
37.94	127.2	0.0948	16.98	187.9	0.1456
75.19	142.2	0.0946	55.09	213.0	0.1453
147.01	171.5	0.0943	100.79	242.9	0.1450
228.05	204.2	0.0938	155.22	278.1	0.1445
295.45	230.8	0.0935	204.62	309.8	0.1441
370.89	260.3	0.0931	303.07	372.0	0.1434
425.17	281.1	0.0928	392.88	427.6	0.1426
13.99	117.2	0.0949	450.95	463.2	0.1422
58.62	135.6	0.0947	9.65	183.0	0.1456
104.55	154.3	0.0945	54.44	212.6	0.1453
160.07	176.9	0.0942	109.32	248.4	0.1449
204.60	194.7	0.0940	154.23	277.5	0.1446
300.07	232.7	0.0935	206.13	310.8	0.1441
396.03	270.0	0.0930	296.52	367.9	0.1434
453.62	292.2	0.0927	398.34	431.0	0.1426
			451.88	463.8	0.1422
40.91	157.2	0.1145			
54.06	163.8	0.1145	55.03	229.2	0.1554
108.49	190.8	0.1141	73.73	242.5	0.1552
155.71	214.4	0.1138	133.45	284.5	0.1548
205.87	239.2	0.1136	180.80	317.1	0.1544
297.53	283.7	0.1130	221.45	345.2	0.1540
390.35	328.0	0.1124	339.54	424.8	0.1530
445.11	353.8	0.1121	414.68	474.3	0.1524
15.74	144.7	0.1147			
89.46	181.3	0.1142	10.15	200.2	0.1579
151.64	212.3	0.1139	20.54	207.9	0.1578
221.61	246.8	0.1135	64.03	239.1	0.1575
293.10	281.5	0.1130	105.98	269.1	0.1571
369.81	318.2	0.1125	157.45	305.5	0.1567
418.64	341.4	0.1122	153.99	302.8	0.1567
			201.79	336.3	0.1563
11.96	164.8	0.1312	290.84	397.5	0.1556
55.12	190.2	0.1309	383.84	460.7	0.1547
116.61	226.0	0.1305	440.75	499.2	0.1543

TABLE 14. Experimental Data on Compressibility of Argon (t, °C; p, atm; ρ, g/cm^3)

t	p	ρ	t	p	ρ
28.85	103.1	0.1735	29.36	165.8	0.2799
51.12	112.5	0.1733	48.72	180.6	0.2797
106.96	135.8	0.1728	99.52	218.6	0.2789
154.77	155.5	0.1723	147.15	253.8	0.2782
254.76	196.2	0.1714	201.47	293.6	0.2774
322.99	223.2	0.1708	294.27	360.0	0.2760
371.31	242.2	0.1703	399.02	433.2	0.2744
15.37	99.0	0.1769	22.77	170.8	0.2973
53.29	115.4	0.1765	56.71	198.6	0.2967
103.80	137.1	0.1760	112.98	244.2	0.2958
161.07	161.2	0.1755	157.87	279.6	0.2951
193.40	174.8	0.1752	207.71	318.6	0.2943
293.35	215.9	0.1743	296.44	386.9	0.2929
398.71	258.2	0.1733	395.39	460.5	0.2912
			491.32	529.3	0.2898
23.11	125.1	0.2170			
55.67	142.5	0.2166	18.53	170.8	0.3032
110.45	172.4	0.2160	62.49	207.6	0.3026
163.74	201.2	0.2154	106.05	243.6	0.3018
206.57	223.8	0.2149	152.57	281.7	0.3011
297.99	272.0	0.2138	220.21	336.0	0.2999
398.44	322.6	0.2126	245.69	356.3	0.2996
500.50	372.5	0.2115	340.88	430.5	0.2980
			447.27	510.9	0.2959
20.08	146.5	0.2584			
66.70	178.1	0.2578	13.74	186.3	0.3383
108.90	206.6	0.2572	47.64	219.4	0.3377
145.79	231.3	0.2567	104.14	273.6	0.3367
208.15	272.3	0.2558	150.69	317.8	0.3358
249.41	299.2	0.2552	202.23	365.6	0.3349
339.53	356.6	0.2539	297.54	452.2	0.3331
449.37	424.6	0.2525	365.79	514.7	0.3317
			400.42	546.3	0.3312

[2.27] for neon and [2.50] for argon. For illustration of reproducibility of the results of measurements, Table 13 lists two series of experiments along five quasi-isochores.

In the first study [2.83], in order to represent these data at round values of temperature, the results of direct measurements were worked up graphically. To attain high accuracy the quasi-isochores were constructed in coordinates of p/T vs T, and the values of p/T at round values of temperature were determined from smooth curves. In substance, once this was done, the results can be represented along isotherms. However, we wished to check the agreement of data along isothermal sections and to obtain "true" [rather than "quasi"] isochores. To do this, we constructed the curves in coordinates (pv, ρ) which are convenient not only because it is possible to select a large scale but also because then there exists a limiting point ($pv \to RT$ at $\rho \to 0$). In addition, isotherms in coordinates (pv, ρ) are smooth curves, the curvatures of which are relatively small over the range of variables under study. This permits reliable determination of derivatives $(\partial p/\partial v)_T$ by the graphic method. The derivatives are needed for reducing the experimental data on pressure to values along isochores, using the equation

$$p = p_{\text{exptl}} + \left(\frac{\partial p}{\partial v}\right)_T \Delta v \tag{2.9}$$

where p_{exptl} is the absolute pressure measured in the experiments, and Δv is the change in the specific volume of the initial value along the given quasi-isochore relative to the actual at the temperature and pressure in the experimental point under study.

The data points for neon and argon were worked up and the results were used for compiling a network of reference values of pressure along isochores at round values of temperature (Table 15). To estimate the accuracy let us analyze the possible measuring error and compare the results with reliable data of others.

In accordance with the theory of errors, the error of determining the pressure is calculated as the sum of errors of determining the gauge, atmospheric and hydrostatic pressures.

$$\overline{\delta p} = \delta p + \frac{100}{p}\left[\left(\frac{\partial p}{\partial v}\right)_T \Delta v + \left(\frac{\partial p}{\partial T}\right)_v \Delta T\right] \tag{2.10}$$

It should be added to this that some inaccuracy in determining the pressure can be due to the impossibility of holding the separating membrane in the zero position. The readoff from the null indicator scale was obtained with an accuracy of 0.5 division which corresponds to an error of 0.015 atm in determining

TABLE 15. Analyzed Experimental Data on the Pressure p, atm

Neon					
	p at v, cm³/g				
t, °C	11.553	10.536	8.718	7.624	6.867
25	110.6	121.9	149.3	172.5	193.3
50	120.0	132.2	162.0	187.4	210.0
75	129.4	142.7	174.8	202.3	226.6
100	138.9	153.1	187.5	217.2	243.2
125	148.3	163.5	200.3	232.0	259.8
150	157.7	174.0	213.2	246.7	276.8
200	176.5	194.7	238.6	276.3	310.0
250	195.3	215.4	264.1	305.8	343.2
300	214.0	236.1	289.5	335.3	376.4
350	232.8	256.8	314.8	364.6	409.6
400	251.5	277.5	340.2	394.1	442.4
450	270.2	298.2	365.5	423.5	475.4

TABLE 15. Analyzed Experimental Data on the Pressure *p*, atm (*Continued*)

Argon					
t, °C	p at v, cm³/g				
	5.657	4.607	3.869	3.569	3.297
25	103.2	126.2	149.7	162.3	176.1
50	114.0	139.9	167.3	181.6	197.5
75	125.0	154.1	184.5	200.8	218.7
100	136.0	167.8	201.6	219.8	239.8
125	146.7	181.6	218.6	238.6	260.9
150	157.7	195.7	235.9	257.6	281.6
200	179.4	222.8	269.4	295.1	323.2
250	200.7	250.0	303.3	332.5	364.8
300	221.7	277.1	336.8	369.4	405.9
350	242.9	304.2	370.0	406.8	446.2
400	263.7	330.6	403.5	442.8	486.8

the pressure. If it is remembered that pressures between 50 and 600 atm were measured by a precision class 0.05 dead-weight pressure gauge, the absolute error in calculating the gauge pressure ranged respectively from 0.025 to 0.3 atm. This means that the overall absolute error at minimum and maximum pressures is $\Delta p_{min} = 0.040$ atm and $\Delta p_{max} = 0.315$ atm.

The corresponding relative errors are $\delta p_{min} = 0.82\%$ and $\delta p_{max} = 0.053\%$.

As previously noted, the compressibility of gases was investigated by the relative method. For this reason, the principal contribution to the error in determining the specific volume is made by errors of the results of the Amsterdam laboratory which, in our opinion, do not exceed 0.05%. In addition, an inaccuracy may be committed when correcting for the increase in the piezometer volume with rising temperature and pressure. Given the low accuracy of functions $\alpha = \alpha(t)$ and $E = E(t)$, and also the approximate nature of Eq. (2.7), the overall error in determining corrections ΔV_t and ΔV_p can be estimated at 0.03%. This means that the relative error of determining the specific volume is 0.08%.

Careful analysis showed that the error in determining the temperature ranges from 0.02 to 0.05 K.

Derivatives $\left(\frac{\partial p}{\partial v}\right)_T$ and $\left(\frac{\partial p}{\partial T}\right)_v$ contained in Eq. (2.10) were found graphically at individual control points along isotherms and isochores. Then Eq. (2.10) was used for calculating errors which, on the average, comprised 0.12% and nowhere exceeded 0.21%.

An additional check on the accuracy of our data involved comparing them with experimental results of other authors (Figs. 19 and 20). The relative deviations from data of the Amsterdam laboratory lie in the majority of cases within ±0.1% and do not exceed the value of ±0.2% anywhere. This is in agreement with the conclusion drawn by us upon analysis of errors of measurement and workup. Note that our values for argon lie systematically 0.08% higher on the

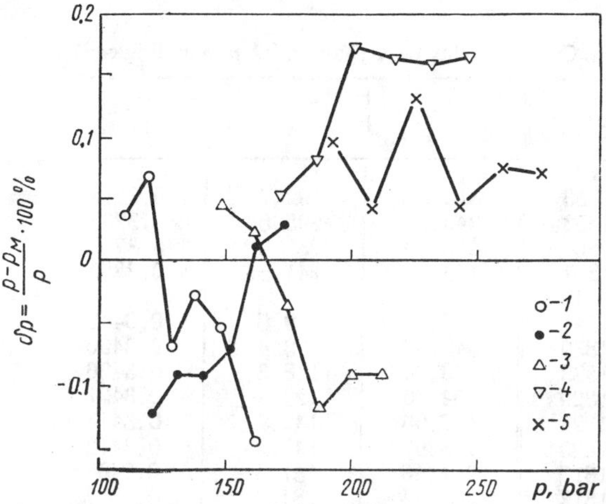

FIGURE 19. Comparison of error in pressure data obtained by the authors for neon relative to those of Michels et al. [2.27] at different specific volumes. 1) v = 11.553 cm³/g; 2) 10.536; 3) 8.718; 4) 7.624; 5) 6.867.

average than the results of Michels [2.50] and Lecocq [2.54]. However, matching results with an allowance for possible experimental error supports the results of Lecocq [2.54] and indicates that the doubts expressed by Altunin and Gvozdkov [2.80] are not too justified.

Experimental data on the compressibility of krypton and xenon are listed in

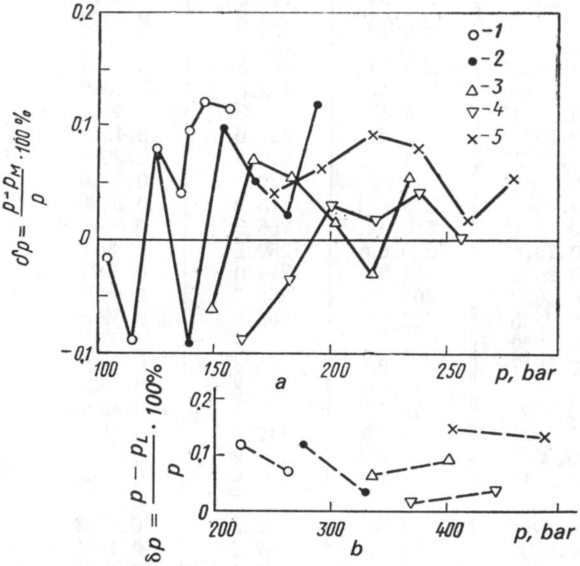

FIGURE 20. Comparison of error in pressure data obtained by the authors for argon relative to those of Michels et al. [2.50] (a) and Lecocq [2.54] (b) at different specific volumes. 1) v = 5.657 cm³/g; 2) 4.607; 3) 3.869; 4) 3.569; 5) 3.297.

TABLE 16. Experimental Data on the Compressibility of Krypton (t, °C; p, atm; ρ, g/cm³)

t	p	ρ	t	p	ρ
17.32	47.1	0.1853	312.01	192.7	0.3274
18.90	47.4	0.1853	346.70	205.5	0.3268
46.13	52.9	0.1850	416.28	231.5	0.3255
80.10	59.8	0.1847	443.66	241.5	0.3250
100.97	64.0	0.1845			
139.57	71.7	0.1841	15.52	79.5	0.3442
159.69	75.7	0.1839	48.37	93.4	0.3436
217.93	87.3	0.1833	108.08	118.3	0.3426
262.14	95.7	0.1829	134.50	129.1	0.3421
308.26	104.6	0.1824	167.95	142.7	0.3415
347.07	111.9	0.1820	183.97	149.2	0.3412
382.26	118.5	0.1817	204.48	157.4	0.3408
439.50	129.4	0.1812	265.74	181.7	0.3397
			312.62	200.0	0.3390
17.79	58.7	0.2374	346.99	213.3	0.3382
57.01	69.3	0.2369	388.40	229.3	0.3376
92.97	79.0	0.2365	397.71	232.9	0.3373
106.37	82.5	0.2362	446.62	251.3	0.3366
163.15	97.5	0.2356			
196.82	106.3	0.2351	15.71	90.8	0.4052
202.79	107.8	0.2351	16.66	91.3	0.4052
253.33	120.9	0.2344	57.59	112.3	0.4044
300.17	132.8	0.2338	79.73	123.5	0.4039
345.18	144.2	0.2332	103.47	135.5	0.4034
385.37	154.2	0.2327	118.83	143.1	0.4031
434.29	166.4	0.2321	129.10	148.3	0.4027
			151.48	159.6	0.4022
16.00	69.4	0.2912	176.39	172.1	0.4018
21.09	71.1	0.2911	200.80	184.2	0.4013
60.02	84.6	0.2905	217.40	192.3	0.4008
82.45	92.3	0.2903	224.35	195.7	0.4008
130.35	108.5	0.2895	248.73	207.6	0.4002
145.01	113.4	0.2893	274.43	219.8	0.3997
173.44	122.9	0.2888	300.30	232.3	0.3990
182.22	125.8	0.2887	323.77	243.9	0.3986
217.41	137.3	0.2881	350.02	256.3	0.3979
253.26	149.1	0.2876	356.69	259.5	0.3978
305.37	166.0	0.2868	387.91	274.0	0.3971
355.76	182.1	0.2859	406.78	282.7	0.3967
386.77	191.7	0.2855	455.65	305.7	0.3956
419.11	201.5	0.2850			
435.67	206.6	0.2847	17.00	96.3	0.4307
			19.96	97.8	0.4307
16.18	77.4	0.3327	22.58	99.4	0.4305
52.43	92.3	0.3321	58.08	119.1	0.4297
75.89	101.6	0.3316	76.65	129.2	0.4294
103.96	112.7	0.3312	91.05	136.9	0.4290
162.62	135.9	0.3301	101.85	142.7	0.4288
185.65	144.9	0.3297	124.95	155.3	0.4283
214.37	155.9	0.3292	129.30	157.6	0.4281
254.28	171.2	0.3285	132.14	159.2	0.4281

TABLE 16. Experimental Data on the Compressibility of Krypton (*t*, °C; *p*, atm; ρ, g/cm³) (*Continued*)

t	*p*	ρ	*t*	*p*	ρ
173.44	181.4	0.4272	407.78	324.0	0.4476
176.92	183.2	0.4270	454.54	349.0	0.4466
203.41	197.3	0.4264			
226.06	209.5	0.4259	17.95	103.1	0.4666
247.47	220.5	0.4253	62.15	130.3	0.4653
275.50	235.4	0.4248	74.35	137.6	0.4653
322.31	259.4	0.4235	106.16	156.8	0.4645
344.29	270.6	0.4232	164.06	191.5	0.4630
379.14	288.3	0.4223	176.91	198.8	0.4627
402.89	300.1	0.4218	206.23	216.1	0.4619
427.77	312.9	0.4210	253.50	243.5	0.4608
450.64	324.1	0.4205	318.95	280.9	0.4591
			346.86	297.1	0.4583
17.30	101.2	0.4575	395.59	324.3	0.4572
18.22	101.8	0.4575	465.16	362.8	0.4554
41.61	115.8	0.4568			
52.96	122.6	0.4566	23.66	118.2	0.5285
106.78	154.1	0.4554	59.88	144.4	0.5275
109.45	155.8	0.4552	71.59	152.7	0.5272
110.94	156.8	0.4552	121.81	187.8	0.5258
138.23	172.7	0.4545	133.48	196.1	0.5254
153.37	181.3	0.4541	165.59	218.4	0.5247
179.86	196.5	0.4535	205.97	246.7	0.5236
206.59	212.0	0.4527	206.43	247.1	0.5236
250.71	237.0	0.4517	256.64	281.1	0.5219
271.28	248.8	0.4513	303.51	312.7	0.5208
298.13	263.8	0.4506	307.33	315.2	0.5205
300.10	265.2	0.4506	357.75	348.8	0.5192
324.68	278.6	0.4498	412.72	384.4	0.5172
352.07	293.5	0.4492	453.44	410.8	0.5162

Tables 16 and 17. To reduce the experimental results to integral values of temperature and specific volume, we used in the more recent study [2.83] a special program for finding the coefficients of the virial equation of state by computer [2.86]. This allows the processing of data distributed arbitrarily over the thermodynamic plane. Then we calculated deviations at each data point. The mean for 153 points for krypton and 172 for xenon was respectively 0.1 and 0.15%. Only at three points for krypton and two for xenon did the error reach 0.21 and 0.25% respectively. This points to the absence of random errors which might go past the expected limits.

Table 18 lists analyzed experimental data on pressure for integral values of temperature and specific volumes. These were then compared with the results of others in Figs. 21 and 22. The mean deviation from data of the Amsterdam laboratory for both substances is 0.1% and the maximum does not exceed 0.17%. Larger differences exist with results of Beattie et al. [2.69, 2.75]. They are as high as 0.6% for krypton and 0.8% for xenon.

At low pressures ($p < 80$ bar) our isotherms, when extrapolated to the

TABLE 17. Experimental Data on the Compressibility of Xenon (t, °C; p, atm; ρ, g/cm³)

t	p	ρ
52.46	91.6	0.9149
69.67	107.6	0.9141
82.20	119.3	0.9135
86.35	123.2	0.9133
133.90	166.7	0.9109
140.35	172.5	0.9107
153.30	184.3	0.9100
179.91	208.3	0.9087
205.15	230.6	0.9075
240.01	261.5	0.9056
255.86	275.3	0.9050
262.54	281.1	0.9046
284.63	300.5	0.9035
305.97	318.6	0.9025
315.20	326.8	0.9018
359.85	364.7	0.8996
406.21	403.6	0.8970
455.30	444.0	0.8946
54.62	92.6	0.8960
70.65	107.3	0.8952
108.30	141.0	0.8933
122.88	154.0	0.8926
158.70	185.5	0.8909
167.86	193.4	0.8906
194.16	216.1	0.8893
201.39	222.4	0.8890
207.53	228.0	0.8885
264.45	276.7	0.8858
310.62	315.5	0.8834
357.40	354.2	0.8811
376.81	370.2	0.8802
396.32	386.2	0.8793
429.23	412.9	0.8774
57.45	91.2	0.8255
73.78	104.4	0.8248
100.11	125.8	0.8237
145.72	161.5	0.8216
177.97	186.7	0.8202
197.17	201.4	0.8193
218.03	217.7	0.8183
234.28	230.2	0.8176
255.75	246.2	0.8167
256.15	246.8	0.8167
295.01	275.5	0.8149
350.81	317.3	0.8123
366.75	328.6	0.8117
402.66	354.7	0.8100
462.39	398.2	0.8071

t	p	ρ
54.37	77.7	0.6303
84.17	94.5	0.6293
95.80	101.0	0.6290
121.25	115.0	0.6281
124.12	116.8	0.6280
150.32	130.9	0.6271
175.69	144.6	0.6263
237.10	177.3	0.6241
271.86	195.6	0.6230
312.45	216.3	0.6215
330.63	225.6	0.6209
342.18	231.8	0.6205
359.00	240.3	0.6200
417.50	269.7	0.6178
467.34	294.9	0.6161
35.99	64.3	0.5759
72.97	83.4	0.5748
99.09	96.2	0.5740
125.54	109.3	0.5732
146.97	119.7	0.5726
174.44	133.0	0.5718
194.76	142.7	0.5711
261.39	173.7	0.5690
281.30	183.0	0.5685
302.49	193.0	0.5676
314.62	198.7	0.5673
351.03	215.1	0.5662
381.75	229.0	0.5653
397.48	236.0	0.5649
450.84	259.8	0.5630
57.76	74.4	0.5592
77.28	83.8	0.5586
95.10	92.4	0.5581
122.42	105.3	0.5573
154.00	120.2	0.5563
171.15	128.2	0.5558
176.40	130.7	0.5557
197.50	140.4	0.5551
233.96	157.0	0.5539
236.32	158.2	0.5538
260.94	169.1	0.5531
282.65	178.7	0.5525
313.18	192.5	0.5515
327.63	198.9	0.5510
330.77	200.2	0.5509
355.89	211.0	0.5502
380.17	221.7	0.5495
405.10	232.3	0.5486

TABLE 17. Experimental Data on the Compressibility of Xenon (t, °C; p, atm; ρ, g/cm³) (*Continued*)

t	p	ρ	t	p	ρ
426.56	241.6	0.5479	389.36	165.1	0.4055
452.20	252.7	0.5471	417.92	173.7	0.4052
			444.60	181.6	0.4045
40.35	64.7	0.5447			
66.50	77.3	0.5439	52.97	53.9	0.3560
81.36	84.2	0.5435	68.95	58.4	0.3557
129.36	106.4	0.5421	104.85	68.3	0.3551
147.82	114.8	0.5416	122.97	73.1	0.3547
170.86	125.0	0.5409	158.34	82.6	0.3540
190.73	133.9	0.5404	180.49	88.4	0.3536
208.71	142.0	0.5398	209.39	96.1	0.3531
258.56	163.8	0.5384	254.00	107.6	0.3522
274.45	170.8	0.5379	303.65	120.4	0.3512
310.97	186.4	0.5368	357.29	133.9	0.3503
319.17	189.8	0.5365	391.71	142.5	0.3495
354.96	204.7	0.5355	421.11	150.0	0.3490
404.70	225.9	0.5341	443.49	155.7	0.3487
451.25	245.3	0.5325			
			56.81	50.9	0.3195
61.96	72.4	0.5149	100.15	61.4	0.3188
100.81	89.5	0.5139	140.54	71.0	0.3181
120.09	97.9	0.5133	158.84	75.3	0.3178
177.13	121.8	0.5118	193.47	83.4	9.3171
179.17	122.6	0.5118	204.15	85.9	0.3169
243.68	149.2	0.5099	261.27	98.9	0.3160
277.64	162.8	0.5089	316.92	111.5	0.3150
286.06	166.3	0.5086	356.04	120.2	0.3144
362.94	196.7	0.5066	388.20	127.1	0.3139
392.01	208.1	0.5058	457.84	142.6	0.3126
422.23	219.7	0.5048			
440.62	227.0	0.5045	47.26	43.9	0.2799
			97.77	54.5	0.2791
51.99	59.4	0.4132	156.82	66.5	0.2782
86.37	70.8	0.4125	183.40	71.8	0.2778
96.97	74.3	0.4122	209.30	77.0	0.2775
170.82	98.1	0.4107	264.77	87.8	0.2766
175.16	99.5	0.4105	307.17	96.0	0.2760
201.19	107.8	0.4100	335.46	101.5	0.2756
257.77	125.4	0.4087	388.66	111.6	0.2748
306.05	140.2	0.4077	428.86	119.2	0.2742
348.72	153.0	0.4067	446.69	122.6	0.2738

TABLE 18. Analyzed Experimental Data on the Pressure p, atm

t, °C	Krypton											
	p at v, cm^3/g											
	1.95	2.00	2.10	2.20	2.30	2.40	2.50	3.00	3.40	4.20	5.00	5.40
25	116.3	113.9	109.4	105.3	101.5	98.0	94.7	81.3	73.1	60.8	52.1	48.6
50	133.8	130.8	125.3	120.3	115.7	111.4	107.5	91.5	81.9	67.7	57.7	53.8
75	151.1	147.6	141.0	135.1	129.7	124.8	120.2	101.8	90.7	74.6	63.3	58.9
100	168.3	164.2	156.7	149.9	143.7	138.1	132.8	111.9	99.4	81.4	68.9	64.0
125	185.5	180.9	172.3	164.7	157.7	151.3	145.5	122.1	108.2	88.2	74.5	60.2
150	202.7	197.5	188.0	179.4	171.7	164.6	158.1	132.1	116.9	95.0	80.1	74.2
200	237.1	230.8	219.2	208.9	199.5	190.9	183.1	152.2	134.2	108.5	91.2	84.4
250	271.2	263.8	250.3	238.2	227.2	217.2	208.1	172.1	151.3	121.9	102.2	94.5
300	305.2	296.7	281.2	267.3	254.7	243.3	232.9	191.9	168.3	135.2	113.1	104.6
350	338.8	329.3	311.9	296.2	282.1	269.2	257.5	211.6	185.2	148.5	124.0	114.6
400	372.2	361.7	342.3	324.9	309.2	294.9	281.9	231.1	202.1	161.7	134.9	124.6
450	405.2	393.7	372.5	353.4	336.2	320.5	306.2	250.5	218.8	174.8	145.7	134.5

TABLE 18. Analyzed Experimental Data on the Pressure p, atm (*Continued*)

t, °C	Xenon												
	p at v, cm^3/g												
	1.10	1.20	1.30	1.40	1.50	1.60	1.70	1.80	1.90	2.00	2.50	3.00	3.50
50	89.1	85.7	82.6	79.8	77.2	74.7	72.4	70.1	68.0	66.0	57.5	50.7	45.2
75	112.2	106.3	101.2	96.7	92.6	88.9	85.6	82.4	79.5	76.8	65.6	57.2	50.6
100	135.3	126.7	119.5	113.3	107.8	102.9	98.5	94.5	90.8	87.4	73.6	63.6	56.0
125	158.2	146.9	137.6	129.7	122.9	116.7	111.3	106.4	101.9	97.8	81.5	70.0	61.3
150	181.1	167.1	155.7	146.1	137.8	130.5	124.0	118.2	112.9	108.2	89.4	76.3	66.6
200	226.5	207.2	191.5	178.5	167.3	157.7	149.2	141.6	134.8	128.7	105.0	88.9	77.1
250	271.5	247.0	227.1	210.6	196.6	184.6	174.0	164.7	156.4	148.9	120.5	101.3	87.5
300	316.4	286.5	262.4	242.6	225.8	211.3	198.8	187.7	177.8	169.0	135.8	113.7	97.9
350	360.9	325.8	297.6	274.3	254.7	237.9	223.3	210.5	199.1	189.0	151.0	125.9	108.1
400	405.0	364.8	332.5	305.9	283.5	264.3	247.7	233.2	220.3	208.8	166.1	138.1	118.3
450	447.9	403.7	367.3	337.3	312.1	290.6	272.0	255.7	241.3	228.6	181.1	150.2	128.5

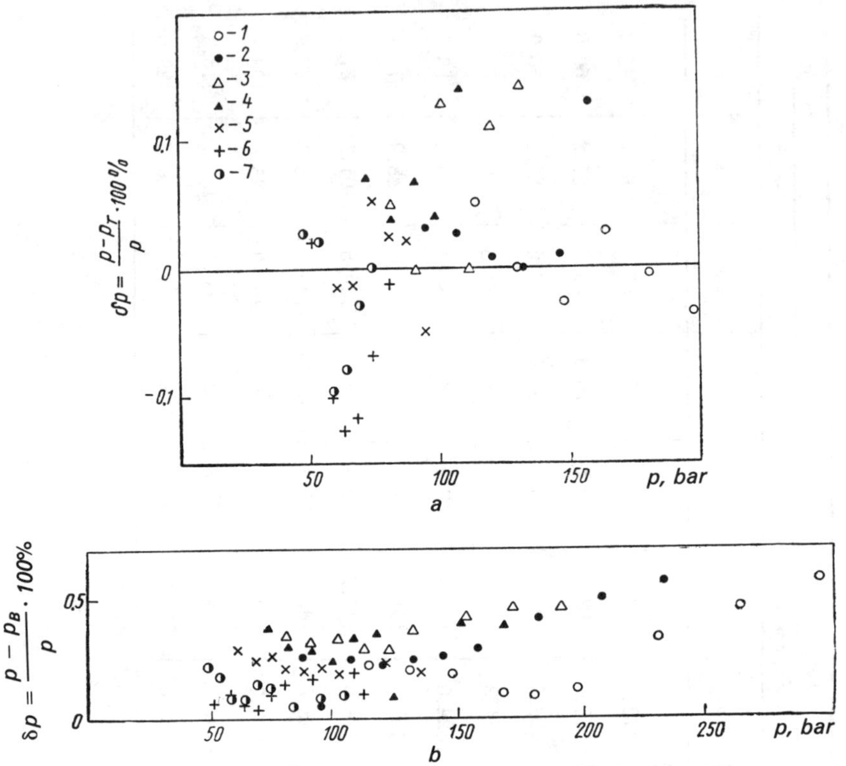

FIGURE 21. Comparison of error in pressure data obtained by the present authors for krypton relative to results of Trappeniers et al. [2.71] (a) and Beattie et al. [2.69] (b) at different specific volumes. 1) v = 2.0 cm^3/g; 2) 2.5; 3) 3.0; 4) 3.4; 5) 4.2; 6) 5.0; 7) 5.4.

point of $pv = RT$, are in satisfactory agreement with data obtained in Canada on the compressibility of the four inert gases [2.26, 2.51, 2.70, 2.78].

2.3. CONCERNING CERTAIN "IDEAL" CURVES OF THE THERMODYNAMIC PLANE OF REAL GASES

The thermodynamic plane of real gases contains a number of curves which exhibit interesting properties and can be useful in obtaining equations of state and in generalizing properties of little-explored substances. For example, "ideal"† curves, which are loci of points for which ideal-gas relationships are valid. In coordinates (Z, ρ, T) these include the ideal-gas curve for which $Z = 1$, the Boyle curve, $(\partial Z/\partial \rho)_T = 0$, the inversion curve, $(\partial Z/\partial T)_p = 0$, and the Joule curve, $(\partial Z/\partial T)_\rho = 0$. The last three curves have in common the fact that they

† The term "ideal" curve is due to Shaber [2.87].

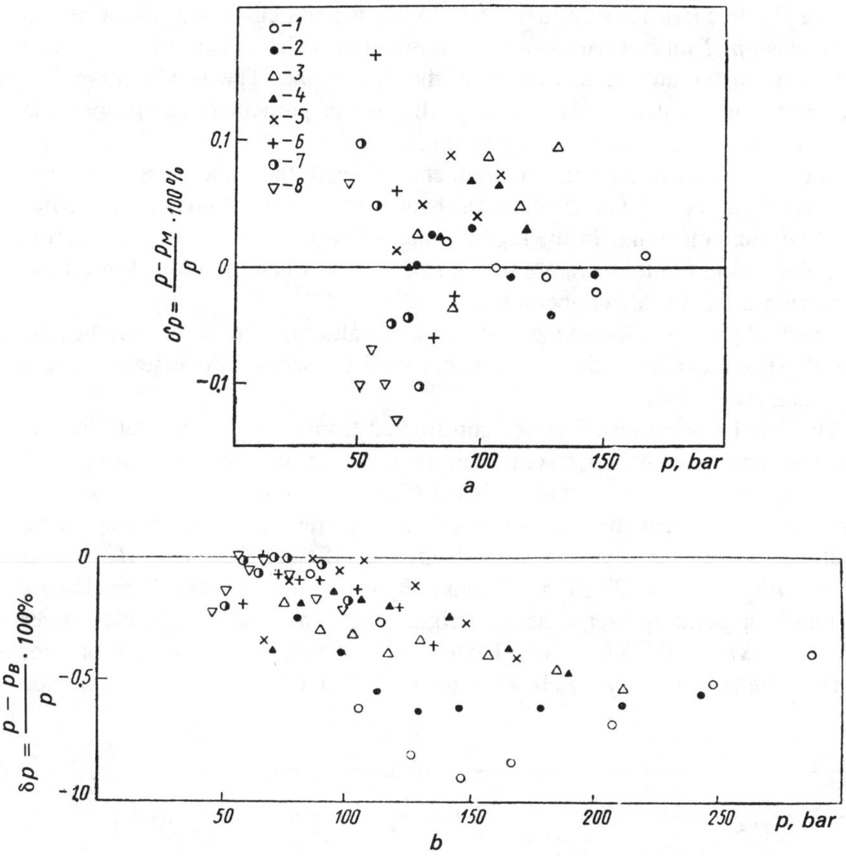

FIGURE 22. Comparison of error in pressure data obtained by the present authors for xenon relative to the results of Michels et al. [2.77] (a) and Beattie et al. [2.75] at different specific volumes. 1) v = 1.2 cm^3/g; 2) 1.4; 3) 1.6; 4) 1.8; 5) 2.0; 6) 2.5; 7) 3.0; 8) 3.5.

are the loci of points of extremal values of Z at isotherms, isobars and isochores, respectively.

To be able to find ideal curves, one must have experimental thermal data over a wide range of temperatures and densities. However, experimental data for the majority of substances are presented for limited ranges of parameters. That makes it impossible to single out, for example, the Joule curve.

By analogy with thermal ideal curves it is possible to single out in the corresponding coordinates the calorific curves of free energy F, enthalpy H and internal energy U, for which $F = F_{id}$, $H = H_{id}$ and $U = U_{id}$. Unfortunately, experimental data on the calorific properties of the majority of substances are simple not available. For that reason, the ideal calorific curves can be determined and analyzed only on the basis of reliable thermal equations of state.

Let us consider the specifics of behavior of ideal curves for the case of argon, with reference to the fact that the situation is similar for other substances.

The Boyle curve in coordinates (ρ, T) has a small curvature. However, in coordinates (p, T) it is a parabola the maximum of which is located in the same point as the maximum of this curve in the Z, p plane. The Boyle curve was investigated in detail by Nedostup [2.88], who in particular, determined the coordinates of curves for argon and xenon.

The inversion curve or, as it is frequently termed, the Joule-Thomson curve has, in (ρ, T) and (p, T) coordinates, a shape similar to the Boyle curve and has a clear physical meaning. In the region lying below the curve, throttling occurs with reduction in the temperature of the substances, whereas above it throttling occurs with a rise in this temperature.

The Joule curve is located at high reduced values of parameters and has the same physical meaning as the inversion curve, but describes processes occurring at constant internal energy.

The relative location of curves constructed from experimental data for hydrogen is shown in Fig. 23, taken from the thesis of Shaber [2.87]. As $\rho \to 0$ they intersect the ordinate axis at Boyle (T_B), inversion (T_{inv}) and Joule (T_J) temperatures, respectively. Each of these temperatures has corresponding to it a singularity on the curve of the second virial coefficient B_2. Thus, $B_2 = 0$ at $T = T_B$, $dB_2/dT = B_2/T$ at $T = T_{inv}$ and $dB_2/dT = 0$ at $T = T_J$. We also note that these temperatures serve as the origin of the three ideal calorific curves ($\Delta F = 0$, $\Delta H = 0$, $\Delta U = 0$). Table 19 lists characteristic conditions for points on the curves analyzed above as $\rho \to 0$, obtained from the virial equation

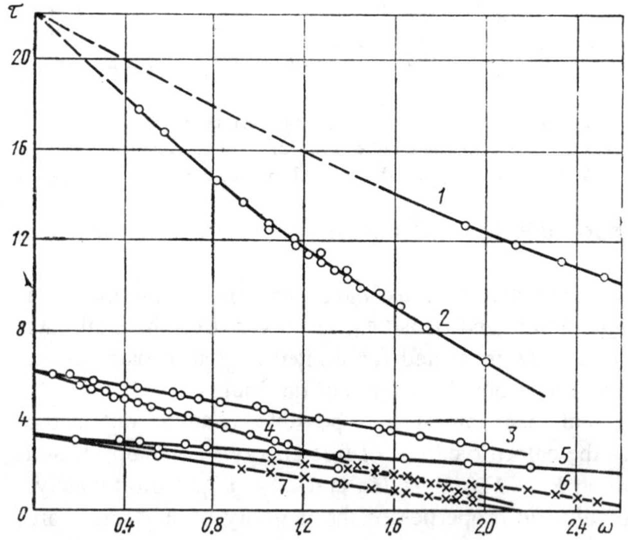

FIGURE 23. "Ideal" curves for hydrogen in dimensionless coordinates (τ, ω). 1) $\Delta U = 0$; 2) $\left(\frac{\partial Z}{\partial T}\right)_v = 0$; 3) $\Delta H = 0$; 4) $\left(\frac{\partial Z}{\partial T}\right)_p = 0$; 5) $\Delta F = 0$; 6) $\Delta Z = 0$; 7) $\left(\frac{\partial Z}{\partial \rho}\right)_T = 0$.

TABLE 19. Limiting Conditions and Slope of Certain "Ideal" Curves in Coordinates (T, ρ) at $\rho = 0$

Curve	Attribute of curve	Temperature at $\rho = 0$	Corresponding point on the curve of B_2	Slope
Boyle	$\left(\frac{dZ}{d\rho}\right)_T = 0$	T_B	$B_2 = 0$	$-\frac{2B_3}{B'_2}$
Ideal gas	$Z = 1$	T_B	$B_2 = 0$	$-\frac{B_3}{B'_2}$
Free energy	$\Delta F = 0$	T_B	$B_2 = 0$	$-\frac{B_3}{2B'_2}$
Inversion	$\left(\frac{dZ}{dT}\right)_p = 0$	T_{inv}	$\frac{dB_2}{dT} = \frac{B_2}{T}$	$-\frac{B'_3 - 2B_3/T}{B''_2}$
Enthalpy	$\Delta H = 0$	T_{inv}	$\frac{dB_2}{dT} = \frac{B_2}{T}$	$-\frac{B'_3 - 2B_3/T}{2B''_2}$
Joule	$\left(\frac{dZ}{dT}\right)_\rho = 0$	T_J	$\frac{dB_2}{dT} = 0$	$-\frac{B'_3}{B''_2}$
Internal energy	$\Delta U = 0$	T_J	$\frac{dB_2}{dT} = 0$	$-\frac{B'_3}{2B''_2}$

of state. Analogous information is presented for the ideal-gas curve which shall be analyzed in more detail, since it will be rather extensively used below.

Morsy [2.89], Nedostup [2.90] and Holleran [2.91] investigated ideal-gas curves for a large group of substances and showed that in coordinates (ρ, T), for substances with a simple molecular structure, the temperature dependence of density at $Z = 1$ becomes linear within the limits of error of experimental thermal data. This was checked out and verified by data on thermal properties of inert gases, air and its principal constituents, light hydrocarbons and other substances which have a relatively simple molecular structure. The linear temperature dependence of ρ, found from experimental data, does not follow directly from the exact equation of state and as of now cannot be explained rigorously from purely theoretical considerations. However, it, does not hold for substances for which orientation and induction effects in molecular interactions (H_2O, NH_3, etc.) or quantum corrections (He, H_2, Ne) are significant. As will be shown, in the case of neon, there exists a significant linear segment up to 70 K, below which the curvature becomes perceptible.

The linearity of the ideal-gas curve in coordinates (ρ, T) causes the curve on the (p, T) plane to be a quadratic equilateral parabola. If

$$\rho = \rho_0 - \frac{\rho_0}{T_B} T \tag{2.11}$$

where ρ_0 is the conditional value of density at the ideal-gas curve at $T = 0$, then

$$p = \rho_0 RT - \frac{\rho_0}{T_B} RT^2 \tag{2.12}$$

from whence we obtain the coordinates of the maximum of the curve ($T_{max} = 0.5\, T_B$, $\rho_{max} = 0.5\, \rho_0$). The point of maximum is interesting because the fact that in it, as shown by Koeppe [2.92], the inversion curve intersects the ideal-gas curve. A similar result was obtained by Nedostup [2.90] in analyzing the derivatives of the values of thermal quantities. The coordinates of this point may be used for forming dimensionless numbers in order to check the law of corresponding states, since the point itself satisfies the requirements formulated by Kazavchinskiy [2.93]. The necessary dimensionless numbers can be obtained in the point of maximum of the ideal-gas curve, where $(dp/dT)_{i.g.c} = 0$ and $Z = 1$ since it belongs to the inversion curve $(dZ/dT)_p = 0$.

It follows from the expressions for T_{max} and ρ_{max} that the reference point is located at the midpoint of the straight-line segment limited by the coordinate axes. This makes it possible to use T_B and ρ_0 as the parameters for reduction. They have a quite definite physical meaning and hence must be analyzed in detail.

The understandable interest in the Boyle temperature has resulted in the appearance of several methods for determining it on the basis of different experimental data. The value of T_B is obtained simplest on the basis of the second virial coefficient. However, the fact that this coefficient has a relatively low accuracy near T_B and that the angle at which the curve of $B_2(T)$ intersects the abscissa is small do not allow finding T_B with sufficient accuracy. A higher accuracy is attained when a reliable function of intermolecular interaction is available which permits transmitting data on the second virial coefficient over a wide temperature range at the experimental accuracy. However, this cannot be attained in practice since the familiar model functions of intermolecular interaction are suitable over narrow temperature ranges. Significantly different values of T_B are obtained, depending on the segment of the curve of $B_2(T)$ at which the potential parameters are determined.

Two-parametric potentials describing rather rigorously the dispersion energy of molecular interaction obey the relationship $\epsilon/k \sim T_B$ which, in particular, for the Lennard Jones (6–12) potential, has the form $\epsilon/k = 0.292\, T_B$. In its turn, the values of ϵ/k are determined with a large tolerance due: (1) to the approximate nature of the Lennard-Jones potential; (2) error of experimental data on the second virial coefficient; and (3) the temperature range in which these data are given. For example, the values of ϵ/k for xenon, found by different methods from the experimental values of Michels et al. [2.77], Beattie et al. [2.75], and Whalley et al. [2.78], range between 216.1 and 235.8 K, which corresponds to Boyle temperatures from 740 to 805 K.

Two more methods are used for determining T_B, based on the use of data on properties of a well-investigated experimental reference substance, with a reliable value of T_B. Both methods employ results at the Boyle curve. In the first of these, one calculates the differences $\Delta T = T_b - T_i$ of the investigated (T_i) and basis (T_b) substance at the same Z, and these differences are then extrapolated using the relationship $\Delta T = f(T_b)$ which is assumed to be linear in the paper by Gyorg and Robert [2.94]. On this assumption satisfactory results are obtained for only a small group of substances which approximately obey the law of corresponding states.

As shown by Kazavchinskiy et al. [2.95], under real conditions, function $T_i = f(T_b)$ is non-linear with small curvature passing through the coordinate origin. It is hence more justified to use an expression in the form (using constants C, D):

$$T_i = CT_b + DT_b^2 \tag{2.13}$$

It was shown by working up a large body of experimental data for a wide circle of substances that the ratio $T_i/T_b = C + DT_b$ can be extrapolated up to T_B for the basis substance. The values of T_B (investigated substance) found in this manner by Nedostup and his coworkers [2.88, 2.95] for many gases have an error of $\pm 2\%$. This error cannot be reduced, since it is difficult to single out Boyle curves on the basis of experimental thermal data.

The most precise method for determining T_B, based on using the experimentally proven fact that the real-gas curve is rectilinear in coordinates (ρ, T), has been developed by Morsy [2.89], Nedostup [2.90], Holleran [2.91], and Kozlov and Spiridonov [2.96]. The latter two authors found the value of T_B for air by singling out the ideal-gas curve from experimental data and by extrapolating it by computer to $\rho = 0$. The error of the results are then commensurable with the error of the original experimental values confirming that this method of determining T_B should be preferred to the preceding.

An equally interesting quantity is the value of ρ_0. Density on the ideal-gas curve is continued to the intersection with the ordinate axis. Naturally, the extrapolation of this curve to $T = 0$ is based on the assumption that it is linear at low temperatures.

As shown already by London [2.97], the main contribution to the interaction between molecules of simple substances (monatomic and diatomic gases, simple hydrocarbons) is made by the dispersion component U_{disp} which exceeds the remaining components by several orders of magnitude. An increase in the dipole moment increases the contribution of orientation forces and, to a lesser extent, also of induction forces. In particular, the orientation and induction components of ammonia are equal. In contrast, the orientation energy of interaction for water exceeds fivefold the dispersion energy. This is apparently due to absence of straight-line segments at the ideal-gas curves in coordinates (ρ, T) for ammonia and water. In addition, the curve for water and steam is steeper.

The orientation energy for the four substances under study is less than

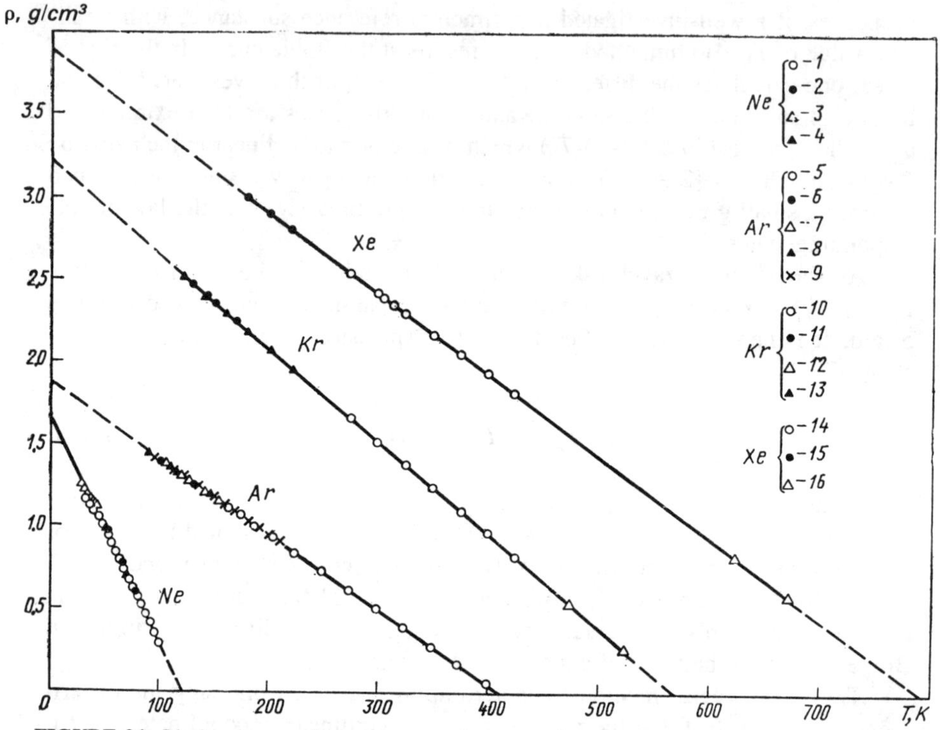

FIGURE 24. Ideal gas curves for neon, argon, krypton and xenon constructed from experimental data. 1) [2.30]; 2) 2.31, 2.32]; 3) [2.98]; 4) [1.94]; 5) [2.50, 2.53]; 6) [2.99]; 7) [1.36]; 8) [1.108]; 9) [1.38]; 10) [2.71]; 11) [2.100]; 12) [2.83]; 13) [2.73]; 14) [2.77]; 15) [2.100]; 16) [2.83].

0.005% of the total potential energy of interaction. For this reason, the ideal gas curves for neon, argon, krypton and xenon are rectilinear over the entire temperature range† where quantum corrections are insignificant (Fig. 24). A linear extrapolation (dashed lines) was drawn outside the limits of data used for constructing the figure. The deviation of experimental data from the straight line for neon starts with $T < 1.6\ T_{cr}$, whereas for argon it starts at $T < T_{cr}$. These systematic deviations do not exceed 0.3% and have a clear meaning. The deviations for heavy gases, krypton and xenon, are smaller which is due to the quantum effects on the dispersion potential energy of interaction.

The natural boundary of the existence of the ideal-gas curve at low temperatures is the point on the crystallization curve which has corresponding to it the liquid-crystal phase transition of the first kind, accompanied by a density jump $\Delta\rho$. At the same due, due to the fact that $Z = 1$, there should occur a pressure

† It should be noted that, according to the Van der Waals equation, the ideal curve in coordinates (ρ, T) is rigorously rectilinear $\left(T = \dfrac{a - ab\rho}{bR}\right)$, and the density is $\rho_0 = b^{-1}$, i.e., equal to the reciprocal of the intrinsic volume of the molecules.

jump proportional to $\Delta\rho$. With argon, for example, the coordinates of the point where $Z = 1$ on the melting curve correspond to the values of 91 K and 310 bar. The ideal-gas curve, shown by a solid line in Fig. 25, was constructed from experimental data on the density of crystalline argon at atmospheric pressure [2.101]. However, raising the pressure to 300 bar decreases the specific volume of the crystal by about 1% [1.29] which is commensurable with the error of experimental data on the density of the crystal. Transition 1–2 corresponds to density jump $\Delta\rho$.

The ideal-gas curve extrapolated from the temperature range of $T \geqslant 91$ K is shown by a dashed line in Fig. 25. The value of ρ_0 cut off on the ordinate axis by this curve is higher than the experimental value of the density of the crystal at $T = 0$. A similar situation prevails also for other inert gases.

Figure 26 shows a curve of $Z = 1$ for argon in coordinates (p, T). One of these was constructed from experimental points [2.101] (solid curve), and the other was calculated from Eq. (2.12) (dashed curve). The jump between points 1–2 is due to a jump in density in the liquid-crystal phase transition. As seen from the figure, both curves intersect the point $p = 0, T = 0$ which is basically entered also by the sublimation curve and other curves of $Z =$ const. For this reason the compressibility factor of the "free" crystal can take any value, including zero. This means that the terminal points of the experimental and analytic curves of $Z = 1$ correspond to the state of the free crystal. Nevertheless, as seen from Fig. 25, the values of density at $T = 0$ differ significantly from one another. This difference can apparently be explained by means of an equation for a crystal, written on the basis of the Clausius virial theorem [2.102],

$$3pv = 2(E_v + E_{z.v}) + \overline{\Sigma r \cdot f(r)} \tag{2.14}$$

where E_v is the kinetic energy of the vibrations of the lattice; $E_{z.v}$ is the energy of zero vibrations; r is the distance between the atoms; here $v \sim r^3$; $f(r)$ is the force of interatomic interaction, equal to $-dU(r)/dr$, where $U(r)$ is the paired interaction potential. We shall assume that

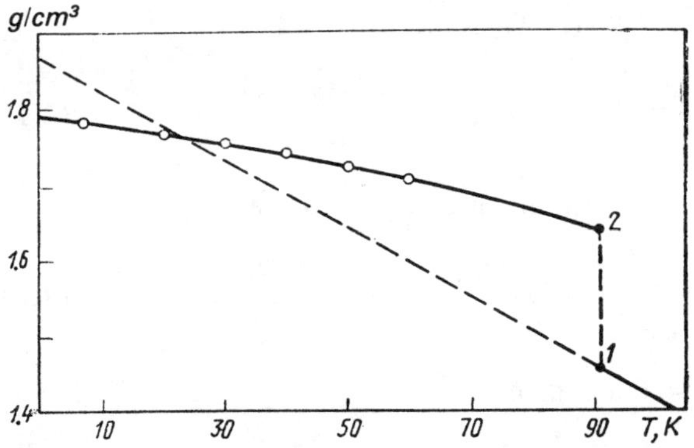

FIGURE 25. Temperature dependence of the density of solid argon at $Z = 1$.

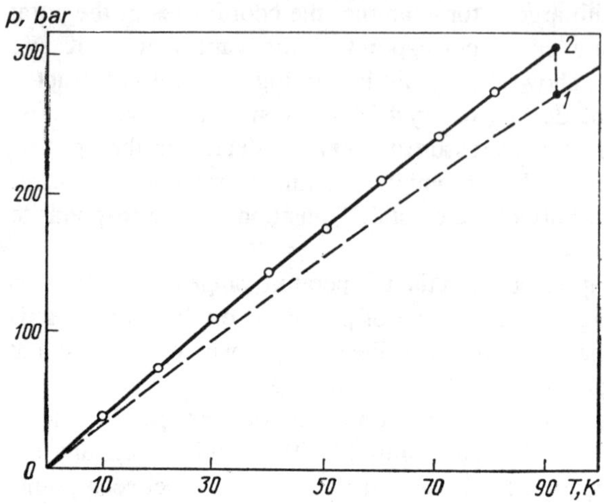

FIGURE 26. Curves of $Z = 1$ for solid argon in coordinates (p, T).

$$U(r) = \epsilon\varphi(r/\sigma) \tag{2.15}$$

which is sufficiently rigorous for the dispersion energy of interaction. On the basis of this, for the static energy of the lattice of one gram-atom of a crystal

$$\Phi(v) = A\Phi_1\left(\frac{B}{v^{1/3}}\right) \tag{2.16}$$

where A and B are constants whose values depend on the nature of interatomic forces and lattice structure. In this case, function $\overline{\Sigma r \cdot f(r)}$ for one gram-atom can be written as

$$\overline{\Sigma r \cdot f(r)} = -v\frac{d\Phi(v)}{dv} \tag{2.17}$$

then the equation of state for the crystal

$$3pv = 2(E_v + E_{z.v}) - v\frac{d\Phi(v)}{dv} \tag{2.18}$$

At $T = 0$ and $E_v = f(T) = 0$, also in our case $p = 0$, i.e.,

$$2E_{z.v} - v\frac{d\Phi(v)}{dv} = 0 \tag{2.19}$$

For the classical case $E_{z.v} = 0$ and then

$$\frac{d\Phi(v)}{dv} = 0 \tag{2.20}$$

TABLE 20. Comparison of Experimental and Analytic Values of Equilibrium Interactomic Distances r_0, Å, in Crystals at $T = 0$

Substance	r_0^{exptl}	$r_0 = \dfrac{1.3284}{\sqrt[3]{\rho_0^{\text{exptl}}/M}}$	r_0^{calc}	r_{01}^{calc}	$r_0 = \dfrac{1.3284}{\sqrt[3]{\rho_0/M}}$
1	2	3	4	5	6
Neon	3.20 [2.106]	3.15	—	—	3.05
Argon	3.83 [2.106] 3.76 [2.107]	3.74	3.64	3.69	3.68
Krypton	4.04 [2.106] 3.99 [2.107]	3.98	3.95	3.98	3.94
Xenon	4.46 [2.106] 4.34 [2.107]	4.33	4.31	4.33	4.29

It can be concluded on the basis of Eqs. (2.18)–(2.20) that the volume of a real crystal is larger than the volume of the crystal considered in the classical approximation. In the latter case, the volume of the crystal is such that the closest interatomic distance corresponds to minimum potential interaction energy. The deviations observed by us are most likely due to the fact that extrapolation of the ideal-gas curve from the high-temperature region to point $T = 0$ does not allow for the effect of quantum phenomena on the volume of a free crystal.

Since the above considerations are only qualitative, it appears best to present certain numerical estimates obtained by means of simple potentials (Lennard-Jones, Kihara) in the Debye approximation, when the energy of zero vibrations is expressed as

$$E_{z.v} = \frac{9}{8} \cdot \frac{R}{\mu} \vartheta$$

Detailed calculations which permit the estimation of the effect of this energy on the density of argon, krypton and xenon at $T = 0$ were performed by Rossi and Danon [2.103]. They assumed additivity of the potential energy†, and they included the energy of zero vibrations in the quasi-harmonic approximation which introduces a negligible error in the calculation of equilibrium states [2.105]. The values of these distances, computed by Rossi and Danon, are listed in Table 20 with the energy of zero vibrations (column 5) and without it (column 4). This

† We note that they also attempted to make allowance for the nonadditivity of the potential energy of interactions. However, their calculations were performed within the framework of the Axilrod-Teller theory [2.104], valid for attraction energy at large distances.

table also lists values of r_0 in crystals of neon, argon, krypton and xenon at $T \to 0$, obtained by X-ray diffraction measurements (column 2), from experimental data on the density of the crystal measured by the piezometer method [2.101] (column 3), and also those determined by us on the basis of values of density on the extrapolation curves of an idea gas at $T = 0$ (column 6).

It is seen from Table 20 that the values of r_0 calculated by Rossi and Danon, and found by us, are in reasonable agreement with experimental results of Gavrilko [2.101], Pollack [2.106] and Shakhparonov [2.107]. Comparison of values of r_0^{calc} and r_{01}^{calc} shows that the energy of zero vibrations increases the static volume ($\sim r^3$) of the argon crystal by 4%, of krypton by 2–4% and of xenon by 2%. These numbers are quite close to the differences between the values of ρ_0 determined by piezometer measurements and obtained by us from the ideal-gas curve.

The above means that the parameters T_B and ρ_0 of the ideal gas curves are related to the characteristic quantities of the classical potential function of interactomic interaction. These are the depth of the potential well and the coordinates of the maximum of the potential energy curve. From this point of view, the use of T_B and ρ_0 appears valid also from physical considerations.

The parameters of the ideal-gas curves, found by us from experimental data on the compressibility of the four inert gases are

	T_B, K	ρ_0, g/cm^3
neon	122.10	1.673
argon	407.76	1.870
krypton	567.50	3.210
xenon	791.00	3.890

When written in the dimensionless coordinate system $\omega = \rho/\rho_0$ and $\Theta = T/T_B$, Eq. (2.11) becomes

$$\omega = 1 - \Theta \tag{2.21}$$

This means that the dimensionless thermodynamic surfaces which are described by an equation in the form $Z = f(\omega, \Theta)$ will coincide (within the experimental error), not at a single point, but along a curve passing over a wide domain on the plane of state. This circumstance leads to the assumption that correlation of properties of similar substances by means of the new reference parameters T_B and ρ_0 will ensure a higher accuracy and over a wider range of temperature and pressure than the use of other reference points, such as the critical or a point located on the Boyle curve.

It can be shown that isochores of two gases at identical values of ω in coordinates (Z, Θ) intersect the ideal-gas curve not only at common points but at the same angles, i.e.,

$$\left(\frac{\partial Z_1}{\partial \Theta}\right)_{\omega=1-\Theta} = \left(\frac{\partial Z_2}{\partial \Theta}\right)_{\omega=1-\Theta} \tag{2.22}$$

For isochores with sufficiently low density, when the equation of state can be limited to the second virial coefficient, Eq. (2.22) becomes

$$\left(\frac{dB^*_{2.1}}{d\Theta}\right)_{\Theta\to 1} = \left(\frac{dB^*_{2.2}}{d\Theta}\right)_{\Theta\to 1} \tag{2.23}$$

where $B^*_2 = B_2\rho_0$. It was found by verifying this equation against experimental values of the second virial coefficient for the four gases under study that the curves of $B^*_2(\Theta)$ agree within the possible error of the original data (see Sec. 2.4).

It was established by Stewart [1.29] that the virial coefficients of the equation of state

$$Z = 1 + B^*_2\omega + B^*_3\omega^2 + B^*_4\omega^3 + \ldots \tag{2.24}$$

are interrelated in a certain fashion due to the linearity of the ideal-gas curve, for which this equation becomes

$$B^*_2 + B^*_3\omega + B^*_4\omega^2 + \ldots = 0 \tag{2.25}$$

We shall differentiate the above, upon substitution of Eq. (2.21):

$$\begin{aligned} B^{*\prime}_2 + B^{*\prime}_3(1 - \Theta) - B^*_3 + B^{*\prime}_4(1 - \Theta)^2 \\ -2B^*_4(1 - \Theta) + \ldots = 0 \end{aligned} \tag{2.26}$$

At $\Theta = 1$

$$B^*_3 = B^{*\prime}_2 \tag{2.27}$$

Differentiating Eq. (2.25), we find analytically that at $\Theta = 1$

$$B^*_4 = B^{*\prime}_3 - \frac{1}{2}B^{*\prime\prime}_2 = \frac{1}{2}B^{*\prime\prime}_2 \tag{2.28}$$

The general expression, relating the virial coefficients at $i \geqslant 2$, is written as

$$B^*_{i+1} = \frac{1}{(i-1)!}B^{*(i-1)}_2 \tag{2.29}$$

This equation is rigorously valid at $T = T_B$. However, detailed calculations performed by Holleran [2.108] showed that at $2.5 \leqslant kt/\epsilon \leqslant 4.5$, the virial coefficients B_2 through B_5, calculated by means of the Lennard-Jones potential [2.109, 2.110], confirm Eq. (2.29) with a high accuracy. It is seen from this that satisfaction of Eq. (2.23) brings about satisfaction of Eq. (2.22), confirmed also by Nedostup [2.111] on the basis of workup of *pvT* data for a number of well-investigated substances.

The results of Holleran and Sinka, who obtained a generalized equation for the slope of isochores when the latter intersect the ideal-gas curve can serve as additional consideration in favor of the validity of Eq. (2.22). The generalized relationship was constructed from experimental data for Ar, Kr, CH_4, CF_4 and a number of binary methane-tetrafluoromethane mixtures. The accuracy with which this relationship is satisfied additionally shows that Eq. (2.22) is valid for the gases under study within the limits of experimental error.

The above shows that the use of parameters T_B and ρ_0 produces satisfactory prerequisites for correlating the properties of a wide circle of substances for which the ideal-gas curve is rectilinear in coordinates (ρ, T). It shall be shown below how this curve can be used for correlating thermodynamic properties of the four inert gases.

2.4. SECOND VIRIAL COEFFICIENTS

Upon certain assumptions, the thermodynamic functions can be calculated from the equation of state

$$Z = 1 + B_2(T)\rho \tag{2.30}$$

containing only the second virial coefficient, which is related to the interparticle potential interaction function by the simplest equation

$$B_2(T) = -2\pi N_A \sigma^3 \int_0^\infty \left(e^{-\frac{U(r)}{kT}} - 1\right) r^2 dr \tag{2.31}$$

At higher temperatures, the range of values of pressure for which Eq. (2.30) is valid broadens since the contribution of terms with higher-order virial coefficients at a given p_{max} drops off rapidly. This circumstance is important in cases when it is necessary to perform an extrapolation with respect to the equation of state at high temperatures for which experimental data are usually lacking. Such an extrapolation is frequently justified on the basis of theoretical values of virial coefficients, the determination of which is based on various interparticle interaction functions.

If the experimental and analytic curves of $B_2(T)$ can be made to be identical to one another over a wide temperature range, then the selected potential is regarded as sufficiently realistic and the extrapolation of the second virial coefficient on this basis to the high-temperature range is fully validated. However, it should be noted that, at low and moderate temperatures, the properties of the gas are basically governed by attractive forces between particles and at high temperatures by the repulsive potential. For this reason, the agreement between curves of $B_2(T)$ at low and moderate temperatures is an insufficient basis for their agreement at high temperatures, and the extrapolation must be validated on the basis of data on other properties.

The values of the second virial coefficient can be determined on the basis of various experimental results (on the speed of sound, enthalpy, specific heat, etc.). However, data on the compressibility of gases are most frequently used [2.113]. The most exact values were obtained in experimental studies of the thermal properties of highly rarefied gases. Then the correction for nonideality is fully determined by paired interactions of molecules.

Table 21 lists studies in which the values of the second virial coefficient for the four monatomic gases were found from experimental data on compressibility at low pressure. The widest temperature range was investigated at the Canadian National Research Laboratory [2.26, 2.51, 2.70, 2.78] where B_2 was determined by a modified Burnett technique, based on measuring the pressure of the gas before and after expanding it by allowing it to flow from a smaller- to a

TABLE 21. Studies Concerned with Experimental Determination of Second Virial Coefficients of Neon, Argon, Krypton and Xenon

Year	Author	Temperature, K	Reference
	Neon		
1949	Oishi	273—373	[2.114]
1954	Nicholson and Schneider	273—973	[2.26]
	Argon		
1949	Oishi	273—373	[2.114]
1953	Whalley et al.	273—873	[2.51]
1956	Cottrell et al.	303—363	[2.115]
1962	Dymond and Smith	273—323	[2.116]
1962	Pool et al.	90	[2.117]
1962	Fender and Halsey	85—124	[2.118]
1967	Weir et al.	80—190	[2.119]
1968	Birne et al.	84—271	[2.120]
	Krypton		
1954	Whalley and Schneider	273—873	[2.70]
1962	Thomaes and Steenwinkel	110—270	[1.119]
1962	Fender and Halsey	108—138	[2.118]
1967	Weir et al.	111—224	[2.119]
1968	Birne et al.	117—252	[2.120]
	Xenon		
1955	Whalley et al.	273—973	[2.78]
1955	Reeves and Whytlaw-Gray	273—313	[2.121]
1966	Greenlief and Constabaris	298	[2.122]

larger-volume vessel. They used direct measurements for determining the coefficients of the equation of state in the form

$$pv^{A} = A + Bp + Cp^{2} \tag{2.32}$$

by means of the method of least squares. The mean values of B and C were selected on the basis of results of their own measurements, which were assigned the greatest weight, and of analogous results of others obtained from compressibility data.

Cottrell and his coworkers [2.115] present values of the second virial coefficients for argon: -0.331 ± 0.045, -0.232 ± 0.032 and -0.108 ± 0.055 cm^3/g at 303, 333, and 363 K respectively. Dymond and Smith [2.116] determined the values of B_2 for argon also at three temperatures (273, 298 and 323 K) with an absolute error of ± 0.05 cm^3/g. Pool et al. [2.117] estimated that the values of B_2 of argon obtained by them at $T = 90$ K also contain the same error which corresponds to a relative error of $\pm 1\%$.

Fender and Halsey [2.118] investigated low-temperature argon using a constant-volume gas thermometer. As a result of these measurements, they obtained a correction for adsorption of the gas by the vessel walls and also estimated the contribution of the three-particle interaction by means of the Lennard-Jones (6–12) potential. The maximum error is estimated by them at 1.5%. However, at $T = 90$ K the difference between the values of B_2 obtained by Pool et al. [2.117] and those obtained by Fender and Halsey [2.118] is more than 4%, which lies outside the limits of the overall error of these experiments.

Studies by Weir et al. [2.119] and by Birne et al. [2.120] contributed to refining the values of B_2 for argon at low temperatures. The data in [2.119] were obtained by the Burnett technique and exhibit a high reproducibility. The final results were corrected for adsorption of the gas and for the contribution of three-particle interactions. The error of the published values of B_2 amounts to 0.25 cm^3/g at 80 K, 0.08 cm^3/g at 82 to 86 K and 0.025 cm^3/g at higher temperatures. Within the above error limits the results in [2.119, 2.120, 2.117] at $T = 90$ K are in agreement. Larger differences (lying outside the limits of the overall measuring error) exist from values of Fender and Halsey [2.118] which apparently are less accurate.

Several of the aforementioned studies were also concerned with krypton. The accuracy of values of B_2 for it are estimated by the authors to be the same as for argon. The values of B_2 for krypton determined by Thomaes and Steenwinkel [1.119] by the relative method are listed at $T = 110$, 134, 174, 235 and 270 K and were compared with analytic values for hydrogen. The values in [1.119] deviate by more than 5% from the results in [2.119, 2.120] which are in good mutual agreement and which can be regarded as the most reliable at above room temperatures.

Five values for xenon were obtained by Reeves and Whytlaw-Gray [2.121] by the relative method using nitrogen as the standard substance. The value of B_2 in [2.122] deviates from data in [2.121] by less than 1%.

When experimental data on the compressibility of gases at low and moderate densities are available, coefficients B_2 and B_3 can be determined by several methods. The isothermal planes of state in coordinates (Z, ρ) are presented by a family of curves with rather well known configurations. Using this together with the virial equation of state (2.1), it is possible to determine coefficient B_2 as $(\partial Z/\partial \rho)_{T,\rho} \to 0$, respectively. Then B_3 will be defined as $(\partial^2 Z/\partial \rho^2)_{T,\rho} \to 0$, etc. In principle, such a method can be used if the error of experimental values of Z does not exceed $\pm 0.001\%$. However, the real error of modern data lies basically within the limits of 0.1–0.2%, and this required new methods for finding the low-order virial coefficients.

We transform Eq. (2.1) to the form

$$y = \frac{Z-1}{\rho} = B_2(T) + B_3(T)\rho + B_4(T)\rho^2 + \ldots \tag{2.33}$$

Now in these new coordinates (y, ρ) we extrapolate the experimental isotherm by graphic construction to $\rho = 0$. Then the segment cut off on the ordinate axis is equal numerically to the value of B_2 and the slope is equal numerically to B_3. Such a simultaneous determination of the two virial coefficients is known as Kay's method. If this method is used with consideration of the possible error of experimental values of Z, then it is easy to establish the optimum and limiting values of the two virial coefficients and the range of values of density past which the straight-line segment of the isotherm becomes curved due to high-order virial coefficients.

We wish to note, however, that in this interpretation Kay's method has its shortcomings. Thus, when B_3 is determined from the slope factor in coordinates (y, ρ), and then from the segment cut off at the ordinate axis when constructing the isotherms $(Z - 1 - B_2\rho)/\rho = f(\rho)$, the results deviate to 10%. Such significant differences are due both to insufficient accuracy of the original data and to change in the range of values of density at which the equation with two virial coefficients is valid upon transition from one isotherm to another. As a result, it is difficult to find in graphical constructions the maximum values of density to which such an equation is valid. As a result of the above circumstances the accuracy of determining the lower-order virial coefficients (particularly the third) depends to a significant measure on the experience and intuition of the investigator.†

In their desire to eliminate these random factors, some investigators made an attempt to establish objective criteria for virial coefficients B_i when the series is truncated. Michels et al. [2.124] showed, for the case of experimental data on the compressibility of carbon dioxide at 49.712°C, that a relation exists between

† Holleran [2.123] suggested an original method for determining B_2 and B_3 based on extrapolating y to the zero value of density along a curve along which the temperature is a function of the density and y = const. The method makes it possible to somewhat improve the accuracy of determining low-order virial coefficients, but it requires highly time consuming workup of original experimental data.

the power of the polynomial and the width of the density interval over which experimental data are represented. Assuming that the values of B_i should be independent of the methods of their determination, the power of the polynomial and the width of the density range, Michels et al. [2.124] found by the method of least squares the coefficients of a polynomial which describe more exactly the original data. It is seen from the graphs of deviations listed in [2.124] that the difference between the experimental and computed values can be minimized by use of a certain power m of the polynomial. However, in this case rather close (in accuracy) values of B_2 and B_3 are obtained and these can apparently be identified with the "real" virial coefficients. The subsequent coefficients are determined with an error which is commensurable with the value of the coefficient itself or even exceeds it due to the limited accuracy of the original data.

An analogous problem was solved by Hal and Canfield [2.125] who determined the coefficient of the approximating polynomial by the method of least squares with the aid of orthonormal functions in order to reduce the volume of computational operations. In selecting a polynomial which would best approach the virial expansion, Hall and Canfield used more rigorous criteria than those used by Michels for truncation of the series and calculated the values of the virial coefficients of carbon dioxide at the same 49.172°C isotherm. The results obtained in [2.124] and [2.125] are virtually identical.

Vukalovich et al. [2.9] assumed that the error of the coefficients of the approximating polynomial depends to a large measure on its power, due to correlation between the coefficients. If the original values of compressibility were determined with an error of 0.1–0.2%, then one can actually determine with acceptable accuracy only B_2 and B_3 by using for this purpose a third or fourth power polynomial to describe the compressibility data with experimental accuracy. A further increase in the power of the polynomial may, in the opinion of Vukalovich et al. [2.9], bring about a steep rise in the error of determining B_2 and B_3.

Vasserman and his coworkers [2.126] determined the lower-order virial coefficients from pvT data by computer by the method of least squares, gradually expanding the density range along isotherms by including additional experimental points. Simultaneously while determining the coefficient they calculated its error resulting from the error of approximation of the original data. The value which was obtained with the lowest error was identified with the virial coefficient. Due to the different lengths of isotherms vs. density, the number of coefficients needed to describe experimental data was selected by the Fisher statistical test.

Vasserman et al. employed a scheme for successive determination and analytic description of virial coefficients. Checks against experimental data on the compressibility of air showed that B_2 and B_3 were determined reliably. However, B_4 and B_5 should be treated as empirical temperature functions since their accuracy is highly affected by errors in determining and approximating the preceding coefficients.

The methods suggested in [2.9, 2.124–2.126] are fully reliable but highly time consuming. For this reason, the virial coefficients for the four inert gases were determined by means of a specially worked out program [2.86] which employed the principle of selecting values of B_2 at the stability segment. The salient feature of the problem consists of finding virial coefficients at once with respect to two variables. This permitted them not only to determine the values of the coefficients but also to find their temperature dependence in analytic form. Note that the above method for determining B_2 permits use of experimental data obtained at an arbitrary temperature in ranges of 52 to 973 K for neon; 80 to 1223 K for argon; 107 to 873 K for krypton; and 273 to 973 K for xenon. These values represent the entire necessary temperature range from the triple point to ~1300 K only for argon. For the three remaining gases, B_2 must be determined by interpolation.

Initially we attempted to solve this problem by means of tabulated values of the analytic virial coefficient $B^*(T^*)$, based on the Lennard-Jones (m − 6), Buckingham (exp-6) and Kihara model functions of intermolecular interaction. However, a check showed that the experimental values of B_2 cannot be described by the analytic function $B^*(T^*)$ over the range of temperatures under study without going past the limits of experimental error. An attempt to obtain agreement at low temperatures ($T < T_B$) increased the discrepancies at high temperatures and conversely. This indicates that the above model potentials do not fully reflect the actual interaction between the atoms of inert gases. Nevertheless, it can be assumed that the interaction between the atoms of the substances under study obey the same law and that their paired potentials conform relative to any two parameters of the transformation. This assumption is supported by calculations performed by Mason et al. [2.127], Rossi and Danon [2.103], and Dymond [2.128].

The above assumption can be checked on the basis of experimental data on the thermophysical properties of rarefied gases, in particular, by overlaying curves of $B_2(T)$ in logarithmic coordinates. When the inert gases have a single function of intermolecular interaction, we will have

$$B_{2i}(T) = b_{0i} B_2^*(T^*) \qquad (2.34)$$

where $B_2^*(T^*)$ is a function universal for the four gases; $b_{0i} = \frac{2}{3}\pi N_A \sigma_i^3$; $T^* = \frac{kT_i}{\epsilon_i}$; σ_i and ϵ_i are the potential parameters of the ith gas relative to which paired potential the gases conform. Since the actual form of $B_2^*(T)$ is unknown, it can be replaced by the exponential curve of $B_2(T)$ of the most fully investigated substance—argon. The fact that the corresponding curves for other gases can be made to match the curve for argon when plotted in log coordinates will mean that the above assumption is correct†.

† To confirm our assumption, we used data on transport coefficients to T = 2100 K (see Chapter 4) for the temperature range above 973 K where no experimental data are available on the second virial coefficients of neon, krypton, and xenon.

It was shown by Rabinovich et al. [2.129] that it is possible to make the curves of $B_2(T)$ of the four gases coincide in logarithmic coordinates by parallel translation of coordinate systems. The only exception was a small part of the curve for neon at $T < 100$ K, where quantum phenomena exert their effect. The coefficients of transition from one substance, taken as the reference, to the other can be determined from the shift of the origins of the coordinate systems relative to one another. However, due to the scatter of experimental data these conversion transitions cannot be uniquelly determined by this technique.

To eliminate this lack of uniqueness, we converted to a dimensionless coordinate system using T_B and ρ_0, the parameters of the ideal-gas curve (see p. 94) which are related to parameters ϵ and σ of the intermolecular-paired interaction function. In these coordinates all the data on the second virial coefficients of the four inert gases formed, within the limits of experimental accuracy, a single general curve. To represent it analytically, we investigated polynomials in the form

$$B_2^* = \sum_{j=0}^{n} \frac{b_j}{\Theta^j} \tag{2.35}$$

where $B_2^* = \rho_0 B_2$ and $\Theta = T/T_B$. Coefficients b_j were determined by the method of least squares:

$b_0 = 1.211472$; $b_3 = 0.5705028$;
$b_1 = -0.1767558$; $b_4 = -0.1183557$;
$b_2 = -1.495367$; $b_5 = 0.008502477$.

A fifth power polymial was found to be optimal since a further increase in polymial order did not reduce the difference between experimental and analytic results.

It is seen from comparison of experimental values of B_2 for argon with the generalized curve calculated from Eq. (2.35) (Fig. 27) that the curve provided an optimum averaging of the original data over the range of temperatures under study.

Figure 28 compares analogous data for neon, krypton, and xenon with the same generalized curve. As in the preceding case, the data are satisfactorily described by Eq. (2.35), which can be used up to $\Theta = 11$ since the generalized curve was extrapolated slightly by making the experimental values of B_2 for neon at $T > 300$ K coincide with the analytic curve based on the Lennard-Jones (12-6) potential. But, at low temperatures, there exist perceptible deviations in the case of neon. This curve segment is shown in Fig. 29 on a larger scale. The systematic deviations start at $T < 100$ K and increase with reduction in temperature. For this reason, when $B_2(T)$ is calculated for neon at low temperatures, it is additionally necessary to make corrections for quantum effects.

We thus see that the generalized equation of state with one virial coefficient

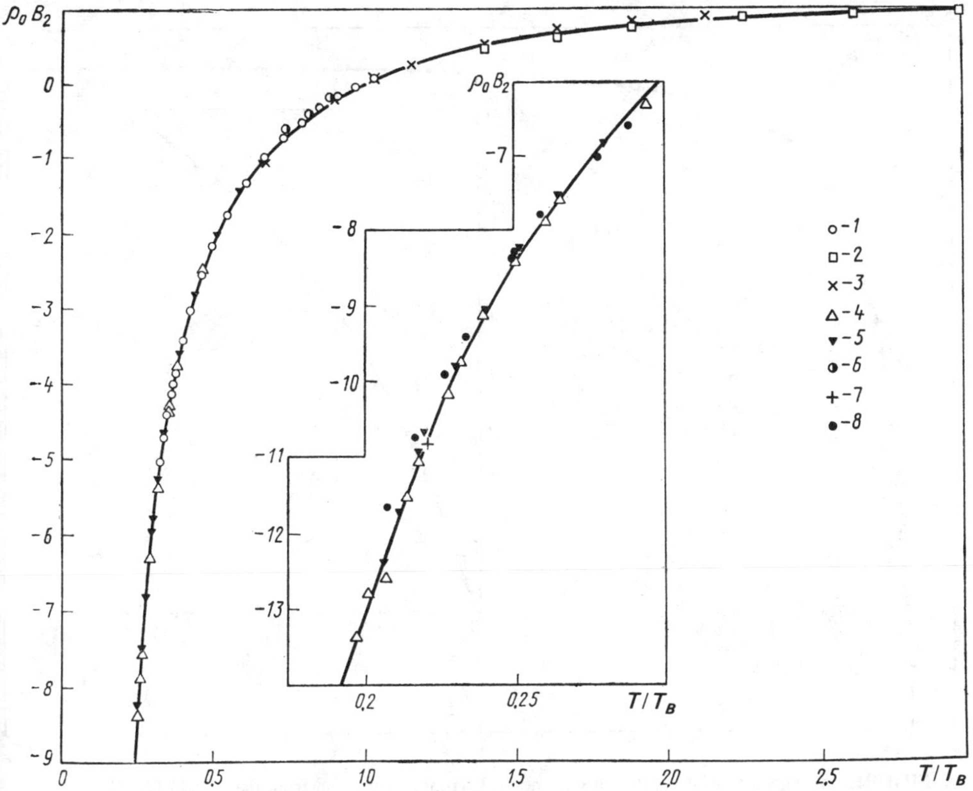

FIGURE 27. Second virial coefficient of argon from data of: 1) [2.50, 2.53]; 2) [2.54]; 3) [2.51]; 4) [2.119]; 5) [2.120]; 6) [2.115]; 7) [2.117]; 8) [2.118]; the solid curve was computed from Eq. (2.35).

obtained above describes the original test data on compressibility of the four inert gases with an error of less than $\pm 0.1\%$ and can be used for calculating thermodynamic properties at low densities not only over the temperature range for which experimental data were obtained but also outside of it.

2.5. ON THE THERMODYNAMIC SIMILARITY OF INERT GASES

The principal formulation of the law of corresponding states first suggested by Van der Waals postulates that, in a given system of dimensionless coordinates, when any two reduced variables of different gases are equal, then the third variables are also equal. This results in a vanishing of the difference between the equations of state represented in dimensionless form, i.e., function

$$\pi = f(\omega, \tau) \tag{2.36}$$

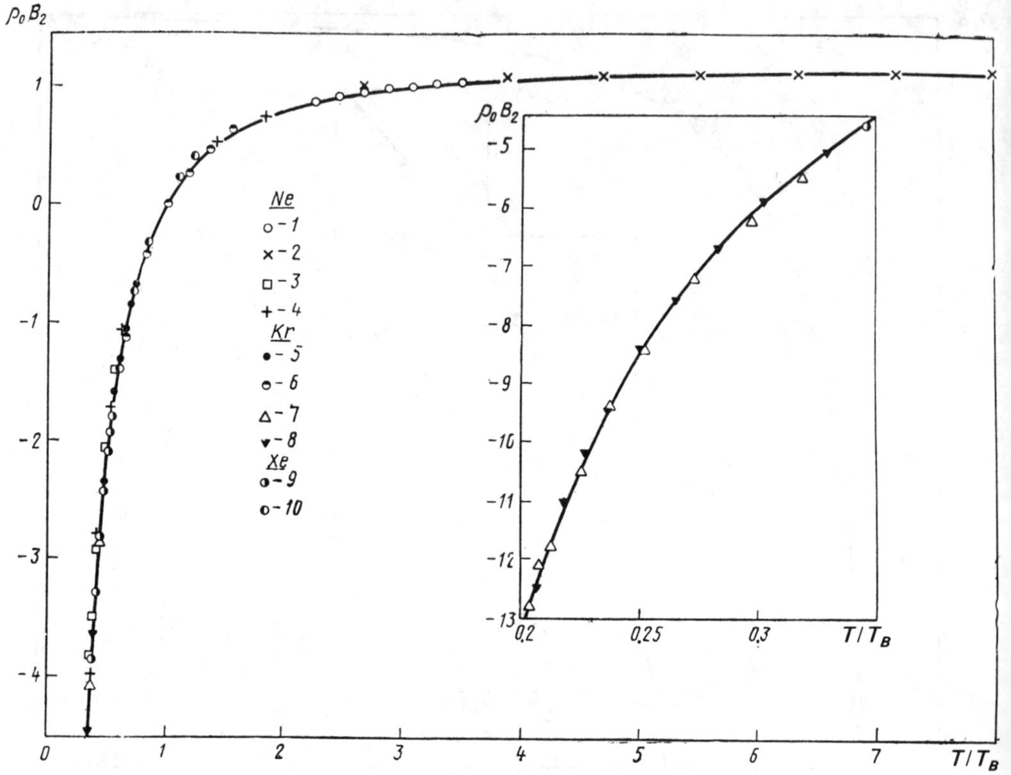

FIGURE 28. Second virial coefficients for neon, krypton, and xenon from data of: 1) [2.27]; 2) [2.26]; 3) [1.94]; 4) [2.31, 2.32]; 5) [2.71]; 6) [2.70]; 7) [2.119]; 8) [2.120]; 9) [2.47]; 10) [2.78]; the solid curve was plotted from Eq. (2.35).

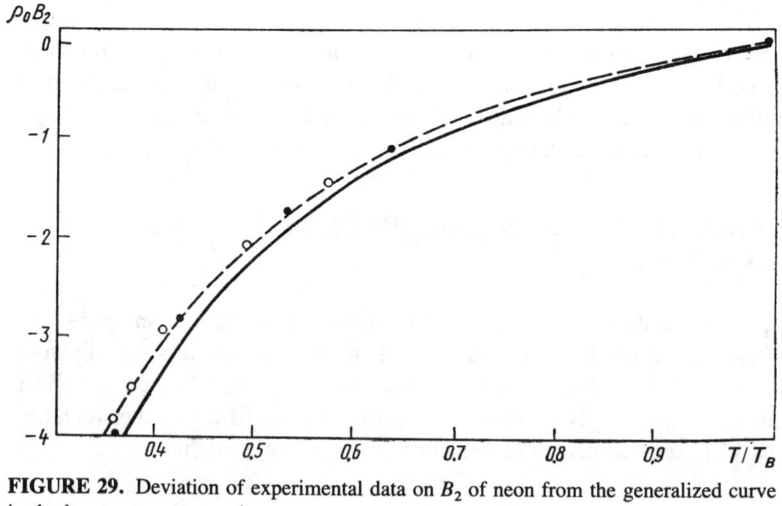

FIGURE 29. Deviation of experimental data on B_2 of neon from the generalized curve in the low-temperature region.

where $\pi = p/p_{ref}$; $\omega = \rho/\rho_{ref}$ and $\tau = T/T_{ref}$ (the subscript ref pertains to coordinates of the reference similarity point), is valid for all substances. The above can be proven rigorously only for gases obeying Van der Waals equation or other equations of state containing two constants.

This law is satisfied approximately for real substances resulting in the appearance of many studies which, in order to investigate the properties of little-explored substances, give various formulations and modifications of the law of corresponding states. The main feature of these studies consists of obtaining empirical correlations which are suitable for individual groups of substances over narrow ranges of parameters. At the same time, the theory permits us to resolve the question of why the law of corresponding states is now universal for all the substances over a wide range of variables.

The law of corresponding states has been validated rigorously by Pitzer [2.130] and Guggenheim [2.131]. They showed that thermodynamically similar gases are those which interact according to a two-parameter potential in the form

$$U(r) = \epsilon\varphi(\sigma/r) \tag{2.37}$$

which is conformal with respect to ϵ and σ. The authors of [2.130] and [2.131] obtained transition from paired to many-particle interaction by postulating additivity of the potential energy of interaction.

If these assumptions are satisfied, then it becomes self-evident that the coordinates of the reference points of similarity of various gases will be uniquely related to the potential parameters: $T_{ref} \sim \epsilon$ and $v_{ref} \sim \sigma^3$. These are an indispensible condition for satisfaction of the law of corresponding states. For example, the coordinates of critical points of thermodynamically similar gases (which are most frequently used as points of reference) are related to parameters of the Lennard-Jones (12-6) potential by the expression

$$T_{cr} = 1.3\frac{\epsilon}{k};\ v_{cr} = 1.678\sigma^3$$

The first assumption on the form of the intermolecular two-particle interaction function made by Pitzer and Guggenheim means that only gases with predominance of the dispersion component of potential energy in their intermolecular interaction can be thermodynamically similar. For this reason, Pitzer [2.130] emphasized that the law of corresponding states can be obeyed only by substances consisting of spherical or pseudospherical molecules (CH_4, CF_4, etc.). The inert gases under study here, whose thermodynamic properties are correlated at low values of density (see Sec. 2.4), where one can restrict himself to analysis of paired interaction, should be included in this class of substances.

In the most general case, the potential energy of interaction is comprised of both the dispersion and orientation components. This has necessitated [2.109] statistical averaging over the distance. As a result of this, the potential energy of interaction becomes a function not only of the interparticle distance, but also of the temperature

$$U(r, T) = \frac{\iint U(r, \Theta) \exp\left[-\frac{U(r, \Theta)}{kT}\right] d\Omega_1 d\Omega_2}{\iint \exp\left[-\frac{U(r, \Theta)}{kT}\right] d\Omega_1 d\Omega_2} \tag{2.38}$$

where potential $U(r, \Theta)$ is a function of distance r and orientation angles Θ, whereas $d\Omega_i$ are differentials of the phase volume.

At high temperatures, where it suffices to retain the first term in the series expansion of the exponent, the temperature dependence of the potential of interaction of polar gases is written as [2.109]

$$U(r, T) = U(r) - \frac{2}{3} \frac{\mu^4}{kT} r^{-6}$$

where μ is the dipole moment.

Another reason for the appearance of temperature dependence of the interaction potential is the elongation effect which is associated with the effect of molecular rotations and vibrations on the paired potentials [2.132]. For the selected form of the potential this influence manifests itself in the form of temperature dependence of the potential parameters. Kessel'man and his coworkers [2.133, 2.134] and Tsykalo [2.135] suggested methods for determining $\epsilon(T)$ and $\sigma(T)$ from experimental data on the thermophysical properties of rarefied gas.

The above means that according to Eq. (2.38), to ensure conditions under which gases will be thermodynamically similar it is necessary to allow, in general, for temperature dependence of the parameters of the reference point.

The deviations from the law of corresponding states at higher pressure (density) are due to the fact that the second assumption of additivity of the potential energy of interaction is not entirely correct due to the presence of nonadditive forces in many-particle interactions. Even when the paired potential is conformal, which ensures thermodynamic similarity of rarefied gases, with increasing pressures the gases become less similar. This occurs because nonadditive components of the potential energy of interaction are not conformal with respect to the force constants ϵ and σ relative to which the paired potentials are conformal. The nonadditive (n.a) component of the potential energy of attraction for an elementary three-particle interaction

$$\Delta U_{\text{n.a}} = c_6 \frac{\alpha}{r^9} (1 + 3\cos\Theta_1 \cos\Theta_2 \cos\Theta_3) \tag{2.39}$$

Here c_6 is the Van der Waals attraction constant; α is the polarizability and Θ_1 through Θ_3 are the angles of the triangle formed by interacting molecules.

If the attraction force is proportional to r^{-6}, then the nonadditive component is proportional to r^{-9}, and it contains an additional parameter. It follows from this that the agreement between the values of α/σ^3 for different gases is a necessary condition for conformality of $\Delta U_{n.a}$. Calculations performed by Donth [2.136] show that the values of this ratio are not the same for all the inert gases and amount to 0.8, 1.0 and 1.21 for argon, krypton, and xenon, respectively.

Extensive use is made lately of the concept of effective paired potential, related to statistical averaging of the energy of many-particle interaction, and consisting of additive and nonadditive components:

$$\overline{U} = \Sigma U_{ad} + \Sigma U_{n.a} = \Sigma U_{eff} \tag{2.40}$$

Calculations of this potential [2.137, 2.138] performed in the approximation of the Perkus Yevick theory for three interacting particles showed that, in general, the effective potential is a function not only of the intermolecular distance, but of the temperature and density. Within the framework of the assumptions made, U_{eff} depends little on temperature but exhibits a strong (linear in the first approximation) density dependence. This means that, in the most general case, the potential energy of two- and many-particle interactions of molecules of different structures can be described by means of an effective paired potential, as expressed by Eq. (2.37), in which parameters ϵ and σ are a function of temperature and density.

If we now extend the above general remarks to the subject under study here, inert gases, we must remember that their properties at low densities can be correlated with the experimental accuracy using parameters T_B and ρ_0. However, at high pressures, when the contribution of many-particle interactions becomes more important, it becomes necessary to use variable parameters. Since otherwise, it is impossible to correlate thermodynamic properties of inert gases over a wide temperature and pressure range with acceptable accuracy.

The above is illustrated by various attempts at constructing a generalized equation of state for elements of the helium group. Thus, Levelt [2.139] correlated the properties of argon and xenon at temperatures between 273 and 423 K at densities between 0 and $2\rho_{cr}$ using the parameters of the Lennard-Jones (12-6) potential and the coordinates of the critical points. In the first case, as seen from the tables given in [2.139], the deviations $\Delta Z = Z_{Xe} - Z_{Ar}$, in the corresponding states, which have a negative sign, increased with the density and temperature from 0.4 to 50% over the above range of variables. Smaller differences are observed when using coordinates of the critical points: behaving in the same fashion, the differences range from 0.3% at low values of the reduced density to 12% at high reduced densities. The reason for this is most likely that use of coordinates of the critical points, located in a region of rather dense gas, incorporate to some measure many-particle interactions.

Rowlinson [2.140] correlated the properties of argon and xenon on the basis of the Kihara potential, which in his opinion is the most realistic of the simple three-parametric potentials. This is apparently the reason why the observed differences ΔZ do not exceed 2.5% over the same range of temperature and density values as in [2.139].

The differences in values of Z of inert gases in corresponding state can be reduced even further by using reference parameters which are a function of temperature and density. The finding of such parameters is the subject matter in [2.141–2.143]. However, in our case this will be best done using the ideal-gas curve.

It was shown above that the use of T_B and ρ_0 as the reference parameters makes it possible to bring about coincidence between the thermodynamic planes of two gases along the entire ideal-gas curve. Parameters T_B and ρ_0 are uniquely related to the potential parameters of some effective potential and are a function of temperature and density. This latter relationship follows from the rectilinearity of the ideal-gas curve in coordinates (ρ, T),

$$T_B = \frac{T}{1 - \dfrac{\rho}{\rho_0}}; \ \rho_0 = \frac{\rho}{1 - \dfrac{T}{T_B}} \tag{2.41}$$

It is also of interest to analyze the property of angles formed upon intersection of the ideal-gas curve by isotherms and isochores. For this we write the expression for the total differential

$$dZ = \left(\frac{\partial Z}{\partial \Theta}\right)_\omega d\Theta + \left(\frac{\partial Z}{\partial \omega}\right)_\Theta d\omega \tag{2.42}$$

If the conditions $Z = 1$; $dZ = 0$; $\omega + \Theta = 1$ and $d\omega = -d\Theta$ are satisfied, then

$$\left(\frac{\partial Z}{\partial \Theta}\right)_{\omega=1-\Theta} = \left(\frac{\partial Z}{\partial \omega}\right)_{\Theta=1-\omega} \tag{2.43}$$

The property of the derivative on the left-hand side was analyzed in Sec. 2.3, where Eq. (2.22) at condition (2.23) has been validated. Similar considerations using Eq. (2.43) lead to the conclusion that the use of parameters T_B and ρ_0 ensures identity of properties with respect to their value and derivative at each corresponding isotherm and isochore of two gases. All this creates good prerequisite conditions for the use of Eq. (2.36) for correlating the properties of inert gases†. We shall illustrate this for the case of argon, krypton, and xenon.

† A similar conclusion was drawn by Holleran [2.123], who showed on the basis of experimental data for argon and xenon that thermodynamic properties of these gases can be correlated with acceptable accuracy using the ideal-gas curve.

Argon is used as a basis substance because it is the inert gas whose thermal properties have been more completely investigated over the entire range of parameters. To justify the extrapolation of thermal properties of krypton and xenon to high temperatures and pressures which were not investigated experimentally, we checked the accuracy with which these gases obey the law of corresponding states using different coordinates of the reference point of similarity. The values of Z calculated from the exact equation of state of gaseous argon on the basis of the coordinates of the reference points†

	argon	krypton	xenon
ϵ/k, K	119.7	164.4	212.73
σ, A°	3.405	3.685	4.22
T_{cr}, K	150.65	209.40	289.74
ρ_{cr}, g/cm³	0.531	0.912	1.100
T_B, K	407.76	567.5	791.0
ρ_0, g/cm³	1.87	3.21	3.89

were compared with experimental data for krypton and xenon. It was found that the greatest differences are obtained when using parameters of the Lennard-Jones (12-6) potential. These differences increase with density and with a reduction in temperature. For example, for xenon they are as high as 87% at $t = 150°C$ and $\rho = 2\rho_{cr}$.

The deviations can be reduced significantly when using parameters of the critical points, whereas the use of T_B and ρ_0 leads to reduction of these differences to values commensurable with the experimental error. These differences do not exceed 0.5% over the entire experimental range of temperatures and pressures, with the exception of the region of minima at transcritical isotherms, where they are 1% and more. Figure 30 shows the deviations of the computed values of Z from the experimental for xenon along two isotherms. Smaller deviations of similar nature are also observed in data for krypton. Thus, it can be claimed that the law of corresponding states is satisfied for monatomic gases within ~1% when using the parameters of the ideal-gas curve.

The above error still does not allow the use of the generalized equation of state for calculating the thermodynamic properties of krypton and xenon for ranges for which experimental data are more exact. However, the values of thermodynamic properties of heavy inert gases obtained from this equation at high temperatures and pressures which were not investigated experimentally can be used together with experimental results in compiling individual equations of state.

Tables 22 and 23 show the computed values of Z of krypton and xenon over temperature and density ranges for which no experimental data have been obtained. In conjunction with this, it is interesting to note that analogous computed data on the compressibility of krypton and xenon obtained by another method by Altunin and Gvozdkov [2.80] for the same range of variables agree with our

† The values of the parameters of the Lennard-Jones (12-6) potential for argon were taken from [2.50], for krypton from [2.71], and for xenon from [2.77].

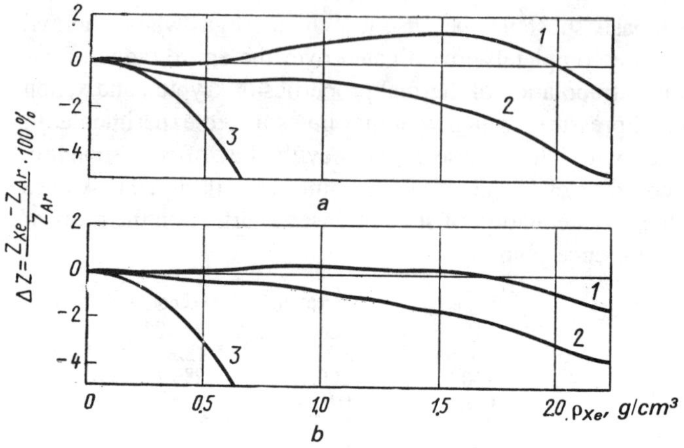

FIGURE 30. Deviations of experimental data on Z of xenon [2.77] for 75°C (a) and 150°C (b) from those calculated from the reduced equation of state for argon in form (2.36) when using parameters of different reference points of similarity. 1) "Ideal"-gas curve; 2) critical points; 3) Lennard-Jones potential.

results to within less than 0.5%. This is apparently the value of the maximum error of data listed in Tables 22 and 23.

Unfortunately, the generalized equation of state cannot be used for calculating the compressibility factor of neon at high reduced temperatures. However, the problem simplifies for $10 < T^* < 25$ since the isotherms at pressures to 1000 bar are limited to the density of from 0.4 ρ_{cr} to 0.2 ρ_{cr}. Under these conditions, extrapolation to the high-temperature range can be performed using an

TABLE 22†. Computed Data on the Compressibility Factor Z of Krypton, Obtained from Generalized Equation of State (2.36)

	Z at t, °C					
g/cm³	200	300	400	600	800	1000
0.1	—	—	—	1.0193	1.0246	1.0278
0.2	—	—	—	1.0424	1.0528	1.0593
0.3	—	—	—	1.0705	1.0861	1.0958
0.4	—	—	—	1.1034	1.1241	1.1370
0.5	—	—	—	1.1406	1.1664	1.1825
0.6	1.0126	1.0822	1.1277	1.1823	1.2131	1.2325
0.7	1.0336	1.1136	1.1662	1.2293	1.2650	1.2876
0.8	1.0613	1.1518	1.2114	1.2831	—	—
0.9	1.0972	1.1983	1.2650	1.3447	—	—
1.0	1.1427	1.2547	1.3282	—	—	—
1.1	1.1995	1.3223	—	—	—	—
1.2	1.2692	1.4016	—	—	—	—
1.3	1.3530	—	—	—	—	—
1.4	1.4519	—	—	—	—	—

† Here and in Tables 23 and 24 the maximum density at each isotherm corresponds to $p \simeq 1000$ bar.

TABLE 23. Computed Data on Compressibility Factor Z of Xenon, Obtained from Generalized Equation of State (2.36)

ρ, g/cm³	Z at t, °C.					
	200	300	400	600	800	1000
0.1	—	—	—	1.0050	1.0123	1.0168
0.2	—	—	—	1.0126	1.0270	1.0359
0.3	—	—	—	1.0236	1.0450	1.0583
0.4	—	—	—	1.0382	1.0665	1.0842
0.5	—	—	—	1.0561	1.0912	1.1131
0.6	—	—	—	1.0772	1.1189	1.1450
0.7	—	—	—	1.1015	1.1496	1.1798
0.8	—	—	—	1.1293	1.1838	1.2181
0.9	—	—	—	1.1613	1.2221	1.2606
1.0	0.8434	0.9850	1.0806	1.1983	1.2656	1.3080
1.1	0.8537	1.0079	1.1125	1.2413	1.3148	—
1.2	0.8702	1.0374	1.1510	1.2911	—	—
1.3	0.8939	1.0743	1.1971	1.3481	—	—
1.4	0.9256	1.1196	1.2517	1.4126	—	—
1.5	0.9666	1.1743	1.3155	—	—	—
1.6	1.0183	1.2396	1.3890	—	—	—
1.7	1.0825	1.3167	1.4727	—	—	—
1.8	1.1609	1.4067	—	—	—	—
1.9	1.2557	1.5112	—	—	—	—
2.0	1.3692	—	—	—	—	—
2.1	1.5037	—	—	—	—	—

equation compiled from elementary functions. The temperature function in this case is the second virial coefficient which ensures reliable extrapolation with respect to temperature. The volume functions are determined by the method of reference isotherms [1.104] on the basis of experimental data on the compressibility of neon [2.27] presented on three isotherms: 0, 75 and 150°C. The results obtained from the auxiliary equation of state for neon are listed in Table 24. The maximum possible error is estimated by us at 0.4%

In our subsequent compilation of individual equations of state for the four gases at temperatures from the triple points to 1300 K at pressures to 1000 bar we shall use the computed values of Z listed in Tables 22 through 24 together with the selected experimental thermal data.

2.6. COMPUTER DETERMINATION OF COEFFICIENTS OF THE VIRIAL EQUATION OF STATE

The experimental pvT data form a thermodynamic plane of state which in general can be described analytically by a function in the form $Z(\rho, T, b_1, b_2, \ldots, b_k)$. Parameters b_i are related to individual properties of the specific substance (molecular structure, intermolecular interaction). However, in solving the purely approximate problem, their number and values are a function of the width of the

TABLE 24. Computed Data of Compressibility Factor Z for Neon Obtained from the Auxiliary Equation of State

ρ, g/cm³	Z at *t*, °C.								
	200	300	400	500	600	700	800	900	1000
0.017996	—	—	—	—	—	—	1.0135	1.0135	1.0136
0.035991	—	—	—	1.0264	1.0266	1.0268	1.0270	1.0271	1.0272
0.071982	—	—	—	1.0534	1.0538	1.0541	1.0543	1.0544	1.0545
0.10797	—	—	—	1.0815	1.0819	1.0822	1.0824	1.0825	1.0826
0.14396	—	—	—	1.1107	1.1111	1.1113	1.1114	1.1115	1.1115
0.17996	1.1367	1.1393	1.1406	1.1412	1.1416	1.1417	1.1417	1.1417	1.1418
0.21595	1.1686	1.1714	1.1727	1.1733	1.1734	1.1734	—	—	—
0.25194	1.2023	1.2052	1.2064	1.2068	1.2069	—	—	—	—
0.28793	1.2378	1.2407	1.2418	1.2420	—	—	—	—	—
0.32392	1.2752	1.2780	1.2789	—	—	—	—	—	—
0.35991	1.3147	1.3173	—	—	—	—	—	—	—
0.39590	1.3562	—	—	—	—	—	—	—	—

range of independent parameters ρ and T over which the original data points are given. For the four gases under study this range was limited to temperatures from the triple point to 1300 K, twice the critical density and the isobar of p = 1000 bar which is shown schematically in Fig. 31 on the basis of data for argon.

The equation of state for neon, argon, krypton and xenon was represented by the virial type expansion:

$$Z(\rho, T) = \frac{p}{\rho RT} = 1 + \sum_{i=1}^{m} B_i(T)\rho^i \tag{2.44}$$

The temperature dependence of coefficients B_i is represented in the form of polynomials in reciprocal powers of $\rho = T/T_{cr}$:

$$B_i = \sum_{j=0}^{j_{max}} b_{ij}\tau^{-j} \tag{2.45}$$

Using this, we reduce the equation of state to the form

$$Z(\rho, T) = 1 + \sum_{i=1}^{m} \sum_{j=0}^{j_{max}} b_{ij}\tau^{-j}\rho^i \tag{2.46}$$

where ρ is in g/cm³.

The assumed form of the equation of state conforms to the theoretically validated Bogolyubov-Mayer equation of state (2.1) when the infinite series in powers of density is replaced by a polynomial to power m. In addition, a polynomial-type equation of state is most convenient for finding coefficients b_{ij} by the method of least squares since the system of equations turns out to be linear relative to the desired parameters.

In mathematical respect, the compilation of an equation of state is reduced to finding a set of constants of Eq. (2.46) so that they would best describe the body of original data on the thermal properties of the gas under study. This

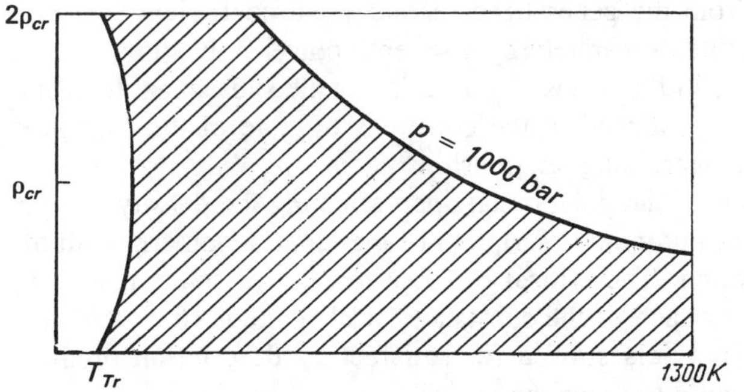

FIGURE 31. Schematic representation of the range of parameters described by equations of state for the four inert gases.

problem was solved by a program which approximated a function of two independent variables by the method of least squares, written for the BESM-4 computer [2.86, 2.144].

In compiling the equations of state for the four inert gases we analyzed the feasibility of complete satisfaction of parameters of the critical point and two critical conditions. However, the calculations showed that the imposition of these conditions significantly distorts the entire thermodynamic surface. It can be assumed that this occurs for two reasons. Firstly, the critical parameters of the majority of substances are known with poor accuracy, with data on the critical density being the least reliable. Even for xenon, the critical density of which was measured in special experiments, $\delta\rho_{cr}$ is not less than 0.5%. At the same time our calculations showed that a $\pm 0.1\%$ change in the critical parameters when introducing critical conditions distorts the configuration of the analytic thermodynamic surface. This results in unacceptable divergences from experimental data. The second reason lies in the virial form of the equation of state which most likely is unable to describe the original thermal data over a wide range of variables with experimental accuracy and to simultaneously ensure satisfaction of the critical conditions. An opinion has been put forward to the effect that both these conditions can be satisfied by combining a virial expansion with scaling-type terms. However, within the framework of current concepts on the nature of the critical point, there is as yet no clarity concerning the form of the equation of state in the transcritical region. Given this, we gave up our attempt to introduce critical conditions into the equation of state for the gas phase.

The program developed for computer determination of coefficients b_{ij} introduced into the equation of state ready-made analytic expressions in the form of Eq. (2.45) as individual temperature functions of B_i by assigning corresponding values of b_{ij}. For argon, krypton and xenon, this made it possible to use the generalized second virial coefficient which we recalculated on the basis of parameters of the ideal-gas curves (see page 109) and values of b_j (see page 102) for each substance to the form corresponding to an expansion with respect to τ. The second virial coefficient of neon is represented over a wider range of reduced temperatures ($\tau_{max} \simeq 30$) than for other inert gases. In the low-temperature region it deviates from the generalized second virial coefficient (see Fig. 29). Hence, it was described separately by a seventh power polynomial in $1/\tau$. The maximum power j_{max} in the series in Eq. 2.45 for the subsequent temperature functions B_i ($i > 1$) contained in the equations of state of the four inert gases was specified by a decreasing relationship.

The maximum power of the polynomial approximating the density was selected so as to provide an optimum description of the longest isotherm with the most complex configuration. The critical or close to critical isotherm was selected for this purpose for each of the substances. Calculations show that seventh-eighth power polynomials suffice for satisfactory description of these curves at densities from zero to twice the critical.

The above means that the specific form of Eq. (2.46) is, to a large measure, determined by the configuration of the thermodynamic surface of the substance under study and by the range of reduced parameters. However, the overall number M of constants b_{ij} and their distribution with respect to temperature functions B_i can be adjusted in order to find the best form of the equation of state which provides for optimum satisfaction of a number of criteria characterizing the quality of the computed thermodynamic surface.

The integral characterization of the quality of description of experimental data in the method of least squares is the residual functional

$$\Phi = \sum_k W_k(Z_k^{\text{exptl}} - Z_k^{\text{calc}})^2 \tag{2.47}$$

where $W_k = 1/\Delta Z_k^2$ is the statistical weight of the kth point; ΔZ_k is the absolute error of measuring the compressibility at this point, defined by us to be equal to $Z_k^{\text{exptl}} \cdot \delta Z$ (here δZ is the mean relative experimental error in each series of measurements). Functional Φ decreases with increasing M to some limit, and then it starts increasing. For this reason, the desired version of the equation of state lies in the region where Φ is at minimum. However, by itself this functional cannot serve as a universal criterion for the quality of the equation of state. The uniformity of approach of the computed thermodynamic surface to experimental data is also determined by the number of points in which the error of the approximation exceeds the experimental and the maximum error of approximation for the given body of points. In addition to the above, we also used as a criterion the degree of symmetry of the histogram of deviations of experimental data from the computed. This made it possible to estimate the closeness of the distribution of deviations to the normal.

For the purpose of obtaining an optimal equation of state for each of the four inert gases we computed a number of versions of equations with the virial component as the first temperature function and with polynomials to powers $m \leqslant 8$. In this case, the computer printout contained all the necessary characteristics of each equation. These were analyzed simultaneously and served as the basis for the final selection.

The accuracy of values of calorific functions calculated from the equation of state is to a large measure determined by the reliability of the thermal data, the extent of their compatibility, and proper formatting of the original body of data. As follows from the brief analysis in Sec. 2.1, both the accuracy and the width of range of parameters of the available experimental data are uneven. In addition, the individual ranges of parameters are not described uniformly. Some have been investigated in extensive detail and contain sufficiently reliable data. Others are less thoroughly investigated or do not contain any experimental information. For this reason, in addition to the previous analysis of experimental studies, our compilation of the starting body of information on the basis of local equations of state included an evaluation of the internal compatibility of the results of measurements of the individual investigators and refined the statistical weight of each test point. Then we checked the degree of mutual agreement of data of individual investigators. Next we assembled in the original body of data the most accurate points, marked by asterisks, in Tables 8, 9, 11 and 12.

However, this body of data was found to be insufficient for compiling the equations of state for neon, krypton and xenon over the necessary range of variables. The part of the thermodynamic surface for which there are no experimental values was complemented by computed values of compressibility listed in Tables 22 through 24.

The range of variables between the saturation curve and the critical isotherm has been investigated insufficiently. It should, however, be remembered that at $\rho < 0.1\ \rho_{cr}$ the accuracy of thermodynamic functions is almost entirely determined by the reliability of the second virial coefficient obtained from the most dependable data (see Sec. 2.4). Thermodynamic functions at higher densities were obtained by analytic interpolation since the values of Z along the saturation curve up to $\sim 1.8\ \rho_{cr}$ were also added to the database. The values of saturated vapor pressure needed for calculating Z_s were determined from Eqs. (1.70) and (1.71) for temperatures at which the corresponding values from the selected original studies were chosen.

The coefficients of the resulting equations of state for the four inert gases are listed in Tables 25 through 28. Each of these equations is analyzed in detail below.

Neon. The body of data includes experimental points at the saturation curve [2.145] up to $\rho_s \simeq 1.8\ \rho_{cr}$. Mathias et al. [2.145] did not give an estimate of the accuracy of their results. However, analysis performed by us [2.146] shows that the data in [2.145] are inaccurate at $\rho_s < 0.04$ g/cm^3, and this is responsible for the anomalous shape of the second virial coefficient at temperatures close to the triple point. As a result of the above, the values of density along the phase equilibrium curve at $\rho_s < 0.04$ g/cm^3 were not included in the body of original data.

At low temperatures, we used data of Gibbons [1.94], Onosovskiy and Moroz [2.31], and Onosovskiy [2.32]. The differences between the results of independent measurements do not exceed the overall experimental error although they exhibit a tendency toward increase with reduction in temperature. The body of original data also included the data of Streett [2.33], although not completely, and the reasons for this should be discussed in more detail.

We did not use these data in the initial compilation of the equation of state of gaseous neon [2.146] since they were not available at the time the work was done. In desiring to obtain more exact values of the thermodynamic functions of gaseous neon at temperatures between 80 and 130 K at pressures above 200 bar we supplemented the body of starting data by information provided by Streett [2.33]. However, this caused a significant increase in the differences between the computed and measured data which had been obtained by other investigators

TABLE 25. Coefficients of Equation of State (2.46) for Neon

i	b_{ij} at j							
	0	1	2	3	4	5	6	7
1	0.701737	$0.504504 \cdot 10^{-1}$	$-0.793757 \cdot 10$	$0.921134 \cdot 10$	$-0.525024 \cdot 10$	$0.106800 \cdot 10$	$0.118040 \cdot 10^{-1}$	$-0.106330 \cdot 10^{-1}$
2	0.240038	$0.181055 \cdot 10$	$-0.917914 \cdot 10$	$0.348565 \cdot 10^2$	$-0.413956 \cdot 10^2$	$0.123012 \cdot 10^2$	—	—
3	$0.346810 \cdot 10$	$-0.331236 \cdot 10$	$-0.247335 \cdot 10^2$	$-0.191754 \cdot 10^2$	$0.775338 \cdot 10^2$	$-0.133045 \cdot 10^2$	—	—
4	$-0.177786 \cdot 10^2$	$0.481756 \cdot 10^2$	$0.839857 \cdot 10^2$	$-0.100423 \cdot 10^3$	$-0.715946 \cdot 10^2$	—	—	—
5	$0.408969 \cdot 10^2$	$-0.152893 \cdot 10^3$	$-0.364078 \cdot 10^2$	$0.188425 \cdot 10^3$	$0.389365 \cdot 10^2$	—	—	—
6	$-0.425484 \cdot 10^2$	$0.194349 \cdot 10^3$	$-0.971971 \cdot 10^2$	$-0.107782 \cdot 10^3$	—	—	—	—
7	$0.165477 \cdot 10^2$	$-0.866427 \cdot 10^2$	$0.846932 \cdot 10^2$	—	—	—	—	—

TABLE 26. Coefficients of Equation of State (2.46) for Argon

i	b_{ij} at j					
	0	1	2	3	4	5
1	0.647846	-0.255839	$-0.585837 \cdot 10$	$0.604954 \cdot 10$	$-0.339695 \cdot 10$	0.660513
2	$-0.288386 \cdot 10^{-1}$	$0.134277 \cdot 10$	$-0.113720 \cdot 10^2$	$0.352681 \cdot 10^2$	$-0.331408 \cdot 10^2$	$0.972579 \cdot 10$
3	$0.341333 \cdot 10$	$0.733959 \cdot 10$	$-0.109504 \cdot 10^2$	$-0.341908 \cdot 10^2$	$0.394560 \cdot 10^2$	$-0.869915 \cdot 10$
4	$-0.121628 \cdot 10^2$	$-0.105136 \cdot 10^2$	$0.397309 \cdot 10^2$	$0.801642 \cdot 10$	$-0.901983 \cdot 10$	—
5	$0.169709 \cdot 10^2$	$0.109696 \cdot 10^2$	$-0.331751 \cdot 10^2$	$-0.234262 \cdot 10^2$	$-0.108524 \cdot 10$	—
6	$-0.653384 \cdot 10$	$-0.247994 \cdot 10^2$	$0.339016 \cdot 10^2$	$0.211293 \cdot 10^2$	—	—
7	$-0.139502 \cdot 10$	$0.193328 \cdot 10^2$	$-0.245424 \cdot 10^2$	—	—	—

TABLE 27. Coefficients of Equation of State (2.46) for Krypton

i	b_{ij} at j: 0	1	2	3	4	5
1	0.377406	—0.148988	—0.341045 · 10	0.352052 · 10	—0.197616 10	0.384117
2	0.142595	—0.485821	—0.134716 · 10	0.774659 · 10	—0.831482 · 10	0.293861 · 10
3	0.358102	0.363255 · 10	—0.670407 · 10	0.709952	0.315334 · 10	—0.226468 · 10
4	—0.142313 · 10	—0.253209 · 10	0.596767 10	—0.393900 · 10	0.452926 · 10	—
5	0.147064 · 10	0.785829	0.433364	—0.423227 · 10	—0.111457 · 10	—
6	—0.378146	—0.166798 · 10	0.150981 · 10	0.173903 · 10	—	—
7	0.957550 · 10^{-2}	0.600554	—0.805498	—	—	—

TABLE 28. Coefficients of Equation of State (2.46) for Xenon

i	b_{ij} at j: 0	1	2	3	4	5
1	0.311432	—0.124048	—0.286506 · 10	0.298409 · 10	—0.169010 · 10	0.331464
2	0.499096 · 10^{-1}	0.660246 · 10^{-1}	—0.863325	0.354419 · 10	—0.358471 · 10	0.119122 · 10
3	0.216360	0.438290	—0.975566	0.766205 · 10^{-1}	0.126497 · 10^{-1}	—0.386903 · 10^{-1}
4	—0.433867	0.123217	—0.519648	0.141667 · 10	0.944493 · 10^{-2}	—
5	0.363104	—0.281127 · 10^{-1}	—0.195138	—0.653019	—	—
6	—0.120343	0.171534	0.114723	—	—	—
7	—0.385342 · 10^{-1}	—0.521881 · 10^{-1}	0.831912 · 10^{-1}	—	—	—
8	0.275163 · 10^{-1}	—0.305918 · 10^{-1}	—	—	—	—

both at low and high temperatures. This necessitated an additional analysis of [2.33].

In the first place, we checked the agreement between these results and data of other investigators located on adjoining segments of the thermodynamic surface. Graphical analysis of isochores, drawn from the results of Michels et al. [2.27] at $T \geqslant 273$ K and of Gibbons [1.94] at $T \leqslant 70$ K was not capable of giving preference to data of Onosovskiy and Moroz [2.31] or Streett [2.33] located in the intermediate temperature range, due to the poor accuracy of graphical constructions. For this reason we compiled two local equations of state. To obtain the first we supplemented the body of original data by results presented in [2.27, 1.94, 2.33] and by experimental points on the saturation curve. The second equation was obtained by replacing data in [2.33] by those in [2.31]. It was found by analyzing these auxiliary equations of state that only the second of these provides a satisfactory description of the entire database. The first equation is not capable of describing experimental data, particularly distorting the saturation curve.

It was noted previously that Streett corrected his density measurements by 0.25%. Our analysis showed that disregarding this correction improves significantly the agreement between data in [2.33] with the results of Onosovskiy and Moroz [2.31] and Holborn and Otto [2.23]. For this reason, we recorrected the data in Streett's article by the same amount and listed the results for four isotherms in Table 29. The values along the 80 K isotherm as before (even after

TABLE 29. Corrected Data of Streett [2.33] Used in Compiling the Equation of State of Neon

At T = 90 K		At T = 100 K		At T = 110 K		At T = 130 K	
p, атм	Z	p, атм	Z	p, атм	Z	p, атм	Z
73.055	0.95785	75.457	0.9868	74.549	1.0059	73.597	1.0263
141.034	0.98777	141.306	1.0213	141.101	1.0434	143.072	1.0704
175.024	1.0256	171.012	1.0498	175.023	1.0741	175.022	1.0963
243.007	1.1271	209.013	1.0944	209.015	1.1100	212.141	1.1305
276.998	1.1858	209.013	1.0944	243.005	1.1507	243.279	1.1614
310.988	1.2478	243.955	1.1416	276.994	1.1945	276.995	1.1962
344.984	1.3110	277.535	1.1909	310.988	1.2407	310.989	1.2349
378.974	1.3753	310.984	1.2427	344.979	1.2889	344.981	1.2742
412.970	1.4406	344.977	1.2978	378.972	1.3386	412.976	1.3552
446.960	1.5064	378.969	1.3539	412.967	1.3890	446.959	1.3967
480.956	1.5718	412.960	1.4106	446.960	1.4402	548.945	1.5237
514.955	1.6369	446.954	1.4679	480.955	1.4915	582.940	1.5668
548.941	1.7023	480.949	1.5259	514.950	1.5431	616.936	1.6103
582.936	1.7680	514.943	1.5839	548.946	1.5949	650.931	1.6543
616.933	1.8325	548.939	1.6422	582.941	1.6470	684.926	1.6974
650.930	1.8973	582.934	1.7003	616.938	1.6991	752.937	1.7839
684.926	1.9613	616.931	1.7582	650.935	1.7514	820.926	1.8701
752.923	2.0890	650.928	1.8158	684.932	1.8030	888.924	1.9552
		684.927	1.8733	752.928	1.9069	956.923	2.0400
		752.923	1.9875	820.922	2.0096	1024.932	2.1248
		820.917	2.1009	888.925	2.1108		
		888.917	2.2133	956.925	2.2116		

correction) deviate significantly from data of others and distort the computed thermodynamic surface of gaseous neon. We note in connection with this that the compressibility at $T = 80$ K was measured by Streett for neon only. Consequently, at this temperature the equipment was not calibrated with either argon or nitrogen. In addition, the 80 K isotherm in all the experiments by Streett and coworkers, including experiments with other substances, is the lowest. For this reason, it is the most doubtful. Therefore, Streett's data along the 80 K isotherm were not included in the body of original data.

At temperatures above 0°C we used data of Michels et al. [2.27] which cover a wide pressure range up to 150°C, data of Nichols and Schneider [2.26] and also our results (see Table 13). The starting body of data for temperature and density ranges for which experimental data are not available was supplemented by computed data listed in Table 24.

Such a body of starting data was used for determining the coefficients of the equation of state of gaseous neon (see Table 25). The equation then served for calculating values of Z which were compared with experimental results.

Table 30 compares the computed values of Z_s with experimental results of Mathias et al. [2.145]. The agreement is quite satisfactory except for the low density region. The reason for these deviations were discussed above. However, it should be mentioned here that the results of McCarty and Stewart [2.40] at $\rho'' < 0.04$ g/cm³ are also lower than those in [2.145]. The computed values of density ρ'' obtained by us and by McCarty and Stewart [2.40] agree with one another within ±0.25%. At the same time significant differences (Table 31) exist between the experimental values of ρ'' [1.94] and those obtained from the

TABLE 30. Comparison of Computed Values of Z_s Along the Saturation Curve of Neon with Experimental Data in [2.145]

T, K†	ρ_s, g/cm³	Values of Z_s	
		experimental	computed
25.13	0.00534	0.96768	0.96640
26.11	0.00711	0.97392	0.95814
27.11	0.00939	0.97155	0.94820
30.09	0.02013	0.91708	0.90702
33.05	0.03831	0.84955	0.84952
36.07	0.06742	0.77419	0.77515
37.79	0.09310	0.71950	0.72091
39.04	0.11591	0.67997	0.67968
41.02	0.16563	0.60401	0.60469
42.98	0.23935	0.52151	0.52022
42.98	0.74866	0.16673	0.16573
41.02	0.85421	0.11711	0.11763

† The temperature scale was corrected in accordance with recommendations in [2.40]:

$$T = T_{\text{exptl}} \frac{273.15}{273.09} - 0.05 \text{ K}.$$

TABLE 31. Comparison of Computed Values of Density of Gaseous Neon Along the Saturation Curve with Experimental Data of Gibbons [1.94]

T, K	ρ'', g/cm³	
	experimental	computed
32.56	0.03489	0.03474
34.97	0.0554	0.0554
37.47	0.0866	0.0876
41.93	0.1936	0.1963
42.94	0.02449†	0.2389
43.93	0.30188	0 2944

† This is apparently a misprint.

equation of state. We also calculated deviations for all the data points obtained in the gas phase. The 14 quasi-isochores selected by us with approximately the same density step and plotted in Fig. 32† do not show all of them since they partially or completely overlap one another. It is seen from Fig. 32 that at temperatures above 50 K the majority of the deviation points do not go above ±0.2%. The deviations increase with reduction in temperature due to internal scatter of the test points themselves and the poorer accuracy of the equation of state in the critical region and at the density limit of its applicability.

Comparison with experimental data of Onosovskiy and his coworkers [2.31, 2.32] along eight isotherms (225 points) showed that the standard deviation‡ is 0.11% with a maximum of 0.54%. Figure 33 gives a graphic idea about the nature of differences between the experimental and computed values. The largest deviations also with opposite sign were noted at the lowest temperatures of 52 and 65 K.

The standard deviation at the four isotherms (82 points) relative to the corrected data of Streett [2.33] was 0.07% with a maximum of 0.20%. It is seen from Fig. 34 that the equation provided a maximum averaging of Streett's corrected data, which results in a normal distribution of deviations. It is also of interest to compare the computed values of Z with the experimental values of Sullivan and Sonntag [2.29] obtained over approximately the same temperature range. It follows from Fig. 35 that the data in [2.29] are systematically higher than the computed. The deviations increase with the density (pressure) at all the isotherms, attaining unacceptably high values.

† In Figs. 32 through 44 $\delta Z = \dfrac{Z_{\text{exptl}} - Z_{\text{calc}}}{Z_{\text{exptl}}} \cdot 100\%$.

‡ Here and henceforth the standard deviation is defined as

$$\sigma_Z = \sqrt{\frac{\sum_{k=1}^{n} \delta Z_k^2}{n}}$$

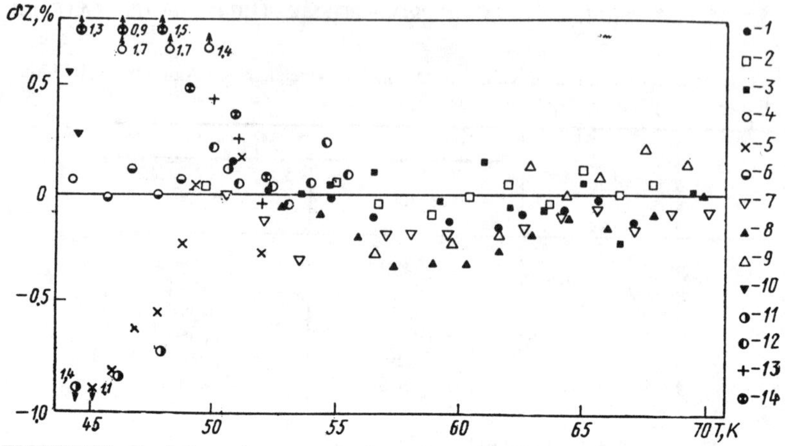

FIGURE 32. Deviations of computed values of Z for neon from experimental data of Gibbons [1.94] at quasi-isochores. 1) ρ = 0.1299 g/cm^3; 2) 0.2042, 3) 0.3012; 4) 0.03485; 5) 0.3656; 6) 0.4680; 7) 0.5635; 8) 0.6169; 9) 0.6804; 10) 0.7469; 11) 0.8403; 12) 0.9048; 13) 0.9396; 14) 0.9625.

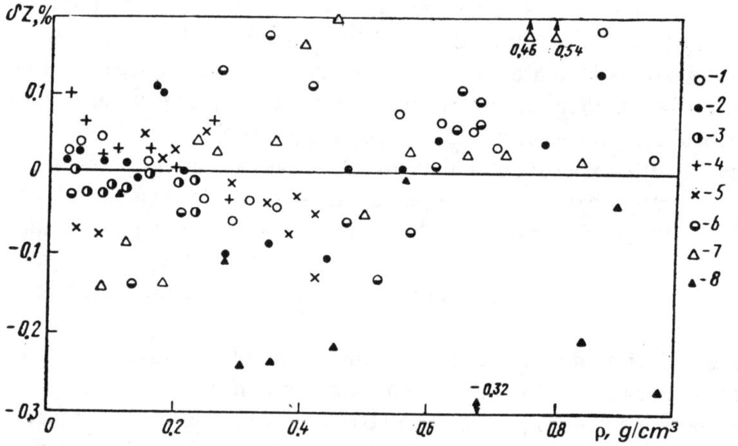

FIGURE 33. Deviations of computed values of Z for neon from experimental data of Onosovskiy and Moroz [2.31] and Onosovskiy [2.32] along isotherms. 1) T = 298.15 K; 2) 273.15; 3) 223.15; 4) 173.15; 5) 123.15; 6) 77.65; 7) 65; 8) 52.

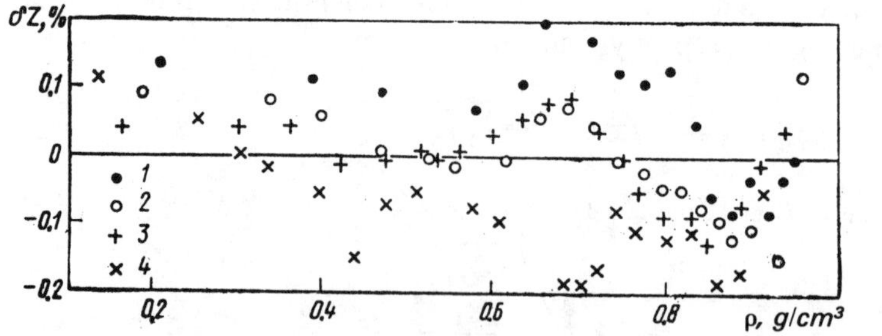

FIGURE 34. Deviations of computed values of Z of neon from Streett's corrected data [2.33] along isotherms. 1) T = 90 K; 2) 100; 3) 110; 4) 130.

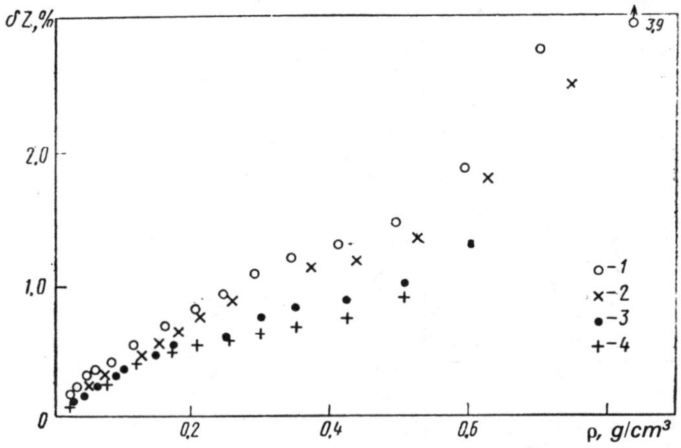

FIGURE 35. Deviations of computed values of Z of neon from the experimental data of Sullivan and Sonntag [2.29] along isotherms. 1) T = 70 K; 2) 80; 3) 100; 4) 120.

Figure 36 presents a comparison with experimental data of Michels et al. [2.27]. The maximum density along the isotherms is limited to a pressure of ~1000 bar. As seen from the figure, the overwhelming part of the deviations lie within the limits of ±0.1%. The calculations in each of the 312 data points showed that the standard deviation is less than 0.08% with a maximum of 0.23% along the 150°C isotherm at density of ~0.7 g/cm³. The satisfactory agreement between computed data and our experimental results (see Table 15) is depicted by Fig. 37 which shows deviations along five isochores: v = 11.553, 10.536, 8.718, 7.624 and 6.867 cm³/g. The standard deviation for 60 points was 0.07%, the maximum being 0.17%. And finally, we note that our equation of state of gaseous neon describes satisfactorily the computed values of Z at high temperatures and pressures (see Table 24), where up to now no experimental data were obtained. The standard deviation for these points was 0.05%.

The above shows that a detailed comparison of the computed and experi-

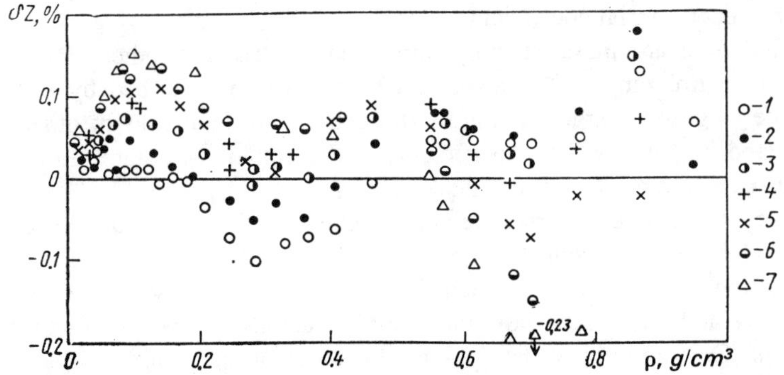

FIGURE 36. Deviations of computed values of Z for neon from experimental data of Michels et al. [2.27] along isotherms. 1) t = 0°C; 2) 25; 3) 50; 4) 75; 5) 100; 6) 125; 7) 150.

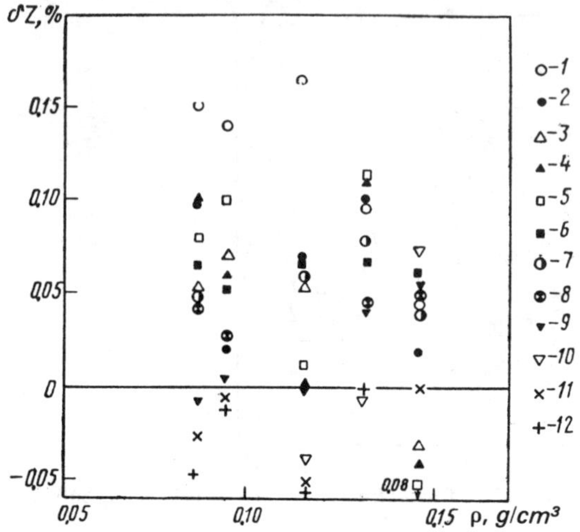

FIGURE 37. Deviations of computed values of Z for neon from experimental data listed in Table 15. 1) t = 25°C; 2) 50; 3) 75; 4) 100; 5) 125; 6) 150; 7) 200; 8) 250; 9) 300; 10) 350; 11) 400; 12) 450.

mental values of the compressibility factor of neon leads to the conclusion that our equation of state, within the limits of compatibility of the original experimental data, describes them with acceptable accuracy. Hence, it can be recommended for calculating the thermodynamic properties of gaseous neon at temperatures from the triple point to 1300 K and density from zero to 2 ρ_{cr} at a pressure of $p \leqslant 1000$ bar.

Argon. The body of starting data included experimental *pvT* data of Michels et al. [2.50, 2.53], Lecocq [2.54], and also our own results (see Table 14). The values of Z_s for T from 120 to 150 K and ρ between 0.06 and 0.99 g/cm³ were calculated from data of Michels et al. [2.53]. At lower temperatures, the accuracy of computed values of Z'' is determined to a significant measure by the reliability of the second virial coefficient.

The selected data are in satisfactory agreement with one another. They cover almost the entire range of parameters which is to be described by the equation of state of gaseous argon and are more exact than the results of other investigators listed in Table 9. The coefficients of the equation of state (2.46) found from this equation are listed in Table 26.

The accuracy of our equation of state was checked by direct comparison of the computed values of Z at each data point. The standard deviations and the maximum deviations on all the experimental isotherms and five isochores are listed in Tables 32 and 33, which show that these deviations do not go past the acceptable limits. Only individual data points deviate from the computed thermodynamic surface more than is permitted by experimental error. On the whole, it is interesting to note that the standard deviation for the entire body of data

TABLE 32. Deviations of Computed Values of Compressibility Z of Argon from Experimental Data

Isotherm, °C	Deviation, %		Isotherm, °C	Deviation, %	
	standard	maximum		standard	maximum
Comparison with studies of Michels et al. [2.50, 2.53]			25	0.07	0.17
			50	0.06	0.17
			75	0.07	0.18
−140	0.09	0.11	100	0.07	0.18
−135	0.07	0.08	125	0.07	0.17
−130	0.05	−0.09	150	0.08	0.19
−125	0.05	−0.22			
−122.5	0.06	−0.20	Comparison with the study by Lecocq [2.54]		
−120	0.07	0.18			
−110	0.08	0.27			
−100	0.07	0.15	300	0.06	−0.35
−85	0.05	−0.15	400	0.05	−0.11
−70	0.05	−0.12	500	0.05	−0.16
−50	0.05	0.09	650	0.13	0.25
−25	0.06	0.13	800	0.02	0.05
0	0.07	0.22	950	0.07	−0.28

points amounts to ~0.07%. This is virtually identical to the result obtained by us [1.104] where the entire gaseous region is described by two equations of state.

The density of argon along the saturation curve was investigated by Mathias and his coworkers [1.113] over the temperature range from −183.15 to −122.44°C. The gas side density at temperatures below −160°C was determined by them from the equation of state. The orthobaric densities of liquid and gaseous argon were investigated in more detail by Michels et al. [2.53] at temperatures from −153 to −122.3°C. Over the entire temperature range the values of density of liquid argon ρ' obtained in these studies are in satisfactory mutual agreement. But, perceptible deviations of 1% and more are observed on the gas side. This makes it impossible to satisfy both groups of data simultaneously.

Table 34 compares the experimental data of Michels et al. [2.53] with those

TABLE 33. Deviation of Computed Values of Z of Argon from Smoothed Experimental Data in Table 15

Isochore, cm³/g	Deviation, %	
	standard	maximum
5.657	0.17	0.35
4.607	0.12	0.26
3.869	0.06	0.17
3.569	0.04	−0.18
3.297	0.10	0.17

TABLE 34. Comparison of Computed Values of Z_s for Argon with Experimental Data in [2.53]*

Density, g/cm^3	Temperature, K	Value of Z_s experimental	Value of Z_s computed	Deviation, %
0.06124	120.33	0.80574	0.80383	0.24
0.08386	126.05	0.75875	0.75801	0.10
0.10418	130.11	0.72192	0.72126	0.09
0.13521	134.98	0.67140	0.67084	0.08
0.17038	139.16	0.62029	0.61994	0.06
0.21157	142.80	0.56668	0.56667	0.03
0.26527	146.13	0.50500	0.50529	—0.06
0.31503	148.19	0.45440	0.45515	—0.17
0.37133	149.66	0.40390	0.40500	—0.27
0.64269	150.41	0.23891	0.23871	0.08
0.70429	149.67	0.21302	0.21303	—0.06
0.72699	149.25	0.20365	0.20383	—0.09
0.77229	148.02	0.18435	0.18411	0.13
0.82226	146.32	0.16392	0.16396	—0.02
0.86410	144.44	0.14663	0.14643	0.14
0.93497	140.43	0.11819	0.11832	—0.11
0.98685	136.77	0.09830	0.09826	0.04

* Since the value of the critical temperature in this study is 0.2 K on the high side, we eliminated three points in the immediate proximity of the critical.

calculated from the equation of state along the saturation curve. The small deviations point to high reliability of the equation over the range of parameters listed in Table 34. Naturally, the computed values of Z_s differ more significantly from data of Mathias et al. [1.113] on the gas side, particularly at $T < 115$ K. Note, however, that it is precisely in this region of temperatures that our computed data are based on a reliable second virial coefficient and are hence more reliable.

When the equation of state for gaseous argon was compiled, new data appeared on the orthobaric densities of argon [2.147] obtained by the method of quasistatic thermograms and estimated by the authors themselves to have an error of 0.15%. Shavandrin et al. [2.147] point out that the sensitivity of the method decreases with increasing distance from the critical point. Table 35 compares experimental values of density along the saturation curve at the gas side with those calculated from the equation of state. The largest deviations are observed at low temperatures ($T < 109$ K). In the remaining region the agreement is satisfactory.

Krypton. The data base is comprised of experimental data of Whalley and Schneider [2.70], Trappeniers et al. [2.71], and Theeuwes and Bearman [2.72] for the gas phase, and also experimental values obtained by us and represented in Table 16. At temperatures and densities for which no experimental data are available the data base was supplemented by computed values from Table 22.

The results of Streett and Staveley [2.73] for the gas phase are shown along

TABLE 35. Comparison of Computed Values of Density of Gaseous Argon Along the Saturation Curve with Experimental Data in [2.147]

T, K	ρ'', g/cm^3		T, K	ρ'', g/cm^3	
	experimental	computed		experimental	computed
102.44	0.02127	0.02204	122.93	0.07092	0.07092
104.39	0.02406	0.02301	126.00	0.08373	0.08368
109.82	0.03315	0.03307	128.88	0.09767	0.09773
114.16	0.04303	0.04310	128.89	0.09804	0.09778
114.87	0.04486	0.04494	129.64	0.1021	0.1018
118.23	0.05447	0.05458	137.06	0.1526	0.1518
119.55	0.05877	0.05882	137.11	0.1526	0.1521
120.72	0.06285	0.06279	138.47	0.1646	0.1641
120.77	0.06286	0.06296	139.55	0.1751	0.1745
121.85	0.06684	0.06683	141.74	0.1993	0.1984
122.44	0.06909	0.06904	145.71	0.2579	0.2567

two isotherms (210 and 220 K) starting with $\rho \simeq 1.3$ g/cm^3. Comparison of these results with data of Theeuwes and Bearman [2.72] showed that they agree within 0.1–0.2% with respect to density, and this corresponds to the overall error of independent measurements. However, the maximum deviations of the same data with respect to compressibility for individual points are as high as 0.4–0.6%, due to the large curvature of isotherms over the range of parameters under study. The simultaneous use of data of Theeuwes and Bearman [2.72] and Streett and Staveley [2.73] for calculating the coefficients of the equation of state resulted in greater discrepancies between experimental and computed values along the saturation curve. For this reason, the results of [2.73] were not included in the data base.

The orthobaric densities of krypton were investigated by Mathias et al. [2.148] and Theeuwes and Bearman [1.118], who obtained results in satisfactory mutual agreement. The basis consisted of the results of [1.118], which were supplemented by data of [2.148] at $\rho < 0.58$ g/cm^3. These experimental data are compared in Table 36 with those computed from equation of state (2.46) whose coefficients are listed in Table 27. The agreement appears to be quite satisfactory.

It is seen from comparison (Fig. 38) of the computed values of Z for krypton and of experimental values of Theeuwes and Bearman [2.72] for the range of T from 208 to 240 K and $\rho = 0.57–1.77$ g/cm^3 that the equation optimally averages these data. As a result, the distribution of deviations becomes normal. The standard deviation was 0.14% with a maximum of 0.37% at $\rho = 1.705$ g/cm^3. Comparison of the analytic values and experimental data [2.73] shows (Fig. 39) that at $\rho < 1.5$ g/cm^3 these data exhibit an internal scatter. At 1.5 to 1.7 g/cm^3 they lie basically above the computed, and at greater density (up to twice the critical) are satisfactorily described by the equation of state.

TABLE 36. Comparison of Computed Values of the Compressibility Factor of Krypton Along Saturation Curve with Experimental Results of Mathias et al. [2.148] and Theeuwes and Bearman [1.118]

Temperature, K	Density, g/cm³	Value of Z_s	
		experimental	computed
		From data of [2.148]	
125.97	0.01333	0.96385†	0.95602
134.13	0.02196	0.93973†	0.93671
144.04	0.03739	0.90503	0.90710
153.34	0.05774	0.87503	0.87391
163.69	0.09004	0.82809	0.82845
170.93	0.12014	0.79161	0.79129
180.83	0.17576	0.73149	0.73072
188.39	0.23501	0.67477	0.67402
193.60	0.29030	0.62570	0.62608
199.64	0.37740	0.55907	0.55912
201.91	0.42170	0.52845	0.52827
		From data of [1.118]	
206.73	0.5793	0.43009	0.43150
208.00	0.6553	0.39132	0.39256
208.76	0.7315	0.35660	0.35763
209.32	0.8075	0.32711	0.32772
209.27	1.0371	0.25410	0.25439
208.82	1.1132	0.23461	0.23446
208.13	1.1893	0.21625	0.21625
206.98	1.2650	0.19808	0.19807
205.38	1.3396	0.18032	0.18020
203.17	1.4177	0.16186	0.16164
200.33	1.4969	0.14331	0.14310
196.94	1.5743	0.12547	0.12541
193.05	1.6494	0.10860	0.10862
192.07	1.6668	0.10479	0.10485

† These values were obtained in [2.148] analytically and were not used in compiling the equation of state.

The overwhelming majority of data points in the study by Trappeniers et al. [2.71] deviated from the computational surface within ±0.1% (Fig. 40). The standard deviation for this cluster of data is somewhat smaller. It should, however, be noted that analytic description of the internally highly compatible experimental results in [2.71] by the local equation of state makes it possible to reduce threefold the standard deviation. However, the nature of the deviations remains the same as shown in Fig. 40.

The differences between our experimental data (see Table 16) and those calculated from the equation of state are shown in Fig. 41. The standard deviation for all the experimental points was 0.18%, the maximum 0.40%. The computed values are basically lower. This is apparently due to the fact that the data

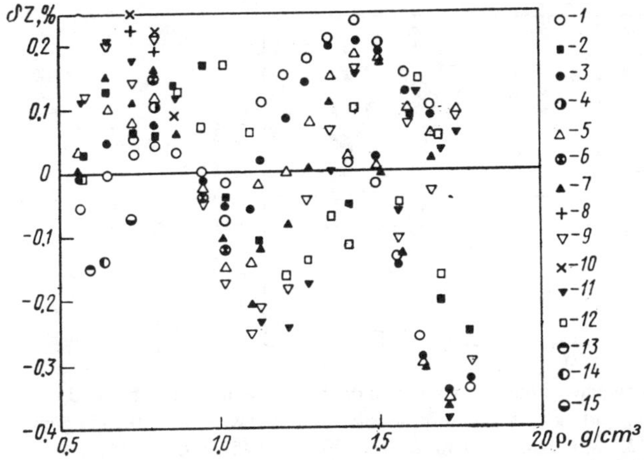

FIGURE 38. Deviations of computed values of Z for krypton from Experimental Values of Theeuwes and Bearman [2.72] 1) T = 240 K; 2) 237; 3) 235; 4) 232; 5) 230; 6) 227; 7) 225; 8) 222; 9) 220; 10) 217; 11) 215; 12) 210; 13) 208; 14) 208.15; 15) 209.

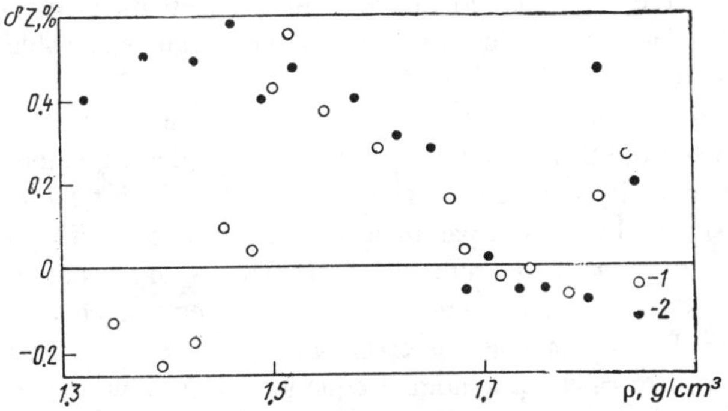

FIGURE 39. Deviations of computed values of Z for krypton from experimental data of Streett and Staveley [2.73] at the 210 K (1) and 220 K (2) isotherms.

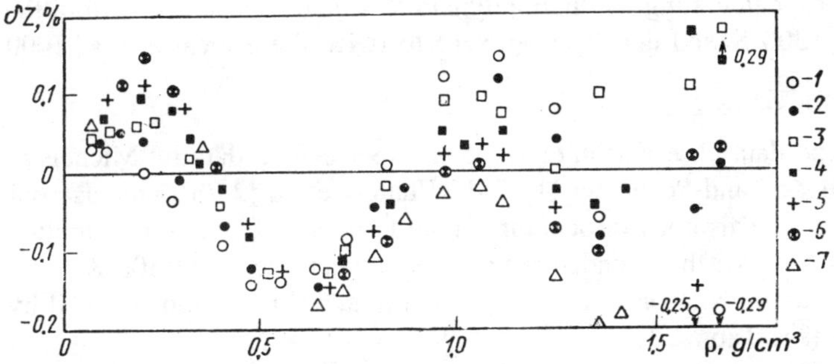

FIGURE 40. Deviations of computed values of Z for krypton from the experimental data of Trappeniers et al. [2.71] along isotherms. 1) t = 0°C; 2) 25; 3) 50; 4) 75; 5) 100; 6) 125; 7) 150.

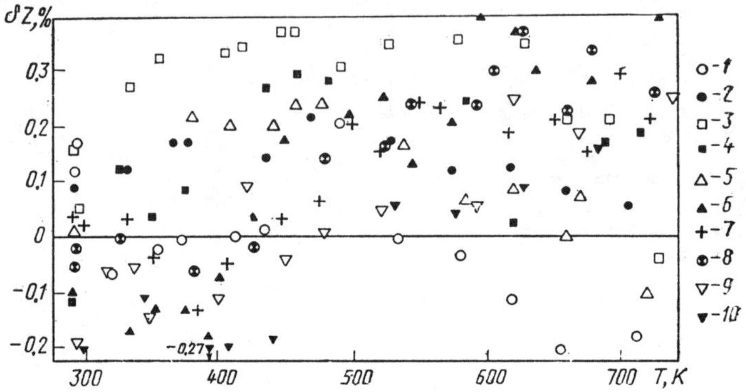

FIGURE 41. Deviations of computed values of Z for krypton from experimental data listed in Table 16 along quasi-isochores starting with values of density listed below. 1) ρ = 0.1853 g/cm^3; 2) 0.2374; 3) 0.2912; 4) 0.3327; 5) 0.3442; 6) 0.4052; 7) 0.4307; 8) 0.4575; 9) 0.4666; 10) 0.5285.

base for the high temperature and pressure region included data computed by the method of similarity (see Table 22) which contained a small systematic error.

The standard deviation for all the 65 points of extrapolated values in Table 22 was 0.11% with a maximum of 0.23%.

Detailed comparison of computed and experimental data on the compressibility of gaseous krypton showed that the local equations of state describe more accurately the internally mutually compatible data of individual investigators over the experimentally explored ranges of parameters. The desire to describe a wide temperature and density range with a number of coefficients larger than in the local equation, involves, as a rule, an increase in individual errors. This, in our opinion, is due to small but systematic inter-laboratory errors of various experimental data and specifics of the polynomial equation. This remarks applies also to equations of state obtained for the other three inert gases.

It must, nevertheless, be admitted that the equation of state being analyzed reflects quite acceptably the original experimental material, and hence it can be recommended for calculating the thermodynamic properties of krypton from the triple point to 1300 K and density from zero to twice the critical at $p \leqslant 1000$ bar.

Xenon. The data base was comprised of experimental data by Michels et al. [2.77], Habgood and Schneider [1.124], Whalley et al. [2.78], and also our own results [2.83]. This body of data turned out to be insufficient for compiling the equation of state for the intended range of parameters (to T = 1300 K and p = 1000 bar). For this reason, the experimental material was supplemented by computed data (see Table 23).

Analysis of experimental data on the density of gaseous xenon along the saturation curve, given only in a single publication [2.149], performed by us [1.89] showed that these data suffer from an error of several percent. The values of ρ″ were corrected by using the entire body of information on gaseous xenon. At low temperatures, we used data on the second virial coefficient, at higher temperatures, the data of Michels et al. [2.77], and the results of studies of compressibility of transcritical xenon [1.12]. The tabulated data on the density of xenon along the saturation curve at round values of temperature obtained as a result are listed in [1.89]. Their maximum error is estimated at 0.6%, and these data were taken as the reference.

The coefficients of the equation of state of gaseous xenon (2.46) found as a result are listed in Table 28. The experimental points of [2.77, 1.124, 2.83] deviate from the computed thermodynamic surface (Figs. 42 through 44) within the limits of the permissible experimental error including systematic interlaboratory errors. The distribution of deviations by sign and magnitude is close to normal. The deviations by different groups of data making up the data base are:

	standard deviation, %	maximum deviation, %
Michels et al. [2.77]	0.10	—0.32
Habgood and Schneider [1.124]	0.08	0.18
Whalley et al. [2.78]	0.04	0.12
Rabinovich et al. [2.83]	0.16	0.38
Extrapolated values of Table 23.	0.11	0.33

The deviations were cut almost in half when each group of data was described by a local equation of state. Apparently the same general considerations put forward in analyzing data and equations of state for gaseous krypton also apply here.

We wish to draw attention to the special nature of deviations of the computed from experimental data in [1.124] obtained in the transcritical region. As seen from Fig. 43, the deviations change sign upon transition through the critical density. This is possible due to the fact that the equation of state does not satisfy critical conditions. This results in the appearance on the computed critical

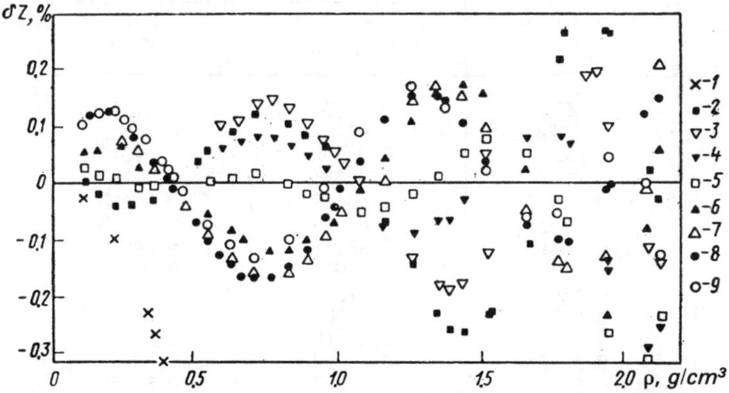

FIGURE 42. Deviations of computed values of Z for xenon from experimental data in [2.77] along isotherms. 1) t = 0°C; 2) 25; 3) 30; 4) 40; 5) 50; 6) 75; 7) 100; 8) 125; 9) 150.

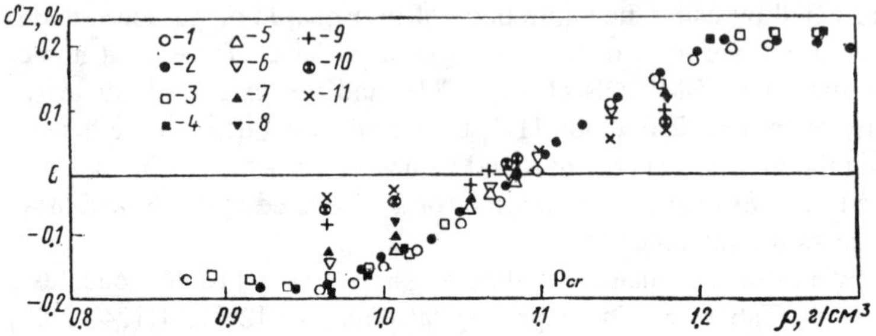

FIGURE 43. Deviations of computed values of *Z* for xenon from experimental data in [1.124] in the transcritical region. 1) *t* = 16.620°C; 2) 16.600; 3) 16.585; 4) 16.550; 5) 16.690; 6) 16,790; 7) 16.990; 8) 17.190; 9) 17.590; 10) 17.990; 11) 18.390.

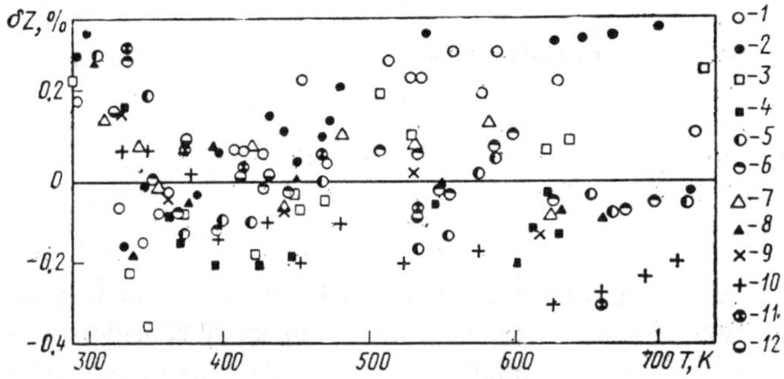

FIGURE 44. Deviations of computed values of *Z* for xenon from experimental data in [2.83] along quasi-isochores starting with different values of density. 1) ρ = 0.9166 g/cm³; 2) 0.8975; 3) 0.8272; 4) 0.6303; 5) 0.5759; 6) 0.5592; 7) 0.5447; 8) 0.5157; 9) 0.4142; 10) 0.3565; 11) 0.3195; 12) 0.2799.

TABLE 37. Comparison of Computed Values of *Z* for Xenon Along the Saturation Curve with Tabulated Data in [1.89]

Temperature, K	Density, g/cm³	Value of Z	
		Rabinovich and Veksler [1.89]	Computed
161.36	0.00820	0.97390	0.97350
170	0.01290	0.96278	0.96238
180	0.02059	0.94592	0.94646
190	0.03124	0.92635	0.92722
200	0.04551	0.90427	0.90453
210	0.06417	0.87973	0.87825
220	0.08812	0.85124	0.84825
230	0.1186	0.81918	0.81415
240	0.1585	0.77820	0.77382
250	0.2109	0.72602	0.72625
260	0.2793	0.67110	0.67148
270	0.3768	0.60072	0.60261
280	0.5274	0.51213	0.51354
289.74	1.100	0.28892	0.28928
280	1.764	0.15312	0.15425
270	1.970	0.11489	0.11553

isotherm of an S-shaped segment, inherent to subcritical isotherms which pass through the two-phase region. Note that these deviations do not go past specified limits.

Table 37 compares the computed values of the compressibility factor of xenon along the saturation curve with tabulated data in [1.89]. The agreement is highly satisfactory. The maximum deviation in only one point is 0.61%, which corresponds to preliminary estimates made in [1.89].

The equation of state of gaseous xenon was used for calculating its thermodynamic properties at temperatures from the triple point to 1300 K and density from zero to twice the critical at pressures of $p \leqslant 1000$ bar.

CHAPTER

THREE

EQUATIONS OF STATE FOR LIQUID NEON, ARGON, KRYPTON, AND XENON

The liquid region of the thermodynamic surface of a substance occupies an intermediate position between the solid and gaseous phases. At close to critical temperatures the properties of the liquid differ little from those of the gas, and for this reason there is no physical criterion which would allow a clear differentiation between these two phases. At the same time, the thermodynamic properties of the liquid near the solidification curve differ significantly from those of a gas since, as shown by X-ray diffraction analysis, the thermal motion of molecules in a liquid is close to that motion in a crystal. This peculiar ''dualism'' highly hampered the development of the theory of the liquid state. In spite of some success achieved during the past several years the problem of the form of the equation of state for a liquid has not been satisfactorily solved. Nevertheless, the use of certain theoretical information permits conclusions to be drawn on the form of the equation of state.

3.1. CONCISE INFORMATION FROM THE THEORY OF LIQUID STATE

The statistical theory of equilibrium properties of substances is based on the fundamental equation

$$F = -kT \ln Q_N \tag{3.1}$$

where F is the free energy; Q_N is a statistical integral (N is the particle population of the system) defined as

$$Q_N = \left[2\pi mkT/h^2\right]^{\frac{3}{2}N} Z_N \tag{3.2}$$

here h is Planck's constant; m is the mass of a particle; Z_N is the configuration integral,

$$Z_N = \frac{1}{N!}\int_V \cdots \int_V \exp\left[-\frac{U_N(q_1, q_2, \ldots q_N)}{kT}\right] dq_1 \ldots dq_N \tag{3.3}$$

where U_N is the potential energy of the system, q_i is the coordinate of the ith molecule and V is the volume of the system.

For low and moderate density gases the configuration integral can be calculated from a series expansion of Z_N in powers of density [1.19]. At values corresponding to the density of the liquid, when the expansion of Z_N may diverge, use is made of other methods. These can be conditionally classified as follows:

a. based on model theories;
b. based on solving integral equations for the radial distribution function;
c. numerical.

The model theories postulate the structure of the liquid and the nature of the thermal motion of molecules. The most popular or the so-called lattice theories (cell and hole) [1.8, 2.109, 3.1–3.3] are based on the free-volume concept. Their underlying concept is that of quasicrystal structure of a liquid in the sense that each particle is permitted to move in a cell near the node of some virtual crystal lattice. The cell is formed by adjoining molecules which limit the motion of the particle under study and may contain one or several particles. According to the hole theory† the number of nodes of such a lattice exceeds the number of particles. The degree of order in this case is lower than for a cell, and such a model conforms closer to a real liquid.

An important assumption which is made in evaluating the configuration integral on the basis of lattice theories consists of replacing the actual interaction of particles of system U_N by some self-consistent external field acting on each particle in its cell.

The Lennard-Jones and Devonshire theory of the "self-consistent" field [3.5] was developed for the case when the cell holds a single particle. Volume $v = V/N$ delegated to a single particle is taken to be equal to the volume of a cell, and the geometric centers of the cell form some regular lattice with the most probable structure. Distance l between the centers is related to the mean distance r_i between the particles and the volume, $v \sim r_i^3$, by the expression:

† The hole model of a liquid was postulated by Frenkel [3.4] long before the corresponding statistical theory was developed.

$$l^3 = \gamma v \tag{3.4}$$

where γ is the structure factor.

For the motion of each molecule in the potential field of its cell

$$U_N = \frac{1}{2} NE(0) - \sum_{i=1}^{N} [E(r_i) - E(0)] \tag{3.5}$$

where the first term is the statistical energy of a lattice comprised of molecules situated in the centers of the cells; $E(r_i)$ is the potential energy of the ith molecule at distance r_i from the center of a cell, calculated on the condition that all the remaining molecules remain in their centers. Under these conditions

$$Z_N = \exp\left[-\frac{NE(0)}{2kT}\right](v_f)^N \tag{3.6}$$

$$v_f = \int_{\Delta} \exp\left\{-\frac{1}{kT} E(r_i) - E(0)]\right\} dr \tag{3.7}$$

Integration in Eq. (3.7) is performed over the entire volume of the cell, and it can be performed if function $E(r_i)$ is known.

Quantity v_f is termed the "free volume" of the liquid and in general is a function of volume and temperature. It is interesting that in a somewhat simplified interpretation of the theory of the self-consistent field (suggested by Frenkel [3.4]) the statistical energy of the lattice is zero and the size of the cell is $\frac{4}{3}\pi l^3$. In addition, Frenkel suggested that Eq. (3.7) be written as:

$$v_f = \int_{\Delta} \exp\left[-\frac{E(r_i)}{kT}\right] dr = v \exp\left[-\frac{\chi(v)}{kT}\right] \tag{3.8}$$

where $\chi(v)$ is a function of the cell volume, similar in its meaning to the mean (statistical) value of the energy of the self-consistent field. On the assumption that the Lennard-Jones potential is valid, the function can be represented in the form

$$\chi(v) = -\frac{c}{v^{\mu}} + \frac{d}{v^{\nu}} \tag{3.9}$$

Dropping simple transformations, we write the equation of state obtained on these assumptions,

$$p = \frac{kT}{v} + \frac{\mu c}{v^{\mu+1}} - \frac{\nu d}{v^{\nu+1}} \tag{3.10}$$

It can be easily seen in this simplified version that the rough approximation is done by replacing the self-consistent potential $E(r_i)$, a function of temperature and volume, by the two-parameter Lennard-Jones potential. Such a replacement conforms to the concept that the potential energy of the particles forming the system is equal to the sum of energies of the paired interaction between particles situated closest to one another, with all the particles assumed to be rigidly held in their nodes. Due to neglect of the thermal motion of the surrounding particles (even in an elementary cell) function $E(r_i)$ and function $\chi(v)$ considered here depend only on volume v by way of the dimension of the elementary cell.

The theoretical possibility of finding the actual self-consistent potential was validated by Kirkwood [3.6], who obtained an integral equation for $E(r, v, T)$. However, the solution to this equation can be obtained only at the cost of simplifications which significantly distort the results [3.7].

An intuitive attempt at refining Eq. (3.10) was taken up by Vasserman and Rabinovich [3.8]. There a temperature dependence was introduced for all the terms of the equation and exponents μ and ν were selected so as to best satisfy the experimental data. In the final analysis they suggested the equation

$$p=\frac{A(T)}{v}+\frac{B(T)}{v^n}+\frac{C(T)}{v^m} \tag{3.11}$$

The equation, when written in this form, while retaining certain corollaries of the cell model, is predominantly empirical since no insight is provided into the structure of temperature functions. Strictly speaking, Eq. (3.11) can be obtained by using the concept of an effective potential which is temperature and volume dependent. However, at the present stage the form of this potential can be determined upon significant simplifications for the case of three-particle interaction. For this reason, the approach based on using experimental data for finding the constants of the equation of state, the form of which is fully validated, is the most preferable in attempting exact approximation of the input data.

The next step in developing the cell theory was made by Barker [1.8, 3.10] Taylor [3.7], and Dahler and Hirschfelder [3.9], who used the Lennard-Jones/Devonshire theory as the first approximation, with a correction in the form of an expansion analogous to the expansion of group integrals of Mayer [3.11] in the statistical mechanics of non-ideal gases. The authors of [3.7, 3.10] did not succeed in finding exact values of correction factors, and simplifications which followed from the independence of binary correlations caused their theories to be suitable only at high densities. There the mean displacements of particles from the equilibrium positions are very small.

The cell-cluster theory developed by de Boer [3.12, 3.13] allows motion of the liquid molecules not only in their own cells, but within the limits of a cluster of cells. On this assumption, in addition to evaluating the cluster integral, it is necessary to incorporate a factor which defines the number m of ways in which clusters of n molecules can be arranged within a lattice of N molecules. The complexity of the problem's solution in general required simplifications. The result was that one had to deal with clusters consisting of not more than two molecules on the assumption that they are statistically independent. This makes this theory suitable only at high densities where the probability of cell exchange and double filling is small.

A result which is important from the practical point of view and is based on the cell-cluster model was obtained by Rabinovich and Abovskiy [1.6, 3.14]. They retained the general computational scheme characteristic of the cell-cluster theory, with each cell holding only one molecule. For this case, the presence of more than one particle in a cell is a weak effect which can be easily accounted for phenomenologically. Their analytic equations were derived on the assumption that the repulsion branch of the two-particle potential function ($R < R_e$) is approximated by the Born-Mayer expression

$$U_{\text{rep}} = U_0 \exp\left(-\frac{\alpha R}{R_e}\right) \tag{3.12}$$

where $\alpha \cong 15$. The attraction branch is described by a significantly smoother relationship, for example, the Lennard-Jones potential. The free cell is a sphere of some diameter d, corresponding to high energies of repulsive interaction of the central molecule with molecules forming the cell. Under these conditions the configurational entropy is determined by the equation

$$S_{\text{conf}} = k \ln Z_{\text{sph}}^{(1)} \left[1 - O\left(\frac{1}{\alpha}\right)\right] \tag{3.13}$$

where $Z_{\text{sph}}^{(1)}$ is the group integral for a system of effective spheres whose diameter is a function of temperature and is equal to

$$d = \frac{R_e}{\alpha} \ln\left(\frac{\Phi_0}{LkT}\right) \tag{3.14}$$

Here L is a fitting parameter.

The final result for the entropy of a system upon allowance for the presence of more than one particle in a cell is

$$S = \frac{3}{2} R\,[1 + \ln(\lambda T r_f^2)] + R\,(m - n \ln T + q v^{\frac{1}{3}}) \tag{3.15}$$

where $$\lambda = \frac{2\pi m k}{h^2}; \quad r_f = v^{\frac{1}{3}} - C + D \ln T$$

C, D, m, n and q are fitting parameters,

$$C = \frac{R_e}{\alpha\gamma^{\frac{1}{3}}} \ln \frac{\Phi_0}{kL};\ D = \frac{R_e}{\alpha\gamma^{\frac{1}{3}}}$$

γ is the structure factor. The exact values of the fitting parameters are unknown. For that reason Eq. (3.15) can be used only when these are determined from experimental data.

A check performed by Rabinovich and Abovskiy [1.6, 3.14] on the basis of experimental data on the properties of liquid argon, krypton, and xenon showed that equations for calculating thermal and calorific properties, obtained from Eq. (3.15) describe experimental results over a wide temperature range at density $\rho \geqq 2.2\ \rho_{cr}$ with an error commensurable with the experimental error. Such high accuracy for Eq. (3.15) is due to the fact that it can be derived without making assumptions:

a. on the structure of the cell and the coordination number
b. on the feasibility of spherical symmetry of the cell potential
c. on the shape of the two-particle potential over the entire range of significant internuclear distances and numerical values of potential parameters
d. on additivity of the potential energy of the system.

The main approximation in deriving Eq. (3.15) consists of statistical independence of binary correlations. This limits the applicability of Eq. (3.15), as in the preceding cases, to $\rho \geqq 2.2\ \rho_{cr}$.

Certain advances were achieved in developing the "hole" theory of the liquid state of matter [1.8, 2.109, 3.1, 3.2]. The principal model concepts of the theory are based on the experimentally established fact that the coordination number of a liquid is smaller than that of a solid, and it decreases with increasing density. This phenomenon causes the appearance of vacant cells ("holes"), the number of which is a function of the parameters of state. However, it can be represented in explicit form only upon making rather rough assumptions [3.15]:

a. on the structure of the cells
b. on the location of nearest neighbors on the surface of the coordination sphere
c. on the dependence of the "free" volume on the fraction of nearest unoccupied cells.

In addition, use is made of other particular assumptions which, together with the basic assumptions, reduce the accuracy of calculations to such a degree that the results of theory cannot be put to practical use. The above pertains also to the modification of the "hole" theory, called the tunnel theory due to Barker [1.8, 3.16] and Chung and Dahler [3.17].

The multistructure Eyring theory [3.18, 3.19] is currently in rather extensive use. Its main idea consists of the assumption that a part of the molecules in a liquid participates in vibrational motion, analogous to that occurring in a solid, whereas another performs translational jumps from one equilibrium position to another. The motion of these particles is here assumed to be close to the motion of particles in a real gas. In addition, a number representing the content of solid-like and gas-like particles in the model is preselected. It is also assumed that the vibrations of solid-like particles are harmonic and independent, and the effect of the potential field in calculating the statistical sum of gas particles is eliminated. As a result, the differences between the experimental and computed data on compressibility are as high as 20–30% [3.18, 3.19].

More rigorous theories are based on the formalism of distribution functions, related to the probability that two liquid particles will find themselves at distance r from one another. Using the radial distribution function $g(r, v, T)$ the equation of state in the approximation of paired additivity of the potential energy of multiparticle interaction takes the form

$$p = \frac{RT}{v} - \frac{2}{3v^2} \int_0^{\infty} \left[\frac{\partial}{\partial r} U(r) \right] g(r, v, T)\, r^3 dr \tag{3.16}$$

It is very difficult to determine theoretically the radial distribution function even on ordinary assumptions. For this reason various methods were suggested for an approximate solution of the problem.

The superposition approximation suggested by Kirkwood [3.20] assumes independence of the paired probability densities. It neglects the effect of simultaneous residence of two particles at points 1 and 2 on the probability of residence of a third particle at point 3. The validity of this approximation is evident only at low densities. For this reason, the computed results for the liquid region deviate from experimental data on pv/RT by 50 to 100% [3.21].

The hyperchain approximation [3.22, 3.23] and the Percus-Yevick approximation [3.24, 3.25] are based on the Ornstein-Zernike relationship [3.26] between the correlation functions. Each of these approximations yields a certain relationship between the direct correlation function and the radial distribution function, as in the case of the superposition approximation, only for low-density gas. Comparison of computed results with experiments showed that the Percus-Yevick approximation is better than the hyperchain approximation at high temperature and allows exact solution of the equation for the system of solid spheres. However, the resulting equation of state does not yield the liquid-crystal transition at high densities [2.137].

For the gas phase, both theories were checked using the Lennard-Jones potential [3.27]. It was found that the virial coefficients, up to the fourth, are in satisfactory agreement with their exact values. The divergence for the fifth coefficient is quite significant. It is apparently the result of the use of the superposition approximation which introduces significant errors at high densities.

A feature peculiar to the Percus-Yevick theory is the use of two different equations of state for the compressibility factor and for pressure. Rowlinson [2.137] was able to avoid this using a self-consistent approximation, the effectiveness of which manifested itself in better agreement of virial coefficients with their exact values [3.28].

The thermodynamic theory of perturbations [1.19], where the initial approximation is the model of solid spheres, is a further development of the theory based on the formalism of distribution functions. This theory was developed for liquids by Zwanzig [3.29] and others [3.30–3.34]. The parameters of the expansion in these studies consisted of quantities which were more realistic potentials than the potential of solid spheres. Thus, Barker and Henderson [3.31–3.33] used two parameters: one defined the steepness of the potential in the repulsion region, and the other the location and depth of the potential well. Then the statistical sum was expanded with respect to these parameters in a double Taylor series. This approach is significantly limited by the "truncation" parameter (the largest interatomic distance to which the analysis is applied) and the feasibility of determining only the lower-order terms of the expansion. Nevertheless, when the results of the Percus-Yevick theory employing a temperature-dependent diameter of the solid spheres are used as the first approximation, satisfactory agreement is observed with experimental data for argon in the high-temperature gas region and in the region of dense liquid at low temperatures. This is due to the fact that the truncation parameter does not play an important role in these regions with the result that the contribution of interaction to the energy of the system at large distances is insignificant. In addition, the rather rapid convergence of the series is due to the fact that at low interatomic distances the realistic potential is close to the potential of solid spheres with temperature dependent diameter.

It appears that the most coherent theory is that of Mansoori and Canfield [3.34], who showed that retention of two or more terms in the expansion for the free energy makes it possible to write an inequality of a certain form. They suggest that the diameter of the effective spheres be sought by way of minimization of the right-hand side of the equality. Such an approach involves additional computational difficulties, but it yields results which are more accurate than those from the previous theory.

It should, however, be noted that the different versions of the theory of perturbation as a rule yield final results which are insufficiently accurate, due to the aforementioned simplifications. It is estimated by Barker et al. [3.33] and Mansoori and Canfield [3.34] that the error in describing experimental data on the properties of argon is 3–5%.

The ever-increasing use of computers has resulted in extensive utilization of the so-called numerical Monte Carlo and molecular dynamics methods [3.35–3.39]. These allow direct calculation of the configuration integral. The methods have significant shortcomings in the assumption of paired additivity of multiparticle interaction and limitations on the number of particles which comprise the system. This assumption results in significant errors at high densities. However, errors can be reduced somewhat by including a nonadditive term to the dispersion three-particle interaction. Unfortunately, the current theory is not capable of defining more complex multiparticle interactions†. For this reason it is preferable to use a method based on employing computers to find the potential from experimental data [2.137].

The fact that only a limited number of particles can be included in the analysis results in the distortion of results basically due to surface effects. This error is reduced by imposition of periodic boundary conditions [3.3, 3.36].

Calculations of thermodynamic functions of argon by the Monte Carlo [3.37] and molecular dynamics [3.38, 3.39] methods showed that the differences between experimental and computed values of pressure are as high as 10%. Although this error is still rather high, numerical methods hold a promise of better accuracy. At present they are used for obtaining unusual "experimental data" of certain idealized systems in order to check the theory of the liquid state.

It can be concluded, in summing up the above brief survey of modern methods of analytic determination of thermodynamic properties of liquids, that not a single one of the existing theories is capable of yielding information with accuracy commensurable with experimental results over a wide range of parameters. The reason lies in: (1) assumptions on the nature of molecular motion and of paired additivity of multiparticle interaction; (2) in the use of approximate two-particle functions; and (3) other approximations, the validity of which cannot always be estimated analytically. For this reason the main task in obtaining an exact approximation of experimental data on thermodynamic properties of liquids over a wide range of density and temperature is the selection of the form of the equation of state with satisfactory analytic validation.

Equations (3.11) and (3.15), the physical meaning of which follows from fully defined model concepts, are of interest from this point of view. Both equations employ approximately the same number of fitting parameters which are found from well worked out methods. However, preference in compiling and analyzing these forms for the purpose of final selection of one of them for calculating thermodynamic properties of liquid neon, argon, krypton, and xenon was given to Eq. (3.11). It is capable of describing these properties starting with $\rho_{min} \geqq 1.8\ \rho_{cr}$ (instead of $\rho_{min} \geqq 2.2\ \rho_{cr}$) with a somewhat better accuracy due to varying the form of temperature functions, and of the number of fitting parameters which are determined from experimental data.

3.2. EXPERIMENTAL DATA ON THERMODYNAMIC PROPERTIES

Experimental *pvT* data for liquid neon, argon, krypton, and xenon were only obtained recently. Argon was the first substance to be investigated. Michels et

† An opinion is current (see, among others, the paper by Toxvaerd and Praestgard [3.40]) that inclusion of three-particle interactions is sufficient for the dense gas and liquid regions. More complex interactions are of opposite signs and partially balance one another.

al. [2.53] obtained, together with data on compressibility of gaseous argon, results for the liquid phase at the −122.5, −125, −130, −135 and −140°C isotherms (from 3 to 11 points at each). Michels et al. [3.41] used a constant-volume piezometer held in a cryostat, connected by a capillary with a differential dead-weight pressure gauge. A diaphragm separates the test substance from the oil of the pressure gauge. The quantity of substance held by the piezometer was determined from results of measurements on the 0 and 25°C isotherms [2.50] obtained in the same laboratory. When the cryostat is cooled, a part of the gas flows from the differential pressure gauge to the piezometer. For this reason, the density of the substance in the latter during the time of the experiment was calculated from the pressure and ratio of volumes of both units.

Soon thereafter, Rogovaya with Kaganer [2.55] investigated the compressibility of argon between −183 and 0°C at pressures to 200 bar using the test facility described in [3.42] employing a constant-volume piezometer. The quantity of substance in the piezometer was determined by the volumetric method using three calibrated vessels. The temperature was measured by a reference-grade platinum resistance thermometer. The cryostat temperature during the experiments was maintained constant with an error of ±0.01 K using a photoelectric thermoregulator. The pressure was measured by a dead-weight pressure gauge. The oil contacting the gas was separated from the test substance by interposing a membrane-type differential pressure gauge and a mercury equalizer. The temperature in the constant-temperature bath was maintained to within ±0.05 K. The pressure of the gas in the calibrated vessels was measured by a mercury-filled manometer to within 0.05 mm Hg. Rogovaya and Kaganer used four piezometers with capacities from 7 to 400 cm^3. The error in determining the piezometer volumes was 0.02%, and of the volume of the gas meter, 0.03%. The correction for the dead space volume was made with an error of 0.05% (relative to the volume of the smallest piezometer). The tested argon contained less than 0.02% of nitrogen and oxygen as impurities. The experiments were performed over a wide range of density, and experimental data were obtained along 11 isotherms. However, only three of them (−135.03, −150.08 and −183.02°C) are for the liquid region. From 6 to 8 data points were obtained for each isotherm. They assumed that the error of the specific volume obtained by them does not exceed 0.05%. This estimate appears to us to be on the low side. When the error of volume determination was added, the error could have reached 0.1%, and the error of pressure measurement is not listed in their paper.

Van Itterbeek and Verbeke [3.43] were the first to determine the density of liquid argon over a narrow temperature range along four isotherms (86.637, 87.915, 89.13 and 90.55 K) at pressures to 150 kg/cm^2 using the standard technique of their laboratory. Subsequently, Van Itterbeek, Verbeke and Staes [1.108] investigated the density of liquid argon over a wide range of parameters with an improved test arrangement. The spring pressure gauge was replaced by a dead-weight unit, and a membrane-type differential manometer was used for separating the oil from the test substance. The pressure difference was determined with an error of 0.013 kg/cm² at pressures to 300 kg/cm². The temperature was measured with an error of less than 0.01 K by a platinum resistance thermometer calibrated in Leiden. An electronic device was used for maintaining the temperature constant within ±0.01 K. Van Itterbeek et al. [1.108] obtained *pvT* data for liquid argon along 11 isotherms at 90.13 to 148.15 K. There were 7 to 16 data points for each isotherm. The maximum pressure was 251–299 kg/cm².

Van Itterbeek et al. [1.108, 3.43] suggested two equations of state for liquid argon. The first, written as

$$\rho = A(T) + B(T)p + Cp^2 \tag{3.17}$$

describes experimental data at 86.6–90.6 K and pressures to 150 kg/cm². The second equation

$$v = A_0(T) + A_1(T)p + A_2(T)p^2 + A_3(T)p^3 + \\ + A_4(T)p^4 + A_5(T)p^5 \tag{3.18}$$

is valid, judging by the results of calculations [1.108] at 90–130 K at pressures to 300 kg/cm². It contains 36 constants, which for this relatively narrow range of parameters may cause an error in determining the derivatives. No mention is made in [1.108, 3.43] of the accuracy of results.

Streett [3.44] investigated the phase equilibrium and density of the neon-argon system at high pressure and obtained data on the density of pure argon along the 101.94, 110.78 and 121.37 K isotherms for pressures between 4 and 552 bar. First, he investigated the phase equilibrium of the mixtures and the density. To measure the latter, a mixture with a known composition was condensed in a piezometer with known volume, maintained at constant temperature, and then the pressure of the mixture (which was above the saturation pressure) was determined. The quantity of gas in the piezometer was determined by discharging it into a gas meter maintained at constant temperature. He performed a series of experiments along isotherms at different pressures for each mixture composition, and then the liquid density was extrapolated to the saturation pressure. The density of pure argon was most likely determined by the same method. The experimental data are given in tables and graphs. The maximum error in determining the density of mixtures is estimated by the author at 1%. However, the accuracy of data for argon is much higher since they agree with the results of the majority of investigators within ±0.3%.

Blagoy and Sorokin [3.45] constructed a facility for determining the density and isothermal compressibility of liquids at low temperatures. The main component of this facility is a variable-volume bellows-type piezometer with a graduated rod connected to the piezometer cover. When the bellows are filled with a certain quantity of substance and compressed, the rod moves in a steel tube from which the piezometer is suspended. The rod travel is determined through a viewing port by means of a KM-6 cathetometer.

The change of the bellows volume is obtained from the expression

$$\Delta V = F_{\text{eff}} \Delta x \qquad (3.19)$$

where F_{eff} is the effective cross section of the bellows which is independent of the compression Δx. The value of F_{eff} was determined by measuring the volume of distilled water displaced from the bellows when compressing it. It was verified in the course of these measurements that F_{eff} is constant within the limits of experimental error. Special measurements were performed to correct for the travel of the zero graduation of the rod due to expansion of the tube and of the piezometer casing as well as compression of the rod and bellows under pressure. A correction was made also for the stiffness of the piezometer itself.

The pressure dependence of the molar volume of the liquid at constant temperature is

$$V = V_0 + \frac{\Delta V}{n} = V_0 + \frac{F_{\text{eff}} \Delta x}{n} \qquad (3.20)$$

where V_0 is the molar volume of liquid at the saturation curve at some initial pressure and n is the number of moles of substance in the piezometer determined by weighing the steel tank into which the test substance was recondensed after the termination of the series of experiments. Two corrections were made here: (1) for the residual gas in the bellows; and (2) for the change in the density of air over the time elapsed between weighing the empty and filled tank.

The pressure in the experiments was measured by an MP-600 (class 0.05) dead-weight pressure gauge, and the temperature by a reference-type platinum resistance thermometer with an error of ± 0.01 K. The operating temperature of the piezometer was maintained automatically with the same accuracy.

As seen from Eq. (3.20), the accuracy of determining the molar volume depends basically on the reliability of determining the initial volume V_0, which makes the main contribution to V. The overall error of measuring the three other quantities (F_{eff}, Δx and n), including errors of reference, should not, in our opinion, exceed 0.1% of V.

Using the above facility, Blagoy and Sorokin [3.45] performed control measurements of the compressibility of liquid argon at 89.94 and 99.90 K to pressures of 250 and 400 kg/cm^2, respectively. The initial data consisted of values of density of the saturated liquid from data by Van Itterbeek et al. [1.108]. Comparison of data in [1.108] and [3.45] along the ~90 K isotherm showed that the disagreement does not exceed 0.1%. Subsequently Sorokin and Blagoy [3.46] used the same equipment to investigate the compressibility of liquid argon on the 89.94, 99.90, 108.23, 119.97 and 128.92 K isotherms from the saturation pressure to 498 kg/cm^2.

Van Witzenburg and Stryland [1.36] determined the compressibility of liquid and solid argon at temperatures between 96.41 and 153.94 K at pressures between 68 and 2022 kg/cm^2. They designed an original experimental apparatus which permitted simultaneous measurement of ρ, p and T.

A ~ 50 cm^3 volume metal vessel, weighing ~3 kg was suspended from a steel capillary (d_{in} = 0.5 mm, d_{out} = 1.5 mm) in the center of a cryostat. The vessel is suspended by way of the vertical part of the capillary to one of the pans of an equal-arm beam balance. The horizontal part of the capillary (1 m long) introduced an error of 3 to 5 mg into the results of weighing. The small diameter of the capillary reduced the correction for the dead space volume to a minimum and limited significantly the undesirable influx of heat. However, when investigating properties of solid argon the capillary frequently became plugged up. This necessitated placing a small electric heater at its end.

The volume of the vessel was calibrated by mercury and, at different temperature and pressure, was calculated from the formula

$$V = 48.303(1 + 1.53 \cdot 10^{-6}p)(1 - 3.8 \cdot 10^{-5}\Delta T - 1.8 \cdot 10^{-8}\Delta T^2) \tag{3.21}$$

where V is in cm^3, p is in kg/cm^2 and $\Delta T = (273 - T)$ K. The formula was obtained on the basis of available data on the compressibility of mercury and on the coefficient of thermal expansion of steel from which the vessel was made. It was found from Eq. (3.21) that at maximum pressure the volume of the vessel increased by 0.3% and at the lowest experimental temperature decreased by 0.8%. The weight of argon in the vessel was determined from three independent measurements: (1) of the evacuated vessel; (2) the vessel with the tested substance; and (3) a metal mockup of the vessel. All the weighings were performed with the capillary connected.

The temperature was determined to within 0.1 K using a platinum resistance thermometer located within the body of the vessel. The pressure in the vessel was measured by a dead-weight pressure gauge with a correction for the hydrostatic pressure of the oil, mercury and gaseous argon. The pressure measurement error is estimated at 0.1%. Weighing was performed under steady conditions and at specified values of p and T which allowed the determination of the behavior of $\rho = f(p, T)$. The argon used in these experiments was 99.995% pure.

The experimental results were published in tabulated form for 12 isotherms. To present these results as well as the coefficients of isothermal compressibility and thermal expansion all at round values of temperature and pressure Van Witzenburg and Stryland [1.36] compiled the empirical equation

$$\rho = \sum_{i=0}^{m} \sum_{j=0}^{n} a_{ij} p^i T^j \tag{3.22}$$

At pressures to 600 kg/cm^2 this equation with 20 constants (m = 4 and n = 3) describes the input data on the density of liquid argon (41 data points) with a standard deviation of 0.07%. Thirty constants ($m = 5$ and $n = 4$) were needed to describe 91 points with a somewhat higher accuracy ($\delta\rho_m = 0.04\%$) at pressures above 450 kg/cm^2. As a result, Van Witzenburg and Stryland described the density over the range of parameters under study by means of two empirical equations containing a total of 50 constants.

Using data of Michels and Prins [3.47] on the melting curve, Van Witzenburg and Stryland determined, on the basis of their experimental results, the density of liquid and crystal along the phase equilibrium curve at six temperatures between 96.41 and 120.08 K ($p = 540$–1669 kg/cm^2).

Streett and Staveley [2.99] investigated the *pvT* behavior of pure (99.98%) liquid argon at temperatures from 100.94 to 143.11 K at pressures from saturation to 680 bar. They constructed a test facility which was described in detail in their previous paper [3.48] and concerned with the compressibility of liquid nitrogen. The technique used in [3.48] was first suggested by Holborn and Otto [2.22, 2.24]. The main unit of the facility is a thick-walled piezometer with a volume of 3.66 cm^3 at room temperature. The piezometer was placed in a cryostat and connected by a capillary to a gas meter and charging line. The temperature in the cryostat was maintained within ± 0.01 K and measured with the same accuracy by a platinum resistance thermometer. The pressure was determined by two reference-grade spring pressure gauges with an error of 0.1%. The experiments were performed along isotherms. At fixed values of pressure and temperature a part of the substance was passed into a gas meter and maintained at constant temperature of $t = 25$°C. The pressure that set on in the gas meter was measured by constant-volume, mercury-filled manometers, and then the specific volume corresponding to the experimental conditions was calculated.

The final accuracy of experimental data depends primarily on: (1) the accuracy of calibration of the piezometer and gas meter volumes; (2) proper correction for the dead space volume; and (3) for the temperature and pressure dependence of the piezometer volume. All these problems are analyzed in detail in [3.48], where the accuracy of the assumed values is validated. The most difficult task consisted of determining the specific volume of the gas in the capillary connecting the piezometer to the charging line since the temperature varies over the length of the capillary and the liquid-gas interface is not known. For this reason, in order to determine the quantity of substance contained in the capillary, Streett and Staveley used another capillary of the same material and dimensions. Its lower end was silver soldered and the upper was connected to a separate calibrated vessel, located in the constant-temperature bath of the gas meter.

In the opinion of Streett and Staveley, corrections for a number of errors and determination of the volume of vessels forming the test facility are fraught with danger of accumulation of systematic errors. This necessitated additional calibration of the facility on the basis of the density of saturated liquid argon [1.113], which was confirmed by new experimental studies of Terry et al.

[3.49] at temperatures from 85 to 118 K. For this purpose, experiments in which the pressure exceeded somewhat the saturation pressure were performed at four values of temperature within this range. The values of v_s were determined by slight extrapolation of isotherms and then these values were compared with data in [1.113, 3.49]. Comparison showed that the measured values of the specific volume are, on the average, 0.25% lower than this. That required correcting all the experimental data in another study by Streett and Staveley [2.99] by this difference. The final error in determining the specific volume according to these authors should not exceed 0.1%.

The experimental results are presented in [2.99] in tabulated form along ten isotherms and are compared with studies of others. At saturation, satisfactory agreement is observed with data of Mathias et al. [1.113] and Michels et al. [2.53] at $T < 130$ K. At higher temperatures, the differences between results in [1.113] and [2.99] become 0.5% at $T = 143$ K. The results in [2.53] and [2.99] are in good mutual agreement in the single-phase region. The values of molar volumes, listed by Van Itterbeek et al. [1.108], are at individual points up to 0.5% lower, as compared with the data in [2.99]. The results of Van Witzenburg and Stryland [1.36] along the 200 atm isobar are lower by the same amount. These differences decrease with rising pressure and do not go past the overall experimental error to a temperature of ~130 K.

To calculate thermodynamic functions at round values of temperature and pressure, Streett and Staveley compiled, on the basis of their experimental data, an equation of state for liquid argon in the form

$$V = \sum_{i=0}^{m} \sum_{j=0}^{n} B_{ij} p'^{i} T'^{j} \tag{3.23}$$

where V is the molar volume

$$p' = \frac{2p - (p_{\max} + p_s)}{p_{\max} - p_s}; \quad T' = \frac{2T - (T_{\max} + T_{\min})}{T_{\max} - T_{\min}}$$

$P_{\max} = 680$ atm, $T_{\min}$ and $T_{\max}$ are the minimum and maximum experimental temperatures. The coefficients of Eq. (3.23) were determined by computer regression from experimental data. The best conformance to original data is attained at $m = 7$ and $n = 5$.

Crawford and Daniels [1.37] investigated the pvT behavior of pure (99.995%) argon at temperatures between 94.72 and 210.16 K and at pressures between 198 and 6464 bar. The experimental technique is basically the same used by Van Witzenburg and Stryland [1.36]. A 9 cm^3 capacity thin-walled piezometer was made of tool steel and designed for pressures to 10 kbar. The piezometer cavity is connected to a charging and pressure maintenance line by a long thin capillary ($d_{\text{out}} = 0.82$ mm, $d_{\text{in}} = 0.25$ mm). The piezometer and a part of the capillary are in the cryostat in which the specified temperature is maintained automatically within ±0.02 K. The uniformity of the temperature field over the height of the piezometer was monitored by a differential copper-constantan thermocouple. The experimental temperature was determined by a platinum resistance thermometer with an error of ±0.02 K. The low pressure was measured within ±1 bar and the high pressure to within ±4 bar by a manganin pressure gauge.

The experiments were performed along isotherms. The piezometer with argon was weighed at specified pressure and temperatures. Upon attaining the minimum pressure (~200 bar) the substance was removed from the piezometer and the latter was weighed at the experimental temperature. This made it possible to calculate the weight of the argon at each experimental point. The density of liquid argon was determined by additional experiments over the same range of temperature and pressure. In this case a steel cylinder was placed in the piezometer so as to take up a greater part of the latter. The difference in the weights of the liquid argon in the two experiments (with and without the steel cylinder) was used along with the precisely known volume of the cylinder.

Crawford and Daniels represented their experimental data on the compressibility of argon in the form of tables and on a graph for 13 values of temperature. The error in the specific volume is estimated by the authors at ±0.02 cm^3/mol which corresponds to a relative error of 0.05–0.10%. At certain temperatures they obtained the values of the molar volume of liquid and solid argon on the melting curve. These results are also given in an earlier study by Crawford and Daniels [1.38] and agree, to within 0.3%, with analogous data of Van Witzenburg and Stryland [1.36]. However, significant differences exist between the results in [1.38] and those of Bridgman [1.77] and Lahr and Eversole [1.82]. These differences increase with the temperature and amount to about 10% at T = 200 K. Apparently the more recent data in [1.36, 1.38], which are in satisfactory agreement with one another and with results for the single-phase region are more dependable.

Lippold [2.98] investigated the isothermal compressibility and density of liquid argon at pressures to 980 bar at temperatures between 96.73 and 149.2 K. The experimental liquid was condensed in a nonrelieved piezometer having a volume known with a high precision. Then the liquid was compressed by a thermocompressor to a high pressure. Then the pressure was reduced at constant temperature by Δp by discharging a certain amount Δm of substance from the piezometer into calibrated vessels maintained at room temperature. If the change in the weight of the substance and its corresponding change in pressure are sufficiently small, then we have

$$\left(\frac{\partial \rho}{\partial p}\right)_T = \frac{1}{V_0}\left(\frac{\Delta m}{\Delta p}\right)_T \tag{3.24}$$

from which expression one can easily calculate the isothermal compressibility.

Particular attention was paid by Lippold to reducing the dead space volume.

For this purpose, the volume of the tube connecting the cavity of the piezometer with a shutoff valve was made as small as possible, and a manganin pressure gauge was used. For this reason, the maximum correction did not exceed 12%. The temperature was measured by a lead resistance thermometer with an error not exceeding 0.01 K. The argon was 99.95% pure. The density of argon was not measured in the experiments. But, it was calculated from measurements of isothermal compressibility and values of density obtained at low pressures (50–100 bar) by fully recondensing the liquid into the calibration vessels.

The values of density of liquid argon are given in a table along seven isotherms. At p = 98 bar the values of density obtained by Lippold are 0.8% lower than values of Van Itterbeek et al. [1.108] at T = 96.73 K and 1.5% higher at 149.2 K. The differences between these two sets of data decrease with increasing pressure. In the opinion of Lippold [2.98] the internal inconsistency of his data does not exceed 0.1%. However, our check showed that the scatter of individual points is as high as 0.3–0.5% which most likely is the magnitude of the random error of Lippold's measurements.

Lippold used his results for checking Tait's equation. It was found to describe the majority of data with an error of up to 1% (for density). However, at $T > 0.9\,T_{cr}$ and pressures below 200 bar quite significant discrepancies occur.

Verbeke et al. [1.107] presented new pvT data for argon at temperature from 87 to 202 K and pressure from saturation to 149 atm (six points at pressures between 1416 and 2299 atm were obtained at the $v \simeq 26.9$ cm^3/mol isochore). They used a constant-volume piezometer and obtained data along quasi-isochores. The quantity of gas in the piezometer was determined by discharging it into a gas meter. The argon used was 99.996% pure. The temperature was measured by a platinum thermometer with an error of 0.003 K. The pressure was determined with an error of 0.01 atm at $p \leqslant 150$ atm. Verbeke et al. list about 320 data points of which about half are for the liquid region and another 100 are for the critical region.

Verbeke et al. represented new experimental data at $V \leqslant 47.1$ cm^3/mol by the equation

$$V = V_1 + (A_0 + A_1T + A_2T^2)\lg \frac{B_{00} + B_{11}T + B_{22}T^2 + p}{B_{00} + B_{11}T + B_{22}T^2 + p_{00} + p_{01}T} + $$
$$+ \sum_{i=0}^{m}\sum_{j=0}^{n} A_{ij}p^i(T - T_0)^j \tag{3.25}$$

When retaining the first two terms in Eq. (3.25) the standard deviation at V between 28.3 and 47.1 cm^3/mol and T from 87 to 169 K amounts to 0.029 cm^3/mol, with the maximum deviation at 0.2%. The error is reduce four-five-fold by adding 25 or 36 coefficients A_{ij}. Particular attention was paid to analytic description of data at V = 40–132 cm^3/mol which includes the critical region. For this region they compiled a new, highly complicated equation of state which contains terms with fractional power exponents which does not contradict the scaling theory [3.50, 3.51].

One's attention is attracted by the fact that Verbeke et al. did not compare their results with experimental data in [1.108, 3.43] obtained earlier in the same laboratory with Verbeke's participation, and pertaining to the range of parameters investigated in the currently discussed article by Verbeke et al. Our comparison showed that the values of molar volumes listed in [1.107] are 0.2–0.4% lower than the experimental data in [1.108, 3.43] as well as the results of the majority of other studies pertaining to the above range of variables. A possible reason for this may be the error committed by Verbeke et al. when calibrating their piezometer.

Dobrovol'skiy and Golubev [3.52] measured the density of argon at temperatures between −180 and −117°C at pressures from 10 to 500 atm gauge. Their studies were performed by hydrostatic weighing, using the facility described in [3.53]. The pressure was measured by class 0.05 dead-weight MP-600 and MP-60 pressure gauges, the temperature by a reference grade platinum resistance thermometer with an error of 0.01 K. The argon contained 0.01% nitrogen and 0.003% oxygen. The experimental data (179 points along 22 isobars) are shown in the table and graphs. The overwhelming majority of points is for the liquid region. The authors estimate the error of experimental data on density at ±0.1% for the liquid and ±0.2% for the gas. However, their comparison with data in [1.36, 1.108] showed that in a number of points the differences amount to 0.3–0.4%.

Dobrovosl'skiy and Golubev performed a graphical analysis of the experimental data in [2.53, 1.108, 3.44, 3.52] and obtained smoothed values of density of liquid argon from −188 to −115°C and pressures from 2 to 485 bar, and on the saturation and melting curves at temperature intervals from −189.38 to −122.29°C and −189.38 to −177°C, respectively.

The thermal properties of the three other monatomic liquids analyzed by us here were investigated in less detail. Gladun [3.54], who measured the specific heat C_v of liquid neon at 25–44 K to a pressure of 100 bar, also obtained values of the density of saturated liquid neon. The technique of determining C_v also provided for the determination of the thermal pressure coefficient γ_v along each of the 34 isochores under study. Using these data Gladun was subsequently able [3.55] to calculate the density of liquid neon at temperatures between 26 and 43 K and pressures of 5 to 100 bar.

Lippold [2.98] investigated the isothermal compressibility and density of liquid neon (99.99% pure) at 28.37–43.11 K and 98–980 bar. The experimental technique was the same as for argon. The values of the density of liquid neon along seven isotherms are given in a table. Graphs present the pressure behavior of isothermal compressibility and density at different temperatures. The experimental results of Lippold for neon at 98 bar are in agreement with data of Gladun [3.55] in the majority of points with a deviation within the limits of 0.1–0.2%. But, at $T = 41$ K the divergences are as high as 0.5%, which lie within the limits of the overall error of two independent determinations.

Unfortunately, there is lack of agreement in [2.98] between the tabulated data and graphs for the 35.07, 37.90, 41.00 and 43.11 K isotherms. Apparently the values of thermodynamic properties at these temperatures should be referred to the pressure scale at which the results are given for argon. Then the tabulated values will agree with those in the graphs. For this reason, we subsequently used data corrected in this manner.

It was noted in the preceding chapter that Gibbons [1.94] investigated the *pvT* behavior of neon at temperatures from 27 to 70 K to a pressure of 204 atm by a constant-volume piezometer. Of the 268 data points given along the quasi-isochores about one-quarter pertain to the liquid phase. He also lists six values of density of the liquid and seven of the density of vapor at saturation. The data on the density of saturated liquid at temperatures of 27.04–34.51 K agree with the results of Mathias et al. [2.145] with deviations within 0.25%. At higher temperatures (40.27 and 44.08 K) the deviations amount to 0.7 and 1.3%.

Onosovskiy [2.32] graphically worked up his experimental *pvT* data for gaseous neon at low temperatures [2.31, 2.32] and determined the thermal properties along the critical isotherm on the saturation curve [1.51, 2.145]. In [2.32] he lists the values of the compressibility factor of neon at 44.4 K and 7.88–104.57 bar (ρ = 48.35–967.0 kg/m^3), the possible error of which is estimated by Onosovskiy at $\pm 0.3\%$.

Data on thermal properties of liquid krypton were obtained for the first time by Theeuwes and Bearman [1.118, 2.72] at 130–240 K and density between 0.6 and 2.6 of the critical (at pressures to 284 bar) by a constant-volume piezometer method. The error of pressure measurement was $\pm 0.01\%$, of density $\pm 0.06\%$ and of temperature 0.01 K. The experimental data in [2.72] are represented in tables and graphically along 47 quasi-isochores, of which 41 are transcritical. The quantity of gas for 34 quasi-isochores was determined by discharging into a volume manometer. For the remaining, they first determined the density relative to one of the quasi-isochores. Then, in its turn, the values of density were found from data for the saturation curve. The error of temperature regulation during the experiments was 0.001 K. For this reason, it is possible to select for the large body of results a number of points, obtained at $T \simeq$ const. which facilitates comparison with data of others.

Blagoy and Sorokin [3.56] investigated the density of liquid krypton (99.91% pure) with nitrogen and oxygen as the impurities at temperatures of 119.98, 129.31, 141.58, 149.95 and 169.94 K at pressures from saturation to 498 kg/cm^2 using the same facility as for liquid argon. The values of density of saturated liquid krypton needed for determining the density of liquid at elevated pressure were taken by Blagoy and Sorokin from data in [2.20]. Using the experimental data on density given in the table, Blagoy and Sorokin determined by graphical differentiation the coefficient of isothermal compressibility and presented values in graphical form along isotherms as a function of pressure.

Subsequently Streett and Staveley [2.73] measured the density of krypton at 120–220 K and 5.3–3679 atm. The experimental facility was basically the same as used for liquid argon. The pressure in the piezometer was measured by a dead-weight pressure gauge with an error of 0.01% (at $p < 2720$ atm) and by a manganin resistance pressure gauge with an error of 0.1% (at higher pressures). The tested krypton was 99.98% pure as confirmed by chromatographic analysis. Streett and Staveley estimate the maximum error of their values of density as 0.1% and the reproducibility at 0.03% deviation. The experimental *pvT* data (385 points) are listed in a table along 11 isotherms (spaced at 10 K). The density of liquid saturated krypton at 120–200 K was determined by the extrapolation of isotherms.

Streett and Staveley described their *pvT* data, including data at the saturation curve, using an equation of state suggested by Strobridge. The standard deviation of the computed values of density from the experimental was 0.03% with a maximum of 0.2%. They emphasized that this equation of state is valid only over the range of parameters investigated by them experimentally, i.e., at $T = 120$–220 K and $\rho/\rho_{cr} = 1.5$–3.0.

The above investigators compared their results with the experimental data of Theeuwes and Bearman [1.118, 2.72] and noted that the values of density of liquid krypton at saturation are 0.10–0.18% higher than the new data, and the results for the single-phase region are systematically higher by 0.1%, on the average. To additionally check the accuracy, Streett and Staveley measured the density of krypton along the 0°C isotherm at five values of pressure (from 127 to 312 atm). The results were found to be 0.05% lower than the data of Trappeniers et al. [2.44]. They attribute the small systematic differences between the results of independent experiments to possible errors of calibration of the volumes of the measuring instrumentation.

Experimental *pvT* data for liquid xenon were first obtained by Michels et al. [2.77] in their study of compressibility at $T = 273.15$–423.15 K over a wide density range. The data points were obtained along two isotherms of the liquid phase: 273.15 and 286.65 K to $p = 1654$ and 1214 atm, respectively.

The *pvT* behavior of liquid xenon was investigated in detail by Theeuwes and Bearman [1.121, 3.57] at 175–240 K and reduced density of 2.16–2.73 (at $p < 291$ atm). They used the same technique as in investigating the properties of krypton. The error of measurement of pressure, density and temperature amounted to 0.01%, 0.06% and 0.01 K, respectively. The experimental data in [3.57] on the density of liquid xenon were obtained along 24 quasi-isochores and are presented in a table and graphically. Four series of experiments were performed for four groups of isochores. The change in density when moving to the next isochore was determined by discharging the gas into the gas meter, and the absolute value of density along the last isochore of a given series was determined when all the gas was discharged. The residual amount of gas in the piezometer was calculated from an equation of state obtained by Michels et al. [2.77].

Blagoy and Sorokin [3.56] obtained data on the density of liquid xenon at the 171.70, 179.45, 199.84 and 219.67 K isotherms from the saturation pressure to 498 bar. The xenon they used contained 0.08% impurities in the form of nitrogen and oxygen. They worked up the results of their measurements using data of Leadbetter and Thomaes [3.58] on the density of saturated liquid xenon. Blagoy and Sorokina also showed graphically the values of the coefficient of isothermal compressibility of liquid xenon along the four isotherms under study to a pressure of 500 bar.

Up to now we have discussed studies concerned with experimental determination of the thermal properties of monatomic liquids. However, a number of studies where other properties of these substances were investigated have also been published.

Gladun [3.54] performed a thorough study of the specific heat of liquid neon (99.99% pure) at constant volume at $T = 25.2–43.14$ K and p to 100 bar. The main component of the test equipment was an adiabatic stainless steel calorimeter with a resistance thermometer soldered in its lower part. The volume of the calorimeter, determined at $T = 25$ and 27 K from data on the density of parahydrogen, was 74.73 cm^3. The calorimeter volume changed due to pressure and temperature during the experiments. This necessitated making the pertinent corrections. The thermometer was calibrated over the 24–44 K range with an error of 0.01 K by measuring the vapor pressure of liquid neon in the calorimeter and using Grilly's equation [1.51].

The experiments were performed along isochores, starting with the two-phase range, and the temperature was gradually raised to values at which the pressure did not exceed 100 bar. The results of measurements were used for calculating the total heat capacity of the system (calorimeter and neon). For this reason, the heat capacity of the calorimeter itself was also investigated over the experimental temperature range. At $T = 30$ and 40 K, its heat capacity was 12 and 23% of the heat capacity over the experimental temperature range, respectively. The temperature behavior of C_v was investigated along 34 isochores ($\rho = 0.03632–0.06139$ mol/cm^3). The results showed that at $\rho > 0.048$ the value of C_v is independent of the temperature over the range of parameters under study, while at lower densities it decreases linearly with rising temperature. These circumstances made it possible for Gladun to obtain values of C'_v for liquid neon at saturation with maximum error estimated at $\pm 1\%$. However, it is not obvious from the graph of C_v vs. temperature and density given in [3.54] that the straight lines for which extrapolation was performed can be drawn uniquely.

Gladun [3.54] used her experimental data on the density, specific heat at constant volume, and thermal pressure coefficient for calculating the coefficients of thermal expansion α_{exp}, isothermal compression β_T and specific heats C_v, C_σ and C_p along the saturation curve at temperatures of 25–43 K. Gladun compared data on C'_v near the triple point in the summary table with data of Clusius [3.59] and found deviations to 30%. She attributes such large differences to the fact that the neon used in the study by Clusius was contaminated by helium. The values of C_σ along the saturation curve agree within 2% with results of Clusius, given in a later study [3.60], but they differ by up to 25% from data in [3.59].

Gladun [3.61] investigated C_v of liquid argon at 87.75–150.5 K at pressures to 100 bar with a facility used previously for investigating C_v of liquid neon. The internal volume of the adiabatic calorimeter was 74.85 cm^3 at 120 K. Before starting her measurements, she measured the heat capacity of the empty calorimeter at $T = 80-155$ K. The variation in this quantity was described by a polynomial with an error of less than 0.05%. At $T = 85$ K the heat capacity of the empty calorimeter was 50% of that of the system, and at 150 K it was 78% of it. The experiments were performed along 12 isochores spaced 28 to 54 cm^3/mol apart. Each experiment was started from the two-phase region and then the temperature was gradually raised. The heat capacity changed jumpwise upon transition through the phase equilibrium curve making it possible to determine the transition temperature. At $V < 34$ cm^3/mol the specific heat C_v of liquid argon is independent of the temperature, whereas at larger specific volumes it decreases linearly with increasing temperature over the range of variables under study. Gladun presents experimental data on the specific heat along seven isochores selected by her. There it is seen that the molar volumes change by less than 0.1%. She also lists values of C_v' along the saturation curve and analytic data on C_σ and C_p'. She estimates the error of values of C_v at $\pm 1\%$ and of C_σ and C_p' at $\pm 2\%$.

Gladun compared her experimental results with those of others and found that the mean differences in her values of C_v and those by Jones and Walker [3.62] are 3%, with the exception of the region of $V < 29.8$ cm^3/mol, where the differences are greater. The data of Eucken and Hauck [3.63] are 10% higher than those of Gladun [3.61] at low temperatures and do not exhibit a minimum or an increase in C_v' as the critical point is approached. The deviation from data in [3.62] on C_p' at pressures of 10, 20, 30 and 40 bar lies within the limits of the overall errors of independent measurements, with the exception of one point at $T = 130$ K ($p = 20$ bar). The values of C_σ are in satisfactory agreement with the data of Flubacher et al. [1.35], given over the range of 86 to 89 K and data in [3.63] at $T \leqq 120$ K. At higher temperatures, the values in [3.63] are significantly lower, as are the results of Clusius at $T < 87$ K.

Gladun and Menzel [3.64] measured C_v of liquid krypton at 116–200 K at pressures to 100 bar using the adiabatic calorimeter described above. Its internal volume was 74.91 cm^3 at $T = 150$ K. The heat capacity of the empty calorimeter was measured at 115–210 K. Its behavior is described by a polynomial with a standard deviation of 0.15%. At 120 K the heat capacity of the empty calorimeter amounted to 67% of the heat capacity of the calorimeter filled with liquid krypton, and at 200 K it was 82%. The temperature was measured by a platinum resistance thermometer which was precalibrated from the results of the measurements of vapor pressure of liquid krypton, and data of Garside and Smith [3.65].

The behavior of C_v of liquid krypton as a function of temperature was investigated along 18 isochores at V between 34 and 54 cm^3/mol. The experimental procedure was the same as in their preceding studies [3.54, 3.61]. They [3.64] found that at $V < 49$ cm^3/mol the value of C_v of liquid krypton is independent of the temperature, and at higher volumes it decreases linearly with rising temperature. In their summary table they list experimental values of V', C_v' and γ_v along the saturation curve, and also the computed values of α_{exp}, β_T, C_σ and C_p'. In their opinion, the maximum error of values of specific heats may be 4% and that of γ_v 2%. The comparison of data on C_σ with results of Clusius [3.60] at 118.3–120.5 K performed by Gladun and Menzel [3.64] showed that the differences do not exceed 0.5%. At the same time, differences of up to 5% were observed between the values of β_T and those calculated by Lim and Aziz [3.66] from results on the speed of sound at T = 116–120 K.

Shenball [3.67] measured the speed of sound in liquid neon at temperatures from 24.6 to 27 K at pressures close to the saturated vapor pressure. The neon, which was 99.9% pure, contained nitrogen, helium and hydrogen as impurities. His experimental values of w, m/s, were approximated by the expression

$$w = 989.0 - 14.602T \tag{3.26}$$

whose coefficients were determined by the method of least squares. The standard deviation of the computed from the experimental data is 0.18% with a maximum of 0.38%. As noted by the author, his absolute values of the speed of sound should be treated as preliminary results, since small subsequent changes may be possible upon making certain corrections.

Gusewell with his coworkers [3.68] investigated the speed of sound in neon-hydrogen and neon-deuterium systems at temperatures of 25–31 K. They also list data on the speed of sound in pure neon at saturation pressure.

Larson et al. [3.69] measured the speed of sound in particularly pure (99.998%) liquid neon at 25–37 K and pressures to 28 kg/cm². The experiments were performed at constant temperature with the pressure lowered gradually to values close to the saturation pressure. Minor extrapolation of isotherms to their intersection with the vapor pressure curve allowed determining the speed of sound in neon along the saturation curve. The pressure was measured with an error of ±0.02 kg/cm², the temperature with an error of 0.01 K. A total of 32 isotherms are presented.

Subsequently, Naugle [370] described these data on the speed of sound in saturated neon by the expression

$$w = 786.65 + 1.9754T - 0.33342T^2 \tag{3.27}$$

with a standard deviation of 0.2%. The author ascribes the same error to the original data. In Eq. (3.27) w is in m/s.

Comparison of the results in [3.67–3.69] shows that at T = 24.6–27 K the values in [3.67] are up to 0.6% lower than those in [3.68, 3.69] which agree with one another within 0.1%, and at higher temperatures within 0.3%.

The speed of sound in liquid argon, including the saturation curve, was investigated by many authors over various ranges of parameters (Table 38).

TABLE 38. Experimental Investigations of the Velocity of Sound in Liquid Argon

Year	Author	Temperature, K	Pressure, bar	Reference
1959	Van Itterbeek et al.	90.3	p_s— 75	[3.71]
1959	Van Itterbeek et al.	84.8—90.3	p_s— 75	[3.72]
1960	Dobbs and Finegold	87.2—90.1	p_s—135	[3.73]
1961	Van Itterbeek and Van Dael	87.3—90.3	p_s—200	[3.74]
1964	Radovskiy	138—150.7	p_s— 60	[2.58]
1966	Blagoy et al.	87.5—119.8	p_s	[3.75]
1966	Van Dael et al.	84.76—148.97	p_s	[2.61]
1967	Aziz et al.	83.83—145.75	p_s	[3.76]
1968	Carome et al.	90.3—140.0	5—100	[2.62]
1968	Bowman et al.	86.2—146.6	p_s— 65	[3.77]
1969	Thoen et al.	100—150.1	p_s—500	[2.63]

The widest range was explored by Thoen and Van Dael [2.63]. Their measurements were performed at 1.2 MHz in 99.996% pure argon. The temperature was measured within ±0.002 K and the pressure within 0.002 atm. The reproducibility of data in all the experiments was within 0.5 m/s.

The results of other investigators, obtained over narrower ranges of parameters, are basically in satisfactory agreement with the data of Thoen et al. [2.63]. Thus, at $p < 65$ bar the differences between the results in [2.58, 2.63, 3.77] do not go past ±1 m/s. Only the data of Carome et al. [2.62] are systematically 0.5% higher at low and 2% at high temperatures. The maximum deviation of 5% is observed along the 140 K isotherm at 40 bar. Table 39 illustrates the agreement between data on the speed of sound in saturated liquid argon.

Aziz et al. [3.76] also present data on the speed of sound in liquid krypton and xenon along the phase equilibrium curve. The experimental results for each of the substances are described by the polynomials

$$w' = a_0 + a_1T + a_2T^2 + a_3T^3 + a_4T^4 + a_5T^5 + a_6T^6 \qquad (3.28)$$

TABLE 39. Speed of Sound in Saturated Liquid Argon

T, K	w', m/s, from data in					
	[3.72]	[2.58]	[3.75]	[2.61]	[3.76]	[2.63]
90	819.5	822	822	819	818.55	819.0
95	—	786	784	783	783.07	
100	—	748	747	747	746.46	747.0
110	—	671	668	670	668.90	669.0
120	—	584	586	584	583.77	585.0
130	—	—	—	487	487.70	488.0
140	—	—	—	371	369.96	371.0
145	—	—	—	296	297.0	295.0
150	—	—	—	—	—	172.0

with an average error of 0.063 and 0.107 m/s for krypton and xenon, respectively.

Blagoy et al. [3.78] performed similar studies of these substances over approximately the same temperature range as in [3.76]. The results of these two studies are given in Table 40. Here it is seen that the differences are higher than for argon.

3.3. ANALYTIC DESCRIPTION OF $p\rho T$ DATA

We described the experimental data on the density of the four substances under study in the liquid state by an equation of state in the form [3.8]

$$p = A(T)\rho^n + B(T)\rho^{n+2} + C(T)\rho^{n+4} + \ldots \tag{3.29}$$

It was used previously for described experimental $p\rho T$ data for nitrogen [3.79], argon [3.80], oxygen [3.81], methane [3.82] and carbon dioxide [3.83] to pressures of 500–700 bar over a temperature range from the triple point to 1.1 T_{cr}. Subsequently, an equation of state was compiled for carbon dioxide in the form of Eq. (3.29) containing four temperature functions which reliably represented the experimental data to 2500 bar [3.84]. In all cases the equation was capable of describing the most reliable data without going past the limits of experimental error.

Vasserman and Rabinovich [3.8, 3.85] described a graphical method for compiling an equation such as (3.29), based on successive finding of tempera-

TABLE 40. Speed of Sound in Liquid Krypton and Xenon

T, K	w', m/s, in krypton, from data in		T, K	w', m/s, in xenon, from data in	
	[3.76]	[3.78]		[3.76]	[3.78]
116	699.96	—	165	645.25	—
118	692.03	—	170	631.53	631
120	684.02	690	180	603.46	602
130	642.91	648	190	574.51	571
140	599.98	605	200	544.49	542
150	—	561	210	513.21	512
160	506.56	517	220	480.43	481
170	—	469	230	445.87	447
180	397.69	410	240	409.13	412
190	—	353	250	369.49	346‡
200	253.25†	292	260	325.79	337
208	172.15†	—	270	276.14	278
			280	217.59†	—
			285	183.69†	—

† These data were obtained in [3.76] by extrapolation.
‡ Apparently there is a printing error in [3.78].

ture functions by the isotherm straightening technique. The use of this method is accompanied by allowance for possible errors of experimental results. This makes it possible to determine the permissible deviations of the temperature functions from the optimum values and to obtain for these functions smooth curves which lend themselves easily to analytic description.

Although the graphic technique for compiling the equation of state for a liquid is rather simple, it is still best to use a computer since it allows considerable speeding up of this process. Vasserman and Kreyzerova [3.86, 3.87] developed two methods for compiling an equation of state such as Eq. (3.29) by computer. The algorithm of the first method provides the successive determination of the optimum and of the limiting values of temperature functions, their analytic description, and comparison of analytic values of density with the experimental. All the calculations, performed by the method of least squares are performed with allowance for the weight $G = 1/\Delta p^2$ of the experimental points. Here, Δp is the permissible deviation for pressure, corresponding to the specified relative error of experiments for values of the density. A feature specific to the method involves determining points in which the deviations from experimental values of density exceed the specified, changing the weight of these points and repeated determination of temperature functions. These computational cycles permit, in a number of cases, a reduction in the number of points in which the deviations exceed the permissible, or to reduce the deviations themselves in these points. The method involves the use of $p\rho T$ data plotted along isotherms at arbitrary pressure. Its effectiveness was checked on the basis of data for nitrogen at 65–140 K and pressure from saturation to 709 bar.

The second (more effective) method [3.87] does not require representation of experimental data along isotherms. It is based on simultaneous determination of all the coefficients of equation of state (3.29) by the method of least squares for a specified form of analytic expressions (polynomials of T) and specified number of constants in each of the expressions. The algorithm of this method, described in detail in [3.87], provides for preliminary determination of the weight of each of the data points. Since the accuracy of experimental data is usually assessed together with specific error of the values of density with consideration of reference errors, we can use the simplified formula

$$\Delta p = \rho \delta \rho (\partial p / \partial \rho)_T \tag{3.30}$$

To work up the data presented at arbitrary temperature, at the first stage an approximated equation of state is compiled without consideration of the weight of data points. This equation is then used for calculating Δp at each point from Eq. (3.30). Then two equations of state are compiled with consideration of the weights. These equations have the same number of constants and have power exponents n of 1 and 2. The subsequent calculations are performed using that value of n which has corresponding to it the smaller weighed sum of squares of deviations of the computed values of pressure from the experimental.

The program provides for the use of a different number of experimental points in compiling the approximate equation of state, selection of the optimum value of n, compiling the principal equations, and final comparison with experimental data. Following the selection of n it is necessary to formulate and solve a system of normal equations for determining the coefficients of temperature functions of the sought equation of state. These coefficients are stored in the computer memory. The equation of state is used for calculating the values of density at the data points. The roots of Eq. (3.29) are determined by Newton's method [3.88] with experimental values of density serving as the initial conditions. After each value is calculated, the relative deviation $\delta\rho$ of the experimental from the computed value is determined and entered into the computer's working storage.

The program in [3.87] provides for performing cycles of calculations involving change of the weight of experimental points and compiling a number of versions of equations of state with specified number of coefficients. The refinement cycles are repeated while the sum of the squares $\Sigma\delta\rho^2$ of relative deviations of values of density decreases. The optimum version of the equation is that which has the minimum sum corresponding to it.

The number of points used for compiling all versions of the equation of state are printed out, as are values of $\Sigma\delta\rho^2$ and the standard deviation σ_ρ and the coefficients of temperature functions. For the optimum version, one also obtains a printout of the total number of points, including points used only for comparison, the corresponding values of $\Sigma\delta\rho^2$ and σ_ρ, deviations $\delta\rho$ in each point and data for constructing histograms of deviations. The program for the Minsk-22 computer provides the compilation of an equation of state with either three or four temperature functions and increases the number of analytic expressions of all or several functions to 28, inclusively.

This technique was checked out by Vasserman and Kreyzerova [3.87] against experimental $p\rho T$ data for nitrogen, argon, and carbon dioxide at pressures from the saturation curves to 709, 1186 and 2493 bar, respectively. The optimum versions of equations of state for nitrogen and argon contain three temperature functions (9 and 10 coefficients) each, and those for CO_2 contain four temperature functions (12 coefficients). The standard deviation of experimental data of density from the computed amounts to 0.13% for nitrogen and argon and 0.08% for carbon dioxide.

Analysis of experimental data on the compressibility of the four monatomic gases performed by Rabinovich [3.89] allowed the estimation of the accuracy and the mutual consistency of results. This made it possible, in compiling equations of state by computer, to specify quite objectively for these data the values of $\delta\rho$ needed for calculating the weight from Eq. (3.30). For this reason, the version of the program for finding the constants of Eq. (3.29) used by us, unlike that in [3.87], does not provide for additional calculations associated with changes of the weight of experimental points. At the same time it appeared best to use data on the speed of sound w' and the specific heats c_v' and c_p' for saturated liquid. These thermodynamic quantities are related by the familiar differential equations:

$$w^2 = -v^2 \frac{c_p}{c_v} \left(\frac{\partial p}{\partial v} \right)_T \tag{3.31}$$

$$c_p - c_v = -T \frac{\left(\frac{\partial p}{\partial T} \right)_v^2}{\left(\frac{\partial p}{\partial v} \right)_T} \tag{3.32}$$

which can be used for calculating the derivatives $\left(\frac{\partial p}{\partial v}\right)_T$ and $\left(\frac{\partial p}{\partial T}\right)_v$. It should be emphasized that the starting equation of state (3.29), with separated independent variables, is linear relative to the sought coefficients since the temperature functions are represented by a polynomial in the temperature. As a result of this, the expressions for derivatives $\left(\frac{\partial p}{\partial v}\right)_T$ and $\left(\frac{\partial p}{\partial T}\right)_v$ are also linear relative to these coefficients which significantly simplifies in our case the approximation of quantities of different kinds. When taking the minimum of the functional we arrive at a system of equations which are linear in the sought coefficients. This presents the opportunity to check the degree of consistency of the various data used in data processing. Then difficulties due to the appearance of possible local minima and poor convergence of results in solving the nonlinear approximated problem are eliminated.

Neon. The constants of the equation of state were obtained by the method of least squares on the basis of experimental data of Mathias et al. [2.145], Gladun [3.54] and Gibbons [1.94] on density ρ' at saturation and data of Gladun [3.55], Lippold [2.98], and Gibbons [1.94] for the single-phase region. To bring the values of thermodynamic properties calculated from the equation of state for the liquid into agreement with data for gas, we also used $p\rho T$ data of Onosovskiy [2.32] at 52 K and interpolated values of density along the 44.4 K critical isotherm listed by him. The values of both derivatives along the saturation curve were calculated from data on the speed of sound [3.69] and specific heats c_p' and c_v' [3.54].

The overall database consisted of 219 points, of which 38 pertained to the saturation curve and 9 to the 44.4 and 52 K isotherms, and also on the basis of 13 computed values of derivatives. The isothermal sections of the surface formed by these data are shown in Fig. 45†. For convenience, Lippold's experi-

† The figure does not show a part of the data of Gladun [3.55], values of Gibbons [1.94] obtained by him along quasi-isochores and the results of Onosovskiy [2.32].

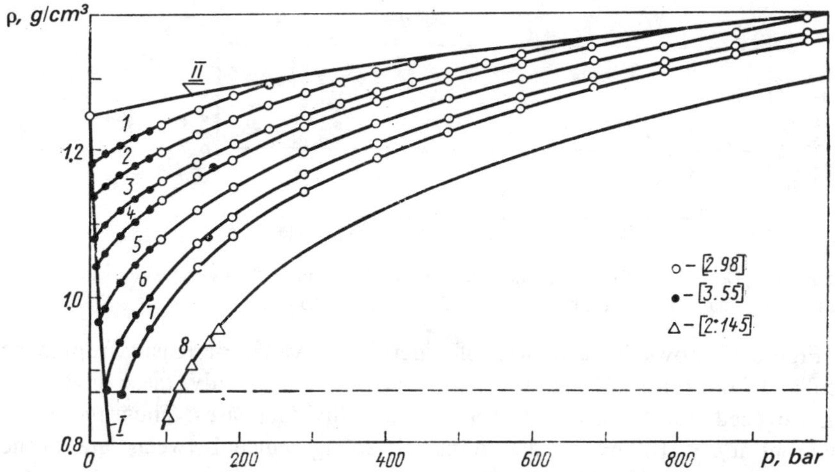

FIGURE 45. Isotherms of liquid neon constructed from data of different investigators. I) Saturation curve; II) solidification curve; 1) 28.37 K isotherm; 2) 30.82; 3) 33.41; 4) 35.07; 5) 39.90; 6) 41.0; 7) 43.11; 8) 50.0; the solid curves were calculated from Eq. (3.33).

mental isotherms [2.98] were supplemented, at the low-pressure range, by points (darkened circles), obtained by us from interpolation of data of Gladun [3.55]. Such a construction is convenient because it made it possible to establish the degree of consistency of both groups of results. Given this, and also preliminary analysis of the initial data, we assumed that $\delta\rho = 0.2\%$ in the single-phase region and 0.1% along the saturation curve.

A series of calculations with different forms of polynomials for the temperature functions showed that optimum conditions (minimum standard deviation at minimum number of constants) has corresponding to it an equation of state in the form

$$p=(-639.184+20.8223T-0.0871395T^2)\rho^2+(-417.945+ \\ +6.69657T-0\ 0673811T^2)\rho^4+(263.293+ \\ +2.0265\cdot 10^{-16}\cdot T^{10})\rho^6 \qquad (3.33)$$

which is valid for $T = 24.55–50$ K and $p = 0.43–1050$ bar at $\rho > 1.8\ \rho_{cr}$ ($\rho > 0.87$ g/cm³). The lower limit of applicability of the equation of state (3.33) for density is given by a dashed curve in Fig. 45.

Figure 46 shows the deviations of experimental values of the density of neon along the saturation curve from data obtained by simultaneous solution of equation of state (3.33) and the equation of the vapor pressure curve [1.51]. The overwhelming majority of experimental points are described with an error not exceeding 0.1%. Greater deviations (to 0.2%) are observed from data of Gibbons [1.94]. Individual points in the study by Gladun [3.54] also higher.

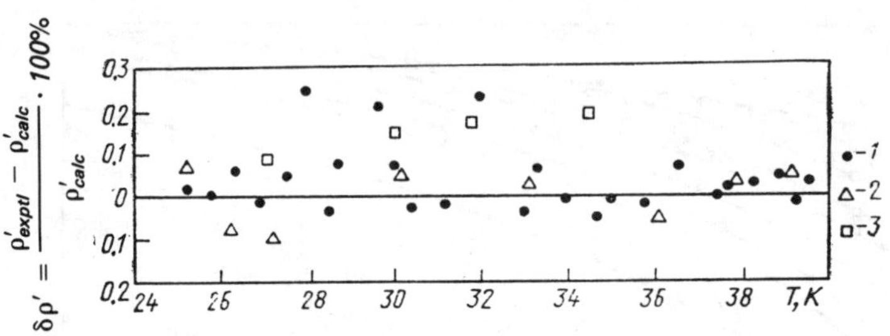

FIGURE 46. Deviations of experimental values of density ρ′ along the saturation curve of neon from those calculated by Eq. (3.33). 1) [3.54]; 2) [1.94]; 3) [2.145].

Figure 47 shows the deviations of experimental values of density from those calculated from Eq. (3.33) at different temperatures. Basically, these deviations do not exceed ±0.2% with the exception of individual cases. Thus, the larger deviations at $p < 100$ bar are due to insufficient agreement between data for the

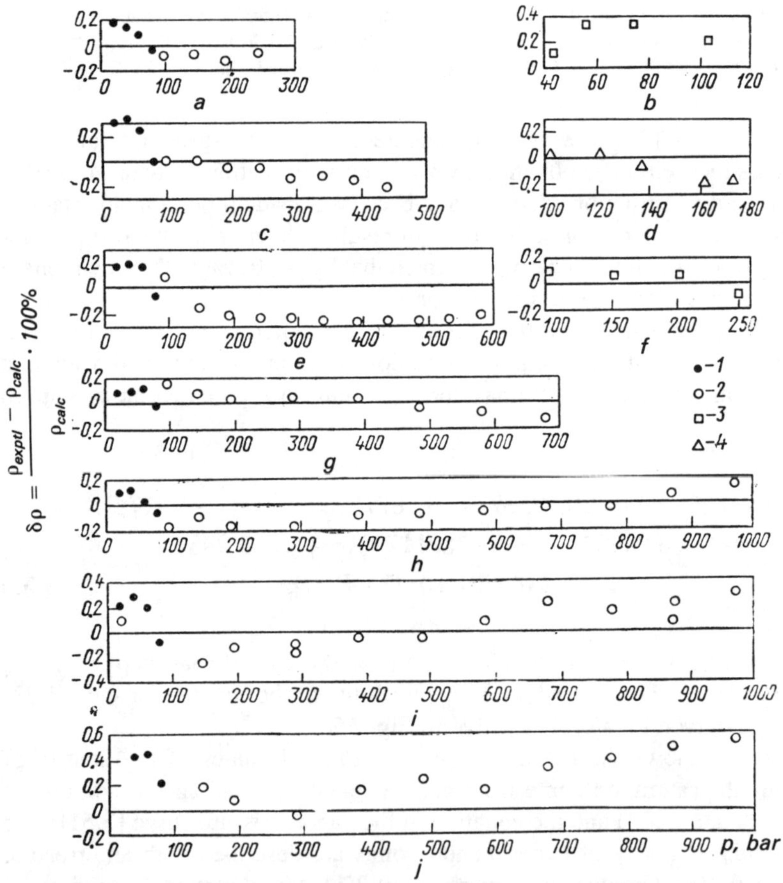

FIGURE 47. Deviations of experimental values of density ρ of neon from those obtained from Eq. (3.33). a) Along the 28.37 K isotherm; b) 44.40; c) 30.82; d) 50.00; e) 33.41; f) 52.00; g) 35.07; h) 37.00; i) 41.00; j) 43.11; 1) [3.55]; 2) [2.98]; 3) [2.32]; 4) [1.94].

single-phase region and those along the saturation curve and values of c_v' and c_p'. The largest deviations are observed along the 43.11 K isotherms, and these could not be reduced even at the expense of making the equation of state more complicated.

Comparison of computed values of density with experimental results of Gibbons [1.94] presented along quasi-isochores showed that the standard deviation for this entire database (73 points) is 0.14%.

Table 41 compares the values of the specific volume of liquid neon along the solidification curve, calculated from equation of state (3.33) using Eq. (1.65) for the p_m, T behavior with analogous data obtained by Abovskiy by another analytic method. The deviations do not exceed 0.1% with the exception of two points where printing mistakes (apparently) occurred. This confirms indirectly the regularity of deviation from Lippold's experimental points [2.98] along 30.82 and 33.41 K isotherms at high pressure (see Fig. 47).

Figure 48 shows the temperature behavior of c_v' and c_p' along the saturation curve obtained from Eq. (3.33) using data on the speed of sound [3.69] (solid curves). The mean deviations from data of Gladun [3.54] do not exceed 1.4%. Better agreement is observed at low temperatures, and then the deviations increase with the temperature to 5 and 3% at $T = 37$ K for c_v' and c_p', respectively. An attempt at reducing these differences resulted in significant deviations of the calculated values of density from mutually agreeable data along the saturation curve.

Argon. Since the range of variables for which tables of thermodynamic properties are to be compiled is limited to a maximum pressure of 1000 bar, we

TABLE 41. Comparison of Computed Values of the Specific Volume of Liquid Neon along the Solidification Curve

T, K	p, bar	$v_l \cdot 10^3$, m³/kg from Eq. (3.33)	$v_l \cdot 10^3$, m³/kg according to Abovskiy [1.6]	δv_l, %
24,55	0.43	0.8012	0.8016	0.05
25	30.45	0.7967	0.7973	0.08
26	98.38	0.7876	0.7900†	—
27	168.0	0.7793	0.7795	0.03
28	239.3	0.7717	0.7718	0.01
29	312.2	0.7645	0.7746†	—
30	386.7	0.7579	0.7579	0
31	462.8	0.7515	0.7516	0.01
32	540.5	0.7455	0.7456	0.01
33	619.7	0.7398	0.7398	0
34	700.5	0.7343	0.7344	0.01
35	782.8	0.7290	0.7292	0.03
36	866.6	0.7240	0.7241	0.01
37	951.9	0.7192	0.7192	0
38	1039	0.7147	0.7145	—0.03

† Apparently printing mistakes at these points.

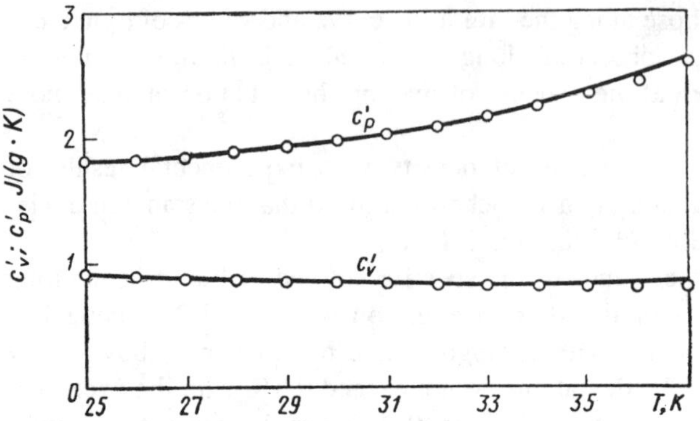

FIGURE 48. Comparison of computed values of specific heats c_v' and c_p' for saturated neon with data of Gladun [3.54].

used only part of the experimental data to a pressure of 1053 bar. The equation was compiled on the basis of experimental results of Van Witzenburg and Stryland [1.36], Crawford and Daniels [1.37] along the solidification curve, Goldman and Scrase [3.90], Terry et al. [3.49], Michels et al. [2.53], Streett and Staveley [2.99] for the saturation curve [2.53, 2.99, 1.37], and the results of Van Itterbeek et al. [1.108, 3.43] for the single-phase region. For the majority of these data it was assumed that $\delta\rho = 0.1\%$. The equation of state was compiled on the basis of 510 data points, of which 92 pertain to the saturation and solidification curve. In addition to $p\rho T$ data, the equation of state contains 12 values of each of derivatives $\left(\frac{\partial p}{\partial T}\right)_v$ and $\left(\frac{\partial p}{\partial v}\right)_T$ along the saturation curve of the liquid at $T = 85\text{–}140$ K, calculated on the basis of experimental values of the speed of sound in [2.58, 2.61, 2.63, 3.76] and of specific heats c_v' and c_p' [3.61].

For comparison with values of density calculated from the equation of state we also used data of Mathias et al. [1.113] for the saturation curve and the results of Streett [3.44], Van Witzenburg and Stryland [1.36], Rogovaya and Kaganer [2.55], Sorokina and Blagoy [3.46], Lippold [2.98], Verbeke et al. [1.107], and Dobrovol'skiy and Golubev [3.52], which were obtained for the range for which this equation is applicable ($T \leqq 173.15$ K, $\rho > 0.925$ g/cm^3) —a total of 501 data points.

A number of versions of the equation of state for liquid argon with a total number of coefficients from 7 to 11 was compiled on a Minsk-22 computer. Due to the unsatisfactory consistency of experimental $p\rho T$ data, increasing the number of coefficients to above 9 did not significantly improve the accuracy of the approximation. Simultaneously we investigated different forms of polynomials for temperature functions and found that when function A(T) of Eq. 3.29 is expressed as

$$A(T) = a_0 + a_1T + a_2T^{-2} + a_4T^{-4} \tag{3.34}$$

the $p\rho T$ data and derivatives $\left(\frac{\partial p}{\partial T}\right)_v$ and $\left(\frac{\partial p}{\partial v}\right)_T$ at the saturation curve for the liquid are described with the optimum accuracy. For this reason we finally selected the equation of state for liquid argon

$$p=\left(-241.8425+3.777497T-\frac{330.3296\cdot 10^4}{T^2}+\frac{54.49258\cdot 10^8}{T^4}\right)\rho^2+(-192.3351-2.816835T+0.01104341T^2)\rho^4+(174.6742+0.8503397T)\rho^6 \quad (3.35)$$

which is valid at $p \leqslant 1053$ bar and $\rho > 0.925$ g/cm³ ($\rho > 1.73\rho_{cr}$) at temperatures from the triple point to 173 K.

The standard deviation of the values of density used in compiling the equation of state (3.35) from the computed are: 0.12% for data along the solidification and saturation curves of [1.36, 1.37, 2.53, 2.99, 3.49, 3.90], 0.14% for [1.108, 3.43], 0.10% for [2.53], 0.07% for [2.99] and 0.11% for [1.37]. The best agreement between results not used for compiling the equation and the analytic are the data of Dobrovol'skiy and Golubev [3.52]—for them $\sigma_\rho = 0.13\%$. For data of Streett [3.44] $\sigma_\rho = 0.21\%$; of Rogovaya and Kaganer [2.55], 0.22%; of Sorokina and Blagoy [3.46], 0.21%, and of Van Witzenburg and Stryland [1.36], 0.24%. The most significant deviations from the computed values of density are those for data of Verbeke et al. [1.107] and Lippold [2.98] where $\sigma = 0.4$ and 0.7%, respectively. These results were apparently obtained with a significant systematic error.

Figure 49 shows deviations of experimental values of density of saturated liquid argon from those calculated from the equation of state (3.55) using equation (1.70) for the vapor pressure curve. It is seen from the figure that Eq. (3.35)

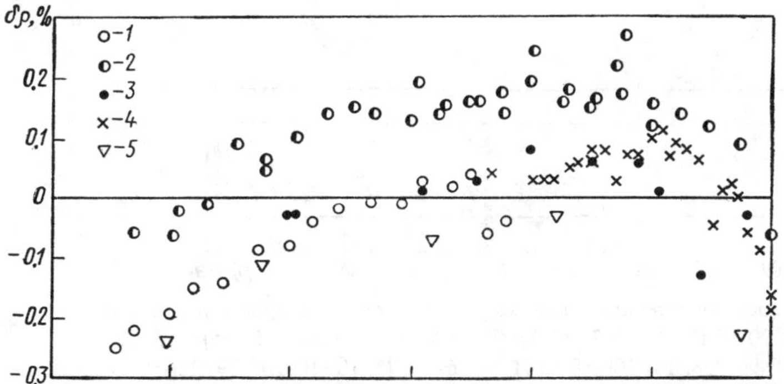

FIGURE 49. Deviations of experimental values of density of saturated liquid argon from those calculated from equation of state (3.55). 1) [3.49]; 2) [3.90]; 3) [2.99] 4) [2.53]; 5) [1.113].

describes the majority of experimental points with an error ±0.15% and only for a part of values in [3.90] and also near the triple and critical points do the deviations rise to 0.20–0.25%.

Figures 50 through 52 show the deviations of the majority of the aforementioned experimental data, including those not used in compiling the equation of state, from the values of density calculated from Eq. (3.35). Deviations for close isotherms are clustered on the graph allowing estimation of the discrepancies between the data of various investigators. The equation of state describes the majority of the results in [2.99] at all the temperatures with an error of ±0.1°C; data of [1.36, 1.37, 2.53, 3.43] and partially also in [3.44] are represented with acceptable accuracy. The results in [1.108] are in satisfactory agreement with

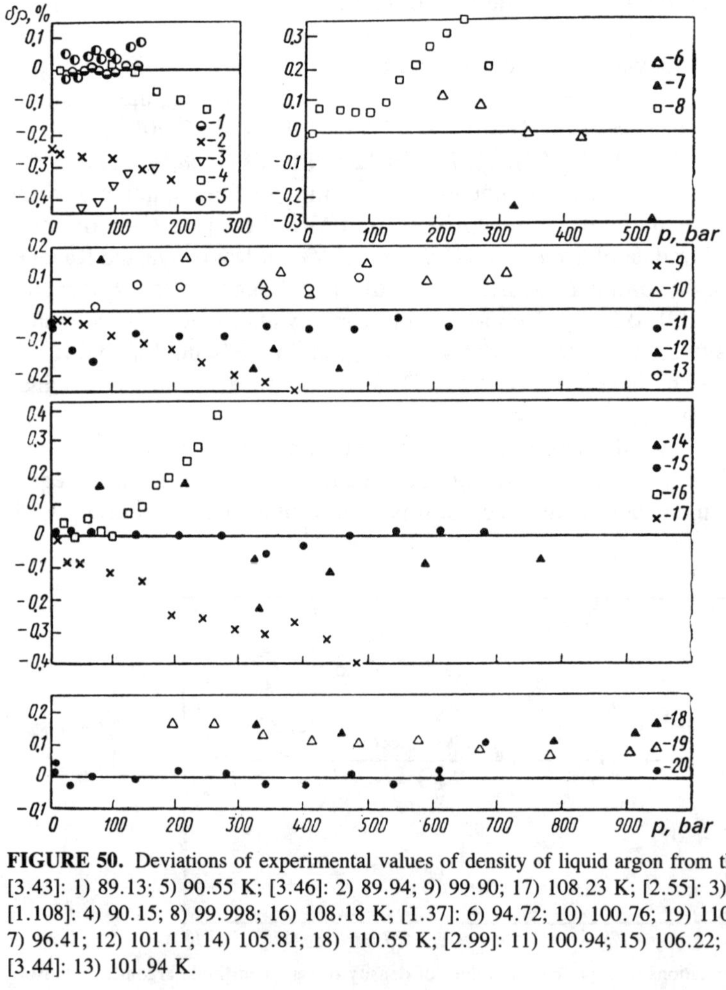

FIGURE 50. Deviations of experimental values of density of liquid argon from the computed. [3.43]: 1) 89.13; 5) 90.55 K; [3.46]: 2) 89.94; 9) 99.90; 17) 108.23 K; [2.55]: 3) 90.13 K; [1.108]: 4) 90.15; 8) 99.998; 16) 108.18 K; [1.37]: 6) 94.72; 10) 100.76; 19) 110.77 K; [1.36]: 7) 96.41; 12) 101.11; 14) 105.81; 18) 110.55 K; [2.99]: 11) 100.94; 15) 106.22; 20) 111.04 K; [3.44]: 13) 101.94 K.

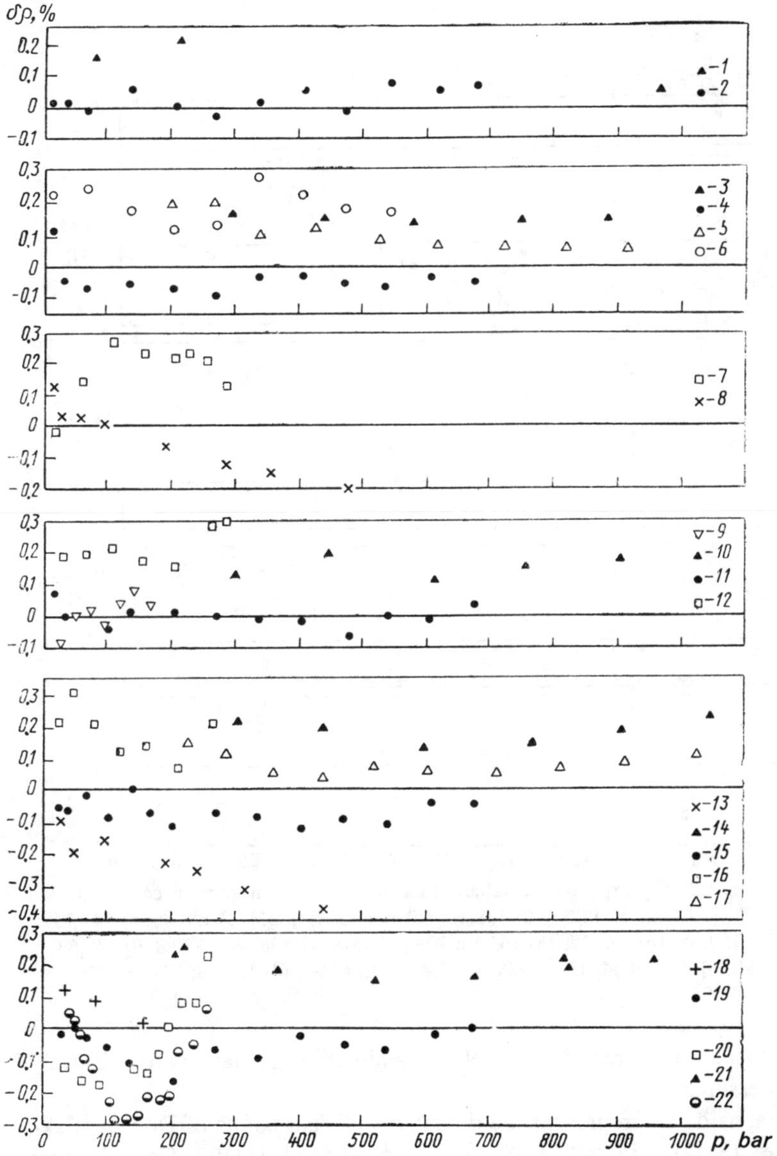

FIGURE 51. Deviations of experimental values of the density of argon from the computed. [1.36]: 1) 115.30; 3) 120.08; 10) 124.87; 14) 129.68; 21) 134.50 K; [2.99]: 2) 115.34; 4) 120.18; 11) 125.18; 15) 130.42; 19) 134.28 K; [1.37]: 5) 120.86; 17) 130.87 K; [3.44]: 6) 121.36 K; [1.108]: 7) 117.10; 12) 127.05; 16) 130.85; 20) 134.40; 22) 136.02 K; [3.46]: 8) 119.97; 13) 128.92 K; [2.55]: 9) 123.07 K; [2.53]: 18) 133.15 K.

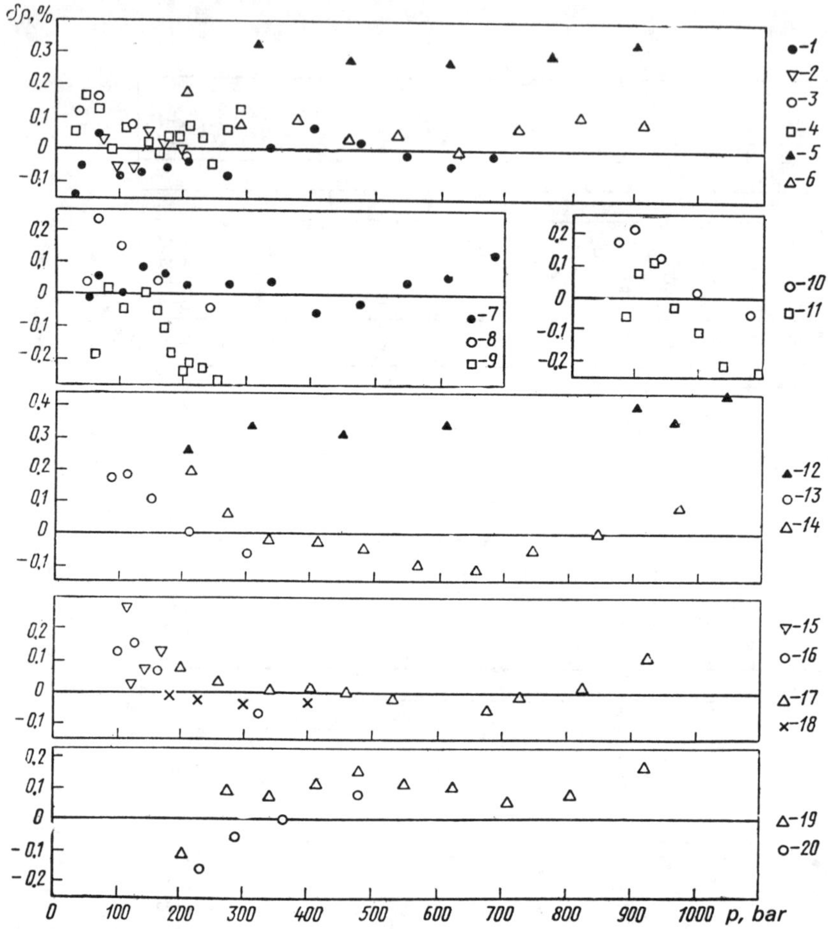

FIGURE 52. Deviations of experimental values of the density of argon from the computed. [2.99]: 1) 138.02; 7) 143.11 K; [2.55]: 2) 138.12; 15) 153.13 K; [2.53]: 3) 138.15; 8) 143.15; 10) 148.15; 13) 150.65; 16) 153.15; 18) 163.15; 20) 173.15 K; [1.108]: 4) 138.98; 9) 146.63; 11) 148.25 K; [1.36]: 5) 139.34; 12) 149.06 K; [1.37]: 6) 140.94; 14) 150.73; 17) 160.47; 19) 170.36 K.

computed data at pressures to 100–150 bar and exhibit deviations to 0.2–0.3% at higher pressures.

The graphs do not represent the deviations of data points of Dobrovol'skiy and Golubev [3.52], Lippold [2.98], and Verbeke et al. [1.107] from the computed. The disagreement with the majority of the results in [3.52] lies within ±0.1%, rising to 0.3–0.4% only in certain points due to scattering of experimental data. The values of density listed in [1.107, 2.98] are in the majority of points 0.3–0.6% higher than those calculated by us apparently due to insufficient reliability. The deviations from data of Lippold [2.98] along the 139 and 149.2 K isotherms at high pressures exceed 1%.

Comparison of experimental data of Crawford and Daniels [1.37] and Van Witzenburg and Stryland [1.36] along the solidification curve with the computed values of ρ (Table 42) shows that the agreement is quite satisfactory in the majority of points except one [1.36].

Figure 53 shows the temperature dependence of c_v' and c_p' along the saturation curve, obtained from Eq. (3.35) using experimental data on the speed of sound from [2.63, 3.76]. At temperatures to 125 K the computed values of c_v' are in agreement with Gladun's experimental data [3.61] with a deviation of 0.3–1.7%, at higher temperatures the discrepancies rise to 3–10%. The maximum disagreements with data in [3.61] on c_p' at T = 87.8 and 135.9 K are 4 and 5%, respectively. These differences exceed the error of values of c_v' and c_p' specified by Gladun.

When experimental values of Gladun [3.61] on c_v' were used for calculating the speed of sound in saturated liquid argon from the equation of state, the values of w' computed in this manner at temperatures from the triple point to 125 K are in satisfactory agreement with the experimental (Fig. 54). At higher temperatures, the computed values are systematically higher than the majority of experimental points. The discrepancies increase with the temperature, becoming as high as 4% at T = 140 K. Only the data points of Carome et al. [2.62] lie higher than the computed curve over the entire temperature range, but at $T > 129$ K deviate perceptibly from data of others.

Unfortunately, an attempt to attain better agreement for specific heats and the speed of sound increased the deviations from the $p\rho T$ data along the saturation curve and close to it. This points to insufficient internal agreement of experimental data on thermal and calorific properties of saturated liquid argon at temperatures above 125 K.

Krypton. The equation of state for liquid krypton was compiled by us on the basis of experimental results of Mathias et al. [2.148], Terry et al. [3.49], Theeuwes and Bearman [1.118], and Streett and Staveley [2.73] for the saturation curve; and data of Theeuwes and Bearman [2.72], Blagoy and Sorokin

TABLE 42. Comparison of Experimental data in [1.36, 1.37] on the Density of Liquid Argon Along the Solidification Curve with the Computed

Author	T, K	p, bar	ρ_l, g/cm³ determined by the cited authors	ρ_l, g/cm³ from Eq. (3.55)	$\delta\rho_l$, %
Crawford and Daniels [1.37]	94.73	451	1.4633	1.4616	0.12
	94.74	459	1.4628	1.4631	−0.02
	100.76	721	1.4856	1.4841	0.10
	108.12	1051	1.5063	1.5068	−0.03
Van Witzenburg and Stryland [1.36]	96.41	529.7	1.465	1.4690	−0.27
	101.11	738.0	1.485	1.4854	−0.03
	105.81	952.1	1.501	1.5007	0.02

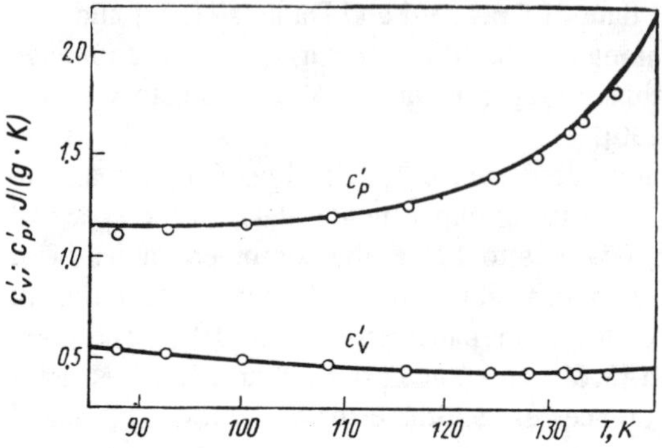

FIGURE 53. Comparison of computed values of c'_v and c'_p of saturated liquid argon with data of Gladun [3.61].

[3.56], and a part of data of Streett and Staveley [2.73] (to 1021 atm) for the single phase region. It was assumed that $\delta\rho = 0.1\%$. The database used for compiling the equation of state contained 592 data points.

Derivatives $\left(\frac{\partial p}{\partial T}\right)_v$ and $\left(\frac{\partial p}{\partial v}\right)_T$ were calculated from data on the speed of

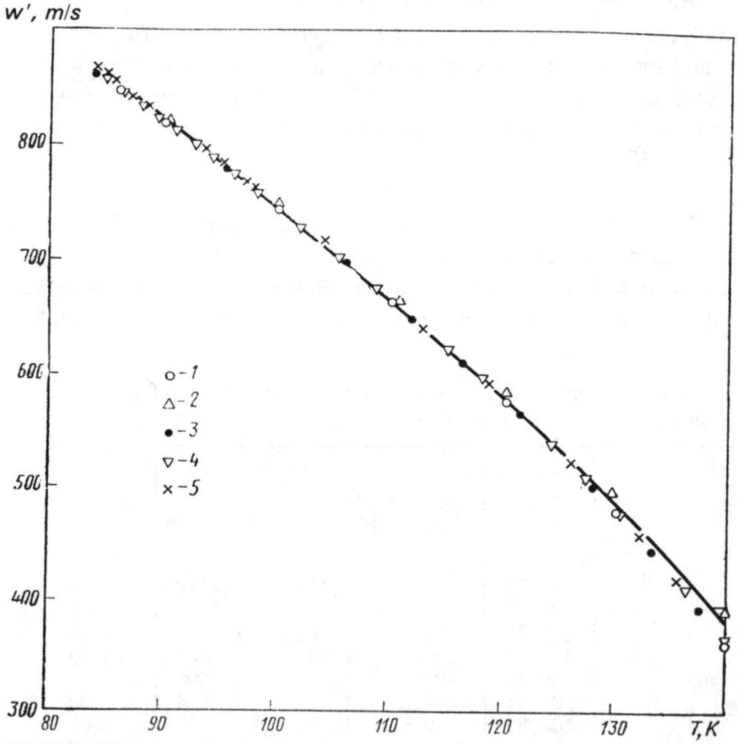

FIGURE 54. Comparison of computed values of the speed of sound in saturated liquid argon with experimental data points. 1) [2.63]; 2) [2.62]; 3) [3.76]; 4) [2.60]; 5) [2.58] in (1964).

sound obtained by Aziz et al. [3.76] and by Blagoy et al. [3.78], and from data of Gladun and Menzel [3.64] on c_v' and c_p' along the saturation curve. The results of calculations were in significant disagreement due to poor consistency of the initial data in [3.76, 3.78]. For this reason, several versions of calculations were performed in order to determine the degree of agreement between the values of derivatives and $p\rho T$ data.

The Minsk-22 computer was used for compiling several versions of the equation of state for liquid krypton with a total number of coefficients from 6 to 11. These were differently redistributed in expressions for temperature functions. Initially we used only $p\rho T$ data. In this case, the standard deviation of experimental values from the computed, found over the entire body of points, decreased insignificantly (from 0.11 to 0.085%) upon increasing the number of coefficients. The inclusion of computed values of both derivatives in all cases increased the rms error, particularly along the saturation curve, due to insufficient agreement of the values of these derivatives and $p\rho T$ data.

For this reason we finally selected an equation of state which describes experimental data on the density of liquid krypton somewhat poorer than the optimum versions, but allows representation of data on the speed of sound w' and specific heats c_v' and c_p' with acceptable accuracy. Equation

$$p=\left(-247.537+1.425643T+\frac{81.64768\cdot10^4}{T^2}-\frac{34.78487\cdot10^8}{T^4}\right)\rho^2+(-117.740+0.505612T-0.000728729T^2)\rho^4+12.69602\rho^6 \qquad (3.36)$$

reliably describes experimental $p\rho T$ data for $p \leqq 1033$ bar and $\rho > 1.78\ \rho_{cr}$ ($\rho > 1.62$ g/cm^3) at temperatures from the triple point to 240 K.

Figure 55 shows deviations of experimental data on density ρ' of liquid krypton along the saturation curve from those calculated from Eq. (3.36) and equation (1.70) of the vapor pressure curve. The standard deviations of experi-

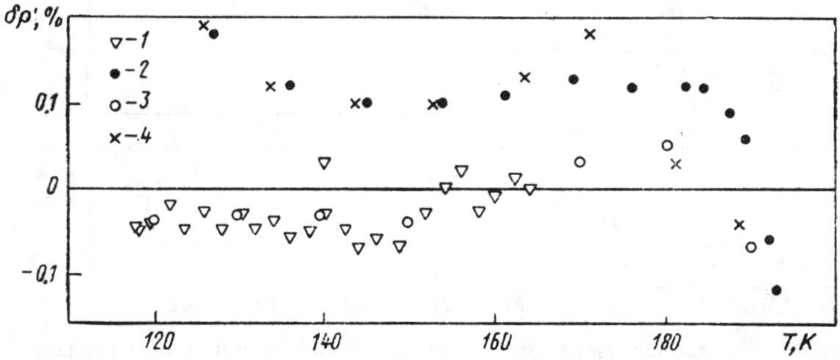

FIGURE 55. Deviations of experimental data on the density of saturated liquid krypton from those calculated from equation of state (3.66). 1) [3.49]; 2) [1.118, 2.72]; 3) [2.73]; 4) [2.148].

mental values of density from the computed are: for [2.148] 0.15%; for [3.49] 0.04%; for [1.118] 0.11%; for [2.73] 0.04%; and for all the 54 points 0.09%.

Figure 56 shows the deviations of experimental data of Theeuwes and Bearman [1.118, 2.72], Blagoy and Sorokin [3.56], and Streett and Staveley [2.73] from those computed by the equation of state. The standard deviation for these studies is 0.1%.

The values of specific volume v_l along the saturation curve, calculated from equation of state (3.36) using the expression for $p_m = f(T)$ from [3.47] are compared with computed data in [1.6, 3.91] in Table 43. The agreement between the results is quite satisfactory, with the exception of the triple point at which the value of density obtained from Eq. (3.36) is more dependable (see Fig. 55).

Equation of state (3.36) and data on the speed of sound in [3.76, 3.78] also yielded values of c_v' and c_p' which differed perceptibly from one another due to insufficient consistency of data on the speed of sound. It was found by comparing the computed results that better agreement with experimental values of c_v' can be attained by using data in [3.76]. In this case, the discrepancies between computed and analytic values of c_v' at 116 to 170 K do not exceed ±4%, which

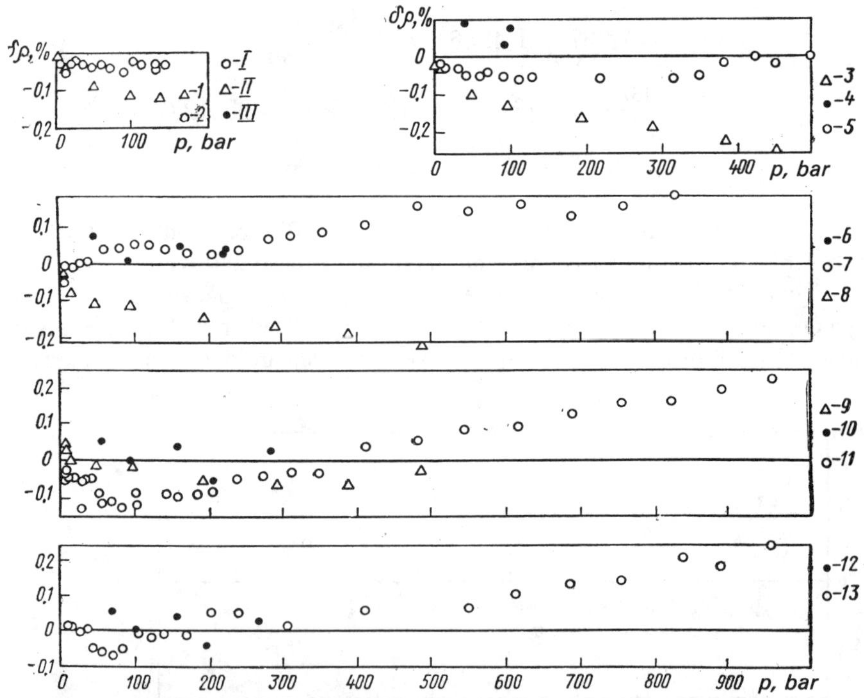

FIGURE 56. Deviations of experimental data on the density of liquid argon from the computed using Eq. (3.36). I) [2.73]; II) [3.56]; III) [1.118, 2.72]; 1) 119.98; 2) 120; 3) 129.31; 4, 5) 130; 6, 7) 140; 8) 141.58; 9) 149.95; 10, 11) 150; 12, 13) 160.

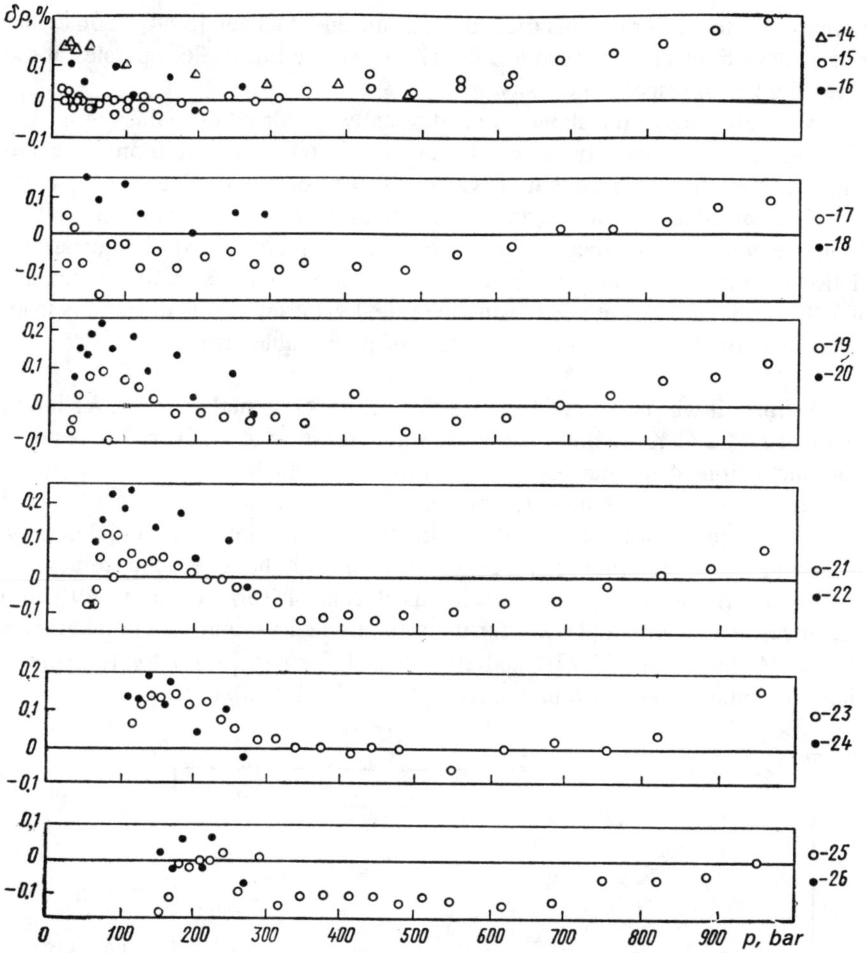

FIGURE 56. 14) 169.94; 15, 16) 170; 17, 18) 180; 19, 20) 190; 21, 22) 200; 23, 24) 210; 25, 26) 220 K. (*Continued*).

TABLE 43. Comparison of Computed Values of the Specific Volume of Liquid Krypton Along the Solidification Curve

T, K	p, bar	v_l, cm³/g from Eq. (3.36)	v_l, cm³/g obtained in [1.6, 3.91]	δv_l, %
115.76	0.73	0.4090	0.4068	—0.54
120	143.5	0.4050	0.4043	—0.17
125	315.5	0.4007	0.4004	—0.07
130	491.7	0.3968	0.3970	0.05
135	672.2	0.3933	0.3936	0.08
140	856.8	0.3901	0.3905	0.10
145	1044	0.3873	0.3876	0.08

corresponds to the error specified by Gladun and Menzel [3.64]. Somewhat greater deviations are observed at $T > 175$ K, where Eq. (3.36) operates at the lower limit with respect to density.

With reference to the above, we additionally calculated the values of w' and c_p' using Eq. (3.36) and experimental data on c_v' [3.64]. The results are shown in Fig. 57. From this it is seen that the values of speed of sound in liquid krypton to 170 K, obtained as a result, deviate from data of Aziz et al. [3.76] within $\pm 1\%$. At higher temperatures they average the results of [3.76, 3.78]. The mean deviation of computed values of c_p' from those in [3.64] is 1.3%, with a maximum of 2.9%. This means that when using measured values of c_v' the deviations from other data are equal to the limiting values of permissible error.

Xenon. It was noted in Chap. 3.2 that the experimental $p\rho T$ data for liquid xenon at $T < 273$ K are limited to a maximum pressure of 500 bar. The result is that calculation of its thermodynamic properties at higher pressure is possible only by extrapolating the equation of state. With reference to this, we attempted to obtain an equation of state of the simplest form describing the initial data on density using the minimum number of constants with the required accuracy.

To determine these constants we used data of Terry et al. [3.49], and Theeuwes and Bearman [1.121] for the saturation curve; and data of Theeuwes [3.57], Michels et al. [2.77], and Blagoy and Sorokin [3.56] for the single-phase region. We obtained an equation of state for liquid xenon

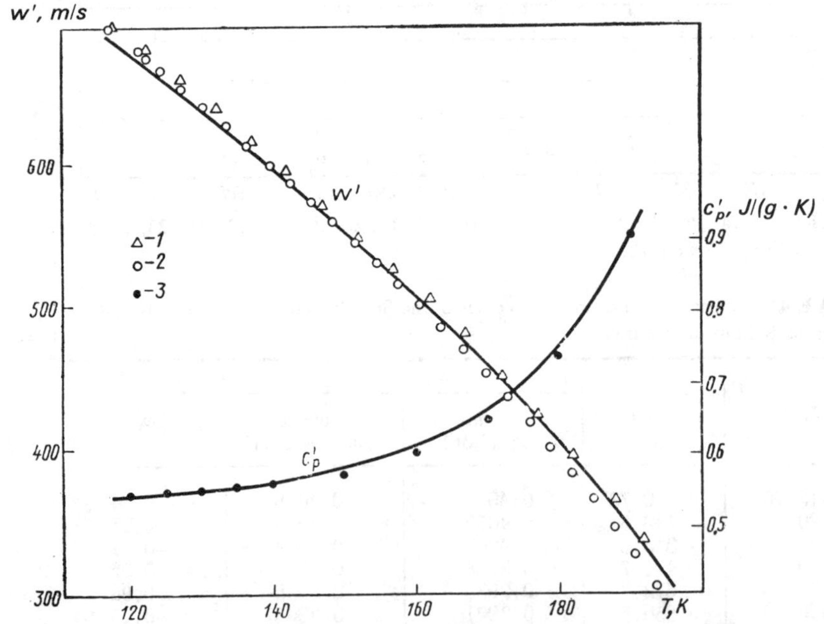

FIGURE 57. Comparison of computed values of the speed of sound w' in liquid krypton and specific heat c_p' along the saturation curve with experimental data. 1) [3.78]; 2) [3.76]; 3) [3.64].

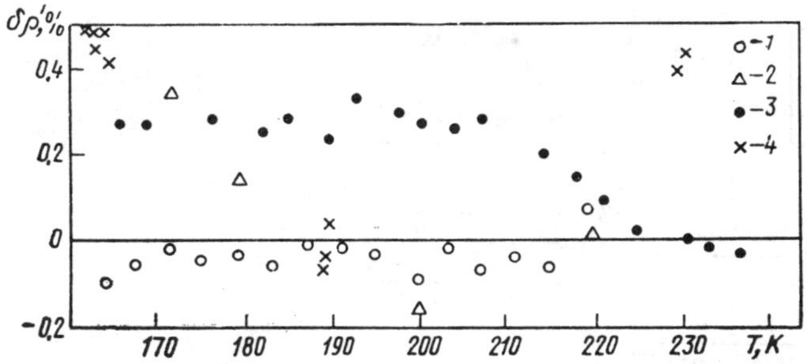

FIGURE 58. Deviations of experimental data on the density of saturated liquid xenon from those calculated from equation of state (3.37). 1) [3.49]; 2) [3.56]; 3) [1.121]; 4) [3.58].

$$p=(-160.945+0.70725\cdot T)\rho^2+(-50.3818+ \\ +0.152241\cdot T-1.58574\cdot 10^{-4}T^2)\rho^4+4.0125\rho^6 \qquad (3.37)$$

where p is in bar, which describes the original data points (365 points) with a standard deviation of 0.095% with respect to density at T = 161.36–300 K and $\rho > 1.8\ \rho_{cr}$ ($\rho > 1.98$ g/cm^3).

Figure 58 shows the deviations of experimental data on density along the saturation curve from values computed from Eq. (3.37) with $p_v(T)$ obtained from Eq. (1.70). The best agreement is observed with data of Terry et al., which were ascribed a higher weight than others. The value of density at the triple point given by Leadbetter and Thomaes [3.58] is up to 0.5% higher. The results in [1.121] are also on the high side. The differences with them decrease at higher temperatures. Data of Blagoy and Sorokin [3.56] used for comparison were obtained by them by working up the results of Leadbetter and Thomaes [3.58].

Figure 59 shows the deviations of experimental values of density of liquid xenon obtained by different investigators from those calculated using Eq. (3.37). The differences from data of Blagoy and Sorokin [3.56] can be reduced to a minimum by using the values of initial volume v' calculated from Eq. (3.37) and not those based on the results of Leadbetter and Thomaes [3.58] (see Fig. 58). The satisfactory agreement with experimental data of Michels et al. [2.77] shows that the equation of state is able to correctly represent the pressure dependence of density to $p \simeq 1000$ bar. This indirectly validates slight extrapolation of this equation with respect to density for $T < 273$ K†.

Table 44 compares computed values of the specific volume of liquid xenon

† The fact that Eq. (3.37) can be extrapolated satisfactorily is also confirmed by comparison with experimental data of Streett et al. [Streett, W. B., Sagan, L. S. and Staveley, L. A. K, J. Chem. Thermodyn, *5*, 633, (1973)] on the density of liquid xenon at 165–289.74 K and pressures to 3815 atm. This study was published recently and was not used for compiling Eq. (3.37). At pressures to 1000 bar the values of density calculated by us agree with data of Streett et al. with a standard deviation of 0.15%. The computed values of density, as well as the experimental data of Michels et al. [2.77] and Theeuwes and Bearman [3.57] are basically higher than the results of Streett et al.

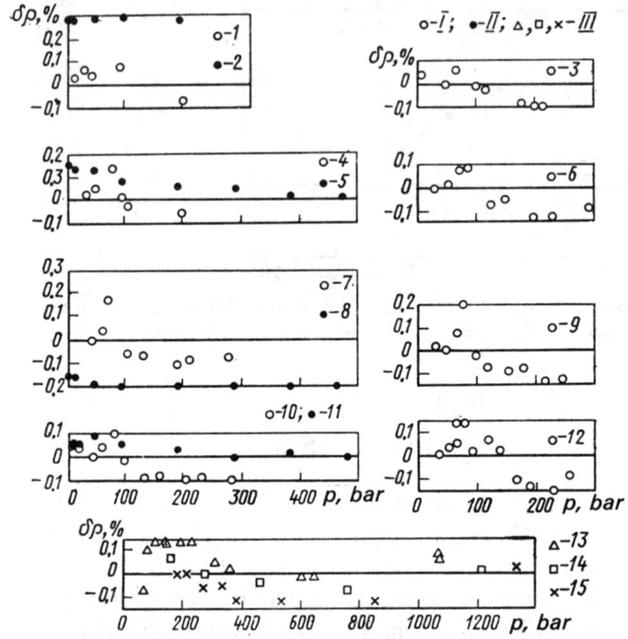

FIGURE 59. Deviations of experimental data on the density of liquid xenon from values obtained from Eq. (3.37). I) [1.121, 3.57]; II) [3.56]; III) [2.77]; 1) T = 170 K; 2) 171.70; 3) 190; 4) 180; 5) 179.45; 6) 210; 7) 200; 8) 199.84; 9) 230; 10) 220; 11) 219.67; 12) 240; 13) 273.15; 14) 286.65; 15) 298.15 K.

along the solidification curve obtained by Abovskiy [1.6] and Abovskiy with Gopanovich [3.91] with values calculated from Eq. (3.37) using the expression for $p_m(T)$ given by Eq. (1.69). The values in [1.6, 3.91] are lower. In these studies the density at the triple point, in the vicinity of which it is possible to expand in a Taylor series in powers of the relative change of density, was as-

TABLE 44. Comparison of Computed Values of the Specific Volume of Xenon Along the Solidification Curve

T, K	p, bar	v_l, cm³/g from Eq. (3.37)	v_l, cm³/g obtained in [1.6, 3.91]	δv_l, %
161.36	0.81	0.3372	0.3351	—0.63
165	93.28	0.3351	0.3329	—0.66
170	224.2	0.3324	0.3301	—0.70
175	357.8	0.3299	0.3274	—0.77
180	493.7	0.3275	0.3250	—0.77
185	631.8	0.3254	0.3226	—0.86
190	772.2	0.3235	0.3203	—1.0
195	914.7	0.3214	0.3181	—1.03
200	1058	0.3195	0.3161	—1.08

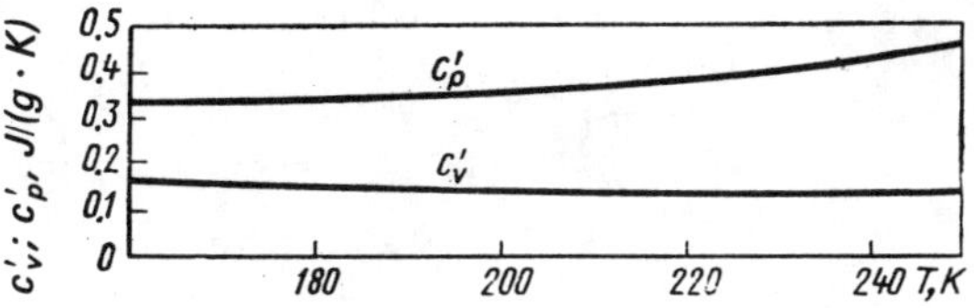

FIGURE 60. Temperature dependence of specific heats c'_v and c'_p of saturated liquid xenon.

sumed on the basis of data of Leadbetter and Thomaes [3.58]. The higher dependability of our data follows from the higher accuracy of description of experimental results of Blagoy and Sorokin [3.56] (including corrections for the initial volume v') which are given for temperatures of 171.70 and 179.45 K up to the solidification curve.

Figure 60 shows the temperature dependence of specific heats c'_v and c'_p obtained from Eq. (3.37) using data in [3.76] on the speed of sound along the saturation curve. The accuracy of the computed values cannot be assessed reliably due to lack of experimental data.

PART TWO

NONEQUILIBRIUM PROPERTIES

CHAPTER
FOUR

DYNAMIC VISCOSITY AND THERMAL CONDUCTIVITY OF MONATOMIC GASES AT ATMOSPHERIC PRESSURE

At present, a large body of information has accumulated on the viscosity and thermal conductivity of inert gases at atmospheric pressure. As these data appeared, a number of investigators checked their mutual compatibility in relation to the kinetic energy of gases. The conclusions they reached were not unique. In certain cases, the ratio η_T/λ_T remained constant and in others it increased or decreased with the temperature.

It was noted that when using the data of Trautz and his coworkers [4.3], the ratio η_T/λ_T increased systematically with the temperature. This made a significant contribution to stimulation of new studies for measuring the viscosity of inert gases at moderate and high temperatures. The results are in better agreement with data on heat conduction. However, the differences between the new values of viscosity frequently exceed the overall error of independent measurements. This necessitated developing a method for correlating and working up most reliable data in order to use them for calculating recommended values of η_T and λ_T for neon, argon, krypton, and xenon at temperatures from the triple points to 1300 K.

4.1. TEMPERATURE BEHAVIOR OF VISCOSITY

Systematic experimental study of the temperature dependence of the viscosity of inert gases at atmospheric pressure started in the 1920s. Since then, these substances have served as the object of investigation in many laboratories all over the world (Table 45). The data of the majority of investigators are given over a limited temperature range. However, certain relatively recent studies by Trautz and his coworkers [4.3] were performed over wider temperature ranges.

There is no need to analyze in detail all the studies listed in the table. The majority of these, published up until 1968, were analyzed by Rabinovich and Kryuchkov [4.26]. Nevertheless, certain features of viscosity data at $T > 400$ K deserve mention.

The viscosity of neon and argon (as of many other substances) at high temperatures was measured for the first time by Trautz and Zink [4.3]†. The data of these investigators were fundamental to a number of experiments. They were subsequently used for calibration of viscosimeters in investigating the viscosity of various gases by the relative method. This was done by Vasilesco [4.14] in measuring the viscosity of argon (and also of CO_2 and nitrogen); by Bonilla and his coworkers [4.15] in investigating the temperature dependence of argon and ordinary-water and heavy-water vapor; by Rigby and Smith [4.18], who determined the viscosity of argon, krypton, and xenon using data on the viscosity of nitrogen. It is therefore little wonder that the values of viscosity of argon are in mutual agreement in the overlapping temperature ranges with an error not exceeding $\pm 1\%$.

At the same time, the viscosity of neon and argon was measured by Kestin and Whitelaw by the oscillating disk technique [4.9]. Systematic deviations are observed between these two sets of results, and they increased with temperature to ~2% at 538 K. Initially, this was attributed to the fundamental difference in the methods and due to noncompensation for errors. However, Guevara et al. [4.21] of the Los Alamos Laboratory obtained, by the capillary method, new values of viscosity of argon at temperatures between 1100 and 2150 K which are 6% higher than those by Trautz et al. [4.3] and 10% higher than those by Vasilesco [4.14] at temperatures of 1100 and 1900 K, respectively. Both sets of data in [4.9, 4.21], when plotted in coordinates (η_T, T), fit on a single smooth curve, although they were obtained by different methods.

It was first noted by Hanley and Childs [4.27], and then it was shown in subsequent experimental studies [4.10–4.12, 4.22, 4.25] that the data of Trautz et al. [4.3] are on the low side at high temperatures.

Guevara et al. [4.21] used the capillary technique for determining the viscosity not only of argon, but also of hydrogen, helium, and nitrogen at T between 1100 and 2150 K and atmospheric pressure. The original experimental facility constructed by them made it possible to obtain relative values of viscosity of gases (relative to their viscosity at $T = 283$ K) without measuring the time of discharge, volumes and absolute velocities of gas flow.

Highly pure gas was discharged from a tank along a connecting tube to a throttling control valve held in a constant-temperature bath. The valve consisted

† This study has been performed faultlessly in the methodological respect and no substantive objections to it were raised up to present.

TABLE 45. Experimental Studies for Determining Viscosity η_T of Monatomic Gases

Year	Author	Temperature, K	Method	Reference
		Neon		
1910	Rankine	283—373	Capillary	[4.1]
1928	Edwards	195—718	»	[4.2]
1929—1935	Trautz et al.	90—1100	»	[4.3]
1940	Van-Itterbeek and Van Paemel	20—293	Oscillating disk	[4.4]
1941	Wobser and Muller	293—371	Falling sphere	[4.5]
1942	Johnston and Grilly	80—300	Oscillating disk	[4.6]
1959	Kestin and Leidenfrost†	293	Same as above	[4.7]
1960	Thornton	~291	Capillary	[4.8]
1963	Kestin and Whitelaw†	296—519	Oscillating disk	[4.9]
1970	Dawe and Smith	293—1600	Capillary	[4.10]
1970	Kestin et al.	298—973	Oscillating disk	[4.11]
1971	Guevara and Stensland	1100—2100	Capillary	[4.12]
		Argon		
1901	Schultze	291—457	Capillary	[4.13]
1910	Rankine	285—373	»	[4.1]
1929—1935	Trautz et al.	293—1100	»	[4.3]
1938	Van-Itterbeek and Van Paemel	55—90	Oscillating disk	[4.4]
1941	Wobser and Muller	293—371	Falling sphere	[4.5]
1942	Johnston and Grilly	80—300	Oscillating disk	[4.6]
1945	Vasilesco	273—1868	Capillary	[4.14]
1956	Bonilla et al.	273—2073	»	[4.15]
1958	Rocco and Halford	211—471	»	[4.16]
1959	Kestin and Leidenfrost†	293	Oscillating disk	[4.7]
1960	Thornton	291	Capillary	[4.8]
1963	Kestin and Whitelaw†	298—538	Oscillating disk	[4.9]
1965	Cronin	140—293	Capillary	[4.17]
1966	Rigby and Smith	293—973	»	[4.18]
1968	Clarke and Smith	114—375	»	[4.19]
1969	Timrot et al.	300—600	Oscillating disk	[4.20]
1969	Guevara et al.	1100—2150	Capillary	[4.21]

TABLE 45. Experimental Studies for Determining Viscosity η_T of Monatomic Gases (*Continued*)

Year	Author	Temperature, K	Method	Reference
1970	Dawe and Smith	293—1600	Capillary	[4.10]
1970	Kalelkar and Kestin	298—1124	Oscillating disk	[4.22]
1970	Kestin et al.	298—973	Same as above	[4.11]
		Krypton		
1910	Rankine	284—373	Capillary	[4.1]
1928	Nasini and Rossi	288—373	»	[4.23]
1929—1935	Trautz et al.	273—373	»	[4.3]
1959	Kestin and Leidenfrost†	293	Oscillating disk	[4.7]
1960	Thornton	291	Capillary	[4.8]
1963	Clifton	297—666	Oscillating disk	[4.24]
1966	Rigby and Smith	293—972	Capillary	[4.18]
1968	Clarke and Smith	134—375	»	[4.19]
1970	Dawe and Smith	293—1600	»	[4.10]
1970	Kalelkar and Kestin	298—1151	Oscillating disk	[4.22]
1970	Kestin et al.	298—973	Same as above	[4.11]
1970	Goldblatt et al.	1100—2000	Capillary	[4.25]
		Xenon		
1910	Rankine	284—373	Capillary	[4.1]
1928	Nasini and Rossi	290—373	»	[4.23]
1929—1935	Trautz et al.	293—550	»	[4.3]
1959	Kestin and Leidenfrost†	298	Oscillating disk	[4.7]
1960	Thornton	291	Capillary	[4.8]
1966	Rigby and Smith	293—972	»	[4.18]
1968	Clarke and Smith	176—375	»	[4.19]
1970	Dawe and Smith	293—1600	»	[4.10]

† The pressure dependence at $p > 1$ atm was also investigated in this study.

of a 55.9 cm long flowmetering stainless steel capillary, with inside diameter d_{in} = 0.071 cm, in the form of a helix (D = 5.7 cm). Then the gas was fed through a connecting tube to a tungsten capillary (L = 8.4 cm, d_{in} = 0.51 mm), and then to a takeup vessel ($v \simeq 100$ liter) which can communicate with the atmosphere or vacuum line for presetting any working pressure and for smoothing out oscillations.

The capillary was placed in a furnace where the temperature can be raised to about 2200 K. Radiation shields of tungsten foil were installed on the tube leading to the capillary in order to ensure that the gas reaches the working temperature before it enters the capillary. Three identical shields were fastened to the heated tube to protect the end ports through which the capillary wall temperature was determined by means of an optical pyrometer. Guevara et al. [4.21] assumed that the accuracy of their data was controlled basically by the accuracy of temperature measurements. For this reason, a great deal of attention was paid to this problem.

The pressure drop Δp_a at the ends of the flowmeter was measured continuously by means of a transducer. The presence in the arrangement of four valves made it possible, by means of another transducer, to measure the pressure drop Δp_b at the ends of the capillary and the overall pressure drop along the line $\Delta p_c = \Delta p_a + \Delta p_b$. The transducers consist of high precision integral digital voltmeters connected to high-sensitivity (10^{-2} mm Hg) recording devices, and capable of rapid indication of the numerical values of pressure differences.

The relative viscosities were calculated from the measured data on the basis of the Poiseuille equation

$$Q = \frac{\pi r^4 \rho}{8L\eta} \Delta p \tag{4.1}$$

in which Q is the mass flowrate of the gas. When the flowmeter and viscosimeter are connected in series and Q = const, we can write the obvious equality

$$Q = \left(\frac{\pi r^4 \rho}{8L\eta}\right)_a \Delta p_a = \left(\frac{\pi r^4 \rho}{8L\eta}\right)_b \Delta p_b \tag{4.2}$$

Since the signal of the pressure difference, expressed in mm of Hg, is transformed into a potential difference in volts, then, setting $\Delta p_a/\Delta p_b = A$, we obtain

$$A = \frac{(\pi r^4 \rho/8L\eta)_a}{(\pi r^4 \rho/8L\eta)_b} = C\left(\frac{\eta L}{r^4 \rho}\right)_b \tag{4.3}$$

Here $C = \dfrac{\rho r^4}{\eta L}$ is a constant since the flowmetering capillary is held in a constant temperature bath and the expression in parentheses depends only on the temperature of the capillary.

Values of $A(T)$ and $A(T_0)$ were determined for each specified experimental temperature, including T_0. It was assumed that T_0 = 283 K and the relative viscosity values were obtained from the expression

$$\frac{\eta(T)}{\eta(T_0)} = \frac{A(T) \cdot \rho(T)}{A(T_0) \cdot \rho(T_0)} = \frac{A(T)}{A(283)} \cdot \frac{283}{T} \tag{4.4}$$

The equation was obtained on the assumption that the gases under study were ideal. Outwardly Eq. (4.4) appears to be rather simple. However, its use for calculating the relative values of viscosity requires incorporation of five corrections: (1) for reducing the readings of the optical pyrometer to the real temperature of the capillary; (2) deviation from Poiseuille's law; (3) thermal expansion of the capillary; (4) additional resistance during the flow of gas through the communicating tubes; and (5) the slip of gas.

Two corrections were made for each reading of the optical pyrometer. One of these is due to the slight loss of visibility through the 0.64 cm thick glass viewing port. This correction was determined by means of a curve obtained from readings of the pyrometer with a reference lamp placed past the viscosimeter port and outside it. The second correction was for the deviation of the readings of the pyrometer used in the experiments from the readings of a calibration-grade pyrometer. The results of a check showed that the error of temperature measurement may be ± 4 K at $T \simeq 1000$ K and doubles at $T \simeq 2200$ K.

The correction for deviation of the real flow of gas from that of an ideal fluid was made as follows. The second transducer yields two readings each of Δp_b and Δp_c for each gas flow mode. These are used for calculating the values of Δp_b^0 and Δp_c^0 using the expressions

$$\Delta p_b^0 = \Delta p_b - \frac{m}{p_{\mathrm{m}}} \Delta p_a^2 \tag{4.5}$$

$$\Delta p_c^0 = \Delta p_c - \frac{m}{p_{\mathrm{m}}} \Delta p_a^2 \tag{4.6}$$

for conformance to the Poiseuille law; here p_{m} is the mean pressure in the capillary whereas m is a coefficient calculated from formulas given in the appendix of [4.21].

An experimental check of Eqs. (4.5) and (4.6) was performed in coordinates $\left(\frac{\Delta p_b}{\Delta p_a}, \Delta p_a\right)$ and $\left(\frac{\Delta p_c}{\Delta p_a}, \Delta p_a\right)$ with the result that the data fit on straight lines. The intersection of the first of them with the ordinate axis as $\Delta p_a \to 0$ yields a value of $\Delta p_b^0/\Delta p_a$ numerically equal to $A(T)$—see Eq. (4.4). At = 283 K the check was performed several times for each gas in order to eliminate changes which could not be predicted or could not be detected in the instrument. In particular, the check showed that the capillary cross section remained unchanged. At high temperatures, such a check was performed only for hydrogen and argon. Guevara et al. [4.21] tend to ascribe the satisfactory agreement between the predicted and experimental values of the slopes of the straight lines to the fact that the transducers are sensitive to very small temperature drops. This makes it possible to perform the experiments at low gas velocities. Thus, for example, in investigating the viscosity of hydrogen the value of Δp_a

$= 0.5\ \Delta p_b$ is equivalent to the flow of gas through the capillary at 35 μg/s at Re = 3. According to the results of comparison, Guevara et al. took as the basis the set of data corresponding to a single determined value of $\Delta p_a = 0.5\ \Delta p_b$ for the entire temperature range. The correction for the thermal expansion of the capillary in the course of heating was made using the expression

$$d = d_0(1 + \alpha\Delta T) \tag{4.7}$$

in which d_0 is the diameter of the capillary at T_0; $\Delta T = T - T_0$ and α is the linear expansion coefficient of tungsten. From this

$$\frac{r^4}{l} = \frac{r_0^4}{l_0}(1 + \alpha\Delta T)^3 \approx \frac{r_0^4}{l_0}(1 + 3\alpha\Delta T) \tag{4.8}$$

The correction factor $f_e = 1 + 3\ \alpha\Delta T$ was calculated from an empirical formula and found to be 1.011 at $T = 1100$ K and 1.029 at 2150 K.

The pressure drop in the flowmeter and piping is included in the pressure drop of the flowmeter and is not included anywhere else. The correction for the pressure drop in the copper tubing (78.7 cm long intake tube and 81.3 cm long discharge tube, $d_{in} = 0.46$ cm) connected to the capillary we obtained from the expression

$$f_1 = 1 + 0.00311\left[1 - \frac{A(283)}{A(T)}\right] \tag{4.9}$$

Its value ranges between 0.28 and 0.30% of $A(T)$ and is independent of the properties of the gas under study. It was taken to be 1.003 in all the experiments.

The correction for the slip of the gas was calculated from the equation

$$f_s = 1 + 4\frac{\xi}{r} = 1 + \frac{4\beta\lambda}{r} \tag{4.10}$$

where λ is the molecular mean free path and the value of $\beta = \xi/\lambda$ as calculated by Albertoni et al. [4.28] was taken to be 1.147;

$$\lambda = 8.589\ \frac{\eta}{p}\left(\frac{T}{M}\right)^{\frac{1}{2}} \tag{4.11}$$

This correction is not made for the flowmeter since the temperature and pressure remain constant for each discharge. For the capillary

$$f_s = \frac{f_s(T)}{f_s(283)} = \frac{1 + 4\beta[\lambda(T)/r]}{1 + 4\beta[\lambda(283)/r]} \approx 1 + 4\beta\left[\frac{\lambda(T)}{r} - \frac{\lambda(283)}{r}\right] \tag{4.12}$$

The feasibility of using semiempirical formulas for correcting for the slip of gas at high temperatures was checked experimentally by Guevara et al. [4.21]. For this purpose they measured A at temperatures of 1594 and 1757 K and a number of values of $p > 1$ atm. In each case, the values of A corresponding to the points of intersection of straight line $A = f(1/p)$ with the coordinate axis were equal to the values of A at $P = 1$ atm, multiplied by f_s. For this reason it was concluded that Eqs. (4.11) and (4.12) can be used for correcting for the slip of gas. It amounts to 1% at 1100 K and 3.9% at 2150 K.

This means that the final formula for analytic determination of the relative viscosity, incorporating the above five corrections, can be written as

$$\frac{\eta(T)}{\eta(283)} = \frac{A(T)}{A(283)} \cdot \frac{283}{T} f_e f_1 f_s \tag{4.13}$$

Guevara et al. [4.21] presented their results in tables. The error in measuring the pressure drop and temperature amounted to 0.1 and 0.4%, respectively. They assume that there are no systematic errors and the random errors can be estimated at ±0.4%.

As previously noted, the experimental data for argon obtained by Guevara et al. [4.21] are systematically higher than those in [4.3, 4.14, 4.15]. Guevara et al. noted that they carefully investigated and analyzed the results in [4.3, 4.14]. Unfortunately, they were not able to find the reason for the systematic differences which increased with the temperature and significantly exceeded the overall error of independent measurements.

Subsequently, new experimental values of viscosity for krypton [4.25] and neon [4.12] were obtained at the Los Alamos laboratory with newly designed equipment. At $T = 1100$ K the results of Trautz and Zink are lower than the data in the latter study by approximately 6%.

Dawe and Smith [4.10] measured the viscosity of nitrogen and the five inert gases at $p < 1$ atm, due to the need to refine the values of viscosity of these gases at high temperatures.

Their experimental facility had two vessels with $V_1 = 1$ liter and $V_2 = 2$ liter, interconnected by supply tubing and a helically wound capillary. It was placed in a bath in which temperature was maintained at a constant level of 298 ± 0.02 K. A mercury filled manometer was fastened to the first vessel. This manometer had soldered in electrical contacts which allow the determination of the time needed for travel of the mercury level which controls the pressure in the vessel. To fix the backpressure, a mercury filled manometer with electrical contacts was also fastened to the second vessel.

The pressure in the vessels can be preset in such a manner that the driving head $\Delta p = p_1 - p_2$ could change between 450 and 50 mm Hg, and the mean pressure in the capillary $\bar{p} = \dfrac{p_1 + p_2}{2}$ from 1000 to 80 mg Hg. The time of discharge of each of the tested gases was determined at 35 different values of Δp and $\bar{p}$.

Three capillaries: platinum, platinum-rhodium, and quartz were used in the experiments. The first two were 100 and 60 cm long and d_{in} of 0.44 and 0.37 mm, respectively. They were in the form of a 47 mm diameter densely wound helix. The intake tubes were made of the same material (L = 26 and 36 cm, respectively, d_{in} = 2 mm) and were smoothly joined to the ends of the capillary. The length was selected so as to allow the gas approaching the capillary to reach the temperature of the latter. The quartz capillary (L = 163 cm, d_{in} = 0.44 mm) was in the form of a 95 mm diameter helix connected to the intake quartz tube (L = 35 cm, d_{in} = 5 mm).

The temperature of the capillary was measured by means of three platinum vs. platinum-rhodium thermocouples, located near the capillary. The possible error of temperature measurement was estimated by the authors at a maximum of 0.2%. The temperature gradient along the capillary was less than 2 K at 1273 K. The quartz capillary was used at T between 300 and 800 K, the metal capillaries at 600–1600 K. The viscosity of the gases (purity of Ne, 99.97%, Ar, 99.995%, Kr, 99.89%, and Xe, 99.89%) was investigated using all the capillaries and the results were compared in the overlapping temperature ranges.

The dynamic viscosity was calculated from the results of measurements using the Poiseuille equation

$$dn/d\tau = -(\pi r^4/8\eta RT)p(dp/dx) \tag{4.14}$$

which should be corrected as necessary, including the most significant correction for the slip. As correctly noted by Dawe and Smith [4.10], the accuracy of calculation of the viscosity coefficient from Eq. (4.14) is in direct proportion to the accuracy of determining all the corrections and of the capillary radius. For this reason, use is frequently made of relative measuring methods. These eliminate the effect of the instrument's geometry and the effect of other factors decreases to some degree. Having decided to use the relative method, Dawe and Smith implemented it in two versions. In accordance with the first version, the viscosity of the test substance was expressed relative to the viscosity of a well-explored reference substance. In accordance with the second, termed the standard to distinguish it from the first, the viscosity of the gas at any temperature is expressed relative to the viscosity of the same gas at the selected temperature (let us say room temperature). Analytic expressions were obtained for these methods.

If Eq. (4.14) is corrected for slip, it can be transformed to

$$dn/d\tau = \frac{\pi r^4}{16\eta RTL}(p^2{}_1 - p^2{}_2)\left(1 + \frac{k}{\bar{p}}\right) \tag{4.15}$$

and then, equating the viscosity of gases a and b under the same conditions, we find

$$\frac{\tau_a}{\tau_b} = \frac{\eta_a}{\eta_b}\left(1 + \frac{k_b - k_a}{\overline{\overline{p}}}\right) \tag{4.16}$$

on the assumption that the effect of slip of gas a is insignificant. Here $\overline{\overline{p}}$ is the arithmetic mean of the mean pressures $\overline{p}$ in the capillary, whereas τ_a and τ_b are the discharge times of the gases measured under the same conditions.

The same procedure was used to obtain an equation for determining the viscosity by the standard method. When comparison was performed at two different values of T_1 and T_0, but at equal remaining conditions,

$$\frac{\tau_0}{\tau_1} = \frac{\eta_0 T_0\,[1 + \alpha(T_1 - T)]^3}{\eta_1 T_1\,[1 + \alpha(T_0 - T)]^3}\left(1 + \frac{k_1 - k_0}{\overline{\overline{p}}}\right) \tag{4.17}$$

where α is the coefficient of linear expansion of the material of the capillary; T is the intermediate temperature.

At pressures from 80 to 1000 mm Hg the viscosity depends little on changes in pressure. The slip correction was made on the basis of experimental data by constructing the linear relationships $\tau_a/\tau_b = f(1/\overline{\overline{p}})$ and $\tau_0/\tau_1 = F(1/\overline{\overline{p}})$ and extrapolating them to $1/\overline{\overline{p}} = 0$. Prior to constructing these graphs, Dawe and Smith made the following corrections in the left-hand sides of Eqs. (4.16) and (4.17). The correction for the kinetic energy of the gas was less than 0.1%, with the exception of several modes of discharge at room temperature. The correction for deviation from ideal-gas behavior was made using second virial coefficients. With the exception of krypton and xenon, it was found to be negligible for gases. For these it amounted to 0.2 and 0.6%, respectively. Analysis of the correction induced by flow of gas through the helically-wound capillary showed that the curvature of the latter has virtually no effect on the experimental results.

A total of 10,000 runs was made over a wide temperature and pressure range at $\mathrm{Re} < 20$. The deviations of the discharge time did not exceed ± 1 sec, which amounted to 0.2%. The height of the mercury columns was measured by cathetometers within ± 0.05 mm. When using Eq. (4.17) the coefficient of linear expansion of the metal capillaries was determined from the equation

$$\alpha = 8.868 \cdot 10^{-6} + 1.234 \cdot 10^{-9}(T - 273) \tag{4.18}$$

whereas for the quartz capillary it was assumed that $\alpha = 0.54 \cdot 10^{-6}$.

Dawe and Smith assume that the errors in their calculations, due to simplifications underlying the derivation of Eqs. (4.16) and (4.17) and also inaccuracies in determining the individual corrections, may result in an overall error in the values of relative viscosity of up to 0.5% and of the standard viscosity to 1%.

The main source of error in the last case is, in their opinion, an insufficiently reliable measurement of the temperature of the capillary.

The disagreement of data on the viscosity of argon in [4.10] and [4.12] is as high as 3% in the overlapping temperature range, which exceeds the overall error of results of independent measurements and of the methods of their workup.

Earlier Clarke and Smith [4.19] investigated the viscosity of nitrogen, argon, krypton, and xenon at lower temperatures and $p \leqslant 1$ atm. The experimental equipment did not basically differ from that considered above. Vessels with capacities of 1 and 5 liters were connected by communicating tubing and a capillary of Pyrex (L = 150 cm, d_{in} = 0.44 mm), wound into a coil (D = 55 mm). The capillary is surrounded by a heat exchanger and placed in a cryostat. There the temperature was maintained within ±0.1 K using a constant-volume gas thermometer. The pressure drop in the first vessel is measured by a mercury manometer which has five platinum contacts. The first serves for presetting the initial pressure, and the following allow determining the time of discharge of the gas through a capillary at three values of Δp. The argon contained less than 0.1% impurities. The impurity content of the krypton and xenon was less than 0.2%.

Clarke and Smith worked up their results by the relative and standard methods using Eqs. (4.16) and (4.17). However, all the corrections, including that for slip, were made analytically. The slip correction, calculated from the equation of Raw and Ellis [4.29] did not exceed 1%. The correction for the departure of the gas from ideal-gas behavior was incorporated by means of second virial coefficients and did not exceed 0.2%. The correction for the kinetic energy of the gas was also the same, even under the most unfavorable conditions. The greatest correction was that due to the coiling of the capillary. To allow for it, Clarke and Smith constructed a graph of the increase in the resistance to the gas due to the curvature of the capillary as a function of the Din number $\mathrm{D} = 2\, r\rho w \sqrt{r/R}/\eta$ and used in their final calculations only those experimental results for which the correction was less than 2%. In addition to the above, they also made a correction for the volume of the gas in the heat exchanger at the experimental temperature. When using the largest heat exchanger (V = 84 cm^3 at T = 114 K), the correction exceeded 1%.

The experimental data on viscosity of gases are given in the tables at 12 values of temperature. In the opinion of the authors, the possible error of their results should not exceed ±0.5%. It is seen from the graphs which compare data of different investigators that, at $T > 170$ K, data for the four substances under study are basically in good mutual agreement. At $T < 170$ K, systematic differences appear between the results of Clarke and Smith [4.19] and Johnston and Grilly [4.6] for argon. These differences increase with reduction of temperature and become ~3% at T = 114 K.

In addition to the above, new studies concerned with determination of the viscosity of inert gases by the capillary method, it is of interest to analyze the experimental data of Timrot et al. [4.20] on the viscosity of argon and of Kallekar and Kestin [4.22] on the viscosity of argon and krypton obtained by the oscillating disk method. They significantly improved the design of the viscosimeter allowing more accurate and satisfactorily reproducible results at high temperatures.

Timrot et al. [4.20] performed their measurements in a high-precision viscosimeter, all the components of which were made of quartz. They calculated the viscosity of the test gas by the Macwood equation [4.30]

$$\eta_T = C\left(\frac{\delta}{\tau} - \frac{\delta_0}{\tau_0}\right) \tag{4.19}$$

where C is the instrument constant; δ, δ_0 and τ, τ_0 are the logarithmic damping decrement and period of oscillations in the test gas and in a vacuum. Since the wire used in the test equipment was close to room temperature, Timrot et al. [4.20] neglected term δ_0/τ_0 due to its smallness. The effect of the possible shift of the zero mark on the scale was eliminated by determining δ from the ratio of sums of successive amplitudes measured in intervals of several oscillations:

$$\delta = -\frac{1}{m}\ln\frac{\Theta_{n+m}}{\Theta_n} \tag{4.20}$$

where m is the number of oscillatory periods; Θ_n is the sum of sequential amplitudes; Θ_{n+m} is the same, following m periods. The instrument constant was determined on the basis of experiments with nitrogen at room temperature using data of Michels and Gibbons [4.31]. The error in the data of the latter two authors is estimated at 0.1%.

The argon used was 99.95% pure. Each experimental point was determined on the basis of five to nine measurements of δ. The error of experimental data was estimated by the authors at 0.8%. The experimental results of Timrot et al. [4.20] over the entire range of temperature are in satisfactory agreement with data of Kestin and Whitelaw [4.9], but 4% higher than the data of Trautz et al. [4.3] at T = 600 K.

The instrument used by Kalelkar and Kestin [4.22] had a special feature in the suspension system of the disk. As in the equipment used by Timrot et al. [4.20] it was made of quartz. Kalelkar and Kestin calculated the viscosity of argon and krypton from the results of their measurements using the rather complex equation due to Newell [4.32], relating η_T to the viscosimeter constant C_N which is independent of the temperature. The analytically calculated values of C_N turned out to be equal to 1.3181 ± 0.0001. It does not include an allowance for the additional resistance of the suspension. For this reason, they had to determine the instrument constant using calibration experiments at a temperature close to 298 K, from well-known values of η_T for nitrogen, helium, argon, and krypton. The value of $C_N = 1.3284 \pm 0.0009$ was used by them for calculating η_T at different temperatures. The error of these data, according to estimates in [4.22], is $\pm 0.2\%$.

Insignificant discrepancies (up to 1%) exist between the results in [4.11] and [4.22] obtained by different methods. The data in [4.22], as new values of viscosity of argon obtained by others [4.10, 4.20, 4.21], are systematically higher than the results of Trautz and Zink [4.3] and Vasilesco [4.14]. The differences increase with temperature and become as high as ~5% at $T = 1124$ K. Kalelkar and Kestin did not analyze the reasons responsible for such significant differences. However, they showed that the data obtained by Kestin and his coworkers in previous studies, together with data of Guevara et al. [4.21] are satisfactorily described by potential function (6–13) at temperatures between 300 and 2150 K and are in good agreement with data obtained in experiments with scattering of molecular beams. These experiments showed that the power exponent for the repulsion branch of the potential function for argon should be between 7 and 14. At the same time, the data in [4.3, 4.14] can be satisfactorily described over approximately the same temperature range only by a potential function (6–40). In addition, the newer experimental data on the viscosity of argon are in better agreement with results on thermal conductivity.

The above circumstances lead one to agree with the statement by Kalelkar and Kestin that the new data on viscosity of inert gases are more exact than those in [4.3, 4.14]. These studies exhibit a systematic error which will most likely require in the future some correction to the large body of experimental data obtained by Trautz and his coworkers for different gases.

The results of still another study by Kestin and his coworkers [4.11] are in agreement with data for argon and krypton given in [4.22] with a deviation within the limits of the overall error ($\pm 0.2\%$) of independent measurements.

Thus, analysis of the principal studies concerned with experimental determination of the viscosity coefficient η_T of the four inert gases points to the existence of systematic differences between data of different investigators, including recently obtained data. It is impossible to determine the reason for this since the methodological refinements of experiments and a number of accompanying effects do not lend themselves to exhaustive analysis and estimate of the degree of their effect on the accuracy of the final results. For this reason, the most reliable values can be selected only on the basis of indirect criteria of their assessment. One of these was used by us in simultaneous workup of data on viscosity and thermal conductivity.

4.2. TEMPERATURE DEPENDENCE OF THE THERMAL CONDUCTIVITY

The bulk of experimental data (Table 46) on the thermal conductivity of the four gases under study at atmospheric pressure was obtained by the heated-wire tech-

TABLE 46. Experimental Studies on Determining the Thermal Conductivity of Monatomic Gases

Year	Author	Temperature, K	Method	Reference
		Neon		
1927	Weber	273	Heated wire	[4.33]
1931	Curie and Lepape	~273	» »	[4.34]
1952	Kannuluik and Carman	90—579	» »	[4.35]
1954	Keyes	91—273	Coaxial cylinders	[4.36]
1959	Zaytseva	413—803	Heated wire	[4.37]
1960—1961	Thornton	291	» »	[4.8]
1964	Sengers et al.	298—348	Planar horizontal layer	[4.38]
1966	Saxena et al.	303—762	Heated wire	[4.39]
1967	Gupta and Saxena	366	» »	]4.40]
1967	Gandhi and Saxena	303—473	» »	[4.41]
1967	Srivastava and Gupta	314—473	» »	[4.42]
1968	V. Saxena and S. Saxena	373—1473	» »	[4.43]
1970	Vargaftik and Yakush	310—1073	» »	[4.44]
		Argon		
1903	Schwarze	~100—200	Heated wire	[4.45]
1911	Eucken	90—273	» »	[4.46]
1917	Weber	~273	» »	[4.47]
1934	Dickins	277—285	» »	[4.48]
1934, 1950, 1952	Kannuluik et al.	90—579	Heated wire	[4.49, 4.50, 4.35]
1952	Schottky	373—773	Heated wire	[4.51]
1954, 1955	Keyes	105—633	Coaxial cylinders	[4.36]
1955	Rothman and Bromley	До 962	Coaxial cylinders	[4.52]
1956, 1957	Schäfer and Reiter	273—1373	Heated wire	[4.53]
1959	Zaytseva	316—788	» »	[4.37]
1960	Vines	533—1173	Coaxial cylinders	[4.54]
1964	Vargaftik and Zimina	311—1201	Heated wire	[4.55]
1965—1966	Timrot and Umanskiy	594—2034	» »	[4.56]
1965—1966	Collins et al.	1500—5000	Shock tube	[4.57, 4.58]
1967	Srivastava and Gupta	313—473	Heated wire	[4.42]
1968	V. Saxena and S. Saxena	400—1500	» »	[4.59]
1968	Bailey and Kellner	90—300	Coaxial cylinders	[4.60]

TABLE 46. Experimental Studies on Determining the Thermal Conductivity of Monatomic Gases (*Continued*)

Year	Author	Temperature, K	Method	Reference
		Krypton		
1931	Curie and Lepape	~ 273	Heated wire	[4.34]
1952	Kannuluik and Carman	195—579	» »	[4.35]
1955	Keyes	131—273	Coaxial cylinders	[4.36]
1956—1957	Schäfer and Reiter	273—1373	Heated wire	[4.54]
1959	Zaytseva	317—798	» »	[4.37]
1966	Grambhir and S. Saxena	308—363	» »	[4.61]
1967	Gupta	312—473	» »	[4.62]
1969	V. Saxena and S. Saxena	373—1473	» »	[4.63]
1970	Umanskiy et al.	500—1200	» »	[4.64]
1970	Vargaftik and Yakush	318—1153	» »	[4.44]
		Xenon		
1931	Curie and Lepape	273	Heated wire	[4.34]
1952	Kannuluik and Carman	195—579	» »	[4.35]
1955	Keyes	155—273	Coaxial cylinders	[4.36]
1959	Zaytseva	306—794	Heated wire	[4.37]
1960—1961	Thornton	291	» »	[4.8]
1967	Gandhi and Saxena	303—363	» »	[4.41]
1969	V. Saxena and S. Saxena	373—1473	» »	[4.63]
1970	Vargaftik and Yakush	306—1179	» »	[4.44]

nique. These data together extend over a rather wide temperature range and hence can serve as a basis for compiling detailed tables of recommended values of λ_T. However, the results of measurements of the different investigators differ significantly from one another requiring analysis of the entire body of experimental data followed by selection of the most reliable of these.

The thermal conductivity of neon at atmospheric pressure was investigated at T between 90 and 1437 K. The data in the earliest study by Weber [4.33] are insufficiently reliable. Over a wide temperature range (90 to 579 K) the thermal conductivity of neon was investigated by Kannuluik and Carman [4.35]. The hot wire was 1.5 mm in diameter. The authors paid a great deal of attention to the elimination of convection. Its absence was established by additional measurements of the thermal conductivity above atmospheric pressure. The error of data in [4.35] increases slightly with the temperature since the radiation correction was incorporated quite roughly. These results are in satisfactory agreement with those of Keyes [4.36] obtained independently by another method (coaxial cylinders). The difference does not exceed 1%.

Zaytseva [4.37] performed a detailed analysis of the correction for the temperature jump between the central heater and the gas.

Some time later, several studies were published in which the thermal conductivity of neon was obtained by a number of versions of the heated-wire method. Thus, V. Saxena and coworkers [4.39] performed their measurements at $p \leqslant 605$ mm Hg in order to make a correction for the temperature jump between the gas and the heater. The maximum difference between λ_T measured at the same temperatures ($T > 600$ K) for pressures from 102 to 605 mm Hg is 8 to 10%. For this reason, the authors themselves assume that their data for the high-temperature region should be regarded as approximate. The averaging curve drawn in their study is apparently based also on data of other investigators. At $T < 600$ K the data of V. Saxena et al. have a smaller error. This is additionally confirmed by experimental results of Gupta and S. Saxena [4.40], who showed that the thermal conductivity of neon at 366 K is virtually independent of pressure between 10 and 492 mm Hg.

Two additional studies were published simultaneously by Gandhi and Saxena [4.41] and Srivastava and Gupta [4.42] (the latter lists six values of λ_T). Finally, in 1968 V. Saxena and S. Saxena [4.43] published their concluding study. It describes in detail the test equipment, analyzes the corrections contained in the analytic equation and lists ±2% as the possible error of the experimental values of λ_T. Within the limits of this error, the results in [4.43] are in agreement with results of preceding studies [4.39–4.42].

Since neon is a light gas, the correction for the temperature jump is very important for it. For this reason, Vargaftik and Yakush [4.44] gave this problem a great deal of attention. To correctly allow for radiation from the heater they used two measuring tubes with different diameters: 4.07 and 2.97 mm. The differences between the results of the two series of experiments, after making the required corrections, did not exceed 1.5%. Experimental data on thermal conductivity of neon over the temperature range in these studies are described by the polynomial

$$\lambda_T = a_0 + a_1 T + a_2 T^2 + a_3 T^3 \tag{4.21}$$

with a standard deviation of 1.1% and maximum deviation of 2%. It is interesting to note that the results of the two fundamental studies [4.43, 4.44], in which the thermal conductivity of neon was investigated over the widest temperature range, agree with one another within the limits of the overall error of independent measurements. However, the values obtained by Vargaftik and Yakush are predominantly higher. For example, at $T = 1073$ K the difference amounts to 3.3%.

Thornton [4.8] and Srivastava and Gupta [4.42] investigated the thermal conductivity of neon in a mixture with other gases. Results of measurement of λ_T of pure neon are also tabulated: one value in [4.8] and six values in [4.42].

The results of Sengers et al. [4.38] were obtained over a wide pressure range from 1 to 2600 bar at T of 298, 323 and 348 K.

The thermal conductivity of argon was investigated experimentally from the triple point to 2100 K. As early as at the start of this century, Schwarze [4.45] determined the values of λ_T at low temperatures. Studies by Eucken [4.46], Weber [4.47], and Dickins [4.48] were performed over a temperature range close to that explored by Schwarze. The most complete studies at low temperatures were performed by Kannuluik with his coworkers [4.50, 4.49, 4.35], of which the last is of the greatest interest. It is concerned with the thermal conductivity of other inert gases in addition to argon. Certain features of this study were noted in analyzing data on the thermal conductivity of neon.

Keyes [4.36] investigated the thermal conductivity of liquid and gaseous argon at atmospheric pessure. The pressure range was significantly broadened (1 to 500 kg/cm^2) by Bailey and Kellner [4.60]†. Their values of λ_T for argon exhibit, in the opinion of the authors, an error of $\pm 2\%$ at low temperatures and of $\pm 1\%$ at $T > 200$ K. To some measure the study by Bailey and Kellner should be regarded as summarizing measurements of thermal conductivity of argon performed over many years by various investigators below room temperature. Comparison of experimental data showed that the results of the four basic studies [4.35, 4.36, 4.46, 4.60], in which the range of temperatures below 300 K has been explored most completely, agree with one another within $\pm 1.5\%$.

More studies are concerned with thermal conductivity of argon at moderate and high temperatures. Rothman and Bromley [4.52], who present their results graphically, point to a significant scatter of points above 773 K.

Schäfer and Reiter [4.53] corrected their measurements for radiation and end losses which at the highest temperatures are as high as 30%. The convection which, in their opinion, arises was eliminated by extrapolating the values of thermal conductivity measured at 250, 500 and 750 mm Hg to $p = 0$. The final results are listed in a table at round values of temperature. Vines [4.54] determined the thermal conductivity of argon at four points (533, 823, 1033 and 1173 K), and he represented the data both in a table and graphically. These values lie above the results obtained in [4.51–4.53].

Vargaftik and Zimina [4.55] performed their investigations at $p \leqslant 1$ atm. The 0.157 mm diameter platinum wire was placed along the axis of a quartz tube ($d_{in} = 4.07$ mm). The temperature difference across the gas layer ranged from 11.3 to 50.3 K. The measured data were corrected for leakage of heat from the ends, radiation, temperature drop across the wall of the quartz tube, and the temperature jump. They compared their data with results of preceding studies and found satisfactory agreement with most of them. The exception are the results in [4.51, 4.53] which are on the low side, starting with temperatures of 573 and 973 K, respectively. They found the reasons for these differences and

† This study will be analyzed in more detail in conjunction with analysis of data on the thermal conductivity of argon at $p > 1$ atm.

showed that when a proper correction is made for the temperature jump, almost all the aforementioned data can be brought into agreement within $\pm 2\%$. In substance, their study correlates experimental data on thermal conductivity of argon at $T > 300$ K.

The experimental results of Srivastava and Gupta [4.42] and Saxena and Saxena [4.59] did not provide significant additional information on the temperature dependence of thermal conductivity of argon. They only validated the results of [4.55] which served as the basis for the correction in [1.104].

Timrot and Umanskiy [4.56] perfected the heated-wire method since the conditions of the study of thermal conductivity of gaseous argon at the high temperatures they selected required reliable elimination of or correction for two accompanying heat transfer mechanisms, convection and radiation. Convection was eliminated by selection of the pertinent dimensions of the measuring tube, in which $\mathrm{Pr} \cdot \mathrm{Gr} = 44$. The radiation correction was 40% and was made analytically with an error of ~5%. The possible error in the values of λ_T due to inaccuracy in determining this correction can be 2%.

Corrections for removal of heat along the measuring wire and for the temperature jump were made experimentally. The first correction was eliminated by compensating the heat losses. The second was corrected by using results of measurements at pressures between 145 and 610 mm Hg. A total of 64 experimental points were obtained and these served to find the coefficients of the approximating quadratic polynomial. The total error of the experimental data is, in the authors' opinion, about 8%.

Although the studies by Collins et al. [4.57, 4.58] were performed over a temperature range not considered here, their results are in satisfactory agreement with data in [4.56]. Thus, the differences at 1500 and 2000 K amount to 0.4 and 4% respectively, the values in [4.56] being lower.

As for the two gases considered above (Arandne), a rather satisfactory body of experimental data is available on the thermal conductivity of krypton at atmospheric pressure. The results of Keyes [4.36], listed at temperatures of 131, 171 and 273 K, are in satisfactory agreement with data in [4.35]. However, the point at 171 K deviates significantly from the smooth curve averaging these data.

At $T > 300$ K all the experimental data on the thermal conductivity of krypton were obtained by various versions of the heated-wire method. Schäfer and Reiter [4.53] published smoothed values at pressures of 250 and 500 mm Hg. Data obtained at higher pressures are somewhat higher. This cannot be easily attributed to incorrect determination of the correction for the temperature jump since for krypton, a heavy gas, such a correction is relatively small even at high temperatures.

The remaining experimental data on the thermal conductivity of krypton are best subdivided into two groups. The first is made up of values obtained by Zaytseva [4.37] and Vargaftik and Yakush [4.44]. Particular attention was paid in these studies to the proper determination of the correction for the temperature jump. The data in these two studies are in good mutual agreement and are described in the latter by a cubic polynomial with an error not exceeding 2%.

The second group is made up of data of Gambhir and S. Saxena [4.61], Gupta [4.62], and V. Saxena and S. Saxena [4.63]. The last two authors estimate the error of their results at ±2%, within which limits they agree with the data in [4.61, 4.62].

Our analysis showed that systematic differences, amounting on the average to 5–6%, exist between the two groups of results, with data in [4.44] being higher. We were not able to determine the reasons for these differences by analyzing the original studies.

Since the thermal conductivity of krypton is lower than that of argon, Umanskiy and his coworkers [4.64] were forced to reduce the diameter of the wire in order to reduce the radiation correction. Their data processing technique was refined relative to that used in previous studies which, in their opinion, led to improvement in the accuracy of the final results. Unfortunately, [4.64] does not list numerical values making it impossible to directly compare the results with data of others. However, the graph plotted there showed that the curve of $\lambda_T = f(T)$ is in good agreement with the results of Kannuluik and Carman [4.35] and lies below the data of Zytseva [4.37]. The error of data not listed in [4.64] are estimated by Umanskiy et al. at 2.5–3%.

Unlike the other three gases, relatively few experimental data are available on the thermal conductivity of xenon at atmospheric pressure. All these are presented in the previously analyzed studies. The results in [4.35, 4.36] are in excellent mutual agreement, in spite of being obtained by different methods.

The experimental data in [4.37, 4.44] agree with one another within ±2%, which is the error of their description by polynomial (4.21).

S. Saxena et al. [4.41, 4.63] estimate the possible error of their data at 2%. The results of the fundamental studies [4.44, 4.63] are in mutual agreement within the overall error of independent measurements. However, the values in [4.63] lie above those in [4.44] at $T \leqslant 800$ K, but are lower at higher temperatures.

The above actually exhaust the list of studies known to us which present experimental data on the temperature dependence of the thermal conductivity of the four inert gases at atmospheric pressure. It can be concluded from the above brief analysis for each of the substances that ranges of parameters exist where the data of different investigators are in rather good agreement with one another, and therefore the recommended values of thermal conductivity have been determined with acceptable accuracy. It is much more difficult to select the more correct values from among the divergent data.

Unfortunately, analysis of original studies does not suffice to determine the reasons for the discrepancies and for giving preference to one set of results over the other. Little is to be gained by using collision integrals, calculated from model functions of intermolecular interaction. Since by selecting the pertinent values of parameters of the potential within limited temperature intervals, it is possible to attain satisfactory agreement of the analytic behavior of $\lambda_T = f(T)$ with any set of experimental results. This necessitates finding an additional objective criterion which would allow the selection of the most reliable data and the use of them for calculating the values of λ_T of the four gases over the temperature range of interest. This is the subject of Sec. 4.3.

4.3. A METHOD OF CORRELATION OF EXPERIMENTAL DATA ON VISCOSITY AND THERMAL CONDUCTIVITY

In accordance with the molecular-kinetic theory of rare gases [2.109] the dynamic viscosity η_T and the thermal conductivity λ_T of monatomic gases can be calculated from the expressions

$$\eta_T \cdot 10^7 = 266.93 f_\eta \frac{\sqrt{TM}}{\sigma^2 \Omega^{(2.2)}(T^*)} \tag{4.22}$$

$$\lambda_T \cdot 10^7 = 1989.1 f_\lambda \frac{\sqrt{T/M}}{\sigma^2 \Omega^{(2.2)}(T^*)} \tag{4.23}$$

where the units of η_T are g/(cm · s) and λ_T are cal/(cm · s · K); M is the atomic weight; $\Omega^{(2.2)}$ is the reduced collision integral which is a function of the reduced temperature $T^* = kT/\epsilon$; f_η and f_λ are also functions of the reduced temperature, which differ little from unity.†

The reduced collision integral depends on the selected function of intermolecular interaction and is usually tabulated. The availability of such tabulated values for the most prevalent potentials and experimental data on the viscosity and thermal conductivity of various gases permits determination of the parameters of the intermolecular interaction function and to check the mutual agreement of theoretical and experimental $\eta_T = f(T)$ and $\lambda_T = f(T)$.

The most frequently used functions of intermolecular interaction for monatomic gases is the Lennard-Jones (6–12) potential which in fact was suggested for these gases

$$U(r) = 4\varepsilon\left[\left(\frac{\sigma}{r}\right)^{12} - \left(\frac{\sigma}{r}\right)^{6}\right] \tag{4.24}$$

To be able to determine ϵ and σ it suffices to have the values of viscosity or thermal conductivity at two temperatures and to proceed in accordance with the recommendations in [2.109]. However, this method has certain shortcomings. It is difficult to calculate ϵ and σ by successive approximations. The use of two different pairs of points at rather large distances from one another yields different values of parameters of the potential. Also, it is impossible to immediately establish the range of satisfactory agreement between the analytic and experimental curves. For this reason, Eqs. (4.22) and (4.23) were checked by a

† It is henceforth assumed that $f_\eta = f_\lambda = 1$.

method permitting determination of the parameters of the potential on the basis of all the reliable experimental data on the viscosity and thermal conductivity of the four gases and to simultaneously establish the range of applicability of these equations. Using Eqs. (4.22) and (4.23) we obtain the expression

$$y = \sigma^2 \Omega^{(2.2)}(T^*) = 266.93 \frac{\sqrt{TM}}{\eta_T} = 1989.1 \frac{\sqrt{T/M}}{\lambda_T} \tag{4.25}$$

from which it follows that the value of y calculated from data on viscosity and thermal conductivity of whatever gas in coordinates (y, T) should coincide with a deviation not exceeding the overall error of η_T and λ_T. This very important circumstance makes it possible to not only check the mutual agreement between the starting experimental data, but to assess their accuracy. It is best to perform the correlation in logarithmic coordinates. As will be shown below, in this case it is possible, in addition to selecting the most reliable data, to determine the parameters of the function of intermolecular interaction and to point out the temperature range over which the collision integral calculated using this function is valid.

Figure 61 presents in these coordinates (lg y, lg T) numerous experimental values of viscosity and thermal conductivity of argon. One can clearly see a set of mutually compatible data on η_T and λ_T over which a single correlating curve can be drawn. However, up to the appearance of studies [4.10, 4.11, 4.21, 4.22] concerned with the viscosity of argon, the data plotted in these coordinates formed two curves. One was from viscosity data of Trautz et al. [4.3] and Vasilesco [4.14] shown by a dashed curve in this figure, and the other from numerous data on the thermal conductivity. The difference between these significantly exceeded the overall errors. This lack of agreement between experimental and analytic results remained unexplained for a long time and apparently was one of the causes for the experimental work performed in [4.10, 4.11, 4.21, 4.22]. Now it can be rather reliably claimed that the values of viscosity of argon in [4.3] and also the results of subsequent studies, where the relative methods were based on the data of Trautz et al., are systematically on the low side at temperatures above 400 K.

The values of η_T and λ_T for argon which are in good agreement were obtained in subsequent studies. The values of η_T obtained by Johnston and Grilly below room temperature [4.6] are in excellent agreement (within $\pm 1\%$) with the thermal conductivity data of Kannuluik and Carman [4.35] and Bailey and Kellner [4.60]. The same agreement with this set of data is exhibited by results of Clarke and Smith [4.19]. However, systematic deviations from the general relationship are observed at $T < 160$ K (see Fig. 61). These deviations increase with reduction in temperature and become as high as $\sim 3\%$ for $T = 114$ K. It should also be pointed out that the thermal conductivity data of Keyes [4.36] at 105 to 160 K confirm the high reliability of the results in [4.6] on the viscosity of argon.

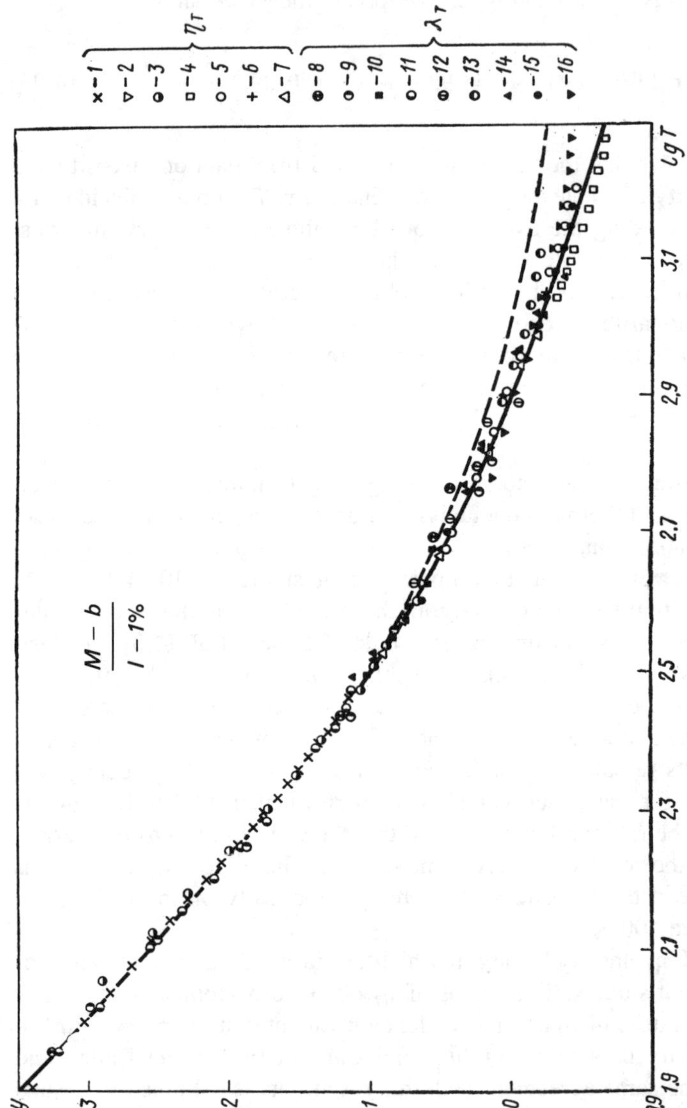

FIGURE 61. Correlation of experimental data on the viscosity and thermal conductivity of argon. 1) [4.6]; 2) [4.9]; 3) [4.19]; 4) [4.21]; 5) [4.10]; 6) [4.22]; 7) [4.11]; 8) [4.35]; 9) [4.36]; 10) [4.42]; 11) [4.60]; 12) [4.51]; 13) [4.53]; 14) [4.55]; 15) [4.59]; 16) [4.56].

The experimental results of Kestin and Whitelaw [4.9], Guevara et al. [4.21], Dowe and Smith [4.10], and Kestin et al. [4.11, 4.22] on the viscosity of argon at $T > 300$ K agree within $\pm 2\%$ with data of Keyes [4.36], Srivastava and Gupta [4.42], Schottky [4.51], Schäfer and Reiter [4.53] (for $T < 1000$ K), Vargaftik and Zimina [4.55], and V. Saxena and S. Saxena [4.59] on thermal conductivity. The values of λ_T at 700 to 1400 K obtained by Timrot and Umanskiy [4.56] agree with this set of data (within the specified limits). However, the differences start increasing rapidly with rising temperature, although they do not anywhere exceed the specified deviation of $\pm 8\%$.

All the above experimental values of viscosity and thermal conductivity of argon were used by us to obtain a function curve of $y = f(T)$ which was expected to provide an exact description of the original data with the smallest number of constants. For this reason we analyzed polynomials of various forms, including those containing exponential terms. Optimal conformance to these requirements was obtained from the polynomial

$$y = a_0 + a_1 \lg T + a_2 T^{-2} + a_3 T^{-3} + \ldots + a_6 T^{-6} \qquad (4.26)$$

The coefficients were determined by the method of least squares with allowance for weights corresponding to the degree of dependability of the input data:

$a_0 = 17.67484$; $a_4 = 39.38774 \cdot 10^8$;
$a_1 = -2.78751$; $a_5 = -16.54629 \cdot 10^{10}$;
$a_2 = 31.14987 \cdot 10^4$; $a_6 = 2.86561 \cdot 10^{12}$
$a_3 = -48.82650 \cdot 10^6$;

At high temperatures the logarithmic term of this equation not only represents correctly the experimental behavior up to $T = 2150$ K, but it describes with acceptable accuracy the results of Collins et al. [4.57, 4.58] to $T = 5000$ K, which were not used in determining coefficients a_i.

Equation (4.26) is suitable at temperatures from the triple point to 5000 K. The curve (Fig. 61) calculated on its basis was subsequently used as the reference curve for checking various model functions of intermolecular interaction and of the collision integrals determined on their basis.

First we checked the Lennard-Jones potential (4.24). For this we plotted $\Omega^{(2.2)}(T^*)$ in logarithmic coordinates on millimeter-graded graph paper on the basis of tabulated data in [2.109] and on a duplicate copy—to the same scale—we constructed the reference curve in coordinates ($\lg y$, $\lg T$). The coincidence of the curves upon parallel translation of coordinate axes should prove the suitability of the selected potential. The parameters of the potential can be easily determined from the relative shift of the origins of the coordinate systems, since

$$\lg y = \lg \sigma^2 + \lg \Omega^{(2.2)}(T^*)$$
$$\lg T^* = -\lg \left(\frac{\epsilon}{k}\right) + \lg T$$

We were not able to bring about coincidence of these curves within the limits of the possible error of experimental data over the temperature range under study. Because of this, we attempted to ensure acceptable agreement over the largest possible temperature range. Almost complete coincidence of the curves was obtained for T between 250 and 1000 K. Outside this range the deviations start becoming significant. The desire to reduce the deviations at low temperatures resulted in an increase in these at high and moderate temperatures, and vice versa. The unacceptable deviation outside the above range can be reduced rather insignificantly by using another function of intermolecular interaction (for example, the Kihara function). It is impossible to make the deviations over the entire temperature under study commensurable with the possible experimental error. The above can be verified on the basis of the thorough computational studies [4.65–4.67] which employ the Lennard-Jones, Buckingham (exp-6) and Kihara potentials to describe the transport coefficients for argon. It was found that the latter potential is valid over a somewhat wider temperature range. However, the computational error below 200 and above 700 K significantly exceeds 1%. Hanley and Childs [4.66] checked the above potentials for neon, krypton, and xenon and selected the Kihara potential for the final calculations. In their subsequent study [4.27], based on new experimental data on the viscosity of argon [4.21], they suggested that the Buckinghan potential be used.

According to our check, it is very difficult to give preference to any of these potentials. Proper selection of the parameters of the potential leads to satisfactory results over any temperature interval, but it is impossible to precisely describe data over the entire experimental range. Apparently this circumstance is responsible for the appearance of empirical potentials such as the three-parametric $(m - 6)$ analyzed in detail in [4.67, 4.68] or the potentials calculated on the basis of experimental data in [4.69–4.71].

The suitability of the $(m - 6)$ potential for calculating the transport coefficients of inert gases at atmospheric pressure was checked in detail by Hanley et al. [4.67, 4.72]. Analysis of their results showed that no significant improvement is obtained over the use of the Kihara potential. Different values of m must be used at high and low temperatures. For this reason, it is impossible to extrapolate to the high temperature range on the basis of experimental data obtained at moderate temperatures.

It is of interest to analyze the rather recent studies concerned with the determination of the potential energy of paired interaction of molecules on the basis of experimental data. Guggenheim and McGlashan [4.70] and Dymond et al. [4.71] calculated the interaction energy of atoms of inert gases on the basis of the values of the second virial coefficient. A common feature of these studies is the preliminary selection of the form of the equation approximating the potential energy $U(r)$, whose coefficients were determined from best satisfaction of values of the second virial coefficient. The function $U(r)$ thus found was used for calculating the collision integrals, in particular $\Omega^{(2.2)}$, which were then used for determining the transport properties. This approach results in large errors because the second virial coefficient is sensitive to the size of the area beneath the curve of potential energy, but it is less sensitive to the configuration of this curve than the collision integrals.

Dymond and Alder [4.69] singled out the potential in numerical form and used a method according to which $U(r)$ was determined also on the basis of data on viscosity and diffusion coefficients. That made it possible to obtain a more reliable configuration of the potential energy curve. This involved rather complex mathematics and in the final analysis they obtained tabulated values of the potential function. This function was then used for calculating the second virial coefficient B^* and the collision integrals $\Omega^{(1.1)*}$ and $\Omega^{(2.2)*}$. The suitability of the resulting values of $\Omega^{(2.2)*}$ were checked against experimental data on the viscosity and thermal conductivity of argon. At temperatures between 100 and 1500 K the most reliable data are described with an error commensurable with the error of measurements. This means that the empirical function $U(r)$ can be used for describing both equilibrium and nonequilibrium properties of argon at low densities.

Guggenheim and McGlashan [4.70] suggested a six parameter function of intermolecular interaction for argon. The coefficients were determined on the basis of the second virial coefficients and data on the properties of the crystal. The technique used in [4.70] was employed by Chakraborti et al. [4.73, 4.74] in determining $U(r)$ for krypton and xenon. The values of $U(r)$ of three substances with close atomic structure differ significantly from one another. The reason for this is most likely not differences in the behavior of each of the substances but the low accuracy and incompatibility of the original experimental data.

In our opinion, empirical functions of intermolecular interaction have the definite advantage of being suitable for correlating data on the thermophysical properties of a group of similar substances. In addition, such an approach allows the extraction of information on the properties of a group of little-explored substances on the basis of experimental results for a well-investigated substance having close molecular structure. The effectiveness of this approach shall be demonstrated for the case of monatomic gases.

We saw in the correlation of the second virial coefficients of the four inert gases (see Chapter 2) that their values can be made to coincide in reduced coordinates. This is due to the existence, for this group of substances, of a single function of intermolecular interaction, the mathematical expression of which is unknown to us. For this reason, there should also exist a single collision integral from which it should be possible to calculate the viscosity and thermal conductivity at atmospheric pressure for the same gases. The above can be checked on the basis of experimental data on η_T and λ_T of krypton, xenon, and neon.

Figure 62 presents the experimental results in logarithmic coordinates. As a whole, the scatter of experimental points is greater for krypton than for argon. At temperatures below 400 K the data of Kannuluik and Carman [4.35], Keyes [4.36], and Clarke and Smith [4.19] are in satisfactory agreement, as before. However, the data of the latter investigators at $T < 200$ K for argon start deviating significantly from the values of thermal conductivity in [4.35, 4.36]. At

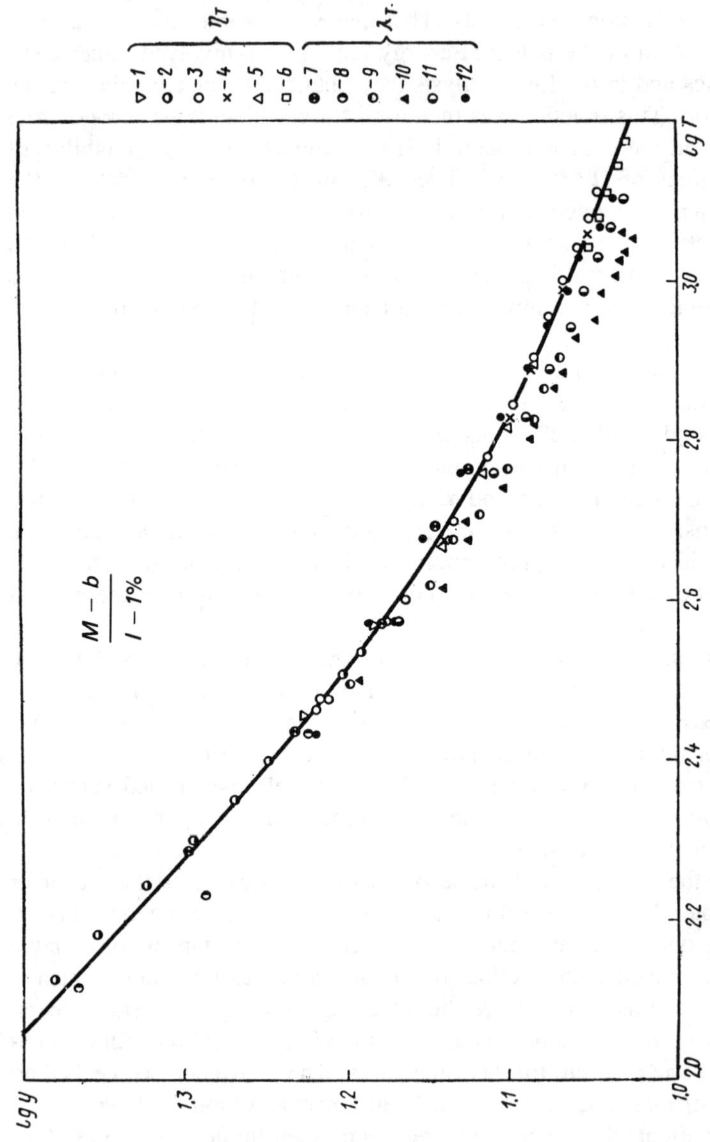

FIGURE 62. Shifted coincidence of the analytic curve obtained from the expression for $y(T)$ for argon for the purpose of correlation of experimental data on viscosity and thermal conductivity of krypton. 1) [4.23]; 2) [4.19]; 3) [4.10]; 4) [4.22]; 5) [4.11]; 6) [4.25]; 7) [4.35]; 8) [4.36]; 9) [4.37]; 10) [4.44]; 11) [4.35]; 12) [4.63].

>400 K the new experimental results [4.10, 4.11, 4.22, 4.25] cluster about a single curve within ±1%. Data of Saxena and Saxena [4.63] are in satisfactory agreement with these results, whereas the data of others [4.37, 4.44] lie significantly higher. It is characteristic that even over the temperature range of 400 to 600 K, which is experimentally the simplest, the discrepancies in the values of thermal conductivity of krypton exceed 5%.

Obviously, given this scatter of points, it is impossible to find the generalized curve by averaging the starting data. Therefore, it is best to correlate data on the transport coefficients of krypton using Eq. (4.26), obtained on the basis of mutually compatible values of viscosity and thermal conductivity of argon.

To eliminate the slight arbitrariness in determining the transformation multipliers and to check for the existence of a single collision integral for the entire group of gases, we used the reduction parameters ρ_0 and T_B, which were found previously and verified in generalizing the second virial coefficients. We next calculated the transformation multipliers of Eq. (4.26)

$$\mu = \frac{T_{BKr}}{T_{BAr}}; \nu = \sqrt[3]{\frac{\rho_{0Ar}^2 \cdot M_{Kr}^2}{\rho_{0Kr}^2 \cdot M_{Ar}^2}} \tag{4.27}$$

shown in Table 47.

Having first recalculated the coefficients of Eq. (4.26) using the values of μ and ν, we used it for correlating experimental data on the viscosity and thermal conductivity of krypton. It is seen from Fig. 62 that the analytic curve averages the data of Nasini and Rossi [4.23], Clarke and Smith [4.19] (for $T > 200$ K), Dawe and Smith [4.10], Kestin et al. [4.11, 4.22], and Goldblatt et al. [4.25] on the viscosity of krypton, and of Kannuluik and Carman [4.35], Keyes [4.36], and Saxena and Saxena [4.63] on thermal conductivity with deviations within the limits of error. The results of Zaytseva [4.37], Vargaftik and Yakush [4.44], and Schafter and Reiter [4.53] lie above the analytic curve, and the deviations exceed the error specified by these investigators. The greatest deviations (6%) are observed with data in [4.44].

Figure 63 shows in logarithmic coordinates the values of y calculated from experimental data on the viscosity and thermal conductivity of xenon. The relative location of the data points remains approximately the same as in Fig. 62. At

TABLE 47. Reduction Parameters and the Transformation Multipliers for Eq. (4.26)

Substance	T_B, K	ρ_0, g/cm³	μ	ν
Neon	122.10	1.673	0.29944	0.68321
Argon	407.76	1.870	1	1
Krypton	567.50	3.210	1.39175	1.14303
Xenon	791.00	3.890	1.93987	1.35657

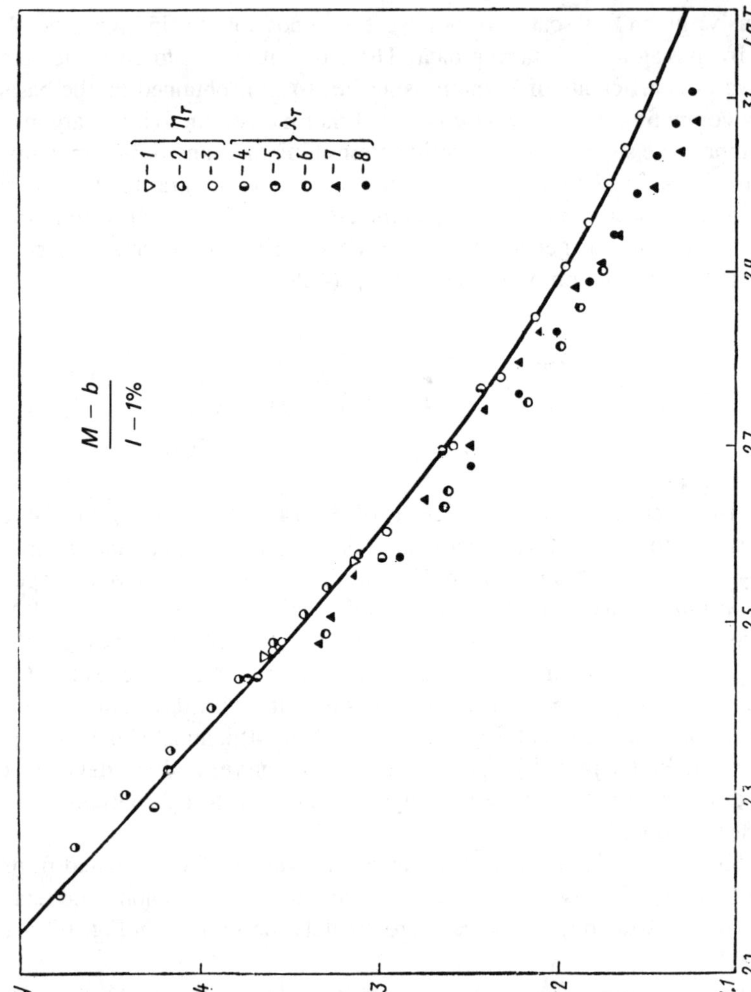

FIGURE 63. Shifted coincidence of the analytic curve for argon with experimental data on the viscosity and thermal conductivity of xenon. 1) [4.23]; 2) [4.19]; 3) [4.10]; 4) [4.35]; 5) [4.36]; 6) [4.37]; 7) [4.44]; 8) [4.63].

temperatures to 600 K the results of [4.10, 4.19, 4.23, 4.35, 4.36] deviate from one another basically within $\pm 2\%$, and a correlating curve can be drawn through them. With increasing temperature this curve separates into two branches and can be drawn either through data on viscosity or the cluster of points on thermal conductivity.

Analyzing these deviations, we noticed that the values of λ_T obtained at $T = 2.73$ K by Kannuluik and Carman [4.35] and Keyes [4.36] by the heated-wire and coaxial cylinders methods, respectively, are in excellent agreement with one another and with viscosity data. At the same time, the results of Zaytseva [4.37] and Vargaftik and Yakush [4.44] at $T \simeq 300$ K are much on the high side (to 5%—see Fig. 63) relative to the data on viscosity of xenon, and this "jump" prevails also at higher temperatures. The results of Saxena and Saxena [4.63] on thermal conductivity of xenon are also on the high side relative to the viscosity data in [4.10]. The correlating curve calculated from Eq. (4.26) using transformation multipliers whose values are listed in Table 47, as seen in Fig. 63, reflects satisfactorily the behavior defined by data in [4.2, 4.10, 4.19, 4.35, 4.36] and yields values of λ_T which are lower (to 6%) than those in [4.37, 4.44, 4.63].

Experimental data on viscosity and thermal conductivity for neon are given for temperatures from 80 to 2100 K. In addition, one value of η_T at $T = 20.4$ K is given by Van Itterbeek and Van Paemel [4.4]. The results of the overwhelming majority of the studies analyzed above are plotted in Fig. 64 in coordinates ($\lg y$, $\lg T$). All these are in satisfactory mutual agreement and form a clearly defined relationship. The only exceptions are data of Zaytseva [4.37] and Vargaftik and Yakush [4.44], which are on the high side at $T > 600$ K and data of Van Itterbeek and Van Paemel [4.4] which are on the average 5% on the high side relative to the results in [4.6, 4.35, 4.36].

The generalized curve calculated from Eq. (4.26) using the transformation multipliers (see Table 47) averages the compatible data on η_T and λ_T within $\pm 1.5\%$. It is important to emphasize that optimum averaging of data by means of an analytic curve at high temperature provides an additional confirmation of the reliability of Eq. (4.26) and on the feasibility of using it for calculating η_T and λ_T for argon, krypton, and xenon to temperatures of 5000, 7000 and 10,000 K, respectively. At $T < T_B$ the analytic curve starts to deviate systematically from the experimental data for neon. Such a deviation was also observed in correlating the second virial coefficients. It is apparently due to quantum-mechanical effects. For this reason, the results at temperatures below 125 K were correlated graphically by a dashed curve which lies on the average 5% higher relative to the results in [4.4] up to $T = 20$ K.

Figure 65 correlates the experimental data (selected by us) on the viscosity and thermal conductivity of the four gases. All the data cluster about the curve calculated from Eq. (4.26) which points to the existence of a single collision integral for these gases. The deviations from the analytic curve do not exceed $\pm 1.5\%$. This is most likely the error of the values of η_T and λ_T for the four

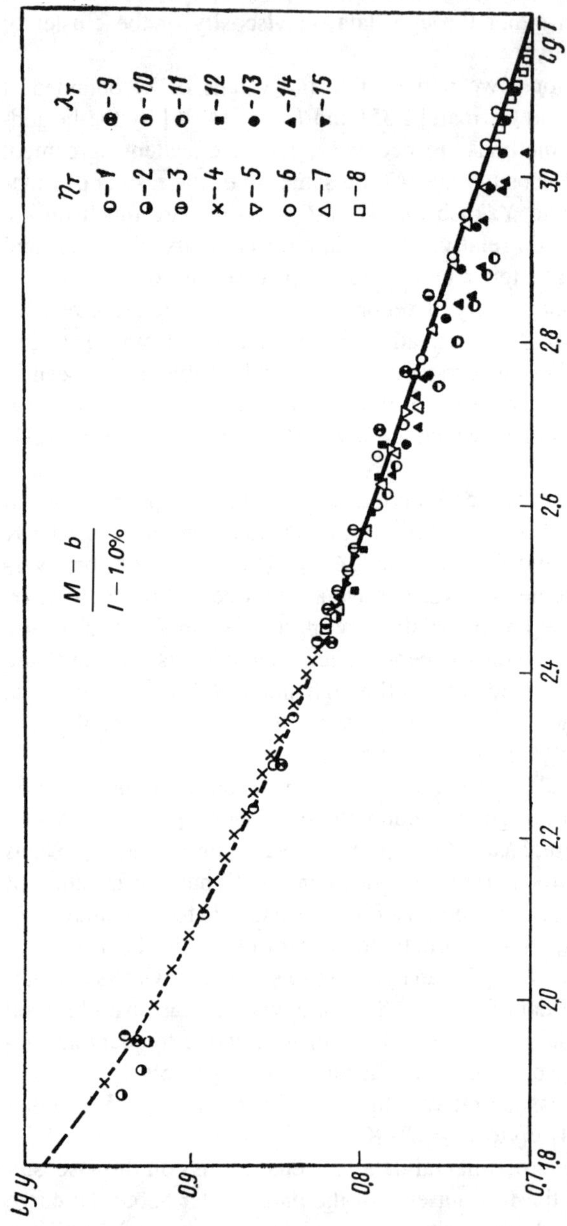

FIGURE 64. Shifted coincidence of the analytic curve for argon with experimental data on the viscosity and thermal conductivity of neon. 1) [4.2]; 2) [4.4]; 3) [4.5]; 4) [4.6]; 5) [4.9]; 6) [4.10]; 7) [4.11]; 8) [4.12]; 9) [4.35]; 10) [4.36]; 11) [4.37]; 12) [4.42]; 13) [4.43]; 14) [4.44]; 15) [4.38].

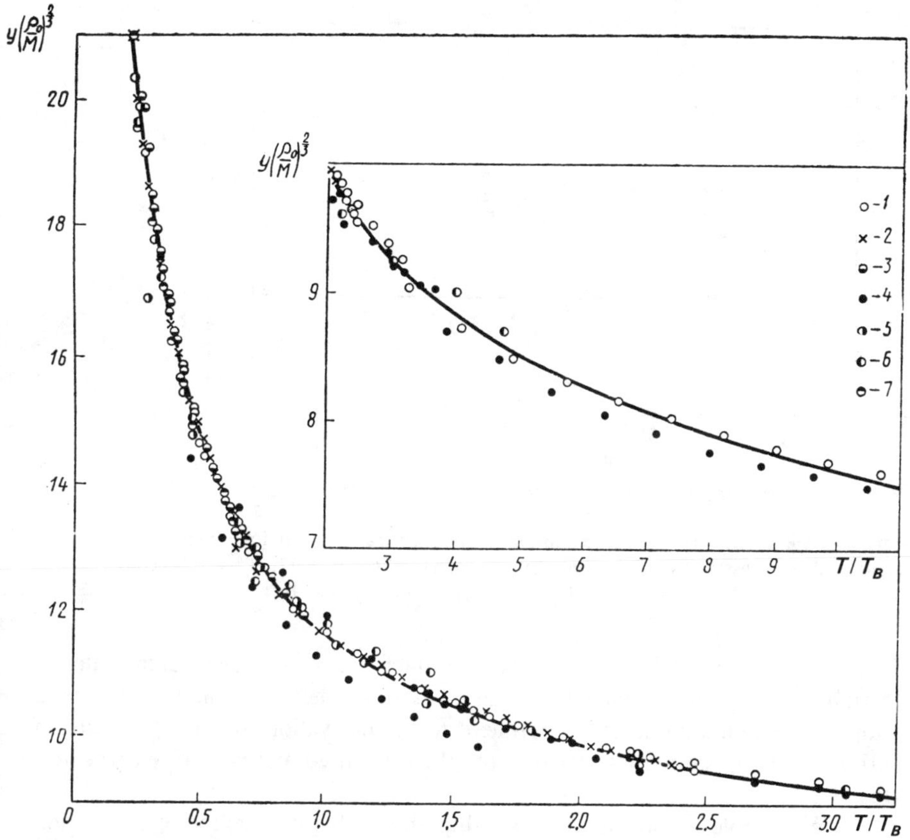

FIGURE 65. Correlation of experimental data on viscosity and thermal conductivity of the four gases. 1) [4.10]; 2) [4.6]; 3) [4.19]; 4) [4.63, 4.43]; 5) [4.36]; 6) [4.35]; 7) [4.60].

gases recommended by us, calculated from Eqs. (4.22) and (4.23), using Eq. (4.26) and listed in Tables VI, VII, XII, XIII, XIX, XX, XXVI and XXVII of Part III. According to our estimate, the error for neon at $T < 125$ K may be as high as 2.5%

It should, however, be noted that the data of V. Saxena and S. Saxena [4.63] on the thermal conductivity of xenon do not fit the correlating curve and deviate from it by up to 5%. It is difficult to explain the reason for this since analogous results of these investigators for neon, argon, and krypton are in excellent agreement with the most recent data on the viscosity of these gases.

The new results on viscosity and thermal conductivity of the four gases at temperatures from the triple points to 1300 K are more exact and exhibit better mutual compatibility than our earlier correlated data on the viscosity [4.26] and thermal conductivity. Further, more thorough, experimental studies of these substances could not yield any unexpected results. Therefore, they will allow reducing the confidence interval of the mutually compatible values of η_T and λ_T.

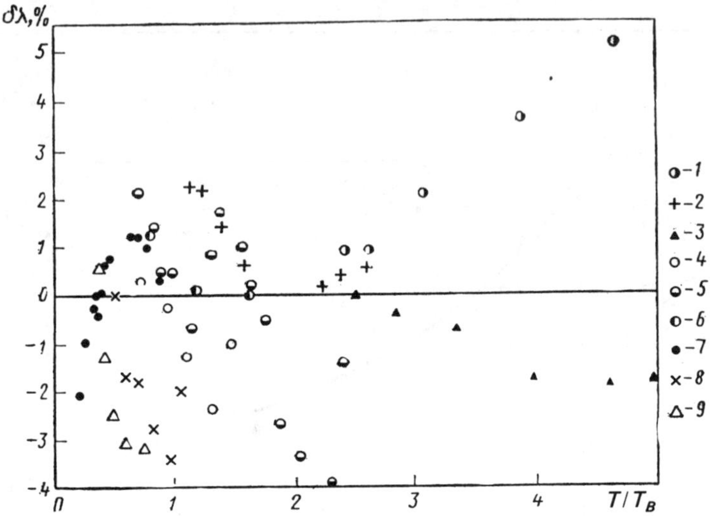

FIGURE 66. Deviation of new experimental data on thermal conductivity of monatomic gases at atmospheric pressure from computational results. 1) [4.76]; 2) [4.77]; 3) [4.78], neon; 4) [4.78]; 5) [4.79]; 6) [4.80]; 7) [4.81], argon; 8) [4.78], krypton; 9) [4.78], xenon.

This claim is definitely confirmed by the fact that new experimental data on thermal conductivity of monatomic gases at atmospheric pressure over a wide temperature range published subsequent to the derivation of Eq. (4.26) are in satisfactory agreement with our results obtained using this expression (Fig. 66)†.

The example of monatomic gases thus shows that the method of correlation of data on viscosity and thermal conductivity at atmospheric pressure used here is quite effective and can be used for other groups of individual substances and gas mixtures.

† After the book was completed we became aware of new experimental data on the thermal conductivity of argon and krypton at temperatures between 900 and 2100 K in the paper by Faubert and Springer [4.82]; of neon, argon, and xenon at T = 900 to 2500 K in Springer and Wingeier [4.83]; and also the analytic data of Hanley [4.84] on the viscosity and thermal conductivity of argon, krypton, and xenon at 0.6 T_{cr} to 2000 K. It was found by comparing these data with the results computed by us that over the above temperature ranges the standard and maximum deviations amounted to: 1.9 and 2.4% for neon according to data in [4.83], 0.7 and 1.2% for argon according to data in [4.82], 0.6 and 0.9% for xenon according to data in [4.83]. The values of viscosity and thermal conductivity coefficients of argon, krypton, and xenon calculated by us and by Hanley agree to T = 2000 K with standard deviations of 0.7 to 0.9% and a maximum of 2.2 to 2.4%.

CHAPTER

FIVE

VISCOSITY OF MONATOMIC GASES AND LIQUIDS AT HIGH PRESSURES

The accumulation of experimental data on the viscosity of substances of engineering importance, including monatomic gases and liquids, over a wide range of variables, made it clear that the surface of the viscosity coefficient in coordinates ($\Delta\eta$, ρ) has a more complex configuration than previously assumed. But, data on the viscosity of monatomic gases obtained by various investigators differ significantly from one another even in that range of parameters, where the simple relationship between excess viscosity and density applies. For this reason, calculation of detailed tables of viscosity of monatomic gases was preceded by analysis of published results and additional experimental studies.

The theory of transport phenomena is far from completion and up to now no equation has been obtained which would be capable of describing the viscosity of both the gas and the liquid within experimental error. As a result, we performed a theoretical analysis which showed that this problem is best solved on the basis of the Enskog equation in which the diameters of the solid spheres are a function of temperature and density. In this chapter, such a relationship is found from experimental data on the viscosity of monatomic substances. It permits the calculation of the viscosity coefficients over a wide range of variables.

5.1. EXPERIMENTAL DATA ON VISCOSITY AT MODERATE AND HIGH PRESSURE

As seen from Table 48, the parameter which has been more extensively investigated both in the gaseous and liquid states, including the saturation curve, is the

TABLE 48. Experimental Studies on the Viscosity of the Four Gases at $p > 1$ atm

Year	Author	Temperature, K	Pressure, atm	Method	Reference
		Neon			
1959	Kestin and Leidenfrost	293	1—28	Oscillating disk	[4.7]
1963	Kestin and Whitelaw	296—519	1—115	Same as above	[4.9]
1963	Flynn et al.	223—373	24—173	Capillary	[5.1]
1963	Förster	25—44	Along the saturation curve	Oscillating disk	[5.2]
1964	Kestin and Nagashima	293 & 303	1—49	Same as above	[5.3]
1964	Trappeniers et al.	298—348	7—1769	Capillary	[5.4]
1966	Reynes and Thodos	373—437	70—820	»	[5.5]
1971	Rabinovich et al.	298—523	25—582	»	[5.6]
1973	Slyusar' et al.	50—300	1—5000	Falling weight	[5.7]
		Argon			
1934	Rudenko and Shubnikov	84—87	Along the saturation curve	Capillary	[5.8]
1941	Van Itterbeek and Van Paemel	84—87	Same as above	Oscillating disk	[5.9]
1953	Golubev and Petrov	273—473	1—475	Capillary	[5.10]
1954	Michels et al.	273—348	9—2000	»	[5.11]
1954	Ivasaki	298—423	1—200	Oscillating disk	[5.12]
1954	Kestin and Pilarczyk	293	1—70	Same as above	[5.13]
1956	Zhdanova	90—270	1—3600	Falling weight	[5.14]
1957	Makita	298—423	1—773	Rolling sphere	[5.15]
1958	Filipova and Ishkin	90—273	1—150	Capillary	[5.16]
1958	Kestin and Wang	298	1—70	Oscillating disk	[5.17]
1959	Kestin and Leidenfrost	293	1—30	Same as above	[4.7]
1962	Van Itterbeek et al.	84—87	Along the saturation curve	»	[5.18]
1963	Flynn et al.	195—373	29—186	Capillary	[5.1]

TABLE 48. Experimental Studies on the Viscosity of the Four Gases at $p > 1$ atm (*Continued*)

Year	Author	Temper-ature, K	Pressure, atm	Method	Reference
1963	Förster	84—116	10	Oscillating disk	[5.2]
1963	Boon and Thomaes	84—89	Along the saturation curve	Capillary	[5.19]
1963	Kestin and Whitelaw	298—538	1—140	Oscillating disk	[4.9]
1964	Kestin and Nagashima	293 & 303	1—51	Same as above	[5.3]
1964	Saji and Kobayashi	84—87	Along the saturation curve	Capillary	[5.20]
1964	Reynes and Thodos	373—473	70—819	»	[5.21]
1964	Lowry et al.	102 & 128	50—500	Torsional vibrations of quartz crystal	[5.22]
1966	Andreyev et al.	293—923	50—500	Capillary	[5.23]
1966	Van Itterbeek et al.	84—90	1—97	Oscillating disk	[5.24]
1967	De Bock et al.	90	1—138	Torsional vibrations of quartz crystal	[5.25]
		86—140	1—193		[5.26]
1967	Boon et al.	84—89	Along the saturation curve	Capillary	[5.27]
1970	Hellemans et al.	104—140	to 100	Oscillating disk	[5.28]
1971	Rabinovich et al.	298—523	25—582	Capillary	[5.6]
1973	Slyusar' et al.	90—300	to 3000	Falling weight	[5.7]
			Krypton		
1959	Kestin and Leidenfrost	293	1—21	Oscillating disk	[4.7]
1963	Boon and Thomaes	116—123	Along the saturation curve	Capillary	[5.19]
1964	Reynes and Thodos	373—473	71—820	»	[5.21]
1965	Trappeniers et al.	298—398	7—2052	»	[5.29]
1967	Boon et al.	116—123	Along the saturation curve	»	[5.27]
1970	Rabinovich	323—523	46—586	»	[3.89]
1973	Slyusar' et al.	120—300	to 2500	Falling weight	[5.7]

TABLE 48. Experimental Studies on the Viscosity of the Four Gases at $p > 1$ atm (*Continued*)

Year	Author	Temperature, K	Pressure, atm	Method	Reference
			Xenon		
1959	Kestin and Leidenfrost	298	1—28	Oscillating disk	[4.7]
1964	Reynes and Thodos	323—473	70—819	Capillary	[5.21]
1965	Trappeniers et al.	273—348	10—875	»	[5.30]
1967	Boon et al.	163—169	Along the saturation curve	»	[5.27]
1970	Rabinovich	323—523	44—590	»	[3.89]
1973	Slyusar' et al.	165—300	to 2000	Falling weight	[5.7]

dynamic viscosity of argon. The majority of studies published up to 1966 were analyzed by Vasserman et al. [1.104, 3.8]. Neon, krypton, and xenon were investigated less completely.

The studies enumerated in Table 48 can be subdivided, for convenience of analysis, into several groups. The first is made up of results obtained in the Amsterdam laboratory [5.4, 5.11, 5.29, 5.30] for the four fluids over the range of parameters traditional to this laboratory. The error is estimated by the experimenters not to exceed 1%. A check on the internal compatibility of results by constructing isotherms in coordinates (η, p) and (η, ρ) showed that the experimental points form smooth curves within the above limit. Only for xenon at 25°C does there exist a perceptible scatter which is apparently due to difficulties of experimentation in the transcritical region.

The second group is made up of results obtained by Reynes and Thodos [5.5, 5.21]. The results for all four fluids in coordinates (η, p) and (η, ρ) exhibit some scatter (to $\pm 1\%$) relative to the averaging isotherms. It is particularly noticeable in the latter case requiring more detailed analysis of values given in [5.21]. It was found that the data on the density of krypton tabulated in [5.21] at experimental parameters differ significantly from the experimental obtained subsequently by Trappeniers et al. [2.71]. Thus, the deviations at the 100°C isotherm are as high as 3%, with data of [5.21] lying below those in [2.71]. Although reliable data on compressibility of xenon were available [2.77], Reynes and Thodos [5.21] list questionable values of the density of xenon. The use of these for calculating viscosity could have resulted in significant error.

Unfortunately, it is impossible to directly compare the data in [5.4, 5.11, 5.29, 5.30] with those in [5.5, 5.21] since their temperature ranges are different. The only exception is the 50°C isotherm, where the viscosity of xenon was investigated both in [5.21] and [5.30]. The data in these two studies are in good agreement at low pressures. However, as the pressures become higher, the discrepancies increase, becoming as high as 9% at 222 atm.

To clarify the compatibility of the two groups of data, we constructed additional graphs in coordinates ($\Delta\eta$, ρ) in which the temperature factor for the majority of gases does not affect the correlation. The deviations of the viscosities of neon and krypton lie within the limits of error of $\pm 2.5\%$. For xenon, these deviations are greater. The results of [5.21] are lower than those in [5.30]. The largest systematic differences were observed for argon where the results in [5.21] are 7% on the high side. An additional check along isobars showed that no smooth curves can be drawn through data of [5.11, 5.21]. Analysis of [5.21] yields no information on the reason for this difference. However, we wish to note that the authors used a long (317.9 cm) helically-wound capillary whose mean diameter was determined by three methods, and assumed to be 0.0248 ± 0.0005 cm. This inaccuracy alone could cause a measuring error of 4%.

Andreyev and his coworkers [5.23] investigated the viscosity of argon by the relative method of the capillary, using a viscosimeter with a ring balance over the widest temperature range. The instrument constant was determined from data for nitrogen at room temperature for pressures from 50 to 500 bar. In their opinion, the measuring error ranged from 1 to 3%, increasing with the temperature. The results of this study are in good agreement with data of Michels et al. [5.11]. Andreyev et al. worked up their results in coordinates ($\Delta\eta$, ρ) and compiled an analytic expression for the correlating curve which agrees within the experimental accuracy with the analytic curve obtained by Rabinovich et al. [1.104].

The results of the Amsterdam laboratory are also confirmed by data of Kestin and Leidenfrost [4.7] at room temperature over a narrow pressure range and of Flynn et al. [5.1] which were obtained over a wider range of parameters. The data of Kestin and Whitelaw [4.9] for neon at room temperature are in satisfactory agreement with the results of Trappeniers et al. [5.4] and Michels et al. [5.11], but they start deviating from the latter with rising temperature and are as much as 1.7% higher at a temperature as low as 75°C. One can judge about further increase in the difference by comparing the data in [5.23] and [4.9]. Hence, it is best to point out the possible reason for the high values obtained in [4.9].

Kestin and Whitelaw [4.9] measured the viscosity of argon and neon by the relative method. The instrument constant was determined from data for nitrogen at $t = 22°C$ and was assumed to remain unchanged over the entire experimentally explored temperature range. This could not be sufficiently validated since the component of the logarithmic damping decrement is a function of the friction in the suspension system and may change with the temperature. Such an assumption is due to differences in the thermal expansion coefficient of materials of the suspension system and the temperature dependence of the internal damping of oscillations of the suspension wire made of a platinum-tungsten alloy.

The third group can be regarded as being made up of studies concerned with the viscosity of the substances under study at high pressure, at saturation, and in the liquid phase. Van Itterbeek and his workers investigated the viscosity of argon first by the relative method [5.9] and then by the absolute method [5.18]. The results of the last study are higher than the preceding by 1.5 to 3.5%.

Förster [5.2] used the oscillating disk technique to determine the viscosity of neon and argon at saturation. The error of experimental data is estimated at ±2%. However, our comparison showed that Förster's assumption of satisfactory agreement between his data on the viscosity of argon and those in [5.18] is not entirely correct, since the differences are as high as 8%.

Boon and Thomaes [5.19] published their article on the viscosity of liquid krypton and argon simultaneously with the paper by Förster. They measured the time needed for a certain volume of liquid to flow and calculated the relative value of the kinematic viscosity. The reference unit of measurement consisted of the viscosity of argon at 88.98 K. The absolute value of viscosity is not given in [5.19]. That made it somewhat difficult to use the results whose error is estimated by the authors at 1%. In a subsequent study [5.27], performed to check for satisfaction of the law of corresponding states with respect to data on density at saturation for a number of simple liquids, Boon et al. presented absolute values of the viscosity of argon, krypton and xenon.

Lately, in addition to traditional methods of determination of viscosity of dense gases and liquids, ever-increasing use is made of a technique based on the torsional oscillations of piezoelectric crystals. Lowry et al. [5.22] were the first to use it to determine the viscosity of liquid argon. Their smoothed data, with error estimated at ±2%, are given along two isotherms for pressures from 50 to 500 atm. This method was developed further by De Bock and his coworkers [5.25, 5.26]. They present in their first article a brief description of the experimental equipment and then list values of viscosity of liquid argon along one isotherm. In the second study, this viscosity is given along eight isotherms. They estimate their maximum error at 3%. Within the limits of this error the isotherms are linear in coordinates (η, p).

The data of Hellemans et al. [5.28], who estimate their error at 2% at the most, were compared with those listed above [5.19, 5.22, 5.26, 5.27]. The differences between the results in [5.28] and the rather reliable data in [5.26, 5.27] are quite large and at times are as high as 5–7%. One's attention is attracted by the fact that, according to the data of [5.28], the pressure dependence of viscosity along the isotherm is smaller than according to the data of others. In particular, along the 147.1 K isotherm the viscosity is virtually independent of pressure. At the same time, the results of De Bock et al. [5.26] are in good agreement with those of Boon et al. [5.27] and Van Itterbeek et al. [5.24]. Poorer agreement is observed with data of Zhdanova [5.14]: the differences are as high as 12% along the 0.98 cm^3/g isochore. Although the same method was used by Lowry et al. [5.22] and De Bock et al. [5.26], their data exhibit the greatest differences, with the results of the former study 18%, on the average, lower than those in the second.

The most thorough investigation of the viscosity of the four inert gases at temperatures from the triple point to 300 K and over a rather wide range of values of density was performed by Slyusar' et al. [5.7, 5.31, 5.32] by the falling weight technique. The values of density needed for determining the viscosity coefficient were available over a narrower range of parameters than the experimental range. For this reason, these studies partially employ values of density obtained by extrapolation of experimental isobars and isotherms.

Estimating the error of their experimental data at 4%, Slyusar' et al. did not allow for the effect of the errors in the values of density in the region not investigated experimentally. Detailed calculations performed by us using equations of state of the four liquids under study (see Chapter 3) showed that the values of temperature and pressure in experimental points listed in [5.7, 5.31, 5.32] do not have corresponding to them the constant values of density listed in the same tables. The differences increase with pressure and are as high as 1 to 1.5%. Consequently, the error of 4% cannot be regarded as representing the entire range of parameters for which the experiments were performed. It is more reasonable to assume that at low temperature and high pressure the error is about 7–8% for Ne, Kr and Xe. Comparison of data by Slyusar' et al. with the results of others shows that the differences do not go past the limits of the overall experimental error. The only exceptions are data in [5.14, 5.22] which, as previously noted, are not too accurate.

The above brief analysis of experimental studies leads to the conclusion that the largest number of measurements were performed at temperatures between 20 and 200°C at rather high pressures. However, the agreement between the results of different investigators in this range, which is of great engineering importance, is rather poor, and no criterion is available for giving preference to a given set of data. This is particularly noticeable for argon, whose viscosity was investigated by Reynes and Thodos [5.21] and by Andreyev et al. [5.23]. New experimental data disagree by up to 9%, with the results in [5.21] confirming the data of Golubev and Petrov [5.10], and those in [5.23], and of Michels et al. [5.11].

These and the previously mentioned circumstances made it necessary to perform additional experiments at temperatures of 25–250°C and pressures to 600 bar in order to obtain most probable values of the dynamic viscosity.

5.2. EXPERIMENTAL STUDY OF THE VISCOSITY OF GASES

The study was performed by the method of the capillary which is theoretically the most valid. In our opinion, the most successful design of experimental equipment for use with the parameters under study is that of Golubev and Petrov [5.33]. For that reason we took it as a basis for our design of the equipment.

The test facility is shown schematically in Fig. 67. High-pressure cylinder *13* (d_{in} = 18 mm, d_{out} = 70 mm) is made of 1Kh18N9T stainless steel. Its

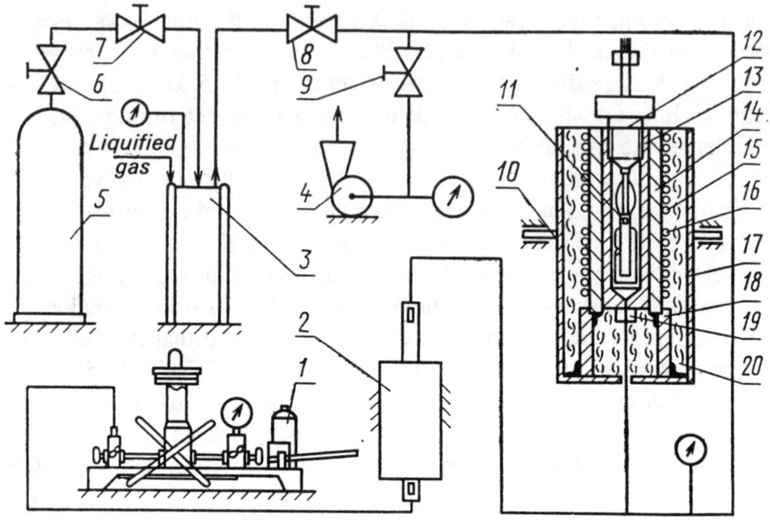

FIGURE 67. Schematic of experimental facility. 1) MP-600 dead-weight 0.05 precision class pressure gauge; 2) booster; 3) thermocompressor; 4) vacuum pump; 5) tank with test gas; 6) through 9) valves; 10) swiveling unit; 11) viscosimeter; 12) stuffing-box seal; 13) high-pressure cylinder; 14) copper sleeve; 15 and 16) heaters; 17) jacket; 18) ring; 19) transition piece; 20) asbestos crumbs.

upper part contains a threaded seat for packing *12*, and the lower has a transition piece *19* for connection to the gas line. To maintain the vessel at constant temperature, it has a copper sleeve *14* (d_{out} = 110 mm) pressed on it. Two heaters, *15* and *16*, consisting of 0.6 mm nichrome wire are wound onto its surface in tubular porcelain insulators. The heaters are supplied from a stabilized power source with power controlled by autotransformers. Each heater is shunted through a rheostat, which allows preselecting the heater power in such a manner as to ensure a uniform temperature field along the high-pressure vessel under steady conditions.

Two holes are drilled in the body of the copper sleeve. A platinum resistance thermometer is placed in one of them, and in the other a differential copper-constantan thermocouple, the ends of which are led out to a miniature voltammeter. The depth of the drillings is selected in accordance with the size of the viscosimeter capillary and its location in the high-pressure vessel.

The entire assembly is fastened in cylindrical jacket *17* by means of asbestos-cement ring *18* and bolt connections. Two shaft halves are welded from the outside to the jacket. These serve for fastening the jacket together with all its contents in special seats to the housing of the unit. One of the seats has a positioning device allowing regulation of the rotation of the unit by 90°. Thermocompressor *3* is made of 1Kh18N9T stainless steel in the form of a thick-walled cylinder and is placed in a metal Dewar. A nichrome heater in porcelain insulation is wound on the outer surface of the cylinder. The pressure is monitored by means of a spring-type pressure gauge. Booster *2*, of stainless steel, is designed in the same manner as in the unit for investigating the *pvT* behavior of the gases.

The most important element of the unit consists of a glass viscosimeter (Fig. 68), the design of which was somewhat modified as compared with that used by Golubev [5.33].

Mercury charging hole *A* is provided on the side surface of the 16 mm diameter lower tank *1*. This location of the charging hole prevents ejection of the mercury from the viscosimeter when the high-pressure vessel is rotated through 90°. The lower skewed end of connecting tube *2* is situated 3–4 mm from the bottom of the tank and the upper end enters the so-called measuring *3, 4* and preliminary *7* vessels. Capillary *5* is welded by means of transition tube *6* to the upper part of the viscosimeter, parallel to the vertical axis. The measuring vessels are in the form of truncated cones. The overall height of the cone and the ratio of cross sectional areas F_1 and F_2 is selected so as to ensure approximately constant rate of lowering of the mercury level on the measuring part.

Four 0.3 mm diameter platinum wire contacts are soldered into the viscosimeter. Three of these (*8* to *10*) are placed on the measuring part and the fourth *11* should be in constant contact with the mercury and be situated in the bottom tank. All the contacts are lead out from the high-pressure vessel through a spe-

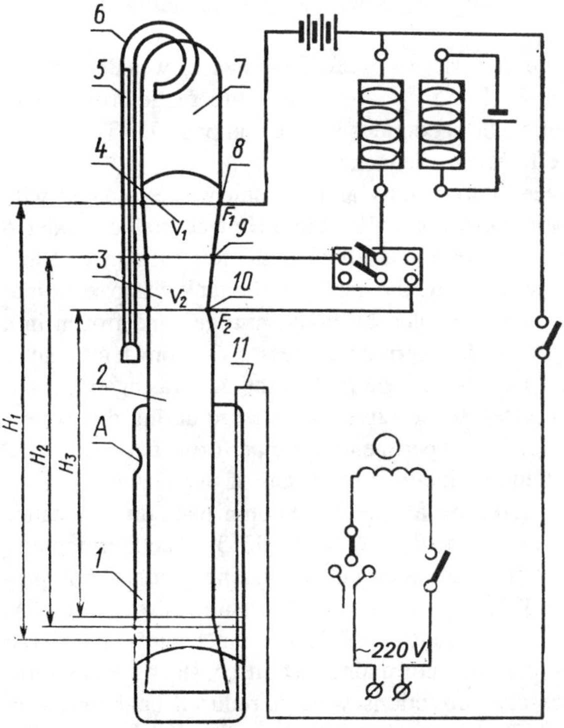

FIGURE 68. Design of viscosimeter and circuit for automatic recording of the gas discharge time.

cial packing and are connected to the system of automatic recording of the discharge time. The operating principle of such a viscosimeter is described in [5.33, 5.34].

First we investigated the viscosity of neon and argon [5.6]. The experimental unit permitted experiments both along isotherms and isobars since any prespecified pressure could be easily preset and maintained by the plunger pump and the handwheel of the deadweight pressure gauge. However, from considerations of convenience and economy of time, the experiments for determining the viscosity of neon and argon were performed primarily along isotherms.

When investigating the viscosity of krypton and xenon [3.89], the system for producing and maintaining the pressure was modified because the density of these substances under the experimental conditions is higher than the density of the organosilicon liquid used as the working fluid. In the new arrangement, the deadweight pressure gauge *1* (see Fig. 67) and booster *2* were disconnected. To measure pressures to 600 bar, a second calibration-grade spring-type pressure gauge (to 250 bar), together with a shutoff valve, were connected parallel to the calibration-grade pressure gauge of the same type. At $p < 250$ bar the pressure was measured from the readings of both pressure gauges. At higher pressures the second gauge was disconnected. Both pressure gauges were calibrated against a 0.02 precision class deadweight MP-600 pressure gauges. The readings of both spring-type pressure gauges were rechecked following the experiments.

The coefficient of dynamic viscosity of the four substances was determined over a temperature range of 25 to 250°C and a pressure range from 25 to 600 bar. The impurity content of the specimens is listed on page 72. The experiments were performed in the following sequence.

Swivel block *10* was placed horizontally and the unit was evacuated with valves *7* through *9* open. When a pressure of 10^{-3} mm Hg was attained, valve *6* was slightly opened and the gas system was filled by the test substance at low pressure. After the unit was reevacuated, the gas from the tank was passed into the thermocompressor. There it was cooled by liquid nitrogen and frozen in a quantity such that the pressure in the thermocompressor at room temperature would not exceed 600 bar. At high pressure a part of the gas was again passed into the tank. If it was again necessary to raise the pressure during the experiment, the heater was started up. The experiments were performed at a specified temperature which was preset and maintained by means of two heaters.

The temperature was measured by a class 1 100 ohm platinum resistance thermometer (R_0 = 100.821 abs. ohm, R_{100}/R_0 = 1.39233). The thermometer was connected to a standard compensation circuit, consisting of a group A precision class potentiometer P – 307, calibration-type 100-ohm resistance foil, M 195/1 galvanometer, class 2 standard cell, two P-33 resistance banks and Ekran type batteries. The resistance coil was not maintained at constant temperature, and for this reason the necessary corrections were introduced analytically in accordance with the unit's certificate. To eliminate the error arising due to the thermal emf, the circuit provides for an inversion current switch. The possible error of temperature measurement cannot, according to our data, exceed 0.05 K.

The uniformity of the temperature field along the capillary was monitored from readings of a differential copper-constantan thermocouple with junctions situated at the level of the start and end of the working length of the capillary. It was assumed that there is no temperature gradient when the temperature difference does not exceed 0.03 K.

The gauge pressure was measured by pressure gauge 1 when investigating the viscosity of neon and argon and by calibration-grade pressure gauges when investigating the viscosity of krypton and xenon. The pressure within the high-pressure vessel was calculated as the sum of the gauge, atmospheric and hydrostatic pressures.

After presetting the required pressure and temperature and holding them unchanged for 20–30 minutes the time of discharge of a certain quantity of the test gas through the capillary was measured. For this, block *10* was returned to the vertical position. The result was that the level of mercury in the connecting tube became higher than the upper contact and then started descending. As yet, all the contacts are blocked by mercury and the current in the circuit bypasses the polarized relay. As soon as the first contact became opened, current passed through the induction coil of the relay starting up an electric timer. When the second contact opened, the timer stopped, which fixed the time of discharge of mercury between two contacts. The circuit has a switch which allows recording the time of discharge of the mercury both between the upper and center contacts (τ_1) and between the upper and lower contacts (τ_2). To reduce random errors in measuring the times of discharge, the experiments were repeated several times at each point and the results were averaged.

The dynamic viscosity η of the gases was calculated from the expression

$$\eta = A_0\left(1 - \frac{\Delta h}{H}\right)(\rho_{Hg} - \rho_g)\tau - B_0(1 + 2\alpha\Delta t)\frac{\rho_g}{\tau} \tag{5.1}$$

which has not been corrected for expansion of the gas in the capillary due to its smallness; here $A_0 = \dfrac{\pi r^4 g}{8Vl_c}H$; $B_0 = \dfrac{mV}{8\pi l_c}$; H is the average difference of mercury levels; ρ_{Hg} is the density of mercury under the experimental conditions; ρ_g is the same for the gas; τ is the time of discharge of gas from the vessel with volume V through the capillary; Δt is the temperature difference equal to $t \simeq 25°C$; α is the coefficient of linear expansion; Δh is the change in the mercury levels due to different temperature expansion of the glass and mercury; Δh was determined as suggested by Golubev and Agayev [5.34]. The values of ρ_g needed for calculations was found for each substance from equations of state (see Chapter 2).

In principle, in order to determine constants A_0 and B_0 one must have two experimental values of η. However, averaging of results obtained using a large number of reliable experimental data yields more exact values. For this reason A_0 and B_0 were calculated from experimental data on the viscosity of neon [5.4] and argon [5.11], which were confirmed by results of a number of investigators at $t = 25°C$ and pressures between 50 and 600 bar. This method also permitted checking for mutual compatibility of the original data. In particular, our experimental and computational results showed that the data in [5.4] and [5.11] are in agreement within less than 1%.

The experiments were performed using three viscosimeters which differed from one another basically by the diameter of the capillaries. In investigating the viscosity of a given substance the diameter of the capillary was selected (between 0.07 and 0.12 mm) in such a manner that the Reynolds number would lie between 400 and 1100.

The first viscosimeter was used for investigating the viscosity of argon at 25 to 250°C and 25 to 600 bar. The second was used for measuring the viscosity of neon over the same range. In addition, experiments with argon were repeated on the second viscosimeter at temperatures of 150, 100 and 25°C. The third viscosimeter served to measure the viscosity of krypton and xenon along the 50, 100, 150, 200 and 250°C isotherms at pressures to 600 bar and the viscosity of argon at $t = 50°C$. The geometric dimensions of the viscosimeters are listed in Table 49.

Experimental data on the viscosity of the four gases are listed in Table 50, where each of the values was determined by averaging the results of five–six measurements of duration $\tau_1 + \tau_2$ of discharge of mercury and the same number of measurements of τ_1. This means that contact *9* (see Fig. 68) was used not only for monitoring the uniformity of motion of the level of the mercury, but also for reducing the random errors.

The error of data listed in Table 50 can be estimated using Eq. (5.1). The accuracy depends primarily on the accuracy of determining constants A_0 and B_0 and the discharge time τ. Since A_0 and B_0 were found from reliable experimental data on the viscosity of neon and argon, the error of these values in our opinion should not exceed 0.7% while the error for τ lies between 0.2 and 0.3%.

The density of mercury was found from reliable data [5.35] which allows for both the temperature and pressure dependence. The maximum possible error

TABLE 49. Description of the Viscosimeter Geometries

Viscosimeter	Difference of mercury levels, cm			Measuring volume, cm³		Length of capillary, cm	Initial height of mercury in the tank, cm
	H_1	H_2	H_3	V_1	V_2		
1	8.552	7.052	5.965	0.450	0.3525	5.650	4.201
2	8.483	7.017	5.705	0.630	0.512	5.627	4.205
3	7.942	6.571	5.344	0.437	0.341	5.641	4.931

TABLE 50. Experimental Values of the Dynamic Viscosities of Neon, Argon, Krypton and Xenon (p, bar; η, Pa · s; ρ, g/cm³)

Neon			Argon		
p	ϱ	$\eta \cdot 10^8$	p	ϱ	$\eta \cdot 10^8$
$t=25°$ C					
25.59	0.0203	3163	25.58	0.0419	2280
50.11	0.0396	3165	50.10	0.0830	2356
99.14	0.0766	3198	99.12	0.1674	2524
148.2	0.1125	3233	148.2	0.2521	2748
197.2	0.1465	3272	197.2	0.3342	3004
295.3	0.2095	3355	295.2	0.4804	3588
393.4	0.2674	3480	393.3	0.5977	4204
491.4	0.3203	3588	491.4	0.6902	4795
589.5	0.3690	3700	589.4	0.7647	5366
$t=50°$ C					
25.60	0.0187	3380	25.60	0.0384	2432
50.12	0.0365	3391	50.11	0.0758	2510
99.16	0.0713	3425	99.14	0.1514	2652
148.2	0.1041	3430	148.2	0.2261	2849
197.2	0.1356	3484	197.2	0.2994	3072
295.3	0.1949	3550	295.2	0.4313	3554
393.4	0.2490	3615	393.3	0.5420	4046
491.4	0.2994	3729	491.4	0.6328	4575
589.5	0.3456	3815	589.4	0.7074	5106
$t = 75°$ C					
—	—	—	25.60	0.0355	2608
—	—	—	50.12	0.0698	2665
99.15	0.0664	3579	99.15	0.1385	2787
148.2	0.0970	3592	148.2	0.2065	2951
197.2	0.1264	3630	197.2	0.2720	3155
295.3	0.1820	3691	295.3	0.3929	3560
393.3	0.2334	3773	393.3	0.4968	4004
491.4	0.2809	3872	491.4	0.5845	4480
589.5	0.3250	3967	589.4	0.6546	4938
$t = 100°$ C					
—	—	—	25.60	0.0330	2744
—	—	—	50.11	0.0648	2791
99.15	0.0619	3743	99.14	0.1278	2907
148.2	0.0906	3759	148.2	0.1899	3059
197.2	0.1184	3782	197.2	0.2501	3237
295.3	0.1708	3838	295.3	0.3613	3593
393.4	0.2195	3911	393.3	0.4589	3982
491.4	0.2646	4007	491.4	0.5428	4399
589.5	0.3070	4071	589.4	0.6149	4790

TABLE 50. Experimental Values of the Dynamic Viscosities of Neon, Argon, Krypton and Xenon (p, bar; η, Pa · s; ρ, g/cm³) (*Continued*)

Neon			Argon		
p	ϱ	$\eta \cdot 10^8$	p	ϱ	$\eta \cdot 10^8$
		$t = 150°$ C			
—	—	—	25.59	0.0291	3006
—	—	—	50.11	0.0566	3051
99.15	0.0545	4053	99.14	0.1113	3149
148.2	0.0804	4061	148.2	0.1650	3264
197.2	0.1051	4081	197.2	0.2162	3402
295.3	0.1522	4111	295.3	0.3129	3672
393.4	0.1961	4186	393.3	0.3997	3988
491.4	0.2374	4243	491.4	0.4762	4303
589.5	0.2762	4301	589.4	0.5440	4641
		$t = 200°$ C			
—	—	—	25.59	0.0258	3253
—	—	—	50.11	0.0504	3282
99.15	0.0492	4373	99.14	0.0988	3364
148.2	0.0723	4374	148.2	0.1457	3460
197.2	0.0948	4381	197.2	0.1913	3564
295.3	0.1375	4415	295.3	0.2768	3797
393.4	0.1778	4451	393.3	0.3545	4057
491.4	0.2164	4515	491.4	0.4259	4327
589.5	0.2516	4569	589.4	0.4900	4599
		$t = 250°$ C			
—	—	—	25.59	0.0228	3494
—	—	—	50.11	0.0456	3527
99.15	0.0446	4660	99.14	0.0880	3611
148.2	0.0657	4661	148.2	0.1313	3698
197.2	0.0862	4672	197.2	0.1721	3792
295.3	0.1254	4694	295.3	0.2492	4000
393.4	0.1625	4730	393.3	0.3202	4224
491.4	0.1981	4776	491.4	0.3858	4445
589.5	0.2309	4815	589.4	0.4452	4679

TABLE 50. Experimental Values of the Dynamic Viscosities of Neon, Argon, Krypton and Xenon (p, bar; η, Pa · s; ρ, g/cm³) (*Continued*)

Krypton			Xenon		
p	ϱ	$\eta \cdot 10^8$	p	ϱ	$\eta \cdot 10^8$
		$t = 50°$ C			
47.5	0.1597	2924	49.0	0.3137	2991
98.1	0.3550	3296	100.4	1.1361	5695
147.3	0.5632	3833	148.1	1.6638	8925
195.4	0.7622	4505	195.0	1.8609	10489
246.4	0.9470	5292	250.3	2.0035	12178
296.0	1.0930	6007	294.4	2.0849	13235
390.7	1.2992	7285	347.4	2.1659	14279
495.2	1.4638	8705	402.3	2.2334	15315
589.0	1.5793	9624	489.9	2.3233	16856
			597.6	2.4123	18895

TABLE 50. Experimental Values of the Dynamic Viscosities of Neon, Argon, Krypton and Xenon (p, bar; η, Pa · s; ρ, g/cm³) (*Continued*)

Krypton			Xenon		
p	ϱ	$\eta \cdot 10^8$	p	ϱ	$\eta \cdot 10^8$
t = 100° C					
47.8	0.1348	3267	49.0	0.2400	3232
98.4	0.2868	3523	97.1	0.5668	3932
148.1	0.4425	3872	146.8	0.9936	5405
197.0	0.5932	4297	195.7	1.3391	7089
295.4	0.8650	5291	246.0	1.5688	8593
392.5	1.0755	6265	353.5	1.8536	10999
491.2	1.2378	7241	441.3	2.0000	12478
593.9	1.3702	8315	490.3	2.0657	13280
			588.8	2.1738	14758
t = 150° C					
46.5	0.1129	3569	50.2	0.2053	3482
98.1	0.2431	3789	98.1	0.4386	3959
147.3	0.3679	4035	122.6	0.5710	4287
196.5	0.4911	4336	197.1	0.9823	5732
294.8	0.7175	5022	294.2	1.3876	7728
392.7	0.9105	5829	392.3	1.6450	9526
491.5	1.0705	6619	491.3	1.8206	10939
592.9	1.2035	7407	591.1	1.9554	12362
t = 200° C					
47.7	0.1029	3888	44.9	0.1580	3697
98.7	0.2138	4063	96.1	0.3579	4037
148.2	0.3202	4258	146.9	0.5680	4537
197.7	0.4250	4489	195.2	0.7753	5170
246.4	0.5241	4726	245.6	0.9760	5925
295.9	0.6200	5025	296.0	1.1475	6711
393.2	0.7936	5628	396.2	1.4174	8199
491.7	0.9486	6289	488.4	1.5950	9371
593.4	1.0899	7024	592.1	1.7570	10669
t = 250° C					
46.8	0.0905	4093	46.8	0.1465	4008
76.5	0.1282	4141	98.1	0.3168	4280
147.1	0.2831	4401	147.3	0.4858	4617
246.1	0.4629	4795	196.1	0.6541	5151
392.2	0.7010	5551	246.2	0.8200	5672
441.3	0.7868	5833	345.3	1.1082	6878
534.7	0.8972	6293	446.4	1.3380	8041
589.4	0.9643	6601	531.6	1.4907	9002

is 0.1%. The same error should also be assigned to data on the density of these substances.

The variation in A_0 over the range of 25 to 250°C does not exceed 2.5%. The difference in H was determined as the logarithmic mean with an error to 0.3%, and that of Δh with an error of ~5%. This gives an overall error of 0.13% in the value of η. The correction to the Hagen-Poiseuille law for isothermal flow of gas through a capillary, given by the second term in Eq. (5.1), was 13% in our experiments. The error due to inaccuracy in calculating the behavior of $B_0(t)$ is estimated at 0.03%, since B_0 goes to 0.4% at t between 25 and 250°C. This means that the overall maximum error of data on the viscosity of the four gases is 1.4%.

5.3. ANALYSIS OF METHODS OF CALCULATING THE VISCOSITIES OF GASES AND LIQUIDS

The rigorous theory of kinetic properties of real gases developed by Bogolyubov [2.6], Born [5.36], Kirkwood [5.37], Yvon [5.38], and others [5.39, 5.41] yields an equation which represents the viscosity coefficient as an expansion in terms of density:

$$\eta_{p,T}/\eta_T = 1 + \beta_\eta(T)\rho + \ldots \tag{5.2}$$

Unlike the equation of state, the first density correction $\beta_\eta(T)\rho$ appears when allowing for the interaction of three rather than two molecules. Dorfman and Cohen [5.42, 5.43], Sengers [5.44] and others showed that allowance for the interaction between four and more molecules requires adding to Eq. (5.2) terms containing ln ρ:

$$\eta_{p,T}/\eta_T = 1 + \beta_\eta(T)\rho + \gamma_\eta(T)\ln\rho \cdot \rho^2 + \ldots \tag{5.3}$$

It is assumed [5.45] that the higher-order terms of this expansion contain the factors ln ρ to the pertinent power. Analytic estimates [5.42, 5.44] and calculations based on experimental data [4.72, 4.78, 5.46] showed that terms containing ln ρ do not have a significant effect on the viscosity.

At present, only the lower-order coefficients of the equation for calculating viscosity can be obtained analytically. It was found on the basis of the physical meaning of the first density correction that its value is most significantly affected by triple collisions and the momentum transfer times. These parameters exert an opposing effect on the numerical values and on the nature of the temperature dependence of the second virial coefficient in the equation for viscosity.

The contribution of triple collisions is predominant at low temperatures, with the result that the second virial coefficient is positive. At high temperatures, the effect of the time of transmission of collisions becomes significant and the virial coefficient changes sign. Various theories allow for these effects in different ways which is responsible for differences in results. For example, the Enskog theory [5.47] makes allowance only for binary collisions. The second virial coefficient β_η obtained as a result turns out to be proportional to $\sqrt{T}$. Allowance for triple collisions was made by Stogryn and Hirschfelder [5.48] and by Kim and Ross [5.49] on the basis of Hill's theory of physical groups [3.2]. In this case, the triple collisions were treated as collisions between doubly connected and free molecules. As a result of this approach, Kim and Ross obtained rather reliable values of β_η for low and moderate temperatures where the effect of the time of momentum transfer can be neglected. Hoffman and Curtiss [5.50] calculated the contribution of three-particle collisions to the viscosity of compressed gases, but then they neglected the effect of connected states. Their results are satisfactory only at high temperatures at negative β_η. These estimates are confirmed by comparing the data obtained from the above three theories with experimental values of β_η (Fig. 69) [5.51]. The coordinates (β_η^*, $1/T^*$) for Fig. 69 are defined as follows: (1) β_η^* is the limitless, density independent, reduced viscosity virial coefficient; (2) $T^* = kT/\epsilon$ based on ϵ from the Lennard-Jones (12-6) potential [5.50].

We wish to mention the attempt to determine the third virial coefficient of viscosity made by Curtiss and his coworkers [5.50, 5.52]. Unfortunately, these studies were performed on the basis of the theory of solid spheres and are only qualitative. Hence, their results cannot be useful in practical calculations.

The above means that the rigorous theory can be used for describing the viscosity of gases only approximately, and for a limited range of densities. For this reason, extensive use is made of various empirical equations describing the experimental data over a wide range of parameters.

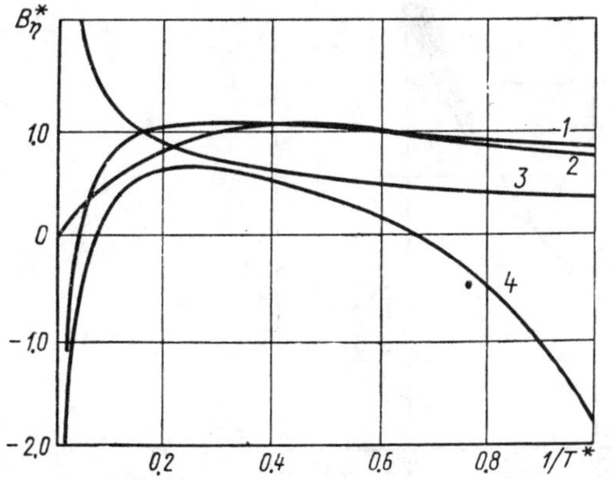

FIGURE 69. Comparison of analytic values of the second virial coefficient of viscosity of inert gases with experimental. 1) [5.49]; 2) experimental curve; 3) [5.47]; 4) [5.50].

Up until recently, the equations most extensively used for calculating the viscosity of compressed gases were those suggested by Vargaftik [5.53] and Golubev [5.33]. The Vargaftik equation expresses the excess viscosity as a function of density

$$\Delta\eta = \eta_{p,T} - \eta_T = f(\rho) \tag{5.4}$$

which is permissible since the separation between excess viscosity isotherms at moderate temperatures and densities in coordinates ($\Delta\eta$, ρ) is small and is fully commensurable with the error of experimental data of various authors. The separation of the isotherms increases with the density and is as high as 30 to 50% for the dense liquid. Figure 70 shows in coordinates ($\Delta\eta$, ρ) the data of Slyusar' et al. [5.7], Michels et al. [5.11], and Hellemans et al. [5.28] on the viscosity of gaseous and liquid argon. It is seen from the figure that the stratification can be neglected to a density of 1.8 ρ_{cr} when $\rho \sim 0.96$ g/cm^3.

Golubev's equation represents the excess viscosity as a function of two vari-

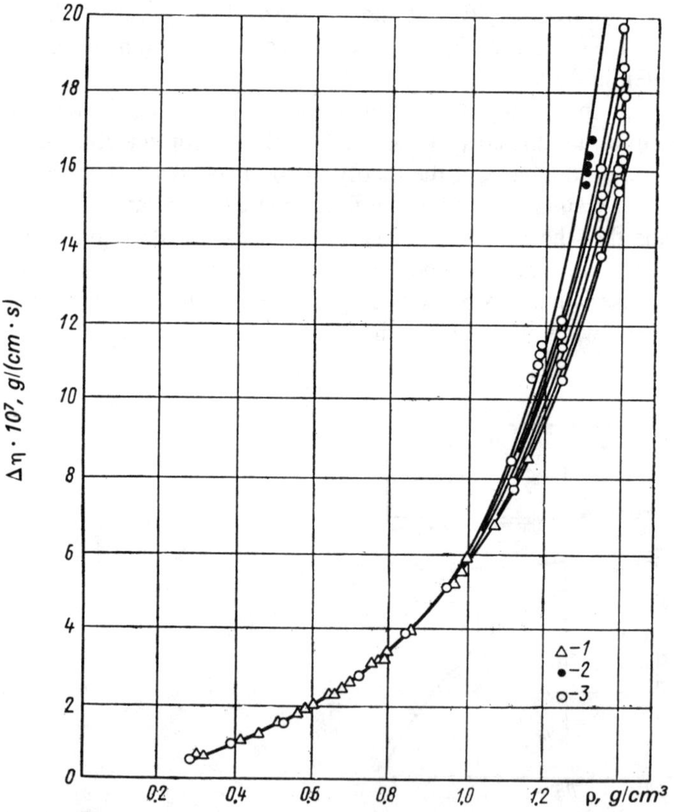

FIGURE 70. Excess viscosity of argon vs. density along isotherms from data in: 1) [5.11]; 2) [5.28]; 3) [5.7].

ables. However, as Eq. (5.4), it is suitable only for describing data on the viscosity of moderately dense gases [3.8, 5.33].

In 1965 Rivkin [5.54] suggested that the viscosity of water and water vapor be described by the equation

$$t = A(\eta) + B(\eta)p \tag{5.5}$$

which follows from the assumption that the constant viscosity curves in coordinates (p, T) are straight lines. In fact, this condition is satisfied over a limited range of parameters and in the vicinity of the critical point. In coordinates (p, T) the graph of the viscosity coefficient with curves of η = const is similar to the familiar diagram of the surface of states in the same coordinates with isochoric sections. Further away from the critical region the lines of η = const cease being rectilinear and acquire a significant curvature. Note that this follows from the existence of a minimum of viscosity coefficients along isobars.

Rivkin et al. [5.55] attempted to validate Eq. (5.5). The scheme used by them for this purpose is not, rigorously speaking, applicable to the critical region for which Eq. (5.5) was suggested. But, it can be referred to the liquid region adjoining the solidification curve. Nevertheless, the results in [5.55] do not contradict experimental data. Over this range of variables it is also possible to single out a region with straight-line, constant-viscosity curves.

Attempts at describing the viscosity of the gas over a wide range of parameters, including the region where the isotherms stratify, has resulted in increasing the complexity of the forms of the equation. Trappeniers et al. [5.30], Kessel'man and Kamenetskiy [5.56] and Vasserman and Nedostup [5.57] suggested equations represented in terms of "elementary functions". These equations have in common the separation of variables T and ρ in their final forms

$$\eta/\eta_T = F_1(\rho) + F_2(\rho)\phi(T) + F_3(\rho)\varphi(T) + \ldots \tag{5.6}$$

Trappeniers et al. [5.30] represent functions $F(\rho)$ by density polynomials, and the temperature functions in the form of lg T to different powers. The equation suggested by Kessel'man and Kamenetskiy [5.56] differs from that by Trappeniers et al. by the fact that the temperature functions contain $1/T$ instead of lg T. Kamenetskiy and Kessel'man [5.56, 5.58] justify this by the feasibility of expanding the virial coefficients of viscosity in a power series in $1/T$. This justification is more intuitive than rigorous, since the form of the higher-order virial coefficients, and due to this, of the equation for viscosity of compressed gas, is not entirely clear. An equation of similar form was suggested by Vasserman and Nedostup [5.57] on the basis of analysis of isochoric sections of the plane of the viscosity coefficient. In connection with the analogy between the temperature dependence of the isochores of viscosity of the gas and the curve of η_T, they suggested using η_T as the temperature function. Then the equation becomes

$$\eta = F_1(\rho) + F_2(\rho)\frac{1}{T} + [1 + F_3(\rho)]\,\eta_T \tag{5.7}$$

Equations for calculating the viscosity of the gas represented by elementary functions are capable of describing experimental data over a wider range of variables than the equations given in [5.33, 5.53, 5.54]. However, they contain a large number of constants which increases significantly when it is necessary to simultaneously describe the viscosity of the gas and the liquid.

As in the case of equilibrium properties, a special position is occupied by theories of transport properties of liquids. This problem has been investigated extensively. In this analysis we restrict ourselves to studies not included in surveys [3.8, 2.84].

One of the most popular theories is the so-called Eyring's activation theory [5.59]. It consists of representing the process of viscous flow as an attempt by a molecule which is in equilibrium with its cell, formed by the surrounding molecules, to overcome a "bottleneck," which prevents it from moving into a free adjoining cell. This process of overcoming the "bottleneck" must be accompanied by overcoming the potential barrier. The frequency of the process, as follows from the "reaction rates theory" [5.59, 2.109], is proportional to $e^{-\frac{\Delta G}{kT}}$, where ΔG is the free energy of activation. The final equation for viscosity has the form

$$\eta = Ae^{-\frac{\Delta G}{kT}} \tag{5.8}$$

where A is a structure parameter depending on the form and energy of the molecular lattice. This is also basically the meaning of Frenkel's viscous flow theory [3.4, 5.60], with the difference that the activation energy is not associated with the heat of vaporization as in the Eyring theory, but the heat of fusion.

The activation model extends the features of the motion of particles in a solid body with lattice vacancies to flow processes. This permits prediction of the distribution of the mean free path l. Such distribution should contain two specific distances: one corresponding to oscillations of the molecule inside the cell, and the other characterizing the travel of the molecule from one cell to another. This means that two peaks should exist in the distribution of the mean free path. The mean free path distribution function simulated by computer [5.61, 5.62] (Fig. 71) is monotonic. It describes, in equal measure, the isochores for different densities from the rarefied gas and to the liquid. It was shown analytically by Alder et al. [5.61, 5.62] that the distribution of the mean free path in rarefied gases is qualitatively the same as in a dense gas. This points to absence of a qualitative jump which would characterize change in the density of the medium. This means that the mean free path distribution function ob-

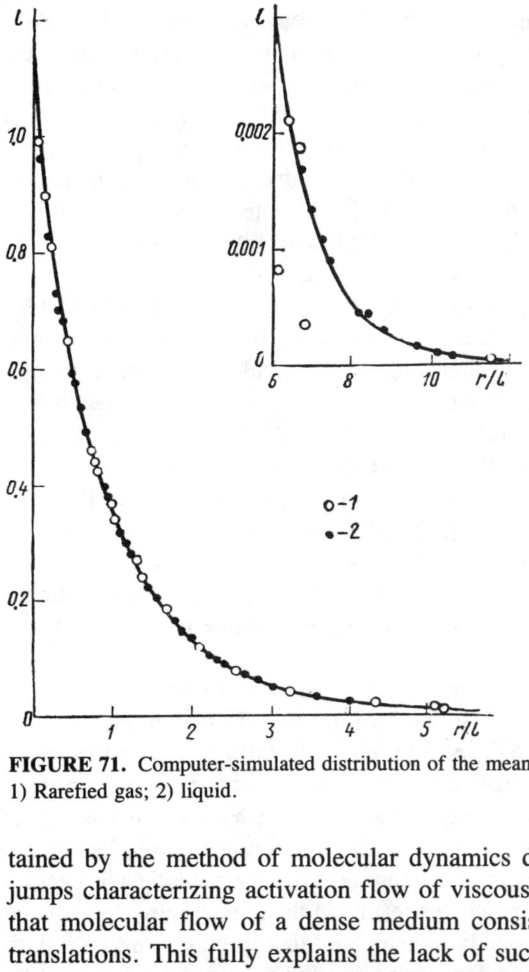

FIGURE 71. Computer-simulated distribution of the mean free path for monatomic substances. 1) Rarefied gas; 2) liquid.

tained by the method of molecular dynamics disproves the existence of large jumps characterizing activation flow of viscous fluids. It follows from Fig. 71 that molecular flow of a dense medium consists of a large number of small translations. This fully explains the lack of success of the activation energy in describing the viscosity coefficient of real liquids.

The "free volume" theories which address to a greater measure the difference between the structure of a real liquid from the structure of a crystal lattice are more promising. The viscosity is represented as a function of the free volume, i.e., that part of the volume into which molecules may move; in this case

$$\eta \approx e^{\frac{b}{v-b}} \tag{5.9}$$

The result of determination of the free volume ($v - b$) depends significantly on the method used. Doolittle et al. [5.63, 5.64] defined the free volume by the Van der Waals equation or by way of the value of the density of the crystal at 0 K. These highly empirical approaches were unsuccessful. Equations such as (5.9) turn out to be suitable only for describing the properties of glasses, greases and other dense media. More rigorous and, at the same time, more elegant is the theory due to Cohen and Turnbull [5.65] in which it is assumed that the molecule may move to a free cell only when the volume of the latter is larger than some minimum. The problem reduces to determination of the necessary free volume which may form as a result of the combination of several free volumes. A check made by Cohen and Turnbull showed that their theory describes the properties of viscous fluids quite satisfactorily.

A more general approach was suggested by Macedo and Litovitz [5.66]. They suggested that in order for molecules to move it is necessary, in addition to the availability of a certain free volume, that the molecule should have enough residual energy to overcome the potential barrier. This theory includes the results of theories of Cohen and Turnbull, and of Eyring. The use of the Macedo-Litovitz hypothesis hinges upon the availability of experimental data which would allow determining the large number of quantities contained in the final expression. In Gray's opinion [2.137], the Macedo-Litovitz theory is not suitable for simple liquids as exemplified by liquefied inert gases since the repulsion potential in such liquids is less steep than in liquids consisting of a large number of molecules. For this reason, the concept of the free volume for them is not as well defined.

Nevertheless, if additional assumptions are made, the free-volume theories can be used rather effectively for describing experimental data on the viscosity of simple liquids. Matheson [5.67] and Lyusternik [5.68] obtained rather satisfactory results by replacing constant b in Eq. (5.9) by a temperature function. This is because the free-volume theory is based on the Van der Waals model which assumes an intermolecular interaction which is simpler than in real gases. As will be shown below, the lack of agreement between the potentials of interaction of real and Van der Waals molecules results in the temperature dependence of b. Matheson [5.67] based his study on the claim that isochores are lines with small curvatures and can be regarded as straight lines over some temperature range. This has resulted in linear temperature dependence of $1/b$ in Eq. (5.9). Comparison with experimental data showed that Eq. (5.9) in the Matheson modification is capable of describing viscosity only over that temperature range where the isochore can be assumed to be linear. Naturally, the greatest error is observed at high densities where the power exponent in Eq. (5.9) is greater than unity, which causes the error to increase exponentially.

From this point of view, it is more preferable to use Lyusternik's [5.68] equation which is a Taylor-series expansion of Eq. (5.9). In fact,

$$\eta = Ae^{\frac{b}{v-b}} = A\left[1 + \frac{b}{v-b} + \frac{1}{2}\left(\frac{b}{v-b}\right)^2 + \ldots\right] \tag{5.10}$$

Constant A is found from the condition that $\eta \rightarrow \eta_T$ as $v \rightarrow \infty$. If the series is truncated after the second term, then

$$\eta = \eta_T + \frac{B}{v - b} \tag{5.11}$$

The temperature dependence of b was determined by Lyusternik on the basis of analysis of experimental data on the viscosity of hydrogen. It was found to be possible to use a rather simple form for b, whereupon Eq. (5.11) became

$$\eta = \eta_T + \frac{B\rho}{1 - \frac{T_0}{T} - b_0\rho} \tag{5.12}$$

The validity of Eq. (5.12) can be checked by constructing isotherms in coordinates ($\rho/\Delta\eta$, ρ). According to this equation, the isotherms should be straight lines. A check of the viscosity of argon on the basis of experimental data showed that at supercritical densities isotherms can be taken to be a straight line with an error not exceeding 3–4%. At lower densities the isotherms curve, indicating that Eq. (5.12) is not capable of describing data on viscosity over the entire range of densities where gas and liquid exist.

For our case, it is desirable to obtain an equation representing the viscosity of the substances under study both in the gaseous and liquid states, and from this point of view, Enskog's theory is of greatest interest [5.47]. Considering only paired collisions of solid spherical molecules with diameter σ, Enskog suggested a theory in which consideration is given to the fact that when the gas is compressed the transport of momentum and the collision energy for 1 sec become significant. It follows from the theory that the frequency of collisions of solid spheres of high-density gas differs from the frequency of collisions of the rarefied gas by a function of $g(\sigma)$, where $g(\sigma)$ is a radial distribution function of solid spheres over distance σ. It is identical to the virial equation of state of a gas made up of solid spheres and according to Ree and Hoover [5.69]

$$g(\sigma) = 1 + 0.625\left(\frac{b}{v}\right) + 0.2869\left(\frac{b}{v}\right)^2 + 0.1103\left(\frac{b}{v}\right)^3 + 0.0386\left(\frac{b}{v}\right)^4 + 0.0138\left(\frac{b}{v}\right)^5 \tag{5.13}$$

where

$$b = \frac{2}{3}\pi N\sigma^3$$

The final equation for density is

$$\eta/\eta_T = \frac{1}{g(\sigma)}\left[1 + 0.8\left(\frac{b}{v}\right)g(\sigma) + 0.761\left(\frac{b}{v}\right)^2 g(\sigma)^2\right] \tag{5.14}$$

In spite of the fact that the assumptions underlying the derivation of Eq. (5.14) make it, rigorously speaking, unsuitable for a real gas, ways were found for using this equation for description of experimental data.

The first of these consists of writing Eq. (5.14) in the somewhat different form:

$$\frac{\eta}{\rho} = \frac{1}{2.545}\left(\frac{\eta}{\rho}\right)_{\min}\left[\frac{1}{y} + 0.8 + 0.761y\right] \tag{5.15}$$

where

$$y = \frac{b}{v} \cdot g(\sigma) = \frac{pv}{RT} - 1 \tag{5.16}$$

When converting to a real gas, pressure p is replaced by thermal pressure $T\left(\frac{\partial p}{\partial T}\right)_v$, it is due to the presence of attraction forces which are absent in the model of solid spheres. Then y is determined from the equation of state of a real gas. A check of this approach, made in [1.104, 5.33, 2.109, 5.70] showed that Eq. (5.15) can be used for describing the viscosity of moderately compressed gases. In certain cases, when experimental viscosity data allow the determination of the location of the minimum of function $\frac{\eta}{\rho} = f(\rho)$, Eq. (5.15) can describe data on the density of dense gases with some loss of accuracy at moderate densities. At the same time, this approach does not permit the use of Enskog's equation for liquids and gases at a density corresponding to that of the liquid.

Another way, suggested by Alder and his coworkers [5.62, 5.71], is based on adding to Eq. (5.14) a term representing the temperature dependence of diameter σ of the solid spheres. They correctly explain their step by the fact that the repulsion energy of molecules of a real gas is less stiff than that of solid spheres. Dymond and Alder [5.71] find this relationship from the Van der Waals equation.

As is known, the Van der Waals gas consists of a system of solid spheres which exhibit an attraction energy proportional to $\sim r^{-6}$. The behavior of a and b in the Van der Waals equation as a function of the parameters of state was determined in [5.71] from satisfaction of experimental data on the compressibility of inert gases. They noted that a is virtually independent of the temperature and density, particularly at high densities while b decreases significantly with increasing temperature and density. At rather high densities ($\rho > 2\ \rho_{cr}$) the behavior of $b(\rho)$ need not be taken into account and diameter σ can be taken to be a function only of the temperature. Analogous results were obtained from computer calculations of constants in the Van der Waals equation for a system of particles interacting according to a potential in the form of a rectangular well [2.137]. This indicates that the principal contribution to the energy of the system at high densities is made by the repulsion energy which can be taken into account with sufficient accuracy by the potential of solid spheres with temperature dependent effective diameter†.

Our calculations showed that the radial distribution function $g(\sigma)$ obtained by computer from the method of molecular dynamics [5.71] is identical to the function calculated from the Ree-Hoover equation [5.69]. As expected, the introduction of term $\sigma(T)$ made possible a more accurate description of the viscosity of dense gases above critical temperatures. At lower temperatures and in the moderate density region the errors increase significantly due to two factors. Firstly, when the model of solid spheres is used for describing the properties of a real gas over a wide range of variables, the density dependence of σ cannot be neglected. This dependence appears due to the presence of attraction forces. The contribution of these forces increases at moderate and low density and also due to nonadditivity of the multiparticle potential energy of interaction. Secondly, the use of the Van der Waals equation for determining the effective diameter may result in inevitable errors in calculating the transport properties due to the absence of complete agreement between the Enskog and Van der Waals theories.

5.4. EQUATION FOR CALCULATING THE VISCOSITY COEFFICIENT OF MONATOMIC GASES AND LIQUIDS

A more effective way is based on determining the density and temperature dependence of the diameter of solid spheres by means of the Enskog equation by requiring satisfaction of experimental viscosity data.

In addition to the previously analyzed grounds for the use of the Enskog theory [5.47] it must be remembered that Eq. (5.14), as will be shown below, satisfies the following fundamental requirements:

(1) the form of the equation is theoretically validated;
(2) the experimental data are used for determining fitting parameters and do not affect the form of the equation as a whole;
(3) the equation is suitable for generalized representation of properties by means of physically validated correlation parameters.

The temperature dependence of σ was determined analytically by a number of investigators on the basis of models of potential functions. Rowlinson [5.72]

† As noted in Sec. 3.1, an analogous result is observed when using the theory of perturbations for describing the properties of liquids.

found for a rarefied gas at high temperature, according to a potential related to $\sim r^{-12}$, that

$$\sigma(T) = a\left(\frac{4}{T^*}\right)^{\frac{1}{12}}\left[1 + \frac{\Phi(T^*)}{12}\right] \tag{5.17}$$

where the values of $\Phi(T^*) = \Phi\left(\frac{kT}{\epsilon}\right)$ were calculated and tabulated in [5.73].

Equation (5.17) is rather complex and its use for other power-law potentials requires rather cumbersome calculations. The relationship used in the theory of perturbations [3.32] has the form

$$\sigma(T) = \int_0^\sigma \left\{1 - \exp\left[-\frac{\varphi(r)}{kT}\right]\right\} dr \tag{5.18}$$

The solution of Eq. (5.18) for any potential can be represented as power series in $1/T$. When potential function $\varphi(r)$ is not available, the equation for $\sigma(T)$ is determined solely from a polynomial with empirical coefficients.

The method of determining $\sigma(T)$ used by Abovskiy [1.6] for the equation of state of high-density inert gases requires special attention. The derivation is rigorous for a system of spherically symmetrical particles interacting according to a repulsion potential and given by the Born-Mayer expression

$$\varphi_{\mathrm{rep}}(r) = \epsilon \exp\left(-\alpha \frac{r}{R_e}\right) \tag{5.19}$$

Restricting himself to high densities, at which the cellular model can be used, Abovskiy employed an equation for the temperature dependence of the diameter of effective spheres [5.74]

$$\sigma(T) = \frac{R_e}{\alpha} \ln\left(\frac{\epsilon}{LkT}\right) \tag{5.20}$$

where L is a fitting parameter; from Abovskiy's estimates for inert gases $\alpha \simeq 15$ and $L \simeq 10$. The reliability of Eq. (5.20) was not checked by Abovskiy, but he did analyze the accuracy of the final equation of state in which Eq. (5.20) was contained in its entirety (together with other quantities) as a part of the fitting parameters.

The accuracy of $\sigma(T)$ is very important for the problem at hand. Because of this, the values of the effective diameter were calculated on the basis of the most compatible experimental data on the viscosity of gases and liquids. For this we calculated the auxiliary function $\eta/\eta_T = f\left(\frac{b}{v}\right)$ using Eqs. (5.13) and (5.14) and then the value of b (and accordingly of σ) was calculated for the experimental values of η/η_T. The resulting data on σ were worked up graphically in coordinates (σ, lg T). Figure 72 shows the result of such workup for the case of argon. Two conclusions can be drawn from the figure: firstly, Eq. (5.20) is valid over a wide temperature range at any density; secondly, the effective diameter of the solid spheres should be a function not only of temperature but also of density. It is remarkable that Eq. (5.20) is suitable not only at high densities, for which this conclusion was rigorously drawn, but also at lower densities. The scatter observed in the latter case is the result of a singularity of the Enskog equation. The numerical value of the derivative $\left(\frac{\partial \eta}{\partial \sigma}\right) \rho$ increases with reduction in density, which is also shown by Fig. 73 [5.71]. From the figure one can see the effect of error in determining the collision diameter (1%) on the error of values of thermophysical properties (η, λ, Z and D—the diffusion coefficients) at different densities. It follows, in part, from the figure, that data at high densities allow determining σ with a very high accuracy. This makes it possible to use such data for checking Eq. (5.20) and also for assessing the agreement between the experimental results of various investigators.

The observed density dependence is the result of various factors. The main one is the increase in the effect of attraction forces with increasing density and nonadditivity of potential energy of multiparticle interaction. The density dependence of σ can be neglected at high densities, whereas at low ρ it plays the same role as the temperature dependence. Due to obvious theoretical diffi-

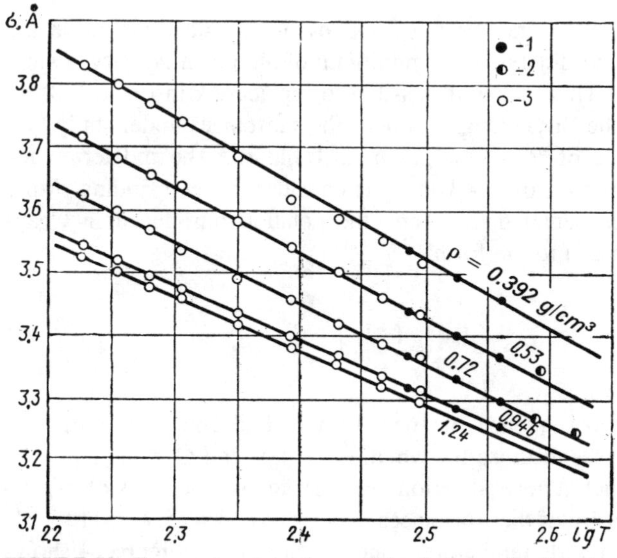

FIGURE 72. Temperature dependence of the diameter of solid spheres of argon at different densities from data of different investigators. 1) [5.11]; 2) [5.23]; 3) [5.7].

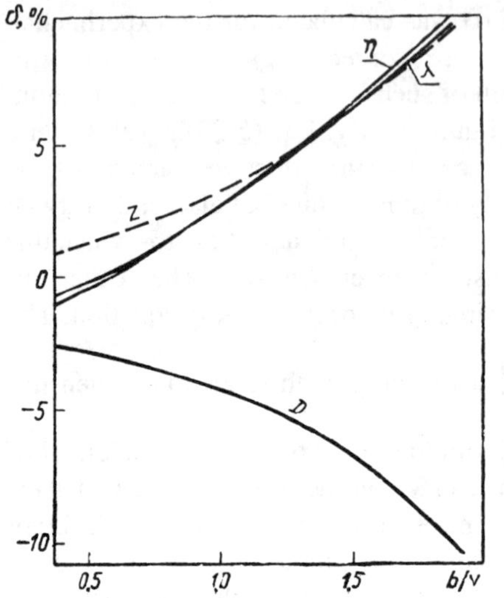

FIGURE 73. Effect of a 1% error in determining the effective diameter of solid spheres on the error in thermophysical properties calculated from the Enskog theory.

culties, the density behavior of σ can be determined only on the basis of experimental results.

An important advantage of Eq. (5.20) consists of the feasibility of correlating data on σ of different substances. For this it is best to rewrite Eq. (5.20) in the dimensionless form

$$\sigma^* = \frac{\sigma}{Re} = \frac{1}{\alpha} \ln \frac{1}{LT^*} \tag{5.21}$$

As shown in Sec. 2.3, the parameters T_B and ρ_0 of the ideal-gas curve are uniquely related to the coordinates of the minimum of the potential curve where ($T_B \approx \epsilon/k$; $\rho_0 \approx 1/R_e^3$). This makes it possible to replace ϵ/k in Eq. (5.20) by experimental values of the Boyle temperature of the substances under study (see page 209) and the value of $R_e = r_0$, given in Table 20. The indeterminacy factor, arising upon transition from ϵ/k to T_B is considered when finding fitting parameter L from experimental data. Using the notation introduced in Chapter 2, we rewrite Eq. (5.21) to the form

$$\sigma^* = \frac{1}{\alpha} \lg \left(\frac{1}{L\Theta} \right) \tag{5.22}$$

In addition, retaining the sequence of analysis, we shall apply the term like isochores of different gases to those for which $\omega = \rho/\rho_0$ and $\Theta = T/T_B$.

The feasibility of such a generalization was checked on the basis of experimental data on the viscosity of the substances under study [5.4, 5.7, 5.11, 5.21, 5.28–5.30, 3.89] over a wide temperature and density range. Figure 74 shows like isochores of the four substances in coordinates (σ^*, lg Θ). It is seen from

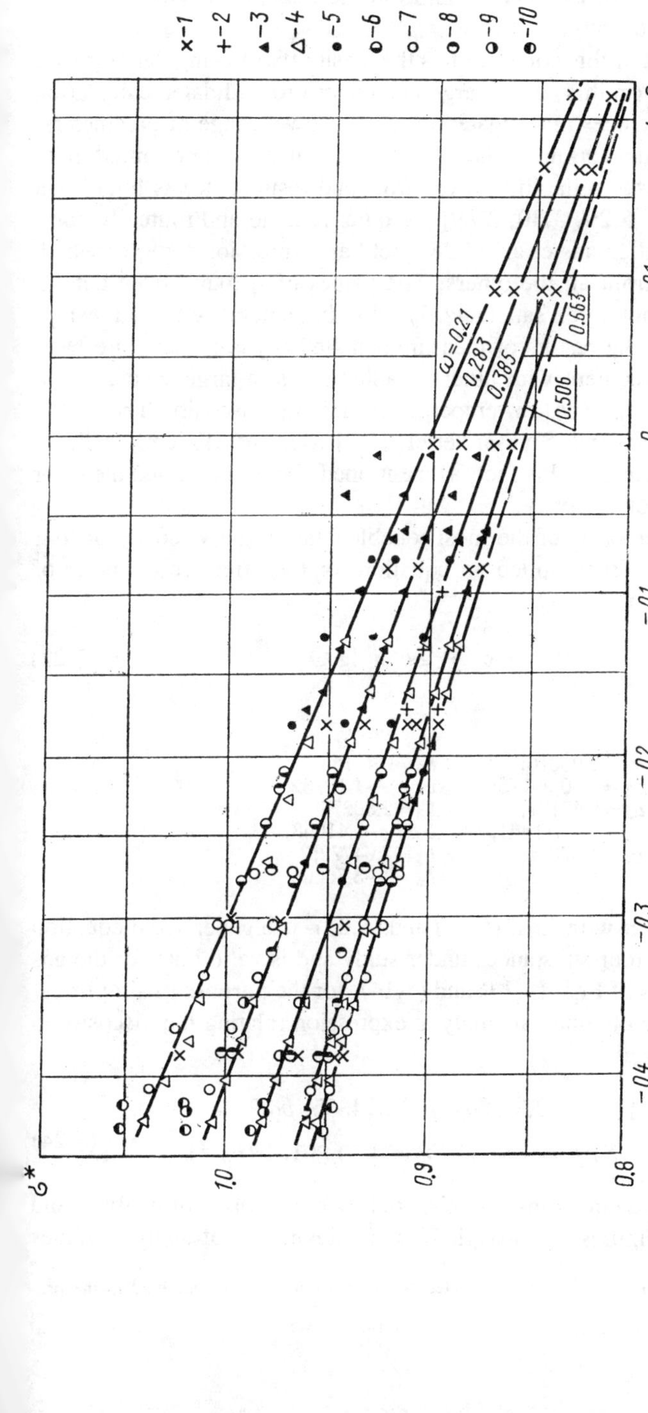

FIGURE 74. Correlation of reduced diameter of solid spheres for inert gases. 1) [5.7] for Ne; 2) [5.11]; 3) [5.21]; 4) [5.7] for Ar; 5) [5.21]; 6) [5.29]; 7) [5.7] for Kr; 8) [5.21]; 9) [5.30]; 10) [5.7] for Xe.

the figure that the majority of data points are described by Eq. (5.22) with a scatter not exceeding that of experimental data over the entire range of parameters, subject to the previously made remark on the relative increase in deviations with reduction in density.

An important result of this correlation is the feasibility of using data for well investigated substances (in this case of argon and neon) for validated calculation of the viscosity of substances investigated over a narrower range of parameters. In addition, simultaneous workup of data for the four substances permitted more dependable analysis of the reliability of experimental results. It was found that data in [5.4, 5.7, 5.11, 5.29, 5.30, 3.89] are quite reliable and mutually compatible. The data of Hellemans et al. [5.28] yield an expression for $\eta(\rho)$ which is somewhat different from all the others. The values of η found from data in [5.21, 5.24, 5.27] do not agree satisfactorily with the generalized relationship. In spite of the fact that the results of measurements by Slyusar' et al. are basically in satisfactory agreement with other reliable values, a large scatter is observed for krypton and xenon at low temperatures and high viscosity. The reason for this scatter is the fact that Slyusar' et al. [5.7] used insufficiently reliable experimental data on density. This fact, as mentioned above, increased the error of the resulting values of viscosity.

Accordingly, on the basis of the most reliable data on the viscosity of four inert gases and liquids we compiled an equation for the effective diameter of solid spheres

$$\sigma^* = \Sigma a_i \omega^i + \Sigma b_i \omega^i \lg \Theta \tag{5.23}$$

whose coefficients are

$$\begin{aligned} &a_0 = 1.03010; && b_0 = 0.48148; \\ &a_1 = -0.99175; && b_1 = -1.18732; \\ &a_2 = 2.47127; && b_2 = 2.80277; \\ &a_3 = -3.11864; && b_3 = -5.41058; \\ &a_4 = 1.57066; && b_4 = 7.04779; \\ & && b_5 = -3.76608. \end{aligned}$$

Equation (5.14), together with Eqs. (5.13) and (5.23), is a generalized equation for the viscosity of the four substances under study and is valid both for the gas and liquid. On the basis of Eqs. (5.13) and (5.14), for the purpose of simplification of calculations, we obtained an analytic expression relating the viscosity to the density†:

$$\begin{aligned} \eta = \eta_T\, [1 &+ 0.27676 b\rho + 0.014355\,(b\rho)^2 + \\ &+ 2.6480\,(b\rho)^3 - 1.9643\,(b\rho)^4 + 0.89161\,(b\rho)^5] \end{aligned} \tag{5.24}$$

The accuracy of the equations was checked by comparison of analytic and experimental values. Figures 75 through 78 show deviations of analytic values

† The needed values of density were calculated from equations of state obtained in the preceding chapters.

of viscosity of the four gases under study from the most reliable experimental data $\delta\eta = \frac{\eta_{calc} - \eta_{exptl}}{\eta_{exptl}} 100\%$. At pressures to 1000 bar the deviations for the gas as well as for the liquid lie within the limits of experimental error. Thus, the deviations from the experimental data of Slyusar' et al. [5.7] basically lie within the limits of 4% which is the error given in [5.7]. A somewhat larger difference

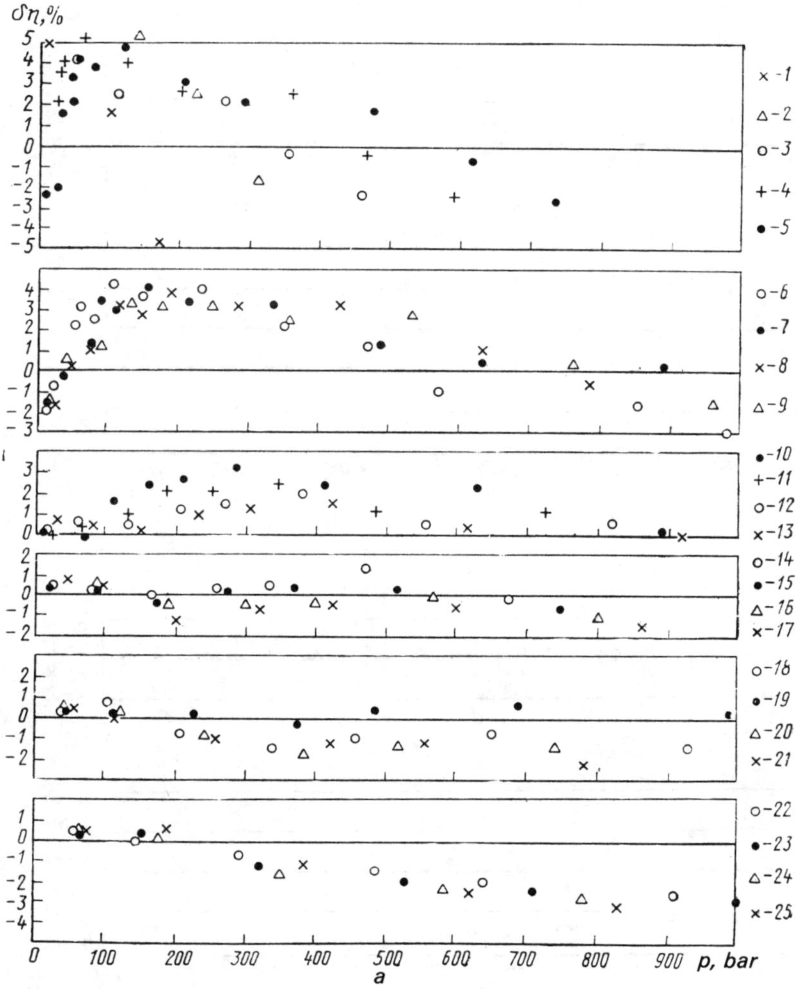

FIGURE 75. Deviations of computed values of the viscosity of neon from the experimental. a) [5.7]; b) [5.4]; c) [5.6]; d) [5.1]; e) [4.9]; 1) T = 30 K; 2) 35; 3) 40; 4) 44.45; 5) 50; 6) 60; 7) 70; 8) 80; 9) 90; 10) 100; 11) 110; 12) 120; 13) 130; 14) 140; 15) 150; 16) 160; 17) 170; 18) 180; 19) 190; 20) 200; 21) 220; 22) 240; 23) 260; 24) 280; 25) 300.

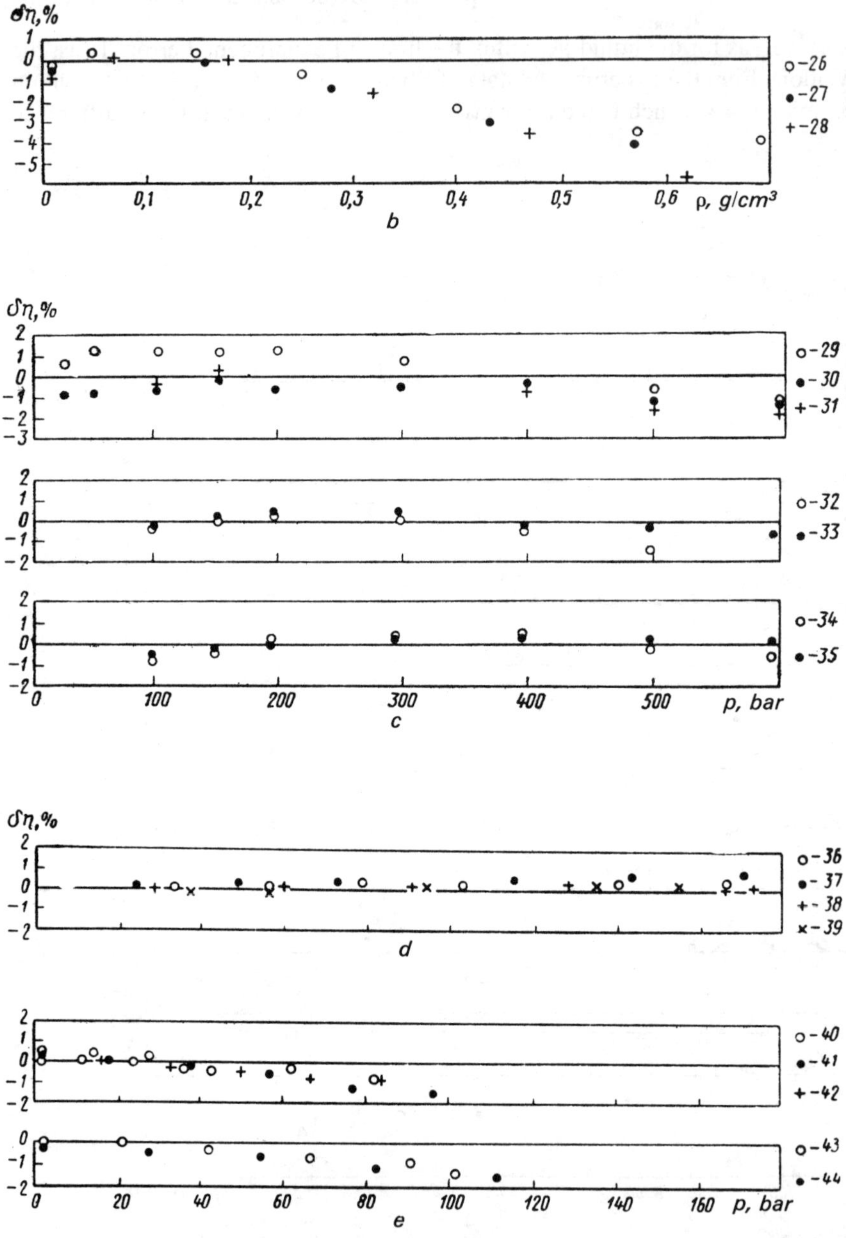

FIGURE 75. Deviations of computed values of the viscosity of neon from the experimental. 26, 29, 38) 298.15; 27, 30) 323.15; 28, 31) 348.15; 32, 39) 373.15; 33) 423.15; 34) 473.15; 35) 523.15; 36) 223.15; 37) 248.15; 40) 296; 41) 371; 42) 424; 43) 468; 44) 519. (*Continued*).

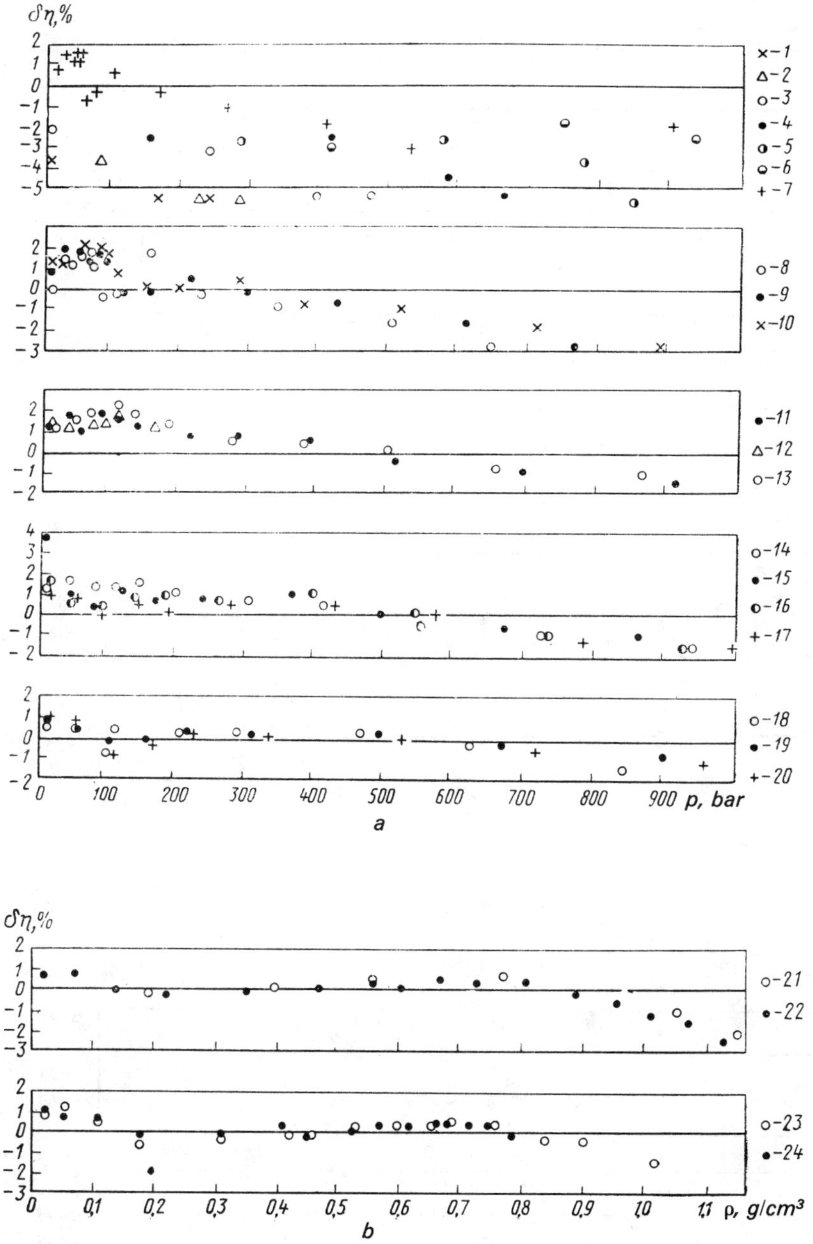

FIGURE 76. Deviation of analytic (computed) values of the viscosity of argon from the experimental. a) [5.7]; b) [5.11]; c) [5.28]; d) [5.6]; e) [4.9]; f) [5.1]; g) [5.10]; h) [5.23]; 1) T = 96.5 K; 2) 100; 3) 109.7; 4) 120; 5) 130; 6) 140; 7) 150.7; 8) 160; 9) 170; 10) 180; 11) 200; 12) 210; 13) 220; 14) 230; 15) 250; 16) 260; 17) 270; 18) 280; 19) 290; 20) 300; 21) 273.15; 22) 298.15; 23) 323.15; 24) 348.15.

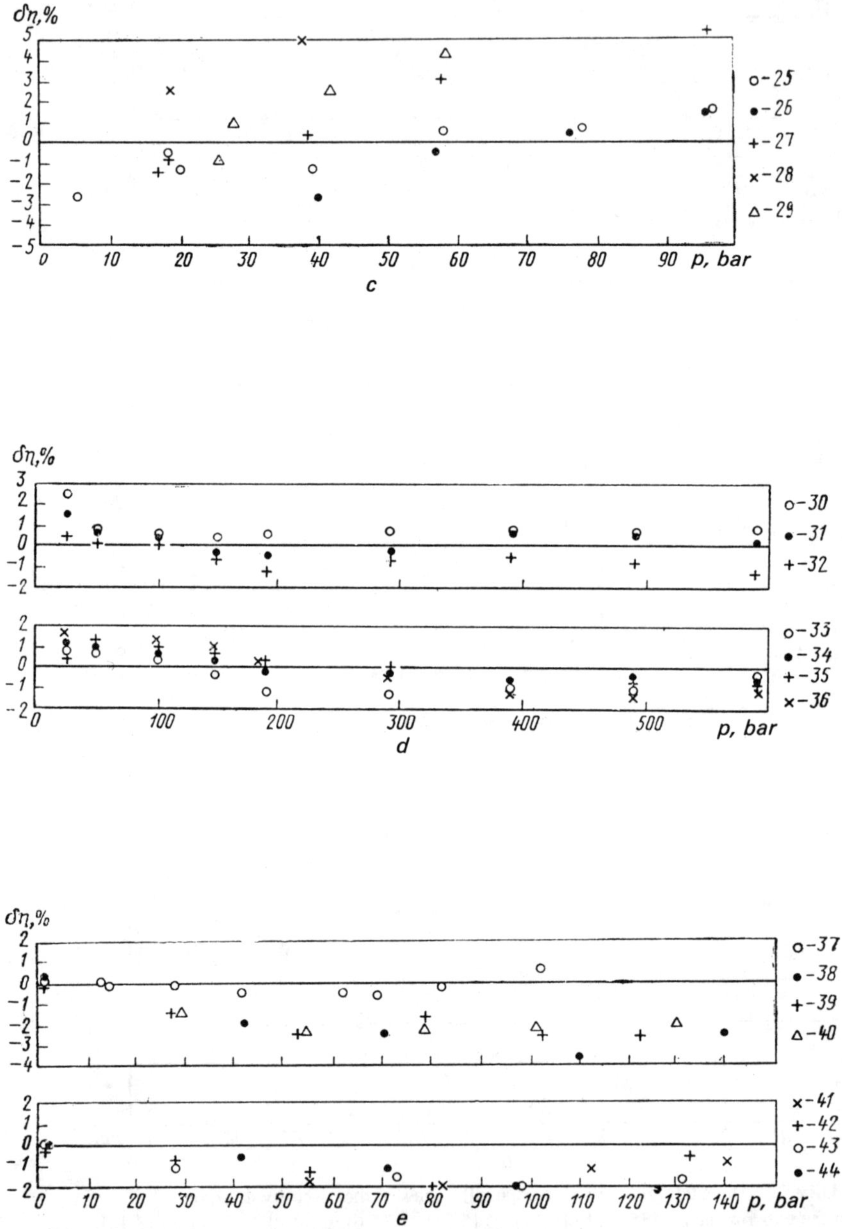

FIGURE 76. Deviation of analytic (computed) values of the viscosity of argon from the experimental. 25) 104.5; 26) 119.6; 27) 125.3; 28) 128.1; 29) 134.5; 30) 298.15; 31) 323.15; 32) 348.15; 33) 373.15; 34) 423.15; 35) 473.15; 36) 523.15; 37) 296; 38) 337; 39) 349; 40) 372; 41) 425; 42) 469; 43) 517; 44) 537. (*Continued*).

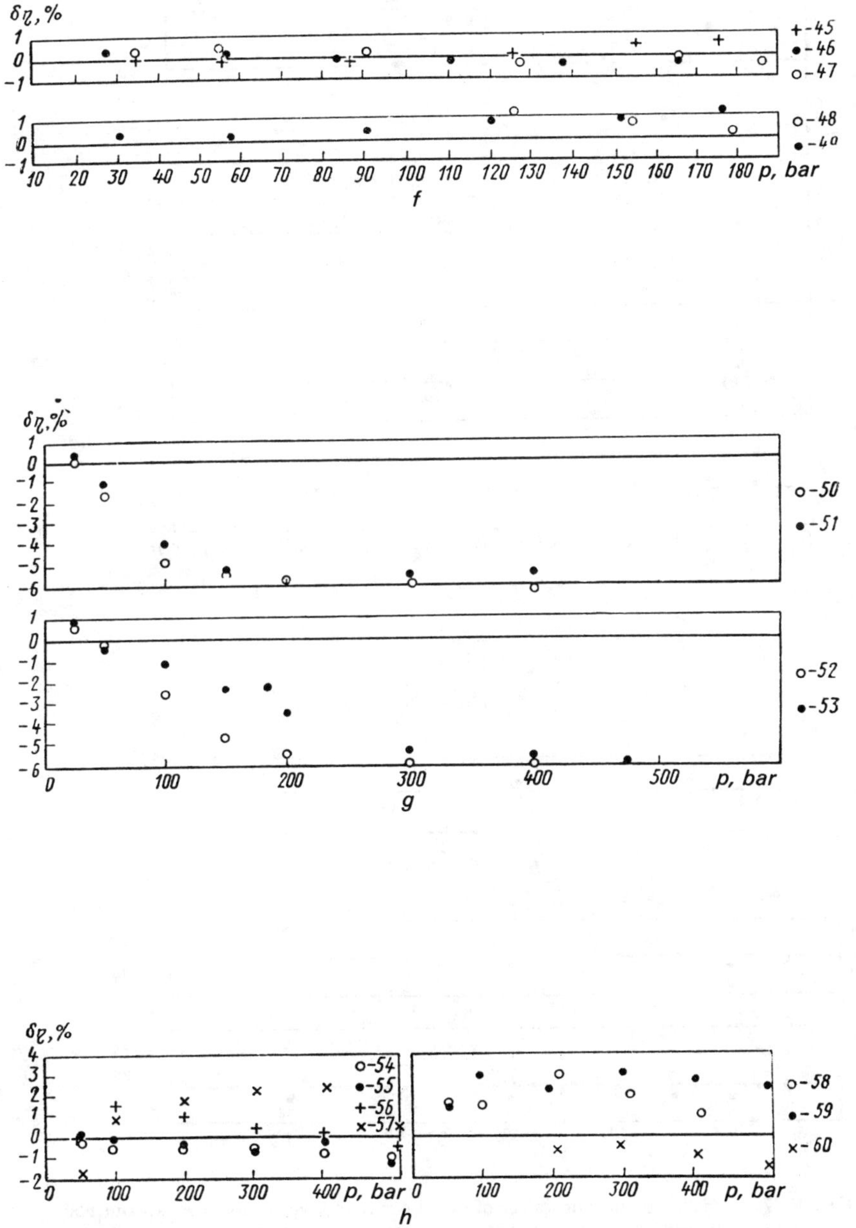

FIGURE 76. Deviation of analytic (computed) values of the viscosity of argon from the experimental. 45) 248; 46) 298; 47) 373; 48) 194; 49) 223; 50) 298.15; 51) 323.15; 52, 55) 373.15; 53, 56) 473.15; 54) 294.15; 57) 573.15; 58) 673.15; 59) 773.15; 60) 923.15. (*Continued*).

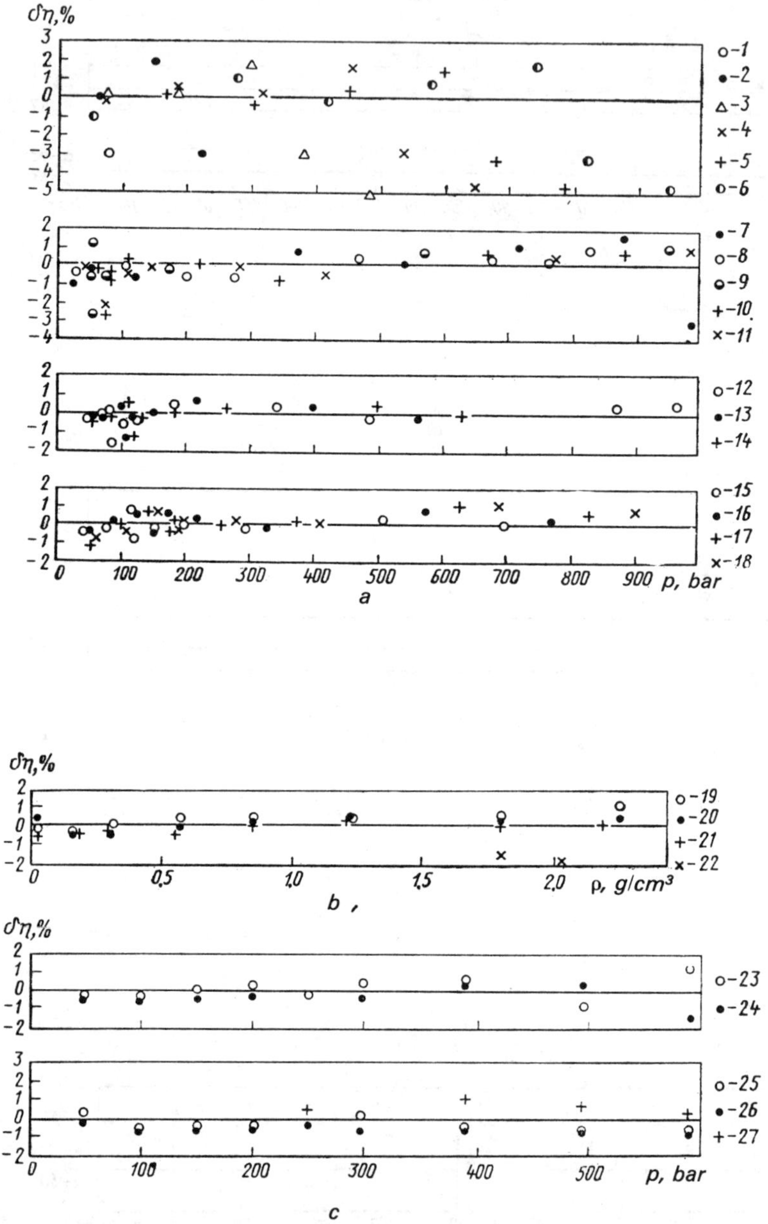

FIGURE 77. Deviation of analytic values of the viscosity of krypton from the experimental. a) [5.7]; b) [5.29]; c) [3.89]; 1) T = 130 K; 2) 140; 3) 150; 4) 160; 5) 170; 6) 180; 7) 190; 8) 200; 9) 210; 10) 220; 11) 230; 12) 240; 13) 250; 14) 260; 15) 270; 16) 280; 17) 290; 18) 300; 19) 298.15; 20) 323.15; 21) 348.15; 22) 398.15; 23) 323.15; 24) 378.15; 25) 423.15; 26) 473.15; 27) 523.15.

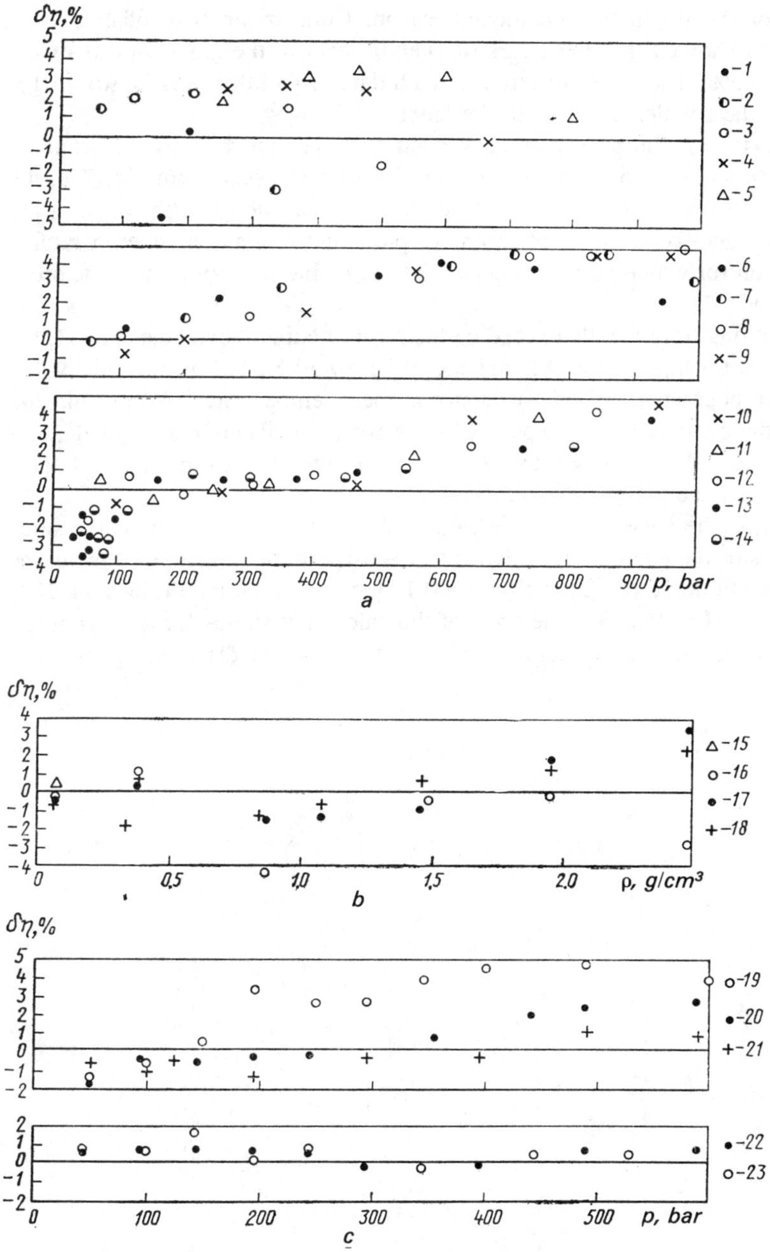

FIGURE 78. Deviations of analytic values of viscosity of xenon from the experimental. a) [5.7]; b) [5.30]; c) [3.89]; 1) T = 170 K; 2) 180; 3) 190; 4) 200; 5) 210; 6) 220; 7) 280; 8) 240; 9) 250; 10) 260; 11) 270; 12) 280; 13) 290; 14) 300; 15) 273.15; 16) 298.15; 17, 19) 323.15; 18) 348.15; 20) 373.15; 21) 423.15; 22) 473.15; 23) 523.15.

to 5–6% is observed over ranges of parameters where the data in [5.7] are less reliable for the previously mentioned reason. Comparison with other experimental data showed that the errors of analytic data in the gas region is of the order of 2–3%. The larger differences with data of Golubev and Petrov [5.10] are due to the smaller accuracy of the latter.

The extrapolation properties of our equation are illustrated by the accuracy of description of data of Rabinovich et al. for neon. As seen from Fig. 75c, the deviations do not exceed 2%. This leads to the conclusion that the error in the calculated values of viscosity of argon, krypton, and xenon at high temperatures in the region for which no experimental data available will be of the same order of accuracy.

Deviations significantly exceeding the error of experimental data (10–15%) are observed for liquid neon at temperatures below 30 K. This is most likely due to quantum phenomena of this substance at these temperatures. As a result, the tables of the coefficient of viscosity of neon were calculated to $T = 40$ K, and the necessary values at lower temperatures were obtained from graphical analysis of the available experimental data.

Thus, on the basis of Eqs. (5.23) and (5.24) we have calculated the summary tables of the coefficient of dynamic viscosity of the four monatomic gases for the gas and liquid ranges to $T = 1300$ K and $p = 1000$ bar (Tables VI, XII, XIX and XXVI of Part 3). The error of the calculated values for the gas phase does not, according to our estimates, exceed 2–3%; for the liquid phase it is ~4%.

CHAPTER

SIX

THERMAL CONDUCTIVITY OF MONATOMIC GASES AND LIQUIDS AT HIGH PRESSURE

Up until recently, the thermal conductivity of gaseous and liquid neon, argon, krypton, and xenon was insufficiently investigated experimentally. The experimental data were obtained for narrow ranges of parameters and were not too accurate. This situation changed radically during the late sixties and early seventies. This necessitated an analysis of experimental data on the thermal conductivity of the substances under study above atmospheric pressure and a correlation of the results on the basis of analytic equations.

6.1. ANALYSIS OF EXPERIMENTAL DATA ABOVE ATMOSPHERIC PRESSURE

As seen from Table 51, argon is the most investigated substance. The main design feature of the equipment used in the first studies [6.2, 6.4] consisted of three coaxial steel tubes between which were held two layers of gas. The test gas filled a 0.203 mm wide annulus between the inner and center tubes, and the reference gas filled the 0.838 mm wide annulus between the middle and outer tubes at atmospheric pressure. Convection was eliminated by maintaining conditions under which $Gr \cdot Pr$ did not exceed 600. During the experiments the pressure was raised to the maximum and then reduced in order to check the reproducibility of results. Data were worked up with correction for end losses and for the increase in the annulus width due to a rise in pressure. Radiation corrections were found to be negligible. The argon was 99.6% pure. The experi-

TABLE 51. Experimental Studies of Thermal Conductivity of the Four Monatomic Gases and Liquids under Pressure

Year	Author	Temperature, K	Pressure, atm	Method	Reference
			Neon		
1964	Sengers et al.	298—348	1—2676	Parallel plate	[6.1]
1971	Tufeu	305—606	1—951	Coaxial cylinders	[4.78]
1971	Golubev and Shpagina	78—321	1—485	Regular mode	[4.77]
1971	Vasil'kovskaya and Golubev	295—566	1—402	Same as above	[4.76]
			Argon		
1951	Lenoir and Comings	314	1—194	Coaxial cylinders	[6.2]
1952	Uhlir	87—194	1—96	Same as above	[6.3]
1953	Lenoir et al.	326	1—217	»	[6.4]
1954	Keyes	363—623	0—19,5	»	[6.5]
1955	Keyes	87—273	0.3—11	»	[6.6]
1956	Michels et al.	273—348	1—1993	Parallel plate	[6.7]
1958	Ziebland and Burton	93—196	1—120	Coaxial cylinders	[6.8]
1960	Tsederberg et al.	203—673	1—484	Heated wire	[6.9]
1963	Michels et al.	273—348	1—2424	Parallel plate	[6.10]
1963	Ikenberry and Rice	91—235	1—532	Coaxial cylinders	[6.11]
1965	Keyes and Vines	413—620	6—146	Same as above	[6.12]
1966	Rosenbaum et al.	279—322	26—704	»	[6.13]
1967	Freud and Rothberg	298	1—3000	»	[6.14]
1968	Bailey and Kellner	89—299	1—484	»	[4.60]
1969	Golubev and Shpagina	97—366	1—485	Regular mode	[4.81]
1969	Le Neindre	298—977	1—987	Coaxial cylinders	[4.79]
1970	Amirkhanov et al.	282—624	97—968	Parallel plate	[6.15]
1972	Amirkhanov et al.	113—253	97—968	Same as above	[6.16]
1971	Tufeu	302—607	1—940	Coaxial cylinders	[4.78]
1971	Tarzimanov and Arslanov	298—653	1—1936	Same as above	[4.80, 6.17]

TABLE 51. Experimental Studies of Thermal Conductivity of the Four Monatomic Gases and Liquids under Pressure (*Continued*)

Year	Author	Temperature, K	Pressure, atm	Method	Reference
		Krypton			
1955	Keyes	123—273	0.3—11	Coaxial cylinders	[6.6]
1963	Ikenberry and Rice	125—235	25—501	Same as above	[6.11]
1971	Tufeu	294—608	1—944		[4.78]
		Xenon			
1955	Keyes	146—273	0.2—11	Coaxial cylinders	[6.6]
1963	Ikenberry and Rice	170—235	25—500	Same as above	[6.11]
1971	Tufeu et al.	303—606	1—945		[4.78, 6.18]

mental data, whose error is estimated by the authors at 1.5–3%, are given in tables.

In the equipment used by Uhlir [6.3] the outer diameter of the inner cylinder was 31.5 mm, the width of the annulus between the cylinders was 0.25 mm and the annulus was 141.7 mm long. The cylinders were made of copper and had helium-filled cavities which served for measuring the temperature difference. A feature of interest in this unit is the use of a helium gas thermometer. As noted by Uhlir, the thermometer is stable to vibrations, sensitive and permits the recording of small temperature differences which allows reducing convective fluxes to a minimum. Due to the small temperature differences, there was no need to use exact compensating devices or to significantly correct the results.

The gaseous and liquid argon contained up to 0.4% of nitrogen as an impurity. The experimental data (66 points) are listed in the table and in the graph in coordinates (λ, T) at nine isobars. Uhlir [6.3] does not exclude the possibility of convection near the critical point, although the experiments were performed under conditions for eliminating it. The error of experimental data is estimated at 2.5%. For the critical region this value appears to us to be too low since the error of temperature measurement could be as high as 0.5 K.

Keyes [6.5] presents experimental data for gaseous argon (10 points) in a table and in a graph. Subsequently, Keyes [6.6] used the same method for determining the thermal conductivity of argon, krypton, and xenon in gaseous and liquid states. Sixteen points were obtained for argon, of which five pertain to the liquid phase. These data are given in a table, but without mentioning their error.

The results of Michels et al. [6.7] are given in a table and graphs. Isotherms in coordinates (λ, p) have the ordinary configuration. But, in coordinates (λ, ρ), instead of being arranged in equidistant fashion in the direction of the ordinate axis, they intersect at $\rho \simeq 200$ and 520 Amagat. Such an anomalous behavior of isotherms was not observed on the basis of analogous data for nitrogen obtained previously with the same equipment.

Michels et al. [6.10], of the same laboratory, investigated for the second time the thermal conductivity of argon. The facility was modified and this eliminated the thermoelectric side effect which caused the appearance of convection and a steep rise in the thermal conductivity at high densities. The space between the plates was 1.28 mm. The temperature difference did not exceed 0.25 to 0.48 K. The measurements were performed with "spectroscopically pure" argon. New experimental data (110 points) are shown along 273.15, 298.15, 323.15 and 348.15 K. The authors estimated the error at 1%. Unlike the results of Michels et al. [6.7] no anomalies were observed in the shape of isotherms in coordinates (λ, ρ). Michels et al. [6.10] calculated the value of $\lambda/\eta C_v$ and found that it increases insignificantly with the temperature. On the isotherms it first increases by 6–12% when the density rises to $\rho \cong \rho_{cr}$, and then decreases.

Ziebland and Burton [6.8] investigated 99.95% pure liquid and gaseous argon. The experimental facility is described in detail in their previous paper [6.19]. The inner cylinder, made of silver, was 100 mm long and had an outer diameter of 32.84 mm. The outer cylinder, made of copper, had an inner diameter of 33.33 mm. Forty three mm long guard heaters, having the same diameter as the inner cylinder, were placed at the ends of the latter. The constant of the facility was determined at $t = 20°C$ and for other temperatures its value was recalculated using familiar coefficients of thermal expansion of the materials of which the cylinders were made. The error in determining the constant did not exceed 0.5%. The temperature difference between the cylinders was measured by copper-constantan thermocouples. It ranged from 0.54 to 6.7 K. In investigating the thermal conductivity of argon, the facility was equipped with an automatic temperature control system which permitted maintaining the temperature of the measuring cell constant over an extended time within several hundredths of a degree.

The experimental data of Ziebland and Burton [6.8] are presented in a table and graph (120 data points for 11 isobars). The thermal conductivity was calculated by averaging the results of several measurements which differ from the mean by not more than 1%. They analyzed the reasons for the onset of convection. They checked for the absence of the latter by measuring the thermal conductivity at unchanged temperature and pressure and different Δt. The deviations of data here did not exceed 1%. As a result of the use of a small temperature difference, convection was virtually absent even at the critical isotherm. Reliable measurements could not be performed at a pressure close to p_{cr} at temperatures from ± 3 K to the critical. Ziebland and Burton assumed that convection in their facility could appear at temperatures exceeding the critical by 5–15 K and for reduced pressures $\pi = p/p_{cr}$ between 1 and 2. In fact, at

$\pi = 1.5$ some points in the temperature range of 158–165 K are on the high side which was taken into account by us in working up the experimental data. However, this may not be due as much to convection as to the presence of a maximum in the thermal conductivity at the transcritical isotherms, discovered during subsequent studies. As a whole, the results of Ziebland and Burton agree with those of Uhlir and Keyes (within 4%).

Tsederberg et al. [6.9] used two glass tubes with d_{in} of 1.07 and 0.52 mm when measuring the thermal conductivity of argon by the heated-wire technique. The first was used to obtain data at temperatures above 273 K and the second at 203 to 305 K. The diameter of the platinum wire was 0.1 mm in both cases. They performed control measurements on the thermal conductivity of air. The differences between their results and the available experimental data did not exceed 1% for the first and 1.9% for the second tube. Since the data obtained with the second tube were systematically on the high side, and the measurements with it were performed by the relative method, the value of the tube constant was reduced by 1.9% relative to that calculated from geometric dimensions.

The experimental data are represented in [6.9] in a graph in coordinates (λ, t) at six isobars. The error is estimated by the authors not to exceed 2.5%. In logarithmic coordinates the results are satisfactorily correlated by a straight line. This allowed them to obtain the following expression for the thermal conductivity of argon

$$\lambda_{p,T} = \lambda_T + 10^{-5} \cdot 0.646\rho^{1.26} \tag{6.1}$$

where the thermal conductivity is in kcal/(m · hr · K) and the density is in kg/m³. Equation (6.1) was used in calculating the thermal conductivity for temperatures between 183 and 873 K and pressures between 1 and 500 kg/cm².

Ikenberry and Rice [6.11] investigated the thermal conductivity of liquid and gaseous argon and krypton and liquid xenon. Sixty-eight points were obtained for argon, of which three along the 91 K isotherm at a pressure above 300 bar pertain to the solid phase. It was shown by Ikenberry and Rice that the method of coaxial cylinders allows the use of a small-diameter measuring cell. It can be easily placed in a casing capable of withstanding high pressures and can be loaded together with the casing into a Dewar. Their experimental unit was smaller than the previously described equipment employing the same method. The outer diameter of the inner cylinder was 9.83 mm. Its length was 76.2 mm and the annulus between the cylinders was 0.38 mm wide. The cylinders were made of copper. A hole with a nichrome elastric heater was drilled along the axis of the inner cylinder. The temperature of the outer cylinder and the difference between the temperatures of the cylinders were measured by copper-constantan thermocouples with errors of 0.1 and 0.01 K, respectively. The temperature difference was approximately 0.4 K which, together with the small width of the annulus, eliminated convection with the exception of the critical region.

The argon used was high purity (99.99%). The experimental data are represented in a table and graph for eight isotherms at maximum pressure of ~500 bar along the majority of isotherms. Ikenberry and Rice noted that their data are in satisfactory agreement with the values of λ obtained by Uhlir [6.3] and Ziebland and Burton [6.8], with the exception of data along the 96 atm isobar at temperatures somewhat above the critical. Unfortunately, they do not present the numerical results of comparison. They estimate the maximum error of their experimental data at ±2%.

The previously performed [3.8] workup of experimental data of Ikenberry and Rice showed that it is possible to draw through them smooth isotherms and isobars on which the majority of points fit within 1–2%. The differences between the results in [6.8] and [6.11] lie basically within ±2%. However, the discrepancies between individual points at temperatures above 135 K may be as high as 6%. The data of Uhlir [6.3] for T of 100 to 130 K agree with those of Ziebland and Burton within 2%. However, at $T = 90$ K the values in [6.3] are up to 4% lower, and at the transcritical isotherms of 140–150 K they are higher by up to 5%. The values of λ obtained by Keyes [6.6] are lower than those in [6.8] by 2–4%. The results of [6.11] frequently are intermediate between the most differing values at the temperatures and pressures where these differences occur. For this reason Vasserman and Rabinovich [3.8] used the experimental results of Ikenberry and Rice as reference values of thermal conductivity of liquid argon, and at pressures to 120 atm also data of Ziebland and Burton.

Keyes and Vines [6.12] represented the values of thermal conductivity of gaseous argon in 21 points along four isotherms in tabulated form and described with satisfactory accuracy their experimental results and data obtained previously by Keyes [6.6] for gaseous argon, using the equation

$$\Delta\lambda = A(p/T) + B(p/T)^2 \tag{6.2}$$

This expression was also checked against the results of Ziebland and Burton. However, even when using new values of A and B, the equation does not satisfactorily represent the pressure dependence of $\Delta\lambda$ at $T < 273$ K. This is seen from the comparison of experimental data in [6.8] and computed results performed by Keyes and Vines [6.12].

Rosenbaum et al. [6.13] investigated the thermal conductivity of argon at 279.35, 293.85, 298.15 and 321.95 K. They described their data (48 points) by the equation

$$\lambda = a_0 + a_1 p + a_2 p^2 + a_3 p^3 + a_4 p^4 \tag{6.3}$$

which represents the pressure dependence of thermal conductivity along isotherms with the experimental error (±1%).

Rosenbaum et al. [6.13] worked up their experimental data and the results of the previously cited studies in coordinates $\Delta\lambda$ vs. ω (excess thermal conductivity vs. reduced density). They established that, within the limits of the overall error of measurements, it is possible to draw a correlating curve over a wide range of densities (from zero to ρ_{tr}). Only close to the critical point does one observe, according to data in [6.3, 6.8], an anomalous increase in $\Delta\lambda$ as compared with the general shape of the correlating curve described by the expression

$$\Delta\lambda = 3.5 \cdot 10^{-5} [\exp(0.846\omega) - 1] \tag{6.4}$$

where $\omega = \rho/\rho_{cr}$ and λ is in cal/(cm · s · K). This means that Rosenbaum et al. and the present authors [1.104] have almost simultaneously proven the possibility of correlating data on the thermal conductivity of liquid and gaseous argon over a wide range of variables using the single curve of $\Delta\lambda = f(\rho)$.

Freud and Rothberg [6.14] used the relative version of the method of coaxial cylinders. Their data are in satisfactory agreement with the results of Michels et al. [6.10].

Bailey and Kellner [4.60] investigated the thermal conductivity of gaseous and liquid argon, paying a great deal of attention to the design of the instrument and accuracy of measurements. The error of their data increased with reduction in temperature from 1.5 to 2% and may be as high as 5% due to smallness of the heat fluxes. The errors were estimated on the basis of careful analysis of the different corrections made in determining the thermal conductivity.

To check for the absence of convective heat transfer, Bailey and Kellner performed a number of experiments at unchanged pressure and temperature, and different heat fluxes with the coaxial cylinders in the vertical and horizontal positions. Even in the critical region the effect of convection did not exceed the measuring error. They also performed several experiments in the region of gas-liquid and liquid-crystal phase transitions. The experimental data on thermal conductivity served for finding the temperatures of phase transitions. They are in satisfactory agreement with published data.

The experimental data of Bailey and Kellner are represented in graphs in coordinates (λ, T) and (λ, P) and in a table (418 points for 19 isobars). Each value of λ is an average of seven individual measurements performed during 30 min. As a whole, these data agree with those of preceding investigators with a deviation within the overall error of measurements in independent experiments. The exception is the results of Tsederberg et al. [6.9] which are up to 15% lower along the 500 kg/cm^2 isobar at $T < 250$ K. The experimental data in [4.60] were correlated by means of the curve of $\Delta\lambda = f(\rho)$, which averages the input data within $\pm 3\%$ and is described by an equation such as (6.4).

Bailey and Kellner paid particular attention to studies of thermal conductivity in the critical region [4.60, 6.20] and were the first to establish that the deviation of values of thermal conductivity of argon in this region from the generalized curve of $\Delta\lambda = f(\rho)$ observed also previously from data in [6.3, 6.8] is not due to convection. But, it is in fact a significant increase in thermal conductivity as the critical point is approached. According to data in

[4.60], at pressures of 49.8 to 53.5 kg/cm^2 and temperature between 150 and 153 K (close to critical density) the thermal conductivity has steep peaks which exceed two- to threefold the values obtained using the generalizing curve of $\Delta\lambda = f(\rho)$. With increasing pressure, the maxima are shifted to higher temperatures and $\Delta\lambda$ decreases. However, even at pressures of 100 kg/cm^2 the values of λ at temperatures of 169 to 181 K exceed by 10% those determined using the curve.

The anomalous rise in thermal conductivity in the critical region was detected for the first time by Guildner [6.21] and Michels et al. [6.22] in investigating the thermal conductivity of carbon dioxide by the method of parallel plates. Guildner took no steps to prevent convection, and the data of both studies near the critical point differ significantly from one another. For this reason, the fact that Bailey and Kellner [4.60, 6.20] confirmed the existence of extremes in thermal conductivity in the critical region by an independent method for another substance is of utmost importance.

Golubev and Shpagina [4.81] investigated the thermal conductivity of gaseous and liquid argon. Golubev's bicalorimeter, which is described in his article [6.23] is cylindrical. The inner and outer cylinders were made of copper and their surfaces in contact with the substance under study were polished and nickel-coated. The 12 mm inner cylinder had a 140 mm long middle (measuring) part, and 50 mm long guard heaters on each end. The layer of the tested substance was 0.3 mm thick. A Nichrome electric heater was placed along the axis of the inner cylinder. The temperature of the outer cylinder was measured by a platinum resistance thermometer within 0.1 K and the difference of temperatures of the cylinders by a three junction copper-constantan differential thermocouple. The pressure was measured by class 0.2 calibration-grade pressure gauges. To check for the absence of convection, experiments were performed at the same parameters of state and differing temperature differences between the cylinder surfaces.

The argon investigated by Golubev and Shpagina contained 0.01% nitrogen and 0.003% oxygen as impurities. The experimental data (423 points along 12 isotherms) are represented in a table and on graphs in coordinates (λ, p) and $(\Delta\lambda, \rho)$. As a rule, two or three values of λ, usually differing from one another by 0.5–1.0%, and less frequently by 2–4%, are given along each isotherm at a given pressure. The error of data, according to Golubev and Shpagina, is $\pm 1.5\%$ and in the critical region it may be as high as $\pm 2.5\%$. The graphs exhibit a steep rise in the thermal conductivity, at temperatures of 143.25, 153.35 and 157.65 K and close to critical pressure (40 to 60 kg/cm^2). This is in qualitative agreement with results obtained by Bailey and Kellner [4.60]. Golubev and Shpagina note that the overwhelming majority of their data agree satisfactorily (within $\pm 5\%$) with data of preceding studies. After working up their experimental results in coordinates $(\Delta\lambda, \rho)$ they obtained a generalized curve about which points cluster the majority of data, with the exception of the critical region, where they deviate significantly upward from this curve. Given the difficulty of performing measurements in the critical region, they obtained a limited body of data in it, and it is difficult to compare them with the results of Bailey and Kellner.

The equipment used in the experiments by Le Neindre [4.79] for determining the thermal conductivity of argon over a wide temperature and pressure range was constructed by Johannin for measuring the thermal conductivity of nitrogen [6.24]. The coaxial cylinders were made of silver and the annulus between them was 0.2 mm wide. The measurements were performed at constant temperatures and different pressures, since equilibrium can be reached more rapidly under these conditions. Le Neindre worked up his result with consideration of numerous corrections and analyzed the possible random and systematic errors which permitted him to estimate the error of experimental data at $\pm 2.5\%$.

The argon used by him was 99.99% pure. The experimental data (154 points along 14 isotherms) are given in a table and in graphs in coordinates (λ, ρ) and ($\Delta\lambda$, ρ). Comparison of these data with the results of Michels et al. [6.10] showed that the difference between them does not exceed 1.5% at $T = 298$ K and 3% at $T = 348$ K, i.e., it lies within the limits of the sum of errors of measurement.

The isotherms constructed from experimental data of [4.79] in coordinates (λ, ρ) lie equidistant in the direction of the ordinate axis which suggested to Le Neindre that the excess thermal conductivity is a unique function of density. It should, however, be noted that the graph of $\Delta\lambda = f(\rho)$ shows data only to 775 K, and the points for the 674 and 775 K isotherms deviate somewhat upward from the correlating curve, drawn over data for lower temperatures. In addition, at a fixed maximum pressure the highest values of density along isotherms decreases significantly with rising temperature (from 972 kg/m^3 at $T = 348$ K and $p = 1000$ bar to 467 kg/m^3 at 775 K and 387 kg/m^3 at 977 K and the same pressure). Therefore, it is impossible to clearly establish on the basis only of Le Neindre's data whether the isotherms in coordinates ($\Delta\lambda$, ρ) at supercritical density agree over a wide temperature range.

Le Neindre made an attempt to additionally analyze the question of the steep rise in the thermal conductivity of substances near the critical point using a special instrument specially designed by him for this purpose. Unfortunately, he was not able to solve the problem. However, the results of several preliminary experiments with carbon dioxide point to a steep rise in λ in the critical region.

Subsequently in the same laboratory Tufeu [4.78] investigated the thermal conductivity of the four gases analyzed here (and also of helium and helium-argon mixtures). The inner cylinder of the measuring cell had an outer diameter of 19.5 mm and was 120 mm long. The thickness of the gas layer was 0.25 mm. Forty mm long guard heaters were placed at each of the inner cylinders. The temperature of the cylinders was measured by platinum—platinum-rhodium thermocouples. The experiments were performed close to constant temperatures. According to Tufeu's estimates, the error of temperature measurement could be 0.3 K, of pressure 0.25 bar at $p < 150$ bar and 2 bar at higher pressure. The error in determining the thermal conductivity in the majority of data points was 1.0–1.5%, but increased with a reduction in pressure, becoming as high as 2.5% at $p \simeq 10$ bar.

The thermal conductivity of argon was measured by Tufeu [4.78] along five isotherms: 303.15, 392.15, 456.15, 540.15 and 606.65 K, with the values of temperature in the test points differing somewhat from the above. But, the deviations basically did not exceed 0.7 K, and were as high as 1.6 K only at $T = 303.15$ K. The experimental data (81 points) are given in tables and in a graph. Tufeu also presents a table of smoothed values of thermal conductivity for round values pressure. Comparison of experimental data of Tufeu and Le Neindre, performed by the former [4.78], showed that the deviations lie within the limits of experimental error. They do not exceed 2% at $p = 900$ bar, with the data of Le Neindre [4.79] basically higher than the results of Tufue [4.78].

Amirkhanov et al. [6.15] investigated the thermal conductivity of argon by the method of parallel plates. The thickness of the layer of the test gas was 0.49 mm and remained virtually unchanged with changes in temperature since the coefficient of linear expansion of the porcelain cylinders used for obtaining the slot between the heater and cooler is quite small. The temperatures of the heater and cooler were measured by copper-constantan thermocouples, the pressure of the gas by deadweight pressure gauge. The correction for radiant heat transfer was insignificant (from 1.3% at 282 K to 2.8% at 623 K). All the experiments were performed at two values of temperature difference ΔT in the layer of gas. The data that were obtained were independent of ΔT. Absence of convective heat transfer is confirmed in the first place by the fact that the product Gr · Pr for all the test points at temperatures from 282 to 473 K and pressures between 100 and 1000 kg/cm^2 was less than 1000. The error in measuring the thermal conductivity with referencing errors are estimated by Amirkhanov et al. at 1.5%.

Amirkhanov et al. [6.15] investigated pure brand A argon, containing 0.01% nitrogen and 0.003% oxygen. They obtained 220 experimental values of λ along 10 isobars and 11 isotherms. The value of λ was measured twice at each pressure. The temperature at these points differs from the round value by not more than ± 1 K. The experimental data are given in a table and graphs in coordinates (λ, t) and $(\Delta\lambda, \rho)$. The curve of $\Delta\lambda = f(\rho)$ is represented by a third-power polynomial. For the majority of points the deviations of analytic values of λ from the experimental do not exceed 1.5%. This equation served to calculate the thermal conductivity for temperatures from 283 to 623°C and pressures of 100 to 1000 kg/cm^2. The results are given in a table.

Subsequently Amirkhanov et al. [6.16] measured the thermal conductivity of gaseous and liquid argon by the same method. In measurements at low temperatures the thickness of the gas layer was reduced to 0.27 mm. The temperature and temperature difference were measured by copper-constantan thermocouples. The pressure was determined by an MP-2500 deadweight pressure gauge. As noted by Amirkhanov et al., they checked for absence of convection by performing experiments at two values of ΔT, differing approximately two-fold from one another. The error of experimental values of λ is estimated in [6.16] at 2.4% for the gas and 3% for the liquid.

The argon investigated by Amirkhanov et al. [6.16] contained 0.01% nitrogen and 0.003% oxygen. They obtained 140 experimental values of λ along 10 isobars and 14 isotherms (values of temperature at points along the isotherms differed from the constant by ± 0.6 K). The data are given in a table and graphs in coordinates (λ, p) and ($\Delta\lambda$, ρ). The density dependence of the excess thermal conductivity is described analytically by a fifth-power polynomial. The equation was constructed based on data from the above study, and also on the basis of results of others [6.2, 6.4, 6.8, 6.9]. The deviations of experimental data of λ in [6.16] from the computed do not exceed 3%. The equation served for calculating the thermal conductivity of argon for the experimentally investigated range of parameters.

Amirkhanov et al. noted that the results of Michels et al. [6.10] lie above the generalized curve, and hence, did not include them in compiling the equation for $\Delta\lambda = f(\rho)$. In our opinion, such a deviation need not be caused by error of data in [6.10] which were obtained by taking pains to prevent convection. But, it points to a separation of isotherms for excess thermal conductivity at supercritical densities.

In the studies by Tarzimanov and Arslanov [4.80, 6.17] the inner cylinder had an outer diameter of 11.36 mm and was 90 mm long. Twenty-nine mm long guard heaters were located at its ends. The thickness of the gas layer under study was 0.33 mm. The cylinders were made of copper and were coated by a thin layer of silver to reduce the radiant heat transfer. The temperature of the cylinders and the temperature difference was measured by a differential three-junction platinum—platinum rhodium thermocouple. The temperature difference ranged between 1 and 3 K. The constant temperature bath was maintained at constant temperature within 0.1 K and measured by a platinum resistance thermometer. The pressure of the gas was measured by a class 0.02 MP-2500 deadweight pressure gauge.

Tarzimanov and Arslanov [4.80, 6.17] used pure, brand A argon. The experimental values (48 points) are shown along eight isobars and can be clustered about eight isotherms (with an error of ± 1.1 K). According to the authors, the error of their data does not exceed 1.5%. They note that their experimental data are in satisfactory agreement with the results of Le Neindre [4.79] and Michels et al. [6.10], and a part of data of Amirkhanov et al. [6.15]. But, they deviate from the data in the latter at temperatures between 298 and 348 K at 1000 bar by up to 10%. Tarzimanov and Arslanov [4.80] correlated the experimental data using a fourth-power polynomial to represent the behavior of $\Delta\lambda = f(\rho)$. Unfortunately, both in this study, and also in the abstract of Tarzimanov's thesis [6.25] no mention was made of the manner in which the density of argon was determined at temperatures above 423 K and pressure above 1000 bar, i.e., over the range of parameters for which experimental pvT data are not available.

The above shows that the numerical experimental results on the thermal conductivity of argon were obtained over a rather wide range of temperatures and pressures and allow compilation of experimentally validated tables of thermal conductivity λ for virtually the entire range of variables analyzed in the present book.

The thermal conductivity of neon at high pressures was investigated for the first time by Sengers et al. [6.1] who obtained data along the 298.15, 323.15 and 348.15 K isotherms at pressures from 1 to 2498, 2620 and 2676 atm, respectively. The experiments were performed with equipment used previously for investigating the thermal conductivity of carbon dioxide in the critical region [6.22]. The thickness of the gas layer was 1.28 mm, the temperature difference across the layer was 0.3–0.4 K.

The neon used was at least 99.99% pure. The experimental data (106 points) with an estimated error of 1% are given in a table and graphs in coordinates (λ, p) and (λ, ρ). The deviations of temperatures in individual points from round values along isotherms lies within the limits of -0.08 to 0.26 K. Sengers et al. [6.1] compiled equations for isotherms in the form of third power polynomials of density and calculated the thermal conductivity at round values of density (in Amagat units). Sengers et al. also calculated the ratio $\lambda/\eta c_v$ and found that it increases by approximately 10% when the density is raised from 1 to 600 Amagat and has a shallow maximum at high values of the latter.

The experimental data obtained by Tufeu [4.78] on the thermal conductivity of neon are given in tables along six isotherms: 305.15, 347.15, 407.15, 483.15, 559.15 and 606.15 K. A total of 115 values of the thermal conductivity coefficient are listed, with deviations of temperature in experimental points from those given above not usually exceeding 1.2 K, with the exception at $T = 559.15$ K, where the deviation is 1.9 K. The isotherms in coordinates (λ, ρ) lie on the graph of experimental data in [4.78] at equal distances in the direction of the ordinate axis. Tufeu also presents a table of smoothed values of λ for neon at round values of pressure (1 to 950 bar) at the five aforementioned temperatures. He notes that his data agree with those of Sengers et al. within 1%.

Golubev and Shpagina [4.77] performed their measurements with equipment described in their previous paper [6.23]. The gas they used contained 0.1% helium and 0.01% nitrogen. They obtained 361 values of the thermal conductivity along 10 isotherms. Most frequently two–three values of λ are given on each isotherm. These, as a rule, agree within 1%, but in certain cases the discrepancy may be as high as 2–3%. The experimental data are given in a table and graphs in coordinates (λ, p) and ($\Delta\lambda$, ρ). The error is estimated by them at $\pm 1.5\%$, apparently on the basis of the scatter of data points. In coordinates ($\Delta\lambda$, ρ) the data points stratify and separate by isotherms. As a result of the graphical workup of their experimental data, they obtained smoothed values of thermal conductivity of neon for T between 78.15 and 323.15 K and p between 1 and 500 atm.

Vasil'kovskaya and Golubev [4.76] measured the thermal conductivity of neon by a cylindrical bicalorimeter, in which the 17 mm OD inner cylinder had a 195 mm long measuring part. The thickness of the test gas layer was 0.4. The purity of the gas was the same as in [4.77]. They present in a table and graph 67 values of λ along seven isotherms in coordinates (λ, *p*). In the majority of cases, two values of λ are given at a given pressure on the isotherm. The deviations between these values of λ range from 1 to 3%. In the opinion of the authors, the error of experimental data is ±1.5%. Within this error margin, the results in [4.76] agree with data of others at atmospheric pressure. Golubev et al. did not compare their results with those of Sengers et al. [6.1], although a part of the latter data was obtained for the range of parameters investigated in [4.76, 4.77].

Of the 15 data points for λ obtained by Keyes [6.6] for krypton and given in a table, three are for the liquid and the other is for the gas.

The special features of the equipment used by Ikenberry and Rice [6.11] were noted above in analyzing data for argon. They present for krypton 37 points along five isotherms, with eight points along the 235.5 K isotherm pertaining to the gaseous phase, and two, at 125.5 K and pressures above 400 bar, to the solid phase. The experimental data, having an estimated error of 2%, are given in a table and a graph in coordinates (λ, *p*).

Tufeu [4.78] also obtained data on the thermal conductivity of gaseous krypton at 294.15, 346.15, 405.75, 482.15, 560.15 and 608.15 K. It is seen from the tables, which list 121 values of λ, that the deviations of temperature in the data points along isotherms from round values (in °C) do not exceed ±1.15 K. The experimental data are also given in a graph in coordinates (λ, ρ), with the 294.15 and 346.15 K isotherms at ω between 0.5 and 1.5 not equidistant from one another and the 405.75 K isotherm, but closer to it. This apparently is a result of the steep rise in thermal conductivity in the critical region. Working up his experimental data, Tufeu obtained smoothed values of thermal conductivity of krypton along the above six isotherms.

Keyes was the first [6.6] to measure the thermal conductivity of xenon, as he was for krypton, and obtained nine values for gaseous and three for liquid xenon.

Subsequently Ikenberry and Rice [6.11] determined the thermal conductivity of liquid xenon along four isotherms and obtained 28 data points, of which four along the 170.2 K isotherm at pressures above 300 bar pertain to the solid phase. The experimental data are given in a table and graph. Their error, as for argon and krypton, is estimated in [6.11] at ≤2%.

The last study known to us on the thermal conductivity of xenon is that by Tufeu [4.78], who obtained results along the 303.15, 346.15, 406.15, 482.15, 560.15 and 606.15 K isotherms. The experimental data (102 points) are given in tables. The deviations of temperature from the above values lie basically between −0.6 and 0.3 K, becoming 1.45 K only along the 303.15 K isotherm. These data are also given in the paper by Tufeu et al. [6.18]. The thermal conductivity along the last mentioned isotherm increases significantly at ω between 0.5 and 1.5, which is due to the closeness of this isotherm to the critical region.

On the graph constructed by Tufeu in coordinates $(\Delta\lambda, \rho)$ this 303.15 K isotherm twice intersects the remaining isotherms. The thermal conductivity over this range of density increases less along the 346.15 K isotherm, which intersects only the 406.15 K isotherm. As for other substances, Tufeu presents also smoothed values of thermal conductivity of xenon at the six temperatures over the above pressure range.

Although the experimental data are given for neon in [4.76–4.78] and for krypton and xenon in [4.78, 6.11] over wide ranges of parameters, they do not extend over the entire temperature and pressure range for which our tables of thermophysical properties are compiled. Up to now, the thermal conductivity of these gases has not been investigated at the following: (1) temperatures above 608 K and elevated pressures; (2) of gaseous neon at temperatures below 77 K (also at 77 to 298 K at pressures above 490 bar); (3) of liquid krypton and xenon above 500 bar and of liquid neon. Hence, it becomes very important to simultaneously work up data on the thermal conductivity of the four substances under study and to compile an equation which would provide an optimum description of the entire accumulated body of experimental data which would be suitable for calculating the thermal conductivity of these substances at pressures to 1000 bar and temperatures to 1300 K.

6.2. EQUATIONS FOR CALCULATING THE THERMAL CONDUCTIVITY OF SUBSTANCES IN THE GASEOUS AND LIQUID STATES

In accordance with the Bogolyubov [2.6] theory, the equation for calculating the thermal conductivity of a real gas can be represented in a form similar to the virial form of the equation of state

$$\lambda_{p,T}/\lambda_T = 1 + B_\lambda \rho + C_\lambda \rho^2 + D_\lambda \rho^3 + \ldots \tag{6.5}$$

where B_λ, C_λ, D_λ, . . . are functions of the temperature. Frequently one uses another form of this equation

$$\lambda_{p,T} = \lambda_T + \lambda_1 \rho + \lambda_2 \rho^2 + \lambda_3 \rho^3 + \ldots \tag{6.6}$$

where the coefficients are related to the corresponding coefficients of Eq. (6.5) by obvious expressions.

Unfortunately, in addition to λ_T only the second virial coefficient λ_1 of Eq. (6.6) can be calculated analytically [5.41, 5.72, 6.26]. In principle, the coefficients of this equation could be determined from experimental data. However, the accuracy of the latter is currently such that only the second virial coefficient can be determined from them with relative reliability [5.46].

Many attempts were made to apply the Enskog theory to determine the thermal conductivity of compressed gases. Some investigators [1.104, 5.70, etc.] computed the thermal conductivity over a wide temperature and density range by calculating the product $b\rho x$ in the Enskog equations on the basis of experimental *pvT* data, or from reliable equations of state describing these data, rather than for the model of solid spheres. However, even under these conditions, differences between the computed values of the thermal conductivity and the experimental results were 10 to 15%.

It was shown in Chapter 5 that when the diameter of solid spheres is defined as a function of temperature and pressure and when this function is determined on the basis of experimental data on viscosity, it becomes possible to describe these data from the Enskog equation with satisfactory accuracy. However, even then the Enskog equation is not suitable to represent the anomalous rise in thermal conductivity in the critical region. If, however, we perform our analysis disregarding this region then, as will be shown below, it becomes possible to describe the data on the thermal conductivity of the four substances by simpler generalized equations within the limits of experimental error.

As noted in Chapter 5, it was shown in a number of new investigations using the kinetic theory of the dense gas [5.42, 5.44, 6.27–6.29] that the transport coefficients are a function not only of the density, but also of the logarithm of its values. For this reason the equation for calculating the thermal conductivity should be

$$\lambda_{p,T} = \lambda_T + \lambda_1\rho + \lambda_2'\rho^2 \ln \rho + \lambda_2''\rho^2 + \ldots \qquad (6.7)$$

The form of the subsequent terms in Eq. (6.7) has not been established, but it is assumed [5.45] that they contain increasing powers of ρ and $\ln \rho$.

Some investigators attempted to investigate the analytic viscosity and density dependence of thermal conductivity on the basis of experimental data in order to check the effectiveness of Eq. (6.7) and of the analogous equation for viscosity. Hanley et al. [5.46] worked up data on the thermal conductivity of neon and showed that the second virial coefficient λ_1 of the equation containing $\ln \rho$ coincides, within the limits of error of its determination, with the corresponding coefficient of the linear equation, compiled from data for the low-density region. This fact, in the opinion of Sengers et al., supports the claim of logarithmic density dependence of transport coefficients, although it does not prove it. Le Neindre [4.79] and Tufeu [4.78] also confirm Eq. (6.7) on the basis of their experimental data on the thermal conductivity of inert gases, since coefficients λ_T, λ_1 and λ_2 constructed as a function of ρ appear for this relationship more stable than for the ordinary polynomial.

On the other hand, Gracki et al. [6.30] claim on the basis of their experimental data on the viscosity of helium, argon, hydrogen, and nitrogen that it is more acceptable to describe data by means of polynomials. Kestin et al. [6.31] specially determined the viscosity of helium, argon, and nitrogen with a high accuracy (error of +0.2%) and investigated the density dependence of their results. They concluded that the contribution of the term $\eta_2'\rho^2 \ln \rho$ is rather small and can be neglected for practical purposes. In addition, it is seen from the table given by them that coefficient η_1 changes significantly upon conversion from the linear to the logarithmic relationship.

It is thus seen that contradictory results are obtained when checking the effectiveness of the logarithmic relationship for analytic description of experimental data on the transport coefficients. Nevertheless, it is seen from the above studies that the accuracy of approximation of experimental results from an equation containing a logarithmic term is virtually the same as when using a polynomial with the same number of terms. This is in agreement with the analytic conclusion of Dorfman and Cohen [5.43], to the effect that the dependence of the transport coefficient on the logarithm of the density is not clearly expressed.

Without analyzing the various empirical equations for calculating the thermal conductivities of gases and liquids discussed in [1.104, 3.8, 6.32], we note that in attempting to analytically describe experimental data over a wide temperature and pressure range it is necessary in the first place to allow for the behavior of the thermal conductivity as a function of the thermal properties. The values of the thermal conductivity in coordinates ($\lambda_{p,T}/\lambda_T$, ρ) and ($\Delta\lambda = \lambda_{p,T} - \lambda_T$, ρ) form a surface. However, analysis of a large number of experimental results shows that in the second case the stratification (separation) of isotherms of "excess" thermal conductivity is basically commensurable with the experimental error and can be detected only at supercritical density for a wide temperature range [6.33] and also in the critical region [4.51, 4.81, 6.20–6.22]. Given the insignificant effect of the temperature factor in coordinates ($\Delta\lambda$, ρ), Eq. (6.6) can be written in the simpler form

$$\lambda_{p,T} = \lambda_T + f(\rho) \tag{6.8}$$

where function $f(\rho)$ is frequently represented by a polynomial. Equation (6.8), suggested by Vargaftik [6.34], has proven itself in describing experimental data on the thermal conductivity of many substances, both in the gaseous and liquid phases, with the exception of the critical region [1.104, 3.8, 4.51, 4.79, 6.13, 6.35]. When using Eq. (6.8) the generalizing curve of $\Delta\lambda = f(\rho)$ is drawn between data points and their deviations from this curve at high densities is attributed to experimental error.

Lately, due to the accumulation of rather reliable data on the thermal conductivity of gases and liquids over a wide range of parameters, equations were suggested for calculating the thermal conductivity which reproduce the separation of isotherms in coordinates ($\Delta\lambda$, ρ). Kessel'man and Kamenetskiy [6.36] suggested an equation for the thermal conductivity of carbon dioxide which has a form analogous to that of an equation obtained by them previously for the viscosity of a real gas [5.56]:

$$\lambda_{p,T}/\lambda_T = 1 + \varphi_0(\omega) + \varphi_1(\omega)\frac{1}{\tau} + \varphi_2(\omega)\frac{1}{\tau^2} \tag{6.9}$$

where functions $\varphi_i(\omega)$ are fourth-power polynomials of the reduced density which do not contain free terms in order to satisfy limiting conditions.

Soon thereafter, Vasserman and Nedostup [6.37] utilized the possibility of factoring the isochoric sections of plane $\lambda = F(\omega, \tau)$ into the linear and curved parts and the similarity of configurations of these sections and of the thermal conductivity curve $\lambda(\tau)$ at atmospheric pressure and suggested an equation for the thermal conductivity gases and liquids in the form

$$\lambda(\omega, \tau) = \alpha(\omega) + \beta(\omega)\tau + \gamma(\omega)\lambda(\tau) \tag{6.10}$$

where $\alpha(\omega)$, $\beta(\omega)$ and $\gamma(\omega)$ are polynomials of the reduced density. To satisfy the limiting functions, functions $\alpha(\omega)$ and $\beta(\omega)$ do not contain free terms, whereas the free term in function $\gamma(\omega)$ is unity. Vasserman and Nedostup [6.37] obtained an equation such as (6.10) for calculating the thermal conductivity of gaseous and liquid nitrogen at T from 77.35 to 973.15 K, which contains 15 coefficients as a function of density. An analogous equation with the same number of coefficients was also compiled for argon [6.38].

Equations (6.9) and (6.10) are of a different form, however. After combining terms with the same powers of density they can be reduced to form (6.5). It can be concluded from analysis of these equations that there is little utility in using them for describing the weakly expressed temperature dependence of the excess thermal conductivity, since the significant number of coefficients in them did not perceptibly improve the accuracy of approximation as compared with Eq. (6.8). It should also be noted that the equations in [6.36–6.38], as Eq. (6.8), do not describe the steep rise in thermal conductivity in the critical region.

It is seen by analyzing experimental data on the thermal conductivity of gaseous and liquid argon, extending over wide temperature and pressure ranges in coordinates $(\Delta\lambda, \rho)$ (Fig. 79), that the overwhelming majority of experimental points cluster about the correlating curve with a scatter not exceeding the experimental error. The isotherms in the critical region exhibit significant stratification due to the steep rise in thermal conductivity as the critical point is approached [4.60, 4.81]. Data for high temperatures at supercritical density lie, as a rule, higher than at low temperatures, although the stratification of isotherms is somewhat masked by the scatter of data points. This nature of stratification of isotherms of excess thermal conductivity at high density was also noted by others [4.77, 5.70, 6.24, 6.33].

Given the above, and also the poorer accuracy of data on transport coefficients as compared with data on thermodynamic properties of substances, it is better to compile the empirical equation for the thermal conductivity of gas and liquid over a wide range of variables with the exception of the critical region on the basis of the rather simple and flexible form of Eq. (6.6). It can be assumed here that the coefficients of lower-order powers of density remain constant, as in Eq. (6.8), whereas the coefficients of higher-order powers of density are a function of temperature. This will allow the reproduction of the stratification of isotherms in coordinates $(\Delta\lambda, \rho)$ at high densities.

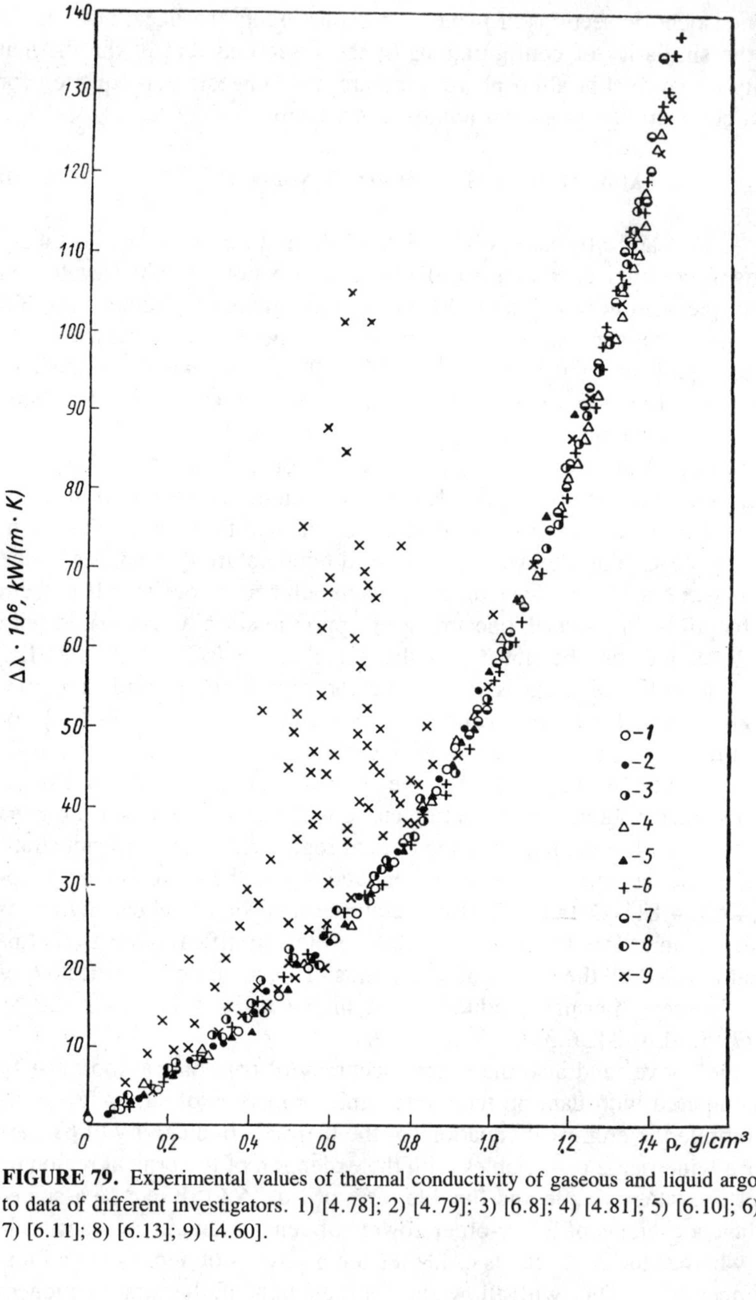

FIGURE 79. Experimental values of thermal conductivity of gaseous and liquid argon according to data of different investigators. 1) [4.78]; 2) [4.79]; 3) [6.8]; 4) [4.81]; 5) [6.10]; 6) [6.16]; 7) [6.11]; 8) [6.13]; 9) [4.60].

A check performed on the basis of experimental data on the thermal conductivity of argon showed that they are quite satisfactorily described by an equation such as (6.6), containing values of density in powers from 1 to 4, in which coefficients λ_1, λ_2, and λ_3 are constants, whereas λ_4 is a linear function of temperature. Most likely an equation in this form is valid also for other inert gases, but their thermal conductivity has not been investigated as well as that of argon. As a result it is difficult to reliably determine the temperature dependence of λ_4 for each of them separately. For this reason, we used a method which permits a generalized equation for calculating the thermal conductivity of the four substances under study in the gaseous and liquid states on the basis of experimental data for these substances.

Many investigators, see among others, [1.104, 2.142, 5.33, 6.32] note the possibility of correlating data on the transport coefficients of gases and liquids at atmospheric as well as above-atmospheric pressures using reduced coordinates. Since monatomic gases have a simple molecular structure, it can be assumed that the dimensionless values of their thermal conductivity in the corresponding states will be identical within the limits of experimental error. This is confirmed by results in [3.89, 4.78, 6.39] and of workup of data on the viscosity and thermal conductivity of inert gases at atmospheric pressure (see Chapter 4).

As noted in Sec. 2.5, satisfactory coincidence of thermodynamic planes of the substances under study in reduced coordinates is attained when using the Boyle temperature T_B and density ρ_0 on the ideal-gas curve at $T = 0$ as the reduction parameters. These parameters were also found to be effective in simultaneously working up data on the transport coefficients of inert gases at $p = 1$ atm—see Sec. 4.3. Hence T_B and ρ_0 were selected as reduction parameters for correlating data on the thermal conductivity of the four substances over a wide temperature and density range. The dimensionless thermal conductivity ratios should be based on value of λ_{T_B} obtained at atmospheric pressure and the Boyle temperature.

If the law of corresponding states is valid for the thermal conductivity, then function $\bar{\lambda} = F(\bar{\omega}, \Theta)$ in reduced coordinates ($\bar{\lambda}_{p,T} = \lambda_{p,T}/\lambda_{T_B}$; $\bar{\omega} = \rho/\rho_0$ and $\Theta = T/T_B$) will be the same for all the monatomic substances. Then their excess thermal conductivity in the above coordinates should also consist of the single function

$$\Delta\bar{\lambda} = \bar{\lambda}_{p,T} - \bar{\lambda}_T = F(\bar{\omega}, \Theta) - f(\Theta) = \Phi(\bar{\omega}, \Theta) \tag{6.11}$$

This assumption was confirmed in working up experimental data for the four substances in coordinates ($\Delta\bar{\lambda}$, $\bar{\omega}$) (Fig. 80). The values of density needed for working up the data were calculated from equations of state obtained in Chapters 2 and 3 and the reduction parameters T_B and ρ_0 were taken from Table 47. The values of λ_{T_B} for argon, krypton, and xenon were determined from Eq. (4.23) using Eq. (4.26) and the transformation multipliers (see Table 47). They were found to be $22.78 \cdot 10^{-3}$, $16.23 \cdot 10^{-3}$ and $12.90 \cdot 10^{-3}$ W/(m · K). Since Eq.

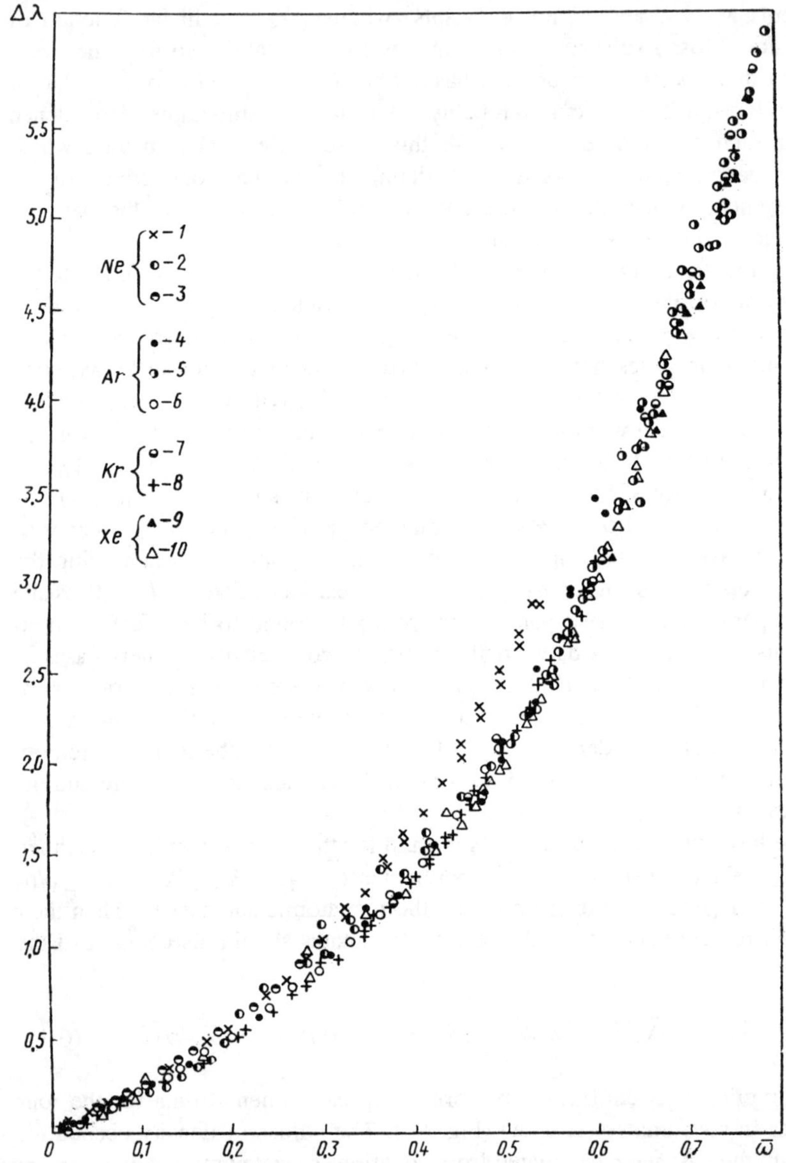

FIGURE 80. Excess thermal conductivity of monatomic gases as a function of reduced density from data of different investigators. 1) [6.1]; 2) [4.77]; 3, 6, 8, 10) [4.78]; 4) [6.10]; 5) [6.15]; 7, 9) [6.11].

(4.26) is not valid for neon at $T < 125$ K, the corresponding value of $\lambda_{T_B} = 25.75 \cdot 10^{-3}$ W/(m · K) was obtained by interpolating data given in Table VII (Part III). It is seen from Fig. 80 that the experimental points for argon, krypton, and xenon at $\overline{\omega} < 0.4$ ($\rho/\rho_{cr} < 1.4$) form a generalized curve, whereas at higher densities they become somewhat stratified (separated) along isotherms. Since a part of data on the thermal conductivity of neon pertains to high reduced temperatures, the corresponding values of $\Delta\overline{\lambda}$ are higher than for other gases with the same density already at $\overline{\omega} > 0.2$. The stratification of isotherms of excess thermal conductivity can be reproduced by assuming a linear temperature dependence of λ_4. Hence function $\Delta\overline{\lambda} = \Phi(\overline{\omega}, \Theta)$ is written analytically in the form

$$\Delta\overline{\lambda} = 1.549\overline{\omega} + 7.364\overline{\omega^2} - 15.54\overline{\omega^3} + (19.6 + 4.75\Theta)\overline{\omega^4} \tag{6.12}$$

To obtain a detailed check on the accuracy of Eq. (6.12) and to additionally evaluate the mutual compatibility of data of various authors, we calculated the thermal conductivity of the substances under study using the above equation and compared them with experimental results. In these calculations the thermal conductivity of the monatomic gases at atmospheric pressure was determined from Eqs. (4.23) and (4.26) using the transformation multipliers, and for neon at temperatures below 125 K from data given in Part III.

Figures 81 through 85 represent the deviations

$$\delta\lambda = \frac{\lambda_{\text{exptl}} - \lambda_{\text{calc}}}{\lambda_{\text{calc}}} \cdot 100\%$$

of the majority of experimental values along different isotherms as a function of pressure. In constructing the graphs we clustered isotherms with close values of reduced temperature. It is seen from the figures that Eq. (6.12) describes experimental data basically with an error of ±3%, and that in many cases the deviations do not exceed 1 to 2%. In a number of cases the deviations are commensurable with the difference between experimental data of different investigators for the given substance. At pressures between 1000 and 2500 bar, Eq. (6.12) describes the experimental data on the thermal conductivity of argon [6.10] and neon [6.1] with the same accuracy as at lower pressures (see Fig. 85).

The figures do not show the results of comparison with experimental data of Uhlir [6.3], Ziebland and Burton [6.8], and Bailey and Kellner [4.60], since these values were obtained along isobars. A detailed comparison performed by us showed that the deviations from data in these studies are mostly within ±4%, being significantly higher only in the critical region. In this region, the experimental data are 20 to 50% higher than those computed without correction for the anomalous rise in thermal conductivity.

As a rule, the experimental values of λ at reduced temperature $\Theta > 1.5$ are higher than the computed (see Figs. 84 and 85). However, the differences basi-

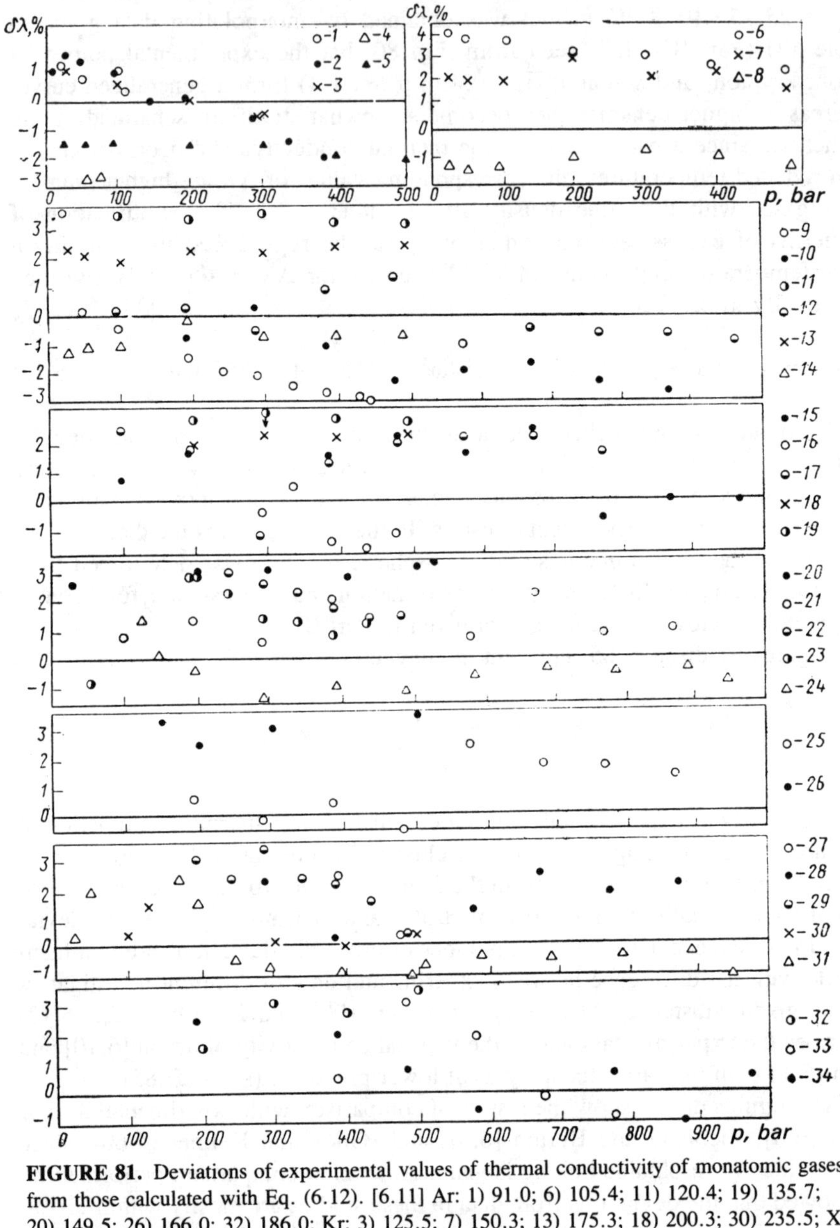

FIGURE 81. Deviations of experimental values of thermal conductivity of monatomic gases from those calculated with Eq. (6.12). [6.11] Ar: 1) 91.0; 6) 105.4; 11) 120.4; 19) 135.7; 20) 149.5; 26) 166.0; 32) 186.0; Kr: 3) 125.5; 7) 150.3; 13) 175.3; 18) 200.3; 30) 235.5; Xe: 4) 170.2; 5) 190.4; 8) 210.2; 14) 235.0 K; [4.81] Ar: 2) 97.45; 9) 111.85; 16) 143.25; 22) 153.35; 23) 157.65; 27) 171.25; 29) 180.85 K; [6.16] Ar: 10) 113.17; 12) 122.84; 15) 133.40; 17) 143.29; 21) 152.56; 25) 163.38; 28) 172.78; 33) 193.10; 34) 203.36 K; [4.78] Xe: 24) 303.45; 31) 346.15 K.

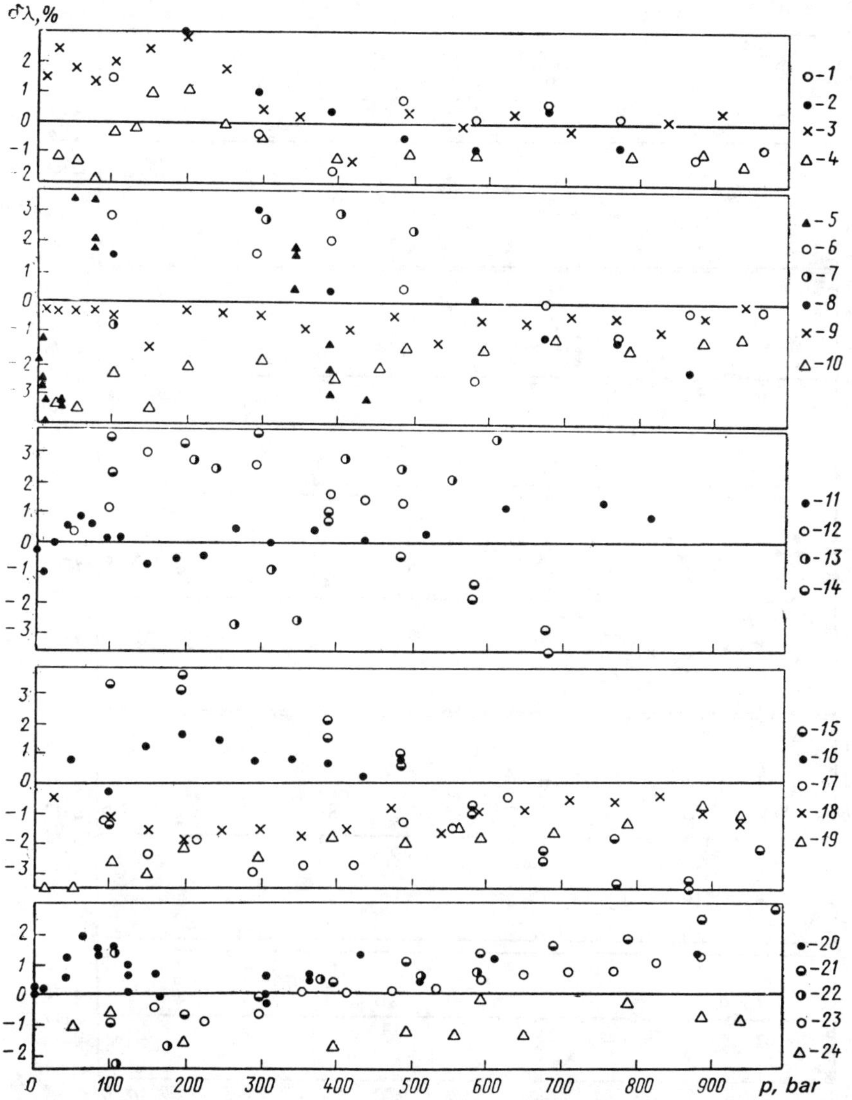

FIGURE 82. Deviations of experimental values of thermal conductivity of monatomic gases from those calculated with Eq. (6.12). [6.16] Ar: 1) 213.44; 2) 223.12; 6) 233.36; 8) 252.96 K; [4.78] Kr: 3) 294.15; 9) 346.15; 18) 405.75; Ar: 23) 303.15; Xe: 4) 406.15; 10) 482.15; 19) 560.15; 24) 605.15 K; [4.77] Ne: 5) 77.85 K; [6.11] Ar: 7) 234.6 K; [6.10] Ar: 11) 273.15; 20) 298.15 K; [4.81] Ar: 12) 273.65; 16) 293.65 K; [6.13] Ar: 13) 279.35; 17) 293.85; 22) 298.15 K; [6.15] Ar: 14) 282.0; 15) 293.05 K; [4.79] Ar: 21) 298.15 K.

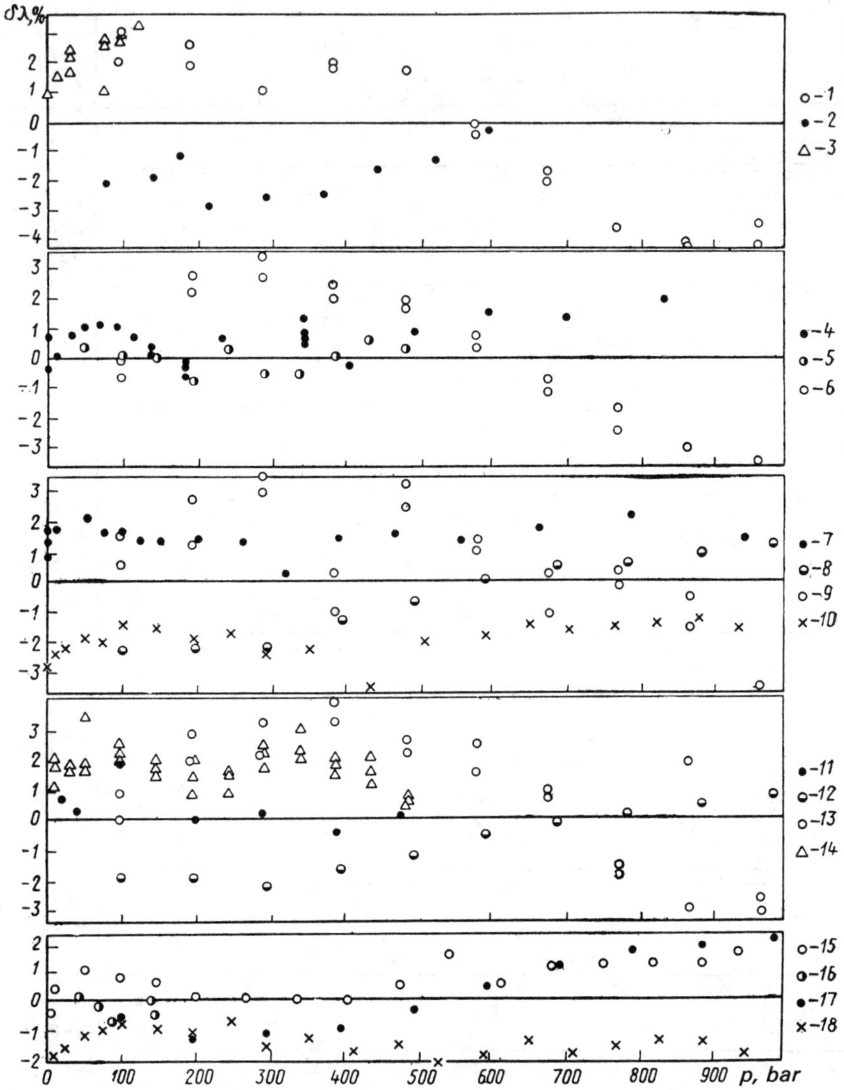

FIGURE 83. Deviations of experimental values of the thermal conductivity of monatomic gases from the computed. [6.15] Ar: 1) 313.56; 6) 333.45; 9) 353.36; 13) 372.98 K; [6.13] Ar: 2) 321.95 K; [4.77] Ne: 3) 96.05; 14) 117.65 K; [6.10] Ar: 4) 323.15; 7) 348.15 K; [4.81] Ar: 5) 323.25; 11) 366.45 K; [4.79] Ar: 8) 348.15; 12) 370.15; 17) 413.15 K; [4.78] Ar: 15) 392.15; Kr: 10) 482.15; 18) 560.15 K; [6.12] Ar: 16) 412.95 K.

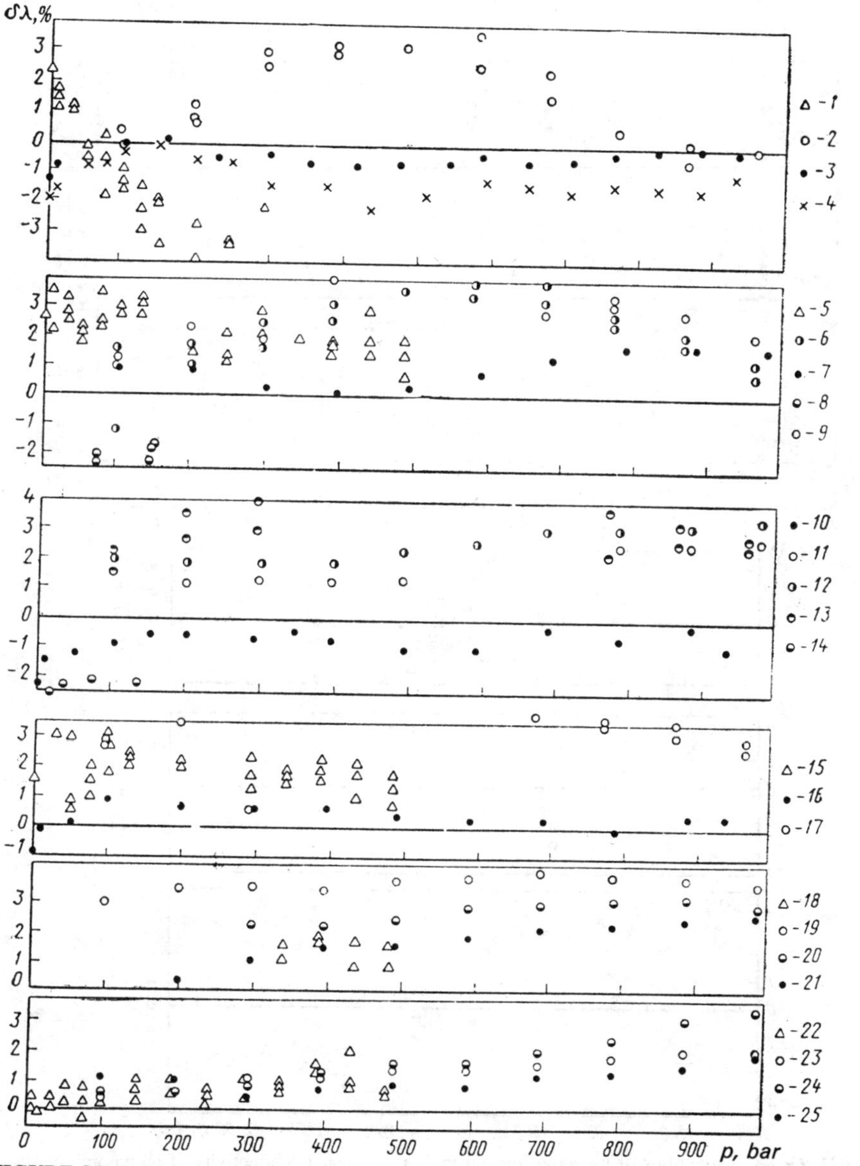

FIGURE 84. Deviations of experimental values of the thermal conductivity of monatomic gases from the computed. [4.77] Ne: 1) 140.2; 5) 153.9; 15) 173.4; 18) 195.65; 22) 273.65 K; [6.15] Ar: 2) 424.7; 6) 472.98; 9) 523.66; 13) 573.47; 17) 623.10 K; [4.78] Ar: 3) 456.15; 10) 540.15; 16) 606.15; Kr: 4) 608.15 K; [4.79] Ar: 7) 475.15; 11) 543.15; 12) 571.15; 19) 646.13; 20) 674.15; 21) 724.15; 23) 775.15; 24) 831.15; 25) 941.15 K; [6.12] Ar: 8) 522.85; 14) 573.55 K.

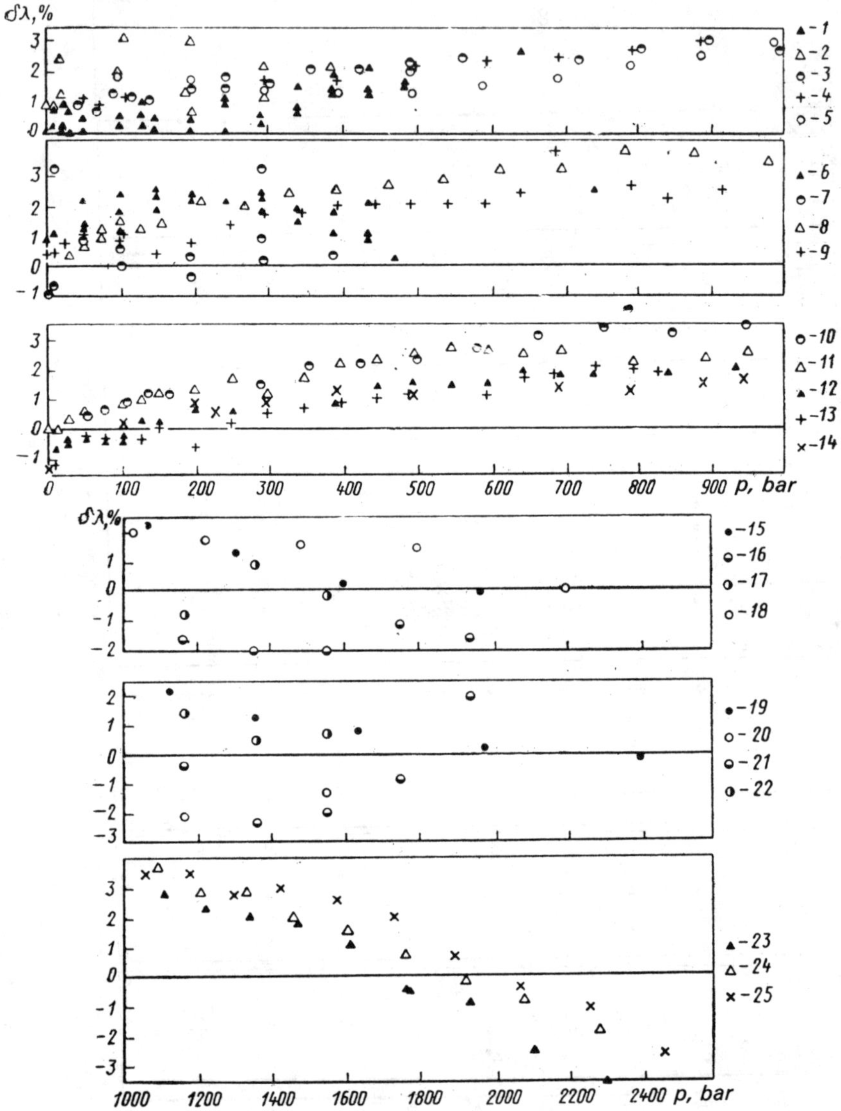

FIGURE 85. Deviations of the experimental values of thermal conductivity of monatomic gases from the computed. [4.77] Ne: 1) 292.85; 2) 295.25; 6) 320.65 K; [6.1] Ne: 3) 298.15; 8) 323.15; 10) 348.15; 23) 298.15; 24) 323.15; 25) 348.15 K; [4.78] Ne: 4) 305.15; 9) 347.15; 11) 407.15; 12) 483.15; 13) 559.15; 14) 606.15 K; [4.79] Ar: 5) 977.15 K; [4.76] Ne: 7) 320.95 K; [6.10] Ar: 15) 298.15; 18) 323.15; 19) 348.15 K; [4.80] Ar: 16) 298.34; 17) 309.68; 20) 349.07; 21) 388.44; 22) 430.69 K.

cally do not exceed 3% and do not tend to increase with the temperature. This permits using Eq. (6.12) for calculating the thermal conductivity of the four monatomic gases from the normal boiling points to a temperature of 1300 K. It is also justified since the corresponding values of $\Delta\lambda$ will decrease with rise in temperature at the fixed maximum pressure of 1000 bar. The computed values of the thermal conductivity are listed in Tables VII, XII, XX and XXVII in Part III. Data for the critical region are identified by dashed lines. It can be assumed, from comparison with experimental results and considerations put forward above, that the error of the computed values of λ does not basically exceed 3–4%. In addition, the computed values will be up to 50% lower only in the critical region.

Equations for the thermal conductivity of gases, written in terms of the independent variables T and p, which can be determined directly in measuring the thermal conductivity are of interest for practical purposes. These equations are most frequently specified in engineering calculations. Equations were suggested recently [6.40, 6.41] which represent the thermal conductivity of superheated steam as a function of pressure and temperature. These equations are convenient for calculations. However, they are valid only over a relatively narrow range of reduced parameters due to the complex configuration of isobars of thermal conductivity.

Rivkin [6.41] also compiled an equation for calculating the thermal conductivity of water and water vapor having the form $t = f_1(\lambda) + p f_2(\lambda)$. It is solved by iterations at the given values of t and p. Although the equation is intended for calculating the thermal conductivity in the critical region, it does not describe the anomalous rise in λ near the critical point. The equation was derived with consideration of the fact that the constant thermal conductivity curves near the vapor pressure curve are virtually rectilinear when plotted in coordinates (p, t). Rivkin noted that this is true only over a limited temperature and pressure range. We performed a check which showed that lines of λ = const curve over a wide range of reduced parameters and pass through a maximum (due to the existence of minima of the thermal conductivity of real gases along isobars).

A somewhat different approach to compiling an equation in the form $\lambda = f(p, T)$ was used by Vasserman [6.42], who analyzed isobaric sections of the plane of the thermal conductivity coefficient. He noted that the isobars of thermal conductivity have a rather simple configuration at temperatures far from the critical ($T > 2.5\ T_{cr}$). At low and moderate pressures these are monotonically rising curves, and at $p > 7\ p_{cr}$ they have a minimum which is shifted with rising pressure to the region of high temperatures. This configuration of isobars of thermal conductivity is similar to the configuration of viscosity isochores. For that reason, the equation for calculating the thermal conductivity of gas at high temperature and pressure can, by analogy with the equation for viscosity [5.57], be compiled in the form [6.42]:

$$\lambda(p, T) = F_1(p) + F_2(p)/T + F_3(p)\lambda_T \tag{6.13}$$

Functions $F_i(p)$ in Eq. (6.13) are represented by polynomials with $p - 1$ as the argument in order to conform to the limiting condition $\lambda(p, T) \underset{p \to 1}{\rightarrow} \lambda_T$. The free terms of the polynomials are respectively 0, 0 and 1. The thermal conductivity λ_T at atmospheric pressure serves as the temperature function in Eq. (6.13). That permits use of this equation for extrapolation to the region of high temperatures, where there are no data on thermal conductivity of compressed gas, but the values of λ_T are known.

We compiled equations such as Eq. (6.13) for calculating the thermal conductivity of neon, argon, krypton, and xenon at temperatures above 140, 298, 346 and 482 K, respectively. The coefficients of these equations were determined from experimental data by the method of least squares, by means of a Minsk-32 computer. This was done by using the weight $G = 1/\lambda^2$ of all the experimental points. The temperature dependence of the thermal conductivity of inert gases at atmospheric pressure used in these calculations is

$$\lambda_T = \sum_{k=0}^{n} a_k \overline{T}^k \tag{6.14}$$

where $\overline{T} = T/1000$. Equation (6.14) was obtained from smoothed values of λ_T given in Tables VII, XIII, XX and XXVII in Part III. Coefficients a_k for the four gases are listed in Table 52. The errors of approximating the reference values of λ_T amount to the following for each of the four substances over the temperature ranges:

	$\delta\lambda_m$, %	$\delta\lambda_{max}$, %
neon (T=42—1300 K)	0.27	0.57
argon (T=90—1300 K)	0.12	0.42
krypton (T=120—1300 K)	0.14	0.31
xenon (T=170—1300 K)	0.13	0.26

Coefficients b_{ij} of functions $F_i(p)$ in Eq. (6.13) for the gases under study are given in Table 53. The coefficients were rounded off with reference to the contribution of the corresponding terms of the equation to the computed values of thermal conductivity. The normed values of the independent variables used in the calculations are $\overline{p} = (p - 1)/1000$ and $\overline{T} = T/1000$, where p is in bar and

TABLE 52. Coefficients a_k of Eq. (6.14)

Substance	a_0	a_1	a_2	a_3	a_4	a_5	a_6
Neon	35.6	26770	—61670	123660	—141920	83400	—19410
Argon	—22.6	7210	—4750	2404	—494	—	—
Krypton	—7.4	3600	—1440	251	58	—	—
Xenon	1.8	1980	—370	—192	119	—	—

TABLE 53. Coefficients of Function $F_i(\bar{p}) = \Sigma b_{ij}\bar{p}^j$ in Eq. (6.13)

Substance	b_{11}	b_{12}	b_{13}	b_{21}	b_{22}
Neon	—1010	567	717	979	—6
Argon	—3136	341	1141	2262	—45
Krypton	—5301	—3900	7271	2529	2682
Xenon	—2428	—14326	15161	1806	7243

TABLE 53. Coefficients of Function $F_i(\bar{p}) = \Sigma b_{ij}\bar{p}^j$ in Eq. (6.13) (*Continued*)

Substance	b_{23}	b_{30}	b_{31}	b_{32}	b_{33}
Neon	—144	1	0.085	—0.026	—0.075
Argon	—312	1	0.490	0.002	—0.238
Krypton	—3332	1	1.382	0.730	—1.679
Xenon	—7053	1	0.862	5.088	—5.466

T is in K. When using Eq. (6.13) the values of thermal conductivity are obtained as $\lambda \cdot 10^8$ kW/(m · K).

The main information on the accuracy of analytic description of experimental data used in compiling Eq. (6.13) is listed in Table 54. There it is seen

TABLE 54. Standard and Maximum Deviations of Experimental Data from the Values of Thermal Conductivity Calculated from Eq. (6.13)

Substance	Author	Number of points	Range of parameters: T, K	Range of parameters: p, bar	$\delta\lambda_m$, %	$\delta\lambda_{max}$, %	Reference
Neon	Sengers et al.	103	298—348	1—2712	0.4	1.5	[6.11]
	Tufeu	115	305—606	1—964	0.4	—1.8	[4.78]
	Golubev and Shpagina	251	140—321	1—491	0.7	3.2	[4.77]
	Vasil'kovskaya and Golubev	67	295—566	1—407	3.1	5.7	[4.76]
Argon	Michels et al.	87	298—348	1—2222	1.6	4.7	[6.10]
	Keyes and Vines	13	413—620	6—148	2.5	—3.4	[6.12]
	Rosenbaum	17	298—322	26—713	3.6	—5.6	[6.13]
	Golubev and Shpagina	56	323—366	1—491	2.0	—3.8	[4.81]
	Le Neindre	143	298—977	1—1000	1.7	—5.7	[4.79]
	Amirkhanov et al.	184	322—624	98—981	2.3	—5.3	[6.15]
	Tufeu	81	302—607	1—952	2.0	—3.6	[4.78]
	Tarzimanov and Arslanov	48	298—653	1—1962	1.9	—4.3	[4.80] [6.17]
Krypton	Tufeu	102	346—608	1—957	1.8	5.8	[4.78]
Xenon	Tufeu	47	482—606	1—953	2.4	5.5	[4.78]

that this equation reproduces satisfactorily the experimental data for each of the gases over a wide temperature and pressure range.

Comparison of thermal conductivity coefficients calculated from Eqs. (6.12) and (6.13) over the range of parameters for which no experimental data are available (to 1300 K and 1000 bar), shows that the differences between both sets of data are 1–3%. Here the values of λ calculated from Eq. (6.13) for neon and argon are systematically on the high side and for krypton and xenon on the low side, as compared with those obtained from Eq. (6.12). The satisfactory agreement of data on λ calculated from different equations is an indirect confirmation of the reliability of extrapolation using these equations.

PART THREE

TABLES OF THERMOPHYSICAL PROPERTIES OF NEON, ARGON, KRYPTON, AND XENON

TABLES OF THERMOPHYSICAL PROPERTIES OF NEON, ARGON, KRYPTON, AND XENON

A. NOMENCLATURE AND UNITS OF QUANTITIES IN TABLES I THROUGH XXVII

T, K	thermodynamic temperature (absolute);
p, bar (1 bar = 10^5 Pa)	pressure;
p_v, bar	saturation vapor pressure along the gas-liquid phase transition curve;
p_m, bar	pressure along the liquid-crystal phase transition curve;
v, m^3/kg	specific volume;
v', m^3/kg	specific volume of saturated liquid;
v'', m^3/kg	specific volume of dry saturated vapor;
v_l, m^3/kg	specific volume of liquid along the crystallization curve;
v_c, m^3/kg	specific volume of crystal along the melting curve;
h, kJ/kg	specific enthalpy;
h', kJ/kg	specific enthalpy of saturated liquid;
h'', kJ/kg	specific enthalpy of dry saturated vapor;
h_l, kJ/kg	specific enthalpy of liquid along the crystallization curve;
h_c, kJ/kg	specific enthalpy of crystal along the melting curve;
s, kJ/(kg · K)	specific entropy;

s', kJ/(kg · K)—specific entropy of saturated liquid;
s'', kJ/(kg · K)—specific entropy of dry saturated vapor;
s_l, kJ/(kg · K)—specific entropy of liquid along the crystallization curve;
s_c, kJ/(kg · K)—specific entropy of crystal along the melting curve;
c_p, kJ/(kg · K)—specific heat at constant pressure;
c'_p, kJ/(kg · K)—specific heat of saturated liquid at constant pressure;
c''_p, kJ/(kg · K)—specific heat of dry saturated vapor at constant pressure;
$c_{p(l)}$, kJ/(kg · K)—specific heat of liquid at constant pressure along the crystallization curve;
$c_{p(c)}$, kJ/(kg · K)—specific heat of crystal at constant pressure along the melting curve;
c_v, kJ/(kg · K)—specific heat at constant volume;
α_{exp}, K^{-1}—coefficient of thermal expansion;
β_T, bar^{-1}—coefficient of isothermal compressibility;
η, Pa · s—dynamic viscosity;
λ, kW/(m · K)—thermal conductivity.

B. METHODS USED IN CALCULATING DATA IN TABLES I–XXVII

The tables of thermophysical properties of the four elements of the helium group given in this part of the book were calculated on the basis of equations obtained in the preceding chapters, over the temperature range from absolute zero to 1300 K and pressures from 1 to 1000 bar. The values of nonequilibrium properties (dynamic viscosity and thermal conductivity) are given for the gaseous and liquid phases at temperatures from the triple point to 1300 K and pressure from 1 to 1000 bar.

The equilibrium properties for each substance are given in the three states of matter, including data along the crystallization and saturation curves. The tables list the values of specific volume, specific enthalpy, specific entropy and specific heat at constant pressure. The values of specific heat at constant volume, and the values of the thermal expansion and isothermal compressibility coefficients for the four substances in the solid state are also presented.

The thermodynamic properties of gaseous neon, krypton, and xenon were calculated from the equations of state listed in Sec. 2.6 which are valid at temperatures from the triple point to 1300 K at $0 < \rho < 2\ \rho_{cr}$. The same properties for the liquid phase were calculated from equations of state obtained in Sec. 3.3 which are valid at $p \leqslant 1000$ bar and $\rho \geqslant 1.8\ \rho_{cr}$ at temperatures from the triple point to 50, 240 and 300 K respectively for neon, krypton, and xenon. The computed values were additionally smoothed graphically in the overlapping ranges of parameters, including the critical.

In order to estimate the accuracy of the computed tabulated data, the data were compared with the most dependable experimental and analytic results obtained by others. In each individual case such an evaluation was made based on consideration of the compatibility of the data being compared, on the accuracy of original data, and on the validity of the extrapolation.

To represent all the numerical values of thermodynamic properties at round values of temperature and pressure it was necessary in the first place to solve the iteration problem. This involved finding the density from the specified values of p and T. To solve this problem by computer we compiled an algorithm according to which the real root of the equation of state in virial form is found at each specified value of p. The value of this root is refined by successive approximations until the specified value of p becomes reproducible with less than 0.01% error.

In accordance with the initial form of the equation of state for the gas we obtained expressions for calculating the enthalpy, entropy and the specific heats c_p and c_v:

$$h = h_T{}^0 + \int_{T_0}^{T} c_p{}^0 dT + RT\left[\left(B_2 - T\frac{dB_2}{dT}\right)\rho + \left(B_3 - \frac{T}{2}\frac{dB_3}{dT}\right)\rho^2 + \right.$$
$$\left. + \left(B_4 - \frac{T}{3}\frac{dB_4}{dT}\right)\rho^3 + \ldots\right] \tag{I}$$

$$s = s_T{}^0 + \int_{T_0}^{T} c_p{}^0 \frac{dT}{T} - R\left[\ln\frac{\rho RT}{p_0} + \left(B_2 + T\frac{dB_2}{dT}\right)\rho + \right.$$
$$\left. + \left(B_3 + T\frac{dB_3}{dT}\right)\frac{\rho^2}{2} + \left(B_4 + T\frac{dB_4}{dT}\right)\frac{\rho^3}{3} + \ldots\right] \tag{II}$$

$$c_v = c_{v\infty} - RT\left[\left(2\frac{dB_2}{dT} + T\frac{d^2B_2}{dT^2}\right)\rho + \left(2\frac{dB_3}{dT} + T\frac{d^2B_3}{dT^2}\right)\rho^2/2 + \right.$$
$$\left. + \left(2\frac{dB_4}{dT} + T\frac{d^2B_4}{dT^2}\right)\frac{\rho^3}{3} + \ldots\right] \tag{III}$$

$$c_p = c_v + R\frac{\left[1 + \left(B_2 + T\frac{dB_2}{dT}\right)\rho + \left(B_3 + T\frac{dB_3}{dT}\right)\rho^2 + \ldots\right]^2}{1 + 2B_2\rho + 3B_3\rho^2 + \ldots} \tag{IV}$$

$$h_T{}^0 - h_0{}^0 = \int_0^{T_{tr}} c_p{}^c dT + \Delta h_s + RT\left(B_2 - T\frac{dB_2}{dT}\right)\rho_{tr} +$$
$$+ \int_{T_{tr}}^{T} c_p{}^0 dT \tag{V}$$

$$s_T{}^0 = \int_0^{T_{tr}} c_p{}^c \frac{dT}{T} + \frac{\Delta h_s}{T_{tr}} + R\left(B_2 + T\frac{dB_2}{dT}\right)\rho_{tr} - -R \ln \rho_{tr} R T_{tr} + \int_{T_{tr}}^{T} c_p{}^0 \frac{dT}{T} \quad \text{(VI)}$$

The first terms in Equations I and II, h_T^0 and s_T^0, are ideal gas properties calculated by Equations V, VI. Subscript tr in Equations V, VI refers to parameters at the triple point. The first terms in Equations V, VI describe the enthalpy and entropy of the crystal at this point. (Their values are listed in Tables I, VIII, XIV and XXI, respectively, for neon, argon, krypton, and xenon.) Δh_s is the heat of sublimation at the triple point, the third terms correct for deviation of the real from the ideal gas, and the term before the last in Eq. (VI) represents the pressure effect change in the ideal entropy upon transition from the triple point to the normal boiling point ($p = 1$ atm).

When the values of Δh_s are available, it is possible to calculate h_T^0 and s_T^0 at any temperature. However, this is most conveniently done at the normal boiling temperature, using reliable spectroscopic data on the standard entropy at $T = 298.15$ K (Sec. 3.1) for each of the four gases. This approach allows a unique determination of Δh_s from Eq. (VI) with deviation within the limits of the experimental error and thus eliminates the incompatibility of values of h_T^0 and s_T^0 at the normal boiling point (nbp). The values used for the standard entropy and the values of Δh_s, $h_{nbp}^0 - h_0$ and s_{nbp}^0 are listed in Table 55.

The specific heat in the ideal-gas state for the substances under study are independent of the temperature and are obtained from the expressions

$$C_p{}^0 = \frac{5}{2}R, \quad C_{v\infty} = \frac{3}{2}R \quad \text{(VII)}$$

where the universal gas constant R is taken to be 8.3143 kJ/(kmol · K).

In computing the tables of thermodynamic properties of gaseous neon, argon, krypton, and xenon, all the isotherms for each substance were divided into two groups. The first was comprised of supercritical isotherms. For these the thermodynamic functions were calculated at p between 1 and 1000 bar (at a

TABLE 55. Standard Entropy and the Computed Values of Δh_s, $h_{nbp}^0 - h_0^0$ and s_{nbp}^0

Substance	$s_{298.15}^0$, kJ/(kg · K)	Δh_s, kJ/kg	$h_{nbp}^0 - h_0^0$, kJ/kg	s_{nbp}^0 kJ/(kg · K)
Neon	7.2323	105.08	120.48	4.7797
Argon	3.8673	193.40	238.60	3.2369
Krypton	1.9539	128.78	163.54	1.7319
Xenon	1.2897	114.38	146.93	1.1988

number of isotherms the maximum pressure was limited by the range of applicability of the equation of state with respect to density). The second group included subcritical isotherms where the calculations were performed only at $p \leqslant p_v$, i.e., the last point for each isotherm was at the saturation curve. For this group, additional use was made of the equations of $p_v = f(T)$.

For the selected form of the equation of state for the liquid (3.29) and using the pertinent differential equations of thermodynamics, we obtained the following expressions for h, s, c_v and c_p:

$$h = h' + \left(2A - T\frac{dA}{dT}\right)(\rho - \rho') + \frac{1}{3}\left(4B - T\frac{dB}{dT}\right)[\rho^3 - (\rho')^3] + \\ + \frac{1}{5}\left(6C - T\frac{dC}{dT}\right)[\rho^5 - (\rho')^5] \quad \text{(VIII)}$$

$$s = s' - \frac{dA}{dT}(\rho - \rho') - \frac{1}{3}\frac{dB}{dT}[\rho^3 - (\rho')^3] - \\ - \frac{1}{5}\frac{dC}{dT}[\rho^5 - (\rho')^5]. \quad \text{(IX)}$$

$$c_v = c'_v - T\frac{d^2A}{dT^2}(\rho - \rho') - \frac{1}{3}T\frac{d^2B}{dT^2}[\rho^3 - (\rho')^3] - \\ - \frac{1}{5}T\frac{d^2C}{dT^2}[\rho^5 - (\rho')^5] \quad \text{(X)}$$

$$c_p = c_v + T\frac{\rho\left(\frac{dA}{dT} + \frac{dB}{dT}\rho^2 + \frac{dC}{dT}\rho^4\right)^2}{2A + 4B\rho^2 + 6C\rho^4} \quad \text{(XI)}$$

As seen from the above equations, the density of saturated liquid $\rho' \geqslant 1.8\ \rho_{cr}$ was the lower limit in integrating the differential equations along the subcritical isotherms. On the transcritical and supercritical isotherms this was the minimum density $\rho_{min} = 1.8\ \rho_{cr}$ at which the original equations of state are still valid. The upper limit is variable and depends on the specified values of $p \leqslant 1000$ bar.

To be able to calculate the enthalpy, entropy, and specific heat at constant volume from Eqs. (VIII), (IX) and (X) one must know the corresponding functions along the saturation curve. These can be calculated from equations of state for the gas and liquid, using equations of the vapor pressure curves (see Chapter 1),

$$h' = h'' - T\frac{dp_v}{dT}(v'' - v') \quad \text{(XII)}$$

$$s' = s'' - \frac{dp_v}{dT}(v'' - v') \quad \text{(XIII)}$$

$$c'_v = c''_v - T\frac{\left[\frac{dp_v}{dT} - \left(\frac{\partial p}{\partial T}\right)_{v=v''}\right]^2}{\left(\frac{\partial p}{\partial v}\right)_{T,\ v=v''}} - T\frac{d^2p_v}{dT^2}(v'' - v') +$$

$$+ T\frac{\left[\frac{dp_v}{dT} - \left(\frac{\partial p}{\partial T}\right)_{v=v'}\right]^2}{\left(\frac{\partial p}{\partial v}\right)_{T,\ v=v'}} \qquad \text{(XIV)}$$

The second and fourth terms in Eq. (XIV) incorporate jumps Δc_v in the specific heat at the gas and liquid sides of the dividing curve, respectively.

The final result obtained from Eq. (XIV) depends to a large measure on the form and accuracy of the vapor pressure curve equations and correctness of determination of the coefficients of equations of state for the gas and liquid. If we add to this that the second and third terms of Eq. (XIV) near the normal boiling point exceed by more than an order of magnitude the values of c'_v, it becomes apparent that good agreement between computed and experimental results cannot be expected. Nevertheless, comparison of values of c'_v calculated from the equation of state and the equations of the vapor pressure curve with experimental data for neon, argon, and krypton showed that the mean discrepancy does not exceed $\pm 4\%$. This result cannot be regarded as fully satisfactory from the quantitative point of view, since the investigators estimate the average error of their data as 1–2%. However, it can be regarded as confirmation of the correctness of selecting the form of the equation for the vapor pressure curve and acceptable accuracy of description of the original thermal data along the phase equilibrium curves and in the adjoining region.

A further improvement in the agreement between experimental and computed results can be attained by using experimental data on the specific heat c'_v in determining the coefficients contained in the equations of state and of the vapor pressure curve. But, in this case the approximating problem becomes highly nonlinear, resulting in significant mathematical difficulties, even if a previously derived equation of the vapor pressure curve is used. In addition, experimental data on c''_v on the gas side of the saturation curve are not available for these substances up to present. For this region we did not deem it possible to correct the equations of state on the basis of data on c'_v we restricted ourselves to the selection of equations which, in addition to satisfactorily describing the thermal properties, ensure good agreement with experimental values of c'_v.

Below are presented results of a comparison judging the accuracy of calorific functions, calculated from equations of state and presented in summary tables.

C. NEON TABLE DISCUSSION

The most complete tables of thermodynamic properties of neon, containing specific values of density, enthalpy, and entropy were calculated by McCarty and Stewart [2.40] from Strobridge's equation at temperatures from 25 to 300 K and pressures between 0.1 and 200 atm. These tables are, to a significant extent, obsolete, since the experimental thermal data on which they are based cover basically the range of 55 to 300 K and 20 to 100 atm and are given to pressures of 240 bar only along the 273.15 and 298.15 K isotherms (a part of the results found in the Leiden laboratory at low temperature is insufficiently reliable). In the remaining range of parameters the starting data were obtained by McCarty and Stewart analytically.

Michels et al. in reference [III.2] used previously obtained experimental data on the compressibility of gaseous neon and calculated a large set of thermodynamic functions at temperatures of 0 to 150°C (six isotherms spaced 25°C apart) and reduced density of 0 to 960 Amagat.

The values of enthalpy, entropy and specific heat c_p calculated by us were compared with the tabulated data of Michels et al. [III.2]. Without presenting these detailed tables here we note that the agreement is satisfactory. Thus, the mean deviation for the entire database for enthalpy turned out to be less than 0.5 kJ/kg. Only for three points (along the 0°C isotherm) were the maximum deviations as high as 1.8 kJ/kg. It should, however, be noted that the values of enthalpy of neon obtained by us are systematically higher (the mean deviation is 0.3% with the maximum never exceeding 1%). The best agreement is observed in values of entropy. In the majority of cases agreement is to within the fourth decimal point with the maximum deviation being 0.005 kJ/(kg · K).

Runge [2.30] used his experimental data on the integral Joule-Thomson effect for computing the enthalpy and entropy and represented them at round values of temperature and pressure at 28 to 100 K and 1 to 240 bar.

Tables 56 and 57 compare the values of enthalpy and entropy obtained by Runge [2.30] with our data. The mean difference for the entire body of points, including those not entered in Table 56, is less than 0.3 kJ/kg. Deviations exceed 1 kJ/kg only in several points in the supercritical region. The mean deviation for the entropy is 0.009 kJ/(kg · K) which cannot be regarded as fully satisfactory. Such a large mean deviation is due to divergences in the supercritical region. The reason for this may be insufficient refinement of the numerical method used by Runge or the limited volume and small accuracy of the starting thermal data.

The tables of thermal properties of data in the three-phase surface over a wide temperature and pressure range calculated by us cover all the previously published data. The equations from which these tables were calculated are based on reliable experimental thermal data. The possible error of calorific functions can be established, in part, by directly comparing their values with the results of others [2.28, 3.54, III.3].

TABLE 56. Comparison of Computed Values of Enthalpy h, kJ/kg, with Data of Runge [2.30]

T, K	p = 1 bar		p = 10 bar		p = 20 bar		p = 30 bar		p = 40 bar	
	$h_{[2.30]}$	h_{calc}	$h_{[2.30]}$	h_{calc}	$h_{[2.30]}$	h_{calc}	$h_{[2.30]}$	h_{calc}	$h_{[2.30]}$	h_{calc}
28	120.0	119.8	—	35.17	—	35.66	35.4	36.16	35.7	36.67
32	124.6	124.4	—	43.29	—	43.66	43.4	44.05	43.7	44.46
36	128.9	128.7	—	52.34	—	52.42	51.9	52.58	52.1	52.79
40	133.1	133.0	124.1	123.8	—	62.83	61.7	62.25	61.3	61.94
44	137.3	137.3	130.7	130.3	117.8	118.1	76.8	78.15	72.8	74.79
48	141.5	141.5	136.0	135.9	127.6	127.6	114.3	115.9	93.1	96.18
60	153.9	154.0	150.3	150.4	145.9	146.1	141.2	141.3	136.3	136.3
80	174.7	174.7	172.6	172.5	170.1	170.1	167.8	167.7	165.4	165.3
100	195.4	195.4	194.1	194.0	192.6	192.4	191.0	190.9	189.7	189.4

TABLE 56. Comparison of Computed Values of Enthalpy h, kJ/kg, with Data of Runge [2.30] *(Continued)*

T, K	p = 60 bar		p = 100 bar		p = 140 bar		p = 200 bar		p = 240 bar	
	$h_{[2.30]}$	h_{calc}	$h_{[2.30]}$	h_{calc}	$h_{[2.30]}$	h_{calc}	$h_{[2.30]}$	h_{calc}	$h_{[2.30]}$	h_{calc}
28	36.9	37.71	40.2	39.84	—	42.04	—	45.42	—	—
32	44.8	45.34	47.6	47.24	50.2	49.28	53.4	52.48	55.3	54.70
36	52.7	53.36	54.9	54.85	57.0	56.62	59.8	59.58	61.6	61.67
40	61.3	61.82	62.5	62.53	64.2	63.87	66.6	66.44	68.0	68.37
44	71.0	71.70	70.7	70.89	71.9	71.51	73.8	73.49	74.9	75.17
48	83.1	83.99	80.0	80.22	80.1	79.76	81.3	80.86	82.1	82.14
60	125.6	126.1	111.4	112.0	106.3	106.8	104.6	105.0	104.6	105.3
80	160.7	160.6	152.7	152.5	147.1	146.8	142.5	142.3	141.3	141.4
100	187.0	186.6	182.2	181.6	178.5	177.8	174.8	174.1	173.5	172.8

TABLE 57. Comparison of Computed Values of Entropy s, kJ/(kg · K), with Data of Runge [2.30]

T, K	p = 1 bar		p = 10 bar		p = 20 bar		p = 30 bar		p = 40 bar	
	$s_{[2.30]}$	s_{calc}	$s_{[2.30]}$	s_{calc}	$s_{[2.30]}$	s_{calc}	$s_{[2.30]}$	s_{calc}	$s_{[2.30]}$	s_{calc}
28	4.791	4.781	—	1.623	—	1.611	1.577	1.599	1.562	1.588
32	4.945	4.932	—	1.894	—	1.878	1.846	1.862	1.828	1.848
36	5.073	5.061	—	2.160	—	2.135	2.097	2.113	2.074	2.093
40	5.184	5.175	4.065	4.063	—	2.408	2.355	2.366	2.317	2.332
44	5.283	5.276	4.226	4.218	3.706	3.719	2.711	2.732	2.595	2.625
48	5.373	5.367	4.341	4.339	3.927	3.927	3.532	3.568	3.035	3.097
60	5.605	5.600	4.598	4.610	4.269	4.272	4.039	4.048	3.858	3.867
80	5.904	5.898	4.928	4.929	4.619	4.620	4.424	4.429	4.280	4.287
100	6.134	6.129	5.167	5.168	4.870	4.869	4.684	4.688	4.551	4.556

Table 58 compares smoothed experimental values of temperature obtained by Gladun [2.28] with those calculated from the equation of state at constant values of pressure and enthalpy. It is seen from the table that the agreement on the whole is good and that the mean deviation lies within ±0.06 K. However, larger differences are observed at individual points. Thus, at h = 101.36 kJ/kg for three points (p = 30, 40 and 50 atm) they (δT) become 0.15, 0.28 and 0.18 K, respectively. An additional graphical check along these isobars showed that the experimental points do not rigorously fit the smooth curves. Deviations (to 0.17 K) are also observed for individual points on three isoenthalpy curves before the last at the maximum values of pressure attained in Gladun's experiments.

The deviations of the calculated values of specific heat c_p of neon from the results of Dedikov [III.3] obtained at even values of temperature and pressure are shown in Fig. 86. Deviations for the majority of points do not exceed ±3%. However, they become greater in the critical region. If it is remembered that

TABLE 57. Comparison of Computed Values of Entropy s, kJ/(kg · K), with Data of Runge [2.30] (*Continued*)

T, K	p = 60 bar		p = 100 bar		p = 140 bar		p = 200 bar		p = 240 bar	
	$s_{[2.30]}$	s_{calc}	$s_{[2.30]}$	s_{calc}	$s_{[2.30]}$	s_{calc}	$s_{[2.30]}$	s_{calc}	$s_{[2.30]}$	s_{calc}
28	1.543	1.566	1.533	1.526	—	1.491	—	1.443	—	—
32	1.805	1.821	1.779	1.773	1.750	1.732	1.701	1.679	1.667	1.647
36	2.039	2.057	1.995	1.997	1.949	1.948	1.888	1.887	1.850	1.852
40	2.264	2.278	2.194	2.198	2.139	2.138	2.067	2.067	2.021	2.028
44	2.495	2.515	2.390	2.395	2.323	2.316	2.238	2.236	2.185	2.194
48	2.757	2.780	2.592	2.599	2.501	2.500	2.401	2.400	2.341	2.347
60	3.559	3.570	3.177	3.191	2.987	3.003	2.834	2.844	2.759	2.772
80	4.067	4.073	3.776	3.781	3.578	3.582	3.381	3.382	3.287	3.287
100	4.361	4.363	4.106	4.106	3.930	3.929	3.742	3.739	3.647	3.643

TABLE 58. Comparison of Smoothed Experimental Values of Temperature in [2.28] with those Computed at Constant Pressure and Enthalpy

p, atm	T_{exptl}, K	T_{calc}, K	p, atm	T_{exptl}, K	T_{calc}, K
	h = 95.66 kJ/kg			h = 112.20 kJ/kg	
1	31.44	31.48	1	46.81	46.87
10	39.82	39.93	10	51.58	51.55
20	46.05	46.17	20	55.86	55.90
30	50.67	50.80	30	59.42	59.45
40	54.25	54.22	40	62.44	62.45
50	56.99	57.00	50	64.94	64.95
60	59.20	59.25	60	67.07	67.05
70	61.05	61.05	70	68.86	68.89
80	62.61	62.62	80	70.43	70.45
100	64.97	64.94	100	73.02	73.05
120	66.64	66.71	120	74.88	74.93
140	67.88	67.85	140	76.35	76.39
160	68.76	68.75	160	77.43	77.49
180	69.35	69.39	180	78.27	78.36
	h = 101.36 kJ/kg			h = 115.60 kJ/kg	
1	36.65	36.65	1	50.15	50.15
10	43.35	43.43	10	54.41	54.36
20	48.93	49.02	20	58.36	58.35
30	53.50	53.35	30	61.72	61.70
40	57.03	56.74	40	64.58	64.56
50	59.61	59.43	50	67.00	67.00
60	61.70	61.66	60	69.08	69.06
70	63.50	63.50	70	70.80	70.82
80	65.04	65.05	80	72.35	72.39
100	67.55	67.55	100	74.85	74.92
120	69.33	69.30	120	76.78	76.86
140	70.66	70.65	140	78.27	78.30
160	71.63	71.65	160	79.35	79.54
180	72.29	72.30	180	80.12	80.27
	h = 106.82 kJ/kg			h = 122.52 kJ/kg	
1	41.73	41.81	1	56.74	56.76
10	47.29	47.35	10	60.25	60.23
20	52.17	52.25	20	63.69	63.70
30	56.14	56.18	30	66.65	66.65
40	59.41	59.40	40	69.20	69.20
50	62.05	62.01	50	71.37	71.39
60	64.23	64.25	60	73.32	73.34
70	66.11	66.10	70	74.97	75.00
80	67.66	67.70	80	76.48	76.52
100	70.16	70.18	100	78.88	78.97
120	72.06	72.08	120	80.72	80.86
140	73.47	73.49	140	82.20	82.35
160	74.50	74.53	160	83.33	83.46
180	75.21	75.23	180	84.20	84.36

TABLE 58. Comparison of Smoothed Experimental Values of Temperature in [2.28] with those Computed at Constant Pressure and Enthalpy (*Continued*)

p, atm	T_{exptl}, K	T_{calc}, K	p, atm	T_{exptl}, K	T_{calc}, K
h = 126.63 kJ/kg			50	78.49	78.50
			60	80.18	80.16
1	60.66	60.69	70	81.66	81.75
10	63.85	63.85	80	82.99	83.08
20	67.03	67.01	100	85.26	85.34
30	69.76	69.72	120	87.03	87.16
40	72.13	72.11	140	88.45	88.62
50	74.19	74.20	160	89.59	89.71
60	76.02	76.00	180	90.45	90.62
70	77.63	77.61			
80	79.03	79.08			
100	81.39	81.52	h = 141.39 kJ/kg		
120	83.24	83.39			
140	84.69	84.82			
160	85.81	85.96	1	74.89	74.92
180	86.68	86.85	10	77.16	77.25
			20	79.47	79.53
			30	81.52	81.58
h = 132.67 kJ/kg			40	83.43	83.47
			50	85.12	85.10
1	66.48	66.52	60	86.66	86.65
10	69.22	69.25	70	87.97	88.00
20	71.94	71.98	80	89.14	89.21
30	74.42	74.43	100	—	91.35
40	76.62	76.58			

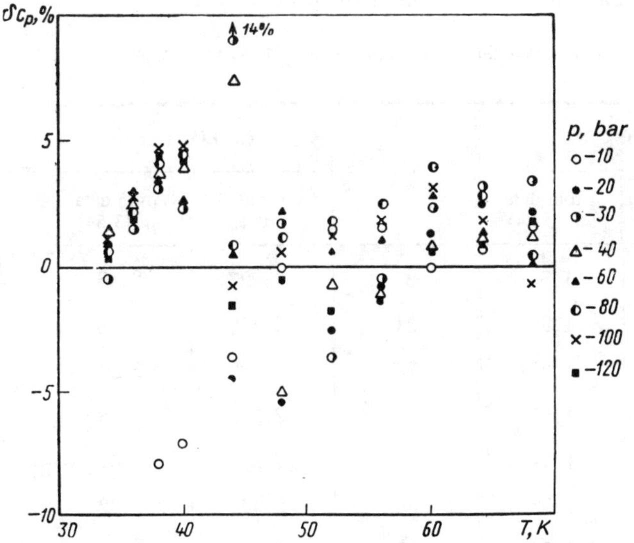

FIGURE 86. Deviations of computed values of specific heat c_p of neon from Dedikov's data [III.3].

Dedikov estimates the error of his data at $\pm 1.2\%$ and near maxima at 5%, then the agreement cannot be regarded as satisfactory. Most likely the main reason is the absence of internal compatibility between the experimental *pvT* data and the values of c_p over the range of variables shown in Fig. 86. This assumption is also reinforced by analysis of deviations of values of enthalpy and entropy given in Tables 56 and 57, particularly in the critical region. For this reason, in order to ensure better accuracy of the computed values of thermodynamic functions of neon it is necessary, in the first place, to perform additional experimental studies of the *pvT* behavior for T between 38 and 55 K and p of 5 to 150 atm.

It is also of interest to compare our data on specific heat c_p of liquid neon at the saturation curve with data of Gladun [3.54] (Table 59). Our data are basically 2–3% higher. The agreement is improved as the critical point is approached. But, this is a result of graphical correction which was used in joining thermodynamic functions calculated from separate equations of state for the liquid and the gas.

Before we evaluate the values given in Tables II through V we must keep in mind that a significant part of the range of parameters presented in these tables is not represented by experimental data (see Table 8). The thermodynamic functions at temperatures from the triple point to 70 K at a pressure of $p \leqslant 200$ bar can be characterized by the following errors: $\delta v = \pm 0.15\%$; $\Delta h = \pm 0.4$ kJ/kg; $\Delta s = \pm 0.005$ kJ/(kg · K) and $\delta c_p = \pm 3\%$. These errors may be two–threefold higher in the critical region. At higher pressures ($p > 200$ bar), where experimental data are not available yet, the errors of the computed values can, by our estimates, exceed $\delta v \simeq 0.25\%$, $\Delta h \simeq 0.6$ kJ/kg, $\Delta s \simeq 0.008$ kJ/(kg · K) and $\delta c_p \simeq 4\%$. The thermodynamic functions given for 70 to 1300 K and 1 to 1000 bar exhibit errors one-half of the above.

TABLE 59. Comparison of Computed Values of Specific Heat c'_p of Neon with Data of Gladun [3.54]

T, K	c'_p, kJ/(kg · K)		T, K	c'_p, kJ/(kg · K)	
	computed value	from data of [3.54]		computed value	from data of [3.54]
25	1.826	1.81	33	2.227	2.18
26	1.868	1.83	34	2.302	2.26
27	1.910	1.86	35	2.392	2.36
28	1.955	1.90	36	2.506	2.48
29	2.003	1.94	37	2.639	2.63
30	2.052	1.98	38	2.825	2.82
31	2.106	2.04	39	3.063	3.07
32	2.163	2.11	40	3.436	3.47

The possible errors of calculated values of thermodynamic properties of solid neon listed in Table I, are given on page 25, and of the values of the dynamic viscosity and thermal conductivity of liquid and gaseous neon (Tables VI and VII) on pages 252 and 279, respectively.

TABLE I. Thermodynamic Properties of Crystalline Neon

T, K	$p = 1$ bar						
	$v \cdot 10^3$	h	s	c_p	c_v	$\alpha_{exp} \cdot 10^5$	$\beta_T \cdot 10^5$
1	0.6636	0	0.000078	0.00024	0.00024	0.09	9.11
2	0.6636	0	0.000639	0.00196	0.00196	0.70	9.11
4	0.6637	0.02	0.00554	0.0177	0.0176	6.06	9.12
6	0.6638	0.10	0.02047	0.0663	0.0661	22.3	9.16
8	0.6644	0.31	0.05126	0.159	0.157	54.4	9.27
10	0.6654	0.75	0.09922	0.278	0.271	99.6	9.48
12	0.6671	1.43	0.1609	0.402	0.383	151.3	9.82
14	0.6696	2.36	0.2325	0.529	0.490	206.1	10.3
16	0.6728	3.55	0.3119	0.662	0.593	264.8	11.0
18	0.6770	5.02	0.3982	0.800	0.688	330.3	11.8
20	0.6823	6.78	0.4906	0.945	0.771	407.9	13.1
22	0.6888	8.84	0.5889	1.103	0.842	505.9	14.9
24	0.6971	11.26	0.6942	1.291	0.902	642.2	17.7
24.56	0.6996	11.96	0.7257	1.345	0.917	690.0	19.3

TABLE I. Thermodynamic Properties of Crystalline Neon (*Continued*)

T, K	$p = 100$ bar						
	$v \cdot 10^3$	h	s	c_p	c_v	$\alpha_{exp} \cdot 10^5$	$\beta_T \cdot 10^5$
1	0.6577	6.54	0.000073	0.00022	0.00022	0.07	8.44
2	0.6577	6.54	0.000595	0.00182	0.00182	0.61	8.44
4	0.6578	6.56	0.00516	0.0164	0.0164	5.25	8.45
6	0.6579	6.63	0.01907	0.0618	0.0617	19.4	8.48
8	0.6584	6.83	0.04784	0.149	0.147	47.4	8.56
10	0.6593	7.24	0.09292	0.262	0.256	87.3	8.73
12	0.6608	7.89	0.1512	0.381	0.366	133.0	8.99
14	0.6629	8.77	0.2193	0.504	0.471	181.2	9.37
16	0.6657	9.91	0.2949	0.630	0.572	231.6	9.87
18	0.6693	11.3	0.3770	0.760	0.666	285.7	10.5
20	0.6737	12.96	0.4643	0.891	0.749	346.1	11.4
22	0.6791	14.90	0.5562	1.027	0.820	416.9	12.6
24	0.6857	17.13	0.6529	1.175	0.880	505.4	14.2
26	0.6939	19.69	0.7555	1.351	0.930	625.3	16.8
26.03	0.6940	19.74	0.7570	1.354	0.931	626.1	16.8

TABLE I. Thermodynamic Properties of Crystalline Neon (*Continued*)

T, K	p = 200 bar						
	$v \cdot 10^3$	h	s	c_p	c_v	$\alpha_{exp} \cdot 10^5$	$\beta_T \cdot 10^5$
1	0.6522	13.09	0.000068	0.00021	0.00021	0.06	7.87
2	0.6522	13.09	0.000557	0.00171	0.00171	0.53	7.87
4	0.6523	13.11	0.00483	0.0154	0.0154	4.61	7.87
6	0.6524	13.17	0.01785	0.0579	0.0578	17.0	7.90
8	0.6528	13.37	0.04484	0.140	0.139	41.8	7.97
10	0.6536	13.75	0.08737	0.248	0.243	77.3	8.10
12	0.6549	14.36	0.1428	0.363	0.350	118.3	8.32
14	0.6568	15.21	0.2076	0.481	0.453	161.3	8.62
16	0.6593	16.30	0.2800	0.603	0.554	205.8	9.01
18	0.6624	17.63	0.3584	0.727	0.647	252.3	9.52
20	0.6662	19.21	0.4417	0.849	0.729	302.3	10.2
22	0.6708	21.05	0.5290	0.972	0.800	358.1	11.0
24	0.6763	23.14	0.6199	1.100	0.860	423.2	12.1
26	0.6830	25.51	0.7146	1.240	0.911	503.5	13.7
27.45	0.6878	27.31	0.7860	1.357	0.943	575.6	15.1

TABLE I. Thermodynamic Properties of Crystalline Neon (*Continued*)

T, K	p = 300 bar						
	$v \cdot 10^3$	h	s	c_p	c_v	$\alpha_{exp} \cdot 10^5$	$\beta_T \cdot 10^5$
1	0.6472	19.59	0.000064	0.00019	0.00019	0.06	7.38
2	0.6472	19.59	0.000524	0.00160	0.00160	0.47	7.38
4	0.6472	19.61	0.00454	0.0145	0.0145	4.08	7.39
6	0.6473	19.67	0.01678	0.0545	0.0543	15.1	7.41
8	0.6476	19.85	0.0422	0.132	0.131	37.2	7.46
10	0.6483	20.21	0.08249	0.235	0.231	69.1	7.50
12	0.6495	20.79	0.1352	0.347	0.335	106.2	7.75
14	0.6512	21.60	0.1973	0.461	0.437	145.2	8.00
16	0.6534	22.64	0.2668	0.580	0.537	185.2	8.33
18	0.6562	23.93	0.3421	0.699	0.629	226.3	8.73
20	0.6596	25.45	0.4221	0.815	0.712	269.4	9.24
22	0.6636	27.20	0.5057	0.930	0.782	315.8	9.88
24	0.6683	29.20	0.5922	1.045	0.843	367.4	10.7
26	0.6739	31.43	0.6815	1.166	0.894	427.6	11.8
28	0.6805	33.94	0.7740	1.299	0.937	501.2	13.2
28.83	0.6836	35.02	0.8135	1.362	0.953	535.9	13.8

TABLE I. Thermodynamic Properties of Crystalline Neon (*Continued*)

T, K	p = 400 bar						
	$v \cdot 10^3$	h	s	c_p	c_v	$\alpha_{exp} \cdot 10^5$	$\beta_T \cdot 10^5$
1	0.6424	26.04	0.000060	0.00018	0.00018	0.05	6.96
2	0.6424	26.04	0.000494	0.00151	0.00151	0.42	6.96
4	0.6424	26.05	0.00428	0.0136	0.0136	3.65	6.96
6	0.6426	26.11	0.01584	0.0514	0.0513	13.5	6.98
8	0.6429	26.28	0.03987	0.125	0.124	33.3	7.03
10	0.6435	26.63	0.07813	0.224	0.221	62.3	7.12
12	0.6445	27.18	0.1285	0.332	0.322	96.2	7.27
14	0.6460	27.96	0.1881	0.444	0.423	131.8	7.48
16	0.6480	28.96	0.2549	0.559	0.521	168.2	7.75
18	0.6505	30.20	0.3276	0.674	0.613	205.3	8.09
20	0.6535	31.67	0.4047	0.786	0.695	243.4	8.50
22	0.6571	33.36	0.4852	0.895	0.766	283.4	9.01
24	0.6613	35.27	0.5682	1.002	0.827	326.5	9.64
26	0.6662	37.40	0.6534	1.111	0.878	374.6	10.4
28	0.6718	39.77	0.7410	1.226	0.922	430.6	11.5
30	0.6784	42.40	0.8313	1.354	0.959	498.7	12.8
30.17	0.6790	42.64	0.8391	1.366	0.962	505.1	12.9

TABLE I. Thermodynamic Properties of Crystalline Neon (*Continued*)

T, K	p = 500 bar						
	$v \cdot 10^3$	h	s	c_p	c_v	$\alpha_{exp} \cdot 10^5$	$\beta_T \cdot 10^5$
1	0.6380	32.44	0.000057	0.00017	0.00017	0.05	6.59
2	0.6380	32.44	0.000468	0.00143	0.00143	0.38	6.59
4	0.6380	32.45	0.00506	0.0129	0.0129	3.29	6.59
6	0.6381	32.51	0.01500	0.0487	0.0486	12.2	6.61
8	0.6384	32.67	0.03779	0.119	0.118	30.1	6.65
10	0.6389	33.0	0.07423	0.214	0.211	56.5	6.73
12	0.6398	33.54	0.1225	0.319	0.310	87.6	6.86
14	0.6412	34.28	0.1798	0.428	0.409	120.5	7.04
16	0.6430	35.25	0.2442	0.540	0.506	153.9	7.27
18	0.6453	36.44	0.3145	0.652	0.598	187.8	7.55
20	0.6480	37.86	0.3891	0.761	0.680	222.2	7.89
22	0.6512	39.50	0.4669	0.865	0.751	257.5	8.31
24	0.6550	41.34	0.5470	0.967	0.812	294.7	8.82
26	0.6593	43.39	0.6290	1.068	0.864	335.0	9.44
28	0.6643	45.65	0.7127	1.171	0.908	380.1	10.2
30	0.6700	48.15	0.7983	1.282	0.946	432.4	11.2
31.48	0.6747	50.31	0.8630	1.370	0.971	475.8	12.1

TABLE I. Thermodynamic Properties of Crystalline Neon (*Continued*)

T, K	p = 600 bar						
	$v \cdot 10^3$	h	s	c_p	c_v	$\alpha_{exp} \cdot 10^5$	$\beta_T \cdot 10^5$
1	0.6338	38.80	0.000054	0.00016	0.00016	0.04	6.26
2	0.6338	38.80	0.000445	0.00136	0.00136	0.34	6.26
4	0.6338	38.81	0.00385	0.0123	0.0123	2.98	6.27
6	0.6339	38.87	0.01424	0.0463	0.0462	11.0	6.28
8	0.6341	39.02	0.03592	0.113	0.112	27.4	6.31
10	0.6346	39.33	0.0707	0.205	0.202	51.6	6.38
12	0.6355	39.84	0.1170	0.306	0.299	80.3	6.49
14	0.6367	40.57	0.1722	0.413	0.396	110.7	6.65
16	0.6384	41.50	0.2345	0.522	0.493	141.7	6.84
18	0.6404	42.66	0.3026	0.632	0.584	172.9	7.09
20	0.6430	44.03	0.3749	0.738	0.666	204.3	7.38
22	0.6459	45.62	0.4503	0.840	0.737	236.2	7.73
24	0.6493	47.40	0.5279	0.937	0.798	269.1	8.15
26	0.6531	49.39	0.6072	1.032	0.851	303.9	8.65
28	0.6575	51.57	0.6878	1.127	0.895	341.7	9.26
30	0.6626	53.95	0.7699	1.227	0.934	384.1	10.0
32	0.6683	56.56	0.8536	1.333	0.967	433.3	11.0
32.75	0.6706	57.50	0.8852	1.373	0.978	451.9	11.4

TABLE I. Thermodynamic Properties of Crystalline Neon (*Continued*)

T, K	p = 700 bar						
	$v \cdot 10^3$	h	s	c_p	c_v	$\alpha_{exp} \cdot 10^5$	$\beta_T \cdot 10^5$
1	0.6298	45.12	0.000052	0.00016	0.00016	0.04	5.97
2	0.6298	45.12	0.000423	0.00130	0.00130	0.31	5.97
4	0.6298	45.13	0.00367	0.0117	0.0117	2.72	5.97
6	0.6299	45.18	0.01356	0.0441	0.0440	10.0	5.99
8	0.6301	45.33	0.03423	0.108	0.107	25.0	6.01
10	0.6306	45.63	0.06751	0.196	0.194	47.3	6.07
12	0.6314	46.12	0.1120	0.295	0.289	73.9	6.17
14	0.6325	46.82	0.1653	0.399	0.384	102.3	6.30
16	0.6340	47.72	0.2256	0.507	0.480	131.1	6.48
18	0.6359	48.84	0.2916	0.614	0.570	160.1	6.69
20	0.6382	50.18	0.3619	0.718	0.652	189.1	6.94
22	0.6409	51.72	0.4353	0.817	0.724	218.2	7.24
24	0.6441	53.46	0.5107	0.910	0.785	247.9	7.59
26	0.6475	55.38	0.5876	1.001	0.838	278.6	8.01
28	0.6515	57.49	0.6656	1.091	0.883	311.3	8.50
30	0.6560	59.79	0.7447	1.182	0.922	346.9	9.10
32	0.6611	62.28	0.8250	1.278	0.955	387.0	9.83
33.99	0.6667	64.90	0.9062	1.378	0.985	431.2	10.7

TABLE I. Thermodynamic Properties of Crystalline Neon (*Continued*)

T, K	p = 800 bar						
	$v \cdot 10^3$	h	s	c_p	c_v	$\alpha_{exp} \cdot 10^5$	$\beta_T \cdot 10^5$
1	0.6261	51.40	0.000049	0.00015	0.00015	0.04	5.71
2	0.6261	51.40	0.000404	0.00124	0.00124	0.29	5.71
4	0.6261	51.41	0.00350	0.0112	0.0112	2.49	5.71
6	0.6262	51.46	0.01295	0.0421	0.0420	9.20	5.72
8	0.6264	51.60	0.03269	0.103	0.103	22.9	5.75
10	0.6268	51.89	0.06458	0.189	0.187	43.6	5.80
12	0.6275	52.36	0.1074	0.285	0.279	68.4	5.88
14	0.6285	53.03	0.1589	0.386	0.373	94.9	6.00
16	0.6299	53.91	0.2174	0.492	0.468	121.9	6.15
18	0.6317	55.00	0.2816	0.598	0.558	149.0	6.34
20	0.6338	56.30	0.3500	0.699	0.640	176.0	6.56
22	0.6363	57.80	0.4214	0.796	0.711	202.8	6.82
24	0.6392	59.49	0.4949	0.887	0.773	229.9	7.12
26	0.6424	61.36	0.5697	0.974	0.826	257.5	7.47
28	0.6460	63.42	0.6455	1.060	0.872	286.3	7.88
30	0.6501	65.64	0.7222	1.145	0.911	317.2	8.37
32	0.6546	68.05	0.7997	1.233	0.945	351.0	8.96
34	0.6598	70.64	0.8782	1.325	0.974	389.0	9.67
35.20	0.6631	72.20	0.9263	1.382	0.991	412.8	10.2

TABLE I. Thermodynamic Properties of Crystalline Neon (*Continued*)

T, K	p = 900 bar						
	$v \cdot 10^3$	h	s	c_p	c_v	$\alpha_{exp} \cdot 10^5$	$\beta_T \cdot 10^5$
1	0.6225	57.64	0.000047	0.00014	0.00014	0.03	5.47
2	0.6225	57.64	0.000387	0.00118	0.00118	0.26	5.47
4	0.6226	57.66	0.00335	0.0107	0.0107	2.29	5.47
6	0.6226	57.70	0.01238	0.0402	0.0402	8.46	5.48
8	0.6228	57.84	0.03129	0.0989	0.0985	21.1	5.50
10	0.6231	58.11	0.06190	0.181	0.180	40.3	5.55
12	0.6238	58.57	0.1032	0.275	0.270	63.5	5.63
14	0.6248	59.22	0.1530	0.375	0.363	88.4	5.73
16	0.6261	60.07	0.2098	0.478	0.456	113.8	5.86
18	0.6279	61.13	0.2723	0.582	0.546	139.2	6.03
20	0.6297	62.40	0.3389	0.682	0.627	164.4	6.22
22	0.6320	63.86	0.4086	0.777	0.699	189.4	6.45
24	0.6346	65.51	0.4803	0.866	0.761	214.4	6.71
26	0.6376	67.34	0.5533	0.951	0.815	239.5	7.01
28	0.6409	69.55	0.6271	1.033	0.861	265.4	7.37
30	0.6447	71.49	0.7017	1.114	0.901	292.7	7.78
32	0.6487	73.83	0.7768	1.195	0.935	321.9	8.26
34	0.6534	76.34	0.8527	1.280	0.965	354.1	8.83
36	0.6585	79.03	0.9294	1.369	0.990	390.3	9.53
36.39	0.6595	79.60	0.9451	1.386	0.995	396.3	9.67

TABLE I. Thermodynamic Properties of Crystalline Neon (*Continued*)

T, K	p = 1000 bar						
	$v \cdot 10^3$	h	s	c_p	c_v	$\alpha_{exp} \cdot 10^5$	$\beta_T \cdot 10^5$
1	0.6191	63.86	0.000045	0.00014	0.00014	0.03	5.26
2	0.6191	63.87	0.000371	0.00113	0.00113	0.24	5.26
4	0.6191	63.86	0.003210	0.0102	0.0102	2.11	5.26
6	0.6192	63.91	0.01187	0.0386	0.0385	7.82	5.27
8	0.6193	64.04	0.03000	0.0949	0.0946	19.6	5.28
10	0.6197	64.30	0.05944	0.175	0.173	37.4	5.33
12	0.6203	64.74	0.09927	0.266	0.261	59.1	5.39
14	0.6212	65.37	0.1476	0.364	0.353	82.6	5.48
16	0.6224	66.20	0.2028	0.465	0.445	106.5	5.60
18	0.6239	67.24	0.2636	0.568	0.534	130.5	5.75
20	0.6258	68.47	0.3287	0.666	0.616	154.3	5.92
22	0.6279	69.90	0.3967	0.759	0.688	177.7	6.12
24	0.6303	71.51	0.4668	0.846	0.750	200.8	6.35
26	0.6331	73.30	0.5381	0.929	0.804	224.0	6.62
28	0.6362	75.24	0.6102	1.008	0.851	247.6	6.92
30	0.6396	77.36	0.6829	1.086	0.891	272.0	7.27
32	0.6435	79.62	0.7560	1.163	0.926	297.8	7.68
34	0.6476	82.06	0.8296	1.242	0.955	325.6	8.15
36	0.6523	84.66	0.9038	1.323	0.981	356.3	8.72
37.55	0.6564	86.70	0.9622	1.388	0.999	381.5	9.19

TABLE II. Thermodynamic Properties of Neon Along the Solidification and Melting Curve

T, K	p_m	$v_l \cdot 10^3$	$v_c \cdot 10^3$	h_l	h_c	s_l	s_c	c_{pl}	c_{pc}
24.55	0.43	0.8012	0.6996	28.22	11.96	1.388	0.7257	1.802	1.344
25	30.45	0.7967	0.6979	30.60	14.25	1.390	0.7350	1.784	1.347
26	98.38	0.7876	0.6942	36.19	19.56	1.396	0.7563	1.756	1.352
27	168.0	0.7793	0.6904	41.88	24.90	1.404	0.7768	1.735	1.356
28	239.3	0.7717	0.6867	47.67	30.35	1.414	0.7970	1.720	1.359
29	312.2	0.7645	0.6831	53.57	36.00	1.426	0.8168	1.709	1.363
30	386.7	0.7579	0.6796	59.63	41.65	1.439	0.8358	1.701	1.366
31	462.8	0.7515	0.6762	65.82	47.42	1.453	0.8543	1.695	1.369
32	540.5	0.7455	0.6730	72.08	53.16	1.468	0.8720	1.690	1.367
33	619.7	0.7398	0.6698	78.50	59.00	1.483	0.8893	1.686	1.374
34	700.5	0.7343	0.6667	84.97	64.95	1.497	0.9064	1.682	1.377
35	782.8	0.7290	0.6637	91.50	70.95	1.511	0.9230	1.678	1.380
36	866.6	0.7240	0.6607	98.08	77.15	1.525	0.9388	1.675	1.384
37	951.9	0.7192	0.6577	104.7	83.34	1.538	0.9542	1.672	1.387
38	1039	0.7147	0.6547	111.4	89.59	1.550	0.9688	1.669	1.390

TABLE III. Thermodynamic Properties of Neon at Saturation

T, K	p_v	$v'' \cdot 10^3$	$v' \cdot 10^3$	h''	h'	s''	s'	c''_p	c'_p
24.55	0.4335	226.6	0.8012	117.0	28.22	5.006	1.388	1.077	1.802
25	0.5103	195.2	0.8060	117.4	29.04	4.954	1.421	1.089	1.826
26	0.7184	142.9	0.8172	118.1	30.90	4.846	1.494	1.118	1.868
27	0.9854	107.0	0.8289	118.7	32.80	4.746	1.564	1.153	1.910
27.09	1.013	104.4	0.8301	118.8	32.98	4.738	1.571	1.156	1.914
28	1.321	81.74	0.8413	119.3	34.75	4.653	1.634	1.194	1.955
29	1.735	63.52	0.8546	119.8	36.75	4.566	1.703	1.242	2.003
30	2.238	50.10	0.8687	120.1	38.80	4.483	1.771	1.298	2.052
31	2.840	40.03	0.8838	120.4	40.90	4.404	1.838	1.365	2.106
32	3.552	32.34	0.9001	120.6	43.06	4.329	1.905	1.444	2.163
33	4.386	26.37	0.9177	120.7	45.28	4.255	1.971	1.538	2.227
34	5.352	21.66	0.9370	120.6	47.57	4.184	2.036	1.652	2.302
35	6.462	17.92	0.9583	120.3	49.92	4.113	2.101	1.792	2.392
36	7.728	14.89	0.9820	119.9	52.34	4.043	2.166	1.964	2.506
37	9.164	12.44	1.009	119.3	54.80	3.973	2.231	2.177	2.639
38	10.78	10.38	1.039	118.4	57.52	3.900	2.297	2.454	2.825
39	12.60	8.685	1.075	117.2	60.30	3.825	2.365	2.806	3.068
40	14.62	7.261	1.116	115.8	63.33	3.749	2.435	3.269	3.436
41	16.88	6.056	1.167	114.0	66.51	3.668	2.507	3.906	4.065
42	19.39	5.038	1.232	111.8	69.82	3.580	2.582	5.025	5.263
43	22.16	4.114	1.329	108.8	73.93	3.478	2.667	7.75	8.06
44	25.22	3.107	1.538	103.0	80.83	3.316	2.812	23.3	25.0
44.40	26.53	2.070		92.5		3.062		∞	

TABLE IV. Thermodynamic Properties of Liquid and Gaseous Neon

T, K	p = 1 bar			p = 2 bar			p = 4 bar		
	$v \cdot 10^3$	h	s	$v \cdot 10^3$	h	s	$v \cdot 10^3$	h	s
26	0.8171	30.92	1.493	0.8167	30.97	1.492	0.8160	31.08	1.490
28	109.9	119.8	4.781	0.8410	34.78	1.633	0.8401	34.88	1.631
30	118.9	122.1	4.860	56.76	120.6	4.539	0.8676	38.87	1.769
32	127.7	124.4	4.932	61.57	123.0	4.619	0.8997	43.08	1.904
34	136.4	126.6	4.999	66.23	125.4	4.691	30.83	122.8	4.351
36	145.0	128.7	5.061	70.78	127.7	4.757	33.44	125.5	4.428
38	153.6	130.9	5.120	75.25	130.0	4.819	35.93	128.1	4.498
40	162.1	133.0	5.175	79.67	132.2	4.876	38.33	130.5	4.561
42	170.6	135.2	4.227	84.03	134.4	4.930	40.68	132.9	4.620
44	179.0	137.3	4.276	88.37	136.6	4.980	42.98	135.2	4.674
46	187.4	139.4	5.323	92.67	138.8	5.028	45.24	137.5	4.724
48	195.8	141.5	5.368	96.96	140.9	5.074	47.48	139.8	4.772
50	204.2	143.6	5.411	101.2	143.1	5.117	49.70	142.0	4.818
55	225.1	148.8	5.509	111.8	148.3	5.218	55.17	147.4	4.922
60	246.0	154.0	5.600	122.4	153.6	5.310	60.57	152.8	5.015
65	266.8	159.2	5.683	132.9	158.8	5.394	65.93	158.2	5.101
70	287.6	164.4	5.760	143.3	164.1	5.471	71.25	163.5	5.179
75	308.3	169.5	5.831	153.8	169.3	5.543	76.54	168.7	5.252
80	329.0	174.7	5.898	164.2	174.5	5.610	81.82	174.0	5.320
85	349.7	179.9	5.961	174.6	179.7	5.673	87.07	179.2	5.384
90	370.4	185.1	6.020	185.0	184.9	5.732	92.32	184.5	5.443
95	391.1	190.2	6.076	195.4	190.1	5.789	97.55	189.7	5.500
100	411.7	195.4	6.129	205.8	195.2	5.842	102.8	194.9	5.553

TABLE IV. Thermodynamic Properties of Liquid and Gaseous Neon (*Continued*)

T, K	p = 1 bar			p = 2 bar			p = 4 bar		
	$v \cdot 10^3$	h	s	$v \cdot 10^3$	h	s	$v \cdot 10^3$	h	s
110	453.0	205.7	6.227	226.5	205.6	5.941	113.2	205.3	5.653
120	494.3	216.1	6.317	247.2	215.9	6.031	123.6	215.7	5.743
130	535.6	226.4	6.400	267.9	226.3	6.113	134.0	226.1	5.826
140	576.9	236.7	6.476	288.5	236.6	6.190	144.3	236.5	5.903
150	618.1	247.0	6.547	309.2	246.9	6.261	154.7	246.8	5.975
160	659.4	257.3	6.614	329.8	257.3	6.328	165.1	257.2	6.041
170	700.6	267.6	6.676	350.5	267.6	6.390	175.4	267.5	6.104
180	741.9	277.9	6.735	371.1	277.9	6.449	185.7	277.8	6.163
190	783.1	288.2	6.791	391.7	288.2	6.505	196.1	288.2	6.219
200	824.3	298.5	6.844	412.4	298.5	6.558	206.4	298.5	6.272
220	906.7	319.2	6.942	453.6	319.2	6.656	227.0	319.2	6.370
240	989.2	339.8	7.032	494.8	339.8	6.746	247.7	339.8	6.460
260	1072	360.4	7.114	536.1	360.4	6.828	268.3	360.4	6.543
280	1154	381.0	7.190	577.3	381.0	6.905	288.9	381.0	6.619
300	1236	401.6	7.262	618.5	401.6	6.976	309.5	401.7	6.690
350	1442	453.1	7.420	721.5	453.1	7.135	361.1	453.2	6.849
400	1648	504.6	7.558	824.5	504.6	7.272	412.6	504.7	6.987
450	1854	556.1	7.679	927.5	556.1	7.394	464.1	556.2	7.108
500	2060	607.6	7.788	1031	607.6	7.502	515.6	607.7	7.217
550	2266	659.1	7.886	1134	659.1	7.600	567.1	659.2	7.315

TABLE IV. Thermodynamic Properties of Liquid and Gaseous Neon (*Continued*)

T, K	p = 1 bar			p = 2 bar			p = 4 bar		
	$v \cdot 10^3$	h	s	$v \cdot 10^3$	h	s	$v \cdot 10^3$	h	s
600	2472	710.6	7.975	1237	710.6	7.690	618.6	710.7	7.404
650	2678	762.1	8.058	1339	762.1	7.772	670.1	762.2	7.487
700	2884	813.5	8.134	1442	813.6	7.849	721.6	813.7	7.563
750	3090	865.0	8.205	1545	865.1	7.920	773.1	865.2	7.634
800	3296	916.5	8.272	1648	916.6	7.986	824.6	916.7	7.701
850	3502	968.0	8.334	1751	968.1	8.049	876.1	968.2	7.763
900	3708	1020	8.393	1854	1020	8.107	927.6	1020	7.822
950	3914	1071	8.449	1957	1071	8.163	979.1	1071	7.878
1000	4120	1123	8.502	2060	1123	8.216	1031	1123	7.930
1050	4326	1174	8.552	2163	1174	8.266	1082	1174	7.981
1100	4532	1225	8.600	2266	1225	8.314	1134	1226	8.029
1150	4738	1277	8.646	2369	1277	8.360	1185	1277	8.074
1200	4944	1328	8.690	2472	1329	8.404	1237	1329	8.118
1250	5150	1380	8.732	2575	1380	8.446	1288	1380	8.160
1300	5356	1431	8.773	2678	1432	8.486	1340	1432	8.201

TABLE IV. Thermodynamic Properties of Liquid and Gaseous Neon (*Continued*)

T, K	p = 6 bar			p = 8 bar			p = 10 bar		
	$v \cdot 10^3$	h	s	$v \cdot 10^3$	h	s	$v \cdot 10^3$	h	s
26	0.8153	31.18	1.488	0.8145	31.28	1.485	0.8138	31.38	1.483
28	0.8392	34.98	1.628	0.8383	35.07	1.626	0.8375	35.17	1.623
30	0.8665	38.96	1.766	0.865_3	39.04	1.763	0.8642	39.13	1.760

TABLE IV. Thermodynamic Properties of Liquid and Gaseous Neon (*Continued*)

T, K	p = 6 bar			p = 8 bar			p = 10 bar		
	$v \cdot 10^3$	h	s	$v \cdot 10^3$	h	s	$v \cdot 10^3$	h	s
32	0.8982	43.15	1.901	0.8967	43.21	1.897	0.8952	43.29	1.894
34	0.9363	47.58	2.035	0.9342	47.62	2.031	0.9322	47.67	2.027
36	20.76	122.8	4.207	0.9816	52.34	2.165	0.9785	52.34	2.160
38	22.67	125.8	4.289	15.89	123.1	4.119	11.66	119.9	3.961
40	24.45	128.6	4.361	17.42	126.4	4.202	13.11	123.8	4.063
42	26.15	131.2	4.425	18.83	129.3	4.275	14.38	127.2	4.146
44	27.80	133.7	4.483	20.16	132.1	4.339	15.55	130.3	4.218
46	29.40	136.1	4.538	21.44	134.7	4.397	16.65	133.2	4.281
48	30.96	138.5	4.588	22.68	137.2	4.451	17.70	135.9	4.339
50	32.51	140.9	4.635	23.89	139.7	4.501	18.71	138.5	4.391
55	36.28	146.5	4.743	26.82	145.5	4.613	21.15	144.6	4.508
60	39.97	152.0	4.839	29.67	151.2	4.711	23.49	150.4	4.610
65	43.61	157.4	4.926	32.46	156.8	4.800	25.76	156.1	4.700
70	47.22	162.8	5.006	35.21	162.2	4.881	28.00	161.6	4.783
75	50.80	168.2	5.080	37.93	167.6	4.956	30.21	167.1	4.858
80	54.36	173.5	5.148	40.63	173.0	5.025	32.40	172.5	4.929
85	57.90	178.8	5.213	43.31	178.4	5.090	34.56	177.9	4.994
90	61.43	184.1	5.273	45.98	183.7	5.151	36.72	183.3	5.055
95	64.94	189.3	5.330	48.64	189.0	5.208	38.86	188.6	5.113
100	68.45	194.6	5.384	51.29	194.3	5.262	40.99	194.0	5.168
110	75.44	205.1	5.483	56.56	204.8	5.363	45.24	204.5	5.269
120	82.41	215.5	5.574	61.81	215.3	5.454	49.46	215.1	5.360
130	89.36	225.9	5.658	67.04	225.7	5.538	53.66	225.6	5.444
140	96.29	236.3	5.735	72.26	236.2	5.615	57.85	236.0	5.522
150	103.2	246.7	5.806	77.47	246.6	5.687	62.02	246.5	5.594

TABLE IV. Thermodynamic Properties of Liquid and Gaseous Neon (*Continued*)

T, K	p = 6 bar			p = 8 bar			p = 10 bar		
	$v \cdot 10^3$	h	s	$v \cdot 10^3$	h	s	$v \cdot 10^3$	h	s
160	110.1	257.1	5.873	82.67	257.0	5.754	66.19	256.9	5.661
170	117.0	267.4	5.936	87.86	267.4	5.817	70.35	267.3	5.724
180	123.9	277.8	5.995	93.04	277.7	5.876	74.51	277.7	5.784
190	130.8	288.1	6.051	98.22	288.1	5.932	78.66	288.1	5.840
200	137.7	298.5	6.104	103.4	298.5	5.985	82.80	298.4	5.893
220	151.5	319.2	6.203	113.7	319.2	6.084	91.09	319.2	5.992
240	165.3	339.8	6.293	124.1	339.8	6.174	99.36	339.9	6.082
260	179.0	360.5	6.375	134.4	360.5	6.257	107.6	360.5	6.164
280	192.8	381.1	6.452	144.7	381.1	6.333	115.9	381.2	6.241
300	206.5	401.7	6.523	155.0	401.8	6.404	124.1	401.8	6.312
350	240.9	453.3	6.682	180.8	453.3	6.563	144.8	453.4	6.471
400	275.3	504.8	6.820	206.6	504.9	6.701	165.4	505.0	6.609
450	309.6	556.3	6.941	232.4	556.4	6.822	186.0	556.5	6.730
500	343.9	607.8	7.049	258.1	607.9	6.931	206.6	608.1	6.839
550	378.3	659.3	7.148	283.9	659.5	7.029	227.2	659.6	6.937
600	412.6	710.8	7.237	309.0	711.0	7.119	247.8	711.1	7.027
650	446.9	762.4	7.320	335.4	762.5	7.201	268.4	762.6	7.109
700	481.3	813.9	7.396	361.1	814.0	7.278	289.0	814.1	7.186
750	515.6	865.4	7.467	386.9	865.5	7.349	309.6	865.6	7.257
800	549.9	916.9	7.534	412.6	917.0	7.415	330.2	917.1	7.323
850	584.3	968.4	7.596	438.4	968.5	7.477	350.8	968.6	7.386
900	618.6	1020	7.655	464.1	1020	7.536	371.4	1020	7.444
950	652.9	1071	7.711	489.9	1071	7.592	392.0	1072	7.500
1000	687.3	1123	7.763	515.6	1123	7.645	412.6	1123	7.553
1050	721.6	1174	7.814	541.4	1174	7.695	433.2	1175	7.603

TABLE IV. Thermodynamic Properties of Liquid and Gaseous Neon (*Continued*)

T, K	p = 6 bar			p = 8 bar			p = 10 bar		
	$v \cdot 10^3$	h	s	$v \cdot 10^3$	h	s	$v \cdot 10^3$	h	s
1100	755.9	1226	7.862	567.1	1226	7.743	453.8	1226	7.651
1150	790.3	1277	7.907	592.9	1277	7.789	474.4	1278	7.697
1200	824.6	1329	7.951	618.6	1329	7.833	495.0	1329	7.741
1250	858.9	1380	7.993	644.4	1380	7.875	515.6	1381	7.783
1300	893.2	1432	8.034	670.1	1432	7.915	536.2	1432	7.823

TABLE IV. Thermodynamic Properties of Liquid and Gaseous Neon (*Continued*)

T, K	p = 12 bar			p = 14 bar			p = 16 bar		
	$v \cdot 10^3$	h	s	$v \cdot 10^3$	h	s	$v \cdot 10^3$	h	s
26	0.8131	31.49	1.481	0.8125	31.59	1.479	0.8118	31.70	1.476
28	0.8366	35.27	1.621	0.8357	35.37	1.618	0.8349	35.48	1.616
30	0.8631	39.22	1.757	0.8620	39.31	1.754	0.8609	39.40	1.751
32	0.8937	43.36	1.891	0.8923	43.43	1.887	0.8909	43.51	1.884
34	0.9302	47.72	2.023	0.9282	47.77	2.019	0.9263	47.82	2.015
36	0.9755	52.35	2.155	0.9726	52.36	2.150	0.9698	52.38	2.145
38	1.036	57.67	2.293	1.031	57.42	2.285	1.027	57.36	2.279
40	10.13	120.8	3.931	7.876	117.2	3.794	1.110	63.19	2.428
42	11.35	124.9	4.030	9.130	122.2	3.917	7.391	119.1	3.803
44	12.43	128.4	4.111	10.17	126.2	4.011	8.441	123.8	3.914
46	13.42	131.5	4.180	11.10	129.7	4.088	9.337	127.8	4.002
48	14.36	134.4	4.242	11.96	132.9	4.155	10.14	131.2	4.075
50	15.25	137.1	4.298	12.77	135.8	4.215	10.89	134.4	4.139

TABLE IV. Thermodynamic Properties of Liquid and Gaseous Neon (*Continued*)

T, K	p = 12 bar			p = 14 bar			p = 16 bar		
	$v \cdot 10^3$	h	s	$v \cdot 10^3$	h	s	$v \cdot 10^3$	h	s
55	17.36	143.5	4.420	14.65	142.5	4.343	12.62	141.4	4.274
60	19.36	149.6	4.525	16.42	148.7	4.451	14.21	147.8	4.386
65	21.30	155.4	4.617	18.12	154.6	4.546	15.73	153.9	4.483
70	23.20	161.0	4.701	19.77	160.4	4.631	17.20	159.7	4.570
75	25.06	166.5	4.778	21.39	166.0	4.709	18.64	165.5	4.648
80	26.91	172.0	4.849	22.99	171.6	4.781	20.05	171.1	4.721
85	28.73	177.5	4.915	24.57	177.1	4.847	21.45	176.6	4.788
90	30.54	182.9	4.977	26.13	182.5	4.910	22.83	182.1	4.851
95	32.34	188.3	5.035	27.69	187.9	4.968	24.20	187.6	4.910
100	34.13	193.7	5.090	29.23	193.3	5.024	25.56	193.0	4.966
110	37.69	204.3	5.191	32.30	204.0	5.126	28.25	203.8	5.068
120	41.22	214.9	5.283	35.34	214.6	5.218	30.93	214.5	5.161
130	44.74	225.4	5.368	38.36	225.2	5.303	33.59	225.1	5.245
140	48.24	235.9	5.446	41.37	235.8	5.381	36.23	235.7	5.323
150	51.73	246.4	5.518	44.37	246.3	5.453	38.86	246.2	5.397
160	55.21	256.8	5.585	47.36	256.7	5.521	41.48	256.7	5.465
170	58.68	267.2	5.648	50.35	267.2	5.584	44.09	267.1	5.528
180	62.15	277.6	5.708	53.32	277.6	5.644	46.70	277.6	5.588
190	65.61	288.0	5.764	56.30	288.0	5.700	49.31	288.0	5.644
200	69.07	298.4	5.817	59.27	298.4	5.753	51.91	298.4	5.698
220	75.98	319.2	5.916	65.19	319.2	5.852	57.10	319.2	5.797
240	82.88	339.9	6.006	71.11	339.9	5.942	62.29	339.9	5.887
260	89.78	360.6	6.089	77.03	360.6	6.025	67.46	360.6	5.970
280	96.67	381.2	6.166	82.93	381.3	6.102	72.64	381.3	6.047
300	103.6	401.9	6.237	88.84	402.0	6.173	77.80	402.0	6.118

TABLE IV. Thermodynamic Properties of Liquid and Gaseous Neon (*Continued*)

T, K	p = 12 bar			p = 14 bar			p = 16 bar		
	$v \cdot 10^3$	h	s	$v \cdot 10^3$	h	s	$v \cdot 10^3$	h	s
350	120.7	453.5	6.396	103.6	453.6	6.332	90.71	453.7	6.277
400	137.9	505.1	6.534	118.3	505.2	6.470	103.6	505.3	6.415
450	155.1	556.6	6.655	133.0	556.7	6.592	116.5	556.8	6.537
500	172.3	608.2	6.764	147.8	608.3	6.700	129.4	608.4	6.645
550	189.5	659.7	6.862	162.5	659.8	6.798	142.3	659.9	6.743
600	206.6	711.2	6.952	177.2	711.3	6.888	155.1	711.4	6.833
650	223.8	762.7	7.034	191.9	762.8	6.971	168.0	763.0	6.916
700	241.0	814.2	7.110	206.6	814.4	7.047	180.9	814.5	6.992
750	258.1	865.7	7.182	221.4	865.9	7.118	193.8	866.0	7.063
800	275.3	917.2	7.248	236.1	917.4	7.184	206.7	917.5	7.129
850	292.5	968.6	7.310	250.8	968.9	7.247	219.5	969.0	7.192
900	309.6	1020	7.369	265.5	1020	7.306	232.4	1021	7.251
950	326.8	1072	7.425	280.2	1072	7.362	245.3	1072	7.306
1000	344.0	1123	7.478	294.9	1123	7.415	258.3	1124	7.359
1050	361.1	1175	7.528	309.6	1175	7.465	271.0	1175	7.410
1100	378.3	1226	7.576	324.4	1226	7.512	283.9	1226	7.457
1150	395.5	1278	7.622	339.1	1278	7.558	296.8	1278	7.503
1200	412.6	1329	7.666	353.8	1329	7.602	309.7	1329	7.547
1250	429.8	1381	7.708	368.5	1381	7.644	322.5	1381	7.589
1300	447.0	1432	7.748	383.2	1432	7.685	335.4	1432	7.630

TABLE IV. Thermodynamic Properties of Liquid and Gaseous Neon (*Continued*)

T, K	p = 18 bar			p = 20 bar			p = 22 bar		
	$v \cdot 10^3$	h	s	$v \cdot 10^3$	h	s	$v \cdot 10^3$	h	s
26	0.8111	31.80	1.474	0.8104	31.91	1.472	0.8097	32.02	1.470
28	0.8341	35.56	1.613	0.8332	35.66	1.611	0.8324	35.76	1.609
30	0.8599	39.48	1.749	0.8588	39.57	1.746	0.8578	39.66	1.743
32	0.8895	43.58	1.881	0.8881	43.66	1.878	0.8868	43.73	1.874
34	0.9244	47.88	2.011	0.9226	47.93	2.007	0.9207	47.99	2.003
36	0.9671	52.40	2.140	0.9645	52.42	2.135	0.9619	52.45	2.130
38	1.023	57.32	2.272	1.018	57.28	2.266	1.014	57.25	2.260
40	1.102	63.00	2.417	1.094	62.83	2.408	1.087	62.68	2.399
42	5.943	115.2	3.680	1.226	69.70	2.577	1.208	69.40	2.563
44	7.055	121.2	3.818	5.899	118.1	3.719	4.891	114.4	3.611
46	7.945	125.7	3.918	6.808	123.4	3.836	5.855	120.8	3.754
48	8.722	129.5	4.000	7.572	127.6	3.927	6.619	125.6	3.856
50	9.431	132.9	4.069	8.253	131.3	4.002	7.283	129.7	3.938
55	11.03	140.3	4.211	9.759	139.2	4.153	8.718	138.0	4.098
60	12.49	147.0	4.327	11.11	146.1	4.272	9.985	145.2	4.222
65	13.87	153.2	4.426	12.38	152.4	4.374	11.16	151.7	4.327
70	15.20	159.1	4.514	13.60	158.5	4.464	12.29	157.9	4.418
75	16.49	164.9	4.594	14.78	164.4	4.545	13.38	163.8	4.501
80	17.77	170.6	4.668	15.94	170.1	4.620	14.45	169.6	4.576
85	19.02	176.2	4.736	17.08	175.8	4.688	15.49	175.3	4.645
90	20.26	181.7	4.799	18.20	181.4	4.752	16.52	181.0	4.709
95	21.48	187.2	4.859	19.31	186.9	4.812	17.54	186.6	4.770
100	22.70	192.7	4.915	20.42	192.4	4.869	18.55	192.1	4.826

TABLE IV. Thermodynamic Properties of Liquid and Gaseous Neon (*Continued*)

T, K	p = 18 bar			p = 20 bar			p = 22 bar		
	$v \cdot 10^3$	h	s	$v \cdot 10^3$	h	s	$v \cdot 10^3$	h	s
110	25.11	203.5	5.018	22.60	203.2	4.972	20.54	203.0	4.931
120	27.50	214.2	5.111	24.75	214.0	5.066	22.51	213.8	5.025
130	29.87	224.9	5.196	26.89	224.7	5.151	24.46	224.5	5.111
140	32.22	235.5	5.275	29.02	235.3	5.230	26.40	235.2	5.190
150	34.57	246.1	5.348	31.14	245.9	5.303	28.33	245.8	5.263
160	36.90	256.6	5.415	33.24	256.5	5.371	30.25	256.4	5.331
170	39.23	267.1	5.479	35.34	267.0	5.435	32.16	266.9	5.395
180	41.56	277.5	5.539	37.44	277.5	5.495	34.07	277.4	5.455
190	43.87	288.0	5.595	39.53	287.9	5.551	35.97	287.9	5.512
200	46.19	298.4	5.649	41.61	298.4	5.605	37.87	298.4	5.565
220	50.81	319.2	5.748	45.78	319.2	5.704	41.66	319.2	5.665
240	55.42	340.0	5.838	49.93	340.0	5.795	45.44	340.0	5.755
260	60.03	360.7	5.921	54.08	360.7	5.878	49.21	360.8	5.838
280	64.63	381.4	5.998	58.22	381.5	5.954	52.98	381.5	5.915
300	69.22	402.1	6.069	62.36	402.2	6.026	56.74	402.2	5.986
350	80.70	453.8	6.229	72.69	453.8	6.185	66.14	453.9	6.146
400	92.16	505.4	6.366	83.01	505.4	6.323	75.52	505.5	6.284
450	103.6	556.9	6.488	93.33	557.0	6.445	84.90	557.1	6.405
500	115.1	608.5	6.597	103.6	608.6	6.553	94.27	608.7	6.514
550	126.5	660.0	6.695	113.9	660.2	6.651	103.6	660.3	6.612
600	138.0	711.6	6.785	124.2	711.7	6.741	113.0	711.8	6.702
650	149.4	763.1	6.867	134.6	763.2	6.824	122.4	763.3	6.784
700	160.9	814.6	6.943	144.9	814.7	6.900	131.7	814.8	6.861
750	172.3	866.1	7.014	155.2	866.2	6.971	141.1	866.4	6.932
800	183.8	917.6	7.081	165.5	917.7	7.038	150.5	917.9	6.998

TABLE IV. Thermodynamic Properties of Liquid and Gaseous Neon (*Continued*)

T, K	p = 18 bar			p = 20 bar			p = 22 bar		
	$v \cdot 10^3$	h	s	$v \cdot 10^3$	h	s	$v \cdot 10^3$	h	s
850	195.2	969.1	7.143	175.8	969.3	7.100	159.8	969.4	7.061
900	206.7	1021	7.202	186.1	1021	7.159	169.2	1021	7.120
950	218.1	1072	7.258	196.4	1072	7.215	178.6	1072	7.175
1000	229.5	1124	7.311	206.7	1124	7.267	187.9	1124	7.228
1050	241.0	1175	7.361	217.0	1175	7.318	197.3	1175	7.278
1100	252.4	1227	7.409	227.3	1227	7.366	206.7	1227	7.326
1150	263.9	1278	7.455	237.6	1278	7.411	216.0	1278	7.372
1200	275.3	1330	7.499	247.9	1330	7.455	225.4	1330	7.416
1250	286.8	1381	7.541	258.2	1381	7.497	234.8	1381	7.458
1300	298.2	1433	7.581	268.5	1433	7.537	244.1	1433	7.498

TABLE IV. Thermodynamic Properties of Liquid and Gaseous Neon (*Continued*)

T, K	p = 24 bar			p = 26 bar			p = 28 bar		
	$v \cdot 10^3$	h	s	$v \cdot 10^3$	h	s	$v \cdot 10^3$	h	s
26	0.8091	32.12	1.468	0.8084	32.23	1.466	0.8078	32.34	1.464
28	0.8316	35.86	1.606	0.8308	35.96	1.604	0.8300	36.06	1.602
30	0.8568	39.75	1.740	0.8558	39.85	1.738	0.8548	39.94	1.735
32	0.8855	43.81	1.871	0.8842	43.89	1.868	0.8829	43.97	1.865
34	0.9190	48.05	2.000	0.9172	48.11	1.996	0.9155	48.17	1.993
36	0.9594	52.47	2.126	0.9570	52.51	2.121	0.9546	52.54	2.117
38	1.011	57.22	2.254	1.007	57.21	2.248	1.003	57.19	2.242
40	1.081	62.55	2.390	1.075	62.44	2.382	1.069	62.34	2.374

TABLE IV. Thermodynamic Properties of Liquid and Gaseous Neon (*Continued*)

T, K	p = 24 bar			p = 26 bar			p = 28 bar		
	$v \cdot 10^3$	h	s	$v \cdot 10^3$	h	s	$v \cdot 10^3$	h	s
42	1.192	69.10	2.550	1.178	68.82	2.537	1.166	68.58	2.525
44	3.948	109.7	3.483	1.510	80.40	2.796	1.456	79.19	2.762
46	5.033	118.0	3.668	4.305	114.7	3.577	3.632	110.8	3.476
48	5.812	123.5	3.786	5.117	121.2	3.715	4.508	118.7	3.643
50	6.467	127.9	3.876	5.771	126.1	3.815	5.168	124.2	3.755
55	7.848	136.8	4.046	7.111	135.6	3.996	6.478	134.3	3.949
60	9.047	144.2	4.175	8.252	143.3	4.130	7.571	142.3	4.088
65	10.15	150.9	4.282	9.295	150.1	4.241	8.5612	149.4	4.201
70	11.20	157.2	4.376	10.28	156.5	4.336	9.492	155.9	4.298
75	12.21	163.3	4.459	11.23	162.7	4.420	10.38	162.2	4.384
80	13.20	169.1	4.535	12.15	168.7	4.497	11.25	168.2	4.462
85	14.17	174.9	4.605	13.05	174.5	4.568	12.09	174.1	4.534
90	15.12	180.6	4.670	13.94	180.2	4.634	12.92	179.8	4.600
95	16.06	186.2	4.731	14.81	185.9	4.695	13.74	185.5	4.661
100	16.99	191.8	4.788	15.67	191.5	4.752	14.55	191.2	4.719
110	18.83	202.8	4.893	17.38	202.5	4.858	16.14	202.3	4.825
120	20.64	213.6	4.987	19.06	213.4	4.952	17.70	213.2	4.920
130	22.44	224.4	5.073	20.72	224.2	5.039	19.25	224.0	5.007
140	24.22	235.1	5.153	22.37	235.0	5.118	20.79	234.8	5.087
150	25.99	245.7	5.226	24.01	245.6	5.192	22.32	245.5	5.161
160	27.75	256.3	5.294	25.64	256.2	5.261	23.83	256.2	5.229
170	29.51	266.9	5.358	27.27	266.8	5.325	25.34	266.8	5.293
180	31.26	277.4	5.418	28.88	277.3	5.385	26.85	277.3	5.354
190	33.01	287.9	5.475	30.50	287.8	5.442	28.35	287.8	5.411
200	34.75	298.3	5.529	32.11	298.3	5.495	29.85	298.3	5.464

TABLE IV. Thermodynamic Properties of Liquid and Gaseous Neon (*Continued*)

T, K	p = 24 bar			p = 26 bar			p = 28 bar		
	$v \cdot 10^3$	h	s	$v \cdot 10^3$	h	s	$v \cdot 10^3$	h	s
220	38.23	319.2	5.628	35.32	319.2	5.595	32.83	319.2	5.564
240	41.69	340.0	5.719	38.52	340.1	5.686	35.81	340.1	5.655
260	45.15	360.8	5.802	41.72	360.9	5.769	38.78	360.9	5.738
280	48.61	381.6	5.879	44.91	381.6	5.846	41.74	381.7	5.815
300	52.06	402.3	5.950	48.10	402.3	5.917	44.70	402.4	5.887
350	60.67	454.0	6.110	56.05	454.0	6.077	52.09	454.1	6.046
400	69.28	505.6	6.248	64.00	505.7	6.215	59.47	505.8	6.184
450	77.88	557.2	6.369	71.94	557.3	6.336	66.84	557.4	6.306
500	86.47	608.8	6.478	79.87	608.9	6.445	74.21	609.0	6.414
550	95.06	660.4	6.576	87.80	660.5	6.543	81.58	660.6	6.513
600	103.7	711.9	6.666	95.73	712.0	6.633	88.94	712.2	6.602
650	112.2	763.5	6.748	103.7	763.6	6.715	96.30	763.7	6.685
700	120.8	815.0	6.825	111.6	815.1	6.792	103.7	815.2	6.761
750	129.4	866.5	6.896	119.5	866.6	6.863	111.0	866.7	6.832
800	138.0	918.0	6.962	127.4	918.1	6.929	118.4	918.3	6.899
850	146.6	969.5	7.025	135.4	969.6	6.992	125.7	969.8	6.961
900	155.2	1021	7.084	143.3	1021	7.051	133.1	1021	7.020
950	163.7	1073	7.139	151.2	1073	7.106	140.5	1073	7.076
1000	172.3	1124	7.192	159.1	1124	7.159	147.8	1124	7.129
1050	180.9	1176	7.243	167.1	1176	7.210	155.2	1176	7.179
1100	189.5	1227	7.290	175.0	1227	7.257	162.5	1227	7.227
1150	198.1	1279	7.336	182.9	1279	7.303	169.9	1279	7.273
1200	206.7	1330	7.380	190.8	1330	7.347	177.2	1330	7.317
1250	215.2	1382	7.422	198.7	1382	7.389	184.6	1382	7.359
1300	223.8	1433	7.463	206.7	1433	7.430	192.0	1433	7.399

TABLE IV. Thermodynamic Properties of Liquid and Gaseous Neon (*Continued*)

T, K	p = 30 bar			p = 32 bar			p = 34 bar		
	$v \cdot 10^3$	h	s	$v \cdot 10^3$	h	s	$v \cdot 10^3$	h	s
26	0.8071	32.44	1.461	0.8065	32.55	1.459	0.8058	32.65	1.457
28	0.8292	36.16	1.599	0.8284	36.26	1.597	0.8276	36.36	1.595
30	0.8538	40.03	1.733	0.8528	40.12	1.730	0.8518	40.22	1.727
32	0.8816	44.05	1.862	0.8804	44.13	1.859	0.8791	44.21	1.856
34	0.9139	48.23	1.989	0.9122	48.30	1.986	0.9106	48.36	1.982
36	0.9523	52.58	2.113	0.9500	52.62	2.109	0.9478	52.66	2.105
38	1.000	57.19	2.237	0.9967	57.18	2.231	0.9936	57.19	2.226
40	1.063	62.25	2.366	1.058	62.17	2.359	1.053	62.10	2.352
42	1.155	68.30	2.514	1.145	68.06	2.503	1.136	67.85	2.493
44	1.410	78.15	2.732	1.370	77.30	2.705	1.334	76.58	2.682
46	2.961	105.7	3.349	1.990	96.10	3.135	1.670	88.35	2.955
48	3.965	115.9	3.568	3.473	112.8	3.488	3.019	109.3	3.402
50	4.640	122.1	3.694	4.173	120.0	3.634	3.756	117.7	3.572
55	5.929	133.0	3.903	5.448	131.7	3.856	5.023	130.3	3.814
60	6.981	141.3	4.048	6.465	140.4	4.009	6.011	139.4	3.972
65	7.926	148.6	4.164	7.370	147.8	4.128	6.881	147.0	4.095
70	8.809	155.3	4.263	8.211	154.6	4.229	7.685	154.0	4.197
75	9.652	161.6	4.350	9.012	161.0	4.318	8.449	160.5	4.287
80	10.47	167.7	4.429	9.785	167.2	4.398	9.184	166.7	4.368
85	11.26	173.6	4.501	10.54	173.2	4.470	9.898	172.8	4.441
90	12.04	179.5	4.568	11.27	179.1	4.538	10.60	178.7	4.509
95	12.81	185.2	4.630	12.00	184.9	4.600	11.28	184.5	4.572
100	13.57	190.9	4.688	12.71	190.6	4.659	11.96	190.3	4.631

TABLE IV. Thermodynamic Properties of Liquid and Gaseous Neon (*Continued*)

T, K	p = 30 bar			p = 32 bar			p = 34 bar		
	$v \cdot 10^3$	h	s	$v \cdot 10^3$	h	s	$v \cdot 10^3$	h	s
110	15.06	202.0	4.794	14.12	201.8	4.766	13.29	201.6	4.739
120	16.53	213.0	4.890	15.50	212.8	4.862	14.60	212.7	4.835
130	17.98	223.9	4.977	16.87	223.8	4.949	15.89	223.6	4.923
140	19.42	234.7	5.057	18.22	234.6	5.029	17.16	234.5	5.003
150	20.85	245.4	5.131	19.56	245.3	5.104	18.43	245.2	5.078
160	22.27	256.1	5.200	20.89	256.0	5.172	19.68	255.9	5.147
170	23.68	266.7	5.264	22.22	266.6	5.237	20.93	266.6	5.211
180	25.08	277.3	5.325	23.54	277.2	5.297	22.18	277.2	5.272
190	26.49	287.8	5.382	24.86	287.8	5.354	23.42	287.7	5.329
200	27.89	298.3	5.435	26.17	298.3	5.408	24.66	298.3	5.383
220	30.67	319.2	5.535	28.79	319.3	5.508	27.12	319.3	5.483
240	33.45	340.1	5.626	31.40	340.1	5.599	29.58	340.2	5.574
260	36.23	360.9	5.709	34.00	361.0	5.683	32.03	361.0	5.657
280	39.00	381.7	5.786	36.59	381.8	5.760	34.47	381.8	5.734
300	41.76	402.5	5.858	39.19	402.5	5.831	36.91	402.6	5.806
350	48.66	454.2	6.018	45.66	454.3	5.991	43.01	454.4	5.966
400	55.55	505.9	6.156	52.12	506.0	6.129	49.09	506.1	6.104
450	62.43	557.6	6.277	58.57	557.7	6.251	55.16	557.8	6.226
500	69.31	609.2	6.386	65.02	609.3	6.359	61.23	609.4	6.334
550	76.18	660.7	6.484	71.46	660.8	6.458	67.30	660.9	6.433
600	83.05	712.3	6.574	77.90	712.4	6.547	73.36	712.5	6.522
650	89.92	763.8	6.656	84.35	763.9	6.630	79.42	764.1	6.605
700	96.79	815.3	6.733	90.79	815.5	6.706	85.49	815.6	6.681
750	103.7	866.9	6.804	97.23	867.0	6.777	91.55	867.1	6.752
800	110.5	918.4	6.870	103.7	918.5	6.844	97.61	918.6	6.819

TABLE IV. Thermodynamic Properties of Liquid and Gaseous Neon (*Continued*)

T, K	p = 30 bar			p = 32 bar			p = 34 bar		
	$v \cdot 10^3$	h	s	$v \cdot 10^3$	h	s	$v \cdot 10^3$	h	s
850	117.4	969.9	6.933	110.1	970.0	6.906	103.7	970.2	6.881
900	124.3	1021	6.992	116.5	1022	6.965	109.7	1022	6.940
950	131.1	1073	7.047	123.0	1073	7.021	115.8	1073	6.996
1000	138.0	1124	7.100	129.4	1125	7.074	121.8	1125	7.049
1050	144.9	1176	7.151	135.9	1176	7.124	127.9	1176	7.099
1100	151.7	1227	7.199	142.3	1228	7.172	134.0	1228	7.147
1150	158.6	1279	7.244	148.7	1279	7.218	140.0	1279	7.193
1200	165.5	1330	7.288	155.2	1331	7.262	146.1	1331	7.237
1250	172.3	1382	7.330	161.6	1382	7.304	152.1	1382	7.279
1300	179.2	1433	7.371	168.0	1434	7.344	158.2	1434	7.319

TABLE IV. Thermodynamic Properties of Liquid and Gaseous Neon (*Continued*)

T, K	p = 36 bar			p = 38 bar			p = 40 bar		
	$v \cdot 10^3$	h	s	$v \cdot 10^3$	h	s	$v \cdot 10^3$	h	s
26	0.8052	32.76	1.455	0.8046	32.87	1.453	0.8040	32.98	1.451
28	0.8269	36.47	1.592	0.8261	36.57	1.590	0.8253	36.67	1.588
30	0.8509	40.31	1.725	0.8499	40.40	1.722	0.8490	40.50	1.720
32	0.8779	44.30	1.853	0.8767	44.38	1.851	0.8755	44.46	1.848
34	0.9090	48.43	1.979	0.9075	48.50	1.975	0.9059	48.57	1.972
36	0.9457	52.70	2.100	0.9436	52.75	2.096	0.9415	52.79	2.093
38	0.9905	57.19	2.221	0.9875	57.20	2.216	0.9846	57.21	2.211
40	1.048	62.04	2.345	1.043	61.99	2.339	1.039	61.94	2.332

TABLE IV. Thermodynamic Properties of Liquid and Gaseous Neon (*Continued*)

T, K	p = 36 bar			p = 38 bar			p = 40 bar		
	$v \cdot 10^3$	h	s	$v \cdot 10^3$	h	s	$v \cdot 10^3$	h	s
42	1.127	67.67	2.484	1.119	67.50	2.474	1.112	67.35	2.466
44	1.303	75.92	2.666	1.275	75.35	2.642	1.251	74.79	2.625
46	1.519	84.80	2.870	1.445	83.23	2.830	1.408	82.19	2.800
48	2.585	105.1	3.305	2.213	100.6	3.197	1.926	96.18	3.097
50	3.382	115.2	3.509	3.044	112.6	3.444	2.740	109.9	3.375
55	4.645	128.9	3.771	4.308	127.5	3.729	4.005	126.1	3.688
60	5.607	138.4	3.936	5.246	137.4	3.901	4.923	136.3	3.867
65	6.446	146.2	4.062	6.058	145.5	4.030	5.709	144.7	4.000
70	7.218	153.3	4.167	6.801	152.7	4.137	6.426	152.0	4.109
75	7.948	159.9	4.258	7.501	159.4	4.230	7.099	158.8	4.20
80	8.650	166.3	4.340	8.173	165.8	4.313	7.744	165.3	4.287
85	9.331	172.3	4.414	8.823	172.0	4.388	8.367	171.5	4.363
90	9.996	178.3	4.482	9.458	178.0	4.456	8.974	177.6	4.432
95	10.65	184.2	4.546	10.08	183.9	4.520	9.569	183.5	4.496
100	11.29	190.0	4.605	10.69	189.7	4.580	10.15	189.4	4.556
110	12.55	201.3	4.713	11.89	201.1	4.689	11.30	200.8	4.665
120	13.79	212.5	4.810	13.07	212.3	4.786	12.42	212.1	4.763
130	15.01	223.5	4.898	14.23	223.3	4.874	13.53	223.2	4.852
140	16.22	234.3	4.979	15.38	234.2	4.955	14.62	234.1	4.933
150	17.42	245.1	5.053	16.52	245.0	5.030	15.70	244.9	5.008
160	18.61	255.8	5.122	17.65	255.8	5.099	16.78	255.7	5.077
170	19.79	266.5	5.187	18.77	266.5	5.164	17.85	266.4	5.142
180	20.97	277.1	5.248	19.89	277.1	5.225	18.91	277.1	5.203
190	22.14	287.7	5.305	21.00	287.7	5.282	19.97	287.7	5.260
200	23.31	298.3	5.359	22.11	298.3	5.336	21.02	298.2	5.315

TABLE IV. Thermodynamic Properties of Liquid and Gaseous Neon (*Continued*)

T, K	p = 36 bar			p = 38 bar			p = 40 bar		
	$v \cdot 10^3$	h	s	$v \cdot 10^3$	h	s	$v \cdot 10^3$	h	s
220	25.64	319.3	5.459	24.32	319.3	5.436	23.12	319.3	5.415
240	27.96	340.2	5.550	26.52	340.2	5.527	25.22	340.3	5.506
260	30.28	361.1	5.634	28.71	361.1	5.611	27.31	361.2	5.590
280	32.59	381.9	5.711	30.90	381.9	5.688	29.39	382.0	5.667
300	34.90	402.7	5.782	33.09	402.7	5.760	31.46	402.8	5.739
350	40.65	454.5	5.942	38.54	454.6	5.920	36.65	454.6	5.899
400	46.39	506.2	6.080	43.98	506.3	6.058	41.82	506.4	6.037
450	52.13	557.9	6.202	49.42	558.0	6.180	46.98	558.1	6.158
500	57.86	609.5	6.311	54.85	609.6	6.288	52.14	609.7	6.267
550	63.59	661.1	6.409	60.28	661.2	6.387	57.30	661.3	6.366
600	69.32	712.6	6.499	65.71	712.7	6.476	62.46	712.9	6.455
650	75.05	764.2	6.581	71.13	764.3	6.559	67.61	764.4	6.538
700	80.77	815.7	6.658	76.56	815.8	6.635	72.76	816.0	6.614
750	86.50	867.2	6.729	81.98	867.4	6.707	77.92	867.5	6.685
800	92.22	918.8	6.795	87.41	918.9	6.773	83.07	919.0	6.752
850	97.95	970.4	6.858	92.83	970.4	6.835	88.22	970.5	6.814
900	103.7	1022	6.917	98.25	1022	6.894	93.37	1022	6.873
950	109.4	1073	6.972	103.7	1073	6.950	98.52	1074	6.929
1000	115.1	1125	7.025	109.1	1125	7.003	103.7	1125	6.982
1050	120.8	1176	7.075	114.5	1176	7.053	108.8	1177	7.032
1100	126.6	1228	7.123	119.9	1228	7.101	114.0	1228	7.080
1150	132.3	1279	7.169	125.4	1279	7.147	119.1	1280	7.126
1200	138.0	1331	7.213	130.8	1331	7.191	124.3	1331	7.170
1250	143.7	1382	7.255	136.2	1382	7.233	129.4	1383	7.212
1300	149.5	1434	7.295	141.6	1434	7.273	134.6	1434	7.252

TABLE IV. Thermodynamic Properties of Liquid and Gaseous Neon (*Continued*)

T, K	p = 42 bar			p = 44 bar			p = 46 bar		
	$v \cdot 10^3$	h	s	$v \cdot 10^3$	h	s	$v \cdot 10^3$	h	s
26	0.8034	33.09	1.449	0.8027	33.19	1.447	0.8021	33.30	1.445
28	0.8246	36.77	1.586	0.8238	36.88	1.583	0.8231	36.98	1.581
30	0.8481	40.59	1.717	0.8471	40.69	1.715	0.8462	40.78	1.712
32	0.8744	44.55	1.845	0.8732	44.64	1.842	0.8721	44.72	1.839
34	0.9044	48.64	1.969	0.9029	48.71	1.966	0.9015	48.78	1.962
36	0.9395	52.84	2.089	0.9375	52.89	2.085	0.9356	52.95	2.081
38	0.9817	57.23	2.207	0.9790	57.25	2.202	0.9763	57.25	2.197
40	1.035	61.91	2.326	1.031	61.88	2.320	1.027	61.85	2.315
42	1.105	67.22	2.457	1.098	67.10	2.449	1.092	67.00	2.442
44	1.229	74.30	2.609	1.209	73.80	2.595	1.193	73.40	2.582
46	1.380	81.38	2.775	1.350	80.70	2.750	1.328	80.10	2.730
48	1.751	93.01	3.023	1.642	90.83	2.971	1.567	89.25	2.931
50	2.474	107.0	3.310	2.245	104.2	3.250	2.059	101.6	3.187
55	3.733	124.6	3.647	3.487	123.1	3.607	3.265	121.6	3.568
60	4.631	135.3	3.834	4.367	134.3	3.802	4.127	133.3	3.770
65	5.395	143.9	3.971	5.109	143.1	3.942	4.850	142.3	3.915
70	6.087	151.4	4.082	5.780	150.7	4.056	5.501	150.1	4.030
75	6.736	158.3	4.178	6.407	157.7	4.153	6.107	157.2	4.129
80	7.356	164.8	4.262	7.004	164.4	4.238	6.683	163.9	4.215
85	7.954	171.1	4.338	7.580	170.7	4.315	7.239	170.3	4.293
90	8.537	177.2	4.408	8.140	176.9	4.386	7.778	176.5	4.364
95	9.107	183.2	4.473	8.687	182.9	4.451	8.305	182.6	4.430
100	9.667	189.1	4.533	9.225	188.8	4.512	8.822	188.5	4.491

TABLE IV. Thermodynamic Properties of Liquid and Gaseous Neon (*Continued*)

T, K	p = 42 bar			p = 44 bar			p = 46 bar		
	$v \cdot 10^3$	h	s	$v \cdot 10^3$	h	s	$v \cdot 10^3$	h	s
110	10.76	200.6	4.643	10.28	200.4	4.622	9.832	200.2	4.602
120	11.84	211.9	4.741	11.31	211.7	4.720	10.82	211.5	4.701
130	12.89	223.0	4.830	12.32	222.9	4.810	11.79	222.7	4.790
140	13.94	234.0	4.912	13.32	233.9	4.891	12.75	233.7	4.872
150	14.97	244.9	4.987	14.30	244.8	4.966	13.70	244.7	4.947
160	16.00	255.6	5.056	15.28	255.6	5.036	14.64	255.5	5.017
170	17.02	266.4	5.121	16.26	266.3	5.101	15.57	266.3	5.082
180	18.03	277.1	5.182	17.23	277.0	5.162	16.50	277.0	5.144
190	19.04	287.7	5.240	18.19	287.6	5.220	17.42	287.6	5.201
200	20.04	298.2	5.294	19.15	298.2	5.274	18.34	298.2	5.256
220	22.05	319.3	5.394	21.07	319.3	5.375	20.17	319.3	5.356
240	24.04	340.3	5.486	22.97	340.3	5.466	22.00	340.4	5.448
260	26.03	361.2	5.569	24.87	361.2	5.550	23.81	361.3	5.531
280	28.01	382.0	5.647	26.77	382.1	5.627	25.62	382.2	5.609
300	29.99	402.9	5.718	28.66	402.9	5.699	27.43	403.0	5.681
350	34.93	454.7	5.878	33.37	454.8	5.859	31.91	454.9	5.841
400	39.86	506.5	6.017	38.07	506.6	5.997	36.44	506.7	5.979
450	44.77	558.2	6.138	42.77	558.3	6.119	40.94	558.4	6.101
500	49.69	609.8	6.247	47.46	609.9	6.228	45.43	610.0	6.210
550	54.60	661.4	6.345	52.15	661.5	6.326	49.91	661.6	6.308
600	59.51	713.0	6.435	56.84	713.1	6.416	54.40	713.2	6.398
650	64.42	764.5	6.518	61.53	764.7	6.499	58.88	764.8	6.480
700	69.33	816.1	6.594	66.21	816.2	6.575	63.36	816.3	6.557
750	74.24	867.6	6.665	70.90	867.8	6.646	67.84	867.9	6.628
800	79.15	919.2	6.732	75.58	919.3	6.713	72.32	919.4	6.694

TABLE IV. Thermodynamic Properties of Liquid and Gaseous Neon (*Continued*)

T, K	p = 42 bar			p = 44 bar			p = 46 bar		
	$v \cdot 10^3$	h	s	$v \cdot 10^3$	h	s	$v \cdot 10^3$	h	s
850	84.05	970.7	6.794	80.26	970.8	6.775	76.80	970.9	6.757
900	88.96	1022	6.853	84.95	1022	6.834	81.28	1022	6.816
950	93.86	1074	6.909	89.63	1074	6.890	85.76	1074	6.871
1000	98.77	1125	6.962	94.31	1125	6.943	90.24	1125	6.924
1050	103.7	1177	7.012	98.99	1177	6.993	94.72	1177	6.974
1100	108.6	1228	7.060	103.7	1228	7.041	99.20	1229	7.022
1150	113.5	1280	7.106	108.4	1280	7.086	103.7	1280	7.068
1200	118.4	1331	7.150	113.0	1331	7.130	108.2	1332	7.112
1250	123.3	1383	7.192	117.7	1383	7.172	112.6	1383	7.154
1300	128.2	1434	7.232	122.4	1434	7.213	117.1	1435	7.194

TABLE IV. Thermodynamic Properties of Liquid and Gaseous Neon (*Continued*)

T, K	p = 48 bar			p = 50 bar			p = 55 bar		
	$v \cdot 10^3$	h	s	$v \cdot 10^3$	h	s	$v \cdot 10^3$	h	s
26	0.8015	33.41	1.443	0.8009	33.52	1.441	0.7994	33.79	1.436
28	0.8224	37.03	1.579	0.8216	37.18	1.577	0.8199	37.44	1.571
30	0.8453	40.88	1.710	0.8445	40.98	1.708	0.8423	41.22	1.701
32	0.8710	44.81	1.837	0.8699	44.90	1.834	0.8672	45.12	1.827
34	0.9000	48.86	1.959	0.8986	48.93	1.956	0.8952	49.12	1.948
36	0.9337	53.00	2.078	0.9318	53.06	2.074	0.9273	53.21	2.065
38	0.9737	57.30	2.193	0.9712	57.33	2.189	0.9651	57.41	2.178
40	1.023	61.83	2.309	1.019	61.82	2.303	1.011	61.81	2.290

TABLE IV. Thermodynamic Properties of Liquid and Gaseous Neon (*Continued*)

T, K	p = 48 bar			p = 50 bar			p = 55 bar		
	$v \cdot 10^3$	h	s	$v \cdot 10^3$	h	s	$v \cdot 10^3$	h	s
42	1.086	66.90	2.434	1.081	66.82	2.425	1.068	66.65	2.409
44	1.172	73.00	2.570	1.161	72.65	2.559	1.141	72.03	2.536
46	1.299	79.54	2.715	1.280	79.11	2.704	1.242	78.22	2.670
48	1.512	88.08	2.900	1.468	87.05	2.873	1.397	85.45	2.820
50	1.919	99.31	3.130	1.803	97.30	3.082	1.620	93.78	2.995
55	3.064	120.1	3.529	2.883	118.6	3.482	2.507	115.0	3.400
60	3.908	132.2	3.740	3.709	131.2	3.710	3.281	128.6	3.639
65	4.613	141.5	3.888	4.396	140.7	3.862	3.927	138.7	3.799
70	5.245	149.4	4.006	5.010	148.8	3.982	4.502	147.2	3.925
75	5.832	156.6	4.106	5.580	156.1	4.083	5.033	154.7	4.030
80	6.390	163.4	4.193	6.120	162.9	4.172	5.534	161.8	4.121
85	6.926	169.9	4.272	6.639	169.5	4.251	6.014	168.5	4.202
90	7.446	176.2	4.343	7.142	175.8	4.323	6.479	174.9	4.275
95	7.954	182.3	4.409	7.632	181.9	4.389	6.931	181.2	4.343
100	8.452	188.2	4.471	8.112	188.0	4.451	7.373	187.3	4.406
110	9.425	199.9	4.582	9.050	199.7	4.663	8.234	199.1	4.519
120	10.37	211.3	4.681	9.965	211.2	4.663	9.073	210.7	4.620
130	11.31	222.6	4.771	10.86	222.4	4.753	9.895	222.1	4.710
140	12.23	233.6	4.853	11.75	233.5	4.835	10.70	233.2	4.793
150	13.14	244.6	4.929	12.62	244.5	4.911	11.50	244.3	4.869
160	14.04	255.4	4.999	13.49	255.4	4.981	12.30	255.2	4.940
170	14.94	266.2	5.064	14.35	266.2	5.047	13.08	266.0	5.006
180	15.83	276.9	5.125	15.21	276.9	5.108	13.86	276.8	5.067
190	16.72	287.6	5.183	16.06	287.6	5.166	14.64	287.5	5.125
200	17.60	298.2	5.238	16.91	298.2	5.220	15.41	298.2	5.180

TABLE IV. Thermodynamic Properties of Liquid and Gaseous Neon (*Continued*)

T, K	p = 48 bar			p = 50 bar			p = 55 bar		
	$v \cdot 10^3$	h	s	$v \cdot 10^3$	h	s	$v \cdot 10^3$	h	s
220	19.35	319.4	5.338	18.60	319.4	5.321	16.95	319.4	5.281
240	21.10	340.4	5.430	20.28	340.4	5.413	18.48	340.5	5.373
260	22.84	361.3	5.514	21.95	361.4	5.497	20.01	361.5	5.457
280	24.58	382.2	5.590	23.62	382.3	5.574	21.52	382.4	5.534
300	26.32	403.1	5.663	25.29	403.1	5.646	23.04	403.3	5.606
350	30.64	455.0	5.823	29.44	455.1	5.806	26.82	455.3	5.767
400	34.95	506.8	5.961	33.58	506.9	5.944	30.58	507.1	5.905
450	39.26	558.5	6.083	37.71	558.6	6.066	34.34	558.9	6.027
500	43.56	610.1	6.192	41.84	610.3	6.175	38.10	610.6	6.136
550	47.86	661.7	6.290	45.97	661.9	6.274	41.85	662.2	6.234
600	52.16	713.3	6.380	50.10	713.5	6.363	45.60	713.8	6.324
650	56.45	764.9	6.463	54.22	765.0	6.446	49.35	765.3	6.407
700	60.75	816.5	6.539	58.35	816.6	6.522	53.10	816.9	6.483
750	65.04	868.0	6.610	62.47	868.1	6.593	56.85	868.4	6.554
800	69.34	919.5	6.677	66.59	919.7	6.660	60.60	920.0	6.621
850	73.63	971.1	6.739	70.71	971.2	6.722	64.35	971.5	6.683
900	77.92	1023	6.798	74.83	1023	6.781	68.09	1023	6.742
950	82.22	1074	6.854	78.96	1074	6.837	71.84	1075	6.798
1000	86.51	1126	6.907	83.08	1126	6.890	75.59	1126	6.851
1050	90.80	1177	6.957	87.20	1177	6.940	79.33	1178	6.901
1100	95.09	1229	7.005	91.32	1229	6.988	83.08	1229	6.949
1150	99.39	1280	7.051	95.44	1280	7.034	86.82	1281	6.995
1200	103.7	1332	7.095	99.56	1332	7.078	90.57	1332	7.038
1250	108.0	1383	7.137	103.7	1383	7.120	94.32	1384	7.080
1300	112.3	1435	7.177	107.8	1435	7.160	98.06	1435	7.121

TABLE IV. Thermodynamic Properties of Liquid and Gaseous Neon (*Continued*)

T, K	p = 60 bar			p = 65 bar			p = 70 bar		
	$v \cdot 10^3$	h	s	$v \cdot 10^3$	h	s	$v \cdot 10^3$	h	s
26	0.7980	34.07	1.431	0.7966	34.34	1.426	0.7952	34.61	1.422
28	0.8181	37.71	1.566	0.8164	37.97	1.561	0.8147	38.23	1.556
30	0.8401	47.47	1.696	0.8381	41.71	1.690	0.8360	41.96	1.684
32	0.8645	45.34	1.821	0.8620	45.57	1.814	0.8595	45.80	1.808
34	0.8919	49.32	1.941	0.8887	49.52	1.934	0.8856	49.72	1.927
36	0.9230	53.36	2.057	0.9189	53.53	2.048	0.9150	53.70	2.040
38	0.9594	57.48	2.168	0.9539	57.62	2.158	0.9488	57.74	2.149
40	1.003	61.82	2.278	0.9953	61.85	2.267	0.9884	61.90	2.257
42	1.056	66.53	2.394	1.046	66.45	2.380	1.036	66.40	2.367
44	1.124	71.70	2.515	1.108	71.45	2.497	1.094	71.25	2.480
46	1.212	77.55	2.642	1.187	77.03	2.619	1.168	76.55	2.598
48	1.334	83.99	2.780	1.296	83.06	2.747	1.259	82.34	2.720
50	1.510	91.48	2.933	1.436	89.87	2.886	1.380	88.67	2.848
55	2.222	111.7	3.318	2.011	108.3	3.246	1.853	106.3	3.182
60	2.936	126.1	3.569	2.657	123.7	3.507	2.431	121.5	3.449
65	3.543	136.7	3.741	3.226	134.9	3.686	2.961	133.0	3.634
70	4.083	145.6	3.872	3.734	144.0	3.822	3.439	142.5	3.775
75	4.581	153.4	3.980	4.202	152.1	3.933	3.880	150.8	3.890
80	5.049	160.6	4.073	4.641	159.5	4.029	4.294	158.4	3.988
85	5.496	167.5	4.156	5.061	166.5	4.114	4.689	165.6	4.074
90	5.928	174.0	4.231	5.465	173.2	4.190	5.069	172.3	4.152
95	6.348	180.4	4.300	5.857	179.6	4.260	5.437	178.8	4.222
100	6.758	186.6	4.363	6.239	185.9	4.324	5.796	185.2	4.288

TABLE IV. Thermodynamic Properties of Liquid and Gaseous Neon (*Continued*)

T, K	p = 60 bar			p = 65 bar			p = 70 bar		
	$v \cdot 10^3$	h	s	$v \cdot 10^3$	h	s	$v \cdot 10^3$	h	s
110	7.556	198.6	4.478	6.982	198.1	4.440	6.493	197.5	4.405
120	8.331	210.3	4.580	7.704	209.9	4.543	7.167	209.4	4.508
130	9.090	221.7	4.671	8.409	221.4	4.635	7.826	221.0	4.601
140	9.836	233.0	4.755	9.102	232.7	4.719	8.473	232.4	4.686
150	10.57	244.1	4.831	9.785	243.9	4.796	9.110	243.7	4.763
160	11.30	255.0	4.902	10.46	254.9	4.867	9.740	254.7	4.835
170	12.02	265.9	4.968	11.13	265.8	4.933	10.36	265.7	4.901
180	12.74	276.7	5.030	11.80	276.7	4.995	10.98	276.6	4.964
190	13.46	287.5	5.088	12.46	287.5	5.054	11.60	287.4	5.022
200	14.17	298.2	5.143	13.11	298.2	5.109	12.21	298.2	5.077
220	15.58	319.5	5.244	14.42	319.5	5.210	13.43	319.6	5.179
240	16.99	340.6	5.336	15.72	340.7	5.302	14.63	340.8	5.271
260	18.38	361.6	5.420	17.01	361.7	5.387	15.84	361.9	5.356
280	19.78	382.6	5.498	18.30	382.7	5.464	17.03	382.9	5.434
300	21.17	403.5	5.570	19.58	403.6	5.537	18.23	403.8	5.506
350	24.63	455.5	5.730	22.78	455.7	5.697	21.20	455.9	5.666
400	28.09	507.4	5.869	25.97	507.6	5.836	24.16	507.8	5.805
450	31.53	559.1	5.991	29.16	559.4	5.958	27.12	559.6	5.927
500	34.98	610.8	6.100	32.34	611.1	6.067	30.07	611.3	6.036
550	38.42	662.5	6.198	35.52	662.7	6.165	33.02	663.0	6.135
600	41.86	714.1	6.288	38.69	714.3	6.255	35.97	714.7	6.224
650	45.30	765.6	6.371	41.86	765.9	6.338	38.92	766.3	6.307
700	48.73	817.2	6.447	45.04	817.5	6.414	41.87	817.8	6.384
750	52.17	868.8	6.518	48.21	869.1	6.485	44.81	869.4	6.455
800	55.61	920.3	6.585	51.38	920.6	6.552	47.76	920.9	6.521

TABLE IV. Thermodynamic Properties of Liquid and Gaseous Neon (*Continued*)

T, K	p = 60 bar			p = 65 bar			p = 70 bar		
	$v \cdot 10^3$	h	s	$v \cdot 10^3$	h	s	$v \cdot 10^3$	h	s
850	59.04	971.8	6.647	54.55	972.2	6.614	50.70	972.5	6.584
900	62.48	1023	6.706	57.72	1024	6.673	53.65	1024	6.643
950	65.91	1075	6.762	60.89	1075	6.729	56.59	1076	6.698
1000	69.34	1126	6.815	64.06	1127	6.782	59.54	1127	6.751
1050	72.78	1178	6.865	67.23	1178	6.832	62.48	1178	6.801
1100	76.21	1229	6.913	70.40	1230	6.880	65.42	1230	6.849
1150	79.65	1281	6.959	73.57	1281	6.926	68.37	1282	6.895
1200	83.08	1332	7.003	76.74	1333	6.970	71.31	1333	6.939
1250	86.51	1384	7.045	79.91	1384	7.012	74.25	1384	6.981
1300	89.95	1435	7.085	83.08	1436	7.052	77.20	1436	7.021

TABLE IV. Thermodynamic Properties of Liquid and Gaseous Neon (*Continued*)

T, K	p = 75 bar			p = 80 bar			p = 85 bar		
	$v \cdot 10^3$	h	s	$v \cdot 10^3$	h	s	$v \cdot 10^3$	h	s
26	0.7938	34.89	1.417	0.7924	35.17	1.412	0.7911	35.44	1.408
28	0.8131	38.50	1.551	0.8115	38.76	1.546	0.8099	39.03	1.541
30	0.8341	42.22	1.679	0.8321	42.47	1.673	0.8302	42.72	1.668
32	0.8571	46.03	1.802	0.8547	46.27	1.796	0.8524	46.51	1.790
34	0.8826	49.94	1.920	0.8797	50.15	1.914	0.8769	50.37	1.907
36	0.9112	53.87	2.033	0.9076	54.06	2.025	0.9041	54.25	2.018
38	0.9439	57.88	2.141	0.9393	58.02	2.132	0.9348	58.17	2.124
40	0.9819	61.98	2.248	0.9758	62.06	2.238	0.9700	62.16	2.229

TABLE IV. Thermodynamic Properties of Liquid and Gaseous Neon (*Continued*)

T, K	p = 75 bar			p = 80 bar			p = 85 bar		
	$v \cdot 10^3$	h	s	$v \cdot 10^3$	h	s	$v \cdot 10^3$	h	s
42	1.027	66.38	2.355	1.019	66.39	2.344	1.011	66.42	2.334
44	1.082	71.11	2.465	1.070	71.00	2.452	1.060	70.93	2.440
46	1.150	76.35	2.579	1.133	76.10	2.561	1.117	75.95	2.546
48	1.232	81.90	2.696	1.207	81.40	2.673	1.186	81.10	2.653
50	1.337	87.74	2.816	1.301	87.01	2.788	1.272	86.40	2.763
55	1.730	104.50	3.129	1.642	102.7	3.084	1.569	101.3	3.046
60	2.247	119.4	3.395	2.097	117.6	3.346	1.973	115.9	3.302
65	2.739	131.3	3.585	2.552	129.7	3.540	2.393	128.1	3.499
70	3.189	141.1	3.730	2.975	139.7	3.688	2.791	138.3	3.649
75	3.605	149.6	3.848	3.369	148.4	3.809	3.163	147.3	3.772
80	3.997	157.4	3.948	3.739	156.3	3.911	3.514	155.3	3.876
85	4.370	164.6	4.036	4.092	163.7	4.001	3.850	162.8	3.967
90	4.728	171.5	4.115	4.432	170.7	4.081	4.172	170.0	4.049
95	5.075	178.2	4.187	4.760	177.5	4.154	4.484	176.8	4.122
100	5.414	184.6	4.253	5.080	184.0	4.221	4.787	183.4	4.190
110	6.069	197.0	4.372	5.699	196.5	4.340	5.374	196.0	4.311
120	6.703	209.0	4.476	6.298	208.6	4.446	5.941	208.2	4.417
130	7.322	220.7	4.570	6.881	220.4	4.540	6.493	220.1	4.512
140	7.928	232.2	4.655	7.453	231.9	4.625	7.033	231.7	4.598
150	8.526	243.5	4.732	8.015	243.3	4.704	7.565	243.1	4.677
160	9.116	254.6	4.804	8.571	254.5	4.776	8.090	254.3	4.749
170	9.701	265.6	4.871	9.120	265.5	4.843	8.609	265.4	4.816
180	10.28	276.6	4.934	9.665	276.5	4.906	9.123	276.4	4.879
190	10.86	287.4	4.992	10.21	287.4	4.965	9.633	287.4	4.938
200	11.43	298.2	5.048	10.74	298.2	5.020	10.14	298.2	4.994

TABLE IV. Thermodynamic Properties of Liquid and Gaseous Neon (*Continued*)

T, K	p = 75 bar			p = 80 bar			p = 85 bar		
	$v \cdot 10^3$	h	s	$v \cdot 10^3$	h	s	$v \cdot 10^3$	h	s
220	12.57	319.6	5.150	11.81	319.6	5.122	11.15	319.7	5.096
240	13.70	340.9	5.242	12.87	340.9	5.215	12.15	341.0	5.189
260	14.82	362.0	5.327	13.93	362.1	5.300	13.14	362.2	5.274
280	15.94	383.0	5.405	14.98	383.1	5.378	14.13	383.3	5.352
300	17.05	404.0	5.477	16.02	404.1	5.450	15.11	404.3	5.425
350	19.83	456.1	5.638	18.63	456.3	5.611	17.57	456.6	5.586
400	22.59	508.1	5.777	21.22	508.3	5.750	20.01	508.6	5.725
450	25.35	559.9	5.899	23.81	560.2	5.872	22.45	560.4	5.847
500	28.11	611.6	6.008	26.40	611.9	5.981	24.88	612.2	5.956
550	30.87	663.3	6.106	28.98	663.6	6.079	27.31	663.9	6.054
600	33.62	715.0	6.196	31.56	715.3	6.169	29.74	715.6	6.144
650	36.37	766.6	6.279	34.14	766.9	6.252	32.17	767.2	6.227
700	39.12	818.1	6.355	36.72	818.5	6.328	34.60	818.8	6.303
750	41.87	869.7	6.426	39.30	870.0	6.400	37.03	870.3	6.375
800	44.62	921.3	6.493	41.87	921.6	6.466	39.45	921.9	6.441
850	47.37	972.8	6.555	44.45	973.2	6.529	41.88	973.5	6.504
900	50.12	1024	6.614	47.03	1025	6.588	44.30	1025	6.563
950	52.87	1076	6.670	49.60	1077	6.643	46.73	1077	6.618
1000	55.61	1127	6.723	52.18	1128	6.696	49.15	1128	6.671
1050	58.36	1179	6.773	54.76	1180	6.746	51.58	1180	6.721
1100	61.11	1230	6.821	57.33	1231	6.794	54.00	1231	6.769
1150	63.86	1282	6.867	59.91	1283	6.840	56.42	1283	6.815
1200	66.60	1333	6.911	62.48	1334	6.884	58.85	1334	6.859
1250	69.35	1385	6.953	65.06	1386	6.926	61.27	1386	6.901
1300	72.10	1436	6.993	67.63	1437	6.966	63.70	1437	6.941

TABLE IV. Thermodynamic Properties of Liquid and Gaseous Neon (*Continued*)

T, K	p = 90 bar			p = 95 bar			p = 100 bar		
	$v \cdot 10^3$	h	s	$v \cdot 10^3$	h	s	$v \cdot 10^3$	h	s
26	0.7898	35.72	1.403	0.7885	36.00	1.399	—	—	—
28	0.8083	39.30	1.536	0.8068	39.57	1.531	0.8053	39.84	1.526
30	0.8284	42.98	1.663	0.8266	43.24	1.658	0.8248	43.50	1.653
32	0.8502	46.75	1.784	0.8480	47.00	1.779	0.8459	47.24	1.773
34	0.8741	50.59	1.901	0.8715	50.81	1.895	0.8689	51.04	1.888
36	0.9007	54.44	2.011	0.8974	54.64	2.003	0.8943	54.85	1.995
38	0.9305	58.33	2.116	0.9264	58.50	2.107	0.9225	58.67	2.099
40	0.9645	62.27	2.220	0.9593	62.40	2.210	0.9544	62.53	2.202
42	1.004	66.46	2.324	0.9971	66.53	2.313	0.9907	66.61	2.304
44	1.050	70.91	2.427	1.041	70.88	2.414	1.033	70.88	2.405
46	1.103	75.79	2.530	1.091	75.63	2.515	1.080	75.51	2.505
48	1.168	80.85	2.634	1.151	80.56	2.616	1.138	80.34	2.604
50	1.246	85.95	2.740	1.224	85.56	2.718	1.205	85.23	2.702
55	1.511	100.2	3.013	1.462	99.31	2.983	1.421	98.54	2.957
60	1.871	114.5	3.261	1.785	113.2	3.224	1.713	112.0	3.191
65	2.258	126.7	3.458	2.142	125.4	3.422	2.042	124.2	3.388
70	2.631	137.1	3.612	2.492	135.9	3.577	2.370	134.8	3.544
75	2.983	146.2	3.737	2.826	145.1	3.703	2.686	144.1	3.672
80	3.317	154.4	3.843	3.143	153.4	3.811	2.988	152.5	3.781
85	3.636	162.0	3.935	3.447	161.2	3.905	3.278	160.4	3.876
90	3.943	169.2	4.018	3.739	168.5	3.988	3.557	167.8	3.960
95	4.239	176.1	4.092	4.022	175.5	4.063	3.828	174.8	4.036
100	4.528	182.8	4.161	4.297	182.2	4.133	4.091	181.6	4.106

TABLE IV. Thermodynamic Properties of Liquid and Gaseous Neon (*Continued*)

T, K	p = 90 bar			p = 95 bar			p = 100 bar		
	$v \cdot 10^3$	h	s	$v \cdot 10^3$	h	s	$v \cdot 10^3$	h	s
110	5.086	195.6	4.283	4.829	195.1	4.256	4.599	194.7	4.231
120	5.625	207.9	4.390	5.342	207.5	4.364	5.089	207.1	4.339
130	6.149	219.8	4.485	5.841	219.5	4.460	5.565	219.2	4.436
140	6.661	231.5	4.572	6.329	231.2	4.547	6.030	231.0	4.523
150	7.165	242.9	4.651	6.808	242.8	4.626	6.487	242.6	4.603
160	7.663	254.2	4.724	7.281	254.1	4.699	6.937	254.0	4.677
170	8.154	265.4	4.791	7.748	265.3	4.767	7.382	265.2	4.745
180	8.641	276.4	4.854	8.210	276.4	4.831	7.823	276.3	4.808
190	9.124	287.4	4.914	8.669	287.3	4.890	8.260	287.3	4.868
200	9.604	298.2	4.969	9.125	298.2	4.946	8.694	298.3	4.924
220	10.56	319.8	5.072	10.03	319.8	5.049	9.554	319.9	5.027
240	11.50	341.1	5.165	10.93	341.2	5.142	10.41	341.3	5.120
260	12.44	362.4	5.250	11.82	362.5	5.227	11.25	362.6	5.206
280	13.38	383.5	5.328	12.70	383.6	5.305	12.10	383.8	5.284
300	14.31	404.5	5.401	13.59	404.7	5.378	12.94	404.9	5.357
350	16.63	456.8	5.562	15.78	457.0	5.539	15.02	457.2	5.518
400	18.93	508.8	5.701	17.97	509.1	5.678	17.10	509.3	5.657
450	21.24	560.7	5.823	20.15	561.0	5.801	19.18	561.2	5.779
500	23.54	612.5	5.932	22.33	612.8	5.910	21.25	613.0	5.889
550	25.83	664.2	6.031	24.51	664.5	6.008	23.32	664.8	6.987
600	28.13	715.9	6.121	26.68	716.2	6.098	25.38	716.5	6.077
650	30.42	767.5	6.203	28.86	767.8	6.181	27.45	768.1	6.160
700	32.71	819.1	6.280	31.03	819.4	6.258	29.51	819.7	6.236
750	35.01	870.7	6.351	33.20	871.0	6.329	31.57	871.3	6.308
800	37.30	922.2	6.418	35.37	922.5	6.395	33.64	922.9	6.374

TABLE IV. Thermodynamic Properties of Liquid and Gaseous Neon (*Continued*)

T, K	p = 90 bar			p = 95 bar			p = 100 bar		
	$v \cdot 10^3$	h	s	$v \cdot 10^3$	h	s	$v \cdot 10^3$	h	s
850	39.59	973.8	6.480	37.54	974.1	6.458	35.70	974.4	6.437
900	41.88	1025	6.539	39.71	1026	6.517	37.76	1026	6.496
950	44.17	1077	6.595	41.88	1077	6.572	39.82	1078	6.551
1000	46.46	1128	6.648	44.05	1129	6.626	41.88	1129	6.604
1050	48.75	1180	6.698	46.22	1180	6.676	43.94	1181	6.654
1100	51.04	1231	6.746	48.39	1232	6.724	46.00	1232	6.702
1150	53.33	1283	6.792	50.56	1283	6.769	48.06	1284	6.748
1200	55.62	1334	6.836	52.73	1335	6.813	50.12	1335	6.792
1250	57.91	1386	6.878	54.89	1386	6.855	52.18	1387	6.834
1300	60.20	1437	6.918	57.06	1438	6.896	54.25	1438	6.875

TABLE IV. Thermodynamic Properties of Liquid and Gaseous Neon (*Continued*)

T, K	p = 120 bar			p = 140 bar			p = 160 bar		
	$v \cdot 10^3$	h	s	$v \cdot 10^3$	h	s	$v \cdot 10^3$	h	s
28	0.7996	40.93	1.508	0.7942	42.04	1.491	0.7892	43.16	1.474
30	0.8180	44.55	1.633	0.8118	45.63	1.614	0.8059	46.71	1.597
32	0.8379	48.24	1.752	0.8305	49.28	1.732	0.8237	50.33	1.713
34	0.8593	51.98	1.866	0.8505	52.96	1.844	0.8426	53.97	1.824
36	0.8826	55.71	1.973	0.8722	56.62	1.949	0.8628	57.57	1.928
38	0.9082	59.43	2.074	0.8956	60.26	2.047	0.8845	61.14	2.025
40	0.9365	63.15	2.173	0.9213	63.87	2.142	0.9080	64.67	2.118

TABLE IV. Thermodynamic Properties of Liquid and Gaseous Neon (*Continued*)

T, K	p = 120 bar			p = 140 bar			p = 160 bar		
	$v \cdot 10^3$	h	s	$v \cdot 10^3$	h	s	$v \cdot 10^3$	h	s
42	0.9683	67.05	2.270	0.9497	67.63	2.235	0.9337	68.33	2.208
44	1.004	71.15	2.365	0.9811	71.51	2.326	0.9619	72.07	2.296
46	1.045	75.47	2.458	1.016	75.56	2.416	0.9931	75.97	2.382
48	1.092	79.91	2.550	1.056	79.79	2.504	1.028	80.02	2.466
50	1.144	84.50	2.641	1.101	84.20	2.590	1.068	84.23	2.548
55	1.304	96.50	2.871	1.229	95.47	2.805	1.177	95.01	2.753
60	1.511	108.8	3.083	1.388	106.8	3.003	1.305	105.8	2.940
65	1.754	120.4	3.270	1.576	117.9	3.180	1.457	116.3	3.109
70	2.011	131.1	3.427	1.781	128.3	3.335	1.624	126.4	3.259
75	2.267	140.6	3.560	1.991	137.9	3.467	1.799	135.9	3.390
80	2.516	149.4	3.673	2.200	146.8	3.582	1.976	144.9	3.506
85	2.759	157.5	3.772	2.405	155.2	3.683	2.153	153.3	3.608
90	2.994	165.2	3.859	2.606	163.1	3.773	2.327	161.3	3.699
95	3.222	172.5	3.938	2.803	170.6	3.855	2.499	169.0	3.782
100	3.445	179.6	4.010	2.995	177.8	3.929	2.667	176.3	3.858
110	3.876	193.0	4.139	3.369	191.6	4.060	2.995	190.4	3.992
120	4.291	205.8	4.251	3.729	204.7	4.175	3.312	203.7	4.108
130	4.694	218.2	4.350	4.078	217.3	4.276	3.621	216.6	4.211
140	5.088	230.2	4.439	4.419	229.6	4.366	3.922	229.0	4.303
150	5.473	242.0	4.520	4.753	241.5	4.449	4.217	241.2	4.387
160	5.853	253.6	4.595	5.082	253.3	4.525	4.506	253.0	4.463
170	6.228	265.0	4.664	5.406	264.8	4.595	4.792	264.7	4.534
180	6.598	276.2	4.728	5.726	276.2	4.660	5.074	276.2	4.600
190	6.965	287.3	4.788	6.043	287.4	4.720	5.353	287.6	4.662
200	7.329	298.4	4.845	6.357	298.6	4.777	5.629	298.8	4.719

TABLE IV. Thermodynamic Properties of Liquid and Gaseous Neon (*Continued*)

T, K	p = 120 bar			p = 140 bar			p = 160 bar		
	$v \cdot 10^3$	h	s	$v \cdot 10^3$	h	s	$v \cdot 10^3$	h	s
220	8.050	320.2	4.949	6.979	320.5	4.882	6.176	320.9	4.824
240	8.765	341.8	5.043	7.593	342.3	4.977	6.716	342.8	4.919
260	9.473	363.2	5.128	8.203	363.8	5.063	7.252	364.4	5.006
280	10.18	384.4	5.207	8.809	385.1	5.142	7.784	385.8	5.085
300	10.88	405.6	5.280	9.412	406.3	5.215	8.313	407.1	5.159
350	12.62	458.1	5.442	10.91	459.0	5.378	9.626	459.9	5.322
400	14.36	510.3	5.581	12.40	511.3	5.517	10.93	512.3	5.462
450	16.09	562.3	5.704	13.88	563.4	5.640	12.23	564.5	5.585
500	17.82	614.2	5.813	15.36	615.3	5.749	13.53	616.4	5.694
550	19.54	665.9	5.912	16.84	667.1	5.848	14.82	668.3	5.793
600	21.26	717.7	6.002	18.32	718.9	5.938	16.11	720.1	5.883
650	22.98	769.3	6.085	19.80	770.6	6.021	17.41	771.8	5.966
700	24.70	821.0	6.161	21.27	822.2	6.097	18.70	823.5	6.042
750	26.42	872.6	6.232	22.75	873.8	6.169	19.99	875.1	6.114
800	28.14	924.1	6.299	24.22	925.4	6.235	21.28	926.7	6.180
850	29.86	975.7	6.361	25.69	977.0	6.298	22.57	978.3	6.243
900	31.58	1027	6.420	27.17	1029	6.357	23.86	1030	6.302
950	33.30	1079	6.476	28.64	1080	6.413	25.14	1081	6.358
1000	35.02	1130	6.529	30.11	1132	6.465	26.43	1133	6.410
1050	36.73	1182	6.579	31.58	1183	6.516	27.72	1185	6.461
1100	38.45	1233	6.627	33.06	1235	6.564	29.01	1236	6.509
1150	40.17	1285	6.673	34.53	1286	6.610	30.30	1288	6.554
1200	41.88	1336	6.717	36.00	1338	6.653	31.59	1339	6.598
1250	43.60	1388	6.759	37.47	1389	6.695	32.87	1391	6.640
1300	45.32	1440	6.799	38.94	1441	6.736	34.16	1442	6.681

TABLE IV. Thermodynamic Properties of Liquid and Gaseous Neon (*Continued*)

T, K	p = 180 bar			p = 200 bar			p = 220 bar		
	$v \cdot 10^3$	h	s	$v \cdot 10^3$	h	s	$v \cdot 10^3$	h	s
28	0.7844	44.28	1.458	0.7799	45.42	1.443	0.7756	46.56	1.428
30	0.8004	47.81	1.580	0.7953	48.93	1.564	0.7904	50.05	1.548
32	0.8173	51.40	1.695	0.8114	52.48	1.679	0.8059	53.58	1.662
34	0.8352	55.00	1.805	0.8284	56.06	1.787	0.8221	57.13	1.769
36	0.8542	58.56	1.907	0.8463	59.58	1.887	0.8391	60.62	1.868
38	0.8745	62.07	2.003	0.8654	63.04	1.981	0.8570	64.04	1.960
40	0.8962	65.53	2.095	0.8857	66.44	2.072	0.8761	67.39	2.049
42	0.9198	69.10	2.183	0.9074	69.94	2.159	0.8964	70.83	2.135
44	0.9454	72.74	2.268	0.9310	73.49	2.243	0.9182	74.34	2.218
46	0.9735	76.49	2.351	0.9566	77.33	2.324	0.9418	77.92	2.298
48	1.004	80.37	2.432	0.9847	81.05	2.403	0.9676	81.59	2.375
50	1.039	84.39	2.511	1.016	84.86	2.480	0.9959	85.35	2.450
55	1.134	94.91	2.709	1.102	95.06	2.672	1.074	95.27	2.638
60	1.244	105.2	2.888	1.197	105.1	2.844	1.159	105.0	2.806
65	1.371	115.3	3.050	1.305	114.8	3.001	1.254	114.5	2.958
70	1.511	125.1	3.196	1.425	124.3	3.142	1.359	123.8	3.095
75	1.659	134.5	3.325	1.554	133.5	3.269	1.471	132.8	3.220
80	1.812	143.4	3.440	1.686	142.3	3.382	1.588	141.5	3.332
85	1.965	151.8	3.542	1.821	150.7	3.485	1.708	149.9	3.434
90	2.118	159.9	3.634	1.956	158.8	3.577	1.828	158.0	3.526
95	2.269	167.6	3.718	2.091	166.6	3.661	1.949	165.8	3.611
100	2.418	175.1	3.795	2.224	174.1	3.739	2.069	173.4	3.688

TABLE IV. Thermodynamic Properties of Liquid and Gaseous Neon (*Continued*)

T, K	p = 180 bar			p = 200 bar			p = 220 bar		
	$v \cdot 10^3$	h	s	$v \cdot 10^3$	h	s	$v \cdot 10^3$	h	s
110	2.710	189.4	3.931	2.486	188.6	3.876	2.306	187.9	3.827
120	2.993	202.9	4.049	2.741	202.3	3.996	2.537	201.8	3.948
130	3.269	216.0	4.153	2.990	215.5	4.101	2.764	215.1	4.055
140	3.538	228.6	4.247	3.233	228.3	4.196	2.986	228.0	4.150
150	3.802	240.9	4.332	3.472	240.7	4.281	3.204	240.6	4.237
160	4.061	252.9	4.409	3.707	252.8	4.360	3.418	252.8	4.316
170	4.316	264.7	4.481	3.938	264.7	4.432	3.629	264.8	4.388
180	4.569	276.3	4.547	4.166	276.4	4.499	3.837	276.6	4.456
190	4.818	287.7	4.609	4.391	288.0	4.562	4.043	288.2	4.519
200	5.065	299.0	4.667	4.614	299.4	4.620	4.247	299.7	4.578
220	5.553	321.3	4.773	5.055	321.8	4.727	4.649	322.3	4.685
240	6.035	343.3	4.869	5.491	343.9	4.823	5.046	344.5	4.782
260	6.513	365.0	4.956	5.922	365.7	4.910	5.439	366.4	4.870
280	6.986	386.5	5.035	6.349	387.3	4.991	5.829	388.1	4.950
300	7.458	407.9	5.109	6.774	408.8	5.065	6.216	409.6	5.024
350	8.627	460.9	5.272	7.829	461.8	5.228	7.176	462.8	5.188
400	9.789	513.4	5.413	8.875	514.4	5.369	8.128	515.5	5.329
450	10.95	565.6	5.536	9.917	566.7	5.492	9.076	567.8	5.452
500	12.10	617.6	5.645	10.96	618.8	5.601	10.02	619.9	5.562
550	13.25	669.5	5.744	11.99	670.7	5.700	10.96	671.9	5.661
600	14.40	721.3	5.834	13.03	722.5	5.791	11.90	723.8	5.751
650	15.55	773.1	5.917	14.06	774.3	5.874	12.84	775.5	5.834
700	16.70	824.8	5.994	15.09	826.0	5.950	13.78	827.3	5.911
750	17.84	876.4	6.065	16.13	877.7	6.021	14.72	878.9	5.982
800	18.99	928.0	6.132	17.16	929.3	6.088	15.66	930.6	6.049

TABLE IV. Thermodynamic Properties of Liquid and Gaseous Neon (*Continued*)

T, K	p = 180 bar			p = 200 bar			p = 220 bar		
	$v \cdot 10^3$	h	s	$v \cdot 10^3$	h	s	$v \cdot 10^3$	h	s
850	20.14	979.6	6.194	18.19	980.9	6.151	16.60	982.2	6.111
900	21.28	1031	6.253	19.22	1033	6.210	17.54	1034	6.170
950	22.43	1083	6.309	20.25	1084	6.265	18.47	1085	6.226
1000	23.57	1134	6.362	21.28	1136	6.318	19.41	1137	6.279
1050	24.72	1186	6.412	22.31	1187	6.369	20.35	1189	6.329
1100	25.86	1237	6.460	23.35	1239	6.417	21.29	1240	6.377
1150	27.01	1289	6.506	24.38	1290	6.462	22.22	1292	6.423
1200	28.15	1341	6.550	25.41	1342	6.506	23.16	1343	6.467
1250	29.30	1392	6.592	26.44	1393	6.548	24.10	1395	6.509
1300	30.44	1442	6.632	27.47	1445	6.589	25.03	1446	6.550

TABLE IV. Thermodynamic Properties of Liquid and Gaseous Neon (*Continued*)

T, K	p = 240 bar			p = 260 bar			p = 280 bar		
	$v \cdot 10^3$	h	s	$v \cdot 10^3$	h	s	$v \cdot 10^3$	h	s
30	0.7858	51.18	1.533	0.7814	52.32	1.519	0.7773	53.47	1.505
32	0.8006	54.70	1.647	0.7957	55.82	1.632	0.7910	56.95	1.618
34	0.8161	58.22	1.754	0.8106	59.32	1.738	0.8053	60.43	1.723
36	0.8323	61.67	1.852	0.8260	62.75	1.836	0.8201	63.84	1.821
38	0.8493	65.07	1.943	0.8422	66.11	1.927	0.8356	67.17	1.911
40	0.8673	68.37	2.031	0.8592	69.38	2.014	0.8517	70.41	1.997

TABLE IV. Thermodynamic Properties of Liquid and Gaseous Neon (*Continued*)

T, K	p = 240 bar			p = 260 bar			p = 280 bar		
	$v \cdot 10^3$	h	s	$v \cdot 10^3$	h	s	$v \cdot 10^3$	h	s
42	0.8864	71.76	2.115	0.8772	72.72	2.097	0.8688	73.71	2.079
44	0.9067	75.26	2.196	0.8963	76.12	2.177	0.8868	77.07	2.158
46	0.9287	78.79	2.274	0.9169	79.60	2.254	0.9062	80.50	2.234
48	0.9526	82.40	2.350	0.9392	83.15	2.328	0.9272	83.99	2.307
50	0.9788	86.10	2.423	0.9636	86.77	2.400	0.9501	87.54	2.378
55	1.051	95.75	2.608	1.029	96.10	2.580	1.012	96.60	2.553
60	1.127	105.3	2.772	1.101	105.5	2.742	1.078	105.8	2.712
65	1.212	114.6	2.920	1.178	114.7	2.886	1.148	114.9	2.856
70	1.305	123.6	3.054	1.261	123.6	3.018	1.224	123.7	2.985
75	1.405	132.4	3.176	1.350	132.3	3.138	1.305	132.3	3.103
80	1.509	141.0	3.287	1.444	140.8	3.247	1.390	140.7	3.210
85	1.616	149.3	3.388	1.541	149.0	3.347	1.478	148.9	3.310
90	1.725	157.4	3.480	1.639	157.0	3.439	1.568	156.8	3.401
95	1.834	165.2	3.565	1.738	164.9	3.523	1.658	164.6	3.485
100	1.943	172.8	3.643	1.838	172.5	3.601	1.750	172.3	3.563
110	2.158	187.5	3.782	2.035	187.2	3.741	1.932	187.0	3.704
120	2.370	201.4	3.904	2.230	201.2	3.864	2.112	201.1	3.826
130	2.578	214.9	4.012	2.422	214.7	3.972	2.289	214.7	3.935
140	2.782	227.9	4.108	2.610	227.8	4.069	2.464	227.9	4.033
150	2.982	240.5	4.195	2.795	240.6	4.157	2.636	240.7	4.122
160	3.179	252.9	4.275	2.977	253.0	4.237	2.805	253.2	4.203
170	3.373	265.0	4.348	3.157	265.2	4.311	2.973	265.5	4.277
180	3.564	276.9	4.416	3.334	277.2	4.380	3.138	277.5	4.346
190	3.754	288.6	4.479	3.509	289.0	4.443	3.301	289.4	4.410
200	3.941	300.1	4.539	3.683	300.6	4.503	3.462	301.0	4.470

TABLE IV. Thermodynamic Properties of Liquid and Gaseous Neon (*Continued*)

T, K	p = 240 bar			p = 260 bar			p = 280 bar		
	$v \cdot 10^3$	h	s	$v \cdot 10^3$	h	s	$v \cdot 10^3$	h	s
220	4.311	322.8	4.647	4.025	323.4	4.612	3.781	324.0	4.579
240	4.676	345.1	4.744	4.363	345.8	4.709	4.095	346.5	4.677
260	5.037	367.1	4.832	4.697	367.9	4.798	4.406	368.6	4.765
280	5.395	388.9	4.913	5.028	389.7	4.878	4.714	390.5	4.847
300	5.750	410.5	4.987	5.357	411.3	4.953	5.020	412.2	4.921
350	6.632	463.8	5.152	6.172	464.8	5.118	5.777	465.8	5.087
400	7.505	516.5	5.292	6.979	517.6	5.259	6.528	518.7	5.228
450	8.375	568.9	5.416	7.782	570.1	5.383	7.274	571.2	5.352
500	9.241	621.1	5.526	8.582	622.3	5.493	8.017	623.5	5.462
550	10.11	673.1	5.625	9.380	674.3	5.592	8.759	675.5	5.561
600	10.97	725.0	5.715	10.18	726.2	5.682	9.499	727.5	5.651
650	11.83	776.8	5.798	10.97	778.1	5.765	10.24	779.3	5.734
700	12.69	828.5	5.875	11.77	829.8	5.842	10.98	831.1	5.811
750	13.55	880.2	5.946	12.56	881.5	5.913	11.71	882.8	5.883
800	14.41	931.9	6.013	13.36	933.2	5.980	12.45	934.5	5.949
850	15.27	983.5	6.075	14.15	984.9	6.042	13.19	986.2	6.012
900	16.13	1035	6.134	14.94	1036	6.101	13.93	1038	6.071
950	16.99	1087	6.190	15.74	1088	6.157	14.66	1089	6.127
1000	17.85	1138	6.243	16.53	1140	6.210	15.40	1141	6.180
1050	18.71	1190	6.294	17.32	1191	6.261	16.14	1193	6.230
1100	19.57	1241	6.341	18.12	1243	6.308	16.87	1244	6.278
1150	20.43	1293	6.387	18.91	1294	6.354	17.61	1296	6.324
1200	21.29	1345	6.431	19.70	1346	6.398	18.34	1347	6.368
1250	22.15	1396	6.473	20.50	1397	6.440	19.08	1399	6.410
1300	23.00	1448	6.514	21.29	1449	6.481	19.82	1450	6.450

TABLE IV. Thermodynamic Properties of Liquid and Gaseous Neon (*Continued*)

T, K	p = 300 bar			p = 350 bar			p = 400 bar		
	$v \cdot 10^3$	h	s	$v \cdot 10^3$	h	s	$v \cdot 10^3$	h	s
30	0.7733	54.62	1.492	0.7641	57.51	1.461	—	—	—
32	0.7866	58.08	1.604	0.7763	60.95	1.572	0.7672	63.85	1.542
34	0.8003	61.55	1.709	0.7890	64.38	1.675	0.7789	67.26	1.645
36	0.8146	64.94	1.806	0.8019	67.74	1.771	0.7908	70.59	1.739
38	0.8293	68.25	1.896	0.8153	71.00	1.860	0.8031	73.81	1.827
40	0.8448	71.46	1.981	0.8292	74.18	1.943	0.8157	76.96	1.909
42	0.8610	74.72	2.062	0.8437	77.40	2.022	0.8289	80.14	1.987
44	0.8781	78.01	2.140	0.8589	80.65	2.098	0.8427	83.35	2.062
46	0.8964	81.36	2.215	0.8752	83.95	2.172	0.8573	86.60	2.135
48	0.9162	84.77	2.288	0.8927	87.29	2.244	0.8731	89.88	2.206
50	0.9379	88.26	2.359	0.9118	90.68	2.314	0.8904	93.24	2.275
55	0.9960	97.09	2.531	0.9615	99.26	2.481	0.9330	101.6	2.436
60	1.057	106.1	2.687	1.015	107.9	2.630	0.9805	110.0	2.581
65	1.122	115.2	2.828	1.070	116.6	2.765	1.030	118.4	2.713
70	1.192	124.0	2.955	1.128	125.1	2.889	1.080	126.7	2.834
75	1.266	132.5	3.071	1.190	133.4	3.002	1.133	134.8	2.944
80	1.344	140.8	3.177	1.254	141.5	3.105	1.188	142.8	3.045
85	1.424	148.9	3.276	1.320	149.4	3.202	1.244	150.6	3.140
90	1.507	156.8	3.366	1.388	157.2	3.291	1.302	158.2	3.227
95	1.590	164.6	3.450	1.458	164.9	3.374	1.361	165.8	3.309
100	1.675	172.2	3.528	1.528	172.5	3.451	1.421	173.3	3.386

TABLE IV. Thermodynamic Properties of Liquid and Gaseous Neon (*Continued*)

T, K	p = 300 bar			p = 350 bar			p = 400 bar		
	$v \cdot 10^3$	h	s	$v \cdot 10^3$	h	s	$v \cdot 10^3$	h	s
110	1.843	186.9	3.669	1.669	187.2	3.592	1.542	188.0	3.526
120	2.010	201.1	3.792	1.811	201.4	3.715	1.664	202.2	3.650
130	2.176	214.7	3.901	1.951	215.2	3.826	1.786	216.1	3.761
140	2.339	228.0	4.000	2.090	228.6	3.925	1.907	229.5	3.861
150	2.499	240.9	4.089	2.227	241.6	4.015	2.026	242.7	3.951
160	2.657	253.5	4.170	2.363	254.4	4.097	2.145	255.5	4.034
170	2.813	265.8	4.245	2.497	266.8	4.173	2.262	268.1	4.111
180	2.968	277.9	4.314	2.630	279.1	4.243	2.378	280.5	4.181
190	3.120	289.8	4.378	2.761	291.1	4.308	2.493	292.6	4.247
200	3.271	301.5	4.438	2.891	303.0	4.369	2.607	304.6	4.308
220	3.570	324.6	4.548	3.148	326.2	4.480	2.832	328.0	3.420
240	3.864	347.2	4.647	3.401	349.0	4.579	3.054	351.0	4.520
260	4.154	369.4	4.736	3.651	371.4	4.669	3.274	373.6	4.610
280	4.442	391.4	4.817	3.899	393.6	4.750	3.491	395.8	4.693
300	4.728	413.1	4.892	4.145	415.4	4.826	3.707	417.8	4.769
350	5.436	466.8	5.057	5.753	469.4	4.992	4.241	472.0	4.936
400	6.137	519.8	5.199	5.355	522.6	5.134	4.769	525.3	5.078
450	6.834	572.4	5.323	5.954	575.3	5.259	5.294	578.2	5.203
500	7.528	624.7	5.433	6.549	627.7	5.369	5.815	630.7	5.313
550	8.220	676.8	5.532	7.143	679.8	5.469	6.335	682.9	5.413
600	8.911	728.7	5.623	7.736	731.9	5.559	6.854	735.0	5.504
650	9.601	780.6	5.706	8.328	783.8	5.642	7.372	787.0	5.587
700	10.29	832.4	5.783	8.919	835.6	5.719	7.890	838.9	5.664
750	10.98	884.1	5.854	9.509	887.4	5.790	8.407	890.7	5.735
800	11.67	935.8	5.921	10.10	939.1	5.857	8.923	942.4	5.802

TABLE IV. Thermodynamic Properties of Liquid and Gaseous Neon (*Continued*)

T, K	p = 300 bar			p = 350 bar			p = 400 bar		
	$v \cdot 10^3$	h	s	$v \cdot 10^3$	h	s	$v \cdot 10^3$	h	s
850	12.36	987.7	5.983	10.69	990.8	5.920	9.440	994.1	5.865
900	13.04	1039	6.042	11.28	1042	5.979	9.956	1046	5.924
950	13.73	1091	6.098	11.87	1094	6.035	10.47	1097	5.980
1000	14.42	1142	6.151	12.46	1146	6.088	10.99	1149	6.032
1050	15.11	1194	6.201	13.05	1197	6.138	11.50	1201	6.083
1100	15.79	1245	6.249	13.64	1249	6.186	12.02	1252	6.131
1150	16.48	1297	6.295	14.23	1300	6.232	12.53	1304	6.177
1200	17.17	1348	6.339	14.81	1352	6.276	13.05	1355	6.221
1250	17.86	1400	6.381	15.40	1404	6.318	13.56	1407	6.263
1300	18.54	1452	6.422	15.99	1455	6.358	14.08	1459	6.303

TABLE IV. Thermodynamic Properties of Liquid and Gaseous Neon (*Continued*)

T, K	p = 450 bar			p = 500 bar			p = 550 bar		
	$v \cdot 10^3$	h	s	$v \cdot 10^3$	h	s	$v \cdot 10^3$	h	s
32	0.7588	66.77	1.514	0.7512	69.71	1.487	—	—	—
34	0.7697	70.16	1.617	0.7615	73.08	1.590	0.7539	76.02	1.565
36	0.7809	73.45	1.711	0.7719	76.34	1.683	0.7637	79.30	1.658
38	0.7922	76.64	1.797	0.7825	79.50	1.770	0.7736	82.42	1.743
40	0.8039	79.74	1.877	0.7933	82.58	1.850	0.7838	85.51	1.822
42	0.8160	82.87	1.953	0.8045	85.69	1.926	0.7942	88.61	1.897
44	0.8286	86.02	2.026	0.8162	88.81	1.999	0.8051	91.70	1.969
46	0.8418	89.21	2.097	0.8286	92.00	2.069	0.8167	94.80	2.038

TABLE IV. Thermodynamic Properties of Liquid and Gaseous Neon (*Continued*)

T, K	p = 450 bar			p = 500 bar			p = 550 bar		
	$v \cdot 10^3$	h	s	$v \cdot 10^3$	h	s	$v \cdot 10^3$	h	s
48	0.8565	92.46	2.172	0.8420	95.23	2.137	0.8292	97.92	2.105
50	0.8705	95.77	2.239	0.8567	98.48	2.203	0.8429	101.1	2.170
55	0.9091	104.0	2.395	0.8910	106.7	2.356	0.8750	109.2	2.319
60	0.9510	112.3	2.537	0.9295	114.9	2.495	0.9090	117.3	2.455
65	0.9955	120.5	2.666	0.9700	122.9	2.623	0.9445	125.3	2.582
70	1.041	128.6	2.785	1.011	130.8	2.741	0.9820	133.1	2.700
75	1.088	136.6	2.893	1.052	138.6	2.849	1.021	140.8	2.808
80	1.136	144.4	2.993	1.095	146.3	2.947	1.060	148.4	2.906
85	1.185	152.1	3.086	1.138	153.9	3.039	1.100	155.9	2.997
90	1.236	159.7	3.173	1.183	161.4	3.125	1.140	163.3	3.082
95	1.287	167.2	3.254	1.229	168.8	3.205	1.181	170.7	3.162
100	1.339	174.6	3.330	1.275	176.2	3.281	1.223	178.0	3.237
110	1.445	189.2	3.469	1.369	190.7	3.419	1.307	192.4	3.374
120	1.552	203.4	3.593	1.464	204.9	3.542	1.392	206.6	3.497
130	1.659	217.3	3.704	1.559	218.7	3.654	1.478	220.5	3.608
140	1.766	230.8	3.804	1.654	232.3	3.754	1.563	234.0	3.709
150	1.871	244.0	3.895	1.749	245.6	3.846	1.649	247.3	3.801
160	1.976	256.9	3.979	1.843	258.6	3.929	1.734	260.4	3.885
170	2.080	269.6	4.056	1.936	271.3	4.007	1.819	273.2	3.962
180	2.183	282.0	4.127	2.029	283.8	4.078	1.903	285.7	4.034
190	2.286	294.3	4.193	2.121	296.1	4.145	1.986	298.1	4.101
200	2.387	306.3	4.255	2.212	308.2	4.207	2.069	310.3	4.163

TABLE IV. Thermodynamic Properties of Liquid and Gaseous Neon (*Continued*)

T, K	p = 450 bar			p = 500 bar			p = 550 bar		
	$v \cdot 10^3$	h	s	$v \cdot 10^3$	h	s	$v \cdot 10^3$	h	s
220	2.587	330.0	4.367	2.392	332.0	4.320	2.233	334.1	4.277
240	2.785	353.1	4.468	2.571	355.2	4.421	2.395	357.5	4.379
260	2.981	375.8	4.559	2.747	378.0	4.513	2.556	380.4	4.471
280	3.175	398.1	4.642	2.922	400.5	4.596	2.715	403.0	4.554
300	3.367	420.2	4.718	3.095	422.7	4.672	2.873	425.3	4.631
350	3.843	474.6	4.886	3.524	477.3	4.841	3.264	480.0	4.800
400	4.313	528.2	5.029	3.949	531.0	4.984	3.650	533.9	4.944
450	4.780	581.1	5.153	4.369	584.1	5.109	4.033	587.1	5.069
500	5.245	633.7	5.264	4.788	636.8	5.220	4.414	639.8	5.180
550	5.707	686.1	5.364	5.205	689.2	5.320	4.794	692.3	5.280
600	6.169	738.2	5.455	5.621	741.4	5.411	5.172	744.6	5.371
650	6.630	790.2	5.538	6.036	793.4	5.494	5.549	796.7	5.455
700	7.090	842.1	5.615	6.450	845.4	5.571	5.926	848.7	5.532
750	7.549	893.9	5.686	6.864	897.2	5.643	6.302	900.5	5.603
800	8.009	945.7	5.753	7.277	949.0	5.710	6.678	952.4	5.670
850	8.468	997.4	5.816	7.690	1001	5.772	7.054	1004	5.733
900	8.926	1049	5.875	8.103	1052	5.832	7.430	1056	5.792
950	9.385	1101	5.930	8.516	1104	5.887	7.805	1108	5.848
1000	9.843	1152	5.984	8.929	1156	5.940	8.180	1159	5.901
1050	10.30	1204	6.034	9.341	1207	5.991	8.555	1211	5.951
1100	10.76	1256	6.082	9.754	1259	6.039	8.930	1262	6.000
1150	11.22	1307	6.128	10.17	1311	6.085	9.305	1314	6.045
1200	11.68	1359	6.172	10.58	1362	6.129	9.680	1366	6.089
1200	12.13	1410	6.214	10.99	1414	6.171	10.05	1417	6.131
1300	12.59	1462	6.255	11.40	1465	6.211	10.43	1469	6.172

TABLE IV. Thermodynamic Properties of Liquid and Gaseous Neon (*Continued*)

T, K	p = 600 bar			p = 650 bar			p = 700 bar		
	$v \cdot 10^3$	h	s	$v \cdot 10^3$	h	s	$v \cdot 10^3$	h	s
34	0.7469	78.97	1.541	0.7404	81.90	1.519	0.7344	84.89	1.498
36	0.7562	82.23	1.634	0.7492	85.08	1.611	0.7427	88.14	1.590
38	0.7655	85.33	1.718	0.7581	88.19	1.695	0.7512	91.26	1.674
40	0.7751	88.37	1.796	0.7671	91.21	1.773	0.7598	94.35	1.751
42	0.7849	91.35	1.870	0.7764	94.22	1.847	0.7686	97.47	1.824
44	0.7952	94.33	1.941	0.7861	97.32	1.918	0.7778	100.4	1.894
46	0.8060	97.42	2.009	0.7963	100.4	1.986	0.7875	103.5	1.962
48	0.8178	100.5	2.075	0.8068	103.5	2.052	0.7978	106.6	2.028
50	0.8307	103.7	2.139	0.8177	106.7	2.116	0.8080	109.7	2.092
55	0.8601	111.7	2.287	0.8465	114.6	2.262	0.8345	117.4	2.236
60	0.8920	119.6	2.422	0.8760	122.4	2.395	0.8616	125.1	2.367
65	0.9245	127.5	2.547	0.9063	130.2	2.516	0.8900	132.8	2.489
70	0.9591	135.3	2.664	0.9385	137.9	2.631	0.9205	140.4	2.602
75	0.9952	143.0	2.772	0.9715	145.5	2.737	0.9514	147.9	2.706
80	1.031	150.6	2.869	1.005	153.0	2.834	0.9821	155.3	2.802
85	1.067	158.1	2.959	1.038	160.4	2.924	1.014	162.7	2.891
90	1.104	165.5	3.043	1.072	167.7	3.008	1.045	170.0	2.975
95	1.141	172.8	3.122	1.106	175.0	3.086	1.077	177.3	3.053
100	1.179	180.0	3.197	1.141	182.2	3.161	1.109	184.5	3.127
110	1.255	194.4	3.334	1.211	196.5	3.297	1.174	198.7	3.263
120	1.332	208.5	3.457	1.282	210.6	3.419	1.239	212.8	3.385
130	1.410	222.4	3.567	1.354	224.4	3.530	1.305	226.6	3.496
140	1.488	235.9	3.668	1.425	238.0	3.631	1.371	240.2	3.596
150	1.566	249.3	3.760	1.497	251.3	3.723	1.437	253.5	3.688

TABLE IV. Thermodynamic Properties of Liquid and Gaseous Neon (*Continued*)

T, K	p = 600 bar			p = 650 bar			p = 700 bar		
	$v \cdot 10^3$	h	s	$v \cdot 10^3$	h	s	$v \cdot 10^3$	h	s
160	1.644	262.3	3.844	1.568	264.4	3.807	1.503	266.6	3.773
170	1.721	275.1	3.922	1.639	277.3	3.885	1.569	279.5	3.851
180	1.798	287.8	3.994	1.710	289.9	3.957	1.634	292.2	3.924
190	1.874	300.2	4.061	1.780	302.4	4.025	1.699	304.6	3.991
200	1.950	312.4	4.124	1.850	314.6	4.088	1.764	316.9	4.054
220	2.100	336.4	4.238	1.989	338.7	4.202	1.893	341.1	4.169
240	2.249	359.8	4.340	2.126	362.2	4.305	2.020	364.7	4.272
260	2.397	382.8	4.432	2.262	385.3	4.397	2.147	387.8	4.365
280	2.543	405.5	4.516	2.397	408.0	4.481	2.272	410.6	4.449
300	2.688	427.8	4.593	2.531	430.5	4.559	2.396	433.1	4.527
350	3.046	482.8	4.763	2.862	485.5	4.729	2.704	488.3	4.697
400	3.401	536.7	4.907	3.190	539.6	4.873	3.009	542.5	4.842
450	3.753	590.0	5.033	3.515	593.0	4.999	3.311	596.0	4.968
500	4.102	642.9	5.144	3.838	646.0	5.110	3.612	649.0	5.079
550	4.451	695.4	5.244	4.160	698.6	5.211	3.911	701.7	5.180
600	4.798	747.8	5.335	4.481	750.9	5.302	4.209	754.1	5.271
650	5.144	799.9	5.419	4.801	803.1	5.385	4.507	806.4	5.355
700	5.490	851.9	5.496	5.120	855.2	5.463	4.803	858.5	5.432
750	5.835	903.8	5.567	5.439	907.1	5.534	5.099	910.4	5.504
800	6.180	955.7	5.634	5.757	959.0	5.610	5.395	962.3	5.571
850	6.524	1007	5.697	6.075	1011	5.664	5.691	1014	5.633
900	6.868	1059	5.756	6.393	1063	5.723	5.986	1066	5.693
950	7.212	1111	5.812	6.711	1114	5.779	6.281	1118	5.749
1000	7.556	1163	5.865	7.029	1166	5.832	6.576	1169	5.802
1050	7.900	1214	5.916	7.346	1218	5.883	6.871	1221	5.852

TABLE IV. Thermodynamic Properties of Liquid and Gaseous Neon (*Continued*)

T, K	p = 600 bar			p = 650 bar			p = 700 bar		
	$v \cdot 10^3$	h	s	$v \cdot 10^3$	h	s	$v \cdot 10^3$	h	s
1100	8.244	1266	5.964	7.663	1269	5.931	7.166	1273	5.900
1150	8.588	1317	6.010	7.981	1321	5.976	7.460	1324	5.946
1200	8.931	1369	6.054	8.298	1372	6.020	7.755	1376	5.990
1250	9.275	1421	6.096	8.615	1424	6.062	8.049	1428	6.032
1300	9.618	1472	6.136	8.932	1476	6.103	8.344	1479	6.072

TABLE IV. Thermodynamic Properties of Liquid and Gaseous Neon (*Continued*)

T, K	p = 750 bar			p = 800 bar			p = 850 bar		
	$v \cdot 10^3$	h	s	$v \cdot 10^3$	h	s	$v \cdot 10^3$	h	s
36	0.7367	91.10	1.569	0.7311	94.06	1.551	0.7257	97.03	1.533
38	0.7448	94.21	1.653	0.7388	97.16	1.634	0.7331	100.1	1.615
40	0.7529	97.31	1.730	0.7466	100.2	1.710	0.7406	103.1	1.691
42	0.7613	100.4	1.803	0.7546	103.2	1.783	0.7483	106.0	1.763
44	0.7701	103.5	1.873	0.7630	106.2	1.853	0.7563	108.9	1.832
46	0.7794	106.6	1.941	0.7718	109.2	1.921	0.7648	111.8	1.899
48	0.7892	109.7	2.007	0.7813	112.3	1.987	0.7741	114.8	1.964
50	0.7991	112.8	2.071	0.7909	115.4	2.051	0.7834	117.8	2.027
55	0.8225	120.4	2.216	0.8129	123.1	2.195	0.8044	125.3	2.172
60	0.8480	128.0	2.344	0.8365	130.7	2.322	0.8268	132.8	2.299
65	0.8750	135.6	2.463	0.8615	138.3	2.440	0.8509	140.3	2.418
70	0.9035	143.2	2.574	0.8880	145.8	2.549	0.8759	147.8	2.526
75	0.9325	150.6	2.676	0.9155	153.2	2.650	0.9014	155.2	2.625

TABLE IV. Thermodynamic Properties of Liquid and Gaseous Neon (*Continued*)

T, K	p = 750 bar			p = 800 bar			p = 850 bar		
	$v \cdot 10^3$	h	s	$v \cdot 10^3$	h	s	$v \cdot 10^3$	h	s
80	0.9618	157.9	2.772	0.9433	160.5	2.745	0.9270	162.6	2.718
85	0.9908	165.2	2.861	0.9706	167.7	2.833	0.9526	170.0	2.805
90	1.020	172.4	2.944	0.9984	174.9	2.915	0.9785	177.3	2.888
95	1.050	179.6	3.022	1.027	182.1	2.993	1.005	184.5	2.966
100	1.080	186.8	3.096	1.055	189.2	3.066	1.032	191.7	3.039
110	1.141	201.1	3.232	1.112	203.5	3.202	1.086	205.9	3.174
120	1.202	215.1	3.354	1.169	217.5	3.324	1.139	219.9	3.296
130	1.263	228.9	3.464	1.226	231.3	3.434	1.193	233.7	3.407
140	1.324	242.4	3.565	1.283	244.8	3.535	1.247	247.3	3.507
150	1.386	255.8	3.657	1.341	258.2	3.627	1.301	260.6	3.599
160	1.447	268.9	3.741	1.398	271.3	3.712	1.355	273.7	3.684
170	1.508	281.8	3.819	1.455	284.2	3.790	1.409	286.6	3.762
180	1.569	294.5	3.892	1.512	296.9	3.863	1.462	299.4	3.835
190	1.630	307.0	3.960	1.569	309.4	3.930	1.515	311.9	3.903
200	1.690	319.3	4.023	1.625	321.8	3.994	1.568	324.3	3.966
220	1.810	343.5	4.138	1.737	346.1	4.109	1.674	348.6	4.082
240	1.929	367.2	4.241	1.849	369.8	4.213	1.778	372.4	4.186
260	2.047	390.4	4.334	1.959	393.0	4.306	1.882	395.7	4.279
280	2.164	413.3	4.419	2.069	415.9	4.391	1.985	418.7	4.364
300	2.280	435.8	4.497	2.178	438.5	4.469	2.088	441.3	4.442
350	2.568	491.1	4.667	2.448	493.9	4.640	2.342	496.8	4.614
400	2.852	545.4	4.812	2.715	548.4	4.785	2.594	551.3	4.759
450	3.135	599.0	4.939	2.980	602.0	4.911	2.843	605.0	4.886
500	3.416	652.1	5.050	3.244	655.2	5.023	3.092	658.3	4.998
550	3.695	704.8	5.151	3.506	708.0	5.124	3.339	711.1	5.099

TABLE IV. Thermodynamic Properties of Liquid and Gaseous Neon (*Continued*)

T, K	p = 750 bar			p = 800 bar			p = 850 bar		
	$v \cdot 10^3$	h	s	$v \cdot 10^3$	h	s	$v \cdot 10^3$	h	s
600	3.974	757.3	5.242	3.767	760.5	5.215	3.585	763.7	5.190
650	4.251	809.6	5.326	4.028	812.8	5.299	3.830	816.0	5.274
700	4.528	861.7	5.403	4.288	865.0	5.376	4.075	868.2	5.351
750	4.805	913.7	5.475	4.547	917.0	5.448	4.320	920.3	5.423
800	5.081	965.7	5.542	4.806	969.0	5.515	4.564	972.3	5.490
850	5.357	1018	5.605	5.065	1021	5.578	4.808	1024	5.553
900	5.633	1069	5.664	5.324	1073	5.637	5.051	1076	5.612
950	5.909	1121	5.720	5.583	1124	5.693	5.295	1628	5.668
1000	6.184	1173	5.773	5.841	1176	5.746	5.538	1180	5.721
1050	6.459	1224	5.824	6.099	1228	5.797	5.781	1231	5.772
1100	6.734	1276	5.872	6.357	1280	5.845	6.024	1283	5.820
1150	7.010	1328	5.917	6.615	1331	5.891	6.266	1335	5.866
1200	7.284	1379	5.961	6.872	1383	5.935	6.509	1386	5.912
1250	7.559	1431	6.004	7.130	1434	6.977	6.752	1438	5.950
1300	7.834	1483	6.044	7.388	1486	6.017	6.994	1489	5.992

TABLE IV. Thermodynamic Properties of Liquid and Gaseous Neon (*Continued*)

T, K	p = 900 bar			p = 950 bar			p = 1000 bar		
	$v \cdot 10^3$	h	s	$v \cdot 10^3$	h	s	$v \cdot 10^3$	h	s
38	0.7278	103.1	1.597	0.7228	106.0	1.579	0.7181	109.0	1.562
40	0.7350	106.1	1.672	0.7298	109.0	1.654	0.7248	112.0	1.637

TABLE IV. Thermodynamic Properties of Liquid and Gaseous Neon (*Continued*)

T, K	p = 900 bar			p = 950 bar			p = 1000 bar		
	$v \cdot 10^3$	h	s	$v \cdot 10^3$	h	s	$v \cdot 10^3$	h	s
42	0.7424	109.1	1.744	0.7369	111.9	1.727	0.7316	115.0	1.710
44	0.7501	112.0	1.814	0.7443	114.9	1.798	0.7388	118.0	1.781
46	0.7583	114.9	1.882	0.7522	117.9	1.867	0.7464	121.0	1.850
48	0.7673	117.9	1.948	0.7606	121.0	1.934	0.7542	124.0	1.916
50	0.7764	120.9	2.012	0.7694	124.1	1.998	0.7621	127.1	1.979
55	0.7969	128.4	2.155	0.7901	131.7	2.140	0.7820	134.6	2.123
60	0.8182	135.9	2.281	0.8111	139.2	2.267	0.8020	142.0	2.250
65	0.8405	143.5	2.398	0.8320	146.6	2.383	0.8222	149.3	2.364
70	0.8635	151.0	2.504	0.8530	153.9	2.488	0.8425	156.5	2.467
75	0.8870	158.4	2.602	0.8742	161.1	2.583	0.8630	163.7	2.561
80	0.9110	165.7	2.692	0.8962	168.3	2.671	0.8838	170.8	2.649
85	0.9352	172.9	2.778	0.9195	175.4	2.756	0.9051	177.9	2.733
90	0.9603	180.0	2.861	0.9434	182.5	2.837	0.9275	185.0	2.814
95	0.9856	187.1	2.939	0.9676	189.6	2.915	0.9509	192.1	2.891
100	1.011	194.2	3.013	0.9920	196.7	2.988	0.9743	199.2	2.964
110	1.062	208.4	3.148	1.041	210.9	3.123	1.021	213.4	3.100
120	1.113	222.4	3.270	1.089	224.9	3.245	1.067	227.5	3.222
130	1.164	236.2	3.381	1.137	238.7	3.356	1.113	241.3	3.332
140	1.215	249.7	3.481	1.185	252.3	3.456	1.159	254.9	3.433
150	1.265	263.1	3.573	1.233	265.6	3.548	1.204	268.2	3.525
160	1.316	276.3	3.658	1.281	278.8	3.633	1.250	281.4	3.610
170	1.366	289.2	3.736	1.329	291.7	3.712	1.295	294.3	3.689
180	1.417	301.9	3.809	1.377	304.5	3.785	1.340	307.1	3.762
190	1.467	314.5	3.877	1.424	317.1	3.853	1.385	319.7	3.830
200	1.517	326.9	3.941	1.472	329.5	3.916	1.430	332.1	3.893

TABLE IV. Thermodynamic Properties of Liquid and Gaseous Neon (*Continued*)

T, K	p = 900 bar			p = 950 bar			p = 1000 bar		
	$v \cdot 10^3$	h	s	$v \cdot 10^3$	h	s	$v \cdot 10^3$	h	s
220	1.617	351.2	4.057	1.566	353.9	4.033	1.520	356.5	4.010
240	1.716	375.0	4.160	1.659	377.7	4.136	1.609	380.4	4.114
260	1.814	398.4	4.245	1.752	401.1	4.230	1.697	403.9	4.208
280	1.911	421.4	4.339	1.845	424.2	4.316	1.785	427.0	4.293
300	2.008	444.1	4.417	1.936	446.9	4.394	1.872	449.7	4.372
350	2.248	499.7	4.589	2.164	502.6	4.566	2.088	505.5	4.544
400	2.486	554.3	4.735	2.389	557.2	4.712	2.302	560.2	4.690
450	2.722	608.1	4.861	2.613	611.1	4.839	2.515	614.1	4.817
500	2.956	661.3	4.974	2.835	664.4	4.951	2.726	667.5	4.929
550	3.190	714.2	5.075	3.057	717.4	5.052	2.937	720.5	5.030
600	3.423	766.8	5.166	3.278	770.0	5.143	3.147	773.2	5.122
650	3.655	819.2	5.250	3.498	822.5	5.227	3.356	825.7	5.206
700	3.886	871.5	5.327	3.717	874.7	5.305	3.565	878.1	5.284
750	4.117	923.6	5.399	3.936	926.9	5.377	3.773	930.1	5.355
800	4.348	975.6	5.466	4.155	978.9	5.444	3.981	982.2	5.423
850	4.579	1028	5.529	4.373	1031	5.507	4.189	1034	5.486
900	4.809	1079	5.589	4.591	1083	5.566	4.396	1086	5.545
950	5.039	1131	5.645	4.809	1135	5.622	4.603	1138	5.601
1000	5.268	1183	5.698	5.027	1186	5.675	4.810	1190	5.654
1050	5.498	1235	5.748	5.245	1238	5.726	5.017	1241	5.705
1100	5.727	1286	5.796	5.462	1290	5.774	5.223	1293	5.753
1150	5.957	1338	5.842	5.680	1341	5.820	5.430	1345	5.799
1200	6.186	1390	5.886	5.897	1393	5.864	5.637	1396	5.843
1250	6.415	1441	5.928	6.114	1445	5.906	5.843	1448	5.885
1300	6.644	1493	5.969	6.331	1496	5.947	6.049	1500	5.925

TABLE V. Isobaric Specific Heat c_p of Liquid and Gaseous Neon

T, K	c_p at p, bar 1	10	20	30	40	50	60	70	80
30	1.129	2.027	1.998	1.972	1.949	1.928	1.910	1.893	1.878
40	1.066	1.818	3.128	2.775	2.545	2.388	2.268	2.180	2.110
50	1.042	1.267	1.741	2.686	4.871	5.509	4.024	3.324	2.949
60	1.039	1.147	1.313	1.534	1.806	2.114	2.407	2.612	2.693
70	1.037	1.103	1.192	1.295	1.408	1.529	1.649	1.760	1.854
80	1.035	1.083	1.140	1.203	1.268	1.335	1.402	1.466	1.525
90	1.034	1.070	1.112	1.156	1.200	1.244	1.288	1.331	1.371
100	1.033	1.062	1.094	1.127	1.160	1.192	1.224	1.255	1.285
120	1.032	1.052	1.073	1.094	1.115	1.135	1.154	1.174	1.192
140	1.031	1.045	1.060	1.075	1.089	1.103	1.117	1.130	1.143
160	1.031	1.041	1.052	1.063	1.073	1.084	1.094	1.103	1.112
180	1.031	1.039	1.047	1.055	1.063	1.071	1.078	1.085	1.092
200	1.031	1.037	1.043	1.049	1.055	1.061	1.067	1.073	1.078
220	1.030	1.035	1.040	1.045	1.050	1.055	1.059	1.064	1.068
240	1.030	1.034	1.038	1.042	1.046	1.050	1.054	1.057	1.061
260	1.030	1.033	1.037	1.040	1.043	1.046	1.049	1.052	1.055
280	1.030	1.033	1.036	1.038	1.041	1.043	1.046	1.048	1.051
300	1.030	1.032	4.035	1.037	1.039	1.041	1.043	1.045	1.047
350	1.030	1.031	1.033	1.035	1.036	1.037	1.039	1.040	1.042
400	1.030	1.031	1.032	1.033	1.034	1.035	1.036	1.037	1.038
450	1.030	1.031	1.031	1.032	1.033	1.034	1.034	1.035	1.036
500	1.030	1.031	1.031	1.032	1.032	1.033	1.033	1.034	1.034
600	1.030	1.030	1.031	1.031	1.031	1.032	1.032	1.032	1.033
700	1.030	1.030	1.030	1.031	1.031	1.031	1.031	1.031	1.032
1000	1.030	1.030	1.030	1.030	1.030	1.030	1.030	1.030	1.031
1300	1.030	1.030	1.030	1.030	1.030	1.030	1.030	1.030	1.030

TABLE V. Isobaric Specific Heat c_p of Liquid and Gaseous Neon (*Continued*)

T, K	c_p at p, bar 100	120	140	160	180	200	220	240	260
30	1.852	1.830	1.811	1.795	1.781	1.768	1.758	1.749	1.740
40	2.030	1.987	1.959	1.934	1.911	1.889	1.868	1.846	1.824
50	2.557	2.354	2.229	2.131	2.054	1.996	1.955	1.915	1.880
60	2.603	2.421	2.262	2.140	2.046	1.978	1.919	1.870	1.832
70	1.973	2.012	2.001	1.964	1.918	1.872	1.828	1.790	1.755
80	1.621	1.684	1.719	1.733	1.733	1.723	1.709	1.691	1.673
90	1.442	1.497	1.537	1.563	1.579	1.588	1.591	1.589	1.584
100	1.339	1.384	1.420	1.448	1.469	1.484	1.494	1.500	1.504

TABLE V. Isobaric Specific Heat c_p of Liquid and Gaseous Neon (*Continued*)

T, K	c_p at p, bar								
	100	120	140	160	180	200	220	240	260
120	1.227	1.258	1.285	1.308	1.328	1.344	1.358	1.369	1.378
140	1.167	1.189	1.210	1.228	1.244	1.258	1.271	1.282	1.291
160	1.130	1.147	1.162	1.176	1.189	1.201	1.212	1.222	1.230
180	1.106	1.119	1.131	1.142	1.152	1.162	1.171	1.179	1.186
200	1.089	1.099	1.108	1.117	1.126	1.134	1.141	1.148	1.154
220	1.077	1.085	1.092	1.099	1.106	1.113	1.119	1.125	1.130
240	1.068	1.074	1.080	1.086	1.092	1.097	1.102	1.107	1.112
260	1.061	1.066	1.071	1.076	1.081	1.085	1.089	1.093	1.097
280	1.055	1.060	1.064	1.068	1.072	1.076	1.079	1.083	1.086
300	1.051	1.055	1.059	1.062	1.065	1.068	1.072	1.074	1.077
350	1.044	1.047	1.049	1.051	1.054	1.056	1.058	1.060	1.062
400	1.040	1.042	1.043	1.045	1.047	1.048	1.050	1.051	1.052
450	1.037	1.038	1.040	1.041	1.042	1.043	1.044	1.045	1.046
500	1.035	1.036	1.037	1.038	1.039	1.040	1.041	1.041	1.042
600	1.033	1.034	1.034	1.035	1.035	1.036	1.036	1.037	1.037
700	1.032	1.032	1.033	1.033	1.033	1.034	1.034	1.034	1.035
1000	1.031	1.031	1.031	1.031	1.031	1.031	1.031	1.032	1.032
1300	1.030	1.030	1.030	1.030	1.031	1.031	1.031	1.031	1.031

TABLE V. Isobaric Specific Heat c_p of Liquid and Gaseous Neon (*Continued*)

T, K	c_p at p, bar								
	280	300	400	500	600	700	800	900	1000
30	1.733	1.727	—	—	—	—	—	—	—
40	1.803	1.781	1.681	1.601	1.547	1.518	1.505	1.502	1.500
50	1.847	1.814	1.682	1.592	1.532	1.506	1.495	1.492	1.490
60	1.801	1.772	1.651	1.568	1.516	1.494	1.484	1.481	1.479
70	1.726	1.700	1.605	1.543	1.501	1.482	1.472	1.468	1.467
80	1.652	1.635	1.566	1.518	1.486	1.469	1.459	1.456	1.454
90	1.578	1.570	1.527	1.493	1.470	1.455	1.446	1.442	1.440
100	1.505	1.504	1.487	1.465	1.449	1.438	1.431	1.427	1.426
120	1.385	1.391	1.404	1.404	1.398	1.393	1.392	1.391	1.390
140	1.299	1.307	1.331	1.341	1.345	1.345	1.345	1.345	1.346
160	1.238	1.245	1.271	1.286	1.294	1.299	1.302	1.303	1.305
180	1.193	1.199	1.224	1.240	1.251	1.258	1.262	1.266	1.268
200	1.160	1.166	1.188	1.203	1.214	1.222	1.228	1.232	1.235
220	1.135	1.140	1.160	1.174	1.185	1.193	1.199	1.203	1.207
240	1.116	1.120	1.138	1.151	1.161	1.168	1.174	1.179	1.183
260	1.101	1.105	1.120	1.132	1.141	1.148	1.154	1.159	1.163
280	1.089	1.092	1.106	1.117	1.125	1.132	1.137	1.141	1.145

TABLE V. Isobaric Specific Heat c_p of Liquid and Gaseous Neon (*Continued*)

T, K	c_p at p, bar								
	280	300	400	500	600	700	800	900	1000
300	1.080	1.083	1.094	1.104	1.112	1.118	1.123	1.127	1.131
350	1.064	1.066	1.074	1.081	1.088	1.093	1.097	1.100	1.103
400	1.054	1.055	1.061	1.067	1.072	1.076	1.080	1.083	1.085
450	1.047	1.048	1.053	1.057	1.061	1.065	1.068	1.070	1.072
500	1.043	1.044	1.047	1.051	1.054	1.057	1.059	1.061	1.063
600	1.038	1.039	1.040	1.042	1.044	1.048	1.048	1.050	1.052
700	1.035	1.035	1.037	1.038	1.039	1.041	1.042	1.043	1.045
1000	1.032	1.032	1.032	1.033	1.033	1.034	1.034	1.035	1.035
1300	1.031	1.031	1.031	1.031	1.031	1.031	1.032	1.032	1.032

TABLE VI. Dynamic Viscosity of Neon

T, K	$\eta \cdot 10^8$, Pa · s at p, bar							
	1	4	8	12	16	20	24	28
26	446	13835	13936	14038	14141	14245	14350	14456
30	507	10080	10180	10280	10380	10480	10580	10690
34	567	588	7209	7328	7446	7562	7677	7791
38	626	644	674	4928	5073	5191	5309	5450
42	684	701	724	761	826	310	3331	3521
46	741	756	776	803	843	902	993	1154
50	798	812	829	851	881	920	972	1040
60	837	949	963	979	997	1019	1045	1076
70	1070	1081	1093	1105	1119	1135	1153	1173
80	1198	1208	1218	1229	1241	1253	1267	1281
90	1319	1328	1337	1347	1357	1367	1378	1390
100	1434	1442	1451	1459	1468	1477	1487	1497
110	1542	1550	1558	1565	1573	1581	1590	1598
120	1645	1652	1659	1667	1674	1681	1688	1696
130	1745	1752	1758	1765	1772	1778	1785	1792
140	1845	1851	1858	1864	1870	1876	1882	1889
150	1944	1950	1956	1962	1967	1973	1979	1985
160	2039	2045	2050	2056	2061	2067	2072	2077
170	2132	2137	2143	2148	2153	2158	2163	2168
180	2223	2228	2233	2238	2243	2248	2253	2257
190	2311	2316	2321	2325	2330	2334	2339	2344
200	2397	2402	2406	2411	2415	2419	2424	2428
220	2564	2568	2572	2576	2580	2584	2588	2592
240	2725	2729	2733	2736	2740	2744	2747	2751
260	2880	2884	2887	2890	2894	2897	2901	2904
280	3029	3032	3035	3039	3042	3045	3048	3051

TABLE VI. Dynamic Viscosity of Neon (*Continued*)

T, K	$\eta \cdot 10^8$, Pa · s at p, bar							
	1	4	8	12	16	20	24	28
300	3175	3178	3181	3184	3187	3190	3193	3196
350	3522	3525	3527	3530	3532	3535	3537	3540
400	3850	3852	3854	3857	3859	3861	3863	3865
450	4163	4165	4167	4169	4171	4173	4174	4176
500	4463	4465	4466	4468	4470	4471	4473	4475
550	4753	4755	4756	4758	4759	4761	4762	4764
600	5033	5034	5036	5037	5038	5040	5041	5042
650	5305	5306	5307	5309	5310	5311	5312	5314
700	5570	5571	5572	5573	5574	5576	5577	5578
750	5828	5829	5830	5831	5832	5833	5834	5835
800	6081	6082	6083	6084	6085	6086	6087	6088
850	6329	6330	6331	6332	6332	6333	6334	6335
900	6572	6573	6574	6574	6575	6576	6577	6578
950	6810	6811	6812	6812	6813	6814	6814	6815
1000	7045	7046	7046	7047	7048	7048	7049	7050
1050	7275	7276	7276	7277	7278	7278	7279	7280
1100	7503	7504	7504	7505	7505	7506	7507	7507
1150	7727	7728	7728	7729	7729	7730	7730	7731
1200	7948	7949	7949	7950	7950	7951	7951	7952
1250	8166	8167	8167	8168	8168	8169	8169	8170
1300	8381	8381	8382	8382	8383	8383	8384	8384

TABLE VI. Dynamic Viscosity of Neon (*Continued*)

T, K	$\eta \cdot 10^8$, Pa · s at p, bar							
	32	36	40	50	60	70	80	100
26	14563	14671	14780	15000	15400	15840	16320	—
30	10790	10900	11000	11280	11560	11860	12180	12800
34	7902	8012	8120	8380	8560	8840	9080	9600
38	5566	5684	5800	6180	6380	6660	6890	7300
42	3690	3848	3990	4280	4760	5080	5360	5680
46	2030	2102	2414	3060	3406	3802	4120	4320
50	1190	1300	1440	1930	2420	2800	3110	3640
60	1130	1170	1278	1348	1515	1710	1910	2280
70	1221	1249	1275	1349	1430	1535	1645	1865
80	1310	1329	1346	1400	1455	1515	1585	1725
90	1393	1412	1427	1464	1505	1553	1603	1714
100	1507	1518	1530	1561	1595	1632	1672	1760
110	1607	1616	1626	1652	1680	1710	1745	1813
120	1704	1712	1720	1743	1766	1792	1819	1877
130	1799	1806	1814	1833	1854	1875	1898	1948

TABLE VI. Dynamic Viscosity of Neon (*Continued*)

T, K	$\eta \cdot 10^8$, Pa · s at p, bar							
	32	36	40	50	60	70	80	100
140	1895	1902	1908	1926	1944	1963	1983	2026
150	1991	1997	2003	2019	2035	2052	2070	2107
160	2083	2088	2094	2108	2123	2139	2155	2188
170	2173	2179	2184	2197	2211	2225	2239	2270
180	2262	2267	2272	2284	2297	2310	2323	2351
190	2348	2353	2357	2369	2381	2393	2405	2430
200	2432	2437	2441	2452	2463	2474	2486	2509
220	2596	2600	2604	2614	2624	2634	2644	2664
240	2755	2758	2762	2771	2780	2789	2798	2816
260	2907	2911	2914	2922	2930	2938	2947	2963
280	3054	3057	3060	3068	3076	3083	3091	3106
300	3199	3201	3204	3211	3218	3225	3232	3246
350	3542	3544	3547	3553	3559	3565	3571	3582
400	3867	3869	3872	3877	3882	3887	3892	3902
450	4178	4180	4182	4186	4191	4196	4200	4209
500	4476	4478	4480	4484	4488	4492	4496	4504
550	4765	4767	4768	4772	4775	4779	4782	4790
600	5044	5045	5047	5050	5053	5056	5060	5066
650	5315	5316	5317	5320	5323	5326	5329	5335
700	5579	5580	5581	5584	5587	5589	5592	5597
750	5836	5837	5838	5841	5843	5846	5848	5853
800	6089	6089	6090	6093	6095	6097	6100	6104
850	6336	6337	6338	6340	6342	6344	6346	6351
900	6578	6579	6580	6582	6584	6586	6588	6592
950	6816	6817	6817	6819	6821	6823	6825	6829
1000	7051	7052	7052	7054	7055	7057	7059	7062
1050	7280	7282	7282	7283	7285	7286	7288	7291
1100	7508	7509	7509	7511	7512	7514	7515	7518
1150	7732	7733	7733	7734	7736	7737	7738	7741
1200	7952	7953	7953	7955	7956	7957	7959	7961
1250	8170	8171	8171	8172	8174	8175	8176	8179
1300	8385	8385	8386	8387	8388	8389	8391	8393

TABLE VI. Dynamic Viscosity of Neon (*Continued*)

T, K	$\eta \cdot 10^8$, Pa · s at p, bar							
	120	140	160	180	200	220	240	260
30	13490	14200	14920	15640	16340	17120	18240	19040
34	10160	10660	11200	11680	12240	12740	12320	13880
38	7700	8220	8800	9160	9520	9960	10320	10760
42	6050	6480	7090	7440	7740	8100	8400	8800

TABLE VI. Dynamic Viscosity of Neon (*Continued*)

T, K	$\eta \cdot 10^8$, Pa · s at p, bar							
	120	140	160	180	200	220	240	260
46	4920	5340	5840	6190	6480	6780	7080	7400
50	4100	4520	4885	5230	5550	5820	6080	6350
60	2620	2940	3238	3523	3798	4060	4320	4568
70	2085	2305	2525	2742	2957	3172	3381	3589
80	1878	2042	2209	2380	2550	2720	2880	3040
90	1835	1965	2099	2231	2363	2492	2627	2759
100	1856	1959	2066	2177	2290	2404	2519	2634
110	1890	1972	2058	2148	2240	2334	2430	2527
120	1940	2008	2079	2153	2230	2309	2390	2471
130	2001	2058	2118	2181	2246	2313	2382	2452
140	2072	2121	2172	2227	2283	2341	2400	2461
150	2148	2191	2236	2283	2332	2383	2435	2488
160	2224	2262	2302	2343	2387	2431	2477	2525
170	2302	2336	2371	2408	2447	2487	2528	2570
180	2380	2411	2443	2476	2511	2546	2583	2611
190	2457	2485	2513	2544	2576	2608	2641	2676
200	2534	2559	2586	2613	2642	2671	2702	2733
220	2686	2707	2730	2754	2778	2802	2828	2854
240	2835	2854	2874	2894	2915	2936	2958	2981
260	2980	2997	3015	3032	3051	3070	3089	3108
280	3121	3136	3152	3168	3184	3201	3218	3235
300	3260	3274	3289	3303	3318	3333	3348	3364
350	3594	3606	3617	3629	3641	3653	3665	3677
400	3912	3922	3932	3942	3952	3962	3972	3983
450	4218	4227	4235	4244	4253	4261	4270	4279
500	4512	4519	4527	4535	4542	4550	4558	4554
550	4797	4804	4811	4818	4824	4831	4838	4845
600	5072	5079	5085	5091	5098	5104	5110	5116
650	5341	5347	5352	5358	5364	5370	5375	5381
700	5603	5608	5613	5619	5624	5629	5634	5639
750	5858	5863	5868	5873	5878	5882	5887	5892
800	6109	6113	6118	6122	6127	6131	6136	6140
850	6355	6359	6363	6367	6371	6376	6380	6384
900	6596	6600	6604	6608	6611	6615	6619	6623
950	6832	6836	6840	6843	6847	6850	6854	6858
1000	7066	7069	7073	7076	7079	7083	7086	7090
1050	7294	7298	7301	7304	7307	7310	7314	7317
1100	7521	7524	7527	7530	7533	7536	7539	7542
1150	7744	7747	7750	7753	7755	7758	7761	7764
1200	7964	7967	7969	7972	7975	7978	7980	7983
1250	8181	8184	8186	8189	8191	8194	8196	8199
1300	8395	8398	8400	8403	8405	8407	8410	8412

TABLE VI. Dynamic Viscosity of Neon (*Continued*)

T, K	$\eta \cdot 10^8$, Pa · s at *p*, bar							
	280	300	350	400	450	500	550	600
30	19800	20670	22080	—	—	—	—	—
34	14400	14780	15960	17720	19640	21440	23360	25400
38	11240	11580	12640	13740	14920	16260	17620	18800
42	9090	9470	10360	11240	12180	13100	14140	14960
46	6640	7960	8700	9440	10240	10980	11760	12520
50	6610	6860	7480	8100	8740	9430	10160	10890
60	4808	5038	5593	6125	6645	7160	7670	8180
70	3789	3980	4435	4855	5275	5675	6064	6445
80	3197	3355	3740	4119	4475	4810	5137	5469
90	2890	3025	3352	3670	3979	4271	4558	4839
100	2748	2861	3141	3410	3670	3918	4162	4392
110	2623	2720	2962	3198	3428	3650	3867	4079
120	2554	2637	2846	3054	3259	3459	3656	3850
130	2524	2595	2778	2961	3145	3326	3503	3680
140	2523	2586	2747	2909	3074	3238	3402	3562
150	2543	2598	2741	2887	3035	3183	3332	3480
160	2573	2622	2749	2880	3015	3149	3286	3421
170	2613	2657	2771	2889	3011	3134	3258	3384
180	2660	2700	2802	2910	3020	3132	3247	3363
190	2711	2747	2840	2937	3038	3142	3256	3354
200	2765	2798	2883	2972	3064	3159	3297	3355
220	2881	2909	2980	3055	3133	3214	3292	3382
240	3004	3027	3088	3152	3219	3289	3361	3434
260	3128	3149	3202	3257	3316	3376	3438	3503
280	3253	3271	3318	3367	3417	3470	3525	3582
300	3379	3395	3437	3480	3525	3572	3621	3671
350	3690	3702	3735	3768	3803	3839	3876	3914
400	3993	4003	4029	4056	4084	4113	4143	4173
450	4287	4296	4318	4341	4364	4388	4412	4437
500	4573	4581	4600	4620	4640	4660	4681	4702
550	4852	4859	4876	4893	4910	4928	4946	4964
600	5122	5128	5144	5159	5175	5191	5206	5223
650	5386	5392	5406	5420	5434	5448	5462	5477
700	5645	5650	5663	5675	5688	5701	5714	5727
750	5897	5901	5913	5925	5937	5949	5961	5972
800	6145	6149	6160	6171	6182	6193	6204	6215
850	6388	6392	6402	6412	6423	6433	6443	6453
900	6627	6631	6640	6650	6659	6669	6678	6688
950	6861	6865	6874	6883	6892	6901	6909	6918
1000	7093	7096	7105	7113	7122	7130	7138	7147
1050	7320	7323	7331	7339	7347	7355	7363	7371
1100	7545	7548	7556	7563	7571	7578	7586	7593
1150	7767	7770	7777	7784	7791	7798	7805	7812
1200	7986	7988	7995	8002	8008	8015	8022	8028

TABLE VI. Dynamic Viscosity of Neon (*Continued*)

T, K	$\eta \cdot 10^8$, Pa · s at p, bar							
	280	300	350	400	450	500	550	600
1250	8201	8204	8210	8217	8223	8229	8236	8242
1300	8415	8417	8423	8429	8435	8441	8447	8453

TABLE VI. Dynamic Viscosity of Neon (*Continued*)

T, K	$\eta \cdot 10^8$, Pa · s at p, bar							
	650	700	750	800	850	900	950	1000
34	19995	20702						
38	17346	18072	18770	19454	20130	20783	21419	22035
42	15087	15774	16457	17121	17779	18426	19057	19692
46	13118	13755	14385	15017	15638	16249	16854	17461
50	11455	12401	12602	13161	13709	14254	14834	15407
60	8549	9062	9556	10039	10483	10908	11284	11748
70	6827	7199	7593	7994	8339	8726	9084	9485
80	5765	6073	6379	6691	6995	7331	7665	8011
90	5091	5345	5609	5864	6124	6388	6658	6952
100	4624	4847	5073	5290	5512	5736	5958	6185
110	4287	4485	4683	4877	5070	5266	5454	5650
120	4036	4219	4396	4572	4750	4920	5091	5262
130	3850	4019	4183	4345	4505	4660	4818	4971
140	3720	3875	4028	4178	4325	4468	4615	4754
150	3625	3770	3911	4051	4189	4326	4461	4593
160	3556	3690	3823	3953	4084	4209	4336	4459
170	3509	3633	3757	3879	4001	4121	4238	4356
180	3478	3595	3710	3825	3937	4051	4162	4275
190	3461	3569	3676	3784	3891	3998	4103	4209
200	3454	3555	3656	3757	3857	3958	4056	4158
220	3468	3556	3645	3734	3822	3912	4001	4091
240	3510	3587	3664	3743	3823	3902	3983	4062
260	3569	3636	3705	3776	3846	3917	3990	4062
280	3640	3700	3761	3823	3887	3951	4015	4080
300	3723	3776	3831	3886	3943	4001	4059	4118
350	3954	3995	4036	4079	4123	4168	4214	4260
400	4204	4237	4270	4304	4339	4375	4411	4449
450	4463	4490	4517	4544	4573	4602	4632	4662
500	4723	4746	4768	4791	4815	4840	4864	4890
550	4983	5002	5021	5041	5061	5082	5103	5125
600	5239	5256	5273	5290	5307	5325	5344	5362
650	5491	5506	5521	5537	5552	5568	5584	5600
650	5491	5506	5521	5537	5552	5568	5584	5600
700	5740	5754	5767	5781	5795	5809	5823	5837

TABLE VI. Dynamic Viscosity of Neon (*Continued*)

T, K	$\eta \cdot 10^8$, Pa · s at p, bar							
	650	700	750	800	850	900	950	1000
750	5985	5997	6009	6021	6034	6047	6059	6072
800	6226	6237	6248	6259	6271	6283	6294	6306
850	6463	6474	6484	6495	6505	6516	6527	6537
900	6697	6707	6717	6726	6736	6746	6756	6766
950	6927	6936	6945	6954	6963	6973	6982	6991
1000	7155	7163	7172	7180	7189	7198	7206	7215
1050	7379	7386	7394	7402	7410	7418	7427	7435
1100	7601	7608	7616	7623	7631	7638	7646	7654
1150	7819	7826	7833	7840	7848	7955	7862	7869
1200	8035	8042	8049	8055	8062	8069	8076	8083
1250	8249	8255	8261	8268	8274	8281	8287	8294
1300	8459	8465	8472	8478	8484	8490	8496	8502

TABLE VII. Thermal Conductivity of Neon

T, K	$\lambda \cdot 10^8$, kW/(m · K) at p, bar							
	1	4	8	12	16	20	24	28
26	689	13340	13406	13473	13531	13599	13659	13721
30	783	11493	11574	11649	11728	11802	11873	11948
34	876	960	9738	9836	9936	10030	10123	10217
38	967	1038	1142	7910	8051	8197	8316	8455
42	1057	1120	1202	1314	1485	5961	6237	6470
46	1145	1201	1270	1357	1468	1618	1830	2177
50	1233	1284	1344	1416	1503	1608	1736	1896
60	1448	1489	1536	1588	1646	1712	1784	1864
70	1654	1689	1727	1769	1814	1863	1915	1971
80	1851	1881	1914	1948	1986	2025	2067	2111
90	2038	2065	2093	2123	2155	2188	2223	2259
100	2216	2240	2265	2292	2319	2348	2378	2409
110	2383	2405	2427	2451	2476	2501	2528	2555
120	2542	2562	2582	2604	2626	2649	2672	2697
130	2697	2715	2734	2754	2774	2795	2816	2838
140	2852	2869	2886	2904	2923	2942	2961	2981
150	3005	3021	3037	3054	3071	3088	3106	3125
160	3152	3167	3182	3197	3213	3230	3246	3263
170	3296	3310	3324	3338	3353	3369	3384	3400
180	3436	3449	3462	3476	3490	3504	3519	3534
190	3573	3585	3598	3611	3624	3637	3651	3665
200	3706	3718	3730	3742	3754	3767	3780	3793
220	3964	3975	3985	3997	4008	4019	4031	4043

TABLE VII. Thermal Conductivity of Neon (*Continued*)

T, K	$\lambda \cdot 10^8$, kW/(m · K) at p, bar							
	1	4	8	12	16	20	24	28
240	4212	4222	4232	4242	4252	4262	4273	4284
260	4451	4460	4469	4478	4488	4497	4507	4517
280	4683	4691	4700	4708	4717	4726	4735	4744
300	4908	4916	4924	4932	4940	4948	4956	4965
350	5445	5452	5458	5465	5472	5479	5486	5493
400	5952	5958	5964	5970	5976	5982	5988	5994
450	6436	6441	6446	6452	6457	6462	6468	6473
500	6900	6905	6909	6914	6919	6924	6929	6933
550	7347	7351	7356	7360	7364	7368	7373	7377
600	7780	7784	7788	7792	7796	7800	7804	7808
650	8200	8204	8207	8211	8214	8218	8222	8226
700	8610	8613	8617	8620	8623	8627	8630	8634
750	9010	9013	9016	9019	9023	9026	9029	9032
800	9400	9403	9406	9409	9412	9415	9418	9421
850	9783	9786	9789	9791	9794	9797	9800	9802
900	10158	10161	10163	10166	10168	10171	10174	10176
950	10527	10529	10532	10534	10537	10539	10542	10544
1000	10889	10891	10894	10896	10898	10901	10903	10905
1050	11246	11248	11250	11253	11255	11257	11259	11262
1100	11598	11600	11602	11604	11606	11609	11611	11613
1150	11944	11946	11948	11950	11952	11954	11956	11958
1200	12285	12287	12289	12291	12293	12295	12297	12299
1250	12623	12625	12627	12629	12630	12632	12634	12636
1300	12956	12958	12960	12961	12963	12965	12967	12969

TABLE VII. Thermal Conductivity of Neon (*Continued*)

T, K	$\lambda \cdot 10^8$, kW/(m · K) at p, bar							
	32	36	40	50	60	70	80	100
26	13784	13840	13907	14060	14206	14352	14498	
30	12022	12092	12165	12337	12510	12679	12840	13152
34	10307	10393	10481	10689	10889	11084	11271	11633
38	8569	8684	8795	9064	9313	9550	9772	10193
42	6675	6863	7030	7406	7746	8044	8318	8824
46		4533	4983	5687	6151	6530	6918	7545
50	2099	2364		3851	4663	5219	5659	6352
60	1951	2061	2152	2454	2805	3187	3572	4286
70	2031	2094	2161	2344	2545	2762	2990	3457
80	2157	2205	2256	2390	2534	2688	2848	3181
90	2297	2336	2377	2484	2598	2718	2842	3101
100	2442	2475	2509	2599	2693	2792	2894	3106
110	2583	2612	2642	2719	2800	2884	2971	3151
120	2722	2748	2774	2842	2913	2986	3062	3219
130	2860	2884	2907	2968	3031	3097	3164	3303

TABLE VII. Thermal Conductivity of Neon (*Continued*)

T, K	$\lambda \cdot 10^8$, kW/(m · K) at p, bar							
	32	36	40	50	60	70	80	100
140	3002	3023	3044	3099	3156	3215	3276	3401
150	3143	3163	3182	3233	3285	3338	3393	3507
160	3281	3298	3316	3363	3411	3460	3510	3614
170	3416	3433	3449	3492	3537	3582	3629	3725
180	3549	3564	3580	3620	3661	3703	3747	3836
190	3679	3694	3708	3746	3784	3824	3864	3948
200	3807	3820	3834	3869	3905	3942	3981	4058
220	4055	4067	4079	4111	4143	4176	4210	4279
240	4295	4306	4317	4345	4374	4404	4435	4497
260	4527	4537	4547	4573	4599	4627	4654	4711
280	4753	4762	4772	4795	4820	4845	4870	4922
300	4973	4982	4990	5012	5035	5058	5081	5129
350	5500	5508	5515	5534	5552	5572	5591	5632
400	6000	6006	6013	6029	6045	6062	6079	6113
450	6479	6484	6490	6504	6518	6533	6548	6578
500	6938	6943	6948	6961	6973	6987	7000	7026
550	7382	7386	7391	7402	7414	7425	7437	7461
600	7812	7816	7820	7830	7841	7851	7862	7884
650	8229	8233	8237	8246	8256	8266	8275	8296
700	8637	8641	8644	8653	8662	8671	8680	8698
750	9035	9039	9042	9050	9058	9067	9075	9092
800	9424	9427	9430	9437	9445	9453	9461	9477
850	9805	9808	9811	9818	9825	9833	9840	9855
900	10179	10182	10184	10191	10198	10205	10212	10226
950	10547	10549	10552	10558	10565	10571	10578	10591
1000	10908	10910	10913	10919	10925	10931	10937	10950
1050	11264	11266	11268	11274	11280	11286	11292	11304
1100	11615	11617	11619	11625	11630	11636	11642	11653
1150	11960	11962	11964	11970	11975	11980	11986	11996
1200	12301	12303	12305	12310	12315	12320	12325	12335
1250	12638	12640	12642	12647	12651	12656	12661	12671
1300	12970	12972	12974	12979	12983	12988	12993	13002

TABLE VII. Thermal Conductivity of Neon (*Continued*)

T, K	$\lambda \cdot 10^8$, kW/(m · K) at p, bar							
	120	140	160	180	200	220	240	260
30	13456	13743	14027	14304	14567	14828	15080	15330
34	11971	12299	12607	12908	13196	13477	13750	14011
38	10582	10949	11293	11620	11934	12242	12532	12816
42	9271	9683	10066	10427	10769	11095	11406	11707

TABLE VII. Thermal Conductivity of Neon (*Continued*)

T, K	λ · 10⁸, kW/(m · K) at p, bar							
	120	140	160	180	200	220	240	260
46	8038	8505	8919	9308	9676	10022	10351	10665
50	6921	7404	7861	8262	8641	9005	9344	9667
60	4899	5429	5894	6314	6697	7055	7396	7706
70	3913	4345	4750	5130	5490	5823	6144	6445
80	3519	3850	4175	4486	4789	5079	5359	5630
90	3366	3632	3894	4152	4404	4651	4890	5126
100	3325	3545	3765	3982	4196	4407	4614	4819
110	3337	3525	3714	3901	4088	4272	4454	4634
120	3381	3545	3710	3876	4040	4204	4365	4526
130	3446	3592	3739	3887	4034	4181	4326	4470
140	3530	3661	3793	3927	4060	4193	4325	4457
150	3624	3743	3864	3985	4107	4229	4350	4471
160	3722	3831	3942	4054	4165	4278	4389	4502
170	3824	3925	4027	4131	4234	4338	4442	4546
180	3928	4022	4117	4213	4309	4407	4504	4601
190	4033	4121	4210	4300	4390	4481	4572	4663
200	4139	4221	4304	4389	4474	4559	4645	4731
220	4351	4423	4498	4573	4649	4725	4802	4879
240	4561	4627	4694	4761	4830	4899	4968	5038
260	4769	4829	4890	4951	5013	5076	5140	5203
280	4975	5030	5086	5142	5199	5257	5315	5374
300	5178	5229	5280	5332	5385	5438	5492	5546
350	5673	5715	5758	5801	5845	5890	5935	5980
400	6148	6184	6221	6258	6296	6335	6373	6412
450	6608	6640	6672	6705	6738	6771	6805	6839
500	7054	7082	7110	7139	7168	7198	7228	7259
550	7486	7511	7536	7562	7589	7615	7642	7670
600	7906	7929	7952	7976	8000	8024	8048	8073
650	8316	8337	8358	8379	8401	8423	8446	8468
700	8717	8736	8756	8775	8796	8816	8836	8857
750	9104	9127	9145	9164	9182	9201	9220	9239
800	9493	9509	9526	9543	9561	9578	9596	9613
850	9870	9886	9901	9917	9933	9950	9966	9983
900	10240	10255	10269	10284	10299	10315	10330	10346
950	10604	10618	10632	10646	10660	10675	10689	10704
1000	10962	10975	10988	11002	11015	11029	11043	11056
1050	11316	11328	11340	11353	11366	11379	11392	11405
1100	11664	11676	11688	11700	11712	11724	11737	11749
1150	12007	12019	12030	12041	12053	12064	12076	12088
1200	12346	12357	12367	12378	12389	12400	12411	12423
1250	12681	12692	12702	12712	12722	12733	12744	12755
1300	13012	13022	13031	13041	13051	13062	13072	13082

TABLE VII. Thermal Conductivity of Neon (*Continued*)

T, K	$\lambda \cdot 10^8$, kW/(m · K) at p, bar							
	280	300	350	400	450	500	550	600
30	15566	15806	16375					
34	14274	14526	15127	15707	16263	16790	17303	17804
38	13086	13357	13994	14598	15175	15724	16258	16768
42	11996	12276	12947	13573	14168	14738	15283	15810
46	10968	11262	11950	12596	13207	13785	14337	14871
50	9974	10271	10968	11617	12227	12804	13358	13885
60	8007	8307	8994	9661	10284	10871	11452	11984
70	6733	7011	7670	8280	8869	9394	9957	10505
80	5891	6144	6746	7309	7853	8363	8868	9347
90	5354	5579	6122	6636	7130	7610	8070	8517
100	5019	5217	5699	6164	6618	7052	7483	7894
110	4810	4986	5417	5836	6246	6643	7036	7425
120	4684	4842	5227	5607	5980	6344	6707	7068
130	4614	4756	5108	5452	5794	6130	6463	6798
140	4587	4717	5039	5355	5669	5981	6292	6600
150	4591	4711	5008	5301	5592	5879	6167	6455
160	4613	4724	4999	5271	5543	5811	6080	6348
170	4649	4753	5010	5265	5519	5771	6021	6275
180	4697	4794	5035	5275	5513	5750	5986	6224
190	4754	4845	5072	5297	5522	5747	5971	6195
200	4817	4903	5117	5330	5543	5755	5967	6180
220	4956	5033	5225	5418	5611	5803	5995	6189
240	5108	5177	5353	5528	5704	5879	6056	6232
260	5267	5331	5492	5653	5814	5976	6138	6300
280	5432	5491	5639	5789	5938	6087	6238	6288
300	5600	5655	5792	5931	6070	6209	6349	6490
350	6026	6072	6188	6306	6424	6543	6662	6782
400	6452	6492	6592	6693	6796	6899	7003	7108
450	6874	6909	6997	7086	7176	7268	7360	7452
500	7289	7320	7399	7478	7559	7640	7722	7805
550	7697	7725	7795	7867	7940	8013	8087	8162
600	8098	8123	8187	8252	8318	8385	8452	8520
650	8491	8514	8572	8632	8692	8753	8815	8878
700	8878	8899	8953	9008	9063	9119	9177	9234
750	9258	9278	9328	9378	9430	9482	9535	9588
800	9631	9650	9696	9743	9791	9839	9889	9938
850	10000	10017	10060	10104	10148	10194	10240	10286
900	10361	10377	10418	10459	10501	10544	10587	10630
950	10719	10734	10772	10811	10850	10890	10931	10972
1000	11070	11085	11120	11157	11194	11232	11270	11309
1050	11418	11431	11465	11500	11535	11571	11607	11644
1100	11762	11774	11806	11839	11873	11907	11941	11976
1150	12100	12112	12143	12174	12206	12238	12271	12304
1200	12434	12445	12475	12504	12535	12565	12597	12628

TABLE VII. Thermal Conductivity of Neon (*Continued*)

T, K	$\lambda \cdot 10^8$, kW/(m · K) at p, bar							
	280	300	350	400	450	500	550	600
1250	12766	12776	12804	12833	12862	12891	12921	12951
1300	13093	13103	13130	13157	13185	13213	13241	13270

TABLE VII. Thermal Conductivity of Neon (*Continued*)

T, K	$\lambda \cdot 10^8$, kW/(m · K) at p, bar							
	650	700	750	800	850	900	950	1000
34	18285	18754						
38	17255	17745	18210	18668	19123	19559	19987	20404
42	16317	16808	17287	17752	18209	18652	19081	19515
46	15385	15879	16359	16836	17291	17735	18170	18604
50	14478	14965	15447	15905	16344	16773	17222	17660
60	12530	13086	13608	14107	14551	14968	15329	15766
70	11006	11504	12017	12526	13030	13427	13850	14317
80	9830	10306	10773	11240	11688	12173	12647	13125
90	8971	9402	9849	10278	10713	11146	11586	12058
100	8312	8714	9124	9518	9919	10323	10724	11129
110	7809	8180	8554	8924	9294	9671	10033	10411
120	7421	7772	8115	8461	8813	9153	9499	9847
130	7122	7452	7775	8099	8424	8742	9070	9392
140	6905	7209	7515	7818	8119	8417	8728	9026
150	6738	7026	7308	7592	7878	8167	8453	8741
160	6616	6884	7151	7418	7691	7956	8229	8497
170	6526	6777	7030	7283	7537	7794	8047	8306
180	6461	6701	6938	7177	7416	7658	7900	8149
190	6419	6645	6869	7096	7325	7555	7787	8021
200	6393	6607	6822	7038	7254	7473	7690	7916
220	6381	6575	6771	6968	7163	7362	7563	7765
240	6409	6587	6765	6945	7127	7308	7495	7678
260	6464	6627	6793	6960	7127	7295	7467	7638
280	6540	6692	6846	7000	7157	7313	7470	7630
300	6631	6774	6916	7060	7206	7352	7501	7649
350	6903	7025	7147	7271	7396	7522	7648	7776
400	7213	7320	7427	7535	7644	7754	7865	7977
450	7546	7640	7735	7831	7928	8025	8123	8223
500	7888	7972	8057	8143	8230	8318	8407	8496
550	8238	8314	8391	8469	8548	8627	8707	8788
600	8590	8659	8729	8801	8872	8944	9017	9091
650	8941	9005	9070	9135	9201	9267	9334	9402
700	9293	9352	9412	9472	9533	9594	9656	9719

TABLE VII. Thermal Conductivity of Neon (*Continued*)

T, K	$\lambda \cdot 10^8$, kW/(m · K) at p, bar							
	650	700	750	800	850	900	950	1000
750	9643	9697	9753	9809	9865	9922	9980	10038
800	9989	10040	10092	10144	10196	10249	10303	10357
850	10334	10381	10430	10478	10528	10577	10628	10678
900	10675	10720	10765	10811	10857	10903	10951	10998
950	11014	11056	11098	11142	11185	11229	11274	11319
1000	11349	11388	11429	11469	11511	11552	11594	11637
1050	11681	11719	11757	11796	11835	11874	11914	11954
1100	12011	12047	12083	12120	12157	12194	12232	12270
1150	12338	12372	12406	12441	12476	12512	12547	12584
1200	12660	12693	12726	12759	12792	12826	12860	12895
1250	12982	13013	13044	13076	13108	13140	13173	13206
1300	13300	13329	13359	13389	13420	13451	13482	13514

D. ARGON TABLE DISCUSSION

As experimental data on thermodynamic properties of argon accumulated, a number of attempts were made to calculate detailed tables of various thermodynamic functions. Unlike other inert gases, rather complete tables containing values of compressibility, density, enthalpy, entropy and specific heat at constant pressure, ratio of specific heats c_p/c_v, and the speed of sound were published for argon as early as 1955 [III.4]. The values of these functions were calculated from the virial equation of state in the form $z = f(p, T)$ and are represented over the temperature range of 70 to 5000 K (the calorific functions at 100 to 3000 K) at pressures to 100 atm.

Din [III.5] worked up experimental *pvT* data of various investigators as well as the available values of calorific quantities by the graphical method and compiled tables of thermodynamic properties of argon (v, h, s, c_p, c_v) for the temperature range from the saturation curve to 600 K at pressures to 5000 atm. Naturally, both studies do not include subsequently published experimental data on the compressibility of argon (see Table 9). The last circumstance served as an important reason for the revision of the aforementioned tables in order to refine the thermodynamic functions of argon and to broaden the range of parameters simultaneously with respect to temperature and pressure.

New tables of thermodynamic properties of gaseous argon [1.104] were calculated by means of two equations of state represented by elementary functions over the temperature range from the saturation curve to 1300 K at pressures to 1000 bar. A separate equation of state was compiled for liquid argon which was then used to calculate its thermodynamic properties from the saturation curve up to the solidification curve [3.8].

Gosman et al. [III.6] of the NBS (USA) also calculated the thermodynamic properties of liquid and gaseous argon from the triple point to 300 K at pressures to 1000 atm using the same original experimental data for finding the coefficients of the equation of state in the form suggested by Strobridge. Their summary tables include molar values of density, enthalpy, internal energy and entropy. The results computed by Vasserman et al. [1.104, 3.8] and Gosman et al. [III.6] from independent equations of state in overlapping ranges of parameters are in satisfactory agreement with one another.

As a result of this, the working group on the properties of atmospheric gases of the IUPAC Commission on Thermodynamics and Thermochemistry recommended issuing International Tables of Thermodynamic Properties of Argon on the basis of the highly compatible results in [1.104, 3.8, III.6]. These tables were compiled and published in London in 1972 [III.7].

Experimental data on the thermodynamic properties of argon which were published following the meetings of the above working group, and which were used by us in chapters 2 and 3 in compiling new equations of state in the virial form for gaseous and liquid argon, agree within the limits of experimental error with the experimental data used for the same purpose in [1.104, 3.8, III.6].

Although the new equations describe the properties of argon somewhat more exactly, and it was possible to obtain a single equation of state for the entire gaseous phase instead of the previous two, we deemed it best not to use them for calculating the thermodynamic properties of argon and to list in Table XI of the present book the data obtained, as in the International Tables [III.7], by averaging the results computed from equations given in [1.104, 3.8, III.6]. The equations of state for gaseous argon and liquid argon given in chapters 2 and 3 are, due to their simple form, convenient to use for local practical calculations. Subsequently these should be utilized when necessary to revise and refine the first International Tables.

Since the data in [III.7] are limited to a maximum temperature of 1100 K, we presented for the 1150–1300 K temperature range computed values of thermodynamic properties of argon [1.104] which are in good agreement with the results in [III.7]. The values of enthalpy of crystalline argon were made compatible with the values for the liquid and gaseous phases by adding the constant value of $\Delta h = 0.67$ kJ/kg to the averaged data. The properties of liquid along the solidification curve, given in Table IX, are determined by graphical extrapolation of isobars constructed from data in [III.7] to solidification temperatures with subsequent interpolation to allow representation at round temperatures.

The possible errors of thermodynamic functions of crystalline argon (Table VIII) are given on page xxx. The coefficients of dynamic viscosity and thermal conductivity (Tables XII and XIII) were calculated from the corresponding equations listed on pages 249, 251, and 278, using data on the specific volumes of argon from Table XI. In these chapters the possible errors of computed values of η and λ are estimated (see pages 252 and 279).

TABLE VIII. Thermodynamic Properties of Crystalline Argon

T, K	p = 1 bar						
	$v \cdot 10^3$	h	s	c_p	c_v	$\alpha_{exp} \cdot 10^5$	$\beta_T \cdot 10^5$
1	0.5645	0	0.000020	0.000060	0.000060	0.01	3.75
2	0.5645	0	0.000163	0.000495	0.000495	0.08	3.75
4	0.5645	0.01	0.001382	0.00435	0.00435	0.72	3.75
6	0.5645	0.02	0.005063	0.0165	0.0165	2.63	3.75
8	0.5645	0.08	0.01298	0.0425	0.0424	6.69	3.75
10	0.5646	0.20	0.02661	0.0833	0.0830	13.3	3.76
12	0.5648	0.42	0.04606	0.132	0.131	21.7	3.78
14	0.5651	0.73	0.07005	0.180	0.178	30.8	3.81
16	0.5655	1.14	0.09700	0.224	0.220	39.6	3.84
18	0.5660	1.63	0.1258	0.266	0.260	47.7	3.87
20	0.5666	2.20	0.1559	0.306	0.297	55.3	3.92
25	0.5684	3.96	0.2342	0.396	0.378	71.72	4.04
30	0.5707	6.12	0.3129	0.466	0.436	85.2	4.18
35	0.5733	8.59	0.3888	0.518	0.475	96.5	4.35
40	0.5763	11.30	0.4610	0.560	0.503	106.4	4.54
45	0.5796	14.19	0.5292	0.596	0.522	115.6	4.74
50	0.5831	17.26	0.5937	0.627	0.536	124.7	4.98
55	0.5870	20.48	0.6551	0.657	0.546	134.1	5.25
60	0.5912	23.85	0.7138	0.687	0.554	144.4	5.57
65	0.5957	27.38	0.7702	0.718	0.559	156.3	5.96
70	0.6008	31.08	0.8250	0.752	0.563	170.4	6.46
75	0.6063	34.96	0.8786	0.791	0.566	187.4	7.08
80	0.6125	39.07	0.9316	0.836	0.567	208.4	7.89
83.81	0.6178	42.34	0.9720	0.877	0.568	228.3	8.68

TABLE VIII. Thermodynamic Properties of Crystalline Argon (*Continued*)

T, K	p = 100 bar						
	$v \cdot 10^3$	h	s	c_p	c_v	$\alpha_{exp} \cdot 10^5$	$\beta_T \cdot 10^5$
1	0.5624	5.58	0.000019	0.000059	0.000059	0.01	3.64
2	0.5624	5.58	0.000158	0.000481	0.000481	0.08	3.64
4	0.5624	5.58	0.001343	0.00423	0.00423	0.68	3.64
6	0.5624	5.60	0.004919	0.0160	0.0160	2.49	3.64
8	0.5624	5.66	0.01262	0.0413	0.0412	6.34	3.65
10	0.5625	5.78	0.02588	0.0812	0.0809	12.6	3.66
12	0.5627	5.99	0.04486	0.129	0.128	20.7	3.67
14	0.5630	6.29	0.06835	0.176	0.175	29.4	3.70
16	0.5634	6.69	0.09482	0.220	0.217	37.9	3.73
18	0.5639	7.17	0.1232	0.262	0.256	45.8	3.76

TABLE VIII. Thermodynamic Properties of Crystalline Argon (*Continued*)

T, K	p = 100 bar						
	$v \cdot 10^3$	h	s	c_p	c_v	$\alpha_{exp} \cdot 10^5$	$\beta_T \cdot 10^5$
20	0.5644	7.73	0.1528	0.301	0.293	53.1	3.80
25	0.5662	9.48	0.2302	0.392	0.375	69.1	3.92
30	0.5684	11.62	0.3081	0.461	0.433	82.1	4.05
35	0.5709	14.06	0.3834	0.514	0.473	93.1	4.21
40	0.5737	16.75	0.4549	0.556	0.500	102.6	4.38
45	0.5768	19.62	0.5225	0.591	0.520	111.3	4.57
50	0.5803	22.66	0.5866	0.622	0.535	119.8	4.79
55	0.5840	25.85	0.6473	0.650	0.545	128.4	5.03
60	0.5879	29.18	0.7053	0.678	0.552	137.7	5.31
65	0.5922	32.66	0.7610	0.707	0.558	148.0	5.65
70	0.5970	36.30	0.8149	0.738	0.562	160.0	6.07
75	0.6021	40.11	0.8674	0.773	0.565	174.2	6.58
80	0.6078	44.11	0.9190	0.812	0.566	191.3	7.23
86.26	0.6160	49.45	0.9812	0.872	0.568	218.5	8.32

TABLE VIII. Thermodynamic Properties of Crystalline Argon (*Continued*)

T, K	p = 200 bar						
	$v \cdot 10^3$	h	s	c_p	c_v	$\alpha_{exp} \cdot 10^5$	$\beta_T \cdot 10^5$
1	0.5603	11.19	0.000019	0.000057	0.000057	0.01	3.54
2	0.5603	11.19	0.000154	0.000467	0.000467	0.08	3.54
4	0.5603	11.20	0.00131	0.00411	0.00411	0.64	3.54
6	0.5603	11.21	0.00478	0.0156	0.0156	2.36	3.55
8	0.5604	11.27	0.0123	0.0402	0.0401	6.01	3.55
10	0.5605	11.38	0.0252	0.0792	0.0789	12.0	3.56
12	0.5607	11.59	0.0437	0.126	0.125	19.7	3.57
14	0.5610	11.89	0.0667	0.173	0.171	28.1	3.59
16	0.5613	12.28	0.0927	0.217	0.213	36.3	3.62
18	0.5618	12.75	0.1206	0.258	0.252	44.0	3.65
20	0.5623	13.31	0.1499	0.298	0.290	51.1	3.69
25	0.5640	15.03	0.2263	0.388	0.371	66.6	3.80
30	0.5661	17.15	0.3035	0.457	0.430	79.3	3.93
35	0.5685	19.58	0.3781	0.510	0.470	89.8	4.07
40	0.5712	22.24	0.4491	0.551	0.498	99.0	4.23
45	0.5742	25.09	0.5162	0.586	0.518	107.3	4.41
50	0.5775	28.10	0.5796	0.616	0.533	115.3	4.61
55	0.5810	31.26	0.6398	0.644	0.543	123.3	4.83
60	0.5848	34.56	0.6973	0.671	0.551	131.7	5.08
65	0.5889	38.00	0.7523	0.698	0.557	140.9	5.38
70	0.5933	41.58	0.8054	0.727	0.561	151.3	5.74
75	0.5982	45.32	0.8569	0.758	0.564	163.4	6.17

TABLE VIII. Thermodynamic Properties of Crystalline Argon (*Continued*)

T, K	p = 200 bar						
	$v \cdot 10^3$	h	s	c_p	c_v	$\alpha_{exp} \cdot 10^5$	$\beta_T \cdot 10^5$
80	0.6035	49.24	0.9074	0.793	0.566	177.7	6.70
88.68	0.6141	56.54	0.9938	0.867	0.567	209.9	7.98

TABLE VIII. Thermodynamic Properties of Crystalline Argon (*Continued*)

T, K	p = 300 bar						
	$v \cdot 10^3$	h	s	c_p	c_v	$\alpha_{exp} \cdot 10^5$	$\beta_T \cdot 10^5$
1	0.5584	16.79	0.000018	0.000055	0.000055	0.01	3.45
2	0.5584	16.79	0.000149	0.000455	0.000455	0.07	3.45
4	0.5584	16.79	0.001270	0.00400	0.00400	0.61	3.45
6	0.5584	16.81	0.004652	0.0151	0.0151	2.24	3.45
8	0.5584	16.86	0.01194	0.0391	0.0391	5.71	3.46
10	0.5585	16.97	0.02453	0.0772	0.0770	11.4	3.46
12	0.5587	17.17	0.04263	0.123	0.123	18.8	3.48
14	0.5590	17.47	0.06517	0.170	0.168	26.9	3.50
16	0.5593	17.85	0.09071	0.213	0.210	34.8	3.52
18	0.5597	18.32	0.1182	0.254	0.249	42.3	3.55
20	0.5603	18.87	0.1470	0.294	0.286	49.2	3.59
25	0.5619	20.57	0.2226	0.384	0.368	64.2	3.69
30	0.5639	22.67	0.2990	0.453	0.427	76.6	3.81
35	0.5662	25.07	0.3730	0.506	0.468	86.8	3.94
40	0.5688	27.71	0.4434	0.547	0.496	95.6	4.09
45	0.5717	30.54	0.5100	0.581	0.517	103.6	4.26
50	0.5748	33.53	0.5730	0.611	0.531	111.2	4.44
55	0.5781	36.66	0.6327	0.638	0.542	118.7	4.65
60	0.5819	39.93	0.6896	0.664	0.550	126.5	4.88
65	0.5857	43.33	0.7440	0.690	0.556	134.7	5.14
70	0.5899	46.87	0.7964	0.717	0.560	143.9	5.45
75	0.5945	50.56	0.8472	0.745	0.563	154.3	5.82
80	0.5995	54.40	0.8967	0.777	0.565	166.5	6.27
90	0.6109	62.66	0.9936	0.853	0.567	198.2	7.53
91.07	0.6123	63.62	1.004	0.863	0.567	202.2	7.69

TABLE VIII. Thermodynamic Properties of Crystalline Argon (*Continued*)

T, K	p = 400 bar						
	$v \cdot 10^3$	h	s	c_p	c_v	$\alpha_{exp} \cdot 10^5$	$\beta_T \cdot 10^5$
1	0.5564	22.36	0.000018	0.000054	0.000054	0.01	3.36
2	0.5564	22.36	0.000146	0.000443	0.000443	0.07	3.36
4	0.5564	22.36	0.001237	0.00390	0.00389	0.58	3.36

TABLE VIII. Thermodynamic Properties of Crystalline Argon (*Continued*)

T, K	p = 400 bar						
	$v \cdot 10^3$	h	s	c_p	c_v	$\alpha_{exp} \cdot 10^5$	$\beta_T \cdot 10^5$
6	0.5565	22.38	0.004529	0.0147	0.0147	2.13	3.36
8	0.5565	22.43	0.01162	0.0381	0.0381	5.43	3.37
10	0.5566	22.54	0.02390	0.0754	0.0752	10.9	3.37
12	0.5567	22.74	0.04160	0.121	0.120	18.0	3.39
14	0.5570	23.02	0.06369	0.167	0.165	25.8	3.41
16	0.5573	23.40	0.08878	0.210	0.207	33.4	3.43
18	0.5578	23.87	0.1158	0.250	0.246	40.7	3.46
20	0.5583	24.40	0.1443	0.290	0.283	47.4	3.49
25	0.5598	26.09	0.2190	0.380	0.365	62.0	3.58
30	0.5617	28.17	0.2947	0.449	0.424	74.0	3.70
35	0.5640	30.55	0.3682	0.502	0.465	83.9	3.82
40	0.5665	33.17	0.4380	0.543	0.494	92.5	3.96
45	0.5693	35.98	0.5041	0.577	0.515	100.1	4.12
50	0.5723	38.95	0.5666	0.607	0.530	107.3	4.29
55	0.5755	42.06	0.6258	0.633	0.541	114.4	4.48
60	0.5790	45.30	0.6822	0.658	0.549	121.6	4.69
65	0.5827	48.66	0.7361	0.683	0.555	129.1	4.93
70	0.5867	52.16	0.7879	0.708	0.559	137.4	5.20
75	0.5910	55.79	0.8380	0.734	0.562	146.7	5.53
80	0.5957	59.57	0.8867	0.763	0.564	157.2	5.91
90	0.6063	67.65	0.9815	0.831	0.566	184.0	6.96
93 41	0.6105	70.60	1.014	0.860	0.566	195.6	7.42

TABLE VIII. Thermodynamic Properties of Crystalline Argon (*Continued*)

T, K	p = 500 bar						
	$v \cdot 10^3$	h	s	c_p	c_v	$\alpha_{exp} \cdot 10^5$	$\beta_T \cdot 10^5$
1	0.5546	27.92	0.000017	0.000052	0.000052	0.01	3.28
2	0.5546	27.91	0.000142	0.000431	0.000431	0.07	3.28
4	0.5546	27.92	0.001205	0.00380	0.00380	0.58	3.28
6	0.5546	27.94	0.004413	0.0144	0.0144	2.03	3.28
8	0.5546	27.99	0.01133	0.0371	0.0371	5.18	3.28
10	0.5547	28.09	0.02331	0.0736	0.0734	10.4	3.29
12	0.5549	28.29	0.04062	0.118	0.118	17.2	3.30
14	0.5551	28.57	0.06228	0.164	0.162	24.7	3.32
16	0.5554	28.94	0.08694	0.206	0.204	32.1	3.34
18	0.5558	29.39	0.1136	0.247	0.242	39.2	3.37
20	0.5563	29.92	0.1417	0.286	0.279	45.7	3.40
25	0.5578	31.65	0.2155	0.376	0.362	60.0	3.49
30	0.5597	33.65	0.2906	0.446	0.422	71.6	3.59
35	0.5618	36.02	0.3634	0.498	0.463	81 3	3.71
40	0.5643	38.62	0.4328	0.539	0.492	89.5	3.84

TABLE VIII. Thermodynamic Properties of Crystalline Argon (*Continued*)

T, K	p = 500 bar						
	$v \cdot 10^3$	h	s	c_p	c_v	$\alpha_{exp} \cdot 10^5$	$\beta_T \cdot 10^5$
45	0.5670	41.40	0.4984	0.573	0.513	96.9	3.99
50	0.5698	44.35	0.5605	0.602	0.528	103.8	4.15
55	0.5730	47.43	0.6193	0.628	0.539	110.5	4.32
60	0.5763	50.65	0.6752	0.653	0.548	117.2	4.52
65	0.5799	53.99	0.7286	0.676	0.554	124.2	4.73
70	0.5837	57.45	0.7798	0.700	0.558	131.7	4.98
75	0.5878	61.03	0.8293	0.725	0.561	140.0	5.27
80	0.5922	64.76	0.8773	0.752	0.563	149.3	5.60
90	0.6021	72.68	0.9703	0.813	0.566	172.4	6.49
95.73	0.6086	77.56	1.023	0.857	0.566	189.4	7.18

TABLE VIII. Thermodynamic Properties of Crystalline Argon (*Continued*)

T, K	p = 600 bar						
	$v \cdot 10^3$	h	s	c_p	c_v	$\alpha_{exp} \cdot 10^5$	$\beta_T \cdot 10^5$
1	0.5528	33.45	0.000017	0.000051	0.000051	0.01	3.20
2	0.5528	33.45	0.000138	0.000421	0.000421	0.06	3.20
4	0.5528	33.45	0.001175	0.00370	0.00370	0.53	3.20
6	0.5528	33.47	0.004302	0.0140	0.0140	1.93	3.20
8	0.5528	33.52	0.01104	0.0362	0.0362	4.94	3.20
10	0.5529	33.63	0.02274	0.0719	0.0718	9.91	3.21
12	0.5531	33.81	0.03968	0.116	0.115	16.5	3.22
14	0.5533	34.09	0.06093	0.161	0.159	23.7	3.24
16	0.5536	34.46	0.08518	0.203	0.200	30.9	3.26
18	0.5540	34.90	0.1115	0.244	0.239	37.7	3.28
20	0.5544	35.43	0.1392	0.283	0.276	44.1	3.31
25	0.5559	37.44	0.2122	0.372	0.359	58.0	3.40
30	0.5577	39.12	0.2866	0.442	0.419	69.4	3.49
35	0.5598	41.47	0.3589	0.494	0.461	78.8	3.61
40	0.5621	44.05	0.4278	0.536	0.490	86.8	3.73
45	0.5647	46.82	0.4930	0.569	0.511	93.9	3.87
50	0.5675	49.74	0.5546	0.598	0.527	100.5	4.02
55	0.5705	52.81	0.6129	0.624	0.538	106.8	4.18
60	0.5737	55.99	0.6684	0.648	0.547	113.2	4.36
65	0.5771	59.30	0.7214	0.670	0.553	119.7	4.56
70	0.5808	62.73	0.7721	0.693	0.557	126.6	4.78
75	0.5847	66.28	0.8210	0.717	0.560	134.1	5.04
80	0.5888	69.95	0.8684	0.742	0.563	142.4	5.34
90	0.5982	77.74	0.9599	0.798	0.565	162.6	6.10
98.01	0.6070	84.48	1.031	0.855	0.566	183.6	6.96

TABLE VIII. Thermodynamic Properties of Crystalline Argon (*Continued*)

T, K	p = 700 bar						
	$v \cdot 10^3$	h	s	c_p	c_v	$\alpha_{exp} \cdot 10^5$	$\beta_T \cdot 10^5$
1	0.5510	38.97	0.000017	0.00005	0.00005	0.01	3.12
2	0.5510	38.97	0.000135	0.00041	0.00041	0.06	3.12
4	0.5510	38.97	0.001146	0.00361	0.00361	0.50	3.12
6	0.5510	38.99	0.004197	0.0137	0.0137	1.85	3.12
8	0.5511	39.04	0.01078	0.0354	0.0353	4.71	3.13
10	0.5511	39.14	0.0222	0.0703	0.0702	9.48	3.13
12	0.5513	39.32	0.0388	0.114	0.113	15.8	3.14
14	0.5515	39.60	0.0596	0.158	0.157	22.8	3.16
16	0.5518	39.95	0.0835	0.200	0.198	29.8	3.18
18	0.5522	40.40	0.1094	0.240	0.236	36.4	3.20
20	0.5526	40.91	0.1367	0.279	0.273	42.6	3.23
25	0.5540	42.54	0.2090	0.369	0.356	56.2	3.31
30	0.5557	44.57	0.2828	0.438	0.416	67.3	3.40
35	0.5578	46.90	0.3545	0.491	0.458	76.4	3.51
40	0.5600	49.46	0.4229	0.532	0.488	84.2	3.62
45	0.5625	52.21	0.4877	0.566	0.510	91.0	3.75
50	0.5652	55.12	0.5489	0.594	0.525	97.4	3.89
55	0.5681	58.16	0.6069	0.620	0.537	103.4	4.05
60	0.5712	61.33	0.6619	0.643	0.545	109.4	4.21
65	0.5745	64.61	0.7144	0.665	0.552	115.5	4.40
70	0.5780	68.01	0.7647	0.687	0.556	121.9	4.60
75	0.5817	71.51	0.8131	0.709	0.560	128.8	4.83
80	0.5857	75.15	0.8600	0.733	0.562	136.4	5.10
90	0.5946	82.81	0.9501	0.785	0.564	154.2	5.77
100	0.6050	91.13	1.037	0.849	0.565	177.7	6.71
100.25	0.6054	91.36	1.039	0.852	0.565	178.4	6.74

TABLE VIII. Thermodynamic Properties of Crystalline Argon (*Continued*)

T, K	p = 800 bar						
	v	h	s	c_p	c_v	$\alpha_{exp} \cdot 10^5$	$\beta_T \cdot 10^5$
1	0.5493	44.47	0.000016	0.000049	0.000049	0.01	3.05
2	0.5493	44.47	0.000132	0.000401	0.000401	0.06	3.05
4	0.5493	44.48	0.001119	0.00352	0.00352	0.48	3.05
6	0.5493	44.49	0.004097	0.0133	0.0133	1.76	3.05
8	0.5494	44.54	0.01052	0.0345	0.0345	4.51	3.05
10	0.5494	44.64	0.02169	0.0688	0.0686	9.08	3.06
12	0.5496	44.82	0.03792	0.111	0.111	15.2	3.07
14	0.5498	45.08	0.0584	0.155	0.154	21.9	3.08
16	0.5500	45.43	0.08187	0.197	0.195	28.7	3.10
18	0.5504	45.87	0.1074	0.237	0.233	35.2	3.13

TABLE VIII. Thermodynamic Properties of Crystalline Argon (*Continued*)

T, K	p = 800 bar						
	v	h	s	c_p	c_v	$\alpha_{exp} \cdot 10^5$	$\beta_T \cdot 10^5$
20	0.5508	46.38	0.1344	0.276	0.270	41.2	3.15
25	0.5522	48.00	0.2059	0.365	0.353	54.4	3.23
30	0.5538	50.01	0.2791	0.435	0.414	65.3	3.31
35	0.5558	52.32	0.3503	0.488	0.456	74.2	3.41
40	0.5580	54.87	0.4182	0.529	0.486	81.7	3.52
45	0.5604	57.60	0.4826	0.562	0.508	88.3	3.65
50	0.5630	60.49	0.5434	0.591	0.524	94.4	3.78
55	0.5658	63.51	0.6010	0.615	0.536	100.2	3.92
60	0.5688	66.65	0.6557	0.638	0.544	105.9	4.08
65	0.5719	69.91	0.7078	0.660	0.551	111.7	4.25
70	0.5753	73.28	0.7577	0.681	0.555	117.7	4.44
75	0.5789	76.75	0.8056	0.702	0.559	124.0	4.65
80	0.5827	80.35	0.8519	0.725	0.561	130.9	4.89
90	0.5911	87.91	0.9408	0.773	0.564	147.0	5.48
100	0.6009	96.06	1.026	0.832	0.565	167.5	6.29
102.48	0.6036	98.20	1.047	0.853	0.565	173.4	6.54

TABLE VIII. Thermodynamic Properties of Crystalline Argon (*Continued*)

T, K	p = 900 bar						
	$v \cdot 10^3$	h	s	c_p	c_v	$\alpha_{exp} \cdot 10^5$	$\beta_T \cdot 10^5$
1	0.5476	49.06	0.000016	0.000048	0.000048	0.01	2.98
2	0.5476	49.06	0.000129	0.000391	0.000391	0.05	2.98
4	0.5476	49.06	0.001093	0.00344	0.00344	0.46	2.98
6	0.5476	49.07	0.004001	0.0130	0.0130	1.69	2.98
8	0.5477	49.12	0.01028	0.0337	0.0337	4.31	2.99
10	0.5478	50.12	0.02120	0.0673	0.0672	8.70	2.99
12	0.5479	50.29	0.0371	0.109	0.109	14.6	3.00
14	0.5481	50.56	0.0572	0.153	0.151	21.1	3.01
16	0.5484	50.90	0.0803	0.194	0.192	27.7	3.03
18	0.5487	51.33	0.1055	0.234	0.230	34.0	3.05
20	0.5491	51.84	0.1322	0.273	0.267	39.9	3.08
25	0.5504	53.43	0.2030	0.362	0.350	52.8	3.15
30	0.5520	55.43	0.2755	0.432	0.411	63.4	3.23
35	0.5539	57.72	0.3462	0.484	0.454	72.1	3.33
40	0.5560	60.26	0.4137	0.525	0.485	79.4	3.43
45	0.5584	62.97	0.4776	0.559	0.507	85.8	3.54
50	0.5609	65.84	0.5381	0.587	0.523	91.7	3.67
55	0.5636	68.84	0.5953	0.612	0.535	97.3	3.80
60	0.5665	71.70	0.6496	0.634	0.543	102.7	3.95
65	0.5695	75.20	0.7014	0.655	0.550	108.2	4.11

TABLE VIII. Thermodynamic Properties of Crystalline Argon (*Continued*)

T, K	p = 900 bar						
	$v \cdot 10^3$	h	s	c_p	c_v	$\alpha_{exp} \cdot 10^5$	$\beta_T \cdot 10^5$
70	0.5727	78.54	0.7509	0.676	0.555	113.8	4.29
75	0.5762	81.99	0.7984	0.696	0.558	119.7	4.48
80	0.5798	85.54	0.8443	0.717	0.561	126.0	4.70
90	0.5877	93.01	0.9321	0.763	0.563	140.6	5.23
100	0.5971	101.03	1.016	0.817	0.564	158.8	5.94
104.67	0.6018	105.00	1.055	0.847	0.564	169.4	6.36

TABLE VIII. Thermodynamic Properties of Crystalline Argon (*Continued*)

T, K	p = 1000 bar						
	$v \cdot 10^3$	h	s	c_p	c_v	$\alpha_{exp} \cdot 10^5$	$\beta_T \cdot 10^5$
1	0.5460	55.43	0.000015	0.000047	0.000047	0.01	2.92
2	0.5460	55.43	0.000126	0.000382	0.000382	0.05	2.92
4	0.5460	55.43	0.001068	0.00336	0.00336	0.44	2.92
6	0.5460	55.44	0.003910	0.0127	0.0127	1.62	2.92
8	0.5461	55.49	0.01004	0.0330	0.0330	4.13	2.92
10	0.5461	55.58	0.02073	0.0659	0.0658	8.35	2.93
12	0.5463	55.76	0.03631	0.107	0.107	14.0	2.93
14	0.5464	56.01	0.05606	0.150	0.149	20.3	2.95
16	0.5467	56.35	0.07881	0.191	0.189	26.7	2.96
18	0.5470	56.78	0.1036	0.231	0.227	32.8	2.98
20	0.5474	57.28	0.1300	0.270	0.264	38.6	3.01
25	0.5487	58.86	0.2001	0.359	0.347	51.2	3.07
30	0.5502	60.83	0.2720	0.428	0.409	61.6	3.15
35	0.5521	63.12	0.3422	0.481	0.452	70.1	3.24
40	0.5541	65.63	0.4093	0.522	0.483	77.2	3.34
45	0.5564	68.33	0.4728	0.556	0.505	83.4	3.45
50	0.5588	71.18	0.5330	0.584	0.521	89.1	3.57
55	0.5615	74.17	0.5898	0.608	0.533	94.5	3.69
60	0.5642	77.27	0.6438	0.630	0.542	99.7	3.83
65	0.5672	80.48	0.6952	0.651	0.549	104.8	3.98
70	0.5703	83.80	0.7443	0.671	0.554	110.1	4.14
75	0.5736	87.21	0.7915	0.690	0.557	115.7	4.32
80	0.5771	90.74	0.8369	0.711	0.560	121.5	4.52
90	0.5848	98.12	0.9237	0.754	0.563	134.8	5.00
100	0.5935	106.01	1.007	0.804	0.564	151.2	5.63
106.84	0.6003	111.76	1.062	0.845	0.564	164.8	6.19

TABLE IX. Thermodynamic Properties of Argon Along the Crystallization and Melting Curves

T, K	p_m	$v_l \cdot 10^3$	$v_c \cdot 10^3$	h_l	h_c	s_l	s_c	c_{pl}	c_{pc}
83.78	0.69	0.7068	0.6178	71.88	42.30	1.333	0.9718	1.050	0.877
84	8.33	0.7062	0.6177	72.48	42.85	1.333	0.9728	1.049	0.876
85	48.75	0.7038	0.6169	75.30	45.75	1.335	0.9773	1.044	0.874
86	89.44	0.7014	0.6162	78.15	48.66	1.336	0.9818	1.040	0.872
87	130.4	0.6991	0.6155	81.05	51.58	1.337	0.9863	1.035	0.870
88	171.7	0.6969	0.6147	84.00	54.52	1.338	0.9907	1.031	0.868
89	213.2	0.6948	0.6139	87.01	57.47	1.340	0.9951	1.027	0.867
90	255.0	0.6927	0.6131	90.07	60.42	1.341	0.9994	1.023	0.865
91	297.1	0.6907	0.6124	93.14	63.37	1.343	1.0036	1.019	0.863
92	339.5	0.6887	0.6116	96.22	66.33	1.344	1.0077	1.015	0.862
93	382.1	0.6868	0.6108	99.30	69.30	1.345	1.0118	1.011	0.860
94	425.0	0.6850	0.6101	102.37	72.29	1.347	1.0158	1.007	0.859
95	468.2	0.6833	0.6093	105.44	75.31	1.348	1.0197	1.003	0.858
96	511.7	0.6816	0.6085	108.51	78.35	1.350	1.0235	1.000	0.857
97	555.4	0.6799	0.6078	111.58	81.40	1.351	1.0273	0.997	0.855
98	599.4	0.6782	0.6070	114.68	84.45	1.352	1.0310	0.993	0.854
99	643.7	0.6766	0.6062	117.81	87.50	1.354	1.0346	0.990	0.853
100	688.2	0.6750	0.6055	120.97	90.57	1.355	1.0382	0.987	0.852
101	733.0	0.6734	0.6047	124.19	93.65	1.356	1.0417	0.984	0.851
102	778.0	0.6718	0.6039	127.42	96.74	1.358	1.0453	0.981	0.850
103	823.3	0.6703	0.6032	130.65	99.84	1.359	1.0489	0.977	0.849
104	868.9	0.6688	0.6024	133.86	102.93	1.361	1.0524	0.974	0.848
105	914.8	0.6673	0.6016	137.09	106.01	1.362	1.0559	0.971	0.847
106	960.9	0.6658	0.6008	140.32	109.10	1.363	1.0593	0.968	0.846
107	1007.2	0.6644	0.6001	143.56	112.24	1.365	1.0626	0.965	0.845

TABLE X. Thermodynamic Properties of Saturated Argon

T, K	p_v	$v'' \cdot 10^3$	$v' \cdot 10^3$	h''	h'	s''	s'	c''_p	c'_p
83.78	0.687	246.95	0.70679	235.06	71.88	3.280	1.333	0.548	1.050
84	0.705	241.51	0.70752	235.15	72.09	3.276	1.335	0.548	1.051
85	0.790	217.65	0.71066	235.55	73.16	3.258	1.348	0.550	1.058
86	0.882	196.68	0.71378	235.89	74.23	3.240	1.360	0.553	1.064
87	0.983	178.14	0.71697	236.28	75.31	3.222	1.372	0.556	1.071
87.28	1.013	173.26	0.71744	236.39	75.61	3.216	1.375	0.556	1.073
88	1.092	161.72	0.72036	236.67	76.38	3.202	1.380	0.559	1.077
89	1.210	147.18	0.72363	237.00	77.46	3.183	1.390	0.562	1.084
90	1.338	134.24	0.72690	237.37	78.55	3.168	1.403	0.565	1.091
91	1.476	122.65	0.73020	237.69	79.63	3.152	1.415	0.568	1.097
92	1.624	112.32	0.73355	238.01	80.77	3.136	1.427	0.572	1.104
93	1.783	103.03	0.73697	238.32	81.86	3.121	1.439	0.575	1.111
94	1.954	94.721	0.74043	238.62	83.00	3.106	1.450	0.579	1.118
95	2.137	87.220	0.74397	238.91	84.15	3.091	1.462	0.583	1.124
96	2.332	80.446	0.74755	239.20	85.29	3.077	1.474	0.587	1.131
97	2.540	74.326	0.75118	239.34	86.39	3.063	1.486	0.591	1.138
98	2.762	68.779	0.75489	239.72	87.54	3.050	1.497	0.596	1.144
99	2.997	63.737	0.75871	239.99	88.70	3.037	1.508	0.601	1.151
100	3.247	59.153	0.76281	240.20	89.85	3.023	1.520	0.606	1.158
101	3.513	54.977	0.76654	240.41	91.01	3.010	1.531	0.611	1.164
102	3.794	51.163	0.77061	240.60	92.22	2.998	1.543	0.617	1.171
103	4.091	47.672	0.77474	240.80	93.39	2.985	1.554	0.623	1.178
104	4.404	44.478	0.77896	240.98	94.55	2.973	1.565	0.629	1.185
105	4.735	41.545	0.78328	241.11	95.77	2.960	1.576	0.636	1.192
106	5.083	38.851	0.78765	241.27	96.98	2.949	1.588	0.643	1.199

TABLE X. Thermodynamic Properties of Saturated Argon (*Continued*)

T, K	p_v	$v'' \cdot 10^3$	$v' \cdot 10^3$	h''	h'	s''	s'	c''_p	c'_p
107	5.450	36.369	0.79222	241.38	98.16	2.937	1.599	0.650	1.206
108	5.836	34.079	0.79686	241.48	99.38	2.926	1.610	0.657	1.214
109	6.241	31.967	0.80157	241.58	100.60	2.914	1.620	0.665	1.221
110	6.665	30.015	0.80635	241.66	101.83	2.903	1.632	0.674	1.229
111	7.111	28.205	0.81128	241.73	103.06	2.892	1.643	0.683	1.237
112	7.577	26.526	0.81632	241.75	104.29	2.881	1.654	0.693	1.246
113	8.065	24.968	0.82148	241.80	105.58	2.870	1.664	0.703	1.255
114	8.574	23.524	0.82672	241.80	106.82	2.859	1.675	0.714	1.264
115	9.107	22.175	0.83216	241.78	108.11	2.848	1.685	0.724	1.274
116	9.663	20.917	0.83776	241.75	109.40	2.837	1.696	0.735	1.284
117	10.243	19.746	0.84354	241.72	110.70	2.829	1.710	0.747	1.295
118	10.847	18.648	0.84946	241.57	112.00	2.815	1.717	0.761	1.307
119	11.476	17.626	0.85552	241.50	113.31	2.806	1.728	0.776	1.321
120	12.131	16.667	0.86178	241.33	114.62	2.794	1.738	0.792	1.336
121	12.812	15.767	0.86830	241.19	115.93	2.784	1.749	0.808	1.352
122	13.520	14.920	0.87503	240.99	117.30	2.774	1.760	0.826	1.369
123	14.256	14.129	0.88190	240.78	118.68	2.763	1.771	0.845	1.387
124	15.019	13.381	0.88907	240.55	120.06	2.753	1.782	0.865	1.407
125	15.812	12.680	0.89648	240.30	121.50	2.743	1.792	0.887	1.427
126	16.634	12.014	0.90417	239.99	122.90	2.732	1.803	0.911	1.448
127	17.487	11.392	0.91219	239.66	124.36	2.721	1.813	0.937	1.470
128	18.370	10.796	0.92053	239.26	125.82	2.711	1.825	0.965	1.494
129	19.286	10.237	0.93266	238.84	127.30	2.700	1.836	0.995	1.521
130	20.233	9.7070	0.93822	238.41	128.79	2.690	1.846	1.026	1.550
131	21.214	9.2036	0.94773	237.90	130.34	2.678	1.857	1.060	1.583

TABLE X. Thermodynamic Properties of Saturated Argon (*Continued*)

T, K	p_v	$v'' \cdot 10^3$	$v' \cdot 10^3$	h''	h'	s''	s'	c''_p	c'_p
132	22.229	8.7263	0.95767	237.36	131.86	2.667	1.868	1.097	1.619
133	23.278	8.2718	0.96808	236.75	133.44	2.656	1.879	1.136	1.660
134	24.364	7.8397	0.97902	236.11	135.09	2.644	1.890	1.176	1.705
135	25.485	7.4270	0.99059	235.40	136.76	2.633	1.902	1.216	1.752
136	26.645	7.0332	1.0028	234.65	138.45	2.621	1.914	1.256	1.799
137	27.842	6.6568	1.0159	233.81	140.16	2.609	1.925	1.298	1.847
138	29.079	6.2958	1.0299	232.88	141.89	2.596	1.937	1.342	1.897
139	30.356	5.9494	1.0453	231.86	143.70	2.583	1.949	1.389	
140	31.675	5.6158	1.0614	230.74	145.58	2.570	1.961	1.438	
141	33.036	5.2940	1.0792	229.55	147.46	2.556	1.974	1.490	
142	34.440	4.9827	1.0982	228.15	149.37	2.541	1.986	1.543	
143	35.889	4.7005	1.1199	226.67	151.34	2.525	1.998	1.597	
144	37.384	4.3842	1.1443	224.94	153.42	2.508	2.012	1.655	
145	38.926	4.0905	1.1721	223.09	155.73	2.490	2.026	1.719	
146	40.516	3.8000	1.2042	220.79	158.24	2.470	2.041		
147	42.156	3.5053	1.2435	218.12	161.10	2.447	2.059		
148	43.847	3.2004	1.2922	214.85	164.47	2.419	2.078		
149	45.591	2.8714	1.3588	210.59	168.64	2.383	2.102		
150	47.388	2.4893	1.4687	204.35	174.64	2.331	2.133		
150.86	48.979	1.8667	1.8667	189.84	189.84	2.201	2.201		

TABLE XI. Thermodynamic Properties of Liquid and Gaseous Argon

T, K	p = 1 bar				p = 2 bar			
	$v \cdot 10^3$	h	s	c_p	$v \cdot 10^3$	h	s	c_p
85	0.7104	73.15	1.351	1.057	0.7102	73.19	1.351	1.057
90	181.57	237.92	3.241	0.553	0.7264	78.57	1.412	1.090
95	192.53	240.42	3.271	0.547	93.528	239.10	3.116	0.578
100	203.45	243.38	3.299	0.542	99.271	242.01	3.146	0.571
105	214.25	246.13	3.326	0.539	104.93	244.87	3.173	0.565
110	225.03	248.82	3.351	0.536	110.51	247.66	3.200	0.559
115	235.73	251.51	3.375	0.533	116.02	250.44	3.225	0.554
120	246.42	254.20	3.397	0.532	121.51	253.22	3.248	0.550
125	257.10	256.88	3.419	0.530	126.99	255.98	3.270	0.546
130	267.71	259.51	3.440	0.530	132.40	258.69	3.292	0.543
135	278.31	262.18	3.460	0.528	137.81	261.39	3.312	0.540
140	288.90	264.81	3.480	0.527	143.20	264.09	3.332	0.539
145	299.49	267.48	3.498	0.527	148.53	266.79	3.351	0.538
150	310.06	270.10	3.516	0.526	153.91	269.48	3.369	0.535
155	320.58	272.72	3.533	0.525	159.28	272.16	3.386	0.532
160	331.10	275.38	3.550	0.524	164.59	274.80	3.403	0.531
170	352.17	280.66	3.582	0.523	175.21	280.11	3.436	0.530
180	373.17	285.89	3.612	0.523	185.81	285.41	3.466	0.529
190	394.10	291.11	3.640	0.522	196.39	290.70	3.494	0.529
200	415.08	296.38	3.667	0.522	206.97	295.99	3.522	0.527
210	436.00	301.60	3.693	0.522	217.48	301.27	3.547	0.526
220	456.91	306.87	3.717	0.521	227.99	306.50	3.572	0.525
230	477.82	312.08	3.740	0.521	238.49	311.78	3.595	0.524
240	498.77	317.29	3.762	0.521	248.99	317.00	3.617	0.523
250	519.67	322.50	3.784	0.521	259.49	322.28	3.639	0.522
260	540.56	327.76	3.804	0.521	269.98	327.49	3.659	0.521
270	561.40	332.97	3.824	0.521	280.41	332.71	3.679	0.521
280	582.28	338.18	3.843	0.521	290.89	337.98	3.698	0.521
290	603.17	343.39	3.861	0.521	301.38	343.19	3.716	0.521
300	624.05	348.60	3.879	0.521	311.80	348.41	3.734	0.521
310	644.85	353.80	3.896	0.521	322.30	353.71	3.751	0.521
320	665.75	359.00	3.912	0.521	332.70	358.91	3.767	0.521
330	686.55	364.30	3.928	0.521	343.20	364.11	3.783	0.521
340	707.45	369.50	3.944	0.521	353.60	369.31	3.799	0.521
350	728.25	374.70	3.959	0.521	364.10	374.51	3.814	0.521
375	780.45	387.70	3.995	0.521	390.20	387.61	3.850	0.521
400	832.55	400.70	4.028	0.521	416.30	400.61	3.884	0.521
425	884.65	413.80	4.060	0.521	442.40	413.71	3.915	0.521
450	936.65	426.80	4.090	0.521	468.40	426.71	3.945	0.521
475	988.75	439.80	4.118	0.521	494.50	439.71	3.974	0.521
500	1040.9	452.80	4.145	0.521	520.50	452.71	4.000	0.521
550	1144.9	478.90	4.194	0.521	572.60	478.81	4.050	0.521
600	1248.9	504.90	4.240	0.521	624.70	504.91	4.095	0.521

TABLE XI. Thermodynamic Properties of Liquid and Gaseous Argon (*Continued*)

T, K	p = 1 bar				p = 2 bar			
	$v \cdot 10^3$	h	s	c_p	$v \cdot 10^3$	h	s	c_p
650	1352.9	530.90	4.281	0.521	676.80	530.91	4.137	0.521
700	1456.9	556.90	4.320	0.521	728.90	557.01	4.176	0.521
750	1560.9	583.00	4.356	0.521	780.90	583.01	4.211	0.521
800	1665.9	609.00	4.389	0.521	833.00	609.01	4.245	0.521
850	1769.9	635.00	4.421	0.521	885.10	635.01	4.277	0.521
900	1873.9	661.00	4.451	0.521	937.10	661.11	4.306	0.521
950	1977.9	687.10	4.479	0.521	989.20	687.11	4.334	0.521
1000	2081.9	713.10	4.506	0.521	1041.0	713.11	4.361	0.521
1050	2185.9	739.10	4.531	0.521	1093.0	739.11	4.387	0.521
1100	2289.9	765.10	4.555	0.521	1145.0	765.11	4.411	0.521
1150	2394	791.35	4.579	0.520	1197	791.35	4.435	0.520
1200	2498	817.35	4.601	0.520	1249	817.35	4.457	0.520
1250	2602	843.35	4.623	0.520	1301	843.35	4.478	0.520
1300	2706	869.35	4.643	0.520	1353	869.35	4.499	0.520

TABLE XI. Thermodynamic Properties of Liquid and Gaseous Argon (*Continued*)

T, K	p = 3 bar				p = 4 bar			
	$v \cdot 10^3$	h	s	c_p	$v \cdot 10^3$	h	s	c_p
85	0.7101	73.24	1.350	1.057	0.7099	73.29	1.350	1.057
90	0.7262	78.62	1.411	1.088	0.7260	78.66	1.411	1.088
95	0.7437	84.14	1.472	1.122	0.7435	84.18	1.471	1.121
100	64.475	240.56	3.051	0.599	0.7626	89.86	1.529	1.156
105	68.408	243.53	3.081	0.590	50.101	242.16	3.012	0.614
110	72.275	246.48	3.108	0.581	53.122	245.22	3.041	0.603
115	76.087	249.36	3.134	0.574	56.084	248.20	3.067	0.593
120	79.858	252.22	3.158	0.567	58.999	251.17	3.092	0.585
125	83.587	255.03	3.181	0.562	61.874	254.06	3.116	0.577
130	87.292	257.82	3.203	0.556	64.720	256.93	3.139	0.571
135	90.970	260.60	3.224	0.552	67.532	259.80	3.160	0.566
140	94.628	263.37	3.244	0.548	70.328	262.60	3.180	0.561
145	98.266	266.09	3.263	0.544	73.100	265.39	3.200	0.557
150	101.89	268.80	3.282	0.541	75.858	268.17	3.219	0.553
155	105.49	271.51	3.299	0.540	78.600	270.95	3.237	0.549
160	109.09	274.21	3.317	0.540	81.330	273.67	3.254	0.544
170	116.27	279.60	3.349	0.537	86.761	279.09	3.287	0.541
180	123.39	284.97	3.380	0.534	92.159	284.49	3.318	0.538
190	130.49	290.29	3.409	0.531	97.529	289.88	3.347	0.536
200	137.58	295.60	3.436	0.529	102.88	295.20	3.375	0.534
210	144.61	300.90	3.462	0.529	108.20	300.57	3.401	0.532
220	151.69	306.19	3.486	0.529	113.51	305.88	3.425	0.531
230	158.70	311.48	3.510	0.527	118.81	311.18	3.449	0.529

TABLE XI. Thermodynamic Properties of Liquid and Gaseous Argon (*Continued*)

T, K	p = 3 bar				p = 4 bar			
	$v \cdot 10^3$	h	s	c_p	$v \cdot 10^3$	h	s	c_p
240	165.72	316.76	3.532	0.526	124.10	316.48	3.471	0.528
250	172.78	322.00	3.554	0.525	129.40	321.77	3.493	0.527
260	179.78	327.27	3.574	0.524	134.68	327.00	3.514	0.526
270	186.78	332.50	3.594	0.523	139.92	332.29	3.534	0.525
280	193.78	337.77	3.613	0.522	145.20	337.52	3.553	0.524
290	200.78	343.00	3.632	0.521	150.48	342.80	3.571	0.523
300	207.77	348.27	3.649	0.521	155.71	348.07	3.589	0.522
310	214.77	353.47	3.666	0.521	161.01	353.27	3.606	0.521
320	221.77	358.67	3.683	0.521	166.21	358.47	3.622	0.521
330	228.67	363.87	3.699	0.521	171.51	363.77	3.638	0.521
340	235.67	369.17	3.714	0.521	176.71	368.97	3.654	0.521
350	242.67	374.37	3.729	0.521	182.01	374.27	3.669	0.521
375	260.07	387.47	3.765	0.521	195.01	387.27	3.705	0.521
400	277.47	400.47	3.799	0.521	208.11	400.37	3.739	0.521
425	294.87	413.57	3.831	0.521	221.21	413.47	3.771	0.521
450	312.27	426.57	3.861	0.521	234.21	426.47	3.801	0.521
475	329.67	439.67	3.889	0.521	247.31	439.57	3.829	0.521
500	347.07	452.67	3.916	0.521	260.31	452.67	3.856	0.521
550	381.77	478.77	3.965	0.521	286.41	478.67	3.905	0.521
600	416.57	504.77	4.011	0.521	312.51	504.77	3.951	0.521
650	451.27	530.87	4.052	0.521	338.61	530.77	3.993	0.521
700	485.97	556.87	4.091	0.521	364.61	556.87	4.031	0.521
750	520.77	582.97	4.127	0.521	390.71	582.97	4.067	0.521
800	555.47	608.97	4.160	0.521	416.71	608.97	4.101	0.521
850	590.17	634.97	4.192	0.521	442.71	634.97	4.132	0.521
900	624.87	661.07	4.222	0.521	468.81	661.07	4.162	0.521
950	659.57	687.07	4.250	0.521	494.81	687.07	4.190	0.521
1000	694.27	713.07	4.277	0.521	520.91	713.07	4.217	0.521
1050	728.97	739.17	4.302	0.521	546.91	739.17	4.242	0.521
1100	763.67	765.17	4.326	0.521	572.91	765.17	4.266	0.521
1150	798.4	791.35	4.351	0.520	598.9	791.35	4.291	0.521
1200	833.1	817.35	4.373	0.520	625.0	817.45	4.313	0.521
1250	867.8	843.45	4.394	0.520	651.0	843.45	4.334	0.520
1300	902.5	869.45	4.414	0.520	677.0	869.45	4.354	0.520

TABLE XI. Thermodynamic Properties of Liquid and Gaseous Argon (*Continued*)

T, K	p = 6 bar				p = 8 bar			
	$v \cdot 10^3$	h	s	c_p	$v \cdot 10^3$	h	s	c_p
85	0.7096	73.38	1.350	1.057	0.7093	73.46	1.349	1.055
90	0.7257	78.74	1.410	1.088	0.7254	78.82	1.410	1.088

TABLE XI. Thermodynamic Properties of Liquid and Gaseous Argon (*Continued*)

T, K	p = 6 bar				p = 8 bar			
	$v \cdot 10^3$	h	s	c_p	$v \cdot 10^3$	h	s	c_p
95	0.7431	84.26	1.470	1.121	0.7427	84.34	1.470	1.121
100	0.7621	89.94	1.528	1.155	0.7616	90.01	1.528	1.155
105	0.7831	95.78	1.585	1.191	0.7825	95.85	1.584	1.189
110	33.886	242.56	2.940	0.655	0.8060	101.85	1.640	1.226
115	36.017	245.88	2.969	0.638	25.904	243.29	2.893	0.690
120	38.093	249.05	2.996	0.623	27.578	246.75	2.923	0.669
125	40.121	252.13	3.021	0.610	29.198	250.04	2.950	0.650
130	42.111	255.14	3.045	0.599	30.776	253.24	2.975	0.634
135	44.071	258.12	3.067	0.589	32.317	256.40	2.999	0.620
140	46.008	261.03	3.089	0.581	33.828	259.48	3.021	0.608
145	47.920	263.92	3.109	0.574	35.317	262.48	3.042	0.597
150	49.818	266.80	3.128	0.569	36.781	265.41	3.062	0.589
155	51.698	269.62	3.147	0.564	38.237	268.37	3.081	0.581
160	53.567	272.47	3.165	0.561	39.671	271.26	3.100	0.575
170	57.268	278.01	3.199	0.553	42.512	276.97	3.134	0.564
180	60.931	283.51	3.230	0.548	45.313	282.58	3.166	0.558
190	64.570	289.00	3.260	0.544	48.090	288.11	3.197	0.553
200	68.183	294.41	3.287	0.541	50.840	293.67	3.225	0.548
210	71.782	299.81	3.314	0.539	53.570	299.10	3.252	0.545
220	75.363	305.20	3.339	0.537	56.283	304.52	3.277	0.541
230	78.932	310.58	3.363	0.534	58.989	309.97	3.301	0.539
240	82.491	315.90	3.385	0.532	61.680	315.32	3.324	0.538
250	86.041	321.21	3.407	0.530	64.361	320.70	3.346	0.536
260	89.587	326.51	3.428	0.530	67.039	326.07	3.367	0.535
270	93.119	331.82	3.448	0.529	69.708	331.39	3.387	0.531
280	96.649	337.11	3.467	0.530	72.369	336.70	3.406	0.530
290	100.18	342.40	3.486	0.527	75.027	342.01	3.425	0.530
300	103.69	347.69	3.504	0.526	77.678	347.31	3.443	0.530
310	107.19	352.89	3.521	0.525	80.328	352.61	3.460	0.530
320	110.69	358.19	3.538	0.524	82.978	357.91	3.477	0.528
330	114.19	363.49	3.554	0.523	85.618	363.21	3.493	0.527
340	117.69	368.69	3.570	0.522	88.258	368.41	3.509	0.526
350	121.19	373.99	3.585	0.521	90.898	373.71	3.524	0.525
375	129.99	387.09	3.621	0.521	97.478	386.91	3.561	0.524
400	138.69	400.19	3.655	0.521	104.09	400.01	3.595	0.523
425	147.49	413.29	3.687	0.521	110.59	413.11	3.627	0.522
450	156.19	426.39	3.717	0.521	117.19	426.21	3.657	0.521
475	164.89	439.39	3.745	0.521	123.69	439.31	3.685	0.521
500	173.59	452.49	3.772	0.521	130.29	452.41	3.712	0.521
550	190.99	478.59	3.822	0.521	143.29	478.51	3.762	0.521
600	208.39	504.69	3.867	0.521	156.39	504.71	3.807	0.521
650	225.79	530.79	3.909	0.521	169.39	530.81	3.849	0.521
700	243.19	556.89	3.948	0.521	182.49	556.91	3.888	0.521

TABLE XI. Thermodynamic Properties of Liquid and Gaseous Argon (*Continued*)

T, K	p = 6 bar				p = 8 bar			
	$v \cdot 10^3$	h	s	c_p	$v \cdot 10^3$	h	s	c_p
750	260.59	582.99	3.984	0.521	195.49	583.01	3.924	0.521
800	277.89	608.99	4.017	0.521	208.59	609.01	3.957	0.521
850	295.29	635.09	4.049	0.521	221.59	635.11	3.989	0.521
900	312.69	661.09	4.079	0.521	234.59	661.11	4.019	0.521
950	329.99	687.09	4.107	0.521	247.59	687.21	4.047	0.521
1000	347.39	713.19	4.134	0.521	260.69	713.21	4.073	0.521
1050	364.79	739.19	4.159	0.521	273.69	739.31	4.098	0.521
1100	382.09	765.29	4.183	0.521	286.69	765.31	4.123	0.521
1150	399.5	791.45	4.206	0.521	299.7	791.45	4.146	0.521
1200	416.7	817.45	4.228	0.521	312.8	817.55	4.168	0.521
1250	434.2	843.45	4.250	0.521	325.8	843.55	4.190	0.521
1300	451.5	869.55	4.270	0.521	338.8	869.65	4.210	0.521

TABLE XI. Thermodynamic Properties of Liquid and Gaseous Argon (*Continued*)

T, K	p = 10 bar				p = 15 bar			
	$v \cdot 10^3$	h	s	c_p	$v \cdot 10^3$	h	s	c_p
85	0.7090	73.55	1.348	1.055	0.7084	73.78	1.346	1.054
90	0.7250	78.91	1.409	1.088	0.7242	79.13	1.407	1.085
95	0.7423	84.42	1.469	1.119	0.7413	84.65	1.467	1.116
100	0.7611	90.08	1.527	1.152	0.7599	90.26	1.524	1.149
105	0.7819	95.91	1.583	1.185	0.7804	96.07	1.580	1.179
110	0.8053	101.90	1.639	1.222	0.8035	102.03	1.636	1.216
115	0.8321	108.09	1.694	1.271	0.8297	108.20	1.690	1.261
120	21.211	244.25	2.862	0.722	0.8605	114.62	1.746	1.326
125	22.603	247.84	2.891	0.696	13.633	241.68	2.769	0.856
130	23.943	251.24	2.918	0.672	14.704	245.70	2.802	0.803
135	25.237	254.59	2.943	0.653	15.708	249.66	2.832	0.759
140	26.499	257.79	2.966	0.636	16.664	253.36	2.859	0.722
145	27.738	260.91	2.988	0.621	17.582	256.87	2.883	0.692
150	28.950	263.99	3.009	0.609	18.472	260.28	2.906	0.668
155	30.149	267.00	3.029	0.599	19.338	263.58	2.928	0.649
160	31.330	269.99	3.048	0.591	20.185	266.80	2.948	0.634
170	33.653	275.87	3.083	0.579	21.836	273.03	2.986	0.614
180	35.942	281.59	3.116	0.571	23.434	279.06	3.021	0.600
190	38.200	287.27	3.147	0.562	24.955	285.03	3.054	0.586
200	40.430	292.87	3.176	0.556	26.494	290.84	3.084	0.576
210	42.641	298.39	3.203	0.550	28.014	296.56	3.112	0.567
220	44.834	303.89	3.228	0.546	29.515	302.20	3.138	0.562
230	47.021	309.32	3.253	0.544	31.004	307.80	3.163	0.557
240	49.192	314.79	3.276	0.543	32.481	313.36	3.187	0.553
250	51.353	320.19	3.298	0.539	33.948	318.87	3.209	0.549

TABLE XI. Thermodynamic Properties of Liquid and Gaseous Argon (*Continued*)

T, K	p = 10 bar				p = 15 bar			
	$v \cdot 10^3$	h	s	c_p	$v \cdot 10^3$	h	s	c_p
260	53.510	325.58	3.319	0.536	35.407	324.35	3.231	0.546
270	55.659	330.92	3.339	0.535	36.859	329.81	3.251	0.543
280	57.800	336.29	3.359	0.534	38.305	335.23	3.271	0.541
290	59.938	341.61	3.377	0.534	39.748	340.63	3.290	0.538
300	62.069	346.98	3.395	0.532	41.183	346.02	3.308	0.537
310	64.199	352.28	3.412	0.530	42.617	351.37	3.326	0.536
320	66.329	357.58	3.429	0.530	44.047	356.75	3.343	0.537
330	68.449	362.88	3.445	0.530	45.476	362.10	3.359	0.535
340	70.569	368.18	3.461	0.527	46.898	367.48	3.375	0.532
350	72.689	373.38	3.476	0.526	48.318	372.78	3.392	0.532
375	77.969	386.58	3.513	0.525	51.856	386.04	3.427	0.531
400	83.239	399.78	3.547	0.524	55.392	399.33	3.461	0.529
425	88.509	412.88	3.579	0.523	58.909	412.55	3.493	0.528
450	93.759	426.08	3.609	0.522	62.423	425.69	3.524	0.527
475	98.999	439.18	3.637	0.521	65.927	438.88	3.553	0.526
500	104.19	452.28	3.664	0.521	69.429	451.99	3.580	0.525
550	114.69	478.38	3.714	0.521	76.412	478.28	3.630	0.524
600	125.19	504.58	3.759	0.521	83.372	504.45	3.675	0.523
650	135.59	530.68	3.801	0.521	90.362	530.66	3.717	0.522
700	146.09	556.88	3.840	0.521	97.305	556.82	3.756	0.521
750	156.49	582.98	3.876	0.521	104.27	582.92	3.792	0.521
800	166.89	608.98	3.910	0.521	111.24	609.04	3.826	0.521
850	177.39	635.08	3.941	0.521	118.15	635.14	3.857	0.521
900	187.79	661.18	3.971	0.521	125.10	661.24	3.887	0.521
950	198.19	687.18	3.999	0.521	132.06	687.33	3.915	0.521
1000	208.59	713.28	4.026	0.521	138.99	713.37	3.942	0.521
1050	219.09	739.28	4.051	0.521	145.91	739.48	3.967	0.521
1100	229.49	765.38	4.076	0.521	152.87	765.53	3.991	0.521
1150	239.9	791.55	4.100	0.521	160.1	791.65	4.015	0.521
1200	250.3	817.55	4.122	0.521	167.0	817.75	4.038	0.521
1250	260.7	843.65	4.143	0.521	174.0	843.75	4.059	0.521
1300	271.1	869.65	4.164	0.521	180.9	869.85	4.080	0.521

TABLE XI. Thermodynamic Properties of Liquid and Gaseous Argon (*Continued*)

T, K	p = 20 bar				p = 25 bar			
	$v \cdot 10^3$	h	s	c_p	$v \cdot 10^3$	h	s	c_p
85	0.7076	73.98	1.345	1.054	0.7069	74.21	1.344	1.052
90	0.7234	79.33	1.406	1.082	0.7225	79.54	1.404	1.079
95	0.7403	84.83	1.466	1.113	0.7393	85.03	1.464	1.109
100	0.7587	90.45	1.523	1.145	0.7575	90.64	1.521	1.139
105	0.7790	96.23	1.579	1.174	0.7776	96.40	1.577	1.167

TABLE XI. Thermodynamic Properties of Liquid and Gaseous Argon (*Continued*)

T, K	p = 20 bar				p = 25 bar			
	$v \cdot 10^3$	h	s	c_p	$v \cdot 10^3$	h	s	c_p
110	0.8016	102.17	1.634	1.209	0.7999	102.30	1.632	1.200
115	0.8274	108.27	1.688	1.250	0.8251	108.37	1.686	1.238
120	0.8574	114.65	1.743	1.307	0.8544	114.69	1.740	1.292
125	0.8935	121.37	1.798	1.398	0.8894	121.34	1.794	1.375
130	9.8822	238.76	2.703	1.010	0.9332	128.57	1.851	1.503
135	10.815	243.78	2.741	0.925	7.6831	236.26	2.652	1.175
140	11.659	248.21	2.773	0.854	8.5379	242.13	2.694	1.053
145	12.443	252.32	2.802	0.797	9.2824	247.10	2.730	0.953
150	13.185	256.15	2.828	0.751	9.9609	251.61	2.760	0.872
155	13.897	259.84	2.852	0.715	10.596	255.74	2.787	0.807
160	14.582	263.36	2.875	0.687	11.198	259.66	2.812	0.757
170	15.903	270.08	2.916	0.652	12.332	266.97	2.856	0.693
180	17.172	276.50	2.952	0.633	13.405	273.85	2.895	0.662
190	18.402	282.71	2.986	0.618	14.435	280.37	2.931	0.647
200	19.603	288.78	3.017	0.595	15.433	286.68	2.963	0.620
210	20.782	294.69	3.046	0.587	16.405	292.80	2.993	0.597
220	21.942	300.51	3.073	0.577	17.362	298.80	3.021	0.594
230	23.084	306.28	3.099	0.570	18.301	304.70	3.047	0.584
240	24.221	311.92	3.123	0.563	19.225	310.51	3.072	0.576
250	25.343	317.53	3.146	0.558	20.143	316.23	3.095	0.569
260	26.460	323.12	3.168	0.555	21.051	321.92	3.118	0.565
270	27.564	328.64	3.188	0.551	21.951	327.53	3.139	0.560
280	28.670	334.15	3.209	0.549	22.843	333.13	3.159	0.555
290	29.763	339.64	3.228	0.547	23.732	338.65	3.179	0.552
300	30.858	345.12	3.246	0.543	24.618	344.16	3.197	0.550
310	31.948	350.52	3.264	0.540	25.498	349.66	3.215	0.550
320	33.028	355.92	3.281	0.539	26.378	355.16	3.232	0.549
330	34.108	361.32	3.298	0.540	27.248	360.66	3.249	0.544
340	35.188	366.72	3.314	0.539	28.118	366.06	3.265	0.542
350	36.268	372.12	3.330	0.537	28.988	371.46	3.281	0.540
375	38.948	385.52	3.367	0.534	31.148	384.96	3.318	0.538
400	41.618	398.82	3.401	0.531	33.298	398.36	3.352	0.535
425	44.278	412.12	3.433	0.529	35.438	411.76	3.384	0.533
450	46.928	425.32	3.463	0.528	37.568	425.06	3.415	0.531
475	49.568	438.52	3.492	0.527	39.688	438.36	3.444	0.530
500	52.208	451.72	3.519	0.526	41.808	451.56	3.471	0.528
550	57.478	478.02	3.569	0.525	46.028	477.96	3.521	0.527
600	62.728	504.32	3.615	0.524	50.238	504.26	3.567	0.526
650	67.968	530.52	3.657	0.523	54.448	530.56	3.609	0.525
700	73.208	556.72	3.696	0.522	58.638	556.76	3.648	0.524
750	78.438	582.92	3.732	0.521	62.828	582.96	3.684	0.523
800	83.668	609.12	3.766	0.521	67.018	609.16	3.718	0.522
850	88.898	635.22	3.798	0.521	71.208	635.36	3.750	0.521

TABLE XI. Thermodynamic Properties of Liquid and Gaseous Argon (*Continued*)

T, K	p = 20 bar				p = 25 bar			
	$v \cdot 10^3$	h	s	c_p	$v \cdot 10^3$	h	s	c_p
900	94.118	661.32	3.828	0.521	75.388	661.46	3.780	0.521
950	99.338	687.42	3.856	0.521	79.568	687.56	3.808	0.521
1000	104.59	713.52	3.883	0.521	83.748	713.66	3.835	0.521
1050	109.79	739.52	3.908	0.521	87.928	739.76	3.860	0.521
1100	114.99	765.62	3.932	0.521	92.098	765.76	3.884	0.521
1150	120.2	791.85	3.955	0.521	96.27	791.95	3.909	0.521
1200	125.4	817.85	3.977	0.521	100.4	818.05	3.931	0.521
1250	130.6	843.95	3.998	0.521	104.6	844.15	3.952	0.521
1300	135.8	870.05	4.019	0.521	108.8	870.15	3.973	0.521

TABLE XI. Thermodynamic Properties of Liquid and Gaseous Argon (*Continued*)

T, K	p = 30 bar				p = 35 bar			
	$v \cdot 10^3$	h	s	c_p	$v \cdot 10^3$	h	s	c_p
85	0.7062	74.44	1.342	1.051	0.7055	74.67	1.341	1.049
90	0.7217	79.76	1.403	1.076	0.7209	79.98	1.401	1.075
95	0.7383	85.23	1.462	1.106	0.7374	85.44	1.460	1.103
100	0.7564	90.83	1.519	1.134	0.7553	91.02	1.517	1.130
105	0.7762	96.57	1.575	1.159	0.7748	96.74	1.573	1.155
110	0.7981	102.45	1.630	1.189	0.7965	102.59	1.627	1.184
115	0.8230	108.48	1.683	1.226	0.8208	108.59	1.680	1.218
120	0.8515	114.75	1.737	1.277	0.8488	114.81	1.734	1.266
125	0.8855	121.32	1.790	1.353	0.8817	121.31	1.786	1.335
130	0.9274	128.39	1.846	1.478	0.9219	128.28	1.841	1.453
135	0.9830	136.34	1.906	1.698	0.9742	135.96	1.900	1.627
140	6.2497	234.19	2.611	1.332	1.0498	144.95	1.964	1.801
145	7.0828	240.91	2.658	1.178	5.3668	232.85	2.581	1.463
150	7.7548	246.44	2.696	1.050	6.1065	240.28	2.632	1.285
155	8.3570	251.28	2.728	0.945	6.7153	246.16	2.671	1.137
160	8.9140	255.70	2.756	0.862	7.2557	251.29	2.703	1.014
170	9.9397	263.73	2.804	0.749	8.2171	260.23	2.757	0.838
180	10.891	271.05	2.846	0.691	9.0853	268.15	2.802	0.736
190	11.786	277.96	2.884	0.667	9.8952	275.46	2.842	0.686
200	12.653	284.51	2.917	0.632	10.664	282.31	2.877	0.647
210	13.492	290.89	2.948	0.611	11.406	288.91	2.910	0.626
220	14.306	297.09	2.977	0.599	12.126	295.31	2.939	0.617
230	15.112	303.11	3.004	0.598	12.833	301.52	2.967	0.600
240	15.901	309.04	3.029	0.588	13.523	307.61	2.993	0.589
250	16.680	314.92	3.053	0.581	14.203	313.55	3.017	0.585
260	17.449	320.66	3.076	0.575	14.873	319.45	3.040	0.580
270	18.210	326.42	3.098	0.568	15.539	325.25	3.062	0.579
280	18.962	332.05	3.118	0.563	16.192	331.03	3.083	0.572

TABLE XI. Thermodynamic Properties of Liquid and Gaseous Argon (*Continued*)

T, K	p = 30 bar				p = 35 bar			
	$v \cdot 10^3$	h	s	c_p	$v \cdot 10^0$	h	s	c_p
290	19.712	337.66	3.138	0.560	16.842	336.73	3.103	0.566
300	20.459	343.26	3.157	0.558	17.490	342.36	3.122	0.562
310	21.199	348.86	3.175	0.554	18.130	347.96	3.141	0.558
320	21.939	354.36	3.193	0.551	18.770	353.56	3.159	0.554
330	22.679	359.86	3.210	0.547	19.410	359.06	3.176	0.551
340	23.409	365.36	3.226	0.544	20.040	364.56	3.192	0.550
350	24.139	370.76	3.242	0.541	20.670	370.06	3.208	0.549
375	25.949	384.26	3.279	0.539	22.240	383.76	3.245	0.546
400	27.749	397.76	3.314	0.541	23.790	397.36	3.280	0.541
425	29.539	411.26	3.347	0.539	25.330	410.86	3.313	0.539
450	31.319	424.76	3.378	0.535	26.870	424.26	3.344	0.535
475	33.099	438.06	3.407	0.530	28.400	437.66	3.373	0.534
500	34.869	451.26	3.431	0.528	29.920	450.96	3.400	0.532
550	38.399	477.66	3.484	0.527	32.950	477.56	3.451	0.529
600	41.919	504.06	3.530	0.526	35.970	503.96	3.497	0.528
650	45.429	530.46	3.572	0.525	38.990	530.36	3.539	0.527
700	48.929	556.76	3.611	0.524	42.000	556.66	3.578	0.526
750	52.429	582.96	3.647	0.523	45.000	582.96	3.614	0.525
800	55.919	609.16	3.681	0.522	48.000	609.16	3.648	0.524
850	59.409	635.36	3.713	0.521	50.990	635.36	3.680	0.523
900	62.899	661.46	3.743	0.521	53.980	661.56	3.710	0.522
950	66.389	687.56	3.771	0.521	56.970	687.76	3.739	0.521
1000	69.869	713.76	3.798	0.521	59.960	713.86	3.766	0.521
1050	73.349	739.86	3.823	0.521	62.950	739.96	3.791	0.521
1100	76.829	765.96	3.847	0.521	65.940	766.06	3.815	0.521
1150	80.31	792.15	3.871	0.522	68.92	792.25	3.838	0.522
1200	83.79	818.25	3.893	0.522	71.90	818.35	3.860	0.522
1250	87.27	844.25	3.914	0.521	74.88	844.45	3.882	0.522
1300	90.75	870.35	3.935	0.521	77.86	870.55	3.903	0.522

TABLE XI. Thermodynamic Properties of Liquid and Gaseous Argon (*Continued*)

T, K	p = 40 bar				p = 45 bar			
	$v \cdot 10^3$	h	s	c_p	$v \cdot 10^3$	h	s	c_p
85	0.7048	74.90	1.339	1.049	—	—	—	—
90	0.7201	80.19	1.399	1.073	0.7194	80.41	1.398	1.073
95	0.7365	85.65	1.459	1.100	0.7355	85.85	1.457	1.098
100	0.7541	91.21	1.515	1.125	0.7531	91.40	1.513	1.122
105	0.7735	96.91	1.571	1.151	0.7722	97.08	1.569	1.146
110	0.7948	102.74	1.625	1.179	0.7932	102.89	1.623	1.173
115	0.8188	108.71	1.678	1.212	0.8168	108.82	1.675	1.204

TABLE XI. Thermodynamic Properties of Liquid and Gaseous Argon (*Continued*)

T, K	p = 40 bar				p = 45 bar			
	$v \cdot 10^3$	h	s	c_p	$v \cdot 10^3$	h	s	c_p
120	0.8461	114.88	1.731	1.252	0.8435	114.95	1.728	1.246
125	0.8781	121.32	1.783	1.319	0.8747	121.33	1.780	1.305
130	0.9168	128.12	1.837	1.411	0.9120	128.08	1.832	1.387
135	0.9662	135.66	1.894	1.597	0.9589	135.38	1.888	1.548
140	1.0348	144.22	1.956	1.869	1.0220	143.61	1.948	1.802
145	1.1525	155.28	2.033	2.459	1.1206	153.66	2.018	2.329
150	4.7617	232.43	2.561	1.581	3.4826	220.22	2.471	2.165
155	5.4323	240.25	2.613	1.381	4.3622	232.89	2.550	1.834
160	5.9815	246.42	2.652	1.212	4.9562	240.79	2.600	1.556
170	6.9131	256.51	2.713	0.960	5.8868	252.50	2.671	1.139
180	7.7261	265.07	2.762	0.801	6.6643	261.93	2.725	0.880
190	8.4720	272.86	2.804	0.715	7.3635	270.18	2.769	0.743
200	9.1745	280.09	2.841	0.669	8.0141	277.79	2.809	0.701
210	9.8464	286.92	2.875	0.647	8.6319	284.92	2.843	0.670
220	10.494	293.52	2.906	0.637	9.2250	291.73	2.875	0.649
230	11.124	299.92	2.934	0.608	9.7998	298.32	2.904	0.629
240	11.743	306.14	2.961	0.596	10.361	304.66	2.932	0.609
250	12.350	312.24	2.986	0.591	10.910	310.92	2.957	0.599
260	12.944	318.24	3.009	0.588	11.449	316.98	2.981	0.594
270	13.533	324.14	3.031	0.586	11.980	322.98	3.004	0.592
280	14.119	329.96	3.053	0.579	12.501	328.89	3.025	0.589
290	14.691	335.75	3.073	0.575	13.021	334.76	3.046	0.582
300	15.262	341.45	3.092	0.569	13.537	340.56	3.065	0.575
310	15.832	347.15	3.111	0.564	14.047	346.26	3.084	0.569
320	16.402	352.75	3.129	0.561	14.557	351.96	3.102	0.564
330	16.962	358.35	3.146	0.558	15.057	357.56	3.119	0.561
340	17.522	363.95	3.162	0.554	15.557	363.16	3.136	0.557
350	18.072	369.45	3.178	0.550	16.047	368.76	3.152	0.556
375	19.452	383.15	3.215	0.546	17.287	382.56	3.189	0.552
400	20.822	396.75	3.250	0.545	18.507	396.36	3.224	0.546
425	22.182	410.35	3.283	0.542	19.717	409.96	3.258	0.543
450	23.532	423.95	3.314	0.540	20.927	423.46	3.290	0.540
475	24.872	437.35	3.344	0.536	22.127	436.96	3.320	0.537
500	26.202	450.75	3.372	0.534	23.307	450.36	3.347	0.535
550	28.862	477.35	3.423	0.530	25.687	477.06	3.397	0.531
600	31.512	503.85	3.469	0.529	28.047	503.56	3.444	0.530
650	34.152	530.25	3.512	0.527	30.397	530.06	3.487	0.529
700	36.792	556.65	3.552	0.526	32.737	556.56	3.526	0.528
750	39.422	582.95	3.588	0.525	35.077	582.96	3.562	0.527
800	42.052	609.25	3.621	0.524	37.417	609.26	3.596	0.526
850	44.672	635.45	3.652	0.523	39.757	635.46	3.628	0.525
900	47.292	661.65	3.682	0.522	42.087	661.66	3.658	0.524
950	49.912	687.85	3.711	0.521	44.417	687.86	3.686	0.523

TABLE XI. Thermodynamic Properties of Liquid and Gaseous Argon (*Continued*)

T, K	p = 40 bar				p = 45 bar			
	$v \cdot 10^3$	h	s	c_p	$v \cdot 10^3$	h	s	c_p
1000	52.532	713.95	3.738	0.521	46.747	714.06	3.713	0.522
1050	55.152	740.05	3.763	0.521	49.077	740.26	3.738	0.521
1100	57.762	766.15	3.787	0.521	51.397	766.36	3.762	0.521
1150	60.37	792.35	3.810	0.522	53.72	792.55	3.786	0.522
1200	62.98	818.45	3.833	0.522	56.04	818.65	3.808	0.522
1250	65.59	844.55	3.854	0.522	58.36	844.75	3.829	0.522
1300	68.20	870.65	3.875	0.522	60.68	870.85	3.850	0.522

TABLE XI. Thermodynamic Properties of Liquid and Gaseous Argon (*Continued*)

T, K	p = 50 bar				p = 60 bar			
	$v \cdot 10^3$	h	s	c_p	$v \cdot 10^3$	h	s	c_p
90	0.7186	80.63	1.396	1.072	0.7171	81.07	1.393	1.069
95	0.7346	86.06	1.455	1.095	0.7329	86.48	1.452	1.091
100	0.7520	91.60	1.512	1.119	0.7499	91.99	1.508	1.113
105	0.7709	97.26	1.567	1.143	0.7683	97.61	1.563	1.134
110	0.7916	103.04	1.621	1.167	0.7886	103.36	1.616	1.158
115	0.8148	108.95	1.673	1.197	0.8110	109.20	1.668	1.183
120	0.8410	115.03	1.725	1.234	0.8363	115.21	1.719	1.218
125	0.8714	121.35	1.776	1.292	0.8652	121.42	1.770	1.265
130	0.9075	128.00	1.828	1.369	0.8991	127.91	1.821	1.332
135	0.9521	135.18	1.883	1.491	0.9401	134.81	1.873	1.409
140	1.0109	143.11	1.940	1.712	0.9919	142.32	1.928	1.539
145	1.0970	152.44	2.006	2.170	1.0622	150.78	1.987	1.789
150	1.3158	170.43	2.202	6.044	1.1701	161.99	2.064	3.967
155	3.3902	222.94	2.473	4.608	1.5277	182.06	2.194	4.295
160	4.0941	234.10	2.541	2.915	2.6401	215.39	2.407	3.520
170	5.0535	248.17	2.630	1.344	3.7729	238.30	2.546	1.970
180	5.8105	258.61	2.689	0.981	4.5205	251.41	2.621	1.210
190	6.4755	267.46	2.737	0.783	5.1411	261.73	2.676	0.898
200	7.0862	275.47	2.778	0.745	5.6944	270.66	2.722	0.810
210	7.6607	282.90	2.814	0.719	6.2061	278.71	2.762	0.774
220	8.2097	289.92	2.847	0.690	6.6899	286.24	2.797	0.731
230	8.7395	296.66	2.877	0.663	7.1522	293.42	2.829	0.698
240	9.2543	303.22	2.905	0.643	7.5989	300.25	2.858	0.671
250	9.7570	309.56	2.931	0.627	8.0328	306.87	2.885	0.650
260	10.250	315.76	2.955	0.614	8.4565	313.29	2.910	0.636
270	10.733	321.87	2.978	0.605	8.8721	319.60	2.934	0.623
280	11.211	327.87	3.000	0.595	9.2807	325.79	2.956	0.611
290	11.682	333.78	3.021	0.586	9.6839	331.82	2.978	0.601
300	12.151	339.61	3.041	0.582	10.080	337.82	2.998	0.594

TABLE XI. Thermodynamic Properties of Liquid and Gaseous Argon (*Continued*)

T, K	p = 50 bar				p = 60 bar			
	$v \cdot 10^3$	h	s	c_p	$v \cdot 10^3$	h	s	c_p
310	12.621	345.41	3.060	0.578	10.480	343.72	3.017	0.590
320	13.081	351.21	3.078	0.575	10.870	349.62	3.036	0.583
330	13.541	356.91	3.095	0.569	11.260	355.42	3.054	0.576
340	13.991	362.61	3.112	0.563	11.640	361.12	3.071	0.568
350	14.441	368.21	3.128	0.560	12.020	366.82	3.087	0.566
375	15.561	382.11	3.167	0.552	12.960	380.82	3.126	0.562
400	16.661	395.91	3.203	0.548	13.900	394.92	3.162	0.556
425	17.761	409.51	3.236	0.544	14.820	408.72	3.196	0.551
450	18.851	423.11	3.267	0.541	15.730	422.42	3.228	0.546
475	19.931	436.61	3.296	0.539	16.640	436.02	3.257	0.543
500	21.001	450.11	3.324	0.536	17.540	449.62	3.284	0.540
550	23.141	476.81	3.375	0.534	19.330	476.52	3.335	0.536
600	25.271	503.51	3.422	0.532	21.110	503.22	3.382	0.533
650	27.391	530.11	3.465	0.531	22.880	529.92	3.425	0.532
700	29.511	556.61	3.504	0.528	24.650	556.52	3.465	0.531
750	31.621	583.01	3.540	0.527	26.410	583.02	3.502	0.528
800	33.721	609.31	3.574	0.525	28.170	609.42	3.536	0.527
850	35.821	635.61	3.606	0.524	29.930	635.72	3.567	0.525
900	37.921	661.81	3.636	0.524	31.680	662.02	3.597	0.524
950	40.021	688.01	3.664	0.523	33.430	688.22	3.626	0.524
1000	42.121	714.21	3.691	0.524	35.180	714.42	3.653	0.523
1050	44.221	740.41	3.716	0.523	36.930	740.62	3.678	0.524
1100	46.311	766.61	3.740	0.523	38.680	766.82	3.702	0.523
1150	48.40	792.75	3.764	0.523	40.42	793.05	3.726	0.523
1200	50.49	818.85	3.786	0.522	42.16	819.15	3.748	0.523
1250	52.58	844.95	3.807	0.522	43.90	845.25	3.769	0.523
1300	54.67	871.05	3.828	0.522	45.64	871.35	3.790	0.522

TABLE XI. Thermodynamic Properties of Liquid and Gaseous Argon (*Continued*)

T, K	p = 70 bar				p = 80 bar			
	$v \cdot 10^3$	h	s	c_p	$v \cdot 10^3$	h	s	c_p
90	0.7156	81.52	1.390	1.065	0.7141	81.96	1.388	1.062
95	0.7311	86.90	1.449	1.087	0.7294	87.33	1.446	1.083
100	0.7478	92.39	1.505	1.108	0.7458	92.79	1.501	1.103
105	0.7659	97.98	1.559	1.128	0.7636	98.35	1.555	1.121
110	0.7857	103.68	1.612	1.148	0.7828	104.02	1.608	1.139
115	0.8074	109.47	1.663	1.172	0.8040	109.76	1.659	1.161
120	0.8318	115.41	1.714	1.200	0.8275	115.62	1.709	1.185
125	0.8594	121.52	1.764	1.241	0.8541	121.65	1.758	1.222
130	0.8915	127.84	1.816	1.298	0.8845	127.85	1.807	1.274

TABLE XI. Thermodynamic Properties of Liquid and Gaseous Argon (*Continued*)

T, K	p = 70 bar				p = 80 bar			
	$v \cdot 10^3$	h	s	c_p	$v \cdot 10^3$	h	s	c_p
135	0.9295	134.55	1.864	1.375	0.9200	134.42	1.856	1.336
140	0.9762	141.73	1.917	1.469	0.9628	141.33	1.906	1.419
145	1.0364	149.63	1.971	1.632	1.0160	148.74	1.958	1.548
150	1.1198	158.82	2.034	2.477	1.0853	157.09	2.015	1.874
155	1.2764	171.45	2.116	3.426	1.1921	167.16	2.082	2.682
160	1.6600	191.77	2.242	3.390	1.3768	180.42	2.165	2.975
170	2.8333	226.58	2.457	2.530	2.1624	213.70	2.367	2.634
180	3.5939	243.56	2.556	1.488	2.9069	235.12	2.490	1.831
190	4.1888	255.69	2.620	1.024	3.4809	249.36	2.567	1.227
200	4.7040	265.66	2.671	0.902	3.9670	260.54	2.624	1.002
210	5.1719	274.48	2.714	0.835	4.4019	270.17	2.671	0.899
220	5.6085	282.53	2.752	0.777	4.8034	278.81	2.712	0.826
230	6.0233	290.13	2.786	0.734	5.1811	286.82	2.747	0.774
240	6.4207	297.28	2.817	0.702	5.5421	294.35	2.779	0.733
250	6.8055	304.18	2.845	0.678	5.8898	301.55	2.809	0.702
260	7.1800	310.88	2.871	0.659	6.2266	308.42	2.836	0.678
270	7.5458	317.38	2.895	0.641	6.5546	315.13	2.861	0.659
280	7.9045	323.71	2.918	0.626	6.8760	321.64	2.885	0.646
290	8.2577	329.93	2.940	0.617	7.1909	328.04	2.907	0.631
300	8.6053	336.04	2.961	0.608	7.5013	334.33	2.928	0.620
310	8.9493	342.14	2.981	0.601	7.8073	340.43	2.948	0.610
320	9.2893	348.04	3.000	0.594	8.1103	346.53	2.967	0.602
330	9.6263	354.04	3.018	0.587	8.4093	352.53	2.986	0.596
340	9.9603	359.84	3.035	0.582	8.7063	358.43	3.004	0.590
350	10.289	365.64	3.051	0.577	9.0003	364.33	3.021	0.584
375	11.109	379.94	3.091	0.567	9.7253	378.83	3.061	0.574
400	11.919	394.04	3.128	0.559	10.439	393.03	3.098	0.565
425	12.719	407.94	3.162	0.554	11.149	407.13	3.132	0.559
450	13.509	421.74	3.193	0.549	11.839	421.03	3.164	0.554
475	14.289	435.44	3.223	0.546	12.529	434.83	3.194	0.549
500	15.069	449.04	3.251	0.543	13.209	448.53	3.222	0.546
550	16.609	476.14	3.302	0.539	14.569	475.73	3.273	0.541
600	18.139	503.04	3.349	0.537	15.919	502.73	3.320	0.537
650	19.669	529.84	3.392	0.534	17.259	529.53	3.363	0.535
700	21.189	556.54	3.432	0.531	18.589	556.23	3.403	0.532
750	22.699	583.04	3.469	0.529	19.919	582.83	3.440	0.531
800	24.209	609.44	3.503	0.527	21.239	609.33	3.475	0.529
850	25.719	635.84	3.535	0.526	22.559	635.83	3.507	0.528
900	27.219	662.14	3.565	0.526	23.879	662.23	3.537	0.528
950	28.719	688.44	3.593	0.524	25.199	688.63	3.565	0.526
1000	30.219	714.64	3.620	0.524	26.519	714.93	3.592	0.525
1050	31.719	740.84	3.646	0.523	27.829	741.13	3.618	0.524
1100	33.219	767.04	3.670	0.523	29.139	767.33	3.642	0.523

TABLE XI. Thermodynamic Properties of Liquid and Gaseous Argon (*Continued*)

T, K	p = 70 bar				p = 80 bar			
	$v \cdot 10^3$	h	s	c_p	$v \cdot 10^3$	h	s	c_p
1150	34.72	793.25	3.693	0.523	30.45	793.55	3.665	0.524
1200	36.22	819.45	3.716	0.523	31.76	819.75	3.688	0.524
1250	37.72	845.65	3.737	0.523	33.07	845.95	3.709	0.523
1300	39.21	871.75	3.758	0.523	34.38	872.05	3.730	0.523

TABLE XI. Thermodynamic Properties of Liquid and Gaseous Argon (*Continued*)

T, K	p = 90 bar				p = 100 bar			
	$v \cdot 10^3$	h	s	c_p	$v \cdot 10^3$	h	s	c_p
90	0.7127	82.41	1.385	1.059	0.7113	82.87	1.382	1.056
95	0.7278	87.76	1.442	1.079	0.7262	88.19	1.439	1.075
100	0.7439	93.20	1.498	1.098	0.7420	93.62	1.494	1.093
105	0.7613	98.73	1.551	1.114	0.7591	99.11	1.548	1.108
110	0.7801	104.36	1.604	1.131	0.7775	104.71	1.600	1.12[illegible]
115	0.8007	110.05	1.654	1.150	0.7976	110.36	1.650	1.139
120	0.8235	115.86	1.704	1.173	0.8197	116.11	1.699	1.161
125	0.8490	121.81	1.752	1.206	0.8443	121.98	1.747	1.189
130	0.8780	127.93	1.800	1.250	0.8720	128.03	1.794	1.228
135	0.9115	134.33	1.849	1.303	0.9037	134.27	1.842	1.273
140	0.9510	141.03	1.897	1.370	0.9405	140.83	1.889	1.330
145	0.9990	148.13	1.947	1.475	0.9844	147.64	1.937	1.418
150	1.0590	155.88	2.000	1.662	1.0378	155.03	1.987	1.549
155	1.1424	164.81	2.059	2.146	1.1072	163.24	2.041	1.826
160	1.2661	175.41	2.126	2.520	1.2013	172.47	2.100	2.155
170	1.7530	202.62	2.291	2.512	1.5293	195.08	2.237	2.314
180	2.3992	226.55	2.428	2.010	2.0371	218.61	2.371	2.052
190	2.9430	242.87	2.516	1.473	2.5319	236.54	2.468	1.628
200	3.4025	255.34	2.580	1.112	2.9625	250.20	2.538	1.233
210	3.8095	265.84	2.631	0.969	3.3447	261.47	2.593	1.038
220	4.1829	275.09	2.674	0.876	3.6933	271.37	2.639	0.929
230	4.5322	283.53	2.712	0.811	4.0180	280.25	2.679	0.853
240	4.8635	291.39	2.746	0.764	4.3253	288.54	2.714	0.798
250	5.1815	298.89	2.776	0.729	4.6192	296.28	2.746	0.754
260	5.4886	306.02	2.804	0.701	4.9024	303.69	2.775	0.721
270	5.7874	312.93	2.830	0.680	5.1771	310.71	2.802	0.698
280	6.0790	319.64	2.854	0.664	5.4449	317.66	2.827	0.679
290	6.3647	326.22	2.877	0.646	5.7066	324.36	2.850	0.662
300	6.6453	332.62	2.899	0.631	5.9633	330.88	2.872	0.645
310	6.9223	338.82	2.920	0.620	6.2163	337.28	2.893	0.634
320	7.1953	345.02	2.939	0.612	6.4663	343.58	2.913	0.623
330	7.4653	351.12	2.958	0.606	6.7123	349.78	2.932	0.616

TABLE XI. Thermodynamic Properties of Liquid and Gaseous Argon (*Continued*)

T, K	p = 90 bar				p = 100 bar			
	$v \cdot 10^3$	h	s	c_p	$v \cdot 10^3$	h	s	c_p
340	7.7323	357.12	2.976	0.597	6.9553	355.88	2.950	0.608
350	7.9973	363.12	2.993	0.591	7.1963	361.98	2.967	0.600
375	8.6493	377.72	3.034	0.583	7.7893	376.78	3.009	0.585
400	9.2903	392.22	3.071	0.577	8.3713	391.28	3.048	0.576
425	9.9223	406.62	3.106	0.571	8.9443	405.58	3.083	0.569
450	10.539	420.82	3.139	0.563	9.5093	419.78	3.116	0.563
475	11.159	434.82	3.169	0.555	10.069	433.78	3.146	0.557
500	11.769	448.62	3.197	0.548	10.619	447.68	3.174	0.552
550	12.979	475.82	3.249	0.541	11.719	475.08	3.226	0.547
600	14.179	502.82	3.296	0.539	12.799	502.38	3.274	0.544
650	15.379	529.72	3.339	0.536	13.879	529.58	3.317	0.540
700	16.569	556.52	3.378	0.533	14.949	556.48	3.356	0.536
750	17.749	583.12	3.415	0.531	16.019	583.18	3.393	0.532
800	18.929	609.62	3.450	0.528	17.079	609.78	3.427	0.530
850	20.109	636.02	3.482	0.528	18.139	636.28	3.459	0.529
900	21.279	662.42	3.512	0.527	19.199	662.68	3.490	0.527
950	22.449	688.82	3.540	0.527	20.259	689.08	3.518	0.526
1000	23.619	715.12	3.567	0.525	21.309	715.38	3.545	0.526
1050	24.789	741.42	3.593	0.524	22.359	741.68	3.571	0.525
1100	25.959	767.62	3.618	0.524	23.409	767.98	3.595	0.524
1150	27.13	793.85	3.641	0.524	24.46	794.15	3.619	0.525
1200	28.29	820.05	3.663	0.524	25.51	820.35	3.641	0.524
1250	29.45	846.25	3.684	0.524	26.56	846.55	3.662	0.524
1300	30.61	872.45	3.705	0.523	27.61	872.75	3.683	0.524

TABLE XI. Thermodynamic Properties of Liquid and Gaseous Argon (*Continued*)

T, K	p = 125 bar				p = 150 bar			
	$v \cdot 10^3$	h	s	c_p	$v \cdot 10^3$	h	s	c_p
90	0.7079	84.00	1.375	1.050	0.7047	85.15	1.368	1.044
95	0.7223	89.29	1.432	1.066	0.7186	90.40	1.424	1.058
100	0.7375	94.66	1.486	1.082	0.7333	95.73	1.479	1.072
105	0.7538	100.10	1.539	1.095	0.7489	101.11	1.531	1.083
110	0.7713	105.61	1.591	1.106	0.7656	106.56	1.582	1.091
115	0.7902	111.17	1.640	1.118	0.7835	112.04	1.630	1.099
120	0.8108	116.80	1.688	1.133	0.8028	117.56	1.678	1.111
125	0.8334	122.52	1.734	1.155	0.8239	123.15	1.723	1.129
130	0.8585	128.36	1.781	1.185	0.8469	128.86	1.768	1.151
135	0.8867	134.37	1.826	1.218	0.8724	134.68	1.812	1.173
140	0.9185	140.56	1.871	1.253	0.9007	140.60	1.855	1.196
145	0.9551	146.94	1.916	1.300	0.9324	146.67	1.897	1.229

TABLE XI. Thermodynamic Properties of Liquid and Gaseous Argon (*Continued*)

T, K	p = 125 bar				p = 150 bar			
	$v \cdot 10^3$	h	s	c_p	$v \cdot 10^3$	h	s	c_p
150	0.9977	153.58	1.961	1.370	0.9683	152.90	1.940	1.279
155	1.0483	160.68	2.008	1.486	1.0093	159.45	1.983	1.343
160	1.1103	168.29	2.056	1.653	1.0568	166.32	2.027	1.421
170	1.2866	185.47	2.159	1.828	1.1794	181.03	2.115	1.555
180	1.5544	204.47	2.268	1.851	1.3499	196.98	2.207	1.623
190	1.8883	222.19	2.367	1.703	1.5687	213.29	2.295	1.589
200	2.2281	238.22	2.446	1.450	1.8166	228.63	2.374	1.461
210	2.5463	251.23	2.510	1.220	2.0696	242.36	2.440	1.292
220	2.8412	262.45	2.562	1.047	2.3142	254.41	2.497	1.141
230	3.1165	272.41	2.607	0.947	2.5478	265.22	2.545	1.027
240	3.3765	281.53	2.645	0.874	2.7705	275.05	2.587	0.942
250	3.6240	289.98	2.680	0.821	2.9834	284.16	2.624	0.876
260	3.8619	297.99	2.711	0.777	3.1877	292.63	2.657	0.827
270	4.0918	305.61	2.740	0.744	3.3855	300.73	2.688	0.788
280	4.3154	312.88	2.766	0.717	3.5774	308.45	2.715	0.753
290	4.5331	319.97	2.791	0.697	3.7642	315.86	2.741	0.727
300	4.7460	326.86	2.814	0.678	3.9463	322.96	2.766	0.709
310	4.9550	333.56	2.836	0.659	4.1253	330.06	2.789	0.691
320	5.1610	340.06	2.857	0.644	4.3013	336.86	2.811	0.671
330	5.3640	346.46	2.877	0.636	4.4743	343.46	2.832	0.657
340	5.5640	352.76	2.896	0.627	4.6453	349.96	2.851	0.646
350	5.7620	359.06	2.914	0.618	4.8143	356.46	2.869	0.637
375	6.2480	374.26	2.956	0.601	5.2273	372.06	2.912	0.618
400	6.7240	389.16	2.994	0.588	5.6313	387.36	2.952	0.602
425	7.1920	403.76	3.030	0.580	6.0273	402.26	2.989	0.590
450	7.6490	418.16	3.064	0.574	6.4153	416.86	3.023	0.580
475	8.1020	432.46	3.095	0.570	6.7973	431.26	3.054	0.574
500	8.5500	446.66	3.124	0.565	7.1743	445.56	3.083	0.570
550	9.4350	474.76	3.177	0.556	7.9183	473.86	3.136	0.561
600	10.309	502.36	3.225	0.547	8.6523	501.76	3.184	0.554
650	11.179	529.56	3.268	0.541	9.3803	529.26	3.228	0.547
700	12.039	556.56	3.308	0.537	10.099	556.56	3.269	0.542
750	12.899	583.36	3.345	0.534	10.819	583.56	3.307	0.538
800	13.749	610.06	3.380	0.531	11.539	610.36	3.342	0.534
850	14.599	636.56	3.412	0.530	12.249	637.06	3.375	0.532
900	15.449	663.06	3.442	0.529	12.959	663.66	3.405	0.531
950	16.299	689.56	3.471	0.529	13.669	690.16	3.433	0.529
1000	17.149	715.96	3.498	0.527	14.379	716.66	3.460	0.528
1050	17.989	742.36	3.524	0.526	15.089	743.06	3.486	0.527
1100	18.829	768.66	3.549	0.526	15.789	769.46	3.511	0.527
1150	19.68	794.95	3.572	0.526	16.49	795.75	3.534	0.527
1200	20.52	821.25	3.594	0.525	17.19	822.05	3.556	0.526
1250	21.36	847.45	3.615	0.525	17.89	848.35	3.577	0.526
1300	22.20	873.65	3.636	0.524	18.59	874.55	3.598	0.525

TABLE XI. Thermodynamic Properties of Liquid and Gaseous Argon (*Continued*)

T, K	p = 175 bar				p = 200 bar			
	$v \cdot 10^3$	h	s	c_p	$v \cdot 10^3$	h	s	c_p
90	0.7017	86.31	1.361	1.038	0.6987	87.47	1.354	1.033
95	0.7151	91.53	1.417	1.051	0.7118	92.67	1.411	1.045
100	0.7293	96.82	1.471	1.064	0.7255	97.93	1.464	1.057
105	0.7443	102.16	1.523	1.072	0.7400	103.22	1.516	1.062
110	0.7603	107.54	1.574	1.078	0.7553	108.55	1.566	1.067
115	0.7773	112.95	1.621	1.085	0.7715	113.90	1.613	1.071
120	0.7956	118.39	1.668	1.090	0.7889	119.26	1.659	1.075
125	0.8153	123.87	1.712	1.104	0.8075	124.66	1.703	1.083
130	0.8367	129.41	1.756	1.121	0.8275	130.11	1.745	1.097
135	0.8600	135.11	1.799	1.140	0.8491	135.64	1.788	1.112
140	0.8856	140.83	1.841	1.160	0.8726	141.25	1.828	1.127
145	0.9139	146.69	1.881	1.178	0.8982	146.91	1.867	1.142
150	0.9453	152.67	1.922	1.214	0.9263	152.68	1.907	1.166
155	0.9802	158.79	1.963	1.254	0.9571	158.57	1.945	1.195
160	1.0195	165.21	2.003	1.301	0.9910	164.63	1.983	1.226
170	1.1153	178.56	2.084	1.390	1.0704	177.15	2.059	1.286
180	1.2400	192.75	2.165	1.451	1.1692	190.21	2.134	1.332
190	1.3960	207.43	2.244	1.458	1.2895	203.68	2.207	1.348
200	1.5781	221.82	2.318	1.401	1.4300	217.10	2.276	1.327
210	1.7750	235.34	2.384	1.299	1.5847	230.13	2.340	1.268
220	1.9746	247.73	2.441	1.184	1.7486	242.43	2.397	1.187
230	2.1711	259.02	2.492	1.078	1.9122	253.84	2.448	1.100
240	2.3617	269.34	2.536	0.991	2.0751	264.44	2.493	1.020
250	2.5457	278.88	2.575	0.921	2.2348	274.28	2.533	0.953
260	2.7236	287.82	2.610	0.866	2.3900	283.52	2.569	0.897
270	2.8958	296.26	2.642	0.822	2.5418	292.26	2.602	0.852
280	3.0635	304.29	2.671	0.787	2.6897	300.59	2.632	0.815
290	3.2266	312.02	2.698	0.758	2.8344	308.60	2.660	0.783
300	3.3863	319.50	2.724	0.733	2.9755	316.30	2.686	0.755
310	3.5433	326.70	2.748	0.709	3.1145	323.70	2.710	0.729
320	3.6963	333.70	2.770	0.689	3.2505	330.90	2.733	0.709
330	3.8473	340.50	2.791	0.675	3.3845	337.90	2.755	0.696
340	3.9963	347.20	2.811	0.665	3.5165	344.80	2.776	0.681
350	4.1443	353.80	2.831	0.652	3.6475	351.60	2.795	0.665
375	4.5043	369.90	2.875	0.631	3.9665	367.90	2.840	0.639
400	4.8563	385.40	2.915	0.615	4.2785	383.60	2.881	0.623
425	5.2003	400.70	2.952	0.604	4.5845	399.10	2.919	0.614
450	5.5373	415.70	2.986	0.596	4.8825	414.40	2.954	0.607
475	5.8683	430.50	3.018	0.588	5.1745	429.50	2.987	0.596
500	6.1953	445.10	3.048	0.580	5.4635	444.30	3.018	0.586
550	6.8383	473.80	3.102	0.565	6.0305	473.20	3.073	0.573
600	7.4713	501.80	3.151	0.554	6.5885	501.60	3.122	0.562
650	8.0983	529.30	3.195	0.547	7.1395	529.50	3.166	0.552
700	8.7203	556.60	3.236	0.543	7.6855	556.90	3.207	0.545

TABLE XI. Thermodynamic Properties of Liquid and Gaseous Argon (*Continued*)

T, K	p = 175 bar				p = 200 bar			
	$v \cdot 10^3$	h	s	c_p	$v \cdot 10^3$	h	s	c_p
750	9,3383	583,70	3,274	0,540	8,2285	584.00	3,244	0.540
800	9,9543	610,60	3,309	0,537	8,7675	611.00	3,279	0,538
850	10,569	637,40	3,341	0,534	9,3045	637,90	3,312	0,536
900	11,179	664,10	3,372	0,532	9,8405	664,60	3,342	0,534
950	11,779	690,70	3,401	0,531	10,359	691,30	3,371	0.532
1000	12,389	717,20	3,428	0,529	10,909	717,90	3,398	0,531
1050	12,999	743,70	3,454	0,528	11,439	744,40	3,424	0,529
1100	13,599	770,10	3,479	0,528	11,969	770,90	3,449	0,528
1150	14,21	796,45	3,501	0,528	12,50	797,25	3,473	0,528
1200	14,81	822,85	3,523	0.528	13,03	823,65	3.495	0,528
1250	15,41	849,15	3,545	0,527	13,56	850,05	3,516	0,527
1300	16,02	875,45	3,565	0,526	14,08	876,45	3,537	0,527

TABLE XI. Thermodynamic Properties of Liquid and Gaseous Argon (*Continued*)

T, K	p = 225 bar				p = 250 bar			
	$v \cdot 10^3$	h	s	c_p	$v \cdot 10^3$	h	s	c_p
90	0,6959	88,65	1,348	1,029	0,6932	89,84	1,342	1,024
95	0,7086	93,82	1,404	1,040	0,7056	94,98	1,398	1,034
100	0,7219	99,04	1,457	1,051	0,7185	100,17	1,450	1,044
105	0,7359	104,30	1,508	1,054	0,7320	105,40	1,501	1,047
110	0,7506	109,58	1,558	1,058	0,7462	110,64	1,550	1,049
115	0,7662	114,89	1,604	1,059	0,7611	115,90	1,597	1,049
120	0,7827	120,18	1,650	1,062	0,7770	121,14	1,642	1,050
125	0,8003	125,51	1,693	1,070	0,7937	126,40	1,684	1,055
130	0,8192	130,88	1,735	1,079	0,8116	131,69	1,726	1,062
135	0,8394	136,32	1,777	1,087	0,8306	137,03	1,766	1,067
140	0,8612	141,76	1,816	1,094	0,8509	142,38	1,805	1,074
145	0,8847	147,26	1,854	1,109	0,8727	147,76	1,843	1,085
150	0,9102	152,85	1,893	1,131	0,8962	153,24	1,880	1,098
155	0,9380	158,58	1,930	1,151	0,9216	158,79	1,916	1,131
160	0,9679	164,38	1,967	1,173	0,9486	164,40	1,952	1,140
170	1,0365	176,29	2,039	1,214	1,0094	176,13	2,022	1,161
180	1,1190	188,61	2,110	1,249	1,0803	187,62	2,089	1,181
190	1,2170	201,20	2,178	1,266	1,1631	199,59	2,154	1,196
200	1,3296	213,87	2 243	1,259	1,2574	211,65	2,215	1,204
210	1,4552	226,33	2,304	1,224	1,3619	223,59	2,274	1,180
220	1,5894	238,33	2,359	1,168	1,4749	235,22	2,328	1,140
230	1,7287	249,66	2,410	1,102	1,5924	246,36	2,378	1,089
240	1,8680	260,35	2,455	1,034	1,7132	256,98	2,423	1,035
250	2,0063	270,37	2,496	0,972	1,8343	267,08	2,464	0,982

TABLE XI. Thermodynamic Properties of Liquid and Gaseous Argon (*Continued*)

T, K	p = 225 bar				p = 250 bar			
	$v \cdot 10^3$	h	s	c_p	$v \cdot 10^3$	h	s	c_p
260	2.1430	279.81	2.533	0.918	1.9551	276.63	2.501	0.933
270	2.2771	288.76	2.567	0.873	2.0741	285.75	2.536	0.889
280	2.4087	297.30	2.598	0.835	2.1918	294.46	2.567	0.852
290	2.5376	305.51	2.627	0.801	2.3076	302.80	2.596	0.817
300	2.6637	313.36	2.653	0.773	2.4209	310.83	2.624	0.786
310	2.7877	320.96	2.678	0.749	2.5329	318.53	2.650	0.759
320	2.9097	328.36	2.701	0.728	2.6439	326.03	2.674	0.739
330	3.0307	335.56	2.724	0.710	2.7529	333.33	2.696	0.724
340	3.1497	342.56	2.744	0.693	2.8599	340.53	2.717	0.704
350	3.2667	349.46	2.764	0.678	2.9659	347.53	2.738	0.689
375	3.5537	366.06	2.810	0.654	3.2259	364.33	2.785	0.664
400	3.8327	382.16	2.852	0.636	3.4799	380.73	2.827	0.646
425	4.1067	397.96	2.890	0.623	3.7289	396.73	2.866	0.632
450	4.3757	413.36	2.925	0.613	3.9719	412.33	2.902	0.620
475	4.6387	428.56	2.958	0.602	4.2109	427.73	2.936	0.607
500	4.8967	443.56	2.989	0.593	4.4449	442.83	2.968	0.597
550	5.4047	472.76	3.045	0.579	4.9049	472.23	3.024	0.583
600	5.9027	501.46	3.096	0.566	5.3559	501.13	3.073	0.572
650	6.3947	529.56	3.141	0.556	5.7999	529.53	3.118	0.561
700	6.8827	557.06	3.182	0.548	6.2409	557.33	3.159	0.552
750	7.3667	584.36	3.219	0.542	6.6779	584.73	3.197	0.545
800	7.8477	611.46	3.254	0.540	7.1119	611.93	3.232	0.541
850	8.3267	638.36	3.286	0.537	7.5439	638.93	3.265	0.539
900	8.8037	665.26	3.317	0.535	7.9739	665.83	3.295	0.536
950	9.2697	691.96	3.346	0.534	8.4029	692.63	3.324	0.534
1000	9.7497	718.66	3.373	0.532	8.8309	719.33	3.352	0.533
1050	10.229	745.26	3.399	0.531	9.2569	745.93	3.378	0.531
1100	10.699	771.76	3.424	0.529	9.6829	772.53	3.403	0.530
1150	11.17	798.15	3.448	0.529	10.11	798.95	3.425	0.530
1200	11.64	824.55	3.470	0.529	10.53	825.45	3.448	0.529
1250	12.11	850.95	3.492	0.528	10.96	851.85	3.469	0.528
1300	12.58	877.35	3.512	0.527	11.38	878.25	3.490	0.528

TABLE XI. Thermodynamic Properties of Liquid and Gaseous Argon (*Continued*)

T, K	p = 275 bar				p = 300 bar			
	$v \cdot 10^3$	h	s	c_p	$v \cdot 10^3$	h	s	c_p
95	0.7026	96.15	1.391	1.029	0.6998	97.33	1.385	1.025
100	0.7152	101.31	1.444	1.039	0.7120	102.46	1.438	1.034
105	0.7283	106.51	1.494	1.040	0.7248	107.63	1.488	1.034
110	0.7420	111.72	1.543	1.041	0.7380	112.81	1.536	1.035

TABLE XI. Thermodynamic Properties of Liquid and Gaseous Argon (*Continued*)

T, K	p = 275 bar				p = 300 bar			
	$v \cdot 10^3$	h	s	c_p	$v \cdot 10^3$	h	s	c_p
115	0.7564	116.93	1.589	1.041	0.7519	117.98	1,582	1,032
120	0,7716	122.13	1.634	1.038	0.7665	123.14	1,626	1.028
125	0,7876	127,33	1.676	1.038	0.7819	128.28	1.668	1.030
130	0,8046	132.51	1.717	1.045	0.7981	133.41	1.709	1.034
135	0,8226	137.77	1.757	1,054	0,8152	138.64	1.748	1.037
140	0.8417	143.08	1.795	1.058	0.8333	143.81	1.785	1.037
145	0,8621	148.37	1.832	1.064	0.8524	149.01	1.822	1,042
150	0,8838	153.68	1.868	1.073	0,8728	154.23	1.857	1.053
155	0.9074	159.13	1.904	1,088	0.8948	159.57	1,892	1,068
160	0.9320	164.56	1.939	1.101	0.9175	164.89	1.927	1.078
170	0.9867	175.72	2.006	1.125	0.9674	175.81	1.992	1.093
180	1.0495	187.04	2.071	1.144	1,0237	186.77	2.055	1.107
190	1.1213	198.56	2.133	1.155	1.0880	197.89	2.115	1.113
200	1.2023	210.12	2.192	1.155	1.1590	209.05	2.173	1.113
210	1.2916	221.64	2.248	1.140	1.2371	220.13	2.227	1.104
220	1.3886	232.90	2.301	1.110	1.3208	231.08	2.278	1.085
230	1.4900	243.83	2.350	1.071	1.4098	241.85	2.325	1.056
240	1.5951	254.30	2.394	1.027	1.5021	252.19	2.370	1.018
250	1.7019	264.39	2.435	0.982	1.5969	262.19	2.410	0.977
260	1.8088	273.95	2.473	0.939	1.6921	271.75	2.448	0.939
270	1.9151	283.16	2.508	0.898	1.7879	280.97	2.482	0.902
280	2.0208	291.97	2.539	0.862	1.8831	289.81	2.514	0.868
290	2.1250	300.41	2.569	0.829	1.9779	298.35	2.544	0.838
300	2.2279	308.56	2.597	0.800	2.0712	306.58	2.572	0.810
310	2.3299	316.46	2.623	0.773	2.1642	314.58	2.598	0.783
320	2.4299	324.06	2.647	0.751	2.2562	322.28	2.623	0.761
330	2.5289	331.46	2.670	0.734	2.3472	329.78	2.646	0.744
340	2.6269	338.76	2.692	0.716	2.4362	337.18	2.668	0.726
350	2.7239	345.86	2.713	0.700	2.5252	344.38	2.689	0.708
375	2.9619	362.96	2.760	0.673	2.7442	361.68	2.736	0.680
400	3.1939	379.56	2.803	0.655	2.9572	378.38	2.780	0.661
425	3.4209	395.76	2.842	0.639	3.1672	394.78	2.820	0.646
450	3.6439	411.56	2.879	0.626	3.3732	410.78	2.857	0.635
475	3.8629	427.06	2.913	0.614	3.5742	426.48	2.892	0.621
500	4.0769	442.36	2.944	0.605	3.7722	441.98	2.924	0.611
550	4.4979	472.16	3.002	0.590	4.1592	471.98	2.981	0.594
600	4.9099	501.36	3.051	0.575	4.5392	501.38	3.031	0.579
650	5.3159	529.86	3.096	0.563	4.9122	530.08	3.076	0.566
700	5.7169	557.66	3.137	0.552	5.2812	557.98	3.118	0.554
750	6.1149	585.16	3.175	0.546	5.6472	585.58	3.156	0.548
800	6.5109	612.36	3.211	0.543	6.0112	612.88	3.191	0.545
850	6.9039	639.46	3.244	0.540	6.3722	640.08	3.224	0.542
900	7.2959	666.46	3.275	0.538	6.7322	667.18	3.255	0.539
950	7.6869	693.26	3.304	0.536	7.0902	694.08	3.284	0.537

TABLE XI. Thermodynamic Properties of Liquid and Gaseous Argon (*Continued*)

T, K	p = 275 bar				p = 300 bar			
	$v \cdot 10^3$	h	s	c_p	$v \cdot 10^3$	h	s	c_p
1000	8.0759	720.06	3.331	0.534	7.4482	720.88	3.312	0.535
1050	8.4639	746.76	3.357	0.533	7.8042	747.58	3.338	0.533
1100	8.8519	773.36	3.382	0.531	8.1602	774.28	3.363	0.532
1150	9.239	799.85	3.405	0.531	8.515	800.75	3.386	0.532
1200	9.625	826.35	3.428	0.530	8.869	827.25	3.409	0.531
1250	10.01	852.85	3.449	0.529	9.223	853.75	3.431	0.530
1300	10.40	879.25	3.470	0.529	9.576	880.25	3.451	0.529

TABLE XI. Thermodynamic Properties of Liquid and Gaseous Argon (*Continued*)

T, K	p = 325 bar				p = 350 bar			
	$v \cdot 10^3$	h	s	c_p	$v \cdot 10^3$	h	s	c_p
95	0.6971	98.52	1.379	1.021	0.6945	99.71	1.374	1.017
100	0.7090	103.62	1.431	1.030	0.7061	104.79	1.425	1.025
105	0.7214	108.77	1.482	1.029	0.7181	109.91	1.475	1.024
110	0.7343	113.92	1.530	1.028	0.7307	115.04	1.523	1.023
115	0.7477	119.05	1.575	1.024	0.7437	120.13	1.568	1.017
120	0.7618	124.17	1.619	1.021	0.7573	125.22	1.612	1.012
125	0.7765	129.27	1.660	1.020	0.7715	130.27	1.653	1.009
130	0.7920	134.38	1.701	1.021	0.7864	135.31	1.693	1.013
135	0.8083	139.48	1.739	1.023	0.8020	140.38	1.731	1.016
140	0.8255	144.62	1.776	1.022	0.8184	145.50	1.767	1.012
145	0.8437	149.73	1.812	1.027	0.8356	150.53	1.803	1.011
150	0.8628	154.86	1.847	1.037	0.8537	155.58	1.837	1.020
155	0.8835	160.11	1.882	1.050	0.8732	160.74	1.872	1.032
160	0.9046	165.38	1.915	1.058	0.8931	165.92	1.904	1.039
170	0.9506	176.03	1.980	1.069	0.9358	176.37	1.968	1.049
180	1.0024	186.75	2.041	1.078	0.9833	186.88	2.028	1.055
190	1.0600	197.57	2.099	1.084	1.0356	197.47	2.085	1.057
200	1.1232	208.41	2.155	1.084	1.0932	208.02	2.139	1.056
210	1.1923	219.24	2.207	1.074	1.1553	218.56	2.191	1.049
220	1.2666	229.90	2.257	1.057	1.2223	229.00	2.239	1.037
230	1.3456	240.35	2.304	1.032	1.2927	239.30	2.285	1.017
240	1.4277	250.57	2.347	1.004	1.3666	249.33	2.328	0.990
250	1.5120	260.42	2.388	0.971	1.4427	259.08	2.367	0.959
260	1.5979	270.00	2.425	0.937	1.5200	268.53	2.404	0.931
270	1.6841	279.17	2.460	0.903	1.5983	277.68	2.439	0.901
280	1.7709	288.07	2.492	0.872	1.6772	286.59	2.471	0.874
290	1.8570	296.63	2.522	0.843	1.7562	295.15	2.501	0.846
300	1.9423	304.95	2.550	0.814	1.8350	303.54	2.530	0.816
310	2.0273	312.95	2.576	0.790	1.9130	311.54	2.557	0.792

TABLE XI. Thermodynamic Properties of Liquid and Gaseous Argon (*Continued*)

T. K	p = 325 bar				p = 350 bar			
	$v \cdot 10^3$	h	s	c_p	$v \cdot 10^3$	h	s	c_p
320	2.1123	320.75	2.601	0.768	1.9910	319.34	2.582	0.773
330	2.1953	328.35	2.624	0.749	2.0680	327.04	2.605	0.757
340	2.2783	335.75	2.646	0.728	2.1440	334.54	2.627	0.740
350	2.3603	342.95	2.667	0.711	2.2200	341.84	2.649	0.724
375	2.5623	360.35	2.715	0.686	2.4070	359.54	2.697	0.692
400	2.7593	377.25	2.759	0.669	2.5910	376.54	2.741	0.667
425	2.9533	393.85	2.799	0.655	2.7720	392.94	2.781	0.643
450	3.1443	410.05	2.836	0.641	2.9490	408.94	2.818	0.643
475	3.3313	425.95	2.870	0.628	3.1230	424.64	2.853	0.637
500	3.5153	441.45	2.902	0.617	3.2950	441.14	2.885	0.627
550	3.8743	471.85	2.959	0.599	3.6300	471.64	2.943	0.604
600	4.2263	501.45	3.010	0.583	3.9580	501.44	2.994	0.585
650	4.5723	530.25	3.057	0.568	4.2800	530.34	3.041	0.572
700	4.9133	558.35	3.100	0.557	4.5980	558.74	3.084	0.561
750	5.2523	585.95	3.139	0.550	4.9130	586.54	3.123	0.553
800	5.5883	613.45	3.174	0.546	5.2260	614.04	3.159	0.547
850	5.9223	640.65	3.207	0.543	5.5370	641.34	3.192	0.544
900	6.2553	667.75	3.238	0.540	5.8460	668.44	3.223	0.541
950	6.5863	694.75	3.267	0.538	6.1540	695.54	3.252	0.539
1000	6.9163	721.65	3.295	0.536	6.4620	722.44	3.280	0.537
1050	7.2463	748.35	3.321	0.534	6.7680	749.24	3.306	0.535
1100	7.5753	775.05	3.346	0.533	7.0730	775.94	3.331	0.533
1150	7.903	801.65	3.369	0.532	7.378	802.55	3.354	0.533
1200	8.230	828.25	3.392	0.531	7.683	829.15	3.376	0.532
1250	8.557	854.75	3.414	0.530	7.987	855.75	3.398	0.531
1300	8.884	881.25	3.431	0.530	8.290	882.25	3.419	0.530

TABLE XI. Thermodynamic Properties of Liquid and Gaseous Argon (*Continued*)

T. K	p = 375 bar				p = 400 bar			
	$v \cdot 10^3$	h	s	c_p	$v \cdot 10^3$	h	s	c_p
95	0.6920	100.90	1.368	1.014	0.6895	102.10	1.363	1.011
100	0.7033	105.96	1.420	1.022	0.7006	107.15	1.414	1.018
105	0.7150	111.07	1.469	1.019	0.7121	112.23	1.463	1.015
110	0.7272	116.17	1.517	1.017	0.7239	117.31	1.511	1.012
115	0.7399	121.23	1.561	1.011	0.7362	122.34	1.555	1.005
120	0.7530	126.28	1.605	1.004	0.7490	127.36	1.598	0.998
125	0.7667	131.29	1.646	1.004	0.7622	132.33	1.639	0.995
130	0.7811	136.30	1.685	1.004	0.7760	137.31	1.678	0.994
135	0.7960	141.36	1.723	1.003	0.7904	142.29	1.716	0.996

TABLE XI. Thermodynamic Properties of Liquid and Gaseous Argon (*Continued*)

T, *K*	*p* = 375 bar				*p* = 400 bar			
	$v \cdot 10^3$	h	s	c_p	$v \cdot 10^3$	h	s	c_p
140	0.8117	146.35	1.759	1.001	0.8055	147.26	1.751	0.994
145	0.8281	151.35	1.794	0.998	0.8212	152.24	1.786	0.990
150	0.8453	156.37	1.829	1.010	0.8376	157.19	1.820	0.999
155	0.8639	161.40	1.862	1.020	0.8554	162.18	1.853	1.010
160	0.8825	166.59	1.894	1.025	0.8730	167.30	1.885	1.011
170	0.9227	176.86	1.957	1.030	0.9107	177.45	1.947	1.012
180	0.9667	187.17	2.016	1.032	0.9519	187.52	2.004	1.014
190	1.0147	197.50	2.072	1.034	0.9967	197.72	2.059	1.017
200	1.0674	207.85	2.125	1.030	1.0453	207.91	2.112	1.017
210	1.1242	218.12	2.175	1.023	1.0973	218.03	2.161	1.009
220	1.1845	228.30	2.223	1.015	1.1525	228.07	2.208	0.998
230	1.2485	238.40	2.268	1.002	1.2112	237.99	2.252	0.983
240	1.3155	248.36	2.310	0.980	1.2718	247.72	2.293	0.963
250	1.3846	258.00	2.349	0.954	1.3349	257.26	2.333	0.941
260	1.4550	267.41	2.386	0.926	1.3998	266.53	2.369	0.917
270	1.5263	276.54	2.420	0.898	1.4653	275.63	2.403	0.893
280	1.5990	285.39	2.452	0.871	1.5321	284.41	2.435	0.870
290	1.6713	293.95	2.482	0.848	1.5991	293.02	2.465	0.846
300	1.7440	302.35	2.511	0.822	1.6662	301.36	2.493	0.823
310	1.8160	310.45	2.538	0.797	1.7332	309.46	2.520	0.798
320	1.8880	318.25	2.563	0.774	1.8002	317.36	2.545	0.783
330	1.9590	325.95	2.587	0.759	1.8662	325.06	2.568	0.765
340	2.0300	333.45	2.609	0.743	1.9322	332.76	2.591	0.747
350	2.1010	340.85	2.630	0.728	1.9982	340.06	2.612	0.729
375	2.2750	358.65	2.679	0.698	2.1612	357.96	2.662	0.700
400	2.4470	375.85	2.724	0.677	2.3222	375.16	2.707	0.681
425	2.6160	392.55	2.765	0.661	2.4802	392.06	2.748	0.666
450	2.7820	408.95	2.802	0.650	2.6362	408.56	2.786	0.657
475	2.9450	425.05	2.837	0.637	2.7892	424.86	2.821	0.642
500	3.1050	440.95	2.870	0.624	2.9402	440.86	2.854	0.629
550	3.4200	471.55	2.928	0.604	3.2362	471.66	2.913	0.609
600	3.7270	501.35	2.981	0.587	3.5252	501.76	2.966	0.594
650	4.0280	530.45	3.028	0.575	3.8082	531.16	3.012	0.578
700	4.3260	558.95	3.070	0.565	4.0882	559.76	3.054	0.564
750	4.6210	587.05	3.108	0.557	4.3652	587.56	3.093	0.554
800	4.9130	614.75	3.144	0.550	4.6402	615.16	3.129	0.550
850	5.2040	642.15	3.177	0.544	4.9122	642.66	3.162	0.548
900	5.4930	669.25	3.208	0.541	5.1832	670.06	3.193	0.545
950	5.7810	696.25	3.237	0.539	5.4542	697.26	3.222	0.542
1000	6.0680	723.25	3.265	0.538	5.7232	724.26	3.250	0.538
1050	6.3540	750.15	3.291	0.536	5.9922	751.16	3.276	0.535
1100	6.6390	776.85	3.316	0.534	6.2602	777.86	3.301	0.534
1150	6.924	803.45	3.339	0.534	6.527	804.45	3.325	0.534
1200	7.208	830.15	3.362	0.532	6.794	831.15	3.348	0.533

TABLE XI. Thermodynamic Properties of Liquid and Gaseous Argon (*Continued*)

T, K	p = 375 bar				p = 400 bar			
	$v \cdot 10^3$	h	s	c_p	$v \cdot 10^3$	h	s	c_p
1250	7.492	856.75	3.383	0.531	7.060	857.75	3.370	0.532
1300	7.776	883.25	3.404	0.531	7.326	884.35	3.390	0.531

TABLE XI. Thermodynamic Properties of Liquid and Gaseous Argon (*Continued*)

T, K	p = 425 bar				p = 450 bar			
	$v \cdot 10^3$	h	s	c_p	$v \cdot 10^3$	h	s	c_p
95	0.6872	103.30	1.357	1.008	0.6849	104.50	1.352	1.005
100	0.6980	108.34	1.408	1.015	0.6955	109.54	1.403	1.012
105	0.7092	113.40	1.458	1.011	0.7064	114.58	1.452	1.007
110	0.7207	118.46	1.505	1.007	0.7177	119.62	1.499	1.003
115	0.7327	123.47	1.549	0.999	0.7293	124.61	1.543	0.994
120	0.7451	128.46	1.592	0.993	0.7414	129.57	1.585	0.987
125	0.7579	133.39	1.632	0.990	0.7538	134.48	1.626	0.982
130	0.7713	138.37	1.671	0.989	0.7668	139.40	1.664	0.983
135	0.7852	143.29	1.709	0.987	0.7802	144.30	1.702	0.982
140	0.7996	148.24	1.744	0.983	0.7941	149.23	1.737	0.975
145	0.8147	153.15	1.778	0.984	0.8086	154.08	1.771	0.972
150	0.8304	158.07	1.812	0.991	0.8237	158.93	1.804	0.982
155	0.8474	163.04	1.845	0.998	0.8400	163.88	1.837	0.991
160	0.8642	168.08	1.877	0.998	0.8560	168.89	1.868	0.990
170	0.8999	178.07	1.937	0.998	0.8899	178.77	1.928	0.988
180	0.9385	188.02	1.994	0.998	0.9265	188.61	1.984	0.985
190	0.9804	198.04	2.048	1.000	0.9658	198.51	2.037	0.987
200	1.0255	208.05	2.099	1.001	1.0078	208.35	2.088	0.985
210	1.0735	218.04	2.148	0.994	1.0526	218.21	2.136	0.979
220	1.1246	227.94	2.194	0.983	1.1003	227.94	2.181	0.968
230	1.1785	237.68	2.238	0.967	1.1498	237.56	2.224	0.954
240	1.2346	247.30	2.279	0.950	1.2018	247.02	2.265	0.939
250	1.2927	256.68	2.317	0.930	1.2557	256.34	2.303	0.923
260	1.3520	265.90	2.353	0.913	1.3109	265.48	2.339	0.903
270	1.4130	274.92	2.387	0.891	1.3671	274.42	2.372	0.881
280	1.4743	283.78	2.419	0.866	1.4243	283.10	2.404	0.861
290	1.5364	292.23	2.449	0.841	1.4822	291.64	2.434	0.842
300	1.5991	300.58	2.477	0.818	1.5404	299.98	2.462	0.821
310	1.6611	308.68	2.503	0.802	1.5984	308.08	2.488	0.801
320	1.7231	316.58	2.528	0.784	1.6574	315.98	2.513	0.784
330	1.7851	324.38	2.552	0.768	1.7154	323.78	2.537	0.767
340	1.8471	331.98	2.575	0.746	1.7734	331.38	2.560	0.749
350	1.9091	339.38	2.596	0.730	1.8314	338.78	2.582	0.734
375	2.0621	357.18	2.646	0.703	1.9754	356.78	2.631	0.709

TABLE XI. Thermodynamic Properties of Liquid and Gaseous Argon (*Continued*)

T, K	p = 425 bar				p = 450 bar			
	$v \cdot 10^3$	h	s	c_p	$v \cdot 10^3$	h	s	c_p
400	2.2131	374.58	2.691	0.687	2.1174	374.28	2.676	0.692
425	2.3611	391.58	2.732	0.674	2.2574	391.38	2.718	0.677
450	2.5071	408.28	2.771	0.661	2.3954	408.18	2.757	0.663
475	2.6521	424.68	2.806	0.647	2.5314	424.58	2.792	0.648
500	2.7951	440.68	2.839	0.635	2.6664	440.68	2.825	0.636
550	3.0741	471.88	2.899	0.615	2.9314	471.88	2.885	0.617
600	3.3471	502.28	2.952	0.597	3.1894	502.38	2.938	0.599
650	3.6151	531.68	2.999	0.579	3.4434	531.98	2.986	0.583
700	3.8791	560.28	3.041	0.565	3.6934	560.68	3.028	0.568
750	4.1401	588.28	3.080	0.556	3.9404	588.88	3.067	0.559
800	4.3991	615.98	3.116	0.551	4.1854	616.68	3.103	0.553
850	4.6561	643.48	3.149	0.548	4.4284	644.28	3.136	0.548
900	4.9111	670.78	3.180	0.544	4.6704	671.58	3.167	0.545
950	5.1661	697.98	3.209	0.542	4.9104	698.78	3.197	0.542
1000	5.4201	724.98	3.237	0.539	5.1504	725.88	3.225	0.541
1050	5.6731	751.98	3.263	0.537	5.3894	752.88	3.251	0.538
1100	5.9251	778.78	3.288	0.536	5.6284	779.78	3.276	0.536
1150	6.177	805.45	3.312	0.535	5.866	806.45	3.300	0.535
1200	6.428	832.15	3.335	0.533	6.103	833.15	3.323	0.534
1250	6.679	858.75	3.357	0.532	6.340	859.75	3.344	0.533
1300	6.929	885.35	3.377	0.531	6.577	886.45	3.365	0.532

TABLE XI. Thermodynamic Properties of Liquid and Gaseous Argon (*Continued*)

T, K	p = 475 bar				p = 500 bar			
	$v \cdot 10^3$	h	s	c_p	$v \cdot 10^3$	h	s	c_p
100	0.6931	110.75	1.398	1.010	0.6907	111.97	1.393	1.007
105	0.7038	115.77	1.446	1.004	0.7012	116.97	1.441	1.001
110	0.7148	120.78	1.493	0.999	0.7119	121.94	1.488	0.995
115	0.7261	125.77	1.537	0.991	0.7230	126.94	1.532	0.988
120	0.7378	130.69	1.579	0.982	0.7344	131.84	1.573	0.979
125	0.7499	135.58	1.619	0.978	0.7462	136.72	1.614	0.970
130	0.7625	140.48	1.658	0.978	0.7584	141.57	1.651	0.696
135	0.7755	145.37	1.695	0.975	0.7710	146.40	1.688	0.968
140	0.7889	150.24	1.730	0.969	0.7840	151.25	1.723	0.963
145	0.8029	155.06	1.763	0.966	0.7975	156.06	1.756	0.960
150	0.8174	159.89	1.797	0.973	0.8114	160.83	1.790	0.964
155	0.8331	164.77	1.829	0.979	0.8267	165.69	1.821	0.972
160	0.8484	169.71	1.860	0.978	0.8413	170.56	1.852	0.971
170	0.8807	179.50	1.920	0.977	0.8722	180.30	1.911	0.969

TABLE XI. Thermodynamic Properties of Liquid and Gaseous Argon (*Continued*)

T, K	p = 475 bar				p = 500 bar			
	$v \cdot 10^3$	h	s	c_p	$v \cdot 10^3$	h	s	c_p
180	0.9155	189.24	1.975	0.975	0.9052	189.94	1.966	0.966
190	0.9525	199.02	2.028	0.975	0.9404	199.62	2.018	0.964
200	0.9921	208.75	2.078	0.972	0.9778	209.24	2.068	0.961
210	1.0343	218.44	2.125	0.965	1.0175	218.83	2.115	0.953
220	1.0785	228.06	2.170	0.956	1.0587	228.33	2.159	0.944
230	1.1247	237.57	2.212	0.943	1.1025	237.69	2.201	0.933
240	1.1734	246.92	2.252	0.929	1.1477	246.99	2.240	0.920
250	1.2230	256.14	2.290	0.912	1.1946	256.12	2.277	0.904
260	1.2748	265.19	2.325	0.896	1.2427	265.07	2.313	0.886
270	1.3271	274.04	2.359	0.877	1.2919	273.84	2.346	0.871
280	1.3804	282.76	2.390	0.859	1.3419	282.48	2.377	0.855
290	1.4344	291.22	2.420	0.840	1.3923	290.98	2.407	0.838
300	1.4892	299.55	2.448	0.820	1.4434	299.25	2.435	0.819
310	1.5432	307.65	2.474	0.799	1.4944	307.35	2.462	0.798
320	1.5982	315.55	2.499	0.779	1.5464	315.25	2.487	0.780
330	1.6532	323.25	2.522	0.766	1.5974	322.95	2.511	0.766
340	1.7072	330.85	2.546	0.751	1.6494	330.55	2.534	0.751
350	1.7612	338.35	2.568	0.738	1.7004	338.05	2.555	0.738
375	1.8972	356.45	2.618	0.714	1.8284	356.15	2.605	0.713
400	2.0312	374.05	2.663	0.694	1.9554	373.75	2.651	0.696
425	2.1632	391.25	2.705	0.677	2.0804	390.95	2.693	0.681
450	2.2942	407.95	2.744	0.662	2.2044	407.85	2.731	0.667
475	2.4232	424.35	2.780	0.649	2.3264	424.35	2.767	0.653
500	2.5512	440.45	2.813	0.640	2.4484	440.55	2.801	0.642
550	2.8032	471.95	2.873	0.623	2.6884	472.15	2.861	0.624
600	3.0482	502.75	2.926	0.604	2.9224	502.95	2.914	0.605
650	3.2892	532.55	2.973	0.585	3.1514	532.85	2.962	0.588
700	3.5272	561.25	3.016	0.569	3.3774	561.75	3.005	0.572
750	3.7612	589.45	3.055	0.559	3.6004	590.15	3.045	0.562
800	3.9932	617.35	3.091	0.554	3.8224	618.05	3.081	0.556
850	4.2242	644.95	3.124	0.550	4.0414	645.75	3.114	0.551
900	4.4542	672.35	3.155	0.546	4.2594	673.25	3.145	0.548
950	4.6822	699.65	3.185	0.543	4.4764	700.55	3.175	0.544
1000	4.9092	726.75	3.213	0.541	4.6924	727.75	3.203	0.541
1050	5.1362	753.75	3.239	0.539	4.9084	754.75	3.229	0.539
1100	5.3622	780.65	3.265	0.537	5.1234	781.65	3.254	0.537
1150	5.587	807.45	3.288	0.536	5.337	808.45	3.277	0.536
1200	5.812	834.15	3.311	0.534	5.551	835.25	3.300	0.535
1250	6.037	860.85	3.333	0.533	5.765	861.95	3.322	0.534
1300	6.261	887.45	3.354	0.532	5.978	888.55	3.343	0.533

TABLE XI. Thermodynamic Properties of Liquid and Gaseous Argon (*Continued*)

T, K	p = 550 bar				p = 600 bar			
	$v \cdot 10^3$	h	s	c_p	$v \cdot 10^3$	h	s	c_p
100	0.6862	114.37	1.382	1.001	0.6819	116.79	1.372	0.996
105	0.6962	119.34	1.430	0.995	0.6916	121.73	1.420	0.990
110	0.7065	124.29	1.477	0.989	0.7015	126.65	1.466	0.983
115	0.7171	129.25	1.520	0.982	0.7116	131.59	1.510	0.976
120	0.7280	134.12	1.562	0.972	0.7220	136.43	1.551	0.966
125	0.7392	138.96	1.602	0.963	0.7327	141.24	1.590	0.956
130	0.7507	143.77	1.639	0.961	0.7436	146.02	1.628	0.954
135	0.7626	148.55	1.676	0.959	0.7549	150.77	1.664	0.951
140	0.7749	153.36	1.710	0.953	0.7665	155.53	1.698	0.945
145	0.7875	158.12	1.743	0.949	0.7784	160.25	1.731	0.940
150	0.8005	162.83	1.776	0.952	0.7907	164.91	1.763	0.942
155	0.8148	167.63	1.807	0.959	0.8041	169.66	1.794	0.948
160	0.8284	172.44	1.838	0.958	0.8168	174.41	1.825	0.947
170	0.8568	182.03	1.896	0.952	0.8433	183.89	1.882	0.940
180	0.8870	191.51	1.950	0.948	0.8711	193.24	1.935	0.934
190	0.9189	200.97	2.001	0.945	0.9004	202.56	1.986	0.929
200	0.9527	210.44	2.050	0.943	0.9312	211.85	2.033	0.925
210	0.9881	219.82	2.095	0.934	0.9634	221.05	2.078	0.919
220	1.0255	229.13	2.139	0.923	0.9971	230.23	2.121	0.909
230	1.0638	238.29	2.180	0.914	1.0318	239.25	2.161	0.898
240	1.1039	247.39	2.218	0.903	1.0679	248.18	2.199	0.887
250	1.1457	256.39	2.255	0.890	1.1051	257.01	2.235	0.876
260	1.1880	265.18	2.290	0.875	1.1432	265.73	2.269	0.863
270	1.2318	273.89	2.322	0.860	1.1823	274.29	2.302	0.850
280	1.2760	282.43	2.353	0.846	1.2222	282.73	2.332	0.837
290	1.3206	290.79	2.383	0.828	1.2625	291.05	2.361	0.824
300	1.3664	299.02	2.411	0.812	1.3035	299.21	2.389	0.807
310	1.4114	307.02	2.438	0.795	1.3445	307.21	2.415	0.789
320	1.4574	314.92	2.463	0.778	1.3855	315.01	2.440	0.775
330	1.5044	322.62	2.487	0.766	1.4275	322.71	2.464	0.764
340	1.5504	330.22	2.509	0.753	1.4695	330.31	2.486	0.754
350	1.5964	337.72	2.531	0.740	1.5115	337.81	2.508	0.743
375	1.7114	355.92	2.581	0.715	1.6155	356.11	2.558	0.718
400	1.8264	373.52	2.627	0.698	1.7185	373.81	2.603	0.699
425	1.9394	390.82	2.669	0.684	1.8205	391.11	2.646	0.685
450	2.0494	407.82	2.708	0.671	1.9215	408.11	2.685	0.676
475	2.1604	424.42	2.744	0.657	2.0225	424.91	2.721	0.663
500	2.2704	440.72	2.777	0.646	2.1235	441.41	2.755	0.652
550	2.4894	472.52	2.838	0.628	2.3245	473.41	2.816	0.632
600	2.7034	503.52	2.891	0.610	2.5215	504.61	2.870	0.613
650	2.9134	533.72	2.939	0.593	2.7155	534.91	2.919	0.595
700	3.1204	562.92	2.983	0.578	2.9055	564.21	2.963	0.580
750	3.3244	591.52	3.023	0.566	3.0935	592.91	3.003	0.568

TABLE XI. Thermodynamic Properties of Liquid and Gaseous Argon (*Continued*)

T, K	p = 550 bar				p = 600 bar			
	$v \cdot 10^3$	h	s	c_p	$v \cdot 10^3$	h	s	c_p
800	3.5254	619.62	3.060	0.558	3.2795	621.21	3.039	0.561
850	3.7254	647.42	3.094	0.553	3.4635	649.11	3.073	0.555
900	3.9244	675.02	3.125	0.549	3.6455	676.81	3.105	0.550
950	4.1224	702.42	3.155	0.546	3.8265	704.21	3.135	0.547
1000	4.3194	729.62	3.183	0.543	4.0075	731.51	3.163	0.544
1050	4.5154	756.72	3.209	0.540	4.1885	758.71	3.189	0.542
1100	4.7104	783.72	3.234	0.538	4.3675	785.81	3.214	0.540
1150	4.905	810.55	3.257	0.537	4.546	812.65	3.238	0.538
1200	5.100	837.35	3.280	0.536	4.725	839.55	3.261	0.536
1250	5.294	864.15	3.302	0.535	4.903	866.35	3.283	0.535
1300	5.488	890.85	3.323	0.534	5.081	893.05	3.304	0.534

TABLE XI. Thermodynamic Properties of Liquid and Gaseous Argon (*Continued*)

T, K	p = 650 bar				p = 700 bar			
	$v \cdot 10^3$	h	s	c_p	$v \cdot 10^3$	h	s	c_p
100	0.6779	119.22	1.362	0.992	—	—	—	—
105	0.6872	124.14	1.410	0.986	0.6831	126.56	1.400	0.982
110	0.6967	129.04	1.456	0.979	0.6922	131.44	1.447	0.975
115	0.7065	133.95	1.499	0.972	0.7016	136.34	1.489	0.968
120	0.7164	138.76	1.540	0.962	0.7112	141.13	1.530	0.957
125	0.7266	143.55	1.580	0.951	0.7210	145.89	1.569	0.946
130	0.7371	148.30	1.617	0.948	0.7310	150.62	1.606	0.943
135	0.7478	153.02	1.653	0.945	0.7412	155.31	1.642	0.939
140	0.7588	157.75	1.687	0.938	0.7517	160.02	1.676	0.932
145	0.7701	162.44	1.719	0.933	0.7625	164.67	1.708	0.926
150	0.7817	167.06	1.751	0.934	0.7735	169.26	1.740	0.927
155	0.7944	171.77	1.782	0.940	0.7856	173.93	1.771	0.932
160	0.8064	176.47	1.812	0.938	0.7969	178.60	1.800	0.931
170	0.8311	185.85	1.869	0.930	0.8202	187.89	1.856	0.920
180	0.8570	195.09	1.922	0.921	0.8445	197.04	1.909	0.910
190	0.8842	204.27	1.971	0.914	0.8698	206.09	1.958	0.903
200	0.9126	213.39	2.018	0.911	0.8962	215.11	2.004	0.900
210	0.9422	222.49	2.062	0.906	0.9236	224.10	2.048	0.896
220	0.9729	231.54	2.104	0.898	0.9519	233.05	2.089	0.887
230	1.0048	240.45	2.144	0.887	0.9813	241.85	2.128	0.877
240	1.0377	249.27	2.182	0.875	1.0117	250.56	2.166	0.864
250	1.0717	257.98	2.217	0.864	1.0422	259.17	2.201	0.854
260	1.1060	266.54	2.251	0.853	1.0741	267.63	2.234	0.843
270	1.1413	275.05	2.283	0.842	1.1064	276.04	2.266	0.832

TABLE XI. Thermodynamic Properties of Liquid and Gaseous Argon (*Continued*)

T. K	p = 650 bar				p = 700 bar			
	$v \cdot 10^3$	h	s	c_p	$v \cdot 10^3$	h	s	c_p
280	1.1774	283.39	2.313	0.830	1.1394	284.30	2.296	0.821
290	1.2143	291.63	2.342	0.814	1.1733	292.45	2.324	0.808
300	1.2514	299.71	2.369	0.799	1.2068	300.49	2.352	0.798
310	1.2884	307.61	2.395	0.785	1.2408	308.39	2.377	0.783
320	1.3264	315.41	2.420	0.774	1.2758	316.19	2.402	0.771
330	1.3644	323.11	2.444	0.765	1.3108	323.79	2.426	0.759
340	1.4024	330.71	2.466	0.752	1.3458	331.39	2.448	0.747
350	1.4404	338.21	2.488	0.740	1.3808	338.79	2.470	0.737
375	1.5344	356.41	2.538	0.717	1.4688	356.99	2.520	0.720
400	1.6274	374.11	2.584	0.701	1.5568	374.79	2.565	0.705
425	1.7194	391.51	2.626	0.689	1.6428	392.29	2.608	0.692
450	1.8124	408.61	2.665	0.680	1.7288	409.39	2.647	0.680
475	1.9054	425.51	2.702	0.667	1.8138	426.29	2.684	0.668
500	1.9984	442.11	2.737	0.656	1.8988	442.89	2.718	0.656
550	2.1844	474.31	2.798	0.635	2.0688	475.19	2.779	0.637
600	2.3684	505.61	2.852	0.615	2.2378	506.59	2.834	0.619
650	2.5484	536.01	2.901	0.598	2.4048	537.19	2.884	0.602
700	2.7244	565.51	2.945	0.583	2.5698	566.89	2.928	0.586
750	2.8984	594.41	2.985	0.572	2.7318	595.89	2.968	0.576
800	3.0704	622.81	3.022	0.564	2.8928	624.49	3.005	0.568
850	3.2414	650.91	3.056	0.557	3.0518	652.89	3.039	0.561
900	3.4104	678.61	3.087	0.553	3.2088	680.69	3.071	0.553
950	3.5784	706.21	3.117	0.549	3.3648	708.19	3.101	0.548
1000	3.7454	733.61	3.145	0.546	3.5198	735.59	3.129	0.546
1050	3.9114	760.81	3.172	0.543	3.6748	762.89	3.156	0.544
1100	4.0774	787.91	3.197	0.540	3.8288	790.09	3.181	0.542
1150	4.242	814.85	3.221	0.539	3.982	817.05	3.205	0.540
1200	4.407	841.75	3.244	0.538	4.136	844.05	3.228	0.538
1250	4.572	868.65	3.266	0.536	4.289	870.95	3.250	0.537
1300	4.736	895.35	3.287	0.535	4.441	897.75	3.271	0.536

TABLE XI. Thermodynamic Properties of Liquid and Gaseous Argon (*Continued*)

T. K	p = 750 bar				p = 800 bar			
	$v \cdot 10^3$	h	s	c_p	$v \cdot 10^3$	h	s	c_p
105	0.6791	128.99	1.391	0.979	0.6753	131.44	1.382	0.976
110	0.6880	133.86	1.437	0.972	0.6839	136.29	1.428	0.969
115	0.6970	138.74	1.480	0.964	0.6927	141.15	1.471	0.961
120	0.7062	143.51	1.521	0.954	0.7015	145.91	1.511	0.951
125	0.7156	148.26	1.560	0.942	0.7106	150.64	1.550	0.939
130	0.7252	152.96	1.596	0.939	0.7198	155.33	1.587	0.935

TABLE XI. Thermodynamic Properties of Liquid and Gaseous Argon (*Continued*)

T, K	p = 750 bar				p = 800 bar			
	$v \cdot 10^3$	h	s	c_p	$v \cdot 10^3$	h	s	c_p
135	0,7351	157,63	1,632	0.935	0.7293	159,98	1.622	0.931
140	0,7451	162,32	1,666	0,927	0,7389	164,65	1.656	0,923
145	0,7554	166,94	1,698	0,921	0,7488	169,25	1,688	0,916
150	0,7659	171,50	1,729	0,922	0,7588	173,78	1,719	0,917
155	0,7774	176,15	1,760	0,926	0,7699	178,40	1,749	0,921
160	0,7881	180,78	1,789	0,924	0,7801	183,01	1,779	0,918
170	0,8102	190,00	1,845	0,912	0,8010	192,17	1,834	0,906
180	0,8331	199,07	1,897	0,902	0,8227	201,17	1,885	0,895
190	0,8568	208.04	1,945	0,895	0.8451	210,07	1,933	0,886
200	0.8815	216,99	1,991	0,891	0.8683	218,92	1,979	0.882
210	0.9071	225.88	2,034	0.884	0.8923	227,71	2,022	0,876
220	0.9335	234,67	2,075	0,876	0.9170	236,47	2,062	0,869
230	0,9607	243.38	2,114	0,867	0,9424	245,08	2,101	0,859
240	0,9886	252,04	2,151	0,858	0.9684	253,65	2,137	0,848
250	1,0171	260,56	2,186	0,846	0.9951	262,07	2,172	0,838
260	1,0463	268,97	2,219	0,834	1,0222	270.40	2,204	0.827
270	1,0763	277,24	2,250	0,823	1,0502	278,63	2,236	0,818
280	1,1066	285,46	2,280	0,814	1,0783	286,77	2,265	0,808
290	1,1376	293,53	2,308	0.804	1,1067	294,80	2,293	0,798
300	1,1688	301,54	2,335	0.789	1,1358	302,74	2,320	0,787
310	1,2008	309,34	2,361	0,776	1,1648	310,54	2,346	0,773
320	1,2328	317,04	2,385	0,763	1,1948	318,24	2,370	0,760
330	1,2648	324,64	2,409	0,757	1,2248	325,74	2,393	0,751
340	1,2968	332,14	2,431	0,746	1,2548	333,24	2,416	0,743
350	1,3288	339,64	2,453	0,736	1,2848	340,64	2,437	0,736
375	1,4098	357,74	2,503	0,718	1,3598	358,84	2,487	0,718
400	1,4908	375,54	2,548	0,705	1,4348	376,64	2,533	0,704
425	1,5708	393,04	2,591	0,693	1,5098	394,04	2,575	0,692
450	1,6508	410,24	2,630	0,681	1,5838	411,24	2,615	0,681
475	1,7318	427,14	2,667	0,669	1,6568	428,14	2,651	0,669
500	1,8118	443,74	2,700	0,658	1,7298	444,74	2,685	0,658
550	1,9698	476,14	2,761	0,638	1,8768	477,14	2,746	0,639
600	2,1258	507,64	2,816	0,621	2,0228	508,74	2,801	0,623
650	2,2798	538,34	2,866	0,605	2,1688	539,54	2,851	0,608
700	2,4338	568,24	2,912	0,590	2,3168	569,64	2,897	0,594
750	2,5868	597,44	2,953	0,578	2,4608	599,04	2,938	0,582
800	2,7378	626,14	2,990	0,569	2,6028	627,84	2,975	0,572
850	2,8868	654,44	3,024	0,562	2,7428	656,34	3,010	0,565
900	3,0338	682,44	3,056	0,556	2,8808	684,44	3,042	0,559
950	3,1798	710,14	3,086	0,552	3,0188	712,24	3,072	0,553
1000	3,3268	737,64	3,114	0,548	3,1558	739,84	3,100	0,550
1050	3,4708	765,04	3,141	0,545	3,2918	767,24	3,127	0,546
1100	3,6148	792,24	3,166	0,543	3,4268	794,54	3,152	0,544

TABLE XI. Thermodynamic Properties of Liquid and Gaseous Argon (*Continued*)

T, K	p = 750 bar				p = 800 bar			
	$v \cdot 10^3$	h	s	c_p	$v \cdot 10^3$	h	s	c_p
1150	3.758	819.35	3.190	0.541	3.561	821.65	3.176	0.542
1200	3.901	846.35	3.213	0.539	3.695	848.75	3.199	0.540
1250	4.044	873.25	3.235	0.537	3.829	875.65	3.221	0.538
1300	4.186	900.15	3.256	0.537	3.963	902.55	3.243	0.537

TABLE XI. Thermodynamic Properties of Liquid and Gaseous Argon (*Continued*)

T, K	p = 850 bar				p = 900 bar			
	$v \cdot 10^3$	h	s	c_p	$v \cdot 10^3$	h	s	c_p
105	0.6718	133.90	1.373	0.973	0.6683	136.36	1.365	0.971
110	0.6801	138.73	1.419	0.966	0.6764	141.18	1.411	0.964
115	0.6885	143.58	1.462	0.959	0.6846	146.03	1.453	0.957
120	0.6971	148.33	1.502	0.948	0.6929	150.76	1.493	0.946
125	0.7059	153.04	1.541	0.936	0.7013	155.46	1.532	0.934
130	0.7148	157.72	1.577	0.932	0.7100	160.13	1.569	0.929
135	0.7238	162.35	1.613	0.928	0.7187	164.75	1.604	0.925
140	0.7331	167.00	1.646	0.919	0.7276	169.38	1.637	0.916
145	0.7426	171.58	1.678	0.912	0.7367	173.94	1.669	0.909
150	0.7522	176.10	1.709	0.912	0.7460	178.44	1.700	0.909
155	0.7628	180.70	1.739	0.916	0.7562	183.02	1.730	0.912
160	0.7726	185.28	1.768	0.913	0.7656	187.58	1.759	0.909
170	0.7926	194.39	1.823	0.901	0.7847	196.65	1.814	0.896
180	0.8132	203.33	1.875	0.888	0.8044	205.53	1.864	0.883
190	0.8344	212.16	1.922	0.880	0.8246	214.31	1.912	0.875
200	0.8564	220.94	1.967	0.876	0.8454	223.04	1.957	0.869
210	0.8789	229.69	2.010	0.869	0.8668	231.72	1.999	0.862
220	0.9022	238.36	2.050	0.860	0.8887	240.29	2.039	0.854
230	0.9260	246.88	2.088	0.851	0.9112	248.78	2.077	0.846
240	0.9504	255.36	2.124	0.841	0.9342	257.23	2.112	0.837
250	0.9754	263.75	2.159	0.832	0.9576	265.54	2.146	0.826
260	1.0010	271.99	2.191	0.821	0.9816	273.75	2.179	0.816
270	1.0265	280.19	2.222	0.812	1.0061	281.87	2.209	0.806
280	1.0532	288.25	2.252	0.804	1.0305	289.90	2.239	0.796
290	1.0797	296.26	2.280	0.792	1.0556	297.78	2.267	0.786
300	1.1068	304.12	2.306	0.779	1.0809	305.62	2.293	0.776
310	1.1338	311.82	2.331	0.764	1.1069	313.32	2.318	0.765
320	1.1608	319.42	2.356	0.755	1.1329	320.92	2.342	0.753
330	1.1888	326.92	2.379	0.750	1.1589	328.42	2.366	0.745
340	1.2168	334.42	2.401	0.741	1.1849	335.82	2.388	0.743
350	1.2448	341.82	2.423	0.733	1.2109	343.22	2.409	0.736

TABLE XI. Thermodynamic Properties of Liquid and Gaseous Argon (*Continued*)

T, K	p = 850 bar				p = 900 bar			
	$v \cdot 10^3$	h	s	c_p	$v \cdot 10^3$	h	s	c_p
375	1,3158	359,92	2,473	0,718	1,2769	361,52	2,459	0,719
400	1,3868	377,72	2,519	0,705	1,3439	379,22	2,505	0,702
425	1,4578	395,22	2,562	0,693	1,4099	396,62	2,548	0,688
450	1,5278	412,42	2,601	0,681	1,4759	413,72	2,587	0,678
475	1,5978	429,32	2,638	0,669	1,5419	430,52	2,624	0,667
500	1,6668	445,92	2,672	0,658	1,6069	447,12	2,658	0,657
550	1,8038	478,32	2,733	0,639	1,7369	479,52	2,719	0,641
600	1,9398	509,92	2,788	0,624	1,8659	511,22	2,774	0,626
650	2,0758	540,82	2,838	0,610	1,9949	542,22	2,824	0,613
700	2,2118	571,02	2,883	0,597	2,1229	572,52	2,870	0,599
750	2,3478	600,62	2,924	0,585	2,2499	602,22	2,911	0,587
800	2,4828	629,62	2,961	0,575	2,3759	631,32	2,948	0,577
850	2,6148	658,12	2,996	0,566	2,5019	660,02	2,983	0,569
900	2,7458	686,32	3,028	0,560	2,6259	688,32	3,016	0,563
950	2,8758	714,22	3,059	0,555	2,7489	716,32	3,046	0,557
1000	3,0048	741,92	3,087	0,552	2,8719	744,12	3,074	0,553
1050	3,1328	769,42	3,114	0,548	2,9929	771,72	3,101	0,549
1100	3,2608	796,82	3,139	0,545	3,1139	799,12	3,127	0,545
1150	3,387	824,05	3,163	0,543	3,233	826,35	3,151	0,544
1200	3,514	851,05	3,186	0,541	3,353	853,45	3,174	0,542
1250	3,640	878,05	3,208	0,539	3,472	880,55	3,196	0,540
1300	3,766	904,95	3,230	0,537	3,591	907,45	3,217	0,538

TABLE XI. Thermodynamic Properties of Liquid and Gaseous Argon (*Continued*)

T, K	p = 950 bar				p = 1000 bar			
	$v \cdot 10^3$	h	s	c_p	$v \cdot 10^3$	h	s	c_p
110	0,6729	143,64	1,402	0,962	0,6695	146,11	1,394	0,961
115	0,6808	148,48	1,445	0,955	0,6772	150,94	1,436	0,953
120	0,6889	153,20	1,485	0,944	0,6850	155,65	1,477	0,942
125	0,6971	157,90	1,524	0,932	0,6930	160,34	1,515	0,930
130	0,7054	162,55	1,560	0,927	0,7010	164,98	1,552	0,925
135	0,7139	167,16	1,595	0,922	0,7092	169,58	1,587	0,920
140	0,7225	171,77	1,628	0,913	0,7176	174,19	1,620	0,911
145	0,7313	176,33	1,660	0,906	0,7261	178,73	1,651	0,904
150	0,7402	180,81	1,691	0,905	0,7347	183,20	1,682	0,903
155	0,7501	185,37	1,721	0,909	0,7442	187,74	1,712	0,906
160	0,7590	189,91	1,750	0,905	0,7529	192,27	1,741	0,902
170	0,7774	198,94	1,804	0,892	0,7705	201,26	1,795	0,888
180	0,7962	207,78	1,855	0,878	0,7885	210,07	1,845	0,874
190	0,8155	216,51	1,902	0,870	0,8070	218,76	1,892	0,865
200	0,8353	225,18	1,946	0,863	0,8260	227,38	1,936	0,858

TABLE XI. Thermodynamic Properties of Liquid and Gaseous Argon (*Continued*)

T, K	p = 950 bar				p = 1000 bar			
	$v\cdot10^3$	h	s	c_p	$v\cdot10^3$	h	s	c_p
210	0.8556	233.81	1.989	0.856	0.8453	235.96	1.979	0.852
220	0.8764	242.31	2.028	0.849	0.8651	244.41	2.018	0.844
230	0.8977	250.77	2.066	0.839	0.8853	252.84	2.055	0.835
240	0.9194	259.14	2.101	0.831	0.9060	261.14	2.090	0.825
250	0.9416	267.37	2.135	0.820	0.9270	269.34	2.124	0.815
260	0.9642	275.55	2.167	0.811	0.9484	277.45	2.156	0.806
270	0.9871	283.60	2.197	0.802	0.9701	285.47	2.186	0.797
280	1.0104	291.60	2.226	0.793	0.9922	293.40	2.215	0.787
290	1.0338	299.46	2.254	0.782	1.0145	301.23	2.243	0.780
300	1.0578	307.28	2.281	0.770	1.0369	308.98	2.269	0.771
310	1.0818	314.88	2.306	0.761	1.0599	316.68	2.294	0.758
320	1.1058	322.48	2.330	0.753	1.0829	324.18	2.318	0.747
330	1.1308	329.98	2.353	0.744	1.1059	331.58	2.341	0.739
340	1.1548	337.38	2.375	0.735	1.1289	338.98	2.363	0.733
350	1.1788	344.68	2.396	0.727	1.1519	346.28	2.384	0.727
375	1.2418	362.68	2.446	0.713	1.2119	364.28	3.434	0.713
400	1.3048	380.38	2.492	0.701	1.2709	381.98	3.480	0.701
425	1.3678	397.78	2.535	0.689	1.3299	399.38	2.522	0.689
450	1.4308	414.88	2.574	0.678	1.3889	416.48	2.561	0.678
475	1.4938	431.68	2.611	0.667	1.4479	433.28	2.598	0.667
500	1.5558	448.28	2.645	0.657	1.5059	449.88	2.633	0.657
550	1.6778	480.68	2.706	0.639	1.6209	482.28	2.695	0.639
600	1.7978	512.28	2.761	0.625	1.7359	513.88	2.750	0.625
650	1.9178	543.28	2.811	0.613	1.8509	544.88	2.800	0.613
700	2.0378	573.68	2.857	0.603	1.9659	575.28	2.845	0.604
750	2.1588	603.58	2.898	0.592	2.0809	605.28	2.886	0.594
800	2.2798	632.98	2.936	0.582	2.1959	634.78	2.924	0.584
850	2.3998	661.88	2.971	0.572	2.3099	663.78	2.959	0.574
900	2.5188	690.28	3.004	0.565	2.4229	692.28	2.992	0.567
950	2.6358	718.38	3.034	0.559	2.5349	720.48	3.023	0.561
1000	2.7518	746.28	3.063	0.555	2.6449	748.48	3.051	0.556
1050	2.8668	773.88	3.090	0.551	2.7539	776.18	3.078	0.552
1100	2.9818	801.38	3.115	0.547	2.8629	803.68	3.104	0.548
1150	3.095	828.75	3.139	0.545	2.971	831.15	3.128	0.546
1200	3.209	855.95	3.163	0.542	3.079	858.35	3.151	0.543
1250	3.322	882.95	3.185	0.540	3.187	885.45	3.174	0.541
1300	3.435	909.95	3.206	0.539	3.295	912.45	3.195	0.539

TABLE XII. Dynamic Viscosity of Argon

T, K	$\eta \cdot 10^8$, Pa · s at p, bar							
	1	10	20	30	40	50	60	70
90	758	24490	24775	25100	25380	25705	26050	26410
100	838	18795	19020	19260	19495	19730	19970	20200
110	917	14280	14510	14720	14940	15160	15380	15595
120	996	1042	10840	11240	11580	11900	12170	12420
130	1074	1116	1204	8709	8994	9259	9513	9754
140	1151	1190	1256	1404	6653	7015	7334	7623
150	1227	1263	1318	1413	1608	4349	5265	5733
160	1303	1337	1384	1457	1576	1781	2235	3303
170	1378	1410	1452	1512	1600	1728	1919	2214
180	1452	1483	1521	1572	1643	1738	1865	2034
190	1525	1554	1589	1635	1694	1771	1867	1988
200	1598	1626	1659	1699	1751	1815	1894	1988
210	1669	1696	1727	1764	1809	1865	1931	2009
220	1740	1766	1795	1829	1870	1919	1976	2043
230	1810	1835	1863	1894	1932	1976	2026	2084
240	1879	1904	1930	1959	1994	2033	2079	2130
250	1947	1971	1996	2024	2056	2092	2133	2179
260	2014	2037	2061	2088	2117	2151	2189	2230
270	2081	2103	2127	2152	2180	2211	2246	2284
280	2146	2168	2190	2214	2241	2270	2302	2337
290	2211	2232	2254	2277	2302	2330	2360	2392
300	2275	2296	2317	2339	2363	2389	2417	2447
310	2337	2357	2378	2399	2422	2446	2473	2502
320	2400	2420	2440	2460	2482	2505	2531	2558
330	2462	2481	2501	2521	2542	2564	2588	2613
340	2523	2542	2561	2580	2600	2622	2644	2669
350	2584	2602	2621	2640	2659	2680	2701	2724
375	2729	2746	2764	2781	2799	2818	2838	2858
400	2873	2890	2906	2922	2939	2956	2974	2993
425	3012	3028	3043	3059	3074	3090	3107	3124
450	3148	3163	3178	3192	3207	3222	3238	3253
475	3279	3293	3307	3321	3335	3349	3364	3379
500	3409	3423	3436	3449	3463	3476	3490	3503
550	3659	3672	3684	3696	3708	3720	3733	3745
600	3899	3911	3922	3933	3944	3955	3967	3978
650	4130	4141	4151	4162	4172	4182	4193	4203
700	4354	4364	4374	4384	4393	4403	4412	4422
750	4570	4579	4589	4598	4607	4616	4624	4633
800	4780	4789	4797	4806	4814	4823	4831	4839
850	4985	4993	5001	5009	5017	5025	5033	5041

TABLE XII. Dynamic Viscosity of Argon (*Continued*)

T, K	η · 10⁸, Pa · s at p, bar							
	1	10	20	30	40	50	60	70
900	5184	5192	5199	5207	5215	5222	5229	5237
950	5379	5386	5394	5401	5408	5415	5422	5429
1000	5570	5577	5584	5591	5597	5604	5611	5617
1050	5757	5764	5770	5777	5783	5789	5796	5802
1100	5940	5946	5953	5959	5965	5971	5977	5983
1150	6120	6126	6132	6138	6144	6149	6155	6161
1200	6297	6303	6308	6314	6320	6325	6331	6336
1250	6471	6476	6482	6487	6493	6498	6503	6508
1300	6642	6647	6652	6658	6663	6668	6675	6678

TABLE XII. Dynamic Viscosity of Argon (*Continued*)

T, K	η · 10⁸, Pa · s at p, bar					
	80	100	125	150	175	200
90	26780	27540	28560	29640	30740	31860
100	20425	20910	21500	22120	22760	23440
110	15820	16260	16820	17360	17925	18500
120	12660	13120	13670	14195	14680	15157
130	9987	10431	10964	11445	11911	12363
140	7889	8379	8928	9428	9899	10343
150	6069	6667	7266	7798	8276	8720
160	4151	5104	5867	6462	6968	7367
170	2658	3706	4654	5336	5884	6357
180	2256	2864	3713	4420	4994	5484
190	2136	2521	3129	3750	4302	4785
200	2100	2378	2818	3313	3801	4250
210	2099	2317	2654	3045	3454	3855
220	2118	2297	2570	2886	3227	3572
230	2149	2301	2529	2792	3080	3379
240	2187	2319	2515	2739	2985	3245
250	2230	2347	2518	2713	2927	3155
260	2276	2380	2532	2704	2893	3095
270	2326	2420	2556	2710	2879	3059
280	2376	2462	2585	2724	2876	3038
290	2428	2507	2619	2745	2884	3031
300	2480	2553	2657	2772	2899	3034
310	2532	2600	2695	2802	2918	3042
320	2586	2649	2738	2837	2945	3059
330	2640	2700	2782	2874	2974	3081

TABLE XII. Dynamic Viscosity of Argon (*Continued*)

T, K	$\eta \cdot 10^8$, Pa · s at *p*, bar					
	80	100	125	150	175	200
340	2694	2750	2827	2913	3006	3106
350	2740	2801	2874	2955	3042	3135
375	2880	2926	2989	3058	3133	3212
400	3012	3053	3109	3170	3235	3304
425	3142	3179	3229	3283	8340	3401
450	3270	3304	3349	3397	3449	3503
475	3394	3426	3466	3510	3557	3606
500	3518	3547	3585	3625	3668	3712
550	3757	3783	3816	3851	3887	3924
600	3989	4012	4041	4072	4103	4135
650	4213	4234	4260	4287	4315	4344
700	4431	4450	4474	4499	4524	4550
750	4642	4660	4682	4704	4727	4750
800	4842	4857	4880	4902	4925	4949
850	5049	5064	5083	5102	5122	5142
900	5244	5258	5277	5295	5313	5331
950	5436	5449	5466	5483	5500	5518
1000	5624	5637	5653	5669	5685	5701
1050	5808	5820	5836	5851	5866	5881
1100	5989	6000	6015	6029	6044	6058
1150	6166	6178	6192	6205	6219	6233
1200	6341	6352	6365	6379	6392	6405
1250	6513	6524	6536	6549	6562	6574
1300	6683	6693	6705	6717	6729	6741

TABLE XII. Dynamic Viscosity of Argon (*Continued*)

T, K	$\eta \cdot 10^8$, Pa · s at *p*, bar						
	225	250	275	300	325	350	375
90	3305	—	—	—	—	—	—
100	24140	24890	25675	26480	27240	28030	28800
110	19065	19340	20210	20770	21325	21900	22480
120	15613	16058	16482	16908	17330	17760	18190
130	12798	13220	13650	14035	14433	14815	15195
140	10765	11176	11569	11952	12328	12691	13051
150	9137	9535	9918	10284	10640	10987	11326
160	7835	8224	8524	8946	9286	9613	9936
170	6779	7167	7532	7877	8207	8524	8827
180	5914	6306	6666	7007	7321	7630	7925
190	5212	5600	5957	6284	6596	6898	7184
200	4661	5036	5383	5703	6007	6295	6571

TABLE XII. Dynamic Viscosity of Argon (*Continued*)

T, K	η · 10⁸, Pa · s at p, bar						
	225	250	275	300	325	350	375
210	4233	4587	4918	5225	5519	5797	6060
220	3914	4240	4549	4844	5123	5386	5642
230	3679	3977	4262	4536	4799	5052	5294
240	3511	3777	4038	4293	4538	4776	5005
250	3391	3629	3866	4100	4329	4551	4763
260	3305	3520	3735	3950	4162	4371	4574
270	3248	3442	3638	3835	4032	4226	4418
280	3209	3385	3564	3746	3927	4108	4286
290	3187	3347	3512	3679	3846	4015	4183
300	3176	3324	3475	3630	3785	3942	4098
310	3173	3309	3449	3592	3738	3883	4030
320	3180	3306	3436	3569	3704	3841	3979
330	3193	3310	3431	3555	3682	3810	3940
340	3210	3320	3433	3549	3667	3788	3910
350	3232	3335	3441	3550	3661	3774	3889
375	3296	3384	3474	3568	3664	3762	3860
400	3377	3453	3532	3614	3698	3783	3871
425	3466	3532	3602	3674	3748	3823	3900
450	3560	3620	3682	3745	3811	3879	3947
475	3657	3711	3766	3823	3882	3943	4004
500	3759	3807	3857	3909	3962	4017	4073
550	3963	4004	4046	4089	4133	4179	4225
600	4169	4204	4239	4276	4314	4353	4392
650	4373	4404	4435	4467	4499	4533	4567
700	4576	4603	4630	4658	4687	4716	4746
750	4774	4798	4823	4848	4873	4900	4926
800	4969	4991	5013	5036	5059	5083	5106
850	5161	5183	5202	5222	5244	5265	5286
900	5349	5368	5387	5406	5425	5444	5464
950	5535	5552	5569	5587	5605	5623	5641
1000	5717	5733	5750	5766	5782	5793	5816
1050	5896	5912	5927	5942	5958	5973	5989
1100	6072	6087	6101	6116	6130	6145	6160
1150	6246	6260	6273	6287	6301	6315	6329
1200	6418	6430	6443	6456	6469	6482	6496
1250	6586	6599	6611	6623	6636	6648	6661
1300	6752	6764	6776	6788	6800	6811	6823

TABLE XII. Dynamic Viscosity of Argon (*Continued*)

T, K	$\eta \cdot 10^8$, Pa · s at p, bar						
	400	425	450	475	500	550	600
100	29600	30400	31230	32060	32880	34550	36170
110	23070	23660	24265	24880	25470	26690	27940
120	18625	19085	19520	19985	20480	21520	22670
130	15572	15936	16298	16657	17012	17711	18398
140	13401	13751	14091	14427	14757	15406	16049
150	11656	11981	12300	12614	12927	13535	14127
160	10244	10547	10847	11140	11427	11989	12542
170	9126	9414	9698	9975	10246	10777	11291
180	8209	9487	8754	9015	9275	9776	10263
190	7455	7720	7977	8228	8473	8949	9408
200	6832	7088	7337	7576	7809	8262	8699
210	6313	6560	6798	7024	7248	7684	8102
220	5885	6121	6346	6567	6785	7163	7592
230	5525	5752	5972	6183	6387	6785	7165
240	5229	5443	5653	5853	6052	6434	6797
250	4980	5184	5383	5579	5765	6129	6481
260	4773	4968	5157	5341	5521	5872	6208
270	4605	4788	4969	5144	5315	5650	5974
280	4462	4137	4807	4975	5138	5456	5768
290	4348	4512	4674	4833	4989	5296	5592
300	4254	4408	4562	4712	4862	5153	5438
310	4176	4322	4467	4611	4753	5033	5303
320	4116	4255	4390	4527	4661	4930	5191
330	4069	4200	4329	4457	4587	4839	5090
340	4032	4155	4277	4401	4522	4764	5004
350	4005	4121	4237	4356	4471	4702	4931
375	3963	4064	4166	4270	4374	4580	4787
400	3959	4049	4140	4232	4324	4509	4698
425	3979	4060	4140	4223	4306	4472	4645
450	4018	4090	4162	4237	4311	4464	4620
475	4068	4132	4198	4265	4333	4471	4614
500	4130	4188	4248	4209	4370	4497	4626
550	4273	4321	4371	4421	4472	4578	4687
600	4432	4474	4516	4559	4602	4691	4784
650	4602	4637	4673	4710	4748	4825	4904
700	4777	4808	4839	4872	4905	4971	5041
750	4953	4981	5009	5037	5066	5125	5186
800	5131	5155	5180	5206	5232	5284	5339
850	5308	5331	5353	5376	5400	5447	5496
900	5484	5505	5525	5546	5567	5610	5654
950	5659	5678	5697	5716	5735	5775	5815
1000	5833	5850	5868	5885	5903	5939	5977
1050	6005	6021	6037	6054	6070	6104	6138
1100	6175	6190	6205	6220	6235	6266	6298

TABLE XII. Dynamic Viscosity of Argon (*Continued*)

T, K	$\eta \cdot 10^8$, Pa · s at p, bar						
	400	425	450	475	500	550	600
1150	6343	6357	6371	6385	6399	6429	6458
1200	6509	6522	6535	6549	6562	6590	6617
1250	6673	6686	6698	6711	6724	6749	6776
1300	6835	6847	6859	6871	6883	6907	6932

TABLE XII. Dynamic Viscosity of Argon (*Continued*)

T, K	$\eta \cdot 10^8$, Pa · s at p, bar							
	650	700	750	800	850	900	950	1000
100	37858	39540	41240	42930	44580	46280	47980	49710
110	29180	30420	31700	32980	34260	35550	36830	38140
120	23900	25020	26100	27200	28370	29590	30860	38100
130	19065	19725	20387	21033	21660	22288	22915	23538
140	16680	17299	17910	18516	19115	19711	20290	20873
150	14714	15288	15855	16419	19976	17528	18073	18616
160	13080	13611	14139	14654	15167	15677	16186	16683
170	11798	12292	12781	13265	13738	14211	14681	15149
180	10738	11199	11656	12107	12549	12989	13426	13863
190	9854	10290	10721	11141	11556	11965	12372	12778
200	9121	9533	9938	10334	10721	11108	11489	11865
210	8498	8889	9270	9644	10013	10373	10733	11087
220	7976	8348	8708	9063	9410	9754	10092	10426
230	7529	7883	8228	8565	8896	9221	9541	9858
240	7145	7482	7815	8137	8451	8760	9066	9365
250	6813	7144	7459	7764	8065	8361	8651	8937
260	6530	6842	7148	7441	7726	8010	8288	8561
270	6284	6585	6876	7157	7439	7705	7974	8236
280	6066	6356	6637	6908	7174	7438	7692	7943
290	5877	6154	6428	6692	6948	7200	7448	7687
300	5712	5983	6244	6498	6745	6989	7227	7460
310	5568	5828	6075	6325	6565	6794	7028	7250
320	5443	5691	5936	6171	6405	6627	6850	7070
330	5334	5572	5806	6036	6267	6478	6697	6908
340	5238	5468	5696	5916	6139	6346	6555	6764
350	5158	5378	5600	5811	6027	6230	6429	6636
375	4995	5191	5395	5591	5785	5976	6163	6345
400	4890	5064	5252	5434	5610	5786	5962	6133
425	4822	4980	5152	5318	5479	5645	5803	5968
450	4781	4926	5084	5240	5387	5541	5692	5842
475	4760	4896	5037	5187	5320	5463	5602	5744
500	4760	4887	5016	5157	5281	5414	5545	5678

TABLE XII. Dynamic Viscosity of Argon (*Continued*)

T, K	$\eta \cdot 10^8$, Pa · s at p, bar							
	650	700	750	800	850	900	950	1000
550	4799	4910	5021	5143	5253	5367	5485	5602
600	4878	4975	5073	5178	5275	5374	5480	5582
650	4985	5069	5155	5245	5332	5419	5515	5603
700	5112	5185	5260	5336	5416	5492	5567	5656
750	5249	5313	5379	5446	5516	5586	5656	5730
800	5394	5451	5510	5570	5631	5694	5758	5822
850	5546	5597	5649	5703	5758	5814	5872	5229
900	5700	5746	5793	5842	5891	5942	5993	6045
950	5856	5898	5942	5985	6030	6076	6123	6170
1000	6014	6053	6092	6132	6174	6215	6258	6301
1050	6173	6208	6244	6281	6319	6357	6397	6436
1100	6330	6363	6397	6431	6466	6501	6537	6574
1150	6488	6519	6550	6582	6614	6647	6680	6714
1200	6646	6674	6703	6733	6763	6794	6825	6856
1250	6802	6829	6856	6884	6912	6941	6970	6999
1300	6957	6982	7068	7034	7061	7087	7114	7142

TABLE XIII. Thermal Conductivity of Argon

T, K	$\lambda \cdot 10^8$, kW/(m · K) at p, bar						
	1	10	20	30	40	50	60
90	592	12038	12114	12193	12272	12344	12418
100	654	10664	10752	10843	10931	11012	11096
110	716	9340	9447	9550	9651	9748	9843
120	778	8771	8154	8283	8404	8523	8634
130	838	925	1073	6981	7145	7298	7442
140	899	977	1093	1304	5766	6007	6216
150	958	1029	1126	1271	1524	4113	4817
160	1017	1082	1167	1282	1444	1700	2208
170	1076	1136	1212	1309	1435	1602	1832
180	1134	1190	1259	1344	1448	1577	1737
190	1191	1244	1307	1382	1472	1579	1706
200	1248	1297	1356	1425	1504	1596	1702
210	1304	1351	1405	1468	1540	1621	1713
220	1359	1403	1454	1513	1578	1651	1732
230	1414	1456	1504	1558	1619	1685	1758
240	1467	1507	1553	1603	1659	1721	1787
250	1521	1560	1603	1650	1703	1759	1821
260	1573	1610	1651	1696	1745	1798	1855

TABLE XIII. Thermal Conductivity of Argon (*Continued*)

T, K	λ · 10⁸, kW/(m · K) at p, bar						
	1	10	20	30	40	50	60
270	1625	1660	1700	1742	1789	1838	1892
280	1676	1710	1747	1788	1832	1879	1929
290	1727	1760	1796	1834	1876	1920	1968
300	1777	1809	1843	1880	1920	1962	2007
310	1826	1856	1890	1925	1963	2003	2045
320	1875	1904	1936	1970	2007	2045	2085
330	1923	1952	1982	2015	2050	2086	2125
340	1970	1998	2027	2059	2092	2128	2164
350	2017	2044	2073	2103	2135	2169	2204
375	2132	2157	2184	2212	2241	2272	2304
400	2244	2267	2292	2318	2345	2373	2403
425	2353	2375	2398	2422	2447	2474	2501
450	2459	2480	2501	2524	2547	2572	2597
475	2562	2582	2602	2623	2645	2668	2692
500	2663	2681	2701	2721	2742	2763	2785
550	2858	2875	2892	2910	2929	2948	2968
600	3045	3060	3076	3092	3110	3127	3145
650	3226	3240	3255	3270	3285	3301	3317
700	3400	3413	3427	3440	3455	3469	3484
750	3569	3581	3594	3607	3620	3633	3647
800	3733	3744	3756	3768	3780	3793	3806
850	3893	3904	3915	3926	3938	3949	3961
900	4049	4059	4070	4080	4091	4102	4113
950	4201	4211	4220	4230	4241	4251	4262
1000	4350	4359	4368	4378	4388	4397	4407
1050	4496	4505	4514	4523	4532	4541	4550
1100	4639	4647	4656	4664	4673	4682	4691
1150	4780	4788	4796	4804	4812	4821	4830
1200	4918	4926	4933	4941	4949	4957	4965
1250	5054	5061	5069	5076	5084	5092	5099
1300	5187	5194	5201	5208	5216	5223	5231

TABLE XIII. Thermal Conductivity of Argon (*Continued*)

T, K	λ · 10⁸, kW/(m · K) at p, bar						
	80	100	125	150	175	200	225
90	12569	12711	12887	13056	13221	13388	13548
100	11262	11420	11612	11798	11979	12156	12326
110	10030	10206	10419	10624	10821	11012	11199
120	8853	9056	9297	9526	9742	9952	10152
130	7707	7949	8230	8488	8730	8960	9176

TABLE XIII. Thermal Conductivity of Argon (*Continued*)

T, K	λ · 10⁸, kW/(m · K) at p, bar						
	80	100	125	150	175	200	225
140	6571	6877	7210	7508	7782	8036	8274
150	5401	5815	6231	6584	6894	7177	7437
160	3957	4709	5284	5717	6076	6390	6673
170	2623	3582	4366	4903	5321	5674	5984
180	2187	2811	3584	4181	4646	5033	5366
190	2028	2450	3051	3612	4084	4484	4830
200	1958	2276	2737	3213	3651	4039	4385
210	1926	2185	2555	2949	3335	3696	4025
220	1919	2135	2445	2777	3113	3436	3744
230	1924	2113	2379	2665	2959	3249	3528
240	1936	2105	2338	2589	2849	3110	3364
250	1957	2109	2318	2542	2774	3009	3242
260	1980	2119	2308	2511	2721	2935	3148
270	2008	2135	2309	2494	2686	2882	3078
280	2037	2156	2316	2486	2663	2844	3025
290	2069	2180	2329	2487	2651	2818	2988
300	2102	2206	2345	2493	2646	2802	2961
310	2136	2234	2365	2503	2646	2793	2942
320	2172	2264	2388	2518	2653	2791	2931
330	2207	2295	2412	2535	2662	2793	2925
340	2243	2326	2437	2554	2675	2799	2924
350	2279	2359	2465	2576	2691	2808	2928
375	2371	2443	2538	2637	2739	2844	2950
400	2464	2530	2615	2705	2797	2892	2988
425	2557	2617	2695	2777	2861	2947	3035
450	2649	2704	2776	2849	2924	3000	3089
475	2740	2792	2858	2926	2996	3067	3148
500	2831	2879	2942	3006	3073	3141	3211
550	3008	3051	3106	3163	3222	3282	3343
600	3181	3219	3269	3320	3372	3426	3480
650	3350	3385	3430	3476	3523	3572	3621
700	3515	3546	3587	3629	3672	3716	3761
750	3675	3704	3742	3780	3820	3861	3902
800	3832	3859	3894	3929	3966	4004	4042
850	3986	4011	4043	4076	4110	4145	4181
900	4136	4160	4190	4221	4253	4285	4318
950	4283	4305	4334	4363	4393	4424	4454
1000	4428	4449	4475	4503	4531	4560	4589
1050	4570	4590	4615	4641	4667	4694	4722
1100	4709	4728	4752	4776	4802	4828	4853
1150	4847	4865	4888	4911	4935	4959	4984
1200	4982	4999	5021	5043	5066	5089	5112
1250	5115	5132	5152	5174	5195	5217	5240
1300	5246	5261	5281	5302	5322	5343	5365

TABLE XIII. Thermal Conductivity of Argon (*Continued*)

T, K	λ · 10⁸, kW/(m · K) at p, bar						
	250	275	300	325	350	375	400
90	13705	—	—	—	—	—	—
100	12493	12657	12821	12977	13132	13286	13433
110	11377	11554	11725	11889	12053	12214	12372
120	10345	10533	10715	10891	11063	11232	11394
130	9385	9594	9779	9968	10148	10325	10500
140	8490	8714	8918	9119	9308	9493	9673
150	7680	7911	8128	8337	8537	8730	8916
160	6933	7177	7405	7624	7830	8030	8221
170	6263	6522	6763	6991	7207	7411	7610
180	5665	5936	6189	6419	6644	6855	7057
190	5139	5420	5675	5914	6145	6361	6563
200	4694	4976	5234	5477	5705	5922	6126
210	4327	4604	4858	5099	5326	5539	5743
220	4029	4294	4545	4779	4999	5211	5412
230	3796	4047	4285	4510	4725	4929	5124
240	3611	3846	4072	4286	4492	4689	4880
250	3469	3688	3900	4104	4299	4488	4672
260	3357	3562	3760	3952	4139	4319	4494
270	3272	3463	3649	3830	4007	4179	4345
280	3206	3384	3559	3730	3897	4060	4219
290	3156	3323	3487	3649	3807	3962	4114
300	3119	3276	3431	3584	3733	3881	4025
310	3091	3238	3385	3530	3673	3813	3951
320	3071	3211	3350	3488	3624	3758	3889
330	3058	3191	3323	3454	3584	3713	3839
340	3050	3177	3303	3428	3552	3675	3796
350	3048	3168	3289	3408	3527	3644	3761
375	3058	3166	3273	3381	3489	3595	3701
400	3085	3182	3280	3378	3476	3573	3669
425	3123	3213	3302	3392	3481	3571	3660
450	3170	3253	3335	3418	3501	3583	3665
475	3223	3300	3376	3452	3528	3606	3683
500	3281	3352	3423	3494	3566	3638	3710
550	3405	3467	3530	3593	3657	3720	3784
600	3535	3591	3647	3704	3761	3818	3875
650	3671	3721	3772	3823	3875	3926	3978
700	3807	3853	3899	3946	3993	4040	4088
750	3944	3986	4029	4072	4115	4159	4202
800	4080	4119	4159	4199	4239	4279	4320
850	4216	4253	4290	4327	4364	4402	4440
900	4352	4386	4420	4455	4490	4524	4560
950	4486	4517	4550	4582	4615	4648	4681
1000	4618	4648	4679	4709	4740	4771	4803

TABLE XIII. Thermal Conductivity of Argon (*Continued*)

T, K	$\lambda \cdot 10^8$, kW/(m · K) at p, bar						
	250	275	300	325	350	375	400
1050	4750	4778	4807	4836	4865	4894	4924
1100	4880	4907	4934	4961	4989	5017	5045
1150	5009	5034	5060	5086	5112	5139	5166
1200	5136	5160	5185	5210	5235	5260	5285
1250	5262	5286	5309	5333	5356	5380	5405
1300	5387	5408	5431	5454	5476	5499	5522

TABLE XIII. Thermal Conductivity of Argon (*Continued*)

T, K	$\lambda \cdot 10^8$, kW/(m · K) at p, bar						
	425	450	475	500	550	600	650
100	13582	13725	13866	14010	14284	14557	14815
110	12527	12676	12823	12972	13260	13535	13809
120	11556	11713	11871	12022	12317	12605	12887
130	10664	10828	10989	11146	11453	11750	12035
140	9848	10018	10185	10346	10659	10963	11258
150	9096	9271	9442	9610	9932	10240	10540
160	8404	8584	8759	8928	9252	9567	9867
170	7798	7982	8162	8333	8666	8981	9287
180	7252	7438	7618	7795	8131	8452	8760
190	6760	6949	7132	7309	7649	7970	8278
200	6324	6515	6697	6874	7212	7536	7844
210	5941	6130	6309	6485	6824	7145	7449
220	5605	5789	5969	6146	6474	6792	7097
230	5314	5497	5673	5842	6171	6483	6779
240	5062	5240	5409	5577	5898	6203	6494
250	4847	5019	5186	5346	5657	5957	6240
260	4664	4828	4988	5145	5448	5739	6017
270	4507	4666	4821	4970	5262	5546	5817
280	4375	4527	4675	4819	5102	5375	5639
290	4263	4408	4551	4690	4964	5226	5480
300	4167	4307	4443	4578	4840	5094	5340
310	4087	4221	4353	4482	4735	4979	5218
320	4021	4147	4274	4397	4643	4880	5108
330	3962	4084	4206	4327	4560	4790	5012
340	3916	4033	4150	4264	4490	4710	4927
350	3876	3988	4102	4212	4430	4643	4852
375	3805	3908	4011	4113	4313	4510	4706
400	3765	3860	3955	4049	4233	4418	4603
425	3749	3836	3925	4011	4183	4356	4532

TABLE XIII. Thermal Conductivity of Argon (*Continued*)

T, K	$\lambda \cdot 10^8$, kW/(m · K) at p, bar						
	425	450	475	500	550	600	650
450	3749	3829	3912	3993	4155	4317	4480
475	3760	3836	3913	3989	4141	4293	4446
500	3782	3854	3926	3997	4140	4283	4427
550	3847	3911	3974	4038	4165	4292	4421
600	3932	3989	4047	4103	4218	4333	4447
650	4030	4082	4134	4186	4290	4394	4498
700	4135	4183	4231	4278	4374	4470	4565
750	4246	4290	4334	4379	4467	4555	4644
800	4361	4401	4443	4483	4566	4648	4730
850	4478	4516	4554	4592	4669	4746	4823
900	4596	4632	4667	4703	4775	4848	4920
950	4715	4748	4782	4816	4884	4952	5020
1000	4834	4866	4898	4930	4994	5058	5122
1050	4954	4984	5014	5044	5105	5166	5227
1100	5073	5101	5130	5159	5216	5274	5332
1150	5192	5220	5247	5274	5329	5384	5440
1200	5311	5337	5368	5389	5441	5494	5546
1250	5429	5454	5478	5503	5553	5604	5654
1300	5546	5569	5592	5617	5664	5713	5761

TABLE XIII. Thermal Conductivity of Argon (*Continued*)

T, K	$\lambda \cdot 10^8$, kW/(m · K) at p, bar						
	700	750	800	850	900	950	1000
100							
110	14077	14332	14590	14836	15080	15320	15560
120	13157	13425	13687	13940	14190	14436	14682
130	12313	12588	12855	13111	13366	13616	13865
140	11541	11818	12090	12353	12612	12862	13109
150	10827	11108	11382	11647	11909	12162	12412
160	10158	10442	10713	10979	11242	11498	11745
170	9579	9864	10140	10408	10671	10925	11178
180	9053	9339	9616	9884	10147	10404	10656
190	8574	8861	9138	9408	9668	9923	10177
200	8138	8424	8702	8969	9231	9486	9733
210	7743	8027	8302	8570	8829	9083	9331
220	7388	7667	7940	8204	8462	8714	8960
230	7066	7342	7611	7873	8128	8377	8621
240	6774	7049	7314	7571	7821	8068	8308
250	6520	6787	7045	7297	7546	7786	8022
260	6286	6550	6803	7047	7290	7528	7760

TABLE XIII. Thermal Conductivity of Argon (*Continued*)

T, K	$\lambda \cdot 10^8$, kW/(m · K) at p, bar						
	700	750	800	850	900	950	1000
270	6080	6335	6581	6827	7059	7295	7522
280	5895	6144	6384	6620	6854	7081	7303
290	5727	5972	6208	6439	6665	6889	7104
300	5583	5818	6048	6272	6494	6711	6925
310	5453	5677	5905	6124	6335	6551	6756
320	5335	5554	5773	5993	6193	6407	6606
330	5230	5445	5656	5869	6066	6270	6469
340	5138	5347	5551	5758	5952	6154	6345
350	5057	5261	5458	5659	5850	6049	6234
375	4891	5082	5267	5450	5632	5813	5984
400	4770	4949	5123	5291	5459	5630	5794
425	4686	4853	5015	5170	5331	5488	5645
450	4626	4782	4936	5081	5231	5377	5528
475	4586	4729	4880	5013	5155	5291	5436
500	4561	4695	4840	4966	5101	5229	5367
550	4544	4666	4797	4914	5034	5152	5279
600	4561	4673	4792	4901	5010	5121	5235
650	4602	4707	4814	4916	5015	5121	5223
700	4660	4757	4852	4950	5042	5141	5235
750	4732	4821	4909	4998	5086	5177	5264
800	4812	4894	4976	5059	5142	5226	5308
850	4899	4976	5053	5131	5208	5286	5363
900	4992	5064	5137	5210	5282	5355	5427
950	5088	5156	5225	5293	5362	5430	5498
1000	5187	5251	5316	5381	5446	5511	5576
1050	5288	5349	5411	5473	5534	5596	5658
1100	5391	5449	5507	5566	5625	5684	5743
1150	5495	5551	5606	5663	5719	5775	5831
1200	5600	5653	5706	5760	5813	5867	5921
1250	5705	5756	5807	5858	5910	5961	6013
1300	5810	5859	5908	5957	6006	6056	6105

E. KRYPTON TABLE DISCUSSION

Tables of thermodynamic properties of krypton in the three phases over such a wide range of parameters (0 to 1300 K and 1 to 1000 bar) were calculated for the first time.

Previously Trappeniers et al. [2.71] used their experimental data on the compressibility of gaseous krypton to calculate tables graphically for its thermodynamic properties at temperatures between 0 and 150°C. Subsequently Vukalovich et al. [III.8] used the *pvT* data in [2.71] for constructing the virial coefficient of state which would be valid over a wider temperature range. For this purpose, Altunin and Gvozdkov [2.80] suggested a method of extrapolation of experimental data to high temperatures, based on the fact that the isolines of $(z - 1)\,v$ = const on the ρ, T diagram are close to a straight line. The latter gave a relatively reliable extrapolation to the second virial coefficient at $\rho \to 0$. The values of B_2 were found by Vukalovich et al. [III.8] using the (18:6) potential with parameters $\epsilon/k = 221.3$ K and $b_0 = 0.6435$ cm³/g. The tables of thermodynamic properties of gaseous krypton at temperatures from 250 to 1300 K and pressures of 1 to 1000 bar calculated from the virial equation of state of Vukalovich et al. are published in Vargaftik's handbook [III.9].

The values of enthalpy, entropy and specific heat c_p calculated by us were compared with analogous data in [2.71, III.8] in Tables 60 through 62. It is seen from Table 60 that on the whole the agreement is rather good, particularly over

TABLE 60†. Comparison of Computed Values of the Enthalpy of Krypton with Tabulated Data in [III.8] (2nd line) and [2.71] (3rd line)

T, K	*h*, kJ/kg, at pressure, bar					
	50	100	200	300	600	1000
250	180.4	157.5	134.0	130.6	133.1	142.4
	180.1	157.6	133.8	130.2	132.3	141.4
300	197.9	186.6	166.8	158.0	155.2	162.3
	197.8	186.3	166.7	157.9	155.0	162.2
	197.9	186.4	166.9	158.1	154.9	162.1
350	212.9	205.2	191.5	182.8	176.6	181.9
	213.0	205.2	191.5	182.8	176.5	182.0
	213.0	205.1	191.5	182.8	176.6	181.9
400	227.1	221.3	211.4	204.1	196.8	200.8
	227.2	221.6	211.3	204.1	197.0	201.2
	227.0	221.2	211.3	204.0	196.8	201.1
500	254.1	250.5	244.5	240.0	234.2	237.4
	254.2	250.7	244.6	240.1	234.4	237.5
600	280.2	277.9	274.2	271.4	268.1	271.7
	280.3	278.0	274.2	271.3	268.0	271.3
800	331.4	330.5	329.3	328.7	329.4	334.6
	331.3	330.4	329.1	328.3	328.5	333.6
1000	381.8	381.8	381.9	382.5	385.9	392.8
	381.7	381.6	381.6	381.9	384.5	391.0
1200	431.9	432.4	433.4	434.7	439.8	448.2
	431.8	432.2	433.0	434.1	438.4	446.1

† Here and in Table 61 the data in [III.8, 2.71] were increased by a scaling constant in order to allow equating the values of enthalpy and entropy at p = 1 bar and T = 300 K.

TABLE 61. Comparison of Computed Values of the Entropy of Krypton with Tabulated Data in [III.8] (2nd line) and [2.71] (3rd line)

T, K	s, kJ/(kg · K), at pressure, bar					
	50	100	200	300	600	1000
250	1,484	1,344	1,213	1,173	1,117	1,074
	1,483	1,345	1,213	1,172	1,114	1,071
300	1,548	1,451	1,333	1,273	1,196	1,149
	1,548	1,450	1,333	1,273	1,196	1,147
	1,548	1,451	1,333	1,273	1,197	1,148
350	1,594	1,509	1,410	1,349	1,262	1,208
	1,595	1,509	1,410	1,350	1,263	1,209
	1,594	1,510	1,410	1,350	1,264	1,209
400	1,632	1,552	1,463	1,407	1,317	1,259
	1,633	1,553	1,463	1,407	1,318	1,260
	1,632	1,552	1,463	1,407	1,318	1,260
500	1,693	1,617	1,537	1,487	1,400	1,340
	1,693	1,618	1,537	1,488	1,401	1,341
600	1,740	1,667	1,591	1,545	1,462	1,403
	1,740	1,667	1,591	1,544	1,462	1,402
800	1,814	1,743	1,671	1,627	1,551	1,494
	1,814	1,743	1,670	1,626	1,549	1,492
1000	1,870	1,800	1,729	1,687	1,614	1,559
	1,871	1,801	1,729	1,687	1,613	1,557
1200	1,916	1,846	1,777	1,735	1,663	1,609
	1,917	1,847	1,776	1,735	1,662	1,607

TABLE 62. Comparison of Computed Values of the Isobaric Specific Heat of Krypton with Tabulated Data in [III.8] (2nd line) and [2.71] (3rd line).

T, K	c_p, kJ/(kg · K), at pressure, bar					
	50	100	200	300	600	1000
250	0,404	0,917	0,685	0,551	0,441	0,402
	0,406	0,889	0,695	0,566	0,471	0,435
300	0,317	0,424	0,580	0,532	0,436	0,404
	0,321	0,428	0,577	0,534	0,440	0,404
	0,318	0,425	0,576	0,533	0,438	0,402
350	0,290	0,340	0,431	0,458	0,417	0,385
	0,292	0,343	0,434	0,461	0,421	0,390
	0,291	0,341	0,433	0,459	0,420	0,389
400	0,277	0,307	0,363	0,397	0,395	0,372
	0,277	0,309	0,366	0,398	0,398	0,376
	0,278	0,308	0,365	0,398	0,397	0,376
500	0,265	0,281	0,309	0,331	0,354	0,353
	0,264	0,280	0,309	0,330	0,353	0,350
600	0,259	0,269	0,287	0,301	0,325	0,332
	0,258	0,268	0,286	0,300	0,321	0,327
800	0,254	0,259	0,268	0,275	0,292	0,301
	0,253	0,258	0,267	0,274	0,288	0,297
1000	0,251	0,254	0,260	0,264	0,275	0,283
	0,251	0,254	0,259	0,264	0,274	0,281
1200	0,250	0,252	0,255	0,258	0,266	0,272
	0,250	0,252	0,256	0,259	0,266	0,272

the range of parameters for which the same starting *pvT* data were used. The differences increase slightly over ranges where the equations of state were constructed on the basis of extrapolated data. Most likely for the high temperature and pressure range this is due to different extrapolation methods used by us and by Vukalovich et al. [III.8]. A comparison could be performed only at $T \geqslant 250$ K. Our values of enthalpy at this temperature are basically higher. But they are more accurate since the equation of state compiled by Vukalovich is not based on compressibility data of krypton at $T < 273.15$ K.

The agreement of values of entropy in Table 61 is quite good. Apparently it is less sensitive to the accuracy of description of the original thermal data.

The values of isobaric specific heat c_p (Table 62) agree basically within $\pm 1\%$. However, the deviations increase with a decrease in temperature and become as high as 8% at $T = 250$ K and $p = 1000$ bar. This is due to the same considerations as for enthalpy.

Table 63 compares the values of c_p' calculated by us for the liquid side of the saturation curve with data of Gladun [3.64]. The agreement as a whole should be regarded as entirely satisfactory.

The lack of direct experimental data on calorific properties of krypton makes it impossible to estimate more precisely the possible error of the thermodynamic functions presented in summary tables XV through XVIII. However, the results of the comparison of calorific quantities presented here, and also of detailed comparison with experimental *pvT* data (see chapters 2 and 3), allow estimation of the confidence intervals of error of the values recommended by us as: $\delta v = +0.1\%$; $\Delta h = \pm 0.15$ kJ/kg; $\Delta s = \pm 0.001$ kJ/(kg · K) and $\delta c_p = \pm 3\%$. These errors may be slightly higher in the transcritical region and for points used in splicing values obtained from separate equations for the gas and liquid.

The possible error of the computed values of thermodynamic functions of crystalline krypton listed in Table XIV are estimated on p. xxx, and the values of coefficients of dynamic viscosity and thermal conductivity of liquid and gaseous krypton (Tables XIX and XX) on pages xxx and xxx, respectively.

TABLE 63. Comparison of Computed Values of the Specific Heat c_p' of Krypton with Data in [3.64]

T, K	c_p, kJ/(kg · K) computed value	c_p, kJ/(kg · K) from data in [3.64]	T, K	c_p, kJ/(kg · K) computed value	c_p, kJ/(kg · K) from data in [3.64]
116	0.547	0.536	160	0.587	0.599
120	0.544	0.537	165	0.610	0.622
125	0.543	0.541	170	0.641	0.649
130	0.542	0.544	175	0.684	0.685
135	0.543	0.549	180	0.734	0.735
140	0.546	0.554	185	0.803	0.803
145	0.551	0.561	190	0.905	0.906
150	0.559	0.570	195	1.100	1.060
155	0.571	0.582	200	1.515	1.301†

† This point drops out from the monotonic curve in coordinates $(1/c_p', T)$.

TABLE XIV. Thermodynamic Properties of Crystalline Krypton

T, K	p = 1 bar						
	$v \cdot 10^3$	h	s	c_p	c_v	$\alpha_{exp} \cdot 10^5$	$\beta_T \cdot 10^5$
1	0.3233	0	0.000021	0.000064	0.000064	0.02	2.95
2	0.3233	0	0.000173	0.000531	0.000531	0.13	2.95
4	0.3233	0.01	0.001509	0.00482	0.00482	1.10	2.95
6	0.3233	0.04	0.005579	0.0180	0.0180	4.03	2.95
8	0.3234	0.10	0.01384	0.0420	0.0420	9.57	2.95
10	0.3235	0.22	0.02564	0.0707	0.0704	16.8	2.96
15	0.3239	0.75	0.06719	0.136	0.134	34.0	3.01
20	0.3246	1.59	0.1141	0.188	0.184	47.0	3.07
25	0.3255	2.64	0.1603	0.223	0.215	56.2	3.14
30	0.3265	3.84	0.2034	0.247	0.235	63.0	3.22
40	0.3288	6.49	0.2791	0.276	0.256	72.7	3.40
50	0.3313	9.37	0.3431	0.295	0.265	80.4	3.61
60	0.3341	12.40	0.3982	0.311	0.270	88.0	3.86
70	0.3372	15.57	0.4471	0.327	0.273	96.9	4.17
80	0.3407	18.97	0.4925	0.345	0.275	108.2	4.59
90	0.3446	22.58	0.5353	0.366	0.275	122.9	5.18
100	0.3492	26.42	0.5765	0.389	0.273	141.7	6.05
110	0.3544	30.52	0.6165	0.414	0.268	165.5	7.30
115.76	0.3579	33.18	0.6390	0.427	0.263	181.7	8.29

TABLE XIV. Thermodynamic Properties of Crystalline Krypton (*Continued*)

T, K	p = 100 bar						
	$v \cdot 10^3$	h	s	c_p	c_v	$\alpha_{exp} \cdot 10^5$	$\beta_T \cdot 10^5$
1	0.3223	3.20	0.000021	0.000062	0.000062	0.01	2.88
2	0.3223	3.20	0.000169	0.000519	0.000519	0.12	2.88
4	0.3223	3.21	0.001474	0.00471	0.00471	1.05	2.88
6	0.3224	3.23	0.005452	0.0176	0.0176	3.86	2.88
8	0.3224	3.30	0.01354	0.0412	0.0411	9.20	2.89
10	0.3225	3.41	0.02511	0.0695	0.0692	16.2	2.90
15	0.3230	3.94	0.06611	0.135	0.133	33.0	2.94
20	0.3237	4.77	0.1126	0.187	0.182	45.7	3.00
25	0.3245	5.81	0.1585	0.222	0.214	54.7	3.07
30	0.3255	7.00	0.2014	0.245	0.234	61.4	3.14
40	0.3277	9.64	0.2768	0.275	0.255	70.8	3.31
50	0.3301	12.51	0.3404	0.294	0.265	78.2	3.50
60	0.3328	15.53	0.3954	0.309	0.270	85.3	3.74
70	0.3358	18.67	0.4438	0.324	0.273	93.5	4.02
80	0.3392	22.04	0.4888	0.341	0.275	103.5	4.39

TABLE XIV. Thermodynamic Properties of Crystalline Krypton (*Continued*)

T. K	p = 100 bar						
	$v \cdot 10^3$	h	s	c_p	c_v	$\alpha_{exp} \cdot 10^5$	$\beta_T \cdot 10^5$
90	0,3429	25,61	0,5310	0,360	0.275	116.1	4,90
100	0,3471	29,38	0,5714	0,381	0,274	132,0	5,61
110	0,3520	33,39	0,6105	0,403	0,269	151,3	6,62
118.7	0,3569	37,18	0,6435	0,420	0,263	171,3	7,85

TABLE XIV. Thermodynamic Properties of Crystalline Krypton (*Continued*)

T, K	p = 200 bar						
	$v \cdot 10^3$	h	s	c_p	c_v	$\alpha_{exp} \cdot 10^5$	$\beta_T \cdot 10^5$
1	0,3214	6.41	0,000020	0.000061	0,000061	0,01	2,82
2	0.3214	6.42	0,000166	0,000507	0,000507	0,12	2,82
4	0.3214	6.43	0,001441	0.00460	0,00460	1,01	2,82
6	0,3215	6.45	0,005330	0,0172	0.0172	3,70	2,82
8	0,3215	6,51	0,01325	0.0404	0.0403	8,84	2,83
10	0,3216	6,63	0,02459	0,0683	0.0681	15,6	2,84
15	0,3220	7,15	0.06505	0,133	0,131	32,0	2,88
20	0.3227	7.97	0,1111	0,185	0.181	44,4	2,93
25	0,3235	9,01	0.1568	0,221	0,213	53,3	2,99
30	0.3245	10,19	0.1994	0,244	0,233	59,8	3,07
40	0,3266	12,82	0,2745	0,274	0,255	69,0	3,22
50	0,3290	15,68	0,3379	0,292	0,265	76,1	3,41
60	0,3316	18,68	0,3925	0,307	0,270	82,8	3,62
70	0,3345	21,81	0,4407	0,322	0,273	90,3	3,88
80	0,3377	25,14	0,4853	0,338	0,275	99,3	4,21
90	0,3413	28,66	0,5269	0,356	0,275	110,4	4,65
100	0,3452	32,38	0,5667	0,375	0,274	124,0	5,25
110	0,3497	36,31	0,6049	0,394	0,270	140,1	6,08
120	0,3549	40,46	0,6419	0,411	0,263	158,6	7,21
121 64	0,3559	41,18	0,6479	0,414	0,262	161,8	7,41

TABLE XIV. Thermodynamic Properties of Crystalline Krypton (*Continued*)

T. K	p = 300 bar						
	$v \cdot 10^3$	h	s	c_p	c_v	$\alpha_{exp} \cdot 10^5$	$\beta_T \cdot 10^5$
1	0,3205	9,63	0.000020	0,000060	0,000060	0,01	2,76
2	0,3205	9,63	0,000162	0,000496	0,000496	0,11	2,76
4	0,3205	9,63	0,001409	0,00450	0,00450	0,97	2,76
6	0,3206	9,66	0,005213	0,0169	0,0168	3,56	2,76
8	0,3206	9,72	0,01297	0,0396	0,0395	8,50	2,77

TABLE XIV. Thermodynamic Properties of Crystalline Krypton (*Continued*)

T, K	p = 300 bar						
	$v \cdot 10^3$	h	s	c_p	c_v	$\alpha_{exp} \cdot 10^5$	$\beta_T \cdot 10^5$
10	0.3207	9.83	0.02409	0.0672	0.0670	15.0	2.78
15	0.3211	10.35	0.06402	0.132	0.130	31.0	2.81
20	0.3218	11.16	0.1097	0.184	0.180	43.2	2.87
25	0.3225	12.19	0.1550	0.219	0.212	52.0	2.93
30	0.3235	13.37	0.1975	0.243	0.232	58.3	2.99
40	0.3255	15.99	0.2722	0.273	0.254	67.3	3.15
50	0.3279	18.83	0.3354	0.291	0.264	74.1	3.32
60	0.3304	21.83	0.3898	0.306	0.269	80.5	3.52
70	0.3332	24.94	0.4378	0.320	0.272	87.5	3.76
80	0.3363	28.24	0.4818	0.335	0.274	95.7	4.06
90	0.3397	31.73	0.5231	0.352	0.275	105.6	4.44
100	0.3435	35.40	0.5622	0.369	0.274	117.4	4.96
110	0.3477	39.26	0.5998	0.387	0.270	131.1	5.64
120	0.3524	43.31	0.6360	0.402	0.264	146.3	6.56
124.58	0.3549	45.35	0.6521	0.408	0.261	153.6	7.03

TABLE XIV. Thermodynamic Properties of Crystalline Krypton (*Continued*)

T, K	p = 400 bar						
	$v \cdot 10^3$	h	s	c_p	c_v	$\alpha_{exp} \cdot 10^5$	$\beta_T \cdot 10^5$
1	0.3196	12.82	0.000019	0.000058	0.000058	0.01	2.71
2	0.3196	12.83	0.000158	0.000485	0.000485	0.11	2.71
4	0.3196	12.83	0.001378	0.00440	0.00440	0.93	2.71
6	0.3197	12.86	0.005101	0.0165	0.0165	3.42	2.71
8	0.3197	12.92	0.01270	0.0388	0.0388	8.19	2.71
10	0.3198	13.03	0.02362	0.0661	0.0659	14.5	2.72
15	0.3202	13.54	0.06303	0.130	0.129	30.1	2.76
20	0.3208	14.34	0.1083	0.183	0.179	42.1	2.81
25	0.3216	15.37	0.1534	0.218	0.211	50.7	2.86
30	0.3225	16.54	0.1956	0.242	0.231	56.9	2.93
40	0.3245	19.15	0.2700	0.272	0.253	65.7	3.07
50	0.3268	21.99	0.3330	0.290	0.264	72.2	3.23
60	0.3293	24.97	0.3872	0.304	0.269	78.3	3.42
70	0.3320	28.07	0.4349	0.318	0.272	84.8	3.64
80	0.3349	31.34	0.4786	0.332	0.274	92.4	3.92
90	0.3382	34.79	0.5194	0.348	0.275	101.3	4.26
100	0.3418	38.42	0.5580	0.364	0.274	111.8	4.71
110	0.3458	42.22	0.5950	0.381	0.271	123.6	5.29
120	0.3502	46.20	0.6305	0.394	0.265	136.5	6.05
127.42	0.3539	49.20	0.6558	0.402	0.260	146.4	6.73

TABLE XIV. Thermodynamic Properties of Crystalline Krypton (*Continued*)

T, K	p = 500 bar						
	$v \cdot 10^3$	h	s	c_p	c_v	$\alpha_{exp} \cdot 10^5$	$\beta_T \cdot 10^5$
1	0,3188	16,02	0.000019	0.000057	0.000057	0.01	2,65
2	0,3188	16,02	0.000155	0.000475	0.000475	0,10	2,65
4	0,3188	16,03	0,001349	0.00431	0,00431	0,89	2,65
6	0,3188	16,05	0,004993	0.0162	0,0162	3,29	2,66
8	0,3189	16,11	0.01244	0,0381	0,0380	7,89	2,66
10	0,3189	16.22	0.02315	0,0651	0.0648	14,0	2,67
15	0.3193	16.72	0,06207	0.129	0.127	29.3	2,70
20	0,3199	17,52	0.1070	0.181	0,177	41.0	2,75
25	0.3207	18.54	0,1517	0,217	0,210	49,5	2.80
30	0,3216	19.70	0,1937	0,241	0,231	55.6	2,86
40	0,3235	22,31	0,2679	0,271	0,253	64.1	3,00
50	0,3257	25.13	0,3306	0,289	0,263	70,5	3,15
60	0,3281	28,10	0,3846	0.303	0,269	76,3	3,33
70	0,3308	31.19	0.4321	0.316	0,272	82.4	3,54
80	0,3336	34,43	0,4754	0,330	0,274	89.4	3,79
90	0,3368	37,85	0,5159	0.345	0.274	97,6	4.10
100	0,3402	41,44	0,5540	0,360	0,274	106,9	4,49
110	0.3440	45.19	0,5905	0,375	0,271	117,4	4,99
120	0,3482	49,11	0,6254	0,388	0,266	128,4	5,64
130	0,3528	53.19	0.6589	0.396	0,258	139,5	6.46
130,24	0,3530	53,19	0,6595	0,396	0.258	139,8	6,48

TABLE XIV. Thermodynamic Properties of Crystalline Krypton (*Continued*)

T, K	p = 600 bar						
	$v \cdot 10^3$	h	s	c_p	c_v	$\alpha_{exp} \cdot 10^5$	$\beta_T \cdot 10^5$
1	0,3179	19,20	0,000019	0,000056	0,000056	0.01	2,60
2	0,3179	19,20	0,000152	0,000465	0,000465	0.10	2,60
4	0.3179	19,21	0,001321	0,00422	0,00422	0.86	2,60
6	0,3180	19,23	0,004890	0.0158	0,0158	3.16	2,60
8	0,3180	19.29	0,01220	0.0374	0,0373	7.61	2,61
10	0.3181	19,40	0,02271	0.0641	0,0638	13.6	2,62
15	0,3185	19.89	0,06113	0.128	0,126	28.5	2,65
20	0,3191	20.69	0,1056	0,180	0,176	40,0	2,69
25	0,3198	21.70	0.1501	0.216	0,209	48,3	2.74
30	0,3206	22,86	0.1919	0.240	0,230	54.3	2,80
40	0,3226	25,45	0,2658	0,270	0,252	62,7	2,93
50	0,3247	28,27	0,3283	0.288	0,263	68.8	3,08
60	0.3271	31.23	0,3821	0.302	0.268	74,4	3.24
70	0,3296	34.30	0,4294	0.315	0,271	80.2	3,44
80	0,3324	37,52	0,4724	0,328	0.273	86.7	3.67

TABLE XIV. Thermodynamic Properties of Crystalline Krypton (*Continued*)

T, K	p = 600 bar						
	$v \cdot 10^3$	h	s	c_p	c_v	$\alpha_{exp} \cdot 10^5$	$\beta_T \cdot 10^5$
90	0.3355	40.91	0.5125	0.342	0.274	94.2	3.95
100	0.3387	44.47	0.5503	0.357	0.274	102.7	4.30
110	0.3423	48.18	0.5862	0.371	0.271	112.0	4.74
120	0.3462	52.04	0.6207	0.383	0.267	121.7	5.30
130	0.3506	56.05	0.6536	0.391	0.260	131.2	5.98
133.05	0.3520	57.30	0.6630	0.392	0.258	133.8	6.21

TABLE XIV. Thermodynamic Properties of Crystalline Krypton (*Continued*)

T, K	p = 700 bar						
	$v \cdot 10^3$	h	s	c_p	c_v	$\alpha_{exp} \cdot 10^5$	$\beta_T \cdot 10^5$
1	0.3171	22.37	0.000018	0.000055	0.000055	0.01	2.55
2	0.3171	22.38	0.000149	0.000456	0.000456	0.09	2.55
4	0.3171	22.39	0.001294	0.00413	0.00413	0.83	2.55
6	0.3171	22.41	0.004791	0.0155	0.0155	3.05	2.56
8	0.3172	22.46	0.01196	0.0367	0.0367	7.34	2.56
10	0.3173	22.57	0.02228	0.0631	0.0629	13.1	2.57
15	0.3176	23.06	0.06023	0.126	0.125	27.7	2.60
20	0.3182	23.85	0.1044	0.179	0.175	39.0	2.64
25	0.3189	24.85	0.1486	0.215	0.208	47.2	2.69
30	0.3198	26.01	0.1902	0.239	0.229	53.1	2.74
40	0.3216	28.59	0.2638	0.269	0.252	61.3	2.87
50	0.3237	31.40	0.3261	0.287	0.263	67.3	3.01
60	0.3260	34.35	0.3797	0.301	0.268	72.6	3.16
70	0.3285	37.41	0.4268	0.313	0.271	78.1	3.35
80	0.3311	40.60	0.4695	0.326	0.273	84.2	3.56
90	0.3341	43.97	0.5092	0.339	0.274	91.1	3.82
100	0.3373	47.49	0.5466	0.353	0.274	98.9	4.14
110	0.3407	51.16	0.5822	0.367	0.272	107.3	4.53
120	0.3445	54.98	0.6162	0.379	0.268	116.0	5.01
130	0.3486	58.93	0.6486	0.386	0.261	124.3	5.59
135.8	0.3511	61.35	0.6664	0.387	0.257	128.5	5.98

TABLE XIV. Thermodynamic Properties of Crystalline Krypton (*Continued*)

T, K	p = 800 bar						
	$v \cdot 10^3$	h	s	c_p	c_v	$\alpha_{exp} \cdot 10^5$	$\beta_T \cdot 10^5$
1	0.3163	25.54	0.000018	0.000054	0.000054	0.01	2.51
2	0.3163	25.54	0.000146	0.000447	0.000447	0.09	2.51
4	0.3163	25.55	0.001269	0.00405	0.00405	0.80	2.51
6	0.3163	25.58	0.004696	0.0152	0.0152	2.94	2.51
8	0.3164	25.63	0.01173	0.0361	0.0360	7.09	2.51

TABLE XIV. Thermodynamic Properties of Crystalline Krypton (*Continued*)

T, K	p = 800 bar						
	$v \cdot 10^3$	h	s	c_p	c_v	$\alpha_{exp} \cdot 10^5$	$\beta_T \cdot 10^5$
10	0,3165	25.73	0.02187	0.0621	0.0619	12.7	2.52
15	0,3168	26.22	0.05935	0.125	0.124	27.0	2,55
20	0.3174	27.00	0.1031	0.178	0,174	38.1	2,59
25	0.3181	28.00	0.1471	0.214	0,207	46.1	2,64
30	0,3189	29.15	0.1885	0.238	0,228	52.0	2.69
40	0.3209	31.72	0,2618	0.268	0,251	60.0	2,81
50	0.3227	34.52	0.3239	0.286	0.262	65.8	2.94
60	0.3250	37.46	0,3773	0,299	0,268	70.9	3.09
70	0.3274	40.51	0.4243	0.312	0,271	76.1	3.26
80	0.3300	43.68	0.4667	0.324	0.273	81.9	3.46
90	0.3328	47.03	0.5061	0.337	0,274	88.3	3.70
100	0.3359	50.52	0.5432	0.350	0.273	95.5	3.99
110	0,3392	54.16	0.5784	0,363	0.272	103,2	4.34
120	0.3428	57.93	0.6119	0.375	0,268	111.0	4.76
130	0.3467	61.83	0.6440	0.382	0,262	118.4	5.27
138.52	0.3502	65.30	0,6696	0.383	0,255	123.8	5.77

TABLE XIV. Thermodynamic Properties of Crystalline Krypton (*Continued*)

T, K	p = 900 bar						
	$v \cdot 10^3$	h	s	c_p	c_v	$\alpha_{exp} \cdot 10^5$	$\beta_T \cdot 10^5$
1	0.3155	28.70	0,000017	0,000053	0.000053	0.01	2.46
2	0.3155	28.70	0.000143	0.000438	0.000438	0.09	2.46
4	0.3155	28.71	0.001244	0.00397	0.00397	0.77	2.46
6	0,3155	28.73	0.004605	0.0149	0.0149	2.83	2.46
8	0.3156	28.79	0.01151	0.0354	0.0354	6.85	2.47
10	0.3157	28.89	0.02147	0.0612	0,0610	12.3	2.47
15	0.3160	29.37	0.05850	0.124	0.123	26.3	2,50
20	0.3166	30.14	0.1019	0.176	0.173	37.2	2,54
25	0.3172	31.14	0,1456	0,212	0.206	45.1	2,59
30	0,3180	32.28	0.1868	0.237	0.228	50.9	2.64
40	0.3198	34.85	0.2599	0.267	0,251	58.7	2.75
50	0.3218	37.64	0,3218	0,285	0.262	64.4	2,87
60	0.3240	40,57	0,3750	0,298	0,267	69.3	3,02
70	0,3263	43.60	0.4218	0,310	0,270	74.3	3.18
80	0,3288	46,76	0.4639	0,322	0.272	79.7	3.37
90	0.3316	50,08	0,5031	0,335	0.273	85.8	3,59
100	0,3345	53.55	0.5399	0.348	0.273	92.4	3.85
110	0,3377	57.15	0.5747	0,360	0,272	100.0	4,17
120	0,3412	60.88	0,6019	0.371	0,268	106.7	4.54
130	0,3449	64.74	0,6396	0,378	0,263	113.3	4.99
140	0,3489	68.70	0,6696	0,380	0.255	118.8	5,51
141.16	0,3493	69.20	0.6730	0.380	0.254	119.4	5.57

TABLE XIV. Thermodynamic Properties of Crystalline Krypton (*Continued*)

T, K	p = 1000 bar						
	$v \cdot 10^3$	h	s	c_p	c_v	$\alpha_{exp} \cdot 10^5$	$\beta_T \cdot 10^5$
1	0.3147	31.85	0.000017	0.000052	0.000052	0.01	2.42
2	0.3147	31.85	0.000140	0.000429	0.000429	0.09	2.42
4	0.3148	31.86	0.001220	0.00389	0.00389	0.74	2.42
6	0.3148	31.88	0.004516	0.0146	0.0146	2.73	2.42
8	0.3148	31.94	0.01129	0.0348	0.0348	6.62	2.42
10	0.3149	32.04	0.02108	0.0603	0.0601	11.9	2.43
15	0.3152	32.51	0.05767	0.123	0.121	25.6	2.46
20	0.3157	33.28	0.1007	0.175	0.172	36.4	2.49
25	0.3164	34.27	0.1442	0.211	0.205	44.2	2.54
30	0.3172	35.41	0.1852	0.236	0.227	49.8	2.59
40	0.3189	37.96	0.2580	0.266	0.250	57.5	2.69
50	0.3209	40.74	0.3197	0.284	0.262	63.0	2.81
60	0.3231	43.67	0.3728	0.297	0.267	67.8	2.95
70	0.3253	46.69	0.4194	0.309	0.270	72.5	3.10
80	0.3277	49.83	0.4613	0.320	0.272	77.7	3.28
90	0.3304	53.13	0.5002	0.332	0.273	83.4	3.48
100	0.3333	56.57	0.5366	0.345	0.273	89.6	3.73
110	0.3363	60.14	0.5712	0.357	0.272	96.2	4.02
120	0.3397	63.84	0.6040	0.368	0.269	102.8	4.35
130	0.3432	67.66	0.6354	0.375	0.264	108.9	4.75
140	0.3470	71.57	0.6651	0.377	0.256	113.8	5.20
143.8	0.3484	73.00	0.6760	0.377	0.253	115.4	5.39

TABLE XV. Thermodynamic Properties of Krypton Along the Crystallization and Melting Curve

T, K	p_m	$v_l \cdot 10^3$	$v_c \cdot 10^3$	h_l	h_c	s_l	s_c	c_{pl}	c_{pc}
115.76†	0.73	0.4090	0.3579	52.78	33.18	0.8095	0.6390	0.547	0.427
116	9.15	0.4087	0.3578	53.30	33.51	0.8118	0.6394	0.546	0.426
117	42.48	0.4077	0.3575	54.66	34.86	0.8207	0.6410	0.540	0.424
118	76.00	0.4068	0.3572	56.04	36.22	0.8294	0.6425	0.535	0.422
119	109.7	0.4059	0.3568	57.44	37.57	0.8380	0.6440	0.530	0.420
120	143.5	0.4050	0.3565	58.87	38.92	0.8465	0.6455	0.526	0.417
121	177.6	0.4041	0.3562	60.30	40.30	0.8548	0.6469	0.522	0.415
122	211.8	0.4032	0.3558	61.75	41.65	0.8631	0.6484	0.518	0.413
123	246.2	0.4024	0.3554	63.21	43.04	0.8713	0.6498	0.514	0.411
124	280.7	0.4016	0.3551	64.68	44.43	0.8795	0.6512	0.510	0.409
125	315.5	0.4007	0.3548	66.16	45.82	0.8875	0.6526	0.506	0.407
126	350.4	0.3999	0.3544	67.64	47.22	0.8953	0.6540	0.502	0.405
127	385.5	0.3991	0.3541	69.13	48.63	0.9032	0.6553	0.498	0.403
128	420.7	0.3983	0.3537	70.62	50.03	0.9111	0.6567	0.495	0.401
129	456.1	0.3975	0.3534	72.11	51.43	0.9190	0.6580	0.492	0.399
130	491.7	0.3968	0.3531	73.61	52.84	0.9268	0.6593	0.488	0.397
131	527.5	0.3961	0.3527	75.11	54.28	0.9346	0.6605	0.484	0.395
132	563.4	0.3954	0.3524	76.61	55.73	0.9423	0.6617	0.480	0.393
133	599.5	0.3947	0.3520	78.11	57.20	0.9500	0.6630	0.477	0.392
134	635.7	0.3940	0.3517	79.61	58.68	0.9577	0.6642	0.474	0.390
135	672.2	0.3933	0.3514	81.11	60.14	0.9653	0.6654	0.471	0.389
136	708.8	0.3926	0.3510	82.62	61.62	0.9729	0.6666	0.468	0.387

TABLE XV. Thermodynamic Properties of Krypton Along the Crystallization and Melting Curve (*Continued*)

T. K	p_m	$v_l \cdot 10^3$	$v_c \cdot 10^3$	h_l	h_c	s_l	s_c	c_{pl}	c_{pc}
137	745.5	0.3920	0.3507	84.13	63.11	0.9805	0.6678	0.465	0.386
138	782.4	0.3914	0.3504	85.64	64.58	0.9880	0.6690	0.462	0.384
139	819.5	0.3907	0.3500	87.16	66.04	0.9955	0.6703	0.460	0.383
140	856.8	0.3901	0.3497	88.68	67.50	1.003	0.6715	0.458	0.382
141	894.2	0.3895	0.3494	90.20	68.95	1.077	0.6727	0.455	0.381
142	931.8	0.3889	0.3490	91.73	70.40	1.151	0.6739	0.453	0.379
143	969.5	0.3883	0.3487	93.26	71.84	1.224	0.6751	0.451	0.378
144	1007	0.3878	0.3484	94.79	73.27	1.297	0.6763	0.449	0.377

† It is found from Eq. (1.65) that at the triple point $p = 1.17$ bar.

TABLE XVI. Thermodynamic Properties of Saturated Krypton

T. K	p_v	$v'' \cdot 10^3$	$v' \cdot 10^3$	h''	h'	s''	s'	c''_p	c'_p
115.76	0.7319	152.9	0.4090	161.8	52.78	1.751	0.8095	0.259	0.547
116	0.7468	150.1	0.4094	161.8	52.90	1.749	0.8107	0.259	0.547
117	0.8116	139.1	0.4107	162.0	53.45	1.743	0.8153	0.260	0.546
118	0.8806	129.1	0.4120	162.2	54.00	1.737	0.8200	0.261	0.546
119	0.9541	120.0	0.4133	162.4	54.55	1.730	0.8246	0.262	0.545
119.76	1.013	113.6	0.4143	162.6	54.99	1.726	0.8279	0.263	0.545
120	1.032	111.6	0.4146	162.6	55.09	1.724	0.8291	0.264	0.544
121	1.115	104.0	0.4159	162.4	55.64	1.718	0.8336	0.265	0.544

TABLE XVI. Thermodynamic Properties of Saturated Krypton (*Continued*)

T, K	p_v	$v'' \cdot 10^3$	$v' \cdot 10^3$	h''	h'	s''	s'	c''_p	c'_p
122	1,203	96,99	0.4172	162.7	56,18	1,713	0.8381	0,266	0,544
123	1,296	90.57	0.4186	163.0	56,73	1.707	0,8425	0.267	0,544
124	1,395	84,68	0,4199	163.2	57,27	1,701	0.8469	0,269	0.543
125	1,499	79,25	0,4213	163.4	57.82	1,696	0,8513	0,270	0.543
126	1,609	74,26	0.4227	163.5	58.37	1,690	0.8555	0,272	0.543
127	1,725	69,66	0.4241	163.7	58,91	1,685	0.8597	0,273	0,543
128	1,848	65,41	0.4255	163,8	59.45	1,679	0,8640	0.275	0.542
129	1,976	61,48	0.4269	164,0	60,00	1.674	0.8682	0,276	0,542
130	2,112	57,84	0,4284	164,1	60,55	1,669	0,8724	0.278	0,542
131	2,254	54,47	0,4299	164.3	61,09	1,664	0,8766	0,280	0,542
132	2,403	51.34	0,4313	164.4	61,64	1,659	0,8806	0,281	0.542
133	2.560	48,44	0,4329	164.5	62,18	1,654	0,8847	0,283	0,543
134	2,724	45.74	0,4344	164,6	62.73	1,649	0.8887	0,285	0.543
135	2,896	43,22	0,4359	164,8	63,28	1,645	0,8928	0,287	0,543
136	3,075	40,87	0,4375	164,9	63,83	1,640	0,8968	0,289	0,544
137	3,263	38,69	0,4391	165,0	64.37	1,635	0,9007	0,291	0,544
138	3.459	36.65	0.4407	165,1	64,92	1,631	0,9046	0,293	0,545
139	3,664	34,74	0,4423	165,2	65,47	1.626	0,9085	0,295	0,545
140	3,878	32,95	0,4440	165,3	66,02	1,622	0.9124	0,297	0,546
141	4.101	31,28	0,4456	165,4	66,57	1,617	0,9162	0,299	0,547
142	4,333	29.72	0,4473	165.5	67,12	1.613	0,9201	0,301	0,548
143	4,574	28,25	0,4491	165,6	67,67	1,609	0,9238	0,304	0,549
144	4,826	26,87	0,4508	165,7	68,23	1,605	0,9276	0,306	0,550
145	5,087	25,57	0,4526	165,8	68,78	1,601	0,9314	0,308	0,551

TABLE XVI. Thermodynamic Properties of Saturated Krypton (*Continued*)

T, K	p_v	$v'' \cdot 10^3$	$v' \cdot 10^3$	h''	h'	s''	s'	c''_p	c'_p
146	5.359	24.35	0.4544	165.9	69.34	1.596	0.9351	0.311	0.552
147	5.641	23.20	0.4562	166.0	69.90	1.592	0.9389	0.314	0.553
148	5.933	22.12	0.4581	166.0	70.46	1.588	0.9426	0.316	0.555
149	6.237	21.10	0.4600	166.1	71.02	1.584	0.9463	0.319	0.557
150	6.552	20.14	0.4619	166.1	71.58	1.580	0.9499	0.322	0.559
151	6.878	19.23	0.4639	166.2	72.15	1.576	0.9536	0.325	0.561
152	7.215	18.37	0.4659	166.2	72.71	1.573	0.9572	0.328	0.563
153	7.565	17.56	0.4679	166.3	73.29	1.569	0.9609	0.331	0.565
154	7.927	16.79	0.4699	166.3	73.86	1.565	0.9645	0.334	0.568
155	8.301	16.06	0.4720	166.4	74.43	1.561	0.9681	0.337	0.571
156	8.688	15.37	0.4742	166.4	75.01	1.557	0.9717	0.341	0.574
157	9.087	14.72	0.4763	166.4	75.59	1.554	0.9753	0.344	0.577
158	9.500	14.10	0.4786	166.4	76.17	1.550	0.9789	0.348	0.580
159	9.926	13.51	0.4808	166.4	76.76	1.546	0.9824	0.352	0.583
160	10.37	12.95	0.4831	166.4	77.34	1.543	0.9859	0.356	0.587
161	10.82	12.42	0.4855	166.4	77.94	1.539	0.9895	0.360	0.591
162	11.29	11.91	0.4879	166.4	78.53	1.535	0.9931	0.364	0.595
163	11.77	11.43	0.4903	166.4	79.13	1.532	0.9966	0.369	0.600
164	12.26	10.97	0.4928	166.4	79.74	1.528	1.000	0.374	0.605
165	12.78	10.53	0.4954	166.3	80.35	1.525	1.004	0.379	0.610
166	13.30	10.12	0.4980	166.3	80.96	1.521	1.007	0.384	0.615
167	13.84	9.719	0.5007	166.2	81.58	1.518	1.011	0.389	0.620
168	14.40	9.339	0.5034	166.2	82.21	1.514	1.014	0.395	0.626
169	14.98	8.976	0.5062	166.1	82.84	1.511	1.018	0.401	0.633
170	15.57	8.628	0.5091	166.0	83.48	1.507	1.022	0.407	0.641

TABLE XVI. Thermodynamic Properties of Saturated Krypton (*Continued*)

T, K	p_v	$v'' \cdot 10^3$	$v' \cdot 10^3$	h''	h'	s''	s'	c''_p	c'_p
171	16.17	8,296	0.5120	165.9	84,12	1.504	1,025	0.414	0.650
172	16.79	7.978	0.5150	165.8	84.77	1.500	1,029	0.421	0.658
173	17.43	7.674	0.5181	165.7	85.43	1.497	1.032	0.429	0.666
174	18.09	7.382	0.5213	165.6	86,09	1.493	1.036	0.436	0,675
175	18.77	7.103	0.5245	165.5	86.76	1,489	1,040	0.445	0,684
176	19.46	6.834	0.5279	165.3	87.44	1.486	1.043	0.454	0.692
177	20.17	6.577	0.5313	165,2	88.13	1.482	1.047	0,463	0.701
178	20.90	6.330	0,5349	165.0	88,83	1.479	1.051	0.474	0.712
179	21.65	6.092	0,5385	164.8	89.54	1.475	1.054	0.485	0.723
180	22.41	5.864	0.5423	164.6	90.26	1.472	1.058	0.496	0.734
181	23.20	5.644	0.5462	164.4	90.99	1.468	1.062	0.509	0.746
182	24.00	5.432	0.5502	164.2	91.73	1.464	1.066	0.523	0.759
183	24.83	5.228	0.5543	164,0	92.48	1.461	1.070	0.537	0.772
184	25.68	5.032	0.5585	163.7	93.25	1.457	1.074	0,553	0.786
185	26.55	4.842	0.5729	163.5	94.03	1.453	1.078	0,571	0.803
186	27.43	4.659	0.5676	163.2	94.82	1.449	1.082	0.590	0.820
187	28.34	4.482	0.5725	162.9	95.63	1.445	1.086	0.611	0.838
188	29.27	4.312	0.5776	162.5	96.47	1.441	1.090	0.634	0.858
189	30.22	4.146	0.5828	162.2	97.32	1.437	1.094	0.659	0.879
190	31.20	3.986	0.5882	161.8	98.19	1.433	1.098	0.687	0.905
191	32.19	3.831	0.5938	161.4	99.08	1.429	1.103	0.718	0.935
192	33.21	3.680	0.5997	161.0	99.99	1.425	1.107	0.754	0.968
193	34.26	3.534	0.6060	160.5	100.9	1.420	1.112	0.793	1.007

TABLE XVI. Thermodynamic Properties of Saturated Krypton (*Continued*)

T, K	p_v	$v'' \cdot 10^3$	$v' \cdot 10^3$	h''	h'	s''	s'	c''_p	c'_p
194	35.33	3.392	0.6127	160.0	101.9	1.416	1.116	0.838	1,050
195	36,41	3.253	0.6199	159.5	102,9	1.411	1.121	0,890	1.100
196	37.53	3.118	0.6275	158.9	103.9	1.407	1,126	0,950	1,155
197	38.67	2,986	0.6357	158,3	105.0	1.402	1.131	1,020	1.221
198	39.83	2.857	0.6444	157.6	106.1	1,397	1.136	1.102	1.302
199	41.02	2.730	0,6539	156.9	107.2	1.391	1.142	1.201	1,399
200	42.23	2.606	0.6641	156.0	108.4	1.386	1.147	1.321	1.515
201	43.47	2,483	0,6755	155.2	109.7	1.380	1.153	1.471	1.667
202	44.74	2.361	0.6881	154.3	111.0	1.373	1.159	1.661	1.855
203	46.03	2.239	0.7022	153.1	112.4	1.366	1,166	1.911	2.105
204	47.36	2.117	0.7185	152.0	113.9	1.359	1.173	2.251	2.445
205	48.70	1.994	0.7377	150.6	115.5	1.351	1.180	2.740	2.944
206	50.08	1.862	0.7606	149.0	117.4	1.342	1.188	3.494	3.718
207	51.49	1.724	0.7880	147.1	119.7	1.330	1.198	5.02	5.21
208	52.92	1.575	0.8278	144.3	122.4	1.315	1.209	8.21	8,70
209	54.38	1.366	0.9132	139.8	126.8	1.291	1.227	28.6	31.3
209,39	54.96	1.098		133.9		1.262		∞	

TABLE XVII. Thermodynamic Properties of Liquid and Gaseous Krypton

T, K	p = 1 bar			p = 5 bar			p = 10 bar		
	$v \cdot 10^3$	h	s	$v \cdot 10^3$	h	s	$v \cdot 10^3$	h	s
120	115.3	162.6	1.728	0.4143	55.19	0.8286	0.4139	55.32	0.8279
125	120.6	163.9	1.738	0.4210	57.89	0.8508	0.4206	58.03	0.8501
130	125.8	165.2	1.749	0.4281	60.62	0.8720	0.4276	60.74	0.8712
135	131.0	166.5	1.759	0.4357	63.33	0.8925	0.4351	63.44	0.8917
140	136.2	167.8	1.768	0.4438	66.04	0.9122	0.4432	66.15	0.9114
145	141.3	169.1	1.777	26.07	165.9	1.603	0.4519	68.87	0.9305
150	146.4	170.4	1.786	27.25	167.4	1.613	0.4613	71.64	0.9492
155	151.5	171.7	1.794	28.41	168.9	1.623	0.4717	74.45	0.9677
160	156.6	172.9	1.802	29.55	170.4	1.632	13.52	166.7	1.548
165	161.7	174.2	1.810	30.68	171.6	1.641	14.18	168.4	1.558
170	166.8	175.5	1.818	31.79	173.2	1.649	14.83	170.1	1.568
175	171.9	176.8	1.826	32.90	174.6	1.657	15.45	171.7	1.577
180	176.9	178.0	1.833	33.99	176.0	1.665	16.06	173.3	1.586
185	182.0	179.3	1.840	35.08	177.4	1.673	16.66	174.8	1.595
190	187.0	180.6	1.847	36.15	178.8	1.680	17.25	176.3	1.603
195	192.0	181.8	1.853	37.22	180.1	1.687	17.83	177.8	1.611
200	197.1	183.1	1.859	38.29	181.5	1.694	18.41	179.3	1.618
205	202.1	184.3	1.865	39.35	182.8	1.701	18.98	180.8	1.625
210	207.1	185.6	1.871	40.40	184.1	1.707	19.54	182.2	1.632
215	212.1	186.9	1.877	41.45	185.3	1.713	20.10	183.6	1.639
220	217.1	188.1	1.883	42.50	186.7	1.719	20.65	185.0	1.645
225	222.1	189.4	1.889	43.54	188.1	1.725	21.20	186.4	1.651
230	227.1	190.6	1.895	44.58	189.4	1.731	21.75	187.8	1.657

TABLE XVII. Thermodynamic Properties of Liquid and Gaseous Krypton (*Continued*)

T, K	p = 1 bar			p = 5 bar			p = 10 bar		
	$v \cdot 10^3$	h	s	$v \cdot 10^3$	h	s	$v \cdot 10^3$	h	s
235	232.2	191.9	1.900	45.62	190.7	1.737	22.29	189.2	1.663
240	237.2	193.1	1.905	46.66	192.0	1.742	22.83	190.5	1.669
245	242.2	194.3	1.910	47.69	193.2	1.747	23.37	191.8	1.675
250	247.2	195.6	1.915	48.72	194.5	1.752	23.90	193.1	1.680
260	257.1	198.1	1.925	50.77	197.1	1.762	24.97	195.8	1.690
270	267.1	200.6	1.934	52.82	199.7	1.772	26.03	198.5	1.700
280	277.1	203.1	1.943	54.86	202.2	1.781	27.08	201.1	1.710
290	287.1	205.6	1.952	56.90	204.8	1.790	28.13	203.7	1.719
300	297.1	208.1	1.961	58.93	207.3	1.799	29.17	206.3	1.728
320	317.0	213.1	1.977	62.99	212.4	1.815	31.24	211.5	1.745
340	336.9	218.1	1.992	67.03	217.4	1.831	33.30	216.6	1.760
360	356.8	223.1	2.006	71.07	222.4	1.845	35.35	221.7	1.775
380	376.7	228.0	2.019	75.09	227.5	1.859	37.39	226.8	1.789
400	396.6	233.0	2.032	79.11	232.5	1.872	39.42	231.9	1.802
420	416.5	237.9	2.044	83.12	237.5	1.884	41.45	237.0	1.814
440	436.4	242.9	2.056	87.13	242.5	1.895	43.48	242.1	1.826
460	456.3	247.9	2.067	91.14	247.5	1.906	45.50	247.1	1.837
480	476.1	252.8	2.078	95.14	252.5	1.917	47.52	252.1	1.848
500	496.0	257.8	2.088	99.14	257.5	1.927	49.53	257.1	1.858
550	545.7	270.2	2.111	109.1	270.0	1.951	54.55	269.7	1.882
600	595.3	282.7	2.133	119.1	282.5	1.973	59.56	282.2	1.904
650	645.0	295.1	2.153	129.1	294.9	1.993	64.56	294.7	1.924
700	694.6	307.5	2.171	139.0	307.3	2.011	69.56	307.2	1.942
750	744.3	319.9	2.188	149.0	319.8	2.028	74.55	319.7	1.959

TABLE XVII. Thermodynamic Properties of Liquid and Gaseous Krypton (*Continued*)

T, K	p = 1 bar			p = 5 bar			p = 10 bar		
	$v \cdot 10^3$	h	s	$v \cdot 10^3$	h	s	$v \cdot 10^3$	h	s
800	793.9	332.3	2.204	158.9	332.2	2.044	79.53	332.2	1.975
850	843.5	344.7	2.219	168.8	344.6	2.059	84.51	344.6	1.990
900	893.1	357.1	2.233	178.8	357.1	2.074	89.49	357.0	2.005
950	942.8	369.5	2.247	188.7	369.5	2.087	94.47	369.5	2.018
1000	992.4	381.9	2.260	198.7	381.9	2.100	99.44	381.9	2.031
1050	1042.0	394.3	2.272	208.6	394.3	2.112	104.4	394.3	2.043
1100	1092.0	406.7	2.283	218.5	406.7	2.124	109.4	406.8	2.055
1150	1141.0	419.1	2.294	228.5	419.2	2.135	114.4	419.2	2.066
1200	1191.0	431.5	2.305	238.4	431.6	2.145	119.3	431.6	2.076
1250	1240.0	443.9	2.315	248.3	444.0	2.155	124.3	444.0	2.086
1300	1290.0	456.3	2.325	258.2	456.4	2.165	129.3	456.5	2.096

TABLE XVII. Thermodynamic Properties of Liquid and Gaseous Krypton (*Continued*)

T, K	p = 15 bar			p = 20 bar			p = 25 bar		
	$v \cdot 10^3$	h	s	$v \cdot 10^3$	h	s	$v \cdot 10^3$	h	s
120	0.4135	55.45	0.8273	0.4132	55.58	0.8266	0.4128	55.71	0.8260
125	0.4201	58.16	0.8494	0.4197	58.28	0.8488	0.4193	58.41	0.8481
130	0.4272	60.86	0.8705	0.4267	60.98	0.8698	0.4262	61.10	0.8691
135	0.4346	63.55	0.8909	0.4341	63.67	0.8901	0.4335	63.78	0.8894
140	0.4426	66.25	0.9105	0.4420	66.36	0.9097	0.4414	66.47	0.9089

TABLE XVII. Thermodynamic Properties of Liquid and Gaseous Krypton (*Continued*)

T, K	p = 15 bar			p = 20 bar			p = 25 bar		
	$v \cdot 10^3$	h	s	$v \cdot 10^3$	h	s	$v \cdot 10^3$	h	s
145	0.4512	68.97	0.9296	0.4505	69.06	0.9287	0.4498	69.16	0.9278
150	0.4605	71.72	0.9482	0.4597	71.80	0.9473	0.4588	71.89	0.9463
155	0.4707	74.52	0.9666	0.4697	74.59	0.9656	0.4688	74.66	0.9645
160	0.4820	77.38	0.9848	0.4808	77.44	0.9836	0.4797	77.49	0.9824
165	0.4947	80.36	1.003	0.4933	80.39	1.002	0.4919	80.42	1.001
170	9.058	166.5	1.513	0.5075	83.47	1.021	0.5057	83.47	1.019
175	9.547	168.4	1.524	0.5240	86.75	1.040	0.5217	86.70	1.038
180	10.02	170.2	1.534	6.910	166.7	1.491	0.5407	90.20	1.057
185	10.47	172.0	1.544	7.308	168.8	1.503	5.329	164.9	1.465
190	10.91	173.7	1.553	7.684	170.8	1.514	5.693	167.3	1.478
195	11.33	175.4	1.562	8.045	172.7	1.524	6.028	169.6	1.490
200	11.75	177.0	1.570	8.393	174.5	1.533	6.344	171.7	1.500
205	12.16	178.6	1.578	8.730	176.2	1.541	6.645	173.7	1.510
210	12.57	180.1	1.585	9.058	177.9	1.549	6.933	175.5	1.519
215	12.96	181.6	1.592	9.379	179.6	1.557	7.212	177.4	1.528
220	13.35	183.1	1.599	9.693	181.2	1.565	7.483	179.1	1.536
225	13.74	184.6	1.606	10.00	182.8	1.572	7.748	180.8	1.544
230	14.13	186.1	1.612	10.31	184.4	1.579	8.006	182.5	1.551
235	14.51	187.5	1.618	10.61	185.9	1.585	8.260	184.1	1.558
240	14.88	188.9	1.624	10.90	187.4	1.591	8 510	185.7	1.565
245	15.26	190.3	1.630	11.20	188.9	1.597	8.756	187.2	1.571
250	15.63	191.7	1.636	11.49	190.3	1.603	8.999	188.7	1.577

TABLE XVII. Thermodynamic Properties of Liquid and Gaseous Krypton (*Continued*)

T, K	p = 15 bar			p = 20 bar			p = 25 bar		
	$v \cdot 10^3$	h	s	$v \cdot 10^3$	h	s	$v \cdot 10^3$	h	s
260	16.37	194.5	1.647	12.06	193.1	1.614	9.476	191.7	1.589
270	17.09	197.2	1.657	12.63	196.0	1.625	9.944	194.7	1.600
280	17.81	199.9	1.667	13.18	198.8	1.635	10.40	197.6	1.610
290	18.53	202.6	1.676	13.74	201.5	1.645	10.86	200.4	1.620
300	19.24	205.3	1.685	14.28	204.2	1.654	11.31	203.2	1.630
320	20.66	210.6	1.702	15.37	209.6	1.672	12.19	208.7	1.647
340	22.06	215.8	1.718	16.43	215.0	1.688	13.06	214.1	1.664
360	23.44	221.0	1.733	17.49	220.3	1.703	13.92	219.5	1.679
380	24.82	226.2	1.747	18.54	225.5	1.717	14.77	224.8	1.694
400	26.20	231.3	1.760	19.58	230.7	1.730	15.62	230.1	1.707
420	27.56	236.4	1.773	20.62	235.9	1.743	16.46	235.4	1.720
440	28.93	241.5	1.785	21.65	241.0	1.755	17.29	240.6	1.732
460	30.29	246.6	1.796	22.68	246.1	1.766	18.12	245.7	1.743
480	31.64	251.7	1.807	23.71	251.2	1.777	18.95	250.8	1.754
500	32.99	256.7	1.817	24.73	256.3	1.788	19.77	255.9	1.765
550	36.36	269.4	1.841	27.27	269.0	1.812	21.81	268.7	1.789
600	39.72	282.0	1.863	29.80	281.7	1.834	23.85	281.5	1.811
650	43.07	294.5	1.883	32.32	294.3	1.854	25.87	294.1	1.832
700	46.41	307.0	1.902	34.83	306.9	1.873	27.89	306.7	1.851
750	49.74	319.5	1.919	37.34	319.4	1.890	29.90	319.3	1.868
800	53.07	332.0	1.935	39.85	331.9	1.906	31.91	331.9	1.884
850	56.40	344.5	1.950	42.35	344.4	1.921	33.91	344.4	1.899
900	59.73	357.0	1.964	44.84	356.9	1.936	35.91	356.9	1.913

TABLE XVII. Thermodynamic Properties of Liquid and Gaseous Krypton (*Continued*)

T. K	p = 15 bar			p = 20 bar			p = 25 bar		
	$v \cdot 10^3$	h	s	$v \cdot 10^3$	h	s	$v \cdot 10^3$	h	s
950	63.05	369.5	1.978	47.34	369.4	1.949	37.91	369.4	1.927
1000	66.37	381.9	1.991	49.83	381.9	1.962	39.91	381.9	1.940
1050	69.69	394.3	2.003	52.33	394.3	1.974	41.91	394.3	1.952
1100	73.01	406.8	2.014	54.82	406.8	1.986	43.90	406.8	1.963
1150	76.32	419.2	2.025	57.31	419.3	1.997	45.90	419.3	1.974
1200	79.64	431.7	2.036	59.79	431.7	2.007	47.89	431.7	1.985
1250	82.95	444.1	2.046	62.28	444.1	2.017	49.88	444.2	1.995
1300	86.27	456.5	2.056	64.77	456.6	2.027	51.87	456.6	2.005

TABLE XVII. Thermodynamic Properties of Liquid and Gaseous Krypton (*Continued*)

T. K	p = 30 bar			p = 35 bar			p = 40 bar		
	$v \cdot 10^3$	h	s	$v \cdot 10^3$	h	s	$v \cdot 10^3$	h	s
120	0.4124	55.84	0.8254	0.4121	55.97	0.8247	0.4117	56.10	0.8241
125	0.4189	58.53	0.8474	0.4185	58.66	0.8468	0.4181	58.79	0.8461
130	0.4258	61.22	0.8684	0.4253	61.34	0.8677	0.4249	61.46	0.8670
135	0.4330	63.90	0.8886	0.4325	64.01	0.8879	0.4320	64.13	0.8872
140	0.4408	66.57	0.9081	0.4402	66.68	0.9073	0.4396	66.79	0.9065
145	0.4491	69.26	0.9270	0.4484	69.36	0.9261	0.4477	69.46	0.9253
150	0.4581	71.98	0.9454	0.4573	72.07	0.9444	0.4565	72.16	0.9435
155	0.4678	74.74	0.9635	0.4669	74.81	0.9625	0.4660	74.89	0.9615
160	0.4786	77.55	0.9813	0.4775	77.61	0.9802	0.4764	77.67	0.9791
165	0.4905	80.45	0.9995	0.4892	80.49	0.9983	0.4879	80.53	0.9970

TABLE XVII. Thermodynamic Properties of Liquid and Gaseous Krypton (*Continued*)

T, K	p = 30 bar			p = 35 bar			p = 40 bar		
	$v \cdot 10^3$	h	s	$v \cdot 10^3$	h	s	$v \cdot 10^3$	h	s
170	0.5040	83.48	1.018	0.5024	83.49	1.016	0.5008	83.50	1.015
175	0.5195	86.66	1.036	0.5175	86.63	1.034	0.5155	86.61	1.033
180	0.5378	90.10	1.055	0.5351	90.02	1.053	0.5325	89.95	1.051
185	0.5602	93.90	1.076	0.5564	93.72	1.074	0.5528	93.57	1.072
190	4.279	163.1	1.442	0.5834	97.95	1.096	0.5782	97.67	1.093
195	4.627	165.9	1.457	3.537	161.2	1.423	0.6118	102.5	1.118
200	4.939	168.5	1.470	3.881	164.6	1.440	2.996	159.4	1.405
205	5.226	170.8	1.482	4.176	167.4	1.454	3.340	163.4	1.425
210	5.495	172.9	1.492	4.443	170.0	1.466	3.622	166.6	1.440
215	5.751	174.9	1.501	4.690	172.3	1.477	3.872	169.3	1.453
220	5.997	176.9	1.510	4.922	174.5	1.487	4.101	171.8	1.465
225	6.235	178.7	1.519	5.143	176.5	1.496	4.314	174.1	1.475
230	6.465	180.5	1.527	5.356	178.4	1.505	4.516	176.2	1.484
235	6.690	182.2	1.534	5.562	180.3	1.513	4.710	178.3	1.493
240	6.909	183.9	1.541	5.761	182.1	1.521	4.896	180.2	1.501
245	7.125	185.6	1.548	5.956	183.9	1.528	5.076	182.1	1.509
250	7.337	187.2	1.554	6.147	185.6	1.535	5.251	183.9	1.516
260	7.751	190.3	1.567	6.517	188.9	1.547	5.590	187.4	1.530
270	8.155	193.4	1.578	6.876	192.1	1.559	5.916	190.7	1.543
280	8.551	196.4	1.589	7.226	195.1	1.571	6.233	193.9	1.554
290	8.940	199.3	1.599	7.569	198.2	1.581	6.541	197.0	1.565
300	9.323	202.2	1.609	7.906	201.1	1.591	6.844	200.0	1.575
320	10.08	207.8	1.627	8.565	206.9	1.610	7.433	205.9	1.595
340	10.81	213.3	1.644	9.210	212.5	1.627	8.007	211.7	1.612
360	11.54	218.7	1.660	9.844	218.0	1.643	8.570	217.3	1.628

TABLE XVII. Thermodynamic Properties of Liquid and Gaseous Krypton (*Continued*)

T, K	p = 30 bar			p = 35 bar			p = 40 bar		
	$v \cdot 10^3$	h	s	$v \cdot 10^3$	h	s	$v \cdot 10^3$	h	s
380	12.26	224.1	1.674	10.47	223.5	1.657	9.124	222.8	1.643
400	12.97	229.5	1.688	11.09	228.9	1.671	9.670	228.3	1.657
420	13.68	234.8	1.701	11.70	234.2	1.684	10.21	233.7	1.670
440	14.38	240.0	1.713	12.31	239.5	1.697	10.75	239.0	1.682
460	15.08	245.2	1.725	12.91	244.8	1.708	11.28	244.3	1.694
480	15.77	250.4	1.736	13.51	250.1	1.719	11.81	249.6	1.705
500	16.46	255.6	1.746	14.10	255.2	1.730	12.33	254.9	1.716
550	18.18	268.5	1.771	15.58	268.1	1.755	13.63	267.8	1.741
600	19.88	281.2	1.793	17.05	281.0	1.777	14.92	280.7	1.763
650	21.57	293.9	1.813	18.50	293.7	1.797	16.20	293.5	1.784
700	23.26	306.5	1.832	19.95	306.4	1.816	17.47	306.2	1.803
750	24.94	319.1	1.849	21.40	319.0	1.834	18.74	318.9	1.820
800	26.62	331.7	1.866	22.84	331.6	1.850	20.00	331.6	1.837
850	28.29	344.3	1.881	24.28	344.2	1.865	21.26	344.2	1.852
900	29.96	356.9	1.895	25.71	356.8	1.880	22.52	356.8	1.866
950	31.63	369.5	1.909	27.14	369.3	1.893	23.78	369.3	1.880
1000	33.30	382.0	1.922	28.57	381.9	1.906	25.03	381.8	1.893
1050	34.96	394.4	1.934	30.00	394.4	1.918	26.28	394.4	1.905
1100	36.63	406.8	1.945	31.43	406.9	1.930	27.53	406.9	1.917
1150	38.29	419.3	1.956	32.86	419.3	1.941	28.78	419.4	1.928
1200	39.95	431.8	1.967	34.28	431.8	1.952	30.03	431.9	1.938
1250	41.61	444.2	1.977	35.71	444.3	1.962	31.28	444.3	1.948
1300	43.27	456.7	1.987	37.13	456.7	1.972	32.52	456.8	1.958

TABLE XVII. Thermodynamic Properties of Liquid and Gaseous Krypton (*Continued*)

T, K	p = 45 bar			p = 50 bar			p = 55 bar		
	$v \cdot 10^3$	h	s	$v \cdot 10^3$	h	s	$v \cdot 10^3$	h	s
120	0,4114	56,23	0,8235	0,4110	56.36	0,8229	0,4107	56,49	0,8222
125	0,4177	58,91	0,8454	0,4173	59,04	0,8448	0,4169	59,17	0,8441
130	0,4244	61,58	0,8663	0,4240	61.71	0,8656	0,4235	61,83	0,8649
135	0,4315	64,24	0,8864	0,4310	64.36	0,8857	0,4305	64,48	0,8850
140	0,4390	66,90	0,9057	0,4385	67,01	0,9050	0,4379	67,12	0,9042
145	0,4471	69,56	0,9244	0,4464	69,67	0,9236	0,4458	69,77	0,9228
150	0,4558	72,25	0,9426	0,4550	72,35	0,9417	0,4543	72,44	0,9408
155	0,4651	74,97	0,9605	0,4642	75,05	0,9595	0,4634	75,14	0,9586
160	0,4754	77,73	0,9780	0,4744	77,80	0,9769	0,4734	77,87	0,9759
165	0,4867	80,58	0,9958	0,4855	80,62	0,9946	0,4843	80,68	0,9935
170	0,4993	83,52	1,014	0,4978	83,55	1,012	0,4964	83,57	1,011
175	0,5136	86,60	1,031	0,5117	86,59	1,030	0,5100	86,59	1,028
180	0,5300	89,88	1,049	0,5277	89,83	1,048	0,5255	89,79	1,046
185	0,5495	93,44	1,069	0,5464	93,33	1,067	0,5435	93,23	1,065
190	0,5734	97,44	1,090	0,5690	97,23	1,087	0,5650	97,05	1,085
195	0,6043	102,1	1,114	0,5977	101,6	1,111	0,5918	101,3	1,108
200	0,6565	107,7	1,144	0,6435	106,8	1,139	0,6310	106,1	1,134
205	2,602	157,8	1,391	0,7280	114,8	1,177	0,6972	112,9	1,166
210	2,941	162,4	1,413	2,321	156,7	1,378	1,576	145,3	1,320
215	3,210	165,9	1,429	2,645	161,7	1,403	2,128	156,2	1,372
220	3,445	169,0	1,443	2,900	165,4	1,420	2,427	161,4	1,396
225	3,657	171,5	1,455	3,119	168,5	1,434	2,663	165,2	1,413
230	3,855	173,9	1,465	3,317	171,4	1,446	2,867	168,5	1,427
235	4,041	176,1	1,475	3,499	173,8	1,457	3,050	171,3	1,439
240	4,218	178,2	1,484	3,671	176,1	1,467	3,220	173,8	1,450

TABLE XVII. Thermodynamic Properties of Liquid and Gaseous Krypton (*Continued*)

T, K	p = 45 bar			p = 50 bar			p = 55 bar		
	$v \cdot 10^3$	h	s	$v \cdot 10^3$	h	s	$v \cdot 10^3$	h	s
245	4.388	180.2	1.492	3.835	178.3	1.476	3.379	176.2	1.460
250	4.553	182.2	1.500	3.991	180.4	1.484	3.530	178.5	1.469
260	4.868	185.8	1.514	4.289	184.2	1.499	3.815	182.6	1.485
270	5.169	189.3	1.527	4.571	187.9	1.513	4.082	186.4	1.500
280	5.460	192.6	1.539	4.842	191.3	1.526	4.336	190.0	1.513
290	5.742	195.8	1.551	5.103	194.7	1.537	4.581	193.5	1.525
300	6.018	199.0	1.561	5.358	197.9	1.548	4.818	196.8	1.536
320	6.553	205.0	1.581	5.850	204.1	1.568	5.275	203.1	1.556
340	7.073	210.9	1.598	6.326	210.0	1.586	5.715	209.2	1.575
360	7.580	216.6	1.615	6.789	215.8	1.603	6.143	215.1	1.592
380	8.079	222.2	1.630	7.244	221.5	1.618	6.561	220.8	1.607
400	8.570	227.7	1.644	7.691	227.1	1.632	6.972	226.5	1.622
420	9.055	233.1	1.657	8.132	232.6	1.646	7.377	232.0	1.635
440	9.536	238.5	1.670	8.568	238.0	1.658	7.776	237.5	1.648
460	10.01	243.9	1.682	9.000	243.4	1.670	8.172	243.0	1.660
440	10.48	249.2	1.693	9.428	248.8	1.682	8.564	248.4	1.672
500	10.95	254.4	1.704	9.853	254.1	1.693	8.953	253.7	1.683
550	12.12	267.5	1.729	10.91	267.2	1.718	9.915	266.9	1.708
600	13.27	280.5	1.751	11.95	280.2	1.740	10.86	280.0	1.730
650	14.41	293.3	1.772	12.98	293.1	1.761	11.81	293.0	1.751
700	15.54	306.1	1.791	14.00	305.9	1.780	12.74	305.8	1.770
750	16.67	318.8	1.808	15.02	318.7	1.798	13.67	318.6	1.788
800	17.80	331.5	1.825	16.04	331.4	1.814	14.59	331.4	1.804
850	18.92	344.1	1.840	17.05	344.0	1.829	15.51	344.0	1.820
900	20.04	356.7	1.854	18.06	356.7	1.844	16.43	356.6	1.834

TABLE XVII. Thermodynamic Properties of Liquid and Gaseous Krypton (*Continued*)

T, K	p = 45 bar			p = 50 bar			p = 55 bar		
	$v \cdot 10^3$	h	s	$v \cdot 10^3$	h	s	$v \cdot 10^3$	h	s
950	21.16	369.3	1.868	19.06	369.3	1.857	17.35	369.2	1.848
1000	22.27	381.8	1.881	20.07	381.8	1.870	18.27	381.8	1.861
1050	23.39	394.4	1.893	21.07	394.4	1.883	19.18	394.4	1.873
1100	24.50	406.9	1.905	22.08	406.9	1.894	20.09	406.9	1.885
1150	25.61	419.4	1.916	23.08	419.4	1.905	21.00	419.5	1.896
1200	26.72	431.9	1.927	24.08	431.9	1.916	21.91	432.0	1.906
1250	27.83	444.4	1.937	25.08	444.4	1.926	22.82	444.5	1.916
1300	28.94	456.9	1.946	26.07	456.9	1.936	23.73	457.0	1.926

TABLE XVII. Thermodynamic Properties of Liquid and Gaseous Krypton (*Continued*)

T, K	p = 60 bar			p = 65 bar			p = 70 bar		
	$v \cdot 10^3$	h	s	$v \cdot 10^3$	h	s	$v \cdot 10^3$	h	s
120	0.4103	56.63	0.8216	0.4100	55.76	0.8210	0.4096	56.89	0.8204
125	0.4166	59.30	0.8435	0.4162	59.42	0.8429	0.4158	59.55	0.8422
130	0.4231	61.95	0.8643	0.4227	62.08	0.8636	0.4223	62.20	0.8629
135	0.4300	64.60	0.8843	0.4296	64.72	0.8835	0.4291	64.84	0.8828
140	0.4374	67.24	0.9034	0.4368	67.35	0.9027	0.4363	67.46	0.9019
145	0.4452	69.88	0.9220	0.4446	69.98	0.9212	0.4440	70.09	0.9204
150	0.4536	72.54	0.9400	0.4529	72.63	0.9391	0.4522	72.73	0.9382
155	0.4626	75.22	0.9576	0.4618	75.31	0.9567	0.4610	75.40	0.9558
160	0.4724	77.94	0.9749	0.4715	78.01	0.9738	0.4705	78.09	0.9728
165	0.4831	80.73	0.9923	0.4820	80.79	0.9912	0.4809	80.85	0.9901

TABLE XVII. Thermodynamic Properties of Liquid and Gaseous Krypton (*Continued*)

T, K	p = 60 bar			p = 65 bar			p = 70 bar		
	$v \cdot 10^3$	h	s	$v \cdot 10^3$	h	s	$v \cdot 10^3$	h	s
170	0.4950	83.61	1.010	0.4937	83.64	1.008	0.4923	83.68	1.007
175	0.5083	86.60	1.027	0.5066	86.61	1.026	0.5050	86.62	1.024
180	0.5233	89.76	1.044	0.5213	89.74	1.043	0.5193	89.72	1.041
185	0.5407	93.15	1.063	0.5381	93.08	1.061	0.5356	93.02	1.059
190	0.5613	96.89	1.083	0.5578	96.75	1.080	0.5545	96.63	1.078
195	0.5864	101.0	1.105	0.5816	100.8	1.102	0.5771	100.6	1.100
200	0.6185	105.6	1.129	0.6112	105.2	1.125	0.6048	104.9	1.122
205	0.6742	111.5	1.157	0.6555	110.5	1.150	0.6440	109.8	1.145
210	0.7767	119.9	1.197	0.7269	117.2	1.182	0.6995	115.7	1.174
215	1.584	147.4	1.326	0.9886	131.1	1.248	0.8209	124.0	1.213
220	1.998	156.3	1.367	1.584	149.3	1.332	1.196	139.9	1.285
225	2.266	161.4	1.390	1.909	156.8	1.365	1.583	151.2	1.337
230	2.481	165.3	1.408	2.145	161.7	1.387	1.846	157.7	1.365
235	2.669	168.6	1.422	2.341	165.6	1.404	2.053	162.3	1.385
240	2.839	171.5	1.434	2.513	168.9	1.417	2.230	166.1	1.401
245	2.996	174.1	1.445	2.670	171.8	1.429	2.388	169.4	1.414
250	3.144	176.5	1.454	2.815	174.4	1.440	2.532	172.3	1.426
260	3.418	180.9	1.472	3.083	179.2	1.459	2.795	177.4	1.446
270	3.674	184.9	1.487	3.328	183.4	1.475	3.032	181.9	1.463
280	3.915	188.7	1.501	3.559	187.3	1.489	3.254	186.0	1.478
290	4.146	192.3	1.513	3.778	191.0	1.502	3.464	189.8	1.492
300	4.369	195.7	1.525	3.989	194.6	1.514	3.664	193.4	1.504
320	4.797	202.2	1.546	4.392	201.2	1.536	4.047	200.3	1.526
340	5.207	208.4	1.564	4.778	207.5	1.556	4.410	206.7	1.546
360	5.605	214.4	1.582	5.150	213.6	1.572	4.761	212.9	1.563

TABLE XVII. Thermodynamic Properties of Liquid and Gaseous Krypton (*Continued*)

T, K	p = 60 bar			p = 65 bar			p = 70 bar		
	$v \cdot 10^3$	h	s	$v \cdot 10^3$	h	s	$v \cdot 10^3$	h	s
380	5.993	220.2	1.597	5.513	219.5	1.588	5.101	218.9	1.579
400	6.373	225.9	1.612	5.868	225.3	1.603	5.434	224.7	1.594
420	6.748	231.5	1.626	6.216	231.0	1.617	5.761	230.5	1.608
440	7.117	237.1	1.639	6.560	236.6	1.630	6.083	236.1	1.621
460	7.483	242.5	1.651	6.900	242.1	1.642	6.401	241.7	1.634
480	7.845	248.0	1.662	7.236	247.6	1.654	6.715	247.2	1.646
500	8.203	253.3	1.673	7.569	253.0	1.665	7.026	252.6	1.657
550	9.090	266.6	1.699	8.392	266.3	1.690	7.794	266.0	1.682
600	9.964	279.7	1.721	9.202	279.5	1.713	8.549	279.3	1.705
650	10 83	292.7	1.742	10.00	292.6	1.734	9.297	292.4	1.726
700	11.69	305.6	1.761	10.80	305.5	1.753	10.04	305.3	1.745
750	12.54	318.4	1.779	11.59	318.3	1.771	10.77	318.2	1.763
800	13.39	331.2	1.795	12.38	331.0	1.787	11.50	331.0	1.780
850	14.24	343.9	1.811	13.16	343.7	1.803	12.23	343.8	1.795
900	15.08	356.6	1.825	13.94	356.4	1.817	12.96	356.5	1.810
950	15.92	369.2	1.839	14.72	369.1	1.831	13.68	369.2	1.823
1000	16.76	381.8	1.852	15.49	381.8	1.844	14.40	381.8	1.836
1050	17.60	394.4	1.864	16.27	394.4	1.856	15.12	394.4	1.849
1100	18.44	406.9	1.876	17.04	407.0	1.868	15.84	407.0	1.861
1150	19.27	419.5	1.887	17.81	419.5	1.879	16.56	419.6	1.872
1200	20.11	432.0	1.898	18.58	432.1	1.890	17.27	432.1	1.882
1250	20.94	444.5	1.908	19.35	444.6	1.900	17.99	444.6	1.892
1300	21.78	457.0	1.918	20.12	457.1	1.910	18.70	457.2	1.902

TABLE XVII. Thermodynamic Properties of Liquid and Gaseous Krypton (*Continued*)

T, K	p = 75 bar			p = 80 bar			p = 85 bar		
	$v \cdot 10^3$	h	s	$v \cdot 10^3$	h	s	$v \cdot 10^3$	h	s
120	0.4093	57.02	0.8198	0.4090	57.15	0.8192	0.4086	57.29	0.8186
125	0.4154	59.68	0.8416	0.4151	59.81	0.8410	0.4147	59.94	0.8404
130	0.4219	62.33	0.8623	0.4214	62.45	0.8616	0.4210	62.58	0.8609
135	0.4286	64.96	0.8821	0.4282	65.08	0.8815	0.4277	65.20	0.8808
140	0.4358	67.58	0.9012	0.4353	67.69	0.9005	0.4347	67.81	0.8997
145	0.4434	70.20	0.9196	0.4428	70.31	0.9188	0.4422	70.41	0.9180
150	0.4515	72.83	0.9374	0.4508	72.93	0.9366	0.4502	73.03	0.9358
155	0.4602	75.49	0.9549	0.4594	75.58	0.9540	0.4587	75.67	0.9531
160	0.4696	78.17	0.9718	0.4687	78.25	0.9709	0.4678	78.33	0.9699
165	0.4799	80.91	0.9890	0.4788	80.98	0.9880	0.4778	81.04	0.9869
170	0.4911	83.73	1.006	0.4898	83.77	1.005	0.4886	83.82	1.004
175	0.5035	86.64	1.023	0.5020	86.67	1.022	0.5006	86.69	1.021
180	0.5174	89.71	1.040	0.5156	89.70	1.039	0.5138	89.71	1.038
185	0.5332	92.97	1.058	0.5309	92.93	1.056	0.5287	92.89	1.055
190	0.5514	96.52	1.077	0.5485	96.43	1.074	0.5457	96.35	1.073
195	0.5729	100.4	1.097	0.5691	100.3	1.094	0.5654	100.1	1.093
200	0.5990	104.6	1.118	0.5936	104.4	1.116	0.5888	104.1	1.114
205	0.6330	109.3	1.141	0.6245	108.9	1.138	0.6175	108.5	1.135
210	0.6805	114.6	1.167	0.6659	113.8	1.162	0.6541	113.2	1.157
215	0.7633	121.2	1.198	0.7304	119.4	1.188	0.7076	118.2	1.181
220	0.9565	131.9	1.247	0.8501	127.4	1.224	0.7942	124.8	1.211
225	1.295	144.7	1.303	1.081	138.4	1.273	0.9480	133.6	1.250
230	1.580	153.0	1.342	1.351	148.0	1.316	1.1679	143.1	1.292
235	1.800	158.7	1.366	1.577	154.9	1.345	1.386	150.8	1.325
240	1.982	163.1	1.384	1.765	160.0	1.367	1.574	156.6	1.350

TABLE XVII. Thermodynamic Properties of Liquid and Gaseous Krypton (*Continued*)

T, K	p = 75 bar			p = 80 bar			p = 85 bar		
	$v \cdot 10^3$	h	s	$v \cdot 10^3$	h	s	$v \cdot 10^3$	h	s
245	2.142	166.8	1.399	1.927	164.1	1.384	1.737	161.3	1.369
250	2.287	170.0	1.412	2.071	167.6	1.398	1.882	165.2	1.385
260	2.545	175.5	1.434	2.326	173.6	1.422	2.134	171.6	1.410
270	2.776	180.3	1.452	2.553	178.7	1.441	2.356	177.0	1.430
280	2.990	184.6	1.468	2.760	183.2	1.457	2.558	181.7	1.448
290	3.192	188.6	1.482	2.954	187.3	1.472	2.746	186.0	1.463
300	3.384	192.4	1.494	3.139	191.2	1.485	2.923	190.0	1.476
320	3.748	199.3	1.517	3.487	198.4	1.508	3.257	197.4	1.500
340	4.093	205.9	1.537	3.815	205.1	1.529	3.571	204.3	1.521
360	4.424	212.2	1.555	4.130	211.5	1.547	3.872	210.7	1.539
380	4.746	218.2	1.571	4.435	217.6	1.563	4.162	217.0	1.556
400	5.060	224.1	1.586	4.732	223.6	1.579	4.444	223.0	1.572
420	5.367	229.9	1.600	5.023	229.4	1.593	4.720	228.9	1.586
440	5.670	235.6	1.614	5.309	235.2	1.606	4.991	234.7	1.599
460	5.969	241.2	1.626	5.591	240.8	1.619	5.258	240.4	1.612
480	6.264	246.8	1.638	5.869	246.4	1.631	5.522	246.0	1.624
500	6.556	252.2	1.649	6.145	251.9	1.642	5.782	251.5	1.635
550	7.276	265.8	1.675	6.823	265.5	1.668	6.424	265.2	1.661
600	7.984	279.1	1.698	7.489	278.8	1.691	7.053	278.6	1.685
650	8.684	292.2	1.719	8.147	292.0	1.712	7.674	291.8	1.706
700	9.377	305.2	1.738	8.799	305.1	1.732	8.289	304.9	1.725
750	10.06	318.1	1.756	9.445	318.0	1.749	8.899	317.9	1.743
800	10.75	330.9	1.773	10.09	330.9	1.766	9.505	330.8	1.760
850	11.43	343.7	1.788	10.73	343.7	1.782	10.11	343.6	1.775
900	12.11	356.4	1.803	11.36	356.4	1.796	10.71	356.4	1.790

TABLE XVII. Thermodynamic Properties of Liquid and Gaseous Krypton (*Continued*)

T, K	p = 75 bar			p = 80 bar			p = 85 bar		
	$v \cdot 10^3$	h	s	$v \cdot 10^3$	h	s	$v \cdot 10^3$	h	s
950	12.78	369.1	1.816	12.00	369.1	1.810	11.31	369.1	1.804
1000	13.46	381.8	1.829	12.63	381.8	1.823	11.90	381.8	1.817
1050	14.13	394.4	1.842	13.26	394.4	1.835	12.50	394.4	1.829
1100	14.80	407.0	1.854	13.89	407.0	1.847	13.09	407.1	1.841
1150	15.47	419.6	1.865	14.52	419.6	1.858	13.68	419.7	1.852
1200	16.14	432.1	1.875	15.15	432.2	1.869	14.27	432.2	1.863
1250	16.81	444.7	1.885	15.78	444.8	1.879	14.86	444.8	1.873
1300	17.48	457.2	1.895	16.40	457.3	1.889	15.45	457.3	1.883

TABLE XVII. Thermodynamic Properties of Liquid and Gaseous Krypton (*Continued*)

T, K	p = 90 bar			p = 95 bar			p = 100 bar		
	$v \cdot 10^3$	h	s	$v \cdot 10^3$	h	s	$v \cdot 10^3$	h	s
120	0.4083	57.42	0.8180	0.4080	57.55	0.8174	0.4077	57.69	0.8169
125	0.4143	60.07	0.8397	0.4140	60.20	0.8391	0.4136	60.34	0.8385
130	0.4206	62.71	0.8603	0.4202	62.83	0.8597	0.4198	62.96	0.8590
135	0.4273	65.32	0.8801	0.4268	65.44	0.8794	0.4264	65.57	0.8787
140	0.4342	67.93	0.8990	0.4337	68.04	0.8983	0.4332	68.16	0.8976
145	0.4416	70.52	0.9173	0.4411	70.63	0.9165	0.4405	70.75	0.9157
150	0.4495	73.13	0.9349	0.4489	73.24	0.9341	0.4482	73.34	0.9333
155	0.4579	75.76	0.9522	0.4572	75.86	0.9513	0.4565	75.95	0.9505
160	0.4670	78.41	0.9690	0.4661	78.49	0.9680	0.4653	78.58	0.9671
165	0.4768	81.11	0.9859	0.4758	81.18	0.9849	0.4748	81.25	0.9839

TABLE XVII. Thermodynamic Properties of Liquid and Gaseous Krypton (*Continued*)

T, K	p = 90 bar			p = 95 bar			p = 100 bar		
	$v \cdot 10^3$	h	s	$v \cdot 10^3$	h	s	$v \cdot 10^3$	h	s
170	0.4874	83.88	1.003	0.4863	83.93	1.002	0.4852	83.99	1.001
175	0.4991	86.73	1.020	0.4978	86.76	1.019	0.4965	86.80	1.017
180	0.5121	89.71	1.037	0.5105	89.73	1.036	0.5089	89.74	1.034
185	0.5266	92.87	1.054	0.5246	92.85	1.053	0.5227	92.84	1.051
190	0.5431	96.28	1.072	0.5406	96.22	1.071	0.5382	96.18	1.069
195	0.5620	100.0	1.091	0.5588	99.89	1.089	0.5558	99.79	1.087
200	0.5842	103.9	1.111	0.5800	103.7	1.109	0.5761	103.6	1.106
205	0.6107	108.1	1.132	0.6050	107.8	1.129	0.5998	107.6	1.126
210	0.6442	112.6	1.153	0.6356	112.1	1.149	0.6280	111.8	1.145
215	0.6904	117.4	1.175	0.6765	116.6	1.170	0.6649	116.1	1.165
220	0.7588	123.0	1.201	0.7335	121.7	1.193	0.7141	120.7	1.187
225	0.8691	130.3	1.234	0.8187	128.0	1.221	0.7834	126.3	1.212
230	1.036	138.8	1.271	0.9449	135.5	1.254	0.8824	132.9	1.241
235	1.228	146.8	1.305	1.105	143.1	1.287	1.013	140.0	1.272
240	1.410	153.2	1.332	1.272	149.9	1.316	1.159	146.7	1.300
245	1.571	158.4	1.354	1.428	155.4	1.339	1.305	152.6	1.324
250	1.715	162.7	1.371	1.568	160.1	1.357	1.441	157.5	1.344
260	1.965	169.6	1.398	1.814	167.6	1.387	1.681	165.6	1.376
270	2 182	175.3	1.420	2.028	173.6	1.410	1.890	171.9	1.400
280	2.379	180.3	1.438	2.219	178.8	1.429	2.077	177.4	1.420
290	2.561	184.8	1.454	2.396	183.5	1.445	2.249	182.2	1.437
300	2.733	188.9	1.468	2.563	187.8	1.459	2.410	186.6	1.451
320	3.054	196.5	1.492	2.873	195.5	1.485	2.710	194.6	1.477
340	3.355	203.5	1.513	3.162	202.6	1.506	2.989	201.8	1.499
360	3.642	210.0	1.532	3.437	209.3	1.525	3.254	208.6	1.519

TABLE XVII. Thermodynamic Properties of Liquid and Gaseous Krypton (*Continued*)

T, K	p = 90 bar			p = 95 bar			p = 100 bar		
	$v \cdot 10^3$	h	s	$v \cdot 10^3$	h	s	$v \cdot 10^3$	h	s
380	3.919	216.3	1.549	3.702	215.7	1.542	3.508	215.1	1.536
400	4.188	222.4	1.565	3.959	221.9	1.558	3.754	221.3	1.552
420	4.451	228.4	1.579	4.211	227.9	1.573	3.995	227.4	1.567
440	4.709	234.2	1.593	4.457	233.8	1.587	4.230	233.3	1.581
460	4.963	240.0	1.606	4.699	239.5	1.600	4.461	239.1	1.594
480	5.213	245.6	1.618	4.937	245.2	1.612	4.689	244.9	1.606
500	5.460	251.2	1.629	5.173	250.9	1.623	4.914	250.5	1.617
550	6.069	264.9	1.655	5.752	264.6	1.649	5.466	264.4	1.644
600	6.666	278.4	1.679	6.319	278.2	1.673	6.007	277.9	1.667
650	7.254	291.6	1.700	6.878	291.5	1.694	6.540	291.3	1.689
700	7.836	304.8	1.719	7.431	304.6	1.714	7.067	304.5	1.708
750	8.414	317.8	1.737	7.979	317.7	1.732	7.589	317.6	1.726
800	8.987	330.7	1.754	8.523	330.6	1.748	8.107	330.5	1.743
850	9.557	343.6	1.770	9.064	343.5	1.764	8.621	343.4	1.759
900	10.12	356.4	1.784	9.603	356.3	1.779	9.133	356.3	1.773
950	10.69	369.1	1.798	10.14	369.1	1.792	9.644	369.1	1.787
1000	11.25	381.8	1.811	10.67	381.8	1.806	10.15	381.8	1.800
1050	11.82	394.4	1.823	11.21	394.5	1.818	10.66	394.5	1.813
1100	12.38	407.1	1.835	11.74	407.1	1.830	11.16	407.1	1.825
1150	12.94	419.7	1.846	12.27	419.7	1.841	11.67	419.8	1.836
1200	13.50	432.3	1.857	12.80	432.3	1.852	12.17	432.4	1.846
1250	14.06	444.9	1.867	13.33	444.9	1.862	12.68	445.0	1.857
1300	14.61	457.4	1.877	13.86	457.5	1.872	13.18	457.5	1.867

TABLE XVII. Thermodynamic Properties of Liquid and Gaseous Krypton (*Continued*)

T, K	p = 110 bar			p = 120 bar			p = 130 bar		
	$v \cdot 10^3$	h	s	$v \cdot 10^3$	h	s	$v \cdot 10^3$	h	s
120	0.4070	57.96	0.8157	0.4064	58.23	0.8145	0.4058	58.50	0.8134
125	0.4129	60.60	0.8373	0.4122	60.86	0.8361	0.4115	61.12	0.8349
130	0.4191	63.22	0.8578	0.4183	63.47	0.8565	0.4175	63.73	0.8553
135	0.4255	65.81	0.8774	0.4246	66.06	0.8761	0.4238	66.31	0.8748
140	0.4323	68.40	0.8962	0.4313	68.63	0.8948	0.4304	68.88	0.8934
145	0.4394	70.97	0.9143	0.4384	71.20	0.9128	0.4373	71.43	0.9114
150	0.4470	73.55	0.9317	0.4458	73.77	0.9302	0.4446	73.98	0.9287
155	0.4551	76.15	0.9488	0.4537	76.34	0.9471	0.4524	76.55	0.9455
160	0.4637	78.75	0.9653	0.4621	78.93	0.9635	0.4606	79.12	0.9618
165	0.4730	81.41	0.9819	0.4712	81.57	0.9800	0.4695	81.73	0.9782
170	0.4830	84.11	0.9988	0.4809	84.25	0.9962	0.4789	84.39	0.9942
175	0.4939	86.89	1.015	0.4915	86.99	1.012	0.4892	87.11	1.010
180	0.5058	89.79	1.031	0.5030	89.85	1.028	0.5003	89.93	1.026
185	0.5190	92.83	1.047	0.5156	92.85	1.045	0.5124	92.88	1.042
190	0.5337	96.10	1.065	0.5295	96.06	1.062	0.5257	96.04	1.059
195	0.5502	99.63	1.083	0.5451	99.52	1.080	0.5404	99.44	1.076
200	0.5690	103.3	1.102	0.5626	103.1	1.098	0.5568	102.9	1.094
205	0.5905	107.1	1.121	0.5824	106.8	1.116	0.5752	106.5	1.111
210	0.6152	111.0	1.139	0.6045	110.6	1.133	0.5953	110.1	1.128
215	0.6463	115.0	1.158	0.6316	114.3	1.151	0.6196	113.7	1.145
220	0.6854	119.2	1.177	0.6645	118.2	1.169	0.6481	117.4	1.162
225	0.7363	124.0	1.198	0.7052	122.4	1.188	0.6823	121.3	1.180
230	0.8039	129.4	1.222	0.7564	127.1	1.209	0.7238	125.5	1.199
235	0.8921	135.4	1.248	0.8209	132.3	1.231	0.7743	130.1	1.219
240	0.9992	141.5	1.274	0.8996	137.7	1.254	0.8349	135.0	1.239

TABLE XVII. Thermodynamic Properties of Liquid and Gaseous Krypton (*Continued*)

T, K	p = 110 bar			p = 120 bar			p = 130 bar		
	$v \cdot 10^3$	h	s	$v \cdot 10^3$	h	s	$v \cdot 10^3$	h	s
245	1.117	147.4	1.298	0.9903	143.2	1.277	0.9056	140.0	1.260
250	1.236	152.7	1.319	1.088	148.4	1.298	0.9842	144.9	1.280
260	1.458	161.5	1.354	1.285	157.6	1.334	1.153	154.1	1.316
270	1.656	168.5	1.381	1.470	165.2	1.362	1.320	161.7	1.346
280	1.834	174.4	1.402	1.638	171.5	1.386	1.477	168.7	1.370
290	1.998	179.6	1.420	1.792	177.1	1.405	1.623	174.6	1.391
300	2.150	184.3	1.436	1.936	182.0	1.422	1.759	179.8	1.408
320	2.431	192.7	1.463	2.202	190.8	1.450	2.010	189.0	1.438
340	2.692	200.2	1.486	2.446	198.6	1.474	2.240	197.1	1.462
360	2.938	207.2	1.506	2.676	205.8	1.495	2.457	204.5	1.484
380	3.173	213.9	1.524	2.896	212.6	1.513	2.663	211.4	1.503
400	3.401	220.2	1.540	3.108	219.1	1.530	2.861	218.1	1.520
420	3.623	226.4	1.556	3.314	225.4	1.545	3.054	224.5	1.535
440	3.839	232.4	1.570	3.515	231.5	1.559	3.242	230.7	1.550
460	4.052	238.3	1.583	3.712	237.5	1.572	3.426	236.8	1.563
480	4.262	244.1	1.595	3.906	243.4	1.585	3.607	242.7	1.576
500	4.468	249.8	1.607	4.097	249.2	1.597	3.784	248.5	1.588
550	4.974	263.8	1.633	4.565	263.3	1.624	4.220	262.8	1.615
600	5.470	277.5	1.657	5.022	277.1	1.648	4.644	276.7	1.639
650	5.957	291.0	1.679	5.471	290.6	1.669	5.061	290.3	1.661
700	6.438	304.2	1.698	5.914	304.0	1.689	5.472	303.7	1.681
750	6.914	317.4	1.716	6.353	317.2	1.707	5.878	317.0	1.699
800	7.387	330.4	1.733	6.787	330.3	1.724	6.280	330.1	1.716
850	7.856	343.3	1.749	7.219	343.2	1.740	6.680	343.1	1.732
900	8.323	356.2	1.764	7.648	356.1	1.755	7.077	356.1	1.747

TABLE XVII. Thermodynamic Properties of Liquid and Gaseous Krypton (*Continued*)

T, K	p = 110 bar			p = 120 bar			p = 130 bar		
	$v \cdot 10^3$	h	s	$v \cdot 10^3$	h	s	$v \cdot 10^3$	h	s
950	8.788	369.0	1.778	8.075	369.0	1.769	7.472	369.0	1.761
1000	9.251	381.8	1.791	8.500	381.8	1.782	7.865	381.8	1.774
1050	9.713	394.5	1.803	8.924	394.5	1.794	8.258	394.6	1.786
1100	10.17	407.2	1.815	9.347	407.2	1.806	8.649	407.3	1.798
1150	10.63	419.9	1.826	9.769	419.9	1.817	9.038	420.0	1.809
1200	11.09	432.5	1.837	10.19	432.6	1.828	9.428	432.7	1.820
1250	11.55	445.1	1.847	10.61	445.2	1.838	9.816	445.3	1.830
1300	12.01	457.7	1.857	11.03	457.8	1.848	10.20	457.9	1.840

TABLE XVII. Thermodynamic Properties of Liquid and Gaseous Krypton (*Continued*)

T, K	p = 140 bar			p = 150 bar			p = 160 bar		
	$v \cdot 10^3$	h	s	$v \cdot 10^3$	h	s	$v \cdot 10^3$	h	s
120	0.4052	58.77	0.8123	—	—	—	—	—	—
125	0.4109	61.39	0.8338	0.4102	61.65	0.8326	0.4096	61.92	0.8315
130	0.4168	63.99	0.8541	0.4161	64.25	0.8528	0.4154	64.51	0.8517
135	0.4230	66.56	0.8735	0.4222	66.82	0.8723	0.4214	67.07	0.8710
140	0.4295	69.12	0.8921	0.4286	69.36	0.8908	0.4277	69.61	0.8895
145	0.4363	71.66	0.9100	0.4353	71.89	0.9086	0.4343	72.13	0.9072
150	0.4435	74.20	0.9272	0.4424	74.43	0.9257	0.4413	74.65	0.9243
155	0.4511	76.75	0.9439	0.4499	76.96	0.9424	0.4486	77.18	0.9409
160	0.4592	79.31	0.9601	0.4578	79.50	0.9585	0.4564	79.70	0.9569
165	0.4678	81.90	0.9764	0.4662	82.08	0.9746	0.4646	82.26	0.9729

TABLE XVII. Thermodynamic Properties of Liquid and Gaseous Krypton (*Continued*)

T. K	p = 140 bar			p = 150 bar			p = 160 bar		
	$v \cdot 10^3$	h	s	$v \cdot 10^3$	h	s	$v \cdot 10^3$	h	s
170	0.4770	84.54	0.9923	0.4752	84.69	0.9904	0.4734	84.85	0.9885
175	0.4870	87.23	1.008	0.4848	87.36	1.006	0.4828	87.50	1.004
180	0.4977	90.02	1.024	0.4952	90.12	1.022	0.4929	90.23	1.019
185	0.5093	92.93	1.040	0.5065	93.00	1.038	0.5037	93.08	1.035
190	0.5221	96.05	1.056	0.5187	96.08	1.054	0.5155	96.12	1.051
195	0.5361	99.39	1.073	0.5320	99.36	1.071	0.5283	99.36	1.068
200	0.5515	102.8	1.091	0.5467	102.7	1.088	0.5422	102.7	1.085
205	0.5687	106.3	1.108	0.5629	106.1	1.104	0.5575	106.1	1.101
210	0.5872	109.8	1.124	0.5801	109.5	1.120	0.5743	109.4	1.117
215	0.6093	113.2	1.140	0.6004	112.8	1.136	0.5929	112.6	1.132
220	0.6347	116.6	1.156	0.6234	116.2	1.151	0.6136	115.8	1.147
225	0.6644	120.3	1.173	0.6497	119.7	1.167	0.6374	119.1	1.162
230	0.6994	124.3	1.190	0.6802	123.4	1.183	0.6645	122.6	1.178
235	0.7409	128.5	1.208	0.7156	127.3	1.200	0.6954	126.3	1.194
240	0.7899	132.9	1.227	0.7566	131.4	1.217	0.7306	130.2	1.210
245	0.8466	137.5	1.246	0.8035	135.7	1.235	0.7706	134.2	1.226
250	0.9104	142.1	1.265	0.8564	140.0	1.253	0.8154	138.3	1.242
260	1.053	151.1	1.300	0.9765	148.5	1.286	0.9178	146.4	1.274
270	1.201	158.8	1.330	1.107	156.5	1.316	1.032	154.2	1.304
280	1.345	166.1	1.355	1.238	163.6	1.342	1.150	161.3	1.330
290	1.482	172.2	1.377	1.365	169.9	1.364	1.267	167.7	1.352
300	1.610	177.6	1.395	1.485	175.5	1.383	1.380	173.5	1.372
320	1.848	187.2	1.426	1.710	185.4	1.415	1.592	183.7	1.405
340	2.066	195.5	1.452	1.916	194.0	1.441	1.788	192.6	1.432
360	2.270	203.2	1.473	2.110	201.9	1.464	1.971	200.6	1.455

TABLE XVII. Thermodynamic Properties of Liquid and Gaseous Krypton (*Continued*)

T, K	p = 140 bar			p = 150 bar			p = 160 bar		
	$v \cdot 10^3$	h	s	$v \cdot 10^3$	h	s	$v \cdot 10^3$	h	s
380	2.465	210.3	1.493	2.294	209.1	1.483	2.146	208.0	1.475
400	2.651	217.1	1.510	2.471	216.0	1.501	2.313	215.1	1.493
420	2.832	223.6	1.526	2.641	222.6	1.517	2.475	221.8	1.509
440	3.009	229.9	1.541	2.808	229.0	1.532	2.632	228.2	1.524
460	3.181	236.0	1.554	2.970	235.3	1.546	2.786	234.5	1.538
480	3.351	242.0	1.567	3.129	241.4	1.559	2.937	240.7	1.551
500	3.517	247.9	1.579	3.286	247.3	1.571	3.085	246.7	1.564
550	3.924	262.3	1.607	3.668	261.8	1.599	3.445	261.2	1.591
600	4.321	276.3	1.631	4.041	275.9	1.623	3.796	275.5	1.616
650	4.709	290.0	1.653	4.405	289.7	1.645	4.140	289.4	1.638
700	5.092	303.5	1.673	4.764	303.2	1.665	4.478	303.0	1.659
750	5.471	316.8	1.691	5.119	316.6	1.684	4.811	316.4	1.677
800	5.846	330.0	1.708	5.470	329.9	1.701	5.141	329.7	1.694
850	6.218	343.0	1.724	5.818	343.1	1.717	5.468	342.9	1.710
900	6.588	356.0	1.739	6.164	356.1	1.732	5.793	356.0	1.725
950	6.955	368.9	1.753	6.508	369.0	1.746	6.116	368.9	1.739
1000	7.321	381.8	1.766	6.850	381.8	1.759	6.438	381.8	1.752
1050	7.686	394.6	1.779	7.191	394.6	1.771	6.758	394.7	1.765
1100	8.050	407.4	1.791	7.531	407.4	1.783	7.077	407.6	1.777
1150	8.412	420.1	1.802	7.870	420.1	1.795	7.395	420.3	1.788
1200	8.774	432.8	1.813	8.208	432.8	1.806	7.713	433.0	1.799
1250	9.135	445.4	1.823	8.545	445.5	1.816	8.029	445.7	1.809
1300	9.495	458.0	1.833	8.882	458.2	1.826	8.345	458.3	1.819

TABLE XVII. Thermodynamic Properties of Liquid and Gaseous Krypton (*Continued*)

T, K	p = 170 bar			p = 180 bar			p = 190 bar		
	$v \cdot 10^3$	h	s	$v \cdot 10^3$	h	s	$v \cdot 10^3$	h	s
125	0.4089	62.19	0.8303	0.4083	62.46	0.8292	0.4077	62.73	0.8281
130	0.4147	64.77	0.8505	0.4140	65.03	0.8493	0.4133	65.30	0.8482
135	0.4206	67.32	0.8698	0.4199	67.58	0.8686	0.4191	67.84	0.8674
140	0.4269	69.86	0.8882	0.4260	70.10	0.8869	0.4252	70.36	0.8857
145	0.4334	72.37	0.9059	0.4325	72.61	0.9045	0.4315	72.85	0.9032
150	0.4402	74.88	0.9229	0.4392	75.11	0.9215	0.4382	75.34	0.9201
155	0.4475	77.39	0.9394	0.4463	77.61	0.9379	0.4452	77.83	0.9365
160	0.4551	79.90	0.9553	0.4538	80.11	0.9537	0.4525	80.32	0.9522
165	0.4631	82.45	0.9712	0.4617	82.64	0.9696	0.4603	82.83	0.9680
170	0.4717	85.02	0.9867	0.4701	85.20	0.9850	0.4685	85.37	0.9833
175	0.4809	87.64	1.002	0.4790	87.80	1.000	0.4772	87.96	0.9982
180	0.4907	90.35	1.017	0.4885	90.48	1.015	0.4865	90.62	1.013
185	0.5012	93.17	1.033	0.4987	93.28	1.031	0.4963	93.39	1.028
190	0.5125	96.18	1.049	0.5096	96.25	1.047	0.5069	96.33	1.044
195	0.5247	99.38	1.065	0.5214	99.41	1.063	0.5183	99.46	1.060
200	0.5381	102.7	1.082	0.5342	102.6	1.079	0.5305	102.6	1.076
205	0.5526	106.0	1.098	0.5480	105.9	1.095	0.5438	105.8	1.093
210	0.5684	109.3	1.113	0.5630	109.1	1.110	0.5580	109.0	1.107
215	0.5858	112.4	1.128	0.5794	112.2	1.125	0.5735	112.1	1.121
220	0.6048	115.5	1.143	0.5971	115.3	1.139	0.5902	115.1	1.135
225	0.6268	118.7	1.158	0.6174	118.4	1.153	0.6091	118.1	1.149
230	0.6512	122.0	1.173	0.6397	121.6	1.167	0.6297	121.2	1.163
235	0.6785	125.5	1.188	0.6650	124.9	1.181	0.6524	124.4	1.177
240	0.7097	129.2	1.203	0.6923	128.4	1.196	0.6776	127.8	1.191

TABLE XVII. Thermodynamic Properties of Liquid and Gaseous Krypton (*Continued*)

T, K	p = 170 bar			p = 180 bar			p = 190 bar		
	$v \cdot 10^3$	h	s	$v \cdot 10^3$	h	s	$v \cdot 10^3$	h	s
245	0,7445	133,0	1,218	0,7232	132,1	1,211	0,7053	131,3	1,205
250	0,7833	136,9	1,234	0,7573	135,7	1,226	0,7357	134,8	1,219
260	0,8718	144,6	1,264	0,8349	143,2	1,255	0,8047	141,9	1,247
270	0,9718	152,2	1,293	0,9232	150,5	1,283	0,8833	149,0	1,274
280	1,078	159,3	1,318	1,018	157,5	1,308	0,9687	155,9	1,299
290	1,185	165,7	1,341	1,116	163,9	1,331	1,058	162,3	1,322
300	1,290	171,7	1,361	1,213	169,9	1,351	1,148	168,3	1,342
320	1,490	182,1	1,394	1,401	180,4	1,385	1,325	179,0	1,376
340	1,676	191,2	1,422	1,578	189,8	1,414	1,492	188,5	1,405
360	1,850	199,4	1,446	1,744	198,2	1,438	1,650	197,0	1,430
380	2,016	206,9	1,466	1,902	205,9	1,458	1,801	204,9	1,452
400	2,176	214,1	1,485	2,054	213,2	1,477	1,946	212,2	1,470
420	2,329	220,9	1,501	2,200	220,1	1,494	2,086	219,2	1,487
440	2,479	227,5	1,517	2,343	226,7	1,509	2,222	226,0	1,503
460	2,624	233,8	1,531	2,481	233,2	1,524	2,354	232,5	1,517
480	2,767	240,0	1,544	2,617	239,4	1,537	2,484	238,8	1,530
500	2,907	246,1	1,557	2,751	245,6	1,550	2,611	245,0	1,543
550	3,249	260,9	1,585	3,075	260,4	1,578	2,920	260,0	1,572
600	3,581	275,2	1,609	3,390	274,8	1,603	3,220	274,5	1,597
650	3,906	289,1	1,632	3,698	288,9	1,625	3,512	288,6	1,619
700	4,225	302,8	1,652	4,000	302,6	1,646	3,800	302,4	1,640
750	4,540	316,3	1,671	4,299	316,1	1,664	4,083	316,0	1,659
800	4,851	329,6	1,688	4,594	329,5	1,681	4,364	329,4	1,676
850	5,160	342,8	1,704	4,886	342,8	1,698	4,641	342,7	1,692
900	5,467	355,9	1,719	5,176	355,9	1,713	4,917	355,8	1,707

TABLE XVII. Thermodynamic Properties of Liquid and Gaseous Krypton (*Continued*)

T, K	p = 170 bar			p = 180 bar			p = 190 bar		
	$v \cdot 10^3$	h	s	$v \cdot 10^3$	h	s	$v \cdot 10^3$	h	s
950	5.771	368.9	1.733	5.465	368.9	1.727	5.190	368.9	1.721
1000	6.074	381.9	1.746	5.751	381.9	1.740	5.462	381.9	1.735
1050	6.376	394.8	1.759	6.037	394.8	1.753	5.733	394.8	1.747
1100	6.677	407.6	1.771	6.321	407.6	1.765	6.003	407.7	1.759
1150	6.977	420.4	1.782	6.605	420.4	1.776	6.272	420.5	1.771
1200	7.276	433.1	1.793	6.887	433.2	1.787	6.540	433.3	1.782
1250	7.574	445.8	1.803	7.169	445.9	1.797	6.807	446.0	1.792
1300	7.871	458.5	1.813	7.451	458.6	1.807	7.074	458.7	1.802

TABLE XVII. Thermodynamic Properties of Liquid and Gaseous Krypton (*Continued*)

T, K	p = 200 bar			p = 220 bar			p = 240 bar		
	$v \cdot 10^3$	h	s	$v \cdot 10^3$	h	s	$v \cdot 10^3$	h	s
125	0.4071	63.00	0.8270	0.4059	63.54	0.8249	0.4048	64.09	0.8227
130	0.4126	65.56	0.8470	0.4113	66.10	0.8448	0.4101	66.63	0.8426
135	0.4184	68.10	0.8662	0.4170	68.62	0.8639	0.4156	69.15	0.8616
140	0.4244	70.61	0.8845	0.4228	71.12	0.8820	0.4213	71.63	0.8797
145	0.4307	73.10	0.9019	0.4289	73.59	0.8994	0.4273	74.09	0.8969
150	0.4372	75.58	0.9187	0.4353	76.06	0.9161	0.4335	76.54	0.9135
155	0.4441	78.06	0.9350	0.4420	78.51	0.9323	0.4399	78.98	0.9296
160	0.4513	80.53	0.9507	0.4489	80.97	0.9478	0.4467	81.41	0.9450
165	0.4589	83.03	0.9664	0.4563	83.44	0.9633	0.4538	83.86	0.9604

TABLE XVII. Thermodynamic Properties of Liquid and Gaseous Krypton (*Continued*)

T, K	p = 200 bar			p = 220 bar			p = 240 bar		
	$v \cdot 10^3$	h	s	$v \cdot 10^3$	h	s	$v \cdot 10^3$	h	s
170	0,4669	85.56	0,9816	0,4640	85.94	0,9784	0,4612	86.34	0,9753
175	0,4754	88.12	0,9964	0,4721	88.47	0,9930	0,4690	88.84	0,9897
180	0,4845	90,77	1,011	0,4807	91,08	1,008	0,4773	91,41	1,004
185	0,4941	93,51	1,026	0,4899	93,78	1,023	0,4859	94,08	1,019
190	0,5044	96,43	1,041	0,4995	96,65	1,038	0,4951	96,91	1,034
195	0,5153	99,53	1,057	0,5098	99,70	1,054	0,5048	99,90	1,050
200	0,5271	102,7	1,073	0,5208	102,8	1,070	0,5151	102,9	1,066
205	0,5398	105,8	1,089	0,5325	105,9	1,085	0,5260	105,9	1,080
210	0,5534	108,9	1,104	0,5450	108,9	1,099	0,5376	108,9	1,094
215	0,5681	112,0	1,119	0,5584	111,8	1,112	0,5499	111,8	1,107
220	0,5839	115,0	1,133	0,5727	114,7	1,126	0,5630	114,6	1,120
225	0,6009	118,0	1,146	0,5879	117,6	1,139	0,5768	117,4	1,133
230	0,6190	121,0	1,159	0,6041	120,5	1,152	0,5915	120,2	1,146
235	0,6415	124,1	1,173	0,6225	123,4	1,165	0,6080	123,1	1,158
240	0,6648	127,3	1,186	0,6430	126,4	1,177	0,6255	126,0	1,170
245	0,6900	130,6	1,199	0,6649	129,6	1,190	0,6445	128,9	1,182
250	0,7175	134,0	1,213	0,6880	132,8	1,203	0,6650	131,9	1,194
260	0,7795	140,9	1,240	0,7396	139,3	1,228	0,7092	138,1	1,218
270	0,8500	147,8	1,266	0,7979	145,8	1,253	0,7587	144,3	1,241
280	0,9272	154,5	1,291	0,8619	152,2	1,276	0,8131	150,4	1,264
290	1,009	160,8	1,313	0,9303	158,4	1,298	0,8715	156,4	1,285
300	1,092	166,8	1,333	1,001	164,3	1,318	0,9327	162,2	1,304
320	1,260	177,4	1,368	1,146	175,2	1,353	1,060	173,0	1,339
340	1,417	187,0	1,397	1,289	184,8	1,383	1,188	182,9	1,369
360	1,567	195,9	1,422	1,426	193,8	1,408	1,312	192,0	1,395

TABLE XVII. Thermodynamic Properties of Liquid and Gaseous Krypton (*Continued*)

T, K	p = 200 bar			p = 220 bar			p = 240 bar		
	$v \cdot 10^3$	h	s	$v \cdot 10^3$	h	s	$v \cdot 10^3$	h	s
380	1.711	203.9	1.444	1.558	202.0	1.430	1.434	200.3	1.418
400	1.850	211.4	1.463	1.686	209.7	1.450	1.551	208.1	1.438
420	1.984	218.5	1.480	1.809	216.9	1.467	1.665	215.5	1.456
440	2.113	225.3	1.496	1.928	223.9	1.484	1.775	222.6	1.472
460	2.240	231.9	1.511	2.044	230.6	1.499	1.883	229.4	1.488
480	2.364	238.3	1.524	2.158	237.1	1.512	1.988	236.1	1.502
500	2.485	244.5	1.537	2.270	243.5	1.525	2.091	242.5	1.515
550	2.780	259.6	1.566	2.541	258.8	1.555	2.342	258.1	1.544
600	3.066	274.2	1.591	2.803	273.5	1.580	2.584	273.0	1.570
650	3.346	288.3	1.614	3.058	287.9	1.603	2.820	287.4	1.594
700	3.620	302.2	1.634	3.309	301.9	1.624	3.051	301.5	1.614
750	3.890	315.9	1.653	3.556	315.6	1.643	3.279	315.4	1.633
800	4.157	329.3	1.671	3.800	329.1	1.660	3.503	329.0	1.651
850	4.421	342.6	1.687	4.041	342.6	1.677	3.725	342.5	1.667
900	4.683	355.8	1.702	4.280	355.8	1.692	3.945	355.8	1.683
950	4.944	368.9	1.716	4.518	369.0	1.706	4.164	369.0	1.697
1000	5.202	381.9	1.729	4.754	382.0	1.719	4.381	382.1	1.710
1050	5.460	394.9	1.742	4.989	395.0	1.732	4.596	395.2	1.723
1100	5.717	407.8	1.754	5.223	407.9	1.744	4.811	408.1	1.735
1150	5.972	420.6	1.766	5.456	420.8	1.756	5.025	421.0	1.747
1200	6.227	433.4	1.777	5.688	433.6	1.767	5.239	433.9	1.758
1250	6.482	446.2	1.787	5.919	446.4	1.777	5.451	446.7	1.768
1300	6.735	458.9	1.797	6.150	459.2	1.787	5.663	459.5	1.778

TABLE XVII. Thermodynamic Properties of Liquid and Gaseous Krypton (*Continued*)

T, K	p = 260 bar			p = 280 bar			p = 300 bar		
	$v \cdot 10^3$	h	s	$v \cdot 10^3$	h	s	$v \cdot 10^3$	h	s
125	0.4036	64.64	0.8207	0.4025	65.19	0.8186	0.4015	65.74	0.8166
130	0.4089	67.17	0.8405	0.4077	67.72	0.8384	0.4065	68.26	0.8363
135	0.4143	69.68	0.8594	0.4130	70.21	0.8572	0.4117	70.75	0.8551
140	0.4199	72.15	0.8774	0.4184	72.67	0.8751	0.4171	73.20	0.8729
145	0.4257	74.60	0.8945	0.4241	75.11	0.8922	0.4226	75.62	0.8899
150	0.4317	77.03	0.9110	0.4300	77.53	0.9086	0.4284	78.03	0.9062
155	0.4380	79.45	0.9270	0.4361	79.94	0.9244	0.4344	80.43	0.9220
160	0.4446	81.87	0.9423	0.4425	82.33	0.9396	0.4406	82.80	0.9370
165	0.4514	84.30	0.9575	0.4492	84.74	0.9547	0.4470	85.20	0.9521
170	0.4586	86.75	0.9723	0.4561	87.17	0.9694	0.4537	87.61	0.9666
175	0.4661	89.23	0.9866	0.4634	89.62	0.9836	0.4607	90.04	0.9806
180	0.4740	91.77	1.001	0.4709	92.15	0.9976	0.4680	92.54	0.9941
185	0.4823	94.40	1.016	0.4789	94.75	1.012	0.4757	95.11	1.008
190	0.4910	97.19	1.031	0.4872	97.51	1.027	0.4837	97.85	1.023
195	0.5002	100.2	1.046	0.4960	100.4	1.042	0.4920	100.7	1.038
200	0.5099	103.2	1.061	0.5052	103.4	1.057	0.5006	103.6	1.053
205	0.5202	106.1	1.076	0.5148	106.3	1.072	0.5099	106.5	1.068
210	0.5310	109.0	1.089	0.5250	109.2	1.085	0.5195	109.3	1.081
215	0 5424	111.9	1.102	0.5357	112.0	1.098	0.5295	112.1	1.094
220	0.5545	114.7	1.115	0.5469	114.8	1.110	0.5401	114.8	1.106
225	0.5672	117.4	1.128	0.5587	117.5	1.122	0.5511	117.4	1.118
230	0.5805	120.1	1.140	0.5711	120.1	1.134	0.5625	120.0	1.130
235	0.5948	122.8	1.152	0.5845	122.7	1.146	0.5745	122.7	1.142
240	0.6100	125.6	1.164	0.5988	125.3	1.158	0.5875	125.4	1.153

TABLE XVII. Thermodynamic Properties of Liquid and Gaseous Krypton (*Continued*)

T, K	p = 260 bar			p = 280 bar			p = 300 bar		
	$v \cdot 10^3$	h	s	$v \cdot 10^3$	h	s	$v \cdot 10^3$	h	s
245	0.6275	128.4	1.175	0.6146	128.0	1.169	0.6020	128.1	1.163
250	0.6464	131.3	1.186	0.6307	130.8	1.180	0.6169	130.6	1.173
260	0.6850	137.2	1.209	0.6652	136.5	1.202	0.6485	136.1	1.195
270	0.7281	143.2	1.232	0.7033	142.3	1.223	0.6827	141.6	1.216
280	0.7752	149.1	1.253	0.7449	148.0	1.244	0.7199	147.2	1.236
290	0.8259	154.9	1.274	0.7895	153.6	1.264	0.7597	152.7	1.255
300	0.8793	160.5	1.293	0.8366	159.1	1.282	0.8017	158.0	1.273
320	0.9935	171.1	1.328	0.9362	169.7	1.316	0.8910	168.4	1.306
340	1.108	181.0	1.358	1.040	179.5	1.346	0.9842	178.2	1.335
360	1.220	190.2	1.383	1.144	188.7	1.372	1.079	187.4	1.362
380	1.331	198.7	1.406	1.246	197.2	1.395	1.173	196.0	1.386
400	1.440	206.6	1.427	1.346	205.3	1.416	1.266	204.1	1.407
420	1.545	214.1	1.445	1.444	213.0	1.435	1.358	211.8	1.426
440	1.648	221.4	1.462	1.540	220.3	1.452	1.448	219.3	1.443
460	1.748	228.4	1.477	1.633	227.3	1.468	1.535	226.4	1.459
480	1.846	235.1	1.492	1.725	234.2	1.482	1.621	233.3	1.473
500	1.942	241.6	1.505	1.814	240.8	1.496	1.705	240.0	1.487
550	2.175	257.4	1.535	2.032	256.7	1.526	1.909	256.1	1.518
600	2.400	272.4	1.561	2.242	271.9	1.553	2.107	271.4	1.545
650	2.619	287.0	1.585	2.447	286.6	1.576	2.299	286.3	1.568
700	2.833	301.2	1.606	2.647	301.0	1.597	2.486	300.7	1.590
750	3.044	315.2	1.625	2.844	315.0	1.617	2.671	314.8	1.609
800	3.252	328.9	1.643	3.038	328.8	1.635	2.852	328.7	1.627
850	3.458	342.4	1.659	3.230	342.4	1.651	3.032	342.4	1.644
900	3.662	355.8	1.674	3.419	355.8	1.666	3.210	355.9	1.659

TABLE XVII. Thermodynamic Properties of Liquid and Gaseous Krypton (*Continued*)

T, K	p = 260 bar			p = 280 bar			p = 300 bar		
	$v \cdot 10^3$	h	s	$v \cdot 10^3$	h	s	$v \cdot 10^3$	h	s
950	3.864	369.1	1.689	3.608	369.2	1.681	3.386	369.3	1.674
1000	4.065	382.2	1.702	3.795	382.4	1.694	3.561	382.5	1.687
1050	4.265	395.3	1.715	3.981	395.5	1.707	3.734	395.7	1.700
1100	4.463	408.3	1.727	4.166	408.5	1.719	3.908	408.8	1.712
1150	4.661	421.3	1.739	4.350	421.5	1.731	4.080	421.8	1.724
1200	4.859	434.2	1.750	4.533	434.4	1.742	4.251	434.7	1.735
1250	5.055	447.0	1.760	4.716	447.3	1.752	4.422	447.6	1.745
1300	5.251	459.8	1.770	4.898	460.1	1.762	4.593	460.5	1.755

TABLE XVII. Thermodynamic Properties of Liquid and Gaseous Krypton (*Continued*)

T, K	p = 320 bar			p = 340 bar			p = 360 bar		
	$v \cdot 10^3$	h	s	$v \cdot 10^3$	h	s	$v \cdot 10^3$	h	s
130	0.4054	68.81	0.8343	0.4043	69.36	0.8323	0.4032	69.92	0.8303
135	0.4105	71.29	0.8530	0.4093	71.83	0.8509	0.4082	72.37	0.8489
140	0.4158	73.73	0.8707	0.4145	74.26	0.8686	0.4132	74.80	0.8666
145	0.4212	76.14	0.8877	0.4198	76.66	0.8855	0.4184	77.19	0.8834
150	0.4268	78.54	0.9039	0.4253	79.05	0.9016	0.4238	79.57	0.8994
155	0.4326	80.92	0.9196	0.4310	81.42	0.9172	0.4294	81.92	0.9149
160	0.4387	83.28	0.9345	0.4369	83.77	0.9321	0.4351	84.26	0.9297
165	0.4450	85.66	0.9495	0.4430	86.13	0.9470	0.4411	86.61	0.9445

TABLE XVII. Thermodynamic Properties of Liquid and Gaseous Krypton (*Continued*)

T, K	p = 320 bar			p = 340 bar			p = 360 bar		
	$v \cdot 10^3$	h	s	$v \cdot 10^3$	h	s	$v \cdot 10^3$	h	s
170	0.4515	88.05	0.9639	0.4493	88.51	0.9613	0.4473	88.97	0.9587
175	0.4583	90.47	0.9778	0.4559	90.90	0.9751	0.4536	91.35	0.9725
180	0.4653	92.94	0.9912	0.4627	93.36	0.9883	0.4603	93.79	0.9860
185	0.4727	95.50	1.005	0.4698	95.89	1.002	0.4671	96.30	1.000
190	0.4803	98.20	1.020	0.4772	98.58	1.017	0.4742	98.97	1.014
195	0.4883	101.0	1.035	0.4849	101.4	1.032	0.4816	101.8	1.029
200	0.4967	103.9	1.050	0.4929	104.3	1.047	0.4893	104.7	1.044
205	0.5054	106.8	1.064	0.5012	107.1	1.061	0.4972	107.5	1.058
210	0.5145	109.6	1.077	0.5098	109.9	1.074	0.5055	110.2	1.071
215	0.5240	112.3	1.089	0.5188	112.6	1.086	0.5141	112.8	1.083
220	0.5339	114.9	1.101	0.5282	115.2	1.097	0.5230	115.3	1.094
225	0.5442	117.5	1.113	0.5379	117.7	1.109	0.5322	117.9	1.105
230	0.5549	120.1	1.125	0.5480	120.2	1.120	0.5418	120.4	1.116
235	0.5661	122.7	1.136	0.5585	122.8	1.131	0.5516	123.0	1.127
240	0.5782	125.4	1.147	0.5695	125.3	1.142	0.5621	125.5	1.138
245	0.5915	128.1	1.157	0.5817	127.8	1.153	0.5735	128.0	1.148
250	0.6050	130.6	1.167	0.5943	130.4	1.163	0.5855	130.5	1.158
260	0.6342	135.9	1.188	0.6217	135.6	1.183	0.6106	135.5	1.178
270	0.6653	141.2	1.209	0.6502	140.8	1.203	0.6370	140.6	1.197
280	0.6988	146.6	1.229	0.6808	146.1	1.222	0.6652	145.7	1.216
290	0.7347	151.9	1.247	0.7135	151.3	1.240	0.6952	150.8	1.234
300	0.7727	157.2	1.265	0.7481	156.5	1.258	0.7269	155.9	1.251
320	0.8534	167.4	1.298	0.8217	166.5	1.290	0.7944	165.8	1.283
340	0.9384	177.0	1.327	0.8994	176.1	1.319	0.8650	175.3	1.312
360	1.025	186.2	1.353	0.9794	185.1	1.345	0.9398	184.3	1.337

TABLE XVII. Thermodynamic Properties of Liquid and Gaseous Krypton (*Continued*)

T, K	p = 320 bar			p = 340 bar			p = 360 bar		
	$v \cdot 10^3$	h	s	$v \cdot 10^3$	h	s	$v \cdot 10^3$	h	s
380	1,112	194.7	1.376	1.060	193,8	1,368	1,015	192.9	1,361
400	1,198	202,9	1,397	1,141	202,0	1,390	1,090	201,1	1,382
420	1,284	210,8	1,417	1,220	209.9	1,409	1,165	209,0	1,401
440	1,368	218,3	1,434	1 299	217.4	1,426	1,239	216,6	1,419
460	1,450	225,5	1,450	1,376	224,7	1,442	1,311	224,0	1,435
480	1,531	232,5	1,465	1,452	231,7	1,457	1,383	231,1	1,450
500	1,610	239,3	1,479	1,527	238.6	1,471	1,454	237,9	1,464
550	1,802	255,5	1,510	1,709	255,0	1,503	1,626	254,5	1,496
600	1,988	271,0	1,537	1,884	270,6	1,530	1,792	270.2	1,523
650	2,169	286,0	1,561	2,055	285.8	1,554	1,954	285.4	1,547
700	2,346	300,5	1,583	2,222	300,3	1,576	2,113	300,1	1,569
750	2,519	314,7	1,602	2,386	314,6	1,595	2,268	314,5	1,589
800	2,690	328,6	1,620	2,548	328,6	1,614	2,421	328,6	1,607
850	2.859	342,4	1,637	2,707	342,4	1,630	2,572	342,4	1,624
900	3,026	355,9	1,652	2,865	356,0	1,646	2,721	356.1	1,640
950	3,192	369.4	1,667	3,021	369,5	1,660	2,869	369,6	1,654
1000	3,356	382,7	1,680	3,176	382,8	1,674	3,016	383,0	1,668
1050	3,520	395,9	1,693	3,330	396,1	1,687	3,162	396,3	1,681
1100	3,682	409,0	1,706	3,483	409,2	1,699	3,307	409,5	1,693
1150	3,844	422,0	1,717	3,636	422,3	1,711	3,451	422,6	1,705
1200	4,005	435,0	1,728	3,787	435,3	1,722	3,594	435,6	1,716
1250	4,165	447,9	1,739	3,939	448,3	1,733	3,737	448,6	1,727
1300	4,325	460,8	1 749	4,089	461,1	1,743	3,880	461,5	1,737

TABLE XVII. Thermodynamic Properties of Liquid and Gaseous Krypton (*Continued*)

T, K	p = 380 bar			p = 400 bar			p = 420 bar		
	$v \cdot 10^3$	h	s	$v \cdot 10^3$	h	s	$v \cdot 10^3$	h	s
130	0.4022	70.47	0.8284	0.4012	71.03	0.8265	0.4002	71.59	0.8246
135	0.4070	72.92	0.8470	0.4059	73.47	0.8450	0.4049	74.03	0.8431
140	0.4120	75.34	0.8645	0.4108	75.88	0.8625	0.4097	76.43	0.8605
145	0.4171	77.72	0.8814	0.4159	78.26	0.8792	0.4146	78.79	0.8772
150	0.4224	80.09	0.8973	0.4210	80.61	0.8951	0.4197	81.14	0.8931
155	0.4279	82.43	0.9127	0.4264	82.95	0.9105	0.4249	83.47	0.9084
160	0.4335	84.76	0.9274	0.4319	85.27	0.9252	0.4303	85.78	0.9230
165	0.4393	87.10	0.9421	0.4375	87.59	0.9398	0.4358	88.09	0.9375
170	0.4453	89.44	0.9563	0.4434	89.92	0.9539	0.4415	90.41	0.9515
175	0.4515	91.81	0.9699	0.4494	92.27	0.9674	0.4474	92.75	0.9650
180	0.4579	94.23	0.9834	0.4556	94.68	0.9809	0.4535	95.14	0.9784
185	0.4645	96.73	0.9972	0.4621	97.16	0.9950	0.4597	97.61	0.9924
190	0.4714	99.37	1.011	0.4687	99.79	1.009	0.4662	100.2	1.006
195	0.4785	102.2	1.026	0.4756	102.6	1.023	0.4729	103.0	1.020
200	0.4859	105.0	1.041	0.4827	105.3	1.037	0.4797	105.7	1.034
205	0.4936	107.8	1.055	0.4901	108.1	1.051	0.4868	108.5	1.047
210	0.5015	110.5	1.068	0.4977	110.8	1.064	0.4941	111.2	1.060
215	0.5097	113.1	1.080	0.5055	113.4	1.076	0.5017	113.8	1.072
220	0.5182	115.6	1.091	0.5137	115.9	1.087	0.5095	116.3	1.084
225	0.5269	118.1	1.102	0.5220	118.4	1.098	0.5175	118.8	1.095
230	0.5360	120.6	1.113	0.5306	120.9	1.109	0.5257	121.2	1.106
235	0.5453	123.1	1.124	0.5395	123.3	1.120	0.5341	123.6	1.117
240	0.5555	125.6	1.135	0.5490	125.8	1.131	0.5430	126.0	1.127

TABLE XVII. Thermodynamic Properties of Liquid and Gaseous Krypton (*Continued*)

T, K	p = 380 bar			p = 400 bar			p = 420 bar		
	$v \cdot 10^3$	h	s	$v \cdot 10^3$	h	s	$v \cdot 10^3$	h	s
245	0,5664	128,1	1,145	0,5595	128,2	1,141	0.5526	128,4	1,137
250	0,5774	130,5	1,155	0,5703	130,6	1,150	0.5629	130,8	1.146
260	0,6008	135,5	1,174	0,5918	135,5	1,168	0,5837	135,6	1,164
270	0,6253	140,5	1,192	0,6148	140,4	1,186	0,6053	140,4	1,182
280	0,6514	145,5	1,210	0,6392	145,3	1,204	0,6282	145,2	1,200
290	0,6792	150,5	1,228	0,6650	150,2	1,222	0,6523	150,1	1,217
300	0,7085	155,5	1,245	0,6922	155,1	1,239	0,6777	154,9	1,234
320	0,7705	165,3	1,277	0,7501	164,7	1,270	0,7318	164,4	1,264
340	0,8369	174,6	1,305	0,8116	174,0	1,298	0,7889	173,6	1,292
360	0,9055	183,5	1,330	0,8750	182,9	1,324	0,8489	182,4	1,317
380	0,9745	192,1	1,353	0,9407	191,5	1,347	0,9099	191,0	1,340
400	1,046	200,3	1,374	1,007	199,7	1,368	0,9709	199,2	1,361
420	1,116	208,3	1,394	1,073	207,6	1,387	1,033	207,1	1,380
440	1,185	215,9	1,412	1,138	215,3	1,405	1,096	214,8	1,398
460	1,254	223,3	1,428	1,203	222,7	1,421	1,158	222,2	1,415
480	1,322	230,4	1,443	1,268	229,9	1,437	1,219	229,4	1,430
500	1,389	237,4	1,457	1,331	236,9	1,451	1,279	236,4	1,445
550	1,552	254,0	1,489	1,486	253,6	1,483	1,427	253,3	1,477
600	1,710	269,9	1,517	1,637	269,6	1,511	1,571	269,3	1,505
650	1,864	285,2	1,541	1,784	284,9	1,535	1,711	284,8	1,530
700	2,015	300,0	1,563	1,927	299,8	1,557	1,848	299,7	1,552
750	2,163	314,4	1,583	2,068	314,3	1,577	1,982	314,3	1,572
800	2,308	328,6	1,601	2,206	328,6	1,596	2,114	328,6	1,590
850	2,451	342,6	1,618	2,343	342,5	1,613	2,245	342,6	1,607
900	2,593	356,2	1,634	2,478	356,3	1,628	2,374	356,4	1,623

TABLE XVII. Thermodynamic Properties of Liquid and Gaseous Krypton (*Continued*)

T, K	p = 380 bar			p = 400 bar			p = 420 bar		
	$v \cdot 10^3$	h	s	$v \cdot 10^3$	h	s	$v \cdot 10^3$	h	s
950	2,733	369.8	1.649	2.611	369,9	1,643	2,501	370,1	1,638
1000	2,873	383,2	1.662	2,744	383,4	1.657	2,628	383,6	1,652
1050	3,011	396,5	1.675	2,876	396,7	1,670	2,753	397,0	1,665
1100	3,149	409,7	1,688	3,007	410,0	1,682	2,878	410,3	1,677
1150	3,286	422,9	1,699	3,137	423,2	1,694	3,002	423,5	1,689
1200	3,422	435,9	1,710	3,266	436,3	1,705	3,126	436,6	1,700
1250	3,557	448,9	1,721	3,395	449,3	1,716	3,249	449,6	1,711
1300	3,693	461,9	1,731	3,524	462,2	1,726	3,372	462,6	1,721

TABLE XVII. Thermodynamic Properties of Liquid and Gaseous Krypton (*Continued*)

T, K	p = 440 bar			p = 460 bar			p = 480 bar		
	$v \cdot 10^3$	h	s	$v \cdot 10^3$	h	s	$v \cdot 10^3$	h	s
130	0,3992	72,15	0,8228	0,3983	72,71	0,8210	0,3974	73,27	0,8192
135	0,4038	74,58	0,8412	0,4028	75,14	0,8394	0,4018	75,70	0,8376
140	0,4086	76,97	0,8586	0,4075	77,52	0,8567	0,4064	78,08	0,8549
145	0,4134	79,33	0,8752	0,4122	79,88	0,8732	0,4111	80,42	0,8713
150	0,4184	81,67	0,8910	0,4171	82,21	0,8890	0,4159	82,75	0,8871
155	0,4235	83,99	0,9063	0,4222	84,52	0,9042	0,4208	85,05	0,9022
160	0,4288	86,29	0,9208	0,4273	86,81	0,9187	0,4259	87,33	0,9166
165	0,4342	88,59	0,9353	0,4326	89,10	0,9331	0,4311	89,61	0,9310

TABLE XVII. Thermodynamic Properties of Liquid and Gaseous Krypton (*Continued*)

T, K	p = 440 bar			p = 460 bar			p = 480 bar		
	$v \cdot 10^3$	h	s	$v \cdot 10^3$	h	s	$v \cdot 10^3$	h	s
170	0.4398	90.90	0.9492	0.4381	91.40	0.9470	0.4364	91.90	0.9448
175	0.4455	93.23	0.9626	0.4437	93.71	0.9603	0.4419	94.21	0.9581
180	0.4514	95.61	0.9760	0.4494	96.08	0.9736	0.4475	96.56	0.9713
185	0.4575	98.06	0.9895	0.4554	98.52	0.9870	0.4533	98.99	0.9847
190	0.4638	100.7	1.003	0.4615	101.1	1.000	0.4592	101.6	0.9980
195	0.4702	103.4	1.017	0.4677	103.8	1.014	0.4653	104.3	1.012
200	0.4769	106.1	1.031	0.4742	106.5	1.028	0.4716	107.0	1.026
205	0.4837	108.9	1.044	0.4808	109.3	1.041	0.4780	109.7	1.039
210	0.4908	111.6	1.057	0.4876	112.0	1.054	0.4845	112.4	1.052
215	0.4980	114.2	1.069	0.4946	114.6	1.066	0.4914	115.0	1.064
220	0.5055	116.7	1.081	0.5018	117.1	1.078	0.4983	117.4	1.075
225	0.5132	119.1	1.092	0.5091	119.5	1.089	0.5054	119.8	1.086
230	0.5211	121.5	1.103	0.5167	121.9	1.100	0.5127	122.2	1.097
235	0.5291	123.9	1.114	0.5246	124.3	1.111	0.5201	124.6	1.107
240	0.5377	126.3	1.124	0.5329	126.7	1.121	0.5282	127.0	1.117
245	0.5468	128.7	1.133	0.5415	129.1	1.130	0.5366	129.3	1.126
250	0.5563	131.1	1.142	0.5505	131.4	1.139	0.5454	131.6	1.135
260	0.5763	135.9	1.160	0.5694	136.1	1.157	0.5627	136.2	1.153
270	0.5967	140.6	1.178	0.5888	140.7	1.174	0.5815	140.8	1.170
280	0.6182	145.3	1.195	0.6092	145.3	1.191	0.6009	145.4	1.187
290	0.6409	150.0	1.212	0.6306	149.9	1.208	0.6211	150.0	1.204
300	0.6647	154.7	1.229	0.6530	154.5	1.224	0.6423	154.6	1.220
320	0.7154	164.1	1.259	0.7007	163.8	1.254	0.6873	163.7	1.249
340	0.7692	173.1	1.286	0.7510	172.8	1.281	0.7350	172.6	1.276
360	0.8253	181.9	1.311	0.8041	181.6	1.306	0.7850	181.3	1.301

TABLE XVII. Thermodynamic Properties of Liquid and Gaseous Krypton (*Continued*)

T. K	p = 440 bar			p = 460 bar			p = 480 bar		
	$v \cdot 10^3$	h	s	$v \cdot 10^3$	h	s	$v \cdot 10^3$	h	s
380	0.8828	190.4	1.334	0.8584	190.0	1.329	0.8365	189.7	1.324
400	0.9413	198.6	1.355	0.9135	198.2	1.350	0.8886	197.8	1.344
420	1.000	206.5	1.374	0.9691	206.1	1.369	0.9415	205.7	1.363
440	1.059	214.2	1.392	1.025	213.8	1.387	0.9940	213.4	1.381
460	1.117	221.7	1.409	1.080	221.3	1.404	1.047	220.9	1.398
480	1.175	228.9	1.424	1.136	228.5	1.419	1.100	228.2	1.414
500	1.233	236.0	1.439	1.190	235.6	1.433	1.152	235.3	1.428
550	1.374	252.9	1.471	1.325	252.6	1.466	1.281	252.4	1.461
600	1.511	269.1	1.499	1.457	268.8	1.494	1.407	268.7	1.489
650	1.645	284.6	1.524	1.585	284.5	1.519	1.530	284.3	1.514
700	1.776	299.6	1.546	1.710	299.6	1.541	1.651	299.5	1.537
750	1.904	314.3	1.567	1.834	314.3	1.562	1.769	314.3	1.557
800	2.031	328.6	1.585	1.955	328.7	1.580	1.885	328.7	1.576
850	2.156	342.7	1.602	2.074	342.8	1.597	2.000	342.9	1.593
900	2.279	356.6	1.618	2.193	356.7	1.613	2.114	356.9	1.609
950	2.401	370.3	1.633	2.310	370.5	1.628	2.226	370.7	1.624
1000	2.522	383.8	1.647	2.426	384.0	1.642	2.337	384.3	1.638
1050	2.642	397.2	1.660	2.541	397.5	1.655	2.448	397.8	1.651
1100	2.762	410.6	1.672	2.655	410.8	1.668	2.558	411.1	1.663
1150	2.880	423.8	1.684	2.769	424.1	1.680	2.667	424.4	1.675
1200	2.998	436.9	1.695	2.882	437.3	1.691	2.775	437.6	1.686
1250	3.116	450.0	1.706	2.995	450.4	1.702	2.883	450.7	1.697
1300	3.233	463.0	1.716	3.107	463.4	1.712	2.991	463.8	1.707

TABLE XVII. Thermodynamic Properties of Liquid and Gaseous Krypton (*Continued*)

T, K	p = 500 bar			p = 550 bar			p = 600 bar		
	$v \cdot 10^3$	h	s	$v \cdot 10^3$	h	s	$v \cdot 10^3$	h	s
135	0,4009	76,26	0,8357	0,3985	77,66	0,8314	0,3963	79,08	0,8271
140	0,4054	78,63	0,8530	0,4029	80,03	0,8485	0,4005	81,43	0,8442
145	0,4100	80,97	0,8694	0,4073	82,35	0,8649	0,4048	83,74	0,8605
150	0,4147	83,29	0,8851	0,4119	84,65	0,8804	0,4092	86,03	0,8759
155	0,4196	85,58	0,9002	0,4165	86,93	0,8954	0,4136	88,29	0,8908
160	0,4245	87,85	0,9146	0,4212	89,18	0,9097	0,4182	90,53	0,9050
165	0,4296	90,13	0,9289	0,4261	91,44	0,9239	0,4228	92,77	0,9191
170	0,4348	92,41	0,9427	0,4311	93,70	0,9375	0,4276	95,01	0,9326
175	0,4402	94,70	0,9559	0,4362	95,97	0,9506	0,4324	97,26	0,9456
180	0,4457	97,05	0,9690	0,4414	98,30	0,9636	0,4374	99,57	0,9586
185	0,4513	99,47	0,9824	0,4467	100,7	0,9768	0,4424	101,9	0,9716
190	0,4571	102,0	0,9956	0,4521	103,2	0,9913	0,4476	104,4	0,9848
195	0,4630	104,7	1,010	0,4577	105,8	1,005	0,4528	107,0	0,9988
200	0,4691	107,4	1,024	0,4634	108,4	1,019	0,4582	109,7	1,013
205	0,4753	110,1	1,037	0,4692	111,1	1,031	0,4636	112,4	1,026
210	0,4817	112,8	1,049	0,4751	113,7	1,043	0,4692	115,0	1,037
215	0,4883	115,4	1,061	0,4812	116,2	1,055	0,4748	117,4	1,048
220	0,4950	117,8	1,072	0,4874	118,6	1,066	0,4806	119,8	1,060
225	0,5018	120,1	1,083	0,4937	121,0	1,077	0,4865	122,1	1,071
230	0,5088	122,4	1,093	0,5001	123,3	1,087	0,4924	124,4	1,081
235	0,5160	124,8	1,103	0,5067	125,6	1,097	0,4985	126,6	1,090
240	0,5238	127,2	1,112	0,5136	127,9	1,106	0,5049	128,8	1,099

TABLE XVII. Thermodynamic Properties of Liquid and Gaseous Krypton (*Continued*)

T, K	p = 500 bar			p = 550 bar			p = 600 bar		
	$v \cdot 10^3$	h	s	$v \cdot 10^3$	h	s	$v \cdot 10^3$	h	s
245	0,5319	129.5	1,121	0,5209	130,1	1,115	0,5115	130,9	1,108
250	0,5403	131,8	1,130	0,5286	132,3	1,124	0,5183	133,1	1,117
260	0,5572	136,4	1,148	0,5440	136,8	1,141	0,5326	137,5	1,134
270	0,5748	140,9	1,166	0,5599	141,3	1,158	0,5472	142,0	1,151
280	0,5932	145,5	1,183	0,5764	145,8	1,174	0,5622	146,4	1,167
290	0,6125	150,1	1,200	0,5935	150,3	1,190	0,5777	150,8	1,182
300	0,6326	154,6	1,215	0,6114	154,8	1,206	0,5937	155,2	1,196
320	0,6752	163,6	1,244	0,6490	163,6	1,234	0,6274	164,0	1,224
340	0,7205	172,5	1,271	0,6890	172,3	1,260	0,6633	172,5	1,250
360	0,7678	181,1	1,296	0,7308	180,7	1,284	0,7007	180,7	1,274
380	0,8165	189,4	1,318	0,7736	189,0	1,306	0,7394	188,9	1,296
400	0,8660	197,5	1,339	0,8179	197,1	1,327	0,7789	196,8	1,317
420	0,9158	205,4	1,358	0,8625	204,9	1,347	0,8190	204,6	1,336
440	0,9665	213,1	1,376	0,9074	212,6	1,365	0,8594	212,3	1,354
460	1,017	220,6	1,393	0,9524	220,0	1,381	0,9000	219,7	1,370
480	1,067	227,9	1,408	0,9974	227,3	1,397	0,9407	227,1	1,386
500	1,117	235,0	1,428	1,042	234,5	1,411	0,9813	234,2	1,400
550	1,241	252,2	1,456	1,154	251,7	1,444	1,082	251,5	1,433
600	1,361	268,5	1,484	1,263	268,2	1,473	1,182	268,1	1,462
650	1,480	284,2	1,509	1,371	284,1	1,498	1,281	284,1	1,488
700	1,596	299,5	1,532	1,476	299,5	1,521	1,377	299,6	1,511
750	1,709	314,3	1,552	1,580	314,4	1,542	1,473	314,7	1,532
800	1,821	328,8	1,571	1,682	329,1	1,560	1,567	329,4	1,551
850	1,932	343,0	1,588	1,783	343,4	1,578	1,659	343,8	1,568
900	2,041	357,1	1,604	1,883	357,5	1,594	1,751	358,0	1,584

TABLE XVII. Thermodynamic Properties of Liquid and Gaseous Krypton (*Continued*)

T, K	p = 500 bar			p = 550 bar			p = 600 bar		
	$v \cdot 10^3$	h	s	$v \cdot 10^3$	h	s	$v \cdot 10^3$	h	s
950	2,149	370,9	1,619	1,981	371.4	1,609	1.842	372.0	1,600
1000	2,256	384.5	1,633	2.079	385,2	1,623	1,932	385.9	1,614
1050	2,362	398,0	1,646	2,176	398,8	1,636	2,021	399.5	1,627
1100	2,468	411.4	1,659	2,272	412.2	1,649	2,109	413.1	1,640
1150	2,573	424.7	1,671	2,368	425.6	1,661	2,197	426,5	1,652
1200	2,677	438.0	1,682	2,463	438.9	1,672	2,285	439,8	1,663
1250	2,781	451,1	1,693	2,558	452.1	1,683	2,372	453.0	1,674
1300	2,884	464,2	1,704	2,652	465.2	1,693	2,458	466.2	1,684

TABLE XVII. Thermodynamic Properties of Liquid and Gaseous Krypton (*Continued*)

T, K	p = 650 bar			p = 700 bar			p = 750 bar		
	$v \cdot 10^3$	h	s	$v \cdot 10^3$	h	s	$v \cdot 10^3$	h	s
135	0.3942	80.50	0,8230	—	—	—	—	—	—
140	0,3983	82.84	0,8401	0,3962	84,26	0,8360	0,3942	85.69	0,8321
145	0,4024	85,14	0,8562	0,4002	86,56	0,8521	0,3980	87,97	0,8481
150	0,4067	87,42	0.8716	0,4043	88,82	0.8674	0,4020	90,23	0,8634
155	0,4109	89.67	0,8864	0,4084	91,06	0,8822	0,4060	92,46	0,8780
160	0,4153	91,89	0,9005	0,4126	93,27	0,8962	0,4101	94,66	0,8920
165	0,4198	94,12	0,9145	0,4169	95.48	0,9100	0,4142	96,86	0,9058

TABLE XVII. Thermodynamic Properties of Liquid and Gaseous Krypton (*Continued*)

T, K	p = 650 bar			p = 700 bar			p = 750 bar		
	$v \cdot 10^3$	h	s	$v \cdot 10^3$	h	s	$v \cdot 10^3$	h	s
170	0.4243	96.34	0.9279	0.4213	97.69	0.9234	0.4184	99.06	0.9191
175	0.4290	98.58	0.9408	0.4257	99.92	0.9363	0.4227	101.3	0.9319
180	0.4337	100.9	0.9536	0.4302	102.2	0.9490	0.4270	103.5	0.9444
185	0.4385	103.2	0.9666	0.4348	104.5	0.9619	0.4314	105.9	0.9574
190	0.4434	105.7	0.9796	0.4395	107.0	0.9748	0.4359	108.3	0.9703
195	0.4483	108.3	0.9930	0.4442	109.6	0.9883	0.4404	110.9	0.9835
200	0.4534	111.0	1.007	0.4490	112.3	1.002	0.4450	113.5	0.9970
205	0.4586	113.6	1.020	0.4539	115.0	1.015	0.4496	116.1	1.010
210	0.4638	116.2	1.032	0.4589	117.5	1.027	0.4543	118.6	1.022
215	0.4691	118.6	1.043	0.4639	119.9	1.038	0.4591	121.0	1.033
220	0.4745	121.0	1.054	0.4690	122.2	1.049	0.4639	123.3	1.044
225	0.4800	123.3	1.064	0.4742	124.5	1.058	0.4688	125.6	1.054
230	0.4856	125.6	1.073	0.4794	126.7	1.067	0.4738	127.9	1.063
235	0.4912	127.9	1.082	0.4847	128.8	1.076	0.4788	130.1	1.072
240	0.4970	130.1	1.091	0.4902	130.9	1.085	0.4839	132.3	1.081
245	0.5030	132.2	1.100	0.4960	133.0	1.094	0.4890	134.4	1.090
250	0.5093	134.3	1.109	0.5020	135.1	1.103	0.4944	136.5	1.098
260	0.5227	138.6	1.126	0.5141	139.4	1.120	0.5059	140.7	1.114
270	0.5362	142.9	1.143	0.5265	143.7	1.136	0.5178	144.9	1.130
280	0.5500	147.2	1.159	0.5393	148.0	1.152	0.5298	149.1	1.146
290	0.5642	151.5	1.164	0.5524	152.3	1.168	0.5420	153.3	1.161
300	0.5788	155.8	1.189	0.5658	156.5	1.182	0.5545	157.4	1.175
320	0.6094	164.3	1.217	0.5938	164.9	1.209	0.5804	165.6	1.202
340	0.6417	172.6	1.242	0.6234	173.1	1.234	0.6076	173.7	1.227
360	0.6757	180.9	1.265	0.6544	181.2	1.257	0.6360	181.7	1.250

TABLE XVII. Thermodynamic Properties of Liquid and Gaseous Krypton (*Continued*)

T, K	p = 650 bar			p = 700 bar			p = 750 bar		
	$v \cdot 10^3$	h	s	$v \cdot 10^3$	h	s	$v \cdot 10^3$	h	s
380	0,7107	188.9	1,287	0,6864	189,2	1.278	0,6656	189.6	1.271
400	0.7466	196.8	1,307	0.7193	197.1	1.299	0.6958	197.4	1.291
420	0.7836	204.6	1,326	0.7526	204.8	1.318	0.7267	2C5.1	1.309
440	0,8206	212.3	1,344	0,7863	212,4	1.335	0.7578	212.6	1.327
460	0,8573	219.8	1,361	0,8203	219,8	1.352	0,7891	220,0	1,344
480	0,8934	227.0	1,376	0,8543	227,1	1,367	0,8206	227,3	1.359
500	0,9308	234,1	1,391	0,8883	234,3	1,382	0.8526	234.4	1.374
550	1,023	251.5	1,424	0.9733	251.6	1.415	0.9307	251.9	1,407
600	1,115	268.2	1,453	1.058	268,3	1,444	1.009	268.7	1.436
650	1,205	284,2	1,479	1,141	284,5	1,470	1.086	284.8	1.462
700	1,294	299.8	1,502	1,224	300.1	1.493	1.163	300.6	1.485
750	1,382	315.0	1,523	1,305	315,4	1,514	1,239	315.9	1,506
800	1,469	329.8	1,542	1,386	330.3	1.533	1,314	330.8	1,526
850	1,555	344.3	1,559	1,466	344.9	1.551	1,389	345.5	1.544
900	1,640	358,6	1,576	1,545	359,3	1.568	1,462	359.9	1,560
950	1,724	372.7	1,591	1,623	373,4	1.583	1,535	374,1	1,575
1000	1,807	386,6	1,605	1,700	387,3	1,597	1,608	388,2	1,590
1050	1,889	400,3	1,619	1,777	401.1	1,611	1,680	402.0	1,603
1100	1,971	413.9	1,631	1,853	414.8	1,623	1.751	415.7	1.616
1150	2,053	427,4	1,643	1,929	428,3	1,635	1.822	429.3	1.628
1200	2.134	440.8	1,655	2,004	441,8	1.647	1.892	442.8	1,640
1250	2.214	454,1	1,666	2,079	455.1	1.658	1.963	456.1	1,651
1300	2,295	467,3	1,676	2.154	468,4	1,668	2,032	469,4	1.661

TABLE XVII. Thermodynamic Properties of Liquid and Gaseous Krypton (*Continued*)

T. K	p = 800 bar			p = 850 bar			p = 900 bar		
	$v \cdot 10^3$	h	s	$v \cdot 10^3$	h	s	$v \cdot 10^3$	h	s
140	0.3922	87.13	0.8283	0.3904	88.56	0.8246	—	—	—
145	0.3960	89.40	0.8443	0.3940	90.83	0.8405	0.3922	92,27	0,8369
150	0.3998	91.65	0,8595	0.3978	93,07	0.8557	0.3958	94.50	0,8520
155	0.4037	93.86	0.8741	0.4015	95,28	0.8702	0.3995	96,70	0,8665
160	0,4076	96,06	0.8879	0,4053	97.46	0.8840	0,4032	98.88	0,8802
165	0,4116	98.25	0.9017	0,4092	99,64	0.8978	0.4069	101,1	0,8940
170	0,4157	100.4	0,9150	0,4131	101.8	0,9109	0,4107	103,2	0,9071
175	0,4198	102,6	0.9277	0,4171	104.0	0,9236	0,4146	105,4	0,9197
180	0,4240	104.9	0,9402	0,4211	106,3	0,9362	0,4184	107,6	0,9322
185	0,4282	107,2	0,9530	0,4252	108,6	0,9488	0,4224	109,9	0,9448
190	0.4325	109.7	0,9658	0,4293	111.0	0,9618	0,4263	112,4	0,9580
195	0,4368	112,3	0.9790	0,4335	113,6	0,9750	0,4304	115,0	0,9712
200	0,4412	114,9	0,9930	0,4377	116,2	0,9890	0,4344	117,6	0,9846
205	0,4457	117,4	1,005	0,4420	118,8	1,001	0,4385	120,2	0,9965
210	0,4502	119,9	1,017	0,4463	121,3	1,013	0,4426	122,7	1,009
215	0,4547	122,3	1,028	0,4506	123,7	1,024	0,4468	125,1	1,020
220	0,4593	124,6	1,039	0,4550	126,0	1,035	0,4510	127,4	1,030
225	0,4639	126,9	1,049	0,4594	128,2	1,045	0,4552	129,6	1,040
230	0,4686	129,2	1,058	0,4639	130,4	1,054	0,4594	131,8	1,049
235	0,4733	131,4	1,067	0,4684	132,6	1,063	0,4637	133,9	1,058
240	0,4781	133,6	1,076	0,4729	134,7	1,071	0,4680	136,0	1,066
245	0,4831	135,7	1,084	0,4774	136,8	1,079	0,4724	138,1	1,074
250	0,4885	137,7	1,092	0,4820	138,9	1,087	0,4767	140,1	1,082
260	0,4995	141,8	1,108	0,4919	143,0	1,103	0,4858	144,1	1,098

TABLE XVII. Thermodynamic Properties of Liquid and Gaseous Krypton (*Continued*)

T, K	p = 800 bar			p = 850 bar			p = 900 bar		
	$v \cdot 10^3$	h	s	$v \cdot 10^3$	h	s	$v \cdot 10^3$	h	s
270	0.5104	145.9	1.124	0.5029	147.1	1.119	0.4964	148.2	1.114
280	0.5214	150.0	1.140	0.5137	151.2	1.135	0.5067	152.3	1.129
290	0.5328	154.1	1.155	0.5245	155.3	1.150	0.5171	156.3	1.144
300	0.5445	158.2	1.169	0.5355	159.3	1.164	0.5274	160.3	1.158
320	0.5685	166.4	1.196	0.5580	167.4	1.190	0.5486	168.3	1.184
340	0.5937	174.4	1.221	0.5816	175.3	1.214	0.5706	176.2	1.208
360	0.6201	182.3	1.244	0.6060	183.1	1.236	0.5935	183.9	1.230
380	0.6474	190.2	1.265	0.6317	190.8	1.257	0.6170	191.5	1.251
400	0.6755	197.9	1.284	0.6577	198.5	1.277	0.6419	199.1	1.271
420	0.7041	205.5	1.302	0.6844	206.1	1.295	0.6669	206.7	1.289
440	0.7331	213.0	1.319	0.7114	213.6	1.312	0.6923	214.2	1.306
460	0.7624	220.3	1.335	0.7387	221.0	1.329	0.7179	221.6	1.323
480	0.7920	227.5	1.350	0.7664	228.3	1.344	0.7437	228.8	1.338
500	0.8221	234.7	1.365	0.7935	235.4	1.359	0.7694	236.0	1.352
550	0.8939	252.3	1.399	0.8619	252.8	1.392	0.8337	253.4	1.385
600	0.9667	269.1	1.428	0.9300	269.6	1.421	0.8977	270.2	1.415
650	1.039	285.3	1.454	0.9976	285.9	1.448	0.9611	286.5	1.441
700	1.111	301.1	1.478	1.065	301.7	1.471	1.024	302.4	1.464
750	1.182	316.4	1.499	1.131	317.1	1.492	1.087	317.8	1.486
800	1.252	331.5	1.518	1.197	332.2	1.512	1.149	332.9	1.505
850	1.322	346.2	1.537	1.263	346.9	1.530	1.210	347.7	1.524
900	1.391	360.7	1.553	1.328	361.5	1.546	1.271	362.3	1.540

TABLE XVII. Thermodynamic Properties of Liquid and Gaseous Krypton (*Continued*)

T. K	p = 800 bar			p = 850 bar			p = 900 bar		
	$v \cdot 10^3$	h	s	$v \cdot 10^3$	h	s	$v \cdot 10^3$	h	s
950	1.459	374.9	1.568	1.392	375.8	1.562	1.332	376.7	1.555
1000	1.527	389.0	1.583	1.456	389.9	1.576	1.393	390.8	1.570
1050	1.594	402.9	1.596	1.519	403.8	1.590	1.453	404.8	1.584
1100	1.661	416.6	1.609	1.582	417.6	1.603	1.513	418.6	1.597
1150	1.728	430.3	1.621	1.645	431.3	1.615	1.572	432.3	1.609
1200	1.794	443.8	1.633	1.708	444.8	1.626	1.631	445.9	1.620
1250	1.860	457.2	1.644	1.770	458.3	1.637	1.690	459.4	1.631
1300	1.926	470.6	1.654	1.832	471.7	1.648	1.748	472.8	1.642

TABLE XVII. Thermodynamic Properties of Liquid and Gaseous Krypton (*Continued*)

T. K	p = 950 bar			p = 1000 bar		
	$v \cdot 10^3$	h	s	$v \cdot 10^3$	h	s
145	0.3904	93.71	0.8333	0.3887	95.16	0.8299
150	0.3939	95.93	0.8484	0.3921	97.37	0.8449
155	0.3975	98.13	0.8628	0.3956	99.57	0.8593
160	0.4011	100.3	0.8766	0.3991	101.7	0.8730
165	0.4047	102.5	0.8902	0.4026	103.9	0.8866
170	0.4084	104.6	0.9033	0.4062	106.1	0.8996
175	0.4121	106.8	0.9159	0.4098	108.2	0.9122
180	0.4159	109.0	0.9284	0.4135	110.4	0.9246
185	0.4197	111.3	0.9409	0.4171	112.7	0.9372
190	0.4235	113.8	0.9538	0.4208	115.2	0.9500

TABLE XVII. Thermodynamic Properties of Liquid and Gaseous Krypton (*Continued*)

T, K	p = 950 bar			p = 1000 bar		
	$v \cdot 10^3$	h	s	$v \cdot 10^3$	h	s
195	0.4274	116.3	0.9672	0.4246	117.7	0.9634
200	0.4313	118.9	0.9804	0.4284	120.3	0.9770
205	0.4352	121.5	0.9935	0.4322	122.9	0.9896
210	0.4392	124.0	1.005	0.4360	125.4	1.001
215	0.4432	126.4	1.015	0.4398	127.8	1.012
220	0.4472	128.7	1.026	0.4437	130.1	1.023
225	0.4513	130.9	1.036	0.4476	132.3	1.033
230	0.4553	133.1	1.045	0.4515	134.4	1.042
235	0.4594	135.2	1.053	0.4554	136.5	1.050
240	0.4636	137.3	1.061	0.4594	138.5	1.058
245	0.4677	139.4	1.069	0.4633	140.5	1.066
250	0.4719	141.4	1.077	0.4674	142.4	1.074
260	0.4805	145.4	1.093	0.4760	146.4	1.090
270	0.4900	149.4	1.109	0.4851	150.3	1.105
280	0.5000	153.4	1.124	0.4945	154.3	1.120
290	0.5102	157.4	1.139	0.5040	158.3	1.135
300	0.5203	161.4	1.153	0.5134	162.3	1.149
320	0.5409	169.4	1.179	0.5324	170.2	1.174
340	0.5618	177.2	1.202	0.5519	178.0	1.197
360	0.5830	184.8	1.224	0.5722	185.7	1.219
380	0.6047	192.4	1.245	0.5928	193.3	1.240
400	0.6279	199.9	1.264	0.6141	200.8	1.259
420	0.6513	207.4	1.282	0.6370	208.2	1.277
440	0.6753	214.9	1.300	0.6600	215.6	1.294
460	0.6994	222.2	1.317	0.6830	222.9	1.310

TABLE XVII. Thermodynamic Properties of Liquid and Gaseous Krypton (*Continued*)

T, K	p = 950 bar			p = 1000 bar		
	$v \cdot 10^3$	h	s	$v \cdot 10^3$	h	s
480	0.7237	229.5	1.333	0.7058	230.2	1.326
500	0.7480	236.8	1.347	0.7285	237.4	1.340
550	0.8087	254.1	1.379	0.7864	254.8	1.373
600	0.8690	270.9	1.409	0.8435	271.7	1.403
650	0.9289	287.3	1.435	0.9001	288.1	1.429
700	0.9883	303.1	1.458	0.9563	304.0	1.453
750	1.047	318.6	1.480	1.012	319.5	1.474
800	1.106	333.8	1.499	1.068	334.6	1.494
850	1.164	348.6	1.518	1.123	349.5	1.512
900	1.222	363.2	1.534	1.178	364.2	1.528
950	1.279	377.6	1.550	1.232	378.6	1.544
1000	1.337	391.8	1.564	1.286	392.8	1.559
1050	1.393	405.8	1.578	1.340	406.8	1.572
1100	1.450	419.7	1.591	1.394	420.7	1.585
1150	1.506	433.4	1.603	1.447	434.5	1.597
1200	1.562	447.0	1.614	1.500	448.2	1.609
1250	1.618	460.6	1.625	1.553	461.7	1.620
1300	1.673	474.0	1.636	1.606	475.2	1.631

TABLE XVIII. Isobaric Specific Heat c_p of Liquid and Gaseous Krypton

T, K	c_p, at p, bar								
	1	10	20	30	40	50	60	70	80
120	0.263	0.543	0.541	0.540	0.538	0.537	0.536	0.535	0.533
150	0.258	0.557	0.552	0.547	0.542	0.538	0.534	0.530	0.526
200	0.252	0.291	0.355	0.481	0.932	1.315	1.034	0.918	0.847
250	0.250	0.268	0.291	0.320	0.357	0.404	0.468	0.552	0.662
300	0.249	0.259	0.272	0.286	0.301	0.317	0.335	0.355	0.377
350	0.249	0.256	0.264	0.272	0.281	0.290	0.300	0.310	0.319
400	0.249	0.254	0.259	0.265	0.271	0.277	0.283	0.289	0.295
450	0.249	0.252	0.256	0.261	0.265	0.269	0.274	0.278	0.282
500	0.248	0.251	0.255	0.258	0.261	0.265	0.268	0.271	0.274
550	0.248	0.251	0.253	0.256	0.259	0.261	0.264	0.267	0.269
600	0.248	0.250	0.252	0.255	0.257	0.259	0.261	0.263	0.265
650	0.248	0.250	0.252	0.254	0.255	0.257	0.259	0.261	0.262
700	0.248	0.250	0.251	0.253	0.254	0.256	0.257	0.259	0.260
750	0.248	0.249	0.251	0.252	0.253	0.254	0.256	0.257	0.258
800	0.248	0.249	0.250	0.251	0.252	0.254	0.255	0.256	0.257
850	0.248	0.249	0.250	0.251	0.252	0.253	0.254	0.255	0.256
900	0.248	0.249	0.250	0.251	0.251	0.252	0.253	0.254	0.255
950	0.248	0.249	0.250	0.250	0.251	0.252	0.252	0.253	0.254
1000	0.248	0.249	0.249	0.250	0.251	0.251	0.252	0.252	0.253
1050	0.248	0.249	0.249	0.250	0.250	0.251	0.251	0.252	0.253
1100	0.248	0.249	0.249	0.250	0.250	0.251	0.251	0.252	0.252
1150	0.248	0.249	0.249	0.250	0.250	0.250	0.251	0.251	0.252
1200	0.248	0.248	0.249	0.249	0.250	0.250	0.250	0.251	0.251
1250	0.248	0.248	0.249	0.249	0.250	0.250	0.250	0.251	0.251
1300	0.248	0.248	0.249	0.249	0.249	0.250	0.250	0.250	0.251

TABLE XVIII. Isobaric Specific Heat c_p of Liquid and Gaseous Krypton (*Continued*)

T, K	c_p, at p, bar								
	90	100	150	200	250	300	350	400	450
120	0.532	0.531	—	—	—	—	—	—	—
150	0.523	0.520	0.507	0.496	0.487	0.480	0.474	0.469	0.465
200	0.806	0.776	0.690	0.640	0.605	0.581	0.564	0.554	0.547
250	0.720	0.917	0.866	0.685	0.597	0.551	0.522	0.500	0.481
300	0.400	0.424	0.538	0.580	0.561	0.532	0.506	0.486	0.470
350	0.330	0.340	0.392	0.431	0.456	0.458	0.444	0.438	0.433
400	0.301	0.307	0.337	0.363	0.383	0.397	0.403	0.404	0.401
450	0.287	0.291	0.311	0.329	0.345	0.357	0.365	0.370	0.373
500	0.278	0.281	0.296	0.309	0.321	0.331	0.339	0.345	0.350
550	0.272	0.274	0.286	0.296	0.306	0.314	0.320	0.327	0.331

TABLE XVIII. Isobaric Specific Heat c_p of Liquid and Gaseous Krypton (*Continued*)

T, K	c_p, at p, bar								
	90	100	150	200	250	300	350	400	450
600	0.267	0.269	0.279	0.287	0.295	0.301	0.309	0.313	0.317
650	0.264	0.266	0.273	0.280	0.287	0.292	0.297	0.302	0.306
700	0.261	0.263	0.269	0.275	0.280	0.285	0.290	0.294	0.298
750	0.259	0.261	0.266	0.271	0.275	0.280	0.284	0.287	0.290
800	0.258	0.259	0.263	0.268	0.272	0.275	0.278	0.282	0.285
850	0.256	0.257	0.261	0.265	0.269	0.272	0.274	0.277	0.280
900	0.255	0.256	0.260	0.263	0.266	0.269	0.271	0.274	0.276
950	0.254	0.255	0.258	0.261	0.263	0.266	0.269	0.271	0.272
1000	0.254	0.254	0.257	0.260	0.262	0.264	0.266	0.268	0.270
1050	0.253	0.254	0.256	0.258	0.260	0.262	0.264	0.266	0.268
1100	0.252	0.253	0.255	0.257	0.259	0.261	0.262	0.264	0.266
1150	0.252	0.252	0.254	0.256	0.258	0.260	0.261	0.263	0.264
1200	0.252	0.252	0.254	0.255	0.257	0.258	0.260	0.261	0.262
1250	0.251	0.252	0.253	0.255	0.256	0.257	0.259	0.260	0.261
1300	0.251	0.251	0.253	0.254	0.255	0.256	0.258	0.259	0.260

TABLE XVIII. Isobaric Specific Heat c_p of Liquid and Gaseous Krypton (*Continued*)

T, K	c_p, at p, bar								
	500	550	600	650	700	750	800	900	1000
150	0.461	0.457	0.455	0.452	0.450	0.448	0.446	0.443	0.441
200	0.541	0.536	0.532	0.528	0.524	0.520	0.516	0.509	0.502
250	0.465	0.452	0.441	0.432	0.425	0.418	0.412	0.406	0.402
300	0.456	0.445	0.436	0.428	0.422	0.416	0.412	0.407	0.404
350	0.428	0.423	0.417	0.411	0.405	0.400	0.396	0.390	0.385
400	0.401	0.398	0.395	0.392	0.390	0.387	0.384	0.378	0.372
450	0.374	0.374	0.373	0.372	0.371	0.370	0.369	0.366	0.362
500	0.352	0.354	0.354	0.355	0.355	0.356	0.356	0.355	0.353
550	0.334	0.337	0.339	0.340	0.341	0.341	0.342	0.342	0.343
600	0.320	0.323	0.325	0.327	0.328	0.329	0.330	0.331	0.332
650	0.309	0.312	0.315	0.316	0.318	0.319	0.320	0.321	0.322
700	0.300	0.303	0.305	0.307	0.309	0.310	0.311	0.313	0.314
750	0.293	0.296	0.298	0.300	0.301	0.303	0.304	0.305	0.307
800	0.287	0.290	0.292	0.293	0.295	0.296	0.297	0.299	0.301
850	0.281	0.283	0.285	0.288	0.290	0.289	0.289	0.291	0.292
900	0.278	0.280	0.282	0.283	0.285	0.286	0.287	0.288	0.290
950	0.275	0.276	0.278	0.280	0.281	0.282	0.283	0.285	0.287
1000	0.272	0.273	0.275	0.276	0.277	0.279	0.280	0.281	0.283
1050	0.269	0.271	0.272	0.273	0.274	0.276	0.277	0.278	0.280
1100	0.267	0.268	0.270	0.271	0.272	0.273	0.274	0.275	0.277
1150	0.265	0.266	0.268	0.269	0.270	0.271	0.271	0.273	0.274
1200	0.264	0.265	0.266	0.267	0.268	0.269	0.269	0.271	0.272
1250	0.262	0.263	0.264	0.265	0.266	0.267	0.268	0.269	0.270
1300	0.261	0.262	0.263	0.264	0.265	0.265	0.266	0.267	0.268

TABLE XIX. Dynamic Viscosity of Krypton

T, K	$\eta \cdot 10^8$, Pa · s at p, bar							
	1	10	20	30	40	50	60	70
120	1088	37510	37793	38109	38388	38670	38953	39239
130	1174	31777	32065	32356	32650	32948	33250	33521
140	1260	26744	27044	27350	27660	27949	28242	28540
150	1345	22322	22694	23003	23319	23622	23910	24203
160	1430	1493	18959	19271	19592	19892	20201	20501
170	1514	1572	15677	16026	16359	16682	16995	17307
180	1597	1651	1748	13136	13508	13863	14205	14531
190	1680	1731	1812	1278	10895	11329	11722	12094
200	1763	1811	1883	2007	2269	8654	9409	9872
210	1844	1890	1954	2055	2227	2586	6370	7515
220	1926	1970	2029	2115	2247	2459	2857	3999
230	2000	2049	2103	2178	2287	2444	2679	3056
240	2087	2128	2179	2247	2339	2466	2638	2877
250	2166	2206	2253	2315	2396	2503	2641	2820
260	2245	2284	2329	2385	2458	2550	2666	2810
270	2323	2361	2403	2456	2522	2604	2704	2825
280	2401	2438	2479	2528	2588	2661	2750	2855
290	2478	2514	2553	2599	2655	2722	2801	2893
300	2554	2589	2627	2670	2722	2783	2855	2938
320	2705	2738	2773	2813	2859	2912	2972	3041
340	2853	2885	2918	2954	2995	3042	3094	3153
360	2998	3028	3060	3094	3131	3173	3219	3270
380	3140	3162	3199	3231	3265	3303	3345	3390
400	3280	3308	3337	3367	3399	3433	3471	3512
420	3418	3445	3472	3501	3531	3563	3598	3635
440	3553	3579	3605	3632	3661	3691	3723	3757
460	3685	3710	3735	3761	3788	3816	3846	3878
480	3815	3832	3864	3888	3914	3941	3969	3998
500	3943	3967	3990	4014	4038	4064	4090	4118
550	4254	4276	4298	4319	4341	4364	4387	4411
600	4553	4573	4593	4613	4634	4654	4675	4696
650	4841	4860	4879	4897	4916	4935	4954	4973
700	5119	5137	5154	5172	5189	5207	5224	5242
750	5389	5406	5422	5439	5455	5471	5487	5504
800	5650	5666	5681	5697	5712	5727	5742	5758
850	5904	5919	5934	5948	5963	5977	5991	6006
900	6152	6166	6180	6194	6208	6221	6235	6248
950	6393	6406	6420	6433	6446	6459	6471	6484
1000	6629	6642	6654	6667	6679	6691	6704	6716
1050	6860	6872	6884	6896	6908	6920	6931	6943

TABLE XIX. Dynamic Viscosity of Krypton (*Continued*)

T, K	$\eta \cdot 10^8$, Pa · s at *p*, bar							
	1	10	20	30	40	50	60	70
1100	7086	7098	7109	7120	7132	7143	7154	7165
1150	7307	7318	7329	7340	7351	7361	7372	7383
1200	7525	7536	7546	7557	7567	7577	7587	7597
1250	7739	7749	7759	7769	7779	7789	7799	7808
1300	7949	7959	7969	7978	7988	7997	8006	8016

TABLE XIX. Dynamic Viscosity of Krypton (*Continued*)

T, K	$\eta \cdot 10^8$, Pa · s at *p*, bar							
	80	100	120	140	160	180	200	220
120	39487	40028	40577	41090	—	—	—	—
130	33830	34387	34921	35465	35982	36508	37044	37550
140	28815	29405	29954	30487	31034	31563	32072	32593
150	24503	25075	25621	26163	26697	27223	27738	28241
160	20792	21363	21924	22454	22986	23499	24010	24518
170	17605	18179	18746	19288	19814	20318	20829	21311
180	14846	15450	16022	16571	17100	17613	18104	18594
190	12443	13093	13698	14258	14798	15316	15802	16288
200	10289	11026	11678	12278	12834	13355	13855	14331
210	8198	9174	9930	10584	11137	11677	12181	12662
220	5712	7342	8305	9051	9683	10253	10768	11251
230	3710	5535	6792	7673	8375	8979	9570	10055
240	3212	4275	5496	6470	7240	7884	8445	8966
250	3049	3706	4600	5502	6281	6944	7522	8040
260	2987	3457	4094	4821	5535	6177	6750	7266
270	2969	3336	3815	4388	4992	5579	6123	6623
280	2977	3279	3661	4119	4621	5136	5633	6104
290	3000	3257	3575	3950	4369	4813	5256	5692
300	3032	3257	3529	3847	4201	4584	4976	5371
320	3118	3296	3507	3749	4018	4321	4625	4935
340	3218	3366	3539	3733	3948	4181	4426	4695
360	3326	3453	3598	3761	3940	4132	4336	4551
380	3439	3550	3675	3814	3966	4130	4303	4485
400	3556	3654	3764	3886	4018	4159	4309	4465
420	3675	3763	3861	3968	4085	4209	4340	4478
440	3794	3873	3961	4058	4162	4272	4389	4512
460	3912	3984	4064	4152	4245	4345	4449	4560
480	4030	4096	4170	4249	4334	4424	4519	4619
500	4147	4209	4277	4349	4427	4510	4596	4687
550	4436	4489	4545	4606	4670	4737	4808	4888
600	4718	4764	4813	4864	4919	4975	5035	5096
650	4993	5034	5077	5121	5168	5217	5268	5321

TABLE XIX. Dynamic Viscosity of Krypton (*Continued*)

T, K	$\eta \cdot 10^8$, Pa · s at p, bar							
	80	100	120	140	160	180	200	220
700	5260	5297	5335	5375	5416	5459	5503	5549
750	5520	5554	5589	5625	5662	5700	5739	5780
800	5773	5804	5836	5869	5902	5937	5972	6008
850	6020	6049	6078	6109	6139	6171	6203	6236
900	6262	6289	6316	6344	6373	6401	6431	6431
950	6497	6523	6548	6574	6601	6627	6655	6682
1000	6728	6752	6776	6801	6825	6850	6876	6901
1050	6954	6977	7000	7023	7046	7070	7094	7118
1100	7176	7198	7220	7241	7263	7286	7308	7330
1150	7393	7414	7435	7455	7476	7497	7518	7539
1200	7607	7627	7647	7667	7687	7707	7727	7747
1250	7818	7837	7856	7875	7894	7913	7932	7951
1300	8025	8043	8062	8080	8098	8116	8134	8152

TABLE XIX. Dynamic Viscosity of Krypton (*Continued*)

T, K	$\eta \cdot 10^8$, Pa · s at p, bar							
	240	260	280	300	320	340	360	380
130	38025	38506	38995	39490	39951	40417	40889	41324
140	33091	33566	34084	34542	35008	35482	35964	36417
150	28730	29232	29718	30186	30665	31124	31592	32028
160	24998	25471	25958	26412	26877	27330	27795	28218
170	21795	22261	22725	23186	23622	24072	24494	24927
180	19054	19519	19975	20419	20849	21277	21686	22109
190	16749	17201	17641	18066	18496	18905	19318	19717
200	14792	15238	15664	16105	16498	16899	17296	17688
210	13120	13559	13984	14399	14799	15195	15576	15948
220	11711	12148	12570	12974	13368	13752	14124	14485
230	10511	10950	11359	11764	12149	12523	12880	13235
240	9444	9920	10301	10721	11097	11477	11824	12153
250	8512	8947	9360	9762	10143	10516	10848	11175
260	7737	8173	8581	8966	9331	9682	10021	10344
270	7084	7512	7913	8292	8650	8994	9324	9642
280	6545	6959	7347	7716	8068	8404	8725	9037
290	6105	6498	6872	7227	7568	7894	8206	8508
300	5753	6121	6475	6816	7141	7455	7757	8049
320	5253	5560	5882	6185	6480	6765	7044	7318
340	4950	5212	5484	5754	6014	6271	6530	6770
360	4774	4999	5227	5462	5693	5920	6146	6369
380	4674	4869	5067	5271	5474	5675	5876	6083
400	4630	4798	4972	5151	5331	5507	5689	5868
420	4622	4771	4924	5080	5239	5401	5561	5724

TABLE XIX. Dynamic Viscosity of Krypton (*Continued*)

T, K	$\eta \cdot 10^8$, Pa · s at p, bar							
	240	260	280	300	320	340	360	380
440	4639	4771	4907	5046	5189	5333	5478	5628
460	4674	4792	4914	5039	5167	5298	5431	5564
480	4722	4829	4939	5052	5168	5287	5407	5529
500	4781	4878	4979	5082	5188	5295	5405	5517
550	4958	5036	5118	5202	5288	5375	5464	5556
600	5160	5225	5294	5363	5435	5508	5584	5658
650	5375	5431	5489	5548	5609	5671	5734	5799
700	5596	5645	5695	5746	5799	5852	5907	5963
750	5821	5864	5908	5953	5999	6046	6094	6142
800	6046	6084	6123	6163	6204	6245	6288	6331
850	6269	6304	6339	6375	6411	6448	6486	6525
900	6492	6523	6555	6587	6620	6654	6688	6723
950	6711	6739	6768	6798	6828	6859	6890	6922
1000	6928	6954	6981	7008	7036	7064	7093	7122
1050	7142	7166	7191	7217	7242	7268	7295	7322
1100	7353	7376	7399	7423	7447	7471	7495	7520
1150	7561	7583	7604	7626	7649	7671	7694	7717
1200	7767	7787	7808	7829	7850	7871	7892	7914
1250	7970	7990	8009	8029	8048	8068	8089	8109
1300	8171	8190	8207	8226	8245	8264	8282	8302

TABLE XIX. Dynamic Viscosity of Krypton (*Continued*)

T, K	$\eta \cdot 10^8$, Pa · s at p, bar							
	400	420	440	460	480	500	550	600
130	41763	42208	42657	43065	43478	—	—	—
140	36876	37304	37738	38177	38624	39034	40084	41122
150	32493	32924	33362	33808	34228	34654	35676	36699
160	28651	29095	29519	29953	30367	30789	31818	32797
170	25350	25785	26185	26594	27013	27418	28389	29358
180	22528	22922	23328	23725	24114	24492	25435	26367
190	20116	20499	20879	21255	21643	22008	22922	23801
200	18072	18446	18809	19172	19533	19892	20758	21609
210	16318	16684	17034	17387	17742	18075	18909	19718
220	14841	15190	15538	15875	16207	16532	17334	18117
230	13585	13910	14249	14580	14895	15216	15984	16732
240	12497	12835	13149	13449	13757	14058	14815	15529
250	11481	11821	12143	12442	12720	13011	13736	14449
260	10663	10970	11269	11566	11871	12135	12825	13496
270	9951	10251	10542	10827	11107	11379	12042	12680
280	9336	9628	9913	10188	10458	10724	11364	11978
290	8800	9085	9361	9628	9892	10147	10770	11362

TABLE XIX. Dynamic Viscosity of Krypton (*Continued*)

T, K	$\eta \cdot 10^8$, Pa · s at p, bar							
	400	420	440	460	480	500	550	600
300	8332	8608	8876	9136	9392	9639	10241	10818
320	7577	7832	8082	8326	8566	8799	9365	9908
340	7010	7249	7478	7708	7928	8144	8673	9179
360	6592	6805	7017	7227	7433	7634	8126	8600
380	6277	6475	6669	6861	7050	7237	7698	8137
400	6048	6236	6407	6584	6759	6932	7356	7768
420	5887	6057	6214	6377	6537	6700	7091	7476
440	5776	5925	6071	6221	6371	6518	6883	7242
460	5699	5834	5971	6109	6245	6381	6720	7056
480	5651	5776	5902	6027	6154	6282	6597	6910
500	5630	5745	5859	5977	6093	6211	6505	6797
550	5648	5742	5836	5933	6030	6127	6374	6626
600	5735	5814	5893	5974	6056	6140	6349	6563
650	5864	5931	5999	6068	6138	6208	6389	6573
700	6020	6077	6136	6196	6256	6317	6474	6636
750	6192	6243	6295	6346	6399	6454	6591	6732
800	6375	6420	6465	6511	6558	6606	6727	6852
850	6564	6604	6645	6686	6728	6770	6879	6991
900	6758	6794	6831	6868	6906	6944	7041	7142
950	6954	6987	7020	7053	7087	7122	7211	7301
1000	7151	7181	7210	7242	7273	7305	7385	7468
1050	7349	7376	7404	7432	7460	7490	7563	7639
1100	7545	7571	7596	7622	7649	7675	7743	7813
1150	7740	7764	7788	7812	7836	7861	7924	7989
1200	7936	7958	7980	8003	8025	8048	8107	8167
1250	8129	8150	8171	8192	8213	8235	8289	8345
1300	8321	8340	8368	8380	8400	8420	8471	8524

TABLE XIX. Dynamic Viscosity of Krypton (*Continued*)

T, K	$\eta \cdot 10^8$, Pa · s at p, bar							
	650	700	750	800	850	900	950	1000
140	42100	43056	43988	44941	45815	—	—	—
150	37680	38653	39614	40561	41444	42349	43229	44082
160	33783	34739	35657	36608	37512	38363	39239	40096
170	30317	31231	32153	33047	33943	34800	35650	36489
180	27278	28187	29060	29918	30783	31624	32433	33239
190	24674	25533	26372	27206	28030	28839	29629	30422
200	22450	23273	24067	24866	25642	26411	27170	27912
210	20516	21292	22070	22808	23551	24295	25016	25729

TABLE XIX. Dynamic Viscosity of Krypton (*Continued*)

T, K	$\eta \cdot 10^8$, Pa · s at p, bar							
	650	700	750	800	850	900	950	1000
220	18879	19620	20357	21068	21774	22470	23169	23849
230	17453	18165	18858	19549	20215	20894	21551	22195
240	16242	16911	17582	18247	18887	19531	20146	20769
250	15140	15754	16449	17033	17726	18335	18923	19509
260	14142	14761	15406	15952	16655	17267	17837	18351
270	13297	13899	14491	15039	15641	16206	16805	17664
280	12570	13146	13709	14253	14795	15327	15877	16360
290	11963	12486	13027	13550	14064	14561	15059	15539
300	11368	11905	12422	12924	13418	13900	14355	14830
320	10424	10928	11411	11885	12343	12790	13185	13655
340	9669	10141	10597	11043	11470	11894	12262	12710
360	9057	9501	9933	10350	10759	11157	11521	11926
380	8565	8982	9386	9781	10160	10550	10908	11284
400	8169	8559	8941	9312	9674	10029	10373	10744
420	7844	8218	8575	8926	9268	9604	9932	10262
440	7586	7939	8275	8605	8929	9246	9556	9861
460	7381	7709	8028	8337	8644	8945	9239	9525
480	7220	7524	7824	8114	8405	8692	8972	9247
500	7088	7376	7655	7926	8212	8483	8725	9015
550	6875	7123	7371	7616	7858	8097	8332	8564
600	6777	6994	7212	7428	7643	7856	8067	8275
650	6762	6952	7144	7334	7525	7716	7905	8094
700	6800	6966	7136	7304	7475	7649	7819	7989
750	6878	7026	7175	7326	7481	7634	7791	7945
800	6981	7113	7248	7383	7522	7660	7801	7940
850	7106	7223	7344	7466	7591	7719	7846	7972
900	7245	7351	7461	7571	7684	7799	7915	8034
950	7395	7491	7591	7691	7794	7899	8005	8112
1000	7553	7641	7731	7823	7917	8012	8109	8208
1050	7718	7798	7880	7965	8051	8138	8228	8318
1100	7885	7959	8035	8113	8191	8272	8354	8438
1150	8055	8124	8193	8265	8338	8412	8489	8566
1200	8229	8298	8357	8423	8490	8560	8631	8703
1250	8403	8462	8522	8584	8647	8711	8777	8844
1300	8578	8633	8689	8746	8805	8865	8927	8989

TABLE XX. Thermal Conductivity of Krypton

T, K	$\lambda \cdot 10^8$, kW/(m · K) at p, bar							
	1	10	20	30	40	50	60	70
120	405	9046	9090	9141	9187	9232	9278	9324
130	437	8319	8368	8417	8467	8517	8569	8615
140	469	7612	7667	7723	7780	7832	7885	7939
150	501	6923	6983	7045	7109	7169	7226	7284
160	532	596	6308	6377	6449	6514	6581	6647

TABLE XX. Thermal Conductivity of Krypton (*Continued*)

T, K	λ · 10⁸, kW/(m · K) at p, bar							
	1	10	20	30	40	50	60	70
170	564	622	5621	5709	5791	5870	5946	6021
180	595	648	731	5016	5119	5216	5309	5396
190	625	674	745	863	4403	4535	4654	4764
200	656	702	765	856	1022	3700	3954	4106
210	687	730	787	864	977	1190	2888	3313
220	717	758	809	877	967	1098	1325	1897
230	747	785	833	893	970	1070	1210	1419
240	777	813	858	912	979	1063	1168	1306
250	806	841	882	932	992	1064	1151	1257
260	836	869	908	955	1009	1073	1148	1235
270	865	897	934	977	1027	1085	1151	1226
280	894	924	960	1000	1047	1099	1158	1225
290	923	952	986	1024	1067	1116	1170	1230
300	951	979	1011	1048	1088	1133	1183	1238
320	1007	1033	1063	1095	1132	1171	1214	1261
340	1062	1086	1114	1144	1176	1212	1250	1291
360	1116	1139	1164	1192	1222	1254	1289	1325
380	1169	1191	1215	1240	1268	1297	1329	1362
400	1221	1242	1264	1288	1314	1341	1370	1400
420	1272	1292	1313	1335	1359	1385	1411	1439
440	1323	1342	1362	1383	1405	1429	1454	1480
460	1372	1390	1409	1429	1450	1472	1496	1520
480	1420	1437	1455	1474	1494	1515	1537	1560
500	1468	1484	1502	1520	1539	1559	1579	1601
550	1584	1599	1614	1631	1647	1665	1683	1702
600	1695	1709	1723	1737	1753	1768	1785	1801
650	1802	1814	1827	1841	1855	1869	1884	1899
700	1906	1918	1929	1942	1955	1968	1981	1995
750	2006	2017	2028	2039	2051	2063	2076	2088
800	2104	2114	2124	2135	2146	2157	2169	2180
850	2198	2207	2217	2227	2237	2248	2259	2269
900	2290	2299	2306	2317	2327	2337	2347	2357
950	2380	2389	2397	2406	2415	2424	2434	2443
1000	2468	2476	2484	2493	2501	2510	2519	2528
1050	2554	2562	2569	2577	2585	2594	2602	2611
1100	2638	2645	2653	2660	2668	2676	2684	2692
1150	2721	2728	2735	2742	2750	2757	2765	2772
1200	2802	2809	2815	2822	2829	2837	2844	2851
1250	2881	2887	2894	2901	2907	2914	2921	2928
1300	2959	2965	2971	2978	2984	2991	2997	3004

TABLE XX. Thermal Conductivity of Krypton (*Continued*)

T, K	λ · 10⁸, kW/(m · K) at p, bar							
	80	100	120	140	160	180	200	220
120	9365	9452	9541	9625				
130	8666	8761	8850	8941	9027	9115	9205	9289
140	7989	8095	8193	8288	8384	8477	8567	8658
150	7343	7454	7560	7664	7765	7866	7962	8057
160	6710	6832	6951	7062	7173	7278	7383	7486
170	6092	6228	6361	6485	6606	6719	6833	6940
180	5480	5638	5786	5925	6057	6184	6304	6422
190	4867	5054	5226	5381	5530	5670	5800	5927
200	4241	4475	4677	4859	5025	5178	5323	5459
210	3557	3896	4150	4365	4543	4714	4871	5020
220	2630	3255	3605	3868	4086	4279	4450	4608
230	1752	2557	3053	3384	3641	3858	4066	4234
240	1488	2010	2564	2929	3226	3468	3675	3864
250	1387	1730	2150	2538	2855	3115	3338	3534
260	1338	1594	1911	2244	2549	2812	3041	3243
270	1312	1517	1766	2042	2314	2565	2790	2992
280	1300	1473	1678	1907	2143	2372	2585	2781
290	1296	1447	1621	1815	2019	2225	2421	2607
300	1297	1431	1584	1753	1931	2114	2293	2467
320	1311	1422	1545	1679	1821	1974	2120	2264
340	1335	1430	1534	1645	1763	1885	2010	2141
360	1364	1447	1538	1634	1735	1840	1947	2056
380	1397	1471	1551	1636	1725	1817	1911	2007
400	1432	1499	1572	1648	1727	1809	1893	1978
420	1468	1530	1596	1665	1737	1811	1887	1964
440	1507	1564	1625	1688	1754	1821	1891	1961
460	1545	1598	1654	1713	1773	1836	1899	1964
480	1584	1633	1685	1740	1796	1854	1913	1973
500	1623	1670	1719	1769	1822	1876	1931	1987
550	1722	1762	1804	1848	1894	1939	1987	2034
600	1819	1854	1892	1930	1970	2010	2052	2093
650	1914	1946	1980	2014	2049	2086	2122	2160
700	2009	2038	2068	2099	2131	2164	2197	2231
750	2101	2128	2155	2184	2213	2243	2273	2304
800	2192	2217	2242	2268	2295	2323	2350	2379
850	2281	2303	2327	2351	2376	2401	2427	2453
900	2367	2389	2411	2433	2456	2480	2504	2528
950	2453	2473	2493	2515	2536	2558	2580	2603
1000	2537	2556	2575	2600	2615	2636	2657	2678
1050	2619	2637	2655	2674	2693	2712	2732	2752
1100	2700	2717	2734	2752	2771	2788	2807	2826
1150	2780	2796	2813	2829	2846	2864	2881	2899
1200	2859	2874	2889	2905	2922	2938	2955	2972
1250	2935	2950	2965	2980	2995	3011	3027	3043
1300	3011	3025	3039	3054	3068	3083	3099	3114

TABLE XX. Thermal Conductivity of Krypton (*Continued*)

T, K	λ · 10⁸, kW/(m · K) at p, bar							
	240	260	280	300	320	340	360	380
130	9369	9449	9530	9613	9690	9768	9847	9920
140	8745	8827	8918	8996	9077	9159	9242	9321
150	8148	8240	8330	8417	8504	8588	8673	8753
160	7582	7677	7774	7863	7954	8043	8133	8215
170	7046	7146	7246	7345	7437	7532	7620	7710
180	6531	6642	6749	6851	6949	7047	7140	7235
190	6047	6163	6274	6382	6488	6590	6690	6786
200	5589	5713	5829	5949	6055	6161	6266	6368
210	5158	5289	5415	5536	5651	5765	5872	5976
220	4757	4896	5029	5155	5277	5394	5505	5614
230	4390	4538	4674	4808	4933	5054	5168	5280
240	4036	4204	4338	4483	4612	4740	4857	4967
250	3710	3870	4020	4165	4301	4434	4550	4664
260	3424	3591	3744	3888	4023	4152	4275	4392
270	3175	3342	3498	3643	3779	3909	4033	4151
280	2961	3127	3281	3425	3562	3692	3815	3934
290	2781	2942	3093	3235	3371	3499	3622	3739
300	2631	2786	2932	3071	3203	3328	3449	3564
320	2408	2542	2681	2809	2933	3051	3165	3277
340	2261	2381	2503	2621	2733	2843	2953	3053
360	2166	2273	2379	2486	2589	2689	2787	2883
380	2103	2200	2295	2391	2485	2576	2665	2756
400	2064	2151	2237	2324	2410	2492	2576	2657
420	2042	2120	2199	2277	2355	2433	2509	2584
440	2032	2103	2175	2247	2319	2390	2461	2532
460	2029	2095	2162	2228	2295	2361	2427	2493
480	2034	2095	2156	2218	2280	2342	2403	2464
500	2044	2100	2158	2216	2273	2331	2389	2446
550	2083	2132	2181	2231	2281	2330	2380	2430
600	2136	2179	2222	2266	2310	2353	2397	2441
650	2198	2236	2274	2313	2352	2391	2431	2470
700	2265	2300	2334	2369	2404	2440	2475	2511
750	2335	2366	2398	2430	2462	2494	2526	2559
800	2407	2436	2465	2494	2524	2554	2583	2613
850	2479	2506	2533	2560	2587	2615	2642	2670
900	2552	2577	2602	2627	2653	2678	2704	2730
950	2626	2649	2672	2696	2720	2744	2768	2792
1000	2699	2721	2743	2765	2788	2810	2833	2855
1050	2772	2793	2814	2835	2856	2877	2898	2919
1100	2845	2864	2884	2904	2923	2943	2964	2984
1150	2918	2936	2954	2973	2992	3011	3030	3049
1200	2989	3006	3024	3042	3060	3078	3096	3114
1250	3060	3076	3093	3110	3127	3144	3161	3179
1300	3130	3146	3162	3178	3194	3210	3227	3243

TABLE XX. Thermal Conductivity of Krypton (*Continued*)

T, K	$\lambda \cdot 10^8$, kW/(m · K) at p, bar							
	400	420	440	460	480	500	550	600
130	9994	10068	10144	10213	10282			
140	9400	9473	9547	9623	9699	9769	9949	10128
150	8835	8913	8991	9071	9146	9222	9404	9584
160	8297	8383	8463	8545	8624	8703	8895	9076
170	7797	7886	7968	8051	8135	8216	8409	8599
180	7328	7415	7503	7591	7674	7755	7956	8151
190	6883	6974	7063	7151	7242	7326	7534	7731
200	6467	6563	6655	6746	6836	6925	7137	7341
210	6079	6180	6275	6370	6465	6553	6772	6979
220	5719	5821	5922	6019	6113	6205	6429	6643
230	5390	5494	5595	5696	5792	5887	6115	6332
240	5080	5190	5292	5388	5487	5582	5818	6037
250	4770	4886	4995	5096	5189	5286	5524	5754
260	4506	4616	4722	4826	4927	5023	5260	5485
270	4265	4376	4482	4586	4687	4784	5020	5243
280	4048	4158	4265	4368	4468	4567	4801	5024
290	3853	3962	4068	4170	4271	4368	4602	4822
300	3677	3785	3890	3991	4090	4186	4417	4638
320	3382	3485	3586	3683	3780	3873	4098	4312
340	3152	3251	3345	3440	3530	3618	3833	4039
360	2978	3067	3157	3244	3330	3414	3618	3814
380	2841	2927	3010	3091	3172	3251	3445	3629
400	2737	2820	2895	2972	3047	3122	3303	3478
420	2659	2736	2806	2878	2948	3020	3189	3355
440	2602	2671	2738	2805	2873	2939	3099	3256
460	2558	2622	2686	2750	2812	2874	3026	3174
480	2525	2586	2646	2705	2764	2824	2968	3108
500	2504	2561	2617	2674	2730	2785	2922	3056
550	2480	2529	2578	2628	2677	2725	2846	2967
600	2485	2529	2573	2616	2660	2704	2812	2919
650	2509	2548	2588	2627	2666	2705	2803	2900
700	2547	2582	2618	2654	2689	2725	2814	2903
750	2591	2624	2657	2689	2722	2755	2836	2918
800	2643	2673	2703	2734	2764	2794	2870	2945
850	2698	2726	2754	2782	2810	2838	2908	2979
900	2756	2782	2808	2834	2860	2886	2952	3018
950	2816	2841	2865	2889	2914	2939	3000	3062
1000	2878	2901	2924	2947	2970	2993	3051	3109
1050	2941	2963	2984	3006	3028	3050	3104	3159
1100	3004	3025	3045	3066	3086	3107	3159	3211
1150	3069	3088	3107	3127	3146	3166	3215	3265
1200	3133	3151	3170	3188	3207	3225	3273	3320
1250	3196	3214	3231	3249	3267	3285	3330	3375
1300	3260	3277	3294	3311	3328	3345	3388	3431

TABLE XX. Thermal Conductivity of Krypton (*Continued*)

T, K	λ · 10^8, kW/(m · K) at p, bar							
	650	700	750	800	850	900	950	1000
140	10296	10461	10622	10787	10940			
150	9758	9928	10098	10264	10418	10576	10732	10883
160	9258	9433	9599	9772	9935	10087	10244	10397
170	8785	8962	9139	9307	9475	9636	9794	9949
180	8340	8526	8702	8874	9045	9211	9368	9524
190	7924	8111	8292	8469	8642	8810	8973	9133
200	7541	7732	7913	8093	8267	8436	8601	8762
210	7181	7373	7563	7740	7916	8091	8258	8420
220	6848	7043	7234	7416	7593	7766	7937	8101
230	6537	6736	6927	7114	7292	7470	7640	7805
240	6251	6449	6644	6834	7014	7193	7361	7529
250	5972	6163	6376	6552	6758	6936	7105	7272
260	5700	5901	6109	6282	6501	6689	6862	7014
270	5456	5661	5860	6041	6238	6420	6610	6763
280	5235	5439	5634	5822	6006	6185	6367	6525
290	5043	5234	5429	5616	5797	5971	6142	6305
300	4846	5047	5238	5424	5603	5777	5940	6107
320	4516	4713	4901	5084	5260	5430	5580	5756
340	4237	4428	4611	4790	4961	5129	5275	5451
360	4002	4185	4364	4535	4704	4867	5016	5182
380	3808	3983	4152	4318	4478	4642	4792	4950
400	3648	3813	3976	4133	4288	4439	4587	4746
420	3513	3673	3826	3978	4125	4270	4412	4556
440	3405	3558	3703	3846	3988	4126	4262	4396
460	3317	3461	3600	3735	3870	4002	4132	4259
480	3246	3381	3513	3641	3769	3896	4021	4144
500	3187	3316	3440	3560	3688	3809	3928	4047
550	3084	3198	3312	3424	3534	3642	3749	3855
600	3024	3128	3231	3333	3434	3532	3630	3726
650	2996	3092	3186	3279	3371	3462	3552	3641
700	2991	3078	3165	3250	3335	3421	3504	3587
750	3000	3080	3161	3240	3320	3398	3477	3554
800	3020	3095	3171	3245	3319	3392	3465	3537
850	3048	3118	3188	3258	3327	3398	3466	3533
900	3083	3149	3215	3280	3345	3410	3476	3541
950	3123	3185	3247	3309	3371	3433	3494	3555
1000	3167	3226	3284	3343	3401	3459	3517	3576
1050	3215	3270	3325	3381	3436	3491	3547	3602
1100	3264	3316	3368	3421	3473	3526	3579	3632
1150	3315	3365	3415	3465	3515	3565	3616	3667
1200	3367	3415	3463	3511	3558	3607	3655	3703
1250	3420	3466	3511	3557	3603	3649	3695	3742
1300	3474	3518	3562	3605	3649	3694	3738	3782

F. XENON TABLE DISCUSSION

The calorific functions of xenon were calculated for the first time by Michels et al. [III.10] graphically from their previous data [2.77] for temperatures of 0 to 150°C at pressures to 2800 bar. Over the same range of variables they list [III.10] at round values of the dimensionless density and pressure, data on the enthalpy, entropy, specific heats c_p and c_v, Helmholtz energy, Gibbs energy, speed of sound and the Joule-Thomson coefficient.

Vukalovich et al. [III.11] suggested a virial equation of state whose coefficients were determined from the *pvT* data in [2.77]. However, in order to cover a wider temperature range, Michels et al. included in their database results obtained by graphical extrapolation on the ρ, *T* diagram of isolines of $(z - 1)\,v =$ const. These were obtained from the same experimental data [2.77] for the corresponding values of the second virial coefficient. The necessary values of the latter at $t > 150$°C were found from the (20:6) potential, the characteristic parameters of which are $\epsilon/k = 328.3$ K and $b_0 = 0.5035$ cm^3/g. The tables of thermodynamic properties of gaseous xenon, calculated from the equation of state in [III.11], which include values of density, enthalpy, entropy, and specific heat c_p at 290 to 1300 K and 1 to 1000 bar, were published in Vargaftik's handbook [III.9].

The results given in [III.9, III.10] and those obtained by us are compared in Tables 64 through 66. It is seen from Table 64 that the deviations for enthalpy

TABLE 64†. Comparison of Computed Values of Enthalpy of Xenon with Tabulated Data in [III.9] (2nd line) and [III.10] (3rd line)

T, K	*h*, kJ/kg, at pressure, bar					
	50	100	200	300	600	1000
300	148.4	106.4	101.7	101.3	104.9	112.8
	148.2	106.3	101.5	101.1	104.8	112.8
	148.1	106.3	101.5	101.2	104.9	112.9
350	163.6	146.2	123.1	119.0	119.7	126.8
	163.7	146.1	123.2	119.0	119.8	126.8
	163.5	146.0	123.0	118.8	119.6	126.6
400	174.7	164.2	145.4	137.4	134.5	140.0
	174.9	164.2	145.4	137.3	134.5	140.2
	174.7	164.1	145.4	137.2	134.4	140.1
500	193.8	187.8	177.1	169.4	162.3	165.7
	194.2	188.0	177.2	169.6	162.6	166.0
600	211.5	207.4	200.3	194.8	187.9	189.9
	211.6	207.6	200.5	195.0	188.2	190.4
800	245.1	243.0	239.5	236.7	233.0	235.0
	245.2	243.1	239.5	236.8	233.0	235.0
1000	277.9	276.8	275.1	273.9	273.0	276.2
	277.8	276.8	275.0	273.7	272.4	275.4
1200	310.2	309.8	309.2	309.0	310.1	314.6
	310.2	309.7	308.9	308.6	309.2	313.3

† Here and in Table 65 scaling constants were added to data in [III.9, III.10] in order to respectively equate the values of enthalpy and entropy at $p = 1$ bar and $T = 300$ K.

TABLE 65. Comparison of Computed Values of Entropy of Xenon with Tabulated Data in [III.9] (2nd line) and [III.10] (3rd line)

T, K	s, kJ/(kg · K), at pressure, bar					
	50	100	200	300	600	1000
300	0.9982	0.8401	0.8073	0.7908	0.7608	0.7370
	0.9977	0.8398	0.8067	0.7903	0.7611	0.7370
	0.9979	0.8396	0.8066	0.7904	0.7612	0.7373
350	1.046	0.9637	0.8735	0.8456	0.8072	0.7797
	1.046	0.9633	0.8735	0.8455	0.8074	0.7801
	1.045	0.9631	0.8731	0.8450	0.8070	0.7795
400	1.075	1.012	0.9330	0.8945	0.8467	0.8162
	1.076	1.012	0.9330	0.8944	0.8466	0.8160
	1.075	1.012	0.9327	0.8940	0.8466	0.8157
500	1.118	1.065	1.004	0.9664	0.9088	0.8731
	1.119	1.066	1.005	0.9668	0.9095	0.8738
600	1.150	1.101	1.047	1.013	0.9555	0.9172
	1.151	1.101	1.047	1.013	0.9562	0.9182
800	1.199	1.152	1.103	1.073	1.021	0.9823
	1.199	1.152	1.103	1.073	1.021	0.9824
1000	1.235	1.190	1.143	1.115	1.065	1.028
	1.236	1.190	1.143	1.115	1.065	1.028
1200	1.265	1.220	1.174	1.147	1.099	1.063
	1.265	1.220	1.174	1.147	1.099	1.062

TABLE 66. Comparison of Computed Values of Isobaric Specific Heat of Xenon with Tabulated Values in [III.9] (2nd line) and [III.10] (3rd line)

T, K	c_p, kJ/(kg · K), at pressure, bar					
	50	100	200	300	600	1000
300	0.429	0.699	0.398	0.354	0.305	0.286
	0.435	0.674	0.403	0.354	0.310	0.294
	0.431	0.669	0.390	0.338	0.289	0.272
350	0.245	0.496	0.463	0.368	0.300	0.278
	0.247	0.494	0.462	0.365	0.295	0.271
	0.246	0.496	0.465	0.371	0.302	0.279
400	0.206	0.285	0.403	0.358	0.289	0.268
	0.207	0.288	0.402	0.360	0.290	0.265
	0.209	0.288	0.401	0.356	0.285	0.259
500	0.181	0.208	0.258	0.282	0.267	0.248
	0.181	0.208	0.260	0.283	0.270	0.252
600	0.173	0.187	0.214	0.232	0.244	0.237
	0.172	0.186	0.214	0.232	0.244	0.236
800	0.166	0.172	0.184	0.194	0.210	0.215
	0.165	0.171	0.183	0.193	0.207	0.211
1000	0.163	0.167	0.174	0.179	0.191	0.198
	0.162	0.166	0.173	0.178	0.189	0.195
1200	0.161	0.164	0.168	0.172	0.181	0.187
	0.161	0.163	0.168	0.171	0.179	0.185

basically do not exceed ±0.1 kJ/kg. However, the deviations over the range of parameters for which experimental *pvT* data are not available exceed 1.3 kJ/kg. Our results, as for krypton, are higher. The reasons apparently are due to different methods of extrapolation of experimental data at high temperatures and pressures. At the same time, good agreement of results for entropy occurs over the same range of parameters (see Table 65), including the extrapolation region.

The agreement of values of c_p at temperatures far from the critical is quite satisfactory. The maximum differences occur along the 300 K isotherm. Our data average results obtained in [III.9, III.10], but it is 4% higher at p = 100 bar.

Approximate calculations from a set of various equations of state for xenon performed by us lead to the following limits of possible errors of thermodynamic functions listed in tables XXII through XXV; $\delta v = \pm 0.15\%$, $\Delta h = \pm 0.15$ kJ/kg; $\Delta s = \pm 0.001$ kJ/(kg · K) and $\delta c_p = \pm 2\%$. The errors may be twice as high at splicing points of the equations of state of gaseous and liquid xenon, also in the critical region, and where up to now no experimental thermal data are available.

The possible errors of thermodynamic functions of crystalline xenon (Table XXI) are listed on page xxx, and of coefficients of dynamic viscosity and thermal conductivity (Tables XXVI and XXVII) on pages xxx and xxx, respectively.

TABLE XXI. Thermodynamic Properties of Crystalline Xenon

T, K	p = 1 bar						
	$v \cdot 10^3$	h	s	c_p	c_v	$\alpha_{exp} \cdot 10^5$	$\beta_T \cdot 10^5$
1	0.2640	0	0.000019	0.000057	0.000057	0.02	2.88
2	0.2640	0	0.000158	0.000490	0.000490	0.13	2.88
4	0.2640	0.01	0.001443	0.00474	0.00474	1.16	2.88
6	0.2641	0.03	0.005491	0.0178	0.0178	4.33	2.89
8	0.2641	0.09	0.01332	0.0381	0.0380	9.74	2.89
10	0.2642	0.19	0.02361	0.0583	0.0581	15.8	2.90
15	0.2645	0.61	0.05595	0.103	0.101	28.2	2.93
20	0.2650	1.21	0.09011	0.133	0.131	36.5	2.98
25	0.2655	1.94	0.1221	0.152	0.148	42.0	3.03
30	0.2661	2.74	0.1510	0.164	0.158	45.8	3.08
40	0.2675	4.47	0.2003	0.178	0.169	50.8	3.19
50	0.2689	6.31	0.2410	0.186	0.174	54.5	3.32
60	0.2704	8.21	0.2755	0.191	0.176	57.7	3.46
70	0.2720	10.16	0.3054	0.196	0.177	61.0	3.62
80	0.2737	12.14	0.3319	0.202	0.177	64.7	3.80
90	0.2756	14.17	0.3558	0.207	0.178	69.0	4.02
100	0.2776	16.30	0.3783	0.214	0.178	74.3	4.29
110	0.2797	18.51	0.3994	0.222	0.178	80.8	4.62
120	0.2820	20.81	0.4197	0.231	0.178	88.6	5.05
130	0.2846	23.19	0.4392	0.241	0.177	97.9	5.60
140	0.2874	25.67	0.4581	0.251	0.175	108.9	6.32
150	0.2905	28.24	0.4765	0.261	0.172	122.0	7.29
160	0.2941	30.94	0.4946	0.270	0.167	137.9	8.65
161.36	0.2946	31.30	0.4969	0.271	0.166	140.1	8.83

TABLE XXI. Thermodynamic Properties of Crystalline Xenon (*Continued*)

T, K	p = 100 bar						
	$v \cdot 10^3$	h	s	c_p	c_v	$\alpha_{exp} \cdot 10^5$	$\beta_T \cdot 10^5$
1	0.2633	2.61	0.000019	0.000056	0.000056	0.01	2.82
2	0.2633	2.61	0.000155	0.000480	0.000480	0.12	2.82
4	0.2633	2.62	0.001413	0.00465	0.00465	1.12	2.82
6	0.2633	2.64	0.005380	0.0174	0.0174	4.17	2.83
8	0.2634	2.70	0.01307	0.0375	0.0374	9.40	2.83
10	0.2634	2.80	0.02320	0.0575	0.0573	15.2	2.84
15	0.2638	3.21	0.05522	0.102	0.101	27.40	2.87
20	0.2642	3.81	0.08916	0.132	0.130	35.6	2.91
25	0.2647	4.53	0.1210	0.151	0.147	41.0	2.96
30	0.2653	5.33	0.1498	0.163	0.158	44.7	3.01

TABLE XXI. Thermodynamic Properties of Crystalline Xenon (*Continued*)

T, K	p = 100 bar						
	$v \cdot 10^3$	h	s	c_p	c_v	$\alpha_{exp} \cdot 10^5$	$\beta_T \cdot 10^5$
40	0.2666	7.06	0.1990	0.177	0.169	49.7	3.12
50	0.2680	8.90	0.2396	0.185	0.173	53.2	3.24
60	0.2695	10.79	0.2740	0.191	0.176	56.2	3.37
70	0.2711	12.73	0.3038	0.196	0.177	59.3	3.52
80	0.2727	14.71	0.3302	0.201	0.177	62.7	3.69
90	0.2745	16.73	0.3539	0.206	0.178	66.7	3.88
100	0.2764	18.84	0.3762	0.212	0.178	71.5	4.12
110	0.2784	21.03	0.3971	0.220	0.178	77.2	4.42
120	0.2807	23.30	0.4171	0.228	0.178	84.0	4.78
130	0.2831	25.65	0.4363	0.236	0.177	92.0	5.25
140	0.2857	28.08	0.4549	0.246	0.176	101.2	5.83
150	0.2885	30.60	0.4729	0.254	0.173	111.5	6.60
160	0.2917	33.21	0.4904	0.262	0.169	123.3	7.60
165.25	0.2934	34.50	0.4994	0.265	0.167	129.5	8.13

TABLE XXI. Thermodynamic Properties of Crystalline Xenon (*Continued*)

T, K	p = 200 bar						
	$v \cdot 10^3$	h	s	c_p	c_v	$\alpha_{exp} \cdot 10^5$	$\beta_T \cdot 10^5$
1	0.2625	5.24	0.000018	0.000055	0.000055	0.01	2.77
2	0.2625	5.24	0.000151	0.000471	0.000471	0.12	2.77
4	0.2626	5.25	0.001385	0.00455	0.00455	1.07	2.77
6	0.2626	5.27	0.005272	0.0171	0.0171	4.01	2.77
8	0.2626	5.33	0.01282	0.0369	0.0368	9.07	2.77
10	0.2627	5.43	0.02280	0.0568	0.0566	14.8	2.78
15	0.2630	5.83	0.05450	0.101	0.0999	26.7	2.81
20	0.2634	6.43	0.08822	0.132	0.129	34.7	2.85
25	0.2640	7.15	0.1199	0.150	0.147	40.0	2.89
30	0.2645	7.95	0.1486	0.162	0.157	43.7	2.94
40	0.2658	9.67	0.1977	0.177	0.168	48.5	3.04
50	0.2672	11.50	0.2382	0.185	0.173	51.9	3.16
60	0.2686	13.39	0.2725	0.190	0.176	54.8	3.28
70	0.2701	15.32	0.3022	0.195	0.177	57.7	3.42
80	0.2717	17.30	0.3285	0.200	0.177	60.9	3.58
90	0.2734	19.30	0.3522	0.205	0.178	64.6	3.76
100	0.2752	21.40	0.3742	0.211	0.178	68.9	3.97
110	0.2772	23.57	0.3949	0.218	0.178	74.1	4.24
120	0.2793	25.81	0.4147	0.225	0.178	80.1	4.56
130	0.2816	28.13	0.4336	0.233	0.177	87.0	4.95
140	0.2841	30.53	0.4518	0.242	0.176	94.8	5.44
150	0.2867	33.00	0.4695	0.249	0.173	103.4	6.06
160	0.2897	35.56	0.4866	0.255	0.170	112.6	6.84
169.05	0.2926	37.92	0.5015	0.259	0.165	121.6	7.75

TABLE XXI. Thermodynamic Properties of Crystalline Xenon (*Continued*)

T, K	p = 300 bar						
	$v \cdot 10^3$	h	s	c_p	c_v	$\alpha_{exp} \cdot 10^5$	$\beta_T \cdot 10^5$
1	0.2618	7.86	0.000018	0.000054	0.000054	0.01	2.71
2	0.2618	7.87	0.000148	0.000461	0.000461	0.11	2.71
4	0.2618	7.87	0.001357	0.00446	0.00446	1.03	2.71
6	0.2619	7.89	0.005169	0.0168	0.0168	3.86	2.71
8	0.2619	7.95	0.01259	0.0363	0.0362	8.76	2.72
10	0.2620	8.04	0.02242	0.0561	0.0559	14.3	2.72
15	0.2623	8.45	0.05381	0.100	0.0991	26.0	2.75
20	0.2627	9.04	0.08731	0.131	0.129	33.9	2.79
25	0.2632	9.76	0.1118	0.150	0.146	39.1	2.83
30	0.2638	10.55	0.1474	0.162	0.157	42.7	2.88
40	0.2650	12.27	0.1964	0.176	0.168	47.4	2.97
50	0.2663	14.10	0.2368	0.184	0.173	50.7	3.08
60	0.2277	15.98	0.2710	0.190	0.175	53.5	3.20
70	0.2692	17.91	0.3007	0.194	0.176	56.3	3.33
80	0.2708	19.88	0.3269	0.199	0.177	59.2	3.47
90	0.2724	21.88	0.3504	0.204	0.177	62.6	3.64
100	0.2742	23.95	0.3723	0.209	0.178	66.6	3.84
110	0.2761	26.10	0.3928	0.216	0.178	71.3	4.07
120	0.2781	28.33	0.4123	0.223	0.178	76.7	4.36
130	0.2803	30.62	0.4310	0.230	0.177	82.8	4.70
140	0.2826	32.99	0.4490	0.238	0.176	89.6	5.12
150	0.2851	35.43	0.4664	0.245	0.174	96.8	5.63
160	0.2878	37.93	0.4832	0.251	0.170	104.4	6.26
170	0.2908	40.50	0.4993	0.254	0.166	112.1	7.04
172.9	0.2917	41.20	0.5040	0.255	0.165	114.3	7.27

TABLE XXI. Thermodynamic Properties of Crystalline Xenon (*Continued*)

T, K	p = 400 bar						
	$v \cdot 10^3$	h	s	c_p	c_v	$\alpha_{exp} \cdot 10^5$	$\beta_T \cdot 10^5$
1	0.2611	10.48	0.000017	0.000053	0.000053	0.01	2.66
2	0.2611	10.48	0.000145	0.000452	0.000452	0.11	2.66
4	0.2611	10.49	0.00133	0.00437	0.00437	0.99	2.66
6	0.2612	10.51	0.00507	0.0165	0.0165	3.72	2.66
8	0.2612	10.56	0.01236	0.0357	0.0357	8.47	2.66
10	0.2613	10.66	0.02204	0.0554	0.0552	13.9	2.67
15	0.2616	11.06	0.05313	0.0993	0.0984	25.3	2.70
20	0.2620	11.65	0.08642	0.130	0.128	33.1	2.73
25	0.2625	12.36	0.1178	0.149	0.146	38.3	2.77
30	0.2630	13.15	0.1463	0.161	0.157	41.8	2.82
40	0.2642	14.87	0.1951	0.176	0.168	46.4	2.91
50	0.2655	16.69	0.2354	0.184	0.173	49.6	3.01
60	0.2669	18.57	0.2696	0.189	0.175	52.3	3.12

TABLE XXI. Thermodynamic Properties of Crystalline Xenon (*Continued*)

T, K	p = 400 bar						
	$v \cdot 10^3$	h	s	c_p	c_v	$\alpha_{exp} \cdot 10^5$	$\beta_T \cdot 10^5$
70	0.2683	20.49	0.2992	0.194	0.176	54.9	3.24
80	0.2698	22.45	0.3253	0.198	0.177	57.7	3.38
90	0.2714	24.45	0.3488	0.203	0.177	60.8	3.54
100	0.2731	26.50	0.3704	0.208	0.178	64.5	3.72
110	0.2749	28.64	0.3908	0.214	0.178	68.8	3.93
120	0.2769	30.85	0.4101	0.221	0.178	73.7	4.18
130	0.2790	33.12	0.4286	0.228	0.177	79.1	4.49
140	0.2812	35.46	0.4464	0.235	0.176	85.1	4.85
160	0.2836	37.86	0.4635	0.241	0.174	91.4	5.28
160	0.2861	40.33	0.4800	0.246	0.171	97.8	5.81
170	0.2888	42.85	0.4959	0.250	0.167	104.1	6.43
176.6	0.2906	44.50	0.5058	0.250	0.164	108.3	6.84

TABLE XXI. Thermodynamic Properties of Crystalline Xenon (*Continued*)

T, K	p = 500 bar						
	$v \cdot 10^3$	h	s	c_p	c_v	$\alpha_{exp} \cdot 10^5$	$\beta_T \cdot 10^5$
1	0.2604	13.09	0.000017	0.000052	0.000052	0.01	2.61
2	0.2604	13.09	0.000143	0.000443	0.000443	0.10	2.61
4	0.2605	13.09	0.001305	0.004289	0.00429	0.96	2.61
6	0.2605	13.12	0.004973	0.0162	0.0162	3.59	2.61
8	0.2605	13.17	0.01214	0.0352	0.0351	8.19	2.61
10	0.2606	13.26	0.02169	0.0547	0.0545	13.5	2.62
15	0.2609	13.66	0.05247	0.0986	0.0977	24.7	2.65
20	0.2613	14.24	0.08555	0.130	0.128	32.4	2.68
25	0.2617	14.95	0.1168	0.149	0.145	37.4	2.72
30	0.2623	15.74	0.1452	0.161	0.156	40.9	2.76
40	0.2634	17.45	0.1939	0.175	0.168	45.4	2.85
50	0.2647	19.27	0.2341	0.183	0.173	48.6	2.94
60	0.2660	21.15	0.2682	0.189	0.175	51.1	3.05
70	0.2674	23.07	0.2977	0.193	0.176	53.6	3.16
80	0.2689	25.02	0.3238	0.197	0.177	56.2	3.29
90	0.2705	27.01	0.3472	0.202	0.177	59.2	3.44
100	0.2721	29.05	0.3686	0.207	0.177	62.6	3.60
110	0.2739	31.17	0.3889	0.213	0.178	66.5	3.80
120	0.2758	33.36	0.4080	0.219	0.178	71.0	4.03
130	0.2777	35.62	0.4263	0.226	0.177	75.9	4.30
140	0.2799	37.93	0.4439	0.232	0.176	81.3	4.62
150	0.2821	40.31	0.4608	0.238	0.174	86.9	4.99
160	0.2845	42.74	0.4770	0.243	0.172	92.4	5.43
170	0.2871	45.22	0.4927	0.246	0.168	97.7	5.95
180	0.2898	47.75	0.5076	0.246	0.163	102.3	6.55
180.25	0.2899	47.80	0.5080	0.246	0.163	102.3	6.99

TABLE XXI. Thermodynamic Properties of Crystalline Xenon (*Continued*)

T, K	p = 600 bar						
	$v \cdot 10^3$	h	s	c_p	c_v	$\alpha_{exp} \cdot 10^5$	$\beta_T \cdot 10^5$
1	0.2598	15.69	0.000017	0.000051	0.000051	0.01	2.56
2	0.2598	15.69	0.00014	0.000435	0.000435	0.10	2.56
4	0.2598	15.69	0.001280	0.00421	0.00421	0.92	2.56
6	0.2598	15.72	0.004881	0.0159	0.0159	3.47	2.56
8	0.2598	15.77	0.01193	0.0346	0.0346	7.93	2.56
10	0.2599	15.86	0.02134	0.0540	0.0539	13.1	2.57
15	0.2602	16.25	0.05183	0.0978	0.0970	24.1	2.60
20	0.2606	16.83	0.08471	0.129	0.127	31.7	2.63
25	0.2610	17.54	0.1158	0.148	0.145	36.6	2.66
30	0.2616	18.33	0.1441	0.160	0.156	40.1	2.70
40	0.2627	20.04	0.1927	0.175	0.167	44.5	2.79
50	0.2639	21.85	0.2328	0.183	0.173	47.5	2.88
60	0.2652	23.72	0.2668	0.188	0.175	50.0	2.98
70	0.2666	25.64	0.2963	0.193	0.176	52.4	3.10
80	0.2680	27.59	0.3223	0.197	0.177	54.9	3.21
90	0.2695	29.57	0.3456	0.201	0.177	57.6	3.34
100	0.2711	31.59	0.3669	0 206	0.177	60.8	3.50
110	0.2728	33.70	0.3870	0.211	0.177	64.4	3.68
120	0.2747	35.88	0.4060	0.217	0.177	68.5	3.89
130	0.2766	38.11	0.4241	0.224	0.177	73.1	4.13
140	0.2786	40.41	0.4415	0.230	0.176	77.9	4.41
150	0.2807	42.76	0.4582	0.236	0.175	82.9	4.74
160	0.2830	45.16	0.4742	0.240	0.172	87.9	5.13
170	0.2854	47.61	0.4897	0.243	0.169	92.5	5.56
180	0.2880	50.10	0.5044	0.244	0.164	96.3	6.06
183.85	0.2890	51.00	0.5101	0.243	0.162	97.8	6.25

TABLE XXI. Thermodynamic Properties of Crystalline Xenon (*Continued*)

T, K	p = 700 bar						
	$v \cdot 10^3$	h	s	c_p	c_v	$\alpha_{exp} \cdot 10^5$	$\beta_T \cdot 10^5$
1	0.2591	18.28	0.000016	0.000050	0.00005	0.01	2.51
2	0.2591	18.28	0.000137	0.000427	0.000427	0.10	2.51
4	0.2591	18.29	0.001257	0.00413	0.00413	0.89	2.51
6	0.2591	18.31	0.004792	0.0156	0.0156	3.35	2.51
8	0.2592	18.36	0.01173	0.0341	0.0341	7.68	2.52
10	0.2592	18.45	0.02100	0.0534	0.0532	12.7	2.52
15	0.2595	18.84	0.05121	0.0971	0.0963	23.5	2.55
20	0.2599	19.42	0.08389	0.128	0.126	31.0	2.58
25	0.2603	20.13	0.1149	0.148	0.144	35.9	2.61
30	0.2609	20.91	0.1431	0.160	0.155	39.2	2.65

TABLE XXI. Thermodynamic Properties of Crystalline Xenon (*Continued*)

T, K	p = 700 bar						
	$v \cdot 10^3$	h	s	c_p	c_v	$\alpha_{exp} \cdot 10^5$	$\beta_T \cdot 10^5$
40	0.2620	22.61	0.1915	0.174	0.167	43.6	2.73
50	0.2632	24.42	0.2316	0.183	0.172	46.6	2.82
60	0.2645	26.29	0.2655	0.188	0.175	49.0	2.91
70	0.2658	28.20	0.2949	0.193	0.176	51.2	3.02
80	0.2672	30.15	0.3208	0.196	0.177	53.6	3.13
90	0.2686	32.12	0.3441	0.200	0.177	56.2	3.26
100	0.2702	34.13	0.3652	0.205	0.177	59.1	3.40
110	0.2718	36.23	0.3852	0.210	0.177	62.5	3.57
120	0.2736	38.39	0.4041	0.216	0.177	66.3	3.76
130	0.2754	40.61	0.4220	0.222	0.177	70.5	3.98
140	0.2774	42.88	0.4392	0.228	0.176	74.9	4.23
150	0.2795	45.21	0.4557	0.233	0.175	79.5	4.53
160	0.2816	47.59	0.4716	0.238	0.173	84.0	4.86
170	0.2839	50.01	0.4868	0.241	0.170	88.0	5.24
180	0.2864	52.47	0.5014	0.241	0.165	91.4	5.66
187.5	0.2883	54.30	0.5117	0.240	0.161	94.0	5.98

TABLE XXI. Thermodynamic Properties of Crystalline Xenon (*Continued*)

T, K	p = 800 bar						
	$v \cdot 10^3$	h	s	c_p	c_v	$\alpha_{exp} \cdot 10^5$	$\beta_T \cdot 10^5$
1	0.2585	20.87	0.000016	0.000049	0.000049	0.01	2.47
2	0.2585	20.87	0.000135	0.000419	0.000419	0.09	2.47
4	0.2585	20.88	0.001234	0.00406	0.00406	0.86	2.47
6	0.2585	20.90	0.004707	0.0153	0.0153	3.24	2.47
8	0.2585	20.95	0.01153	0.0336	0.0336	7.45	2.47
10	0.2586	21.04	0.02067	0.0527	0.0526	12.3	2.48
15	0.2588	21.42	0.05060	0.0964	0.0956	22.9	2.50
20	0.2592	22.00	0.08309	0.128	0.126	30.3	2.53
25	0.2597	22.70	0.1139	0.147	0.144	35.2	2.56
30	0.2602	23.48	0.1420	0.160	0.155	38.5	2.60
40	0.2613	25.18	0.1904	0.174	0.167	42.8	2.68
50	0.2624	26.98	0.2303	0.182	0.172	45.7	2.76
60	0.2637	28.85	0.2642	0.188	0.175	48.0	2.85
70	0.2650	30.76	0.2935	0.192	0.176	50.2	2.95
80	0.2663	32.70	0.3194	0.196	0.176	52.4	3.06
90	0.2678	34.67	0.3426	0.200	0.177	54.8	3.18
100	0.2693	36.67	0.3636	0.204	0.177	57.6	3.31
110	0.2709	38.76	0.3835	0.209	0.177	60.7	3.46
120	0.2726	40.90	0.4022	0.214	0.177	64.3	3.64
130	0.2744	43.10	0.4200	0.220	0.177	68.2	3.84

TABLE XXI. Thermodynamic Properties of Crystalline Xenon (*Continued*)

T, K	p = 800 bar						
	$v \cdot 10^3$	h	s	c_p	c_v	$\alpha_{exp} \cdot 10^5$	$\beta_T \cdot 10^5$
140	0.2763	45.36	0.4370	0.226	0.176	72.3	4.08
150	0.2782	47.67	0.4534	0.231	0.175	76.5	4.34
160	0.2803	50.03	0.4691	0.236	0.173	80.5	4.64
170	0.2825	52.42	0.4842	0.239	0.170	84.2	4.97
180	0.2848	54.85	0.4986	0.239	0.166	87.2	5.33
190	0.2872	57.31	0.5123	0.237	0.162	89.3	5.73
191.0	0.2874	57.50	0.5137	0.238	0.162	89.5	5.77

TABLE XXI. Thermodynamic Properties of Crystalline Xenon (*Continued*)

T, K	p = 900 bar						
	$v \cdot 10^3$	h	s	c_p	c_v	$\alpha_{exp} \cdot 10^5$	$\beta_T \cdot 10^5$
1	0.2578	23.45	0.000016	0.000048	0.000048	0.01	2.42
2	0.2578	23.45	0.000132	0.000412	0.000412	0.09	2.42
4	0.2578	23.46	0.001212	0.00398	0.00398	0.83	2.43
6	0.2579	23.48	0.004624	0.0151	0.0151	3.14	2.43
8	0.2579	23.53	0.01134	0.0331	0.0331	7.22	2.43
10	0.2579	23.62	0.02036	0.0521	0.0520	12.0	2.44
15	0.2582	24.00	0.05001	0.0957	0.0949	22.4	2.46
20	0.2586	24.57	0.08230	0.127	0.125	29.7	2.49
25	0.2590	25.27	0.1130	0.146	0.144	34.5	2.52
30	0.2595	26.05	0.1410	0.159	0.155	37.7	2.55
40	0.2606	27.74	0.1892	0.174	0.167	42.0	2.63
50	0.2617	29.54	0.2291	0.182	0.172	44.8	2.71
60	0.2629	31.41	0.2629	0.187	0.175	47.1	2.79
70	0.2642	33.31	0.2922	0.191	0.176	49.1	2.88
80	0.2655	35.25	0.3180	0.195	0.176	51.2	2.99
90	0.2669	37.22	0.3411	0.199	0.177	53.6	3.10
100	0.2684	39.21	0.3621	0.203	0.177	56.2	3.23
110	0.2699	41.28	0.3818	0.208	0.177	59.1	3.37
120	0.2716	43.41	0.4004	0.213	0.177	62.4	3.53
130	0.2733	45.59	0.4181	0.218	0.177	66.0	3.72
140	0.2751	47.84	0.4349	0.224	0.176	69.9	3.93
150	0.2771	50.13	0.4511	0.229	0.175	73.8	4.17
160	0.2791	52.47	0.4667	0.234	0.173	77.5	4.44
170	0.2812	54.84	0.4816	0.237	0.171	80.9	4.73
180	0.2833	57.25	0.4959	0.238	0.167	83.6	5.05
190	0.2856	59.68	0.5095	0.236	0.163	85.5	5.39
194.5	0.2866	60.61	0.5153	0.235	0.161	86.4	5.54

TABLE XXI. Thermodynamic Properties of Crystalline Xenon (*Continued*)

T, K	$p = 1000$ bar						
	$v \cdot 10^3$	h	s	c_p	c_v	$\alpha_{exp} \cdot 10^5$	$\beta_T \cdot 10^5$
1	0.2572	26.02	0.000016	0.000047	0.000047	0.01	2.38
2	0.2572	26.03	0.000130	0.000405	0.000405	0.09	2.38
4	0.2572	26.03	0.001191	0.00391	0.00391	0.81	2.38
6	0.2572	26.05	0.004545	0.0148	0.0148	3.04	2.39
8	0.2573	26.10	0.01116	0.0326	0.0326	7.01	2.39
10	0.2573	26.19	0.02005	0.0515	0.0514	11.7	2.39
15	0.2576	26.57	0.04943	0.0950	0.0943	21.9	2.41
20	0.2579	27.14	0.08154	0.126	0.125	29.1	2.44
25	0.2584	27.83	0.1121	0.146	0.143	33.8	2.47
30	0.2588	28.61	0.1401	0.159	0.154	37.0	2.51
40	0.2599	30.30	0.1882	0.173	0.167	41.2	2.58
50	0.2610	32.10	0.2280	0.181	0.172	43.9	2.65
60	0.2622	33.96	0.2617	0.187	0.174	46.1	2.73
70	0.2634	35.86	0.2909	0.191	0.176	48.1	2.82
80	0.2648	37.79	0.3166	0.194	0.176	50.2	2.92
90	0.2661	39.75	0.3397	0.198	0.177	52.4	3.03
100	0.2675	41.74	0.3607	0.202	0.177	54.8	3.15
110	0.2690	43.79	0.3802	0.207	0.177	57.6	3.28
120	0.2706	45.91	0.3987	0.212	0.177	60.7	3.43
130	0.2723	48.08	0.4161	0.217	0.177	64.1	3.61
140	0.2741	50.31	0.4329	0.222	0.176	67.7	3.80
150	0.2759	52.59	0.4490	0.227	0.175	71.3	4.02
160	0.2779	54.91	0.4644	0.232	0.173	74.8	4.26
170	0.2799	57.26	0.4792	0.235	0.171	77.9	4.52
180	0.2820	59.65	0.4934	0.236	0.168	80.5	4.81
190	0.2842	62.06	0.5069	0.235	0.164	82.2	5.11
197.9	0.2859	63.7	0.5169	0.233	0.161	83.5	5.35

TABLE XXII. Thermodynamic Properties of Xenon Along the Crystallization and Melting Curve

T, K	p_m	$v_l \cdot 10^3$	$v_c \cdot 10^3$	h_l	h_c	s_l	s_c	c_{pl}	c_{pc}
161.36	0.81	0.3372	0.2946	48.56	31.30	0.6072	0.4969	0.350	0.271
162	16.99	0.3368	0.2944	49.18	31.80	0.6074	0.4973	0.348	0.270
163	42.21	0.3362	0.2941	50.08	32.65	0.6077	0.4979	0.346	0.268
164	67.62	0.3356	0.2938	50.97	33.50	0.6079	0.4986	0.343	0.266
165	93.28	0.3351	0.2936	51.85	34.35	0.6082	0.4992	0.341	0.265
166	119.2	0.3346	0.2933	52.76	35.20	0.6084	0.4999	0.339	0.264
167	145.2	0.3340	0.2931	53.67	36.07	0.6087	0.5005	0.337	0.262
168	171.4	0.3335	0.2928	54.59	36.95	0.6090	0.5011	0.335	0.261
169	197.8	0.3329	0.2925	55.52	37.81	0.6092	0.5017	0.333	0.260
170	224.2	0.3324	0.2923	56.43	38.68	0.6094	0.5023	0.331	0.258

TABLE XXII. Thermodynamic Properties of Xenon Along the Crystallization and Melting Curve (*Continued*)

T, K	p_m	$v_l \cdot 10^3$	$v_c \cdot 10^3$	h_l	h_c	s_l	s_c	c_{pl}	c_{pc}
171	250.7	0,3319	0.2921	57.35	39.57	0.6097	0,5029	0,329	0.257
172	277,4	0.3314	0,2918	58.28	40.45	0.6100	0.5035	0.327	0.256
173	304.1	0.3309	0,2916	59,22	41.32	0.6103	0.5041	0.325	0.255
174	330.9	0.3304	0.2914	60.15	42.20	0,6105	0.5046	0,324	0.253
175	357.8	0.3299	0.2911	61.08	43.08	0.6108	0.5051	0.322	0.252
176	384.8	0.3294	0.2909	62.02	43.95	0.6111	0,5057	0.320	0.251
177	411.9	0.3289	0.2907	62.95	44.83	0,6114	0.5063	0.319	0.250
178	439.1	0.3285	0.2904	63.89	45.72	0.6117	0,5068	0.317	0.249
179	466.3	0.3280	0.2902	64.83	46,61	0,6120	0.5073	0.316	0.248
180	493.7	0.3275	0.2900	65.78	47.50	0,6123	0.5079	0.315	0.247
181	521,1	0.3270	0.2898	66.72	48.41	0.6126	0.5084	0.313	0.246
182	548.7	0.3266	0.2895	67.66	49.33	0.6128	0.5089	0.312	0.245
183	576.3	0.3262	0,2893	68.62	50.25	0,6131	0.5095	0.311	0.244
184	604.0	0.3258	0.2891	69.60	51.18	0,6134	0,5100	0.309	0.243
185	631.8	0.3254	0.2889	70.59	52.10	0.6137	0.5106	0.308	0.242
186	659.7	0.3250	0.2886	71.59	53.01	0.6140	0.5111	0.307	0.241
187	687.7	0.3247	0.2884	72.57	53.90	0,6143	0.5117	0.306	0.240
188	715.8	0.3243	0.2882	73.53	54.80	0.6146	0.5122	0.305	0.239
189	743.9	0.3239	0.2879	74.46	55.70	0.6149	0.5127	0.304	0.239
190	772.2	0.3235	0.2877	75.38	56,60	0.6152	0.5132	0,303	0.238
191	800.5	0.3231	0.2875	76.30	57.50	0,6155	0,5137	0,302	0.237
192	828.9	0.3227	0.2872	77.22	58.40	0.6158	0,5142	0,301	0.237
193	857.4	0.3222	0.2870	78.14	59,30	0.6161	0.5146	0.300	0.236
194	886.0	0.3218	0.2868	79.05	60.20	0.6163	0.5151	0.299	0.235
195	914.7	0.3214	0.2865	79.94	61.10	0,6166	0.5156	0.298	0.235
196	943.5	0.3210	0.2863	80.83	62.00	0,6169	0.5160	0.297	0.234
197	972.3	0.3206	0,2861	81.71	62.90	0,6172	0.5164	0,296	0.234
198	1001	0,3202	0.2859	82,58	63.80	0,6175	0.5169	0.294	0,233

TABLE XXIII. Thermodynamic Properties of Saturated Xenon

T, K	p_v	$v'' \cdot 10^3$	$v' \cdot 10^3$	h''	h'	s''	s'	c''_p	c'_p
161.36	0.8160	121.9	0.3372	145.5	48.98	1.206	0.6072	0.166	0.350
162	0.8480	117.7	0.3377	145.6	49.21	1.204	0.6087	0.166	0.350
163	0.8999	111.4	0.3385	145.7	49.56	1.201	0.6108	0.167	0.350
164	0.9543	105.6	0.3392	145.8	49.91	1.198	0.6129	0.167	0.349
165	1.011	100.1	0.3400	146.0	50.26	1.195	0.6150	0.168	0.349
165.03	1.013	99.96	0.3400	146.0	50.27	1.195	0.6150	0.168	0.349
166	1.071	95.03	0.3408	146.1	50.61	1.192	0.6171	0.168	0.349
167	1.133	90.24	0.3416	146.2	50.96	1.190	0.6192	0.169	0.349
168	1.197	85.75	0.3424	146.3	51.31	1.187	0.6212	0.170	0.349
169	1.265	81.53	0.3431	146.4	51.66	1.184	0.6232	0.170	0.349
170	1.336	77.57	0.3439	146.5	52.01	1.181	0.6253	0.171	0.349
171	1.409	73.85	0.3447	146.6	52.36	1.179	0.6274	0.172	0.349
172	1.486	70.34	0.3456	146.7	52.71	1.176	0.6295	0.172	0.349
173	1.565	67.05	0.3464	146.8	53.06	1.174	0.6315	0.173	0.349
174	1.648	63.94	0.3472	146.9	53.41	1.171	0.6334	0.174	0.349
175	1.734	61.01	0.3480	147.0	53.76	1.168	0.6354	0.175	0.349
176	1.824	58.25	0.3489	147.1	54.11	1.166	0.6374	0.175	0.349
177	1.917	55.64	0.3497	147.2	54.47	1.164	0.6394	0.176	0.349
178	2.014	53.18	0.3506	147.3	54.82	1.161	0.6413	0.177	0.349
179	2.114	50.85	0.3514	147.4	55.17	1.159	0.6433	0.178	0.349
180	2.218	48.65	0.3523	147.5	55.52	1.156	0.6452	0.179	0.349
181	2.325	46.57	0.3532	147.6	55.87	1.154	0.6471	0.179	0.349
182	2.437	44.60	0.3541	147.7	56.22	1.152	0.6490	0.180	0.350
183	2.553	42.73	0.3550	147.8	56.57	1.149	0.6509	0.181	0.350
184	2.672	40.96	0.3559	147.9	56.92	1.147	0.6529	0.182	0.350

TABLE XXIII. Thermodynamic Properties of Saturated Xenon (*Continued*)

T. K	p_v	$v'' \cdot 10^3$	$v' \cdot 10^3$	h''	h'	s''	s'	c''_p	c'_p
185	2.796	39.28	0.3568	147.9	57.28	1.145	0.6548	0.183	0.350
186	2.924	37.68	0.3577	148.0	57.63	1.143	0.6567	0.184	0.350
187	3.056	36.17	0.3586	148.1	57.98	1.140	0.6586	0.185	0.351
188	3.193	34.73	0.3596	148.1	58.34	1.138	0.6604	0.186	0.351
189	3.334	33.36	0.3605	148.2	58.69	1.136	0.6622	0.187	0.352
190	3.480	32.06	0.3615	148.3	59.04	1.134	0.6641	0.188	0.352
191	3.631	30.82	0.3624	148.3	59.40	1.132	0.6659	0.189	0.352
192	3.786	29.64	0.3634	148.4	59.76	1.130	0.6678	0.190	0.353
193	3.946	28.52	0.3644	148.5	60.11	1.128	0.6696	0.191	0.353
194	4.111	27.45	0.3654	148.5	60.47	1.125	0.6714	0.192	0.353
195	4.282	26.43	0.3664	148.6	60.82	1.123	0.6732	0.193	0.354
196	4.457	25.46	0.3674	148.7	61.18	1.121	0.6750	0.194	0.354
197	4.638	24.53	0.3684	148.7	61.54	1.119	0.6768	0.196	0.355
198	4.823	23.64	0.3694	148.8	61.89	1.117	0.6785	0.197	0.355
199	5.015	22.79	0.3705	148.8	62.25	1.115	0.6803	0.198	0.356
200	5.212	21.98	0.3715	148.9	62.61	1.113	0.6820	0.199	0.357
201	5.414	21.21	0.3726	148.9	62.97	1.112	0.6838	0.200	0.357
202	5.622	20.47	0.3737	149.0	63.33	1.110	0.6856	0.202	0.358
203	5.836	19.76	0.3748	149.0	63.70	1.108	0.6873	0.203	0.359
204	6.056	19.08	0.3759	149.0	64.06	1.106	0.6891	0.204	0.360
205	6.282	18.43	0.3770	149.1	64.42	1.104	0.6908	0.206	0.361
206	6.514	17.81	0.3781	149.1	64.78	1.102	0.6925	0.207	0.361
207	6.752	17.21	0.3793	149.2	65.15	1.100	0.6943	9.209	0.362
208	6.996	16.64	0.3804	149.2	65.52	1.098	0.6960	0.210	0.363
209	7.247	16.09	0.3816	149.2	65.88	1.097	0.6977	0.212	0.364

TABLE XXIII. Thermodynamic Properties of Saturated Xenon (*Continued*)

T. K	p_v	$v'' \cdot 10^3$	$v' \cdot 10^3$	h''	h'	s''	s'	c''_p	c'_p
210	7.504	15.56	0.3828	149.2	66.25	1.095	0.6994	0.213	0.365
211	7.768	15.05	0.3840	149.3	66.62	1.093	0.7011	0.215	0.366
212	8.038	14.57	0.3852	149.3	66.99	1.091	0.7028	0.216	0.368
213	8.315	14.10	0.3864	149.3	67.36	1.089	0.7045	0.218	0.369
214	8.599	13.65	0.3877	149.3	67.73	1.088	0.7062	0.220	0.370
215	8.890	13.22	0.3889	149.3	68.10	1.686	0.7079	0.221	0.371
216	9.188	12.81	0.3902	149.3	68.48	1.084	0.7096	0.223	0.372
217	9.494	12.41	0.3915	149.4	68.86	1.082	0.7113	0.225	0.374
218	9.806	12.02	0.3928	149.4	69.24	1.081	0.7129	0.227	0.376
219	10.13	11.65	0.3941	149.4	69.62	1.079	0.7146	0.229	0.377
220	10.45	11.31	0.3955	149.4	70.00	1.077	0.7163	0.231	0.379
221	10.79	10.95	0.3968	149.4	70.38	1.075	0.7180	0.233	0.381
222	11.13	10.62	0.3982	149.3	70.76	1.074	0.7196	0.235	0.382
223	11.48	10.30	0.3996	149.3	71.15	1.072	0.7213	0.237	0.384
224	11.84	9.994	0.4010	149.3	71.54	1.070	0.7230	0.239	0.386
225	12.20	9.697	0.4025	149.3	71.93	1.069	0.7246	0.242	0.388
226	12.58	9.411	0.4039	149.3	72.32	1.067	0.7263	0.244	0.391
227	12.96	9.134	0.4054	149.3	72.71	1.065	0.7279	0.246	0.393
228	13.35	8.867	0.4069	149.3	73.11	1.064	0.7296	0.249	0.395
229	13.75	8.609	0.4085	149.2	73.51	1.062	0.7313	0.252	0.398
230	14.16	8.359	0.4100	149.2	73.91	1.060	0.7330	0.254	0.400
231	14.57	8.118	0.4116	149.1	74.31	1.059	0.7347	0.257	0.403
232	15.00	7.885	0.4132	149.1	74.71	1.057	0.7363	0.260	0.405
233	15.43	7.660	0.4149	149.0	75.12	1.055	0.7380	0.263	0.408
234	15.87	7.442	0.4165	149.0	75.53	1.054	0.7397	0.266	0.411

TABLE XXIII. Thermodynamic Properties of Saturated Xenon (*Continued*)

T, K	p_v	$v''\cdot10^3$	$v'\cdot10^3$	h''	h'	s''	s'	C_p''	C_p'
235	16.33	7.230	0.4182	148.9	75.94	1.052	0.7414	0.269	0.414
236	16.79	7.026	0.4199	148.9	76.36	1.050	0.7431	0.272	0.418
237	17.26	6.828	0.4217	148.8	76.78	1.049	0.7448	0.276	0.421
238	17.74	6.636	0.4234	148.7	77.20	1.047	0.7464	0.279	0.425
239	18.22	6.451	0.4253	148.6	77.62	1.045	0.7481	0.283	0.428
240	18.72	6.271	0.4271	148.5	78.05	1.044	0.7498	0.287	0.432
241	19.23	6.096	0.4290	148.4	78.48	1.042	0.7515	0.291	0.436
242	19.75	5.927	0.4309	148.4	78.92	1.040	0.7532	0.295	0.441
243	20.28	5.763	0.4328	148.3	79.36	1.039	0.7549	0.299	0.445
244	20.81	5.603	0.4348	148.2	79.80	1.037	0.7567	0.304	0.450
245	21.36	5.449	0.4368	148.1	80.24	1.035	0.7584	0.309	0.455
246	21.92	5.299	0.4389	148.0	80.69	1.034	0.7601	0.314	0.460
247	22.49	5.153	0.4410	147.9	81.15	1.032	0.7618	0.319	0.465
248	23.06	5.011	0.4432	147.7	81.61	1.030	0.7636	0.324	0.471
249	23.65	4.873	0.4454	147.6	82.07	1.029	0.7653	0.330	0.476
250	24.25	4.740	0.4476	147.5	82.54	1.027	0.7671	0.336	0.482
251	24.87	4.609	0.4499	147.3	83.01	1.025	0.7689	0.343	0.488
252	25.49	4.483	0.4522	147.2	83.49	1.024	0.7707	0.349	0.495
253	26.12	4.360	0.4546	147.0	83.97	1.022	0.7725	0.357	0.502
254	26.76	4.240	0.4571	146.9	84.46	1.020	0.7743	0.364	0.509
255	27.42	4.123	0.4596	146.7	84.95	1.018	0.7762	0.372	0.517
256	28.09	4.009	0.4622	146.5	85.45	1.017	0.7780	0.381	0.525
257	28.76	3.898	0.4648	146.3	85.96	1.015	0.7798	0.390	0.533
258	29.45	3.790	0.4675	146.1	86.47	1.013	0.7817	0.400	0.542
259	30.16	3.685	0.4702	145.9	86.99	1.011	0.7836	0.410	0.551

TABLE XXIII. Thermodynamic Properties of Saturated Xenon (*Continued*)

T, K	p_v	$v''\cdot 10^3$	$v'\cdot 10^3$	h''	h'	s''	s'	c_p''	c_p'
260	30.87	3.582	0.4730	145.7	87.52	1.009	0.7855	0.420	0.560
261	31.60	3.482	0.4759	145.5	88.06	1.007	0.7874	0.432	0.570
262	32.33	3.384	0.4789	145.2	88.60	1.006	0.7893	0.445	0.580
263	33.08	3.288	0.4820	145.0	89.15	1.004	0.7913	0.459	0.591
264	33.85	3.195	0.4851	144.7	89.71	1.002	0.7932	0.473	0.603
265	34.62	3.103	0.4883	144.4	90.28	1.000	0.7953	0.489	0.615
266	35.41	3.014	0.4916	144.1	90.86	0.9977	0.7973	0.506	0.627
267	36.21	2.926	0.4952	143.8	91.44	0.9956	0.7993	0.525	0.640
268	37.02	2.841	0.4992	143.5	92.03	0.9936	0.8014	0.546	0.654
269	37.85	2.757	0.5034	143.2	92.67	0.9915	0.8036	0.569	0.669
270	38.69	2.674	0.5079	142.8	93.30	0.9893	0.8058	0.593	0.685
271	39.55	2.594	0.5126	142.5	93.95	0.9871	0.8080	0.621	0.703
272	40.41	2.514	0.5175	142.1	94.61	0.9849	0.8104	0.652	0.722
273	41.29	2.436	0.5226	141.7	95.29	0.9826	0.8127	0.686	0.743
274	42.19	2.360	0.5280	141.2	95.99	0.9802	0.8151	0.725	0.768
275	43.10	2.284	0.5337	140.8	96.70	0.9778	0.8175	0.769	0.796
276	44.02	2.210	0.5398	140.3	97.44	0.9753	0.8200	0.818	0.827
277	44.96	2.136	0.5463	139.8	98.20	0.9727	0.8226	0.876	0.860
278	45.91	2.064	0.5532	139.2	98.99	0.9700	0.8252	0.943	0.896
279	46.87	1.992	0.5608	138.6	99.80	0.9672	0.8279	1.021	0.937
280	47.86	1.920	0.5689	138.0	100.6	0.9643	0.8308	1.114	0.995
281	48.85	1.849	0.5779	137.3	101.5	0.9612	0.8337	1.226	1.074
282	49.86	1.778	0.5877	136.6	102.4	0.9580	0.8368	1.363	1.186
283	50.89	1.707	0.5987	135.8	103.4	0.9546	0.8400	1.535	1.343
284	51.93	1.636	0.6113	135.0	104.4	0.9510	0.8433	1.756	1.545

TABLE XXIII. Thermodynamic Properties of Saturated Xenon (*Continued*)

T, K	p_v	$v'' \cdot 10^3$	$v' \cdot 10^3$	h''	h'	s''	s'	c_p''	c_p'
285	52.99	1.564	0.6258	134.0	105.5	0.9471	0.8469	2.046	1.825
286	54.06	1.490	0.6430	132.9	106.7	0.9428	0.8509	2.45	2.31
287	55.15	1.405	0.6645	131.5	108.2	0.9374	0.8557	3.18	3.02
288	56.25	1.305	0.6933	129.8	110.1	0.9304	0.8620	4.72	4.59
289	57.37	1.177	0.7042	127.0	112.8	0.9203	0.8717	10.4	8.75
289.74	58.21	0.9091		120.0		0.8962		∞	

TABLE XXIV. Thermodynamic Properties of Liquid and Gaseous Xenon

T, K	p = 1 bar			p = 5 bar			p = 10 bar		
	$v \cdot 10^3$	h	s	$v \cdot 10^3$	h	s	$v \cdot 10^3$	h	s
165	101.3	146.0	1.196	0.3398	50.34	0.6147	0.3395	50.45	0.6143
170	104.6	146.8	1.201	0.3437	52.09	0.6250	0.3434	52.19	0.6246
175	108.0	147.6	1.206	0.3478	53.83	0.6351	0.3475	53.93	0.6347
180	111.3	148.5	1.210	0.3521	55.57	0.6450	0.3517	55.67	0.6445
185	114.6	149.3	1.215	0.3566	57.32	0.6546	0.3562	57.41	0.6541
190	117.9	150.1	1.219	0.3613	59.07	0.6640	0.3609	59.16	0.6635
195	121.2	151.0	1.224	0.3663	60.83	0.6731	0.3658	60.92	0.6726
200	124.5	151.8	1.228	23.02	149.0	1.117	0.3710	62.68	0.6815
205	127.7	152.6	1.232	23.77	150.0	1.121	0.3765	64.47	0.6904
210	131.0	153.4	1.236	24.52	151.0	1.126	0.3824	66.28	0.6991

TABLE XXIV. Thermodynamic Properties of Liquid and Gaseous Xenon (*Continued*)

T, K	p = 1 bar			p = 5 bar			p = 10 bar		
	$v \cdot 10^3$	h	s	$v \cdot 10^3$	h	s	$v \cdot 10^3$	h	s
215	134.3	154.3	1.240	25.26	151.9	1.131	0.3887	68.11	0.7078
220	137.5	155.1	1.244	25.98	152.9	1.135	11.91	149.7	1.081
225	140.8	155.9	1.247	26.70	153.8	1.139	12.34	150.8	1.086
230	144.0	156.7	1.251	27.42	154.7	1.143	12.75	151.9	1.091
235	147.2	157.5	1.254	28.13	155.6	1.147	13.16	152.9	1.095
240	150.5	158.3	1.258	28.83	156.5	1.151	13.55	154.0	1.100
245	153.7	159.1	1.261	29.53	157.4	1.154	13.94	155.0	1.104
250	156.9	160.0	1.264	30.22	158.3	1.158	14.33	156.0	1.108
255	160.1	160.8	1.268	30.91	159.1	1.161	14.71	157.0	1.112
260	163.3	161.6	1.271	31.60	160.0	1.165	15.09	157.9	1.116
265	166.6	162.4	1.274	32.28	160.9	1.168	15.46	158.9	1.119
270	169.8	163.2	1.277	32.96	161.7	1.171	15.83	159.8	1.123
275	173.0	164.0	1.280	33.64	162.6	1.174	16.19	160.8	1.126
280	176.2	164.8	1.283	34.31	163.4	1.178	16.55	161.7	1.129
285	179.4	165.6	1.285	34.99	164.3	1.181	16.91	162.6	1.133
290	182.6	166.4	1.288	35.66	165.1	1.183	17.27	163.5	1.136
295	185.8	167.2	1.291	36.33	166.0	1.186	17.63	164.4	1.139
300	189.0	168.0	1.294	36.99	166.8	1.189	17.98	165.3	1.142
305	192.2	168.8	1.296	37.66	167.6	1.192	18.33	166.2	1.145
310	195.4	169.6	1.299	38.32	168.5	1.195	18.68	167.0	1.148
315	198.6	170.4	1.301	38.99	169.3	1.197	19.03	167.9	1.150
320	201.8	171.2	1.304	39.65	170.1	1.200	19.37	168.8	1.153
325	205.0	172.0	1.306	40.31	171.0	1.202	19.72	169.7	1.156
330	208.1	172.8	1.309	40.97	171.8	1.205	20.06	170.5	1.158
335	211.3	173.6	1.311	41.63	172.6	1.207	20.40	171.4	1.161

TABLE XXIV. Thermodynamic Properties of Liquid and Gaseous Xenon (*Continued*)

T, K	p = 1 bar			p = 5 bar			p = 10 bar		
	$v \cdot 10^3$	h	s	$v \cdot 10^3$	h	s	$v \cdot 10^3$	h	s
340	214.5	174.4	1.314	42.28	173.4	1.210	20.75	172.2	1.164
345	217.7	175.2	1.316	42.94	174.3	1.212	21.09	173.1	1.166
350	220.9	176.0	1.318	43.59	175.1	1.215	21.43	173.9	1.169
360	227.3	177.6	1.323	44.90	176.7	1.219	22.10	175.6	1.173
370	233.6	179.2	1.327	46.21	178.3	1.224	22.78	177.3	1.178
380	240.0	180.8	1.331	47.51	180.0	1.228	23.45	179.0	1.182
390	246.4	182.3	1.336	48.81	181.6	1.232	24.11	180.7	1.187
400	252.7	183.9	1.340	50.11	183.2	1.236	24.78	182.3	1.191
420	265.5	187.1	1.347	52.70	186.5	1.244	26.11	185.6	1.199
440	278.2	190.3	1.355	55.29	189.7	1.252	27.43	188.9	1.207
460	290.9	193.5	1.362	57.87	192.9	1.259	28.74	192.2	1.214
480	303.6	196.6	1.369	60.44	196.1	1.266	30.05	195.5	1.221
500	316.3	199.8	1.375	63.01	199.3	1.272	31.35	198.7	1.228
550	348.0	207.8	1.390	69.43	207.3	1.288	34.60	206.8	1.243
600	379.8	215.7	1.404	75.83	215.3	1.302	37.83	214.9	1.257
650	411.5	223.6	1.417	82.21	223.3	1.314	41.05	222.9	1.270
700	443.2	231.5	1.428	88.59	231.3	1.326	44.26	231.0	1.282
750	474.9	239.5	1.439	94.96	239.2	1.337	47.48	239.0	1.293
800	506.6	247.4	1.450	101.3	247.2	1.347	50.67	247.0	1.303
850	538.3	255.3	1.459	107.7	255.1	1.357	53.86	254.9	1.313
900	570.0	263.2	1.468	114.0	263.1	1.366	57.05	262.9	1.322
950	601.6	271.1	1.477	120.4	271.0	1.375	60.24	270.9	1.331
1000	633.3	279.1	1.485	126.7	279.0	1.383	63.42	278.8	1.339
1050	665.0	287.0	1.493	133.1	286.9	1.391	66.60	286.8	1.347
1100	696.7	294.9	1.500	139.4	294.8	1.398	69.78	294.8	1.354

TABLE XXIV. Thermodynamic Properties of Liquid and Gaseous Xenon (*Continued*)

T, K	p = 1 bar			p = 5 bar			p = 10 bar		
	$v \cdot 10^3$	h	s	$v \cdot 10^3$	h	s	$v \cdot 10^3$	h	s
1150	728.4	302.8	1.507	145.8	302.8	1.405	72.96	302.7	1.361
1200	760.0	310.7	1.514	152.1	310.7	1.412	76.14	310.6	1.368
1250	791.7	318.7	1.520	158.5	318.6	1.418	79.32	318.6	1.374
1300	823.4	326.6	1.526	164.8	326.6	1.424	82.49	326.5	1.380

TABLE XXIV. Thermodynamic Properties of Liquid and Gaseous Xenon (*Continued*)

T, K	p = 15 bar			p = 20 bar			p = 25 bar		
	$v \cdot 10^3$	h	s	$v \cdot 10^3$	h	s	$v \cdot 10^3$	h	s
165	0.3392	50.56	0.6139	0.3389	50.66	0.6135	0.3387	50.77	0.6132
170	0.3431	52.29	0.6242	0.3428	52.40	0.6238	0.3425	52.50	0.6234
175	0.3472	54.03	0.6343	0.3468	54.13	0.6339	0.3465	54.23	0.6335
180	0.3514	55.77	0.6441	0.3510	55.87	0.6437	0.3507	55.97	0.6432
185	0.3558	57.51	0.6537	0.3554	57.60	0.6532	0.3550	57.70	0.6528
190	0.3604	59.25	0.6630	0.3600	59.34	0.6625	0.3596	59.43	0.6621
195	0.3653	61.00	0.6721	0.3648	61.09	0.6716	0.3644	61.17	0.6711
200	0.3705	62.76	0.6810	0.3699	62.84	0.6804	0.3694	62.92	0.6799
205	0.3759	64.54	0.6898	0.3753	64.62	0.6892	0.3748	64.69	0.6887
210	0.3817	66.35	0.6985	0.3811	66.41	0.6979	0.3804	66.48	0.6973
215	0.3880	68.17	0.7071	0.3872	68.22	0.7065	0.3865	68.28	0.7058
220	0.3947	70.04	0.7157	0.3938	70.08	0.7150	0.3929	70.13	0.7143
225	0.4019	71.95	0.7242	0.4009	71.98	0.7234	0.3999	72.01	0.7227

TABLE XXIV. Thermodynamic Properties of Liquid and Gaseous Xenon (*Continued*)

T, K	p = 15 bar			p = 20 bar			p = 25 bar		
	$v \cdot 10^3$	h	s	$v \cdot 10^3$	h	s	$v \cdot 10^3$	h	s
230	0.4098	73.91	0.7329	0.4087	73.93	0.7320	0.4075	73.95	0.7312
235	8.076	149.8	1.060	0.4172	75.94	0.7407	0.4159	75.94	0.7398
240	8.386	151.1	1.065	0.4267	78.05	0.7495	0.4251	78.02	0.7486
245	8.687	153.2	1.070	5.983	149.1	1.043	0.4354	80.20	0.7576
250	8.979	153.4	1.075	6.243	150.5	1.048	0.4473	82.52	0.7669
255	9.264	154.5	1.079	6.491	151.8	1.054	4.765	148.6	1.030
260	9.543	155.6	1.084	6.730	153.1	1.058	4.993	150.1	1.036
265	9.817	156.7	1.088	6.961	154.3	1.063	5.208	151.6	1.041
270	10.09	157.8	1.092	7.185	155.5	1.068	5.413	152.9	1.046
275	10.35	158.8	1.095	7.404	156.6	1.072	5.610	154.2	1.051
280	10.61	159.8	1.099	7.619	157.7	1.076	5.801	155.5	1.056
285	10.87	160.8	1.103	7.829	158.8	1.080	5.986	156.7	1.060
290	11.12	161.8	1.106	8.035	159.9	1.083	6.166	157.9	1.064
295	11.38	162.7	1.109	8.238	160.9	1.087	6.342	159.0	1.068
300	11.63	163.7	1.113	8.439	162.0	1.090	6.515	160.2	1.072
305	11.87	164.6	1.116	8.637	163.0	1.094	6.684	161.3	1.076
310	12.12	165.5	1.119	8.832	164.0	1.097	6.851	162.3	1.079
315	12.36	166.5	1.122	9.025	165.0	1.100	7.015	163.4	1.082
320	12.61	167.4	1.125	9.217	165.9	1.103	7.177	164.4	1.086
325	12.85	168.3	1.127	9.406	166.9	1.106	7.336	165.4	1.089
330	13.09	169.2	1.130	9.594	167.8	1.109	7.494	166.4	1.092
335	13.33	170.1	1.133	9.781	168.8	1.112	7.652	167.4	1.095
340	13.56	171.0	1.135	9.966	169.7	1.115	7.806	168.4	1.098
345	13.80	171.9	1.138	10.150	170.6	1.117	7.959	169.4	1.101

TABLE XXIV. Thermodynamic Properties of Liquid and Gaseous Xenon (*Continued*)

T, K	p = 15 bar			p = 20 bar			p = 25 bar		
	$v \cdot 10^3$	h	s	$v \cdot 10^3$	h	s	$v \cdot 10^3$	h	s
350	14.03	172.8	1.141	10.33	171.6	1.120	8.111	170.3	1.104
360	14.50	174.5	1.146	10.70	173.4	1.125	8.412	172.2	1.109
370	14.96	176.3	1.150	11.06	175.2	1.130	8.709	174.1	1.114
380	15.42	178.0	1.155	11.41	177.0	1.135	9.003	175.9	1.119
390	15.88	179.7	1.159	11.76	178.7	1.140	9.293	177.7	1.124
400	16.34	181.4	1.164	12.11	180.5	1.144	9.581	179.5	1.128
420	17.24	184.8	1.172	12.81	183.9	1.152	10.15	183.1	1.137
440	18.14	188.1	1.180	13.50	187.4	1.160	10.71	186.6	1.145
460	19.03	191.5	1.187	14.18	190.8	1.168	11.27	190.0	1.153
480	19.92	194.8	1.194	14.85	194.1	1.175	11.82	193.5	1.160
500	20.80	198.1	1.201	15.53	197.5	1.182	12.36	196.9	1.168
550	22.99	206.3	1.217	17.19	205.8	1.198	13.71	205.3	1.183
600	25.17	214.5	1.231	18.84	214.0	1.212	15.04	213.6	1.197
650	27.33	222.6	1.244	20.48	222.2	1.225	16.36	221.8	1.210
700	29.49	230.6	1.256	22.10	230.3	1.237	17.67	230.0	1.223
750	31.64	238.7	1.267	23.72	238.4	1.248	18.97	238.2	1.234
800	33.78	246.7	1.277	25.34	246.5	1.259	20.27	246.3	1.244
850	35.92	254.7	1.287	26.95	254.5	1.269	21.57	254.4	1.254
900	38.05	262.7	1.296	28.55	262.6	1.278	22.86	262.4	1.263
950	40.18	270.7	1.305	30.16	270.6	1.287	24.14	270.5	1.272
1000	42.31	278.7	1.313	31.76	278.6	1.295	25.43	278.5	1.280
1050	44.44	286.7	1.321	33.36	286.6	1.303	26.71	286.5	1.288
1100	46.56	294.7	1.328	34.96	294.6	1.310	27.99	294.5	1.295

TABLE XXIV. Thermodynamic Properties of Liquid and Gaseous Xenon (*Continued*)

T, K	p = 15 bar			p = 20 bar			p = 25 bar		
	$v \cdot 10^3$	h	s	$v \cdot 10^3$	h	s	$v \cdot 10^3$	h	s
1150	48,69	302.6	1.335	36.55	302.6	1,317	29,27	302.5	1,302
1200	50,81	310.6	1.342	38.15	310,5	1.324	30.55	310.5	1,309
1250	52,93	318.5	1,348	39.74	318.5	1,330	31.82	318.5	1,316
1300	55.05	326,5	1,355	41,33	326.5	1,336	33,10	326.4	1,322

TABLE XXIV. Thermodynamic Properties of Liquid and Gaseous Xenon (*Continued*)

T, K	p = 30 bar			p = 35 bar			p = 40 bar		
	$v \cdot 10^3$	h	s	$v \cdot 10^3$	h	s	$v \cdot 10^3$	h	s
165	0.3384	50,88	0.6128	0.3381	50.99	0.6124	0,3379	51,09	0.6120
170	0.3422	52,61	0.6230	0,3419	52,71	0,6226	0,3416	52.82	0,6223
175	0.3462	54,33	0,6331	0.3459	54.43	0.6327	0.3455	54.54	0,6323
180	0,3503	56.06	0,6428	0,3500	56,16	0.6424	0.3496	56.26	0,6420
185	0.3546	57,79	0,6523	0.3543	57.89	0.6519	0,3539	57,99	0.6515
190	0.3592	59,52	0.6616	0.3587	59.61	0.6611	0,3583	59.70	0,6607
195	0.3639	61,26	0.6706	0.3634	61,35	0,6701	0.3630	61.43	0,6697
200	0,3689	63.00	0,6794	0.3684	63,09	0,6789	0,3679	63.17	0,6784
205	0.3742	64,77	0.6881	0,3736	64.84	0,6876	0.3731	64.92	0.6871
210	0,3798	66,55	0,6967	0,3791	66.61	0,6962	0,3785	66.69	0,6956
215	0,3857	68.34	0,7052	0.3850	68.40	0,7046	0.3843	68.46	0.7040
220	0,3921	70,18	0,7136	0.3913	70,23	0.7130	0.3905	70.29	0.7123
225	0.3990	72,05	0.7220	0,3981	72.09	0.7213	0,3972	72,13	0,7206

TABLE XXIV. Thermodynamic Properties of Liquid and Gaseous Xenon (*Continued*)

T, K	p = 30 bar			p = 35 bar			p = 40 bar		
	$v \cdot 10^3$	h	s	$v \cdot 10^3$	h	s	$v \cdot 10^3$	h	s
230	0.4064	73.97	0.7304	0.4054	74.00	0.7297	0.4043	74.03	0.7289
235	0.4146	75.94	0.7390	0.4133	75.95	0.7381	0.4121	75.96	0.7373
240	0.4236	78.00	0.7476	0.4221	78.00	0.7467	0.4207	78.00	0.7458
245	0.4336	80.15	0.7565	0.4318	80.10	0.7555	0.4301	80.07	0.7544
250	0.4450	82.43	0.7657	0.4428	82.35	0.7646	0.4407	82.30	0.7634
255	0.4581	84.89	0.7755	0.4553	84.75	0.7741	0.4527	84.66	0.7728
260	3.766	146.4	1.013	0.4701	87.36	0.7842	0.4667	87.20	0.7828
265	3.989	148.3	1.021	0.4879	90.23	0.7952	0.4834	90.00	0.7933
270	4.193	150.0	1.028	3.266	146.4	1.006	0.5060	93.20	0.8054
275	4.384	151.5	1.033	3.468	148.3	1.014	2.720	144.3	0.9921
280	4.565	153.0	1.038	3.652	150.1	1.020	2.925	146.7	1.001
285	4.738	154.4	1.043	3.823	151.8	1.026	3.108	148.7	1.009
290	4.904	155.7	1.048	3.984	153.3	1.031	3.273	150.5	1.016
295	5.065	157.0	1.052	4.138	154.7	1.036	3.426	152.2	1.021
300	5.221	158.2	1.056	4.285	156.1	1.041	3.570	153.8	1.026
305	5.373	159.4	1.060	4.427	157.4	1.045	3.707	155.3	1.031
310	5.522	160.6	1.064	4.565	158.7	1.049	3.839	156.7	1.036
315	5.668	161.7	1.068	4.699	159.9	1.053	3.966	158.1	1.041
320	5.811	162.8	1.071	4.830	161.1	1.057	4.089	159.4	1.045
325	5.952	163.9	1.074	4.958	162.3	1.061	4.208	160.6	1.049
330	6.091	165.0	1.077	5.083	163.4	1.065	4.325	161.8	1.053
335	6.227	166.0	1.080	5.207	164.5	1.068	4.439	163.0	1.057
340	6.362	167.0	1.083	5.328	165.6	1.071	4.551	164.2	1.060
345	6.496	168.1	1.086	5.448	166.7	1.074	4.660	165.3	1.063
350	6.628	169.1	1.089	5.566	167.8	1.077	4.769	166.4	1.066

TABLE XXIV. Thermodynamic Properties of Liquid and Gaseous Xenon (*Continued*)

T, K	p = 30 bar			p = 35 bar			p = 40 bar		
	$v \cdot 10^3$	h	s	$v \cdot 10^3$	h	s	$v \cdot 10^3$	h	s
360	6.888	171.1	1.095	5.798	169.8	1.083	4.980	168.6	1.072
370	7.144	173.0	1.100	6.026	171.8	1.088	5.186	170.7	1.078
380	7.396	174.9	1.105	6.249	173.8	1.094	5.388	172.7	1.083
390	7.645	176.7	1.110	6.469	175.7	1.099	5.586	174.7	1.088
400	7.892	178.6	1.115	6.685	177.6	1.103	5.780	176.7	1.093
420	8.377	182.2	1.124	7.111	181.3	1.112	6.162	180.5	1.103
440	8.854	185.8	1.132	7.528	185.0	1.121	6.535	184.2	1.111
460	9.325	189.3	1.140	7.939	188.6	1.129	6.901	187.9	1.119
480	9.790	192.8	1.147	8.345	192.2	1.136	7.261	191.5	1.127
500	10.25	196.3	1.154	8.746	195.7	1.144	7.617	195.0	1.134
550	11.39	204.8	1.170	9.732	204.3	1.160	8.490	203.8	1.151
600	12.51	213.2	1.185	10.70	212.7	1.175	9.347	212.3	1.166
650	13.62	221.5	1.198	11.66	221.1	1.188	10.19	220.7	1.179
700	14.72	229.7	1.211	12.61	229.4	1.200	11.03	229.1	1.192
750	15.81	237.9	1.222	13.55	237.6	1.212	11.86	237.4	1.203
800	16.90	246.0	1.232	14.49	245.8	1.222	12.68	245.6	1.214
850	17.98	254.1	1.242	15.42	254.0	1.232	13.50	253.8	1.224
900	19.06	262.2	1.252	16.34	262.1	1.242	14.31	261.9	1.233
950	20.13	270.3	1.260	17.27	270.2	1.251	15.12	270.0	1.242
1000	21.21	278.4	1.268	18.19	278.2	1.259	15.93	278.1	1.250
1050	22.28	286.4	1.276	19.11	286.3	1.267	16.74	286.2	1.258
1100	23.35	294.4	1.284	20.03	294.3	1.274	17.54	294.3	1.265
1150	24.42	302.4	1.291	20.95	302.4	1.281	18.35	302.3	1.272
1200	25.48	310.4	1.298	21.86	310.4	1.288	19.15	310.4	1.279
1250	26.55	318.4	1.305	22.78	318.4	1.295	19.95	318.4	1.286
1300	27.61	326.4	1.311	23.69	326.4	1.301	20.75	326.4	1.292

TABLE XXIV. Thermodynamic Properties of Liquid and Gaseous Xenon (*Continued*)

T, K	p = 45 bar			p = 50 bar			p = 55 bar		
	$v \cdot 10^3$	h	s	$v \cdot 10^3$	h	s	$v \cdot 10^3$	h	s
165	0.3376	51.20	0.6117	0.3373	51.31	0.6113	0.3371	51.42	0.6109
170	0.3413	52.92	0.6219	0.3411	53.03	0.6215	0.3408	53.13	0.6211
175	0.3452	54.64	0.6319	0.3449	54.74	0.6315	0.3446	54.85	0.6311
180	0.3493	56.36	0.6416	0.3489	56.46	0.6411	0.3486	56.56	0.6407
185	0.3535	58.08	0.6510	0.3531	58.18	0.6506	0.3528	58.28	0.6502
190	0.3579	59.79	0.6602	0.3575	59.89	0.6598	0.3571	59.98	0.6593
195	0.3625	61.52	0.6692	0.3621	61.61	0.6687	0.3617	61.70	0.6683
200	0.3674	63.25	0.6779	0.3669	63.34	0.6774	0.3664	63.42	0.6769
205	0.3725	65.00	0.6865	0.3720	65.08	0.6860	0.3714	65.16	0.6855
210	0.3779	66.76	0.6950	0.3773	66.83	0.6945	0.3767	66.90	0.6940
215	0.3836	68.53	0.7034	0.3830	68.59	0.7028	0.3823	68.66	0.7022
220	0.3897	70.34	0.7117	0.3890	70.04	0.7111	0.3882	70.46	0.7104
225	0.3963	72.18	0.7199	0.3954	72.22	0.7192	0.3946	72.27	0.7185
230	0.4033	74.06	0.7282	0.4023	74.09	0.7275	0.4014	74.13	0.7268
235	0.4110	75.98	0.7365	0.4098	76.00	0.7357	0.4087	76.02	0.7350
240	0.4193	78.00	0.7449	0.4180	78.00	0.7440	0.4167	78.00	0.7432
245	0.4285	80.05	0.7534	0.4269	80.03	0.7525	0.4254	80.00	0.7517
250	0.4387	82.25	0.7623	0.4369	82.20	0.7613	0.4351	82.15	0.7605
255	0.4503	84.58	0.7716	0.4480	84.50	0.7704	0.4458	84.42	0.7695
260	0.4636	87.05	0.7812	0.4607	86.90	0.7798	0.4580	86.80	0.7787
265	0.4793	89.80	0.7918	0.4755	89.53	0.7901	0.4721	89.40	0.7888
270	0.4986	92.87	0.8036	0.4932	92.55	0.8017	0.4886	92.25	0.8001
275	0.5289	96.54	0.8166	0.5188	96.06	0.8140	0.5102	95.65	0.8117
280	2.294	142.0	0.9800	0.5616	100.2	0.8292	0.5465	99.52	0.8256
285	2.509	144.8	0.9900	1.925	139.0	0.9665	0.6110	104.8	0.8444

TABLE XXIV. Thermodynamic Properties of Liquid and Gaseous Xenon (*Continued*)

T, K	p = 45 bar			p = 50 bar			p = 55 bar		
	$v \cdot 10^3$	h	s	$v \cdot 10^3$	h	s	$v \cdot 10^3$	h	s
290	2.692	147.3	0.9992	2.186	143.1	0.9802	1.690	137.2	0.9561
295	2.853	149.4	1.007	2.368	146.0	0.9904	1.933	141.6	0.9713
300	3.000	151.2	1.013	2.525	148.4	0.9982	2.114	144.8	0.9822
305	3.136	152.9	1.018	2.667	150.4	1.005	2.268	147.4	0.9911
310	3.266	154.5	1.023	2.797	152.2	1.011	2.404	149.6	0.9981
315	3.389	156.0	1.028	2.920	153.8	1.016	2.528	151.5	1.004
320	3.507	157.5	1.033	3.036	155.4	1.021	2.644	153.2	1.010
325	3.621	158.8	1.037	3.146	156.9	1.026	2.754	154.9	1.015
330	3.731	160.1	1.041	3.253	158.4	1.031	2.858	156.5	1.020
335	3.839	161.4	1.045	3.356	159.8	1.036	2.958	158.0	1.025
340	3.943	162.6	1.049	3.456	161.1	1.038	3.054	159.4	1.029
345	4.046	163.8	1.052	3.553	162.4	1.042	3.148	160.8	1.033
350	4.147	165.0	1.056	3.648	163.6	1.046	3.238	162.1	1.037
360	4.342	167.3	1.062	3.832	166.0	1.053	3.413	164.6	1.044
370	4.533	169.5	1.068	4.009	168.3	1.059	3.581	167.0	1.050
380	4.718	171.6	1.073	4.182	170.5	1.065	3.743	169.3	1.056
390	4.899	173.7	1.079	4.350	172.6	1.070	3.901	171.5	1.062
400	5.077	175.7	1.084	4.514	174.7	1.075	4.054	173.7	1.068
420	5.424	179.6	1.094	4.834	178.7	1.085	4.352	177.8	1.078
440	5.762	183.4	1.103	5.145	182.6	1.094	4.640	181.8	1.087
460	6.093	187.2	1.111	5.448	186.4	1.103	4.921	185.7	1.096
480	6.419	190.8	1.119	5.745	190.2	1.111	5.195	189.5	1.104
500	6.739	194.4	1.126	6.038	193.8	1.118	5.464	193.2	1.111
550	7.525	203.3	1.143	6.753	202.7	1.135	6.122	202.2	1.129
600	8.293	211.9	1.158	7.451	211.5	1.150	6.762	211.0	1.144

TABLE XXIV. Thermodynamic Properties of Liquid and Gaseous Xenon (*Continued*)

T, K	p = 45 bar			p = 50 bar			p = 55 bar		
	$v \cdot 10^3$	h	s	$v \cdot 10^3$	h	s	$v \cdot 10^3$	h	s
650	9.049	220.4	1.171	8.136	220.0	1.164	7.390	219.7	1.158
700	9.797	228.8	1.184	8.813	228.5	1.177	8.009	228.2	1.170
750	10.54	237.1	1.195	9.482	236.8	1.188	8.620	236.6	1.182
800	11.27	245.4	1.206	10.15	245.1	1.199	9.227	244.9	1.193
850	12.00	253.6	1.216	10.81	253.4	1.209	9.829	253.2	1.203
900	12.73	261.8	1.225	11.46	261.6	1.218	10.42	261.4	1.212
950	13.45	269.9	1.234	12.12	269.8	1.227	11.02	269.6	1.221
1000	14.17	278.0	1.242	12.77	277.9	1.235	11.62	277.8	1.229
1050	14.89	286.1	1.250	13.41	286.0	1.243	12.21	285.9	1.237
1100	15.61	294.2	1.258	14.06	294.1	1.251	12.80	294.0	1.245
1150	16.33	302.2	1.265	14.71	302.2	1.258	13.38	302.1	1.252
1200	17.04	310.3	1.272	15.35	310.2	1.265	13.97	310.2	1.259
1250	17.75	318.3	1.279	15.99	318.3	1.272	14.56	318.3	1.266
1300	18.47	326.3	1.285	16.64	326.3	1.278	15.14	326.3	1.272

TABLE XXIV. Thermodynamic Properties of Liquid and Gaseous Xenon (*Continued*)

T, K	p = 60 bar			p = 65 bar			p = 70 bar		
	$v \cdot 10^3$	h	s	$v \cdot 10^3$	h	s	$v \cdot 10^3$	h	s
165	0.3368	51.53	0.6106	0.3366	51.64	0.6102	0.3363	51.75	0.6099
170	0.3405	53.24	0.6207	0.3402	53.35	0.6204	0.3399	53.46	0.6200
175	0.3443	54.95	0.6307	0.3440	55.06	0.6303	0.3437	55.16	0.6299
180	0.3483	56.67	0.6403	0.3480	56.77	0.6399	0.3476	56.87	0.6395
185	0.3524	58.37	0.6498	0.3521	58.47	0.6493	0.3517	58.57	0.6489

TABLE XXIV. Thermodynamic Properties of Liquid and Gaseous Xenon (*Continued*)

T, K	p = 60 bar			p = 65 bar			p = 70 bar		
	$v \cdot 10^3$	h	s	$v \cdot 10^3$	h	s	$v \cdot 10^3$	h	s
190	0.3567	60.08	0.6589	0.3563	60.17	0.6584	0.3559	60.27	0.6580
195	0.3612	61.79	0.6678	0.3608	61.88	0.6673	0.3604	61.97	0.6669
200	0.3659	63.51	0.6764	0.3655	63.60	0.6759	0.3650	63.68	0.6754
205	0.3709	65.24	0.6850	0.3704	65.32	0.6845	0.3699	65.40	0.6840
210	0.3761	66.98	0.6934	0.3756	67.05	0.6929	0.3750	67.13	0.6923
215	0.3817	68.73	0.7017	0.3810	68.80	0.7011	0.3804	68.87	0.7005
220	0.3875	70.52	0.7098	0.3868	70.58	0.7092	0.3861	70.64	0.7086
225	0.3938	72.32	0.7179	0.3930	72.38	0.7173	0.3922	72.43	0.7166
230	0.4004	74.17	0.7261	0.3995	74.21	0.7254	0.3986	74.25	0.7247
235	0.4077	76.05	0.7342	0.4066	76.08	0.7335	0.4056	76.11	0.7327
240	0.4155	78.00	0.7424	0.4142	78.01	0.7416	0.4131	78.02	0.7408
245	0.4240	80.00	0.7507	0.4226	80.00	0.7498	0.4212	80.00	0.7489
250	0.4333	82.13	0.7592	0.4317	82.10	0.7582	0.4301	82.08	0.7572
255	0.4437	84.36	0.7680	0.4418	84.30	0.7669	0.4399	84.25	0.7658
260	0.4555	86.66	0.7771	0.4531	86.57	0.7759	0.4508	86.50	0.7747
265	0.4689	89.20	0.7870	0.4658	89.05	0.7857	0.4630	88.90	0.7841
270	0.4844	92.00	0.7980	0.4806	91.75	0.7962	0.4770	91.55	0.7943
275	0.5035	95.25	0.8092	0.4981	94.86	0.8070	0.4933	94.53	0.8048
280	0.5340	98.85	0.8223	0.5235	98.25	0.8192	0.5154	97.70	0.8163
285	0.5830	103.4	0.8382	0.5612	102.1	0.8330	0.5455	101.0	0.8284
290	0.6832	110.1	0.8612	0.6132	106.4	0.8481	0.5908	104.7	0.8410
295	1.495	135.3	0.9477	0.8627	119.0	0.8903	0.6676	110.5	0.8605
300	1.739	140.4	0.9647	1.364	134.1	0.9411	0.9561	123 5	0.9032
305	1.916	143.8	0.9760	1.593	139.4	0.9587	1.282	133.6	0.9373
310	2.064	146.5	0.9848	1.761	143.0	0.9704	1.485	138.8	0.9542

TABLE XXIV. Thermodynamic Properties of Liquid and Gaseous Xenon (*Continued*)

T, K	p = 60 bar			p = 65 bar			p = 70 bar		
	$v \cdot 10^3$	h	s	$v \cdot 10^3$	h	s	$v \cdot 10^3$	h	s
315	2.194	148.8	0.9922	1.902	145.8	0.9795	1.641	142.4	0.9660
320	2.312	150.9	0.9987	2.025	148.3	0.9872	1.773	145.4	0.9752
325	2.422	152.8	1.005	2.138	150.4	0.9939	1.889	147.9	0.9830
330	2.526	154.5	1.010	2.242	152.4	0.9998	1.995	150.1	0.9897
335	2.624	156.1	1.015	2.339	154.2	1.005	2.093	152.1	0.9958
340	2.718	157.7	1.019	2.432	155.9	1.010	2.185	153.9	1.001
345	2.809	159.2	1.024	2.521	157.5	1.015	2.273	155.7	1.006
350	2.896	160.6	1.028	2.606	159.0	1.019	2.356	157.3	1.011
360	3.064	163.2	1.035	2.768	161.8	1.027	2.514	160.3	1.020
370	3.224	165.7	1.042	2.922	164.5	1.035	2.663	163.1	1.027
380	3.378	168.1	1.049	3.068	167.0	1.041	2.804	165.7	1.034
390	3.526	170.5	1.055	3.210	169.4	1.047	2.939	168.2	1.041
400	3.671	172.7	1.061	3.347	171.6	1.053	3.070	170.6	1.047
420	3.951	176.9	1.071	3.611	176.0	1.064	3.321	175.1	1.058
440	4.220	181.0	1.080	3.865	180.2	1.074	3.561	179.4	1.068
460	4.481	185.0	1.089	4.110	184.2	1.083	3.793	183.5	1.077
480	4.737	188.8	1.097	4.350	188.2	1.091	4.018	187.5	1.085
500	4.988	192.6	1.105	4.584	192.0	1.099	4.238	191.4	1.093
550	5.596	201.7	1.122	5.152	201.2	1.116	4.772	200.8	1.111
600	6.188	210.6	1.138	5.703	210.2	1.132	5.288	209.8	1.127
650	6.768	219.3	1.152	6.242	219.0	1.146	5.792	218.6	1.141
700	7.338	227.9	1.164	6.772	227.6	1.159	6.286	227.3	1.154
750	7.902	236.3	1.176	7.295	236.1	1.171	6.774	235.8	1.166
800	8.460	244.7	1.187	7.812	244.5	1.181	7.256	244.3	1.176
850	9.014	253.0	1.197	8.325	252.8	1.192	7.735	252.7	1.187

TABLE XXIV. Thermodynamic Properties of Liquid and Gaseous Xenon (*Continued*)

T, K	p = 60 bar			p = 65 bar			p = 70 bar		
	$v \cdot 10^3$	h	s	$v \cdot 10^3$	h	s	$v \cdot 10^3$	h	s
900	9.564	261.3	1.206	8.835	260.1	1.201	8,209	261.0	1.196
950	10.11	269.5	1.215	9.341	269.4	1.210	8,681	269.2	1.205
1000	10.66	277.7	1.224	9.846	277.6	1.218	9,151	277.5	1.214
1050	11.20	285.8	1,232	10.35	285.8	1.226	9,618	285.7	1.222
1100	11.74	294.0	1.239	10.85	293.9	1,234	10,08	293,8	1,229
1150	12.28	302.1	1.246	11.35	302.0	1.241	10,55	301.9	1.236
1200	12.82	310.2	1.253	11.85	310.1	1.248	11,01	310.0	1.243
1250	13.36	318.3	1.260	12.34	318.2	1.255	11,47	318.1	1.250
1300	13.89	326.3	1.266	12.84	326.2	1,261	11,93	326.2	1.256

TABLE XXIV. Thermodynamic Properties of Liquid and Gaseous Xenon (*Continued*)

T, K	p = 75 bar			p = 80 bar			p = 85 bar		
	$v \cdot 10^3$	h	s	$v \cdot 10^3$	h	s	$v \cdot 10^3$	h	s
165	0.3360	51.85	0.6095	0.3358	51,96	0.6092	0.3355	52.07	0.6088
170	0.3397	53.56	0,6196	0,3394	53,67	0,6193	0,3391	53,78	0,6189
175	0.3434	55,26	0.6295	0,3432	55.37	0,6291	0.3428	55.48	0.6288
180	0.3473	56,97	0.6391	0.3470	57.07	0.6387	0.3467	57.18	0.6383
185	0.3514	58,67	0.6485	0.3510	58.77	0.6481	0.3507	58.87	0.6477
190	0,3556	60.36	0.6576	0.3552	60.46	0.6571	0.3548	60.56	0.6567
195	0.3600	62,07	0.6664	0.3595	62.16	0.6660	0.3591	62.25	0.6655
200	0,3646	63,77	0.6750	0,3641	63.86	0.6745	0,3637	63.95	0.6740
205	0,3694	65.49	0.6835	0,3689	65.57	0.6830	0,3684	65.65	0.6825
210	0.3744	67.21	0,6918	0.3739	67.29	0.6913	0.3733	67.37	0.6908

TABLE XXIV. Thermodynamic Properties of Liquid and Gaseous Xenon (*Continued*)

T, K	p = 75 bar			p = 80 bar			p = 85 bar		
	$v \cdot 10^3$	h	s	$v \cdot 10^3$	h	s	$v \cdot 10^3$	h	s
215	0.3798	68.94	0.7000	0.3792	69.01	0.6994	0.3786	69.09	0.6989
220	0.3854	70.71	0.7081	0.3847	70.77	0.7075	0.3841	70.84	0.7069
225	0.3914	72.49	0.7160	0.3906	72.54	0.7154	0.3899	72.60	0.7148
230	0.3978	74.30	0.7240	0.3969	74.35	0.7234	0.3961	74.40	0.7227
235	0.4046	76.15	0.7320	0.4036	76.18	0.7313	0.4027	76.22	0.7305
240	0.4119	78.05	0.7400	0.4108	78.05	0.7392	0.4098	78.10	0.7384
245	0.4199	80.00	0.7480	0.4186	80.0	0.7472	0.4174	80.05	0.7464
250	0.4285	82.05	0.7562	0.4271	82.05	0.7554	0.4256	82.06	0.7546
255	0.4380	84.20	0.7647	0.4363	84.15	0.7638	0.4346	84.13	0.7628
260	0.4486	86.40	0.7735	0.4465	86.35	0.7725	0.4445	86.28	0.7714
265	0.4604	88.80	0.7827	0.4579	88.70	0.7815	0.4555	88.60	0.7802
270	0.4737	91.35	0.7926	0.4706	91.20	0.7911	0.4678	91.05	0.7895
275	0.4891	94.20	0.8028	0.4852	93.90	0.8009	0.4816	93.60	0.7989
280	0.5082	97.16	0.8135	0.5025	96.67	0.8109	0.4974	96.25	0.8083
285	0.5340	100.2	0.8245	0.5245	99.55	0.8211	0.5175	99.00	0.8178
290	0.5668	103.6	0.8360	0.5534	102.7	0.8318	0.5429	102.0	0.8282
295	0.6210	108.0	0.8512	0.5950	106.5	0.8450	0.5772	105.4	0.8400
300	0.7344	115.0	0.8743	0.6634	111.5	0.8616	0.6272	109.6	0.8534
305	0.9894	125.8	0.9093	0.7978	118.9	0.8853	0.7092	115.1	0.8720
310	1.230	133.6	0.9354	1.006	127.7	0.9144	0.8466	122.2	0.8954
315	1.406	138.4	0.9506	1.196	134.0	0.9342	1.017	129.3	0.9180
320	1.548	142.2	0.9626	1.348	138.6	0.9491	1.172	134.7	0.9350
325	1.670	145.1	0.9717	1.476	142.1	0.9601	1.304	138.9	0.9480
330	1.779	147.6	0.9795	1.588	145.0	0.9690	1.419	142.3	0.9583
335	1.878	149.9	0.9863	1.689	147.6	0.9767	1.521	145.1	0.9670

TABLE XXIV. Thermodynamic Properties of Liquid and Gaseous Xenon (*Continued*)

T, K	p = 75 bar			p = 80 bar			p = 85 bar		
	$v \cdot 10^3$	h	s	$v \cdot 10^3$	h	s	$v \cdot 10^3$	h	s
340	1.970	151.9	0.9923	1.781	149.8	0.9834	1.615	147.6	0.9744
345	2.057	153.8	0.9978	1.868	151.9	0.9894	1.701	149.9	0.9810
350	2.139	155.6	1.003	1.950	153.8	0.9949	1.782	152.0	0.9870
360	2.294	158.8	1.012	2.101	157.3	1.005	1.932	155.7	0.9974
370	2.438	161.8	1.020	2.242	160.4	1.013	2.069	159.0	1.006
380	2.574	164.5	1.027	2.374	163.3	1.021	2.198	162.0	1.014
390	2.705	167.1	1.034	2.500	166.0	1.028	2.320	164.8	1.022
400	2.830	169.6	1.040	2.621	168.5	1.034	2.437	167.4	1.029
420	3.070	174.2	1.052	2.851	173.3	1.046	2.658	172.4	1.041
440	3.298	178.6	1.062	3.069	177.8	1.057	2.867	177.0	1.051
460	3.518	182.8	1.071	3.278	182.1	1.066	3.067	181.3	1.061
480	3.731	186.8	1.080	3.481	186.2	1.075	3.260	185.5	1.070
500	3.939	190.8	1.088	3.678	190.2	1.083	3.448	189.6	1.078
550	4.443	200.3	1.106	4.154	199.8	1.101	3.901	199.3	1.097
600	4.928	209.4	1.122	4.614	209.0	1.117	4.337	208.6	1.113
650	5.401	218.3	1.136	5.060	217.9	1.131	4.760	217.6	1.127
700	5.866	227.0	1.149	5.498	226.7	1.144	5.174	226.4	1.140
750	6.323	235.6	1.161	5.929	235.3	1.156	5.581	235.1	1.152
800	6.775	244.1	1.172	6.354	243.9	1.167	5.983	243.6	1.163
850	7.223	252.5	1.182	6.776	252.3	1.178	6.381	252.1	1.174
900	7.667	260.8	1.192	7.193	260.7	1.187	6.775	260.5	1.183
950	8.109	269.1	1.201	7.608	269.0	1.196	7.167	268.9	1.192
1000	8.548	277.4	1.209	8.021	277.2	1.205	7.556	277.1	1.201
1050	8.985	285.6	1.217	8.432	285.5	1.213	7.944	285.4	1.209
1100	9.421	293.8	1.225	8.841	293.7	1.221	8.329	293.6	1.217

TABLE XXIV. Thermodynamic Properties of Liquid and Gaseous Xenon (*Continued*)

T, K	p = 75 bar			p = 80 bar			p = 85 bar		
	$v \cdot 10^3$	h	s	$v \cdot 10^3$	h	s	$v \cdot 10^3$	h	s
1150	9.855	301.9	1.232	9.249	301.8	1.228	8.714	301.8	1.224
1200	10.29	310.0	1.239	9.655	310.0	1.235	9.097	309.9	1.231
1250	10.72	318.1	1.246	10.06	318.1	1.241	9.479	318.1	1.237
1300	11.15	326.2	1.252	10.46	326.2	1.248	9.860	326.2	1.244

TABLE XXIV. Thermodynamic Properties of Liquid and Gaseous Xenon (*Continued*)

T, K	p = 90 bar			p = 95 bar			p = 100 bar		
	$v \cdot 10^3$	h	s	$v \cdot 10^3$	h	s	$v \cdot 10^3$	h	s
165	0.3353	52.18	0.6084	—	—	—	—	—	—
170	0.3389	53.88	0.6185	0.3386	53.99	0.6182	0.3383	54.10	0.6178
175	0.3425	55.58	0.6284	0.3423	55.69	0.6280	0.3420	55.79	0.6276
180	0.3464	57.28	0.6379	0.3460	57.38	0.6376	0.3457	57.49	0.6372
185	0.3503	58.97	0.6473	0.3500	59.07	0.6469	0.3497	59.17	0.6465
190	0.3544	60.65	0.6563	0.3541	60.75	0.6559	0.3537	60.85	0.6555
195	0.3587	62.35	0.6651	0.3583	62.44	0.6647	0.3579	62.53	0.6642
200	0.3632	64.04	0.6736	0.3628	64.13	0.6731	0.3623	64.22	0.6727
205	0.3679	65.74	0.6820	0.3674	65.83	0.6816	0.3670	65.91	0.6811
210	0.3728	67.45	0.6903	0.3723	67.53	0.6898	0.3718	67.61	0.6893
215	0.3780	69.16	0.6984	0.3774	69.24	0.6979	0.3768	69.31	0.6973
220	0.3834	70.91	0.7063	0.3828	70.98	0.7058	0.3821	71.05	0.7052
225	0.3892	72.66	0.7142	0.3884	72.73	0.7136	0.3877	72.79	0.7130

TABLE XXIV. Thermodynamic Properties of Liquid and Gaseous Xenon (*Continued*)

T, K	p = 90 bar			p = 95 bar			p = 100 bar		
	$v \cdot 10^3$	h	s	$v \cdot 10^3$	h	s	$v \cdot 10^3$	h	s
230	0.3953	74.45	0.7221	0.3945	74.51	0.7215	0.3937	74.56	0.7208
235	0.4018	76.27	0.7299	0.4009	76.31	0.7293	0.4000	76.36	0.7286
240	0.4087	78.11	0.7378	0.4077	78.15	0.7372	0.4067	78.20	0.7364
245	0.4162	80.05	0.7457	0.4150	80.06	0.7450	0.4139	80.10	0.7442
250	0.4243	82.05	0.7536	0.4229	82.02	0.7528	0.4216	82.03	0.7520
255	0.4330	84.10	0.7617	0.4315	84.05	0.7608	0.4300	84.05	0.7599
260	0.4426	86.20	0.7703	0.4408	86.12	0.7691	0.4391	86.09	0.7681
265	0.4532	88.50	0.7789	0.4511	88.40	0.7776	0.4490	88.34	0.7764
270	0.4650	90.90	0.7877	0.4624	90.75	0.7862	0.4600	90.63	0.7848
275	0.4783	93.35	0.7966	0.4751	93.10	0.7948	0.4722	92.93	0.7933
280	0.4933	95.85	0.8055	0.4894	95.50	0.8035	0.4858	95.25	0.8018
285	0.5118	98.55	0.8148	0.5062	98.11	0.8124	0.5011	97.75	0.8104
290	0.5342	101.4	0.8249	0.5267	100.9	0.8220	0.5202	100.4	0.8195
295	0.5636	104.5	0.8358	0.5527	103.9	0.8324	0.5436	103.2	0.8292
300	0.6036	108.3	0.8478	0.5864	107.3	0.8436	0.5729	106.5	0.8401
305	0.6622	112.9	0.8644	0.6322	111.3	0.8576	0.6108	110.1	0.8524
310	0.7551	118.5	0.8818	0.6980	115.9	0.8722	0.6621	114.1	0.8656
315	0.8818	124.8	0.9024	0.7913	121.3	0.8900	0.7322	118.8	0.8806
320	1.023	131.0	0.9204	0.9075	127.0	0.9078	0.8231	123.9	0.8968
325	1.154	135.5	0.9357	1.028	132.1	0.9236	0.9267	128.9	0.9124
330	1.270	139.3	0.9475	1.141	136.4	0.9366	1.031	133.4	0.9261
335	1.375	142.6	0.9572	1.243	139.9	0.9474	1.130	137.3	0.9377

TABLE XXIV. Thermodynamic Properties of Liquid and Gaseous Xenon (*Continued*)

T, K	p = 90 bar			p = 95 bar			p = 100 bar		
	$v \cdot 10^3$	h	s	$v \cdot 10^3$	h	s	$v \cdot 10^3$	h	s
340	1.467	145.4	0.9655	1.337	143.0	0.9565	1.222	140.6	0.9476
345	1.553	147.8	0.9727	1.422	145.7	0.9644	1.306	143.6	0.9562
350	1.634	150.1	0.9791	1.502	148.1	0.9714	1.385	146.2	0.9637
360	1.782	154.0	0.9903	1.648	152.4	0.9833	1.529	150.7	0.9765
370	1.917	157.5	0.9999	1.781	156.1	0.9935	1.659	154.6	0.9872
380	2.042	160.7	1.0080	1.903	159.4	1.002	1.779	158.1	0.9965
390	2.161	163.6	1.016	2.018	162.5	1.010	1.891	161.3	1.005
400	2.273	166.4	1.023	2.128	165.3	1.017	1.998	164.2	1.012
420	2.487	171.5	1.035	2.334	170.5	1.030	2.198	169.6	1.025
440	2.688	176.2	1.046	2.528	175.4	1.041	2.385	174.6	1.037
460	2.880	180.6	1.056	2.712	179.9	1.052	2.562	179.2	1.047
480	3.065	184.9	1.065	2.890	184.2	1.061	2.733	183.6	1.056
500	3.244	189.0	1.074	3.062	188.4	1.069	2.899	187.8	1.065
550	3.676	198.8	1.092	3.475	198.3	1.088	3.295	197.8	1.084
600	4.090	208.2	1.109	3.871	207.8	1.105	3.673	207.4	1.101
650	4.493	217.3	1.123	4.254	216.9	1.119	4.039	216.6	1.116
700	4.886	226.1	1.136	4.628	225.8	1.133	4.397	225.6	1.129
750	5.272	234.8	1.148	4.996	234.6	1.145	4.747	234.4	1.141
800	5 653	243.4	1.159	5.358	243.2	1.156	5.093	243.0	1.152
850	6.030	251.9	1.170	5.717	251.8	1.166	5.434	251.6	1.163
900	6.404	260.4	1.179	6.072	260.2	1.176	5.773	260.1	1.172
950	6.775	268.7	1.188	6.424	268.6	1.185	6.108	268.5	1.181
1000	7.143	277.0	1.197	6.774	276.9	1.193	6.441	276.8	1.190
1050	7.510	285.3	1.205	7.122	285.2	1.201	6.772	285.1	1.198
1100	7.875	293.5	1.213	7.468	293.5	1.209	7.102	293.4	1.206

TABLE XXIV. Thermodynamic Properties of Liquid and Gaseous Xenon (*Continued*)

T, K	p = 90 bar			p = 95 bar			p = 100 bar		
	$v \cdot 10^3$	h	s	$v \cdot 10^3$	h	s	$v \cdot 10^3$	h	s
1150	8.238	301.7	1.220	7.813	301.7	1.216	7.430	301.6	1.213
1200	8.600	309.9	1.227	8.157	309.8	1.223	7.757	309.8	1.220
1250	8.962	318.0	1.234	8.499	318.0	1.230	8.083	318.0	1.227
1300	9.322	326.1	1.240	8.841	326.1	1.236	8.408	326.1	1.233

TABLE XXIV. Thermodynamic Properties of Liquid and Gaseous Xenon (*Continued*)

T, K	p = 110 bar			p = 120 bar			p = 130 bar		
	$v \cdot 10^3$	h	s	$v \cdot 10^3$	h	s	$v \cdot 10^3$	h	s
170	0.3378	54.32	0.6171	0.3373	54.54	0.6164	0.3368	54.75	0.6157
175	0.3414	56.01	0.6269	0.3409	56.22	0.6262	0.3403	56.43	0.6255
180	0.3451	57.69	0.6364	0.3445	57.90	0.6357	0.3440	58.11	0.6349
185	0.3490	59.37	0.6457	0.3484	59.58	0.6449	0.3477	59.78	0.6441
190	0.3530	61.05	0.6546	0.3523	61.24	0.6538	0.3516	61.44	0.6530
195	0.3572	62.73	0.6634	0.3564	62.92	0.6625	0.3558	63.11	0.6617
200	0.3615	64.40	0.6718	0.3607	64.59	0.6709	0.3599	64.78	0.6700
205	0.3660	66.09	0.6802	0.3651	66.27	0.6792	0.3642	66.45	0.6783
210	0.3708	67.78	0.6883	0.3698	67.95	0.6873	0.3688	68.12	0.6864
215	0.3757	69.47	0.6963	0.3746	69.63	0.6953	0.3736	69.79	0.6943
220	0.3809	71.19	0.7042	0.3797	71.34	0.7031	0.3786	71.49	0.7021
225	0.3864	72.92	0.7119	0.3851	73.06	0.7108	0.3838	73.20	0.7097
230	0.3922	74.68	0.7196	0.3907	74.80	0.7185	0.3893	74.93	0.7173
235	0.3983	76.45	0.7273	0.3966	76.56	0.7261	0.3951	76.67	0.7249

TABLE XXIV. Thermodynamic Properties of Liquid and Gaseous Xenon (*Continued*)

T, K	p = 110 bar			p = 120 bar			p = 130 bar		
	$v \cdot 10^3$	h	s	$v \cdot 10^3$	h	s	$v \cdot 10^3$	h	s
240	0.4048	78.25	0.7351	0.4029	78.35	0.7338	0.4012	78.45	0.7324
245	0.4117	80.14	0.7427	0.4096	80.20	0.7414	0.4077	80.27	0.7399
250	0.4191	82.05	0.7504	0.4168	82.10	0.7490	0.4146	82.13	0.7474
255	0.4271	84.03	0.7582	0.4245	84.05	0.7566	0.4219	84.05	0.7550
260	0.4358	86.03	0.7662	0.4327	86.00	0.7644	0.4298	86.00	0.7626
265	0.4452	88.20	0.7742	0.4416	88.06	0.7722	0.4383	88.00	0.7702
270	0.4554	90.40	0.7824	0.4513	90.19	0.7800	0.4475	90.10	0.7778
275	0.4667	92.62	0.7906	0.4618	92.35	0.7878	0.4574	92.20	0.7854
280	0.4792	94.85	0.7988	0.4734	94.52	0.7958	0.4682	94.31	0.7930
285	0.4931	97.20	0.8072	0.4862	96.80	0.8040	0.4800	96.51	0.8008
290	0.5093	99.65	0.8156	0.5007	99.15	0.8122	0.4930	98.80	0.8088
295	0.5290	102.3	0.8243	0.5176	101.7	0.8207	0.5080	101.2	0.8170
300	0.5525	105.2	0.8343	0.5375	104.4	0.8296	0.5253	103.7	0.8256
305	0.5812	108.3	0.8450	0.5607	107.2	0.8390	0.5452	106.3	0.8344
310	0.6170	111.7	0.8559	0.5886	110.2	0.8490	0.5683	109.0	0.8434
315	0.6625	115.5	0.8678	0.6223	113.4	0.8593	0.5952	111.9	0.8527
320	0.7202	119.5	0.8807	0.6632	116.8	0.8700	0.6268	114.9	0.8622
325	0.7904	123.9	0.8942	0.7124	120.5	0.8815	0.6639	118.2	0.8722
330	0.8700	128.2	0.9075	0.7699	124.3	0.8932	0.7070	121.6	0.8826
335	0.9533	132.3	0.9197	0.8340	128.2	0.9047	0.7558	125.0	0.8930
340	1.036	136.0	0.9306	0.9018	131.9	0.9156	0.8094	128.5	0.9032
345	1.115	139.3	0.9403	0.9705	135.3	0.9257	0.8661	131.9	0.9131
350	1.189	142.2	0.9488	1.038	138.5	0.9348	0.9242	135.1	0.9223
360	1.327	147.3	0.9631	1.166	144.0	0.9504	1.039	140.8	0.9386
370	1.452	151.7	0.9750	1.284	148.7	0.9634	1.149	145.9	0.9524

TABLE XXIV. Thermodynamic Properties of Liquid and Gaseous Xenon (*Continued*)

T. K	p = 110 bar			p = 120 bar			p = 130 bar		
	$v \cdot 10^3$	h	s	$v \cdot 10^3$	h	s	$v \cdot 10^3$	h	s
380	1.567	155.5	0.9851	1.394	152.8	0.9744	1.252	150.3	0.9641
390	1.674	158.9	0.9941	1.495	156.5	0.9839	1.348	154.2	0.9743
400	1.775	162.1	1.002	1.591	159.9	0.9924	1.440	157.8	0.9833
420	1.963	167.8	1.016	1.770	165.9	1.007	1.609	164.1	0.9988
440	2.139	172.9	1.028	1.936	171.3	1.020	1.766	169.8	1.012
460	2.305	177.7	1.039	2.092	176.3	1.031	1.913	174.9	1.023
480	2.464	182.3	1.048	2.241	181.0	1.041	2.053	179.8	1.034
500	2.617	186.6	1.057	2.384	185.5	1.050	2.188	184.3	1.043
550	2.984	196.9	1.077	2.725	196.0	1.070	2.508	195.1	1.064
600	3.332	206.6	1.094	3.049	205.9	1.087	2.811	205.1	1.081
650	3.669	215.9	1.109	3.361	215.3	1.102	3.102	214.7	1.096
700	3.997	225.0	1.122	3.665	224.5	1.116	3.384	223.9	1.110
750	4.319	234.1	1.134	3.962	233.4	1.128	3.661	233.0	1.123
800	4.635	242.6	1.146	4.254	242.3	1.140	3.932	241.9	1.134
850	4.947	251.3	1.156	4.542	250.9	1.150	4.199	250.6	1.145
900	5.257	259.8	1.166	4.827	259.5	1.160	4.464	259.2	1.155
950	5.563	268.2	1.175	5.109	268.0	1.169	4.725	267.8	1.164
1000	5.867	276.6	1.184	5.389	276.4	1.178	4.985	276.3	1.172
1050	6.169	285.0	1.192	5.667	284.8	1.186	5.242	284.7	1.181
1100	6.470	293.3	1.200	5.943	293.1	1.194	5.498	293.0	1.189
1150	6.769	301.5	1.207	6.219	301.4	1.201	5.753	301.3	1.196
1200	7.067	309.7	1.214	6.493	309.7	1.208	6.007	309.6	1.203
1250	7.364	317.9	1.221	6.766	317.9	1.215	6.259	317.8	1.210
1300	7.661	326.1	1.227	7.038	326.1	1.221	6.511	326.0	1.216

TABLE XXIV. Thermodynamic Properties of Liquid and Gaseous Xenon (*Continued*)

T, K	p = 140 bar			p = 150 bar			p = 160 bar		
	$v \cdot 10^3$	h	s	$v \cdot 10^3$	h	s	$v \cdot 10^3$	h	s
170	0.3363	54.97	0.6150	0.3358	55.19	0.6143	0.3353	55.41	0.6136
175	0.3398	56.65	0.6247	0.3393	56.86	0.6240	0.3387	57.08	0.6233
180	0.3434	58.32	0.6342	0.3428	58.54	0.6334	0.3423	58.75	0.6327
185	0.3471	59.99	0.6434	0.3465	60.20	0.6426	0.3459	60.41	0.6419
190	0.3510	61.65	0.6522	0.3503	61.85	0.6515	0.3497	62.05	0.6507
195	0.3549	63.31	0.6609	0.3542	63.50	0.6601	0.3535	63.70	0.6593
200	0.3591	64.97	0.6692	0.3583	65.16	0.6683	0.3576	65.35	0.6675
205	0.3634	66.63	0.6775	0.3626	66.81	0.6766	0.3617	67.00	0.6757
210	0.3679	68.29	0.6855	0.3670	68.47	0.6846	0.3661	68.65	0.6837
215	0.3725	69.96	0.6934	0.3715	70.12	0.6924	0.3706	70.30	0.6915
220	0.3774	71.65	0.7011	0.3763	71.81	0.7001	0.3753	71.97	0.6991
225	0.3825	73.34	0.7086	0.3813	73.49	0.7076	0.3802	73.64	0.7066
230	0.3879	75.06	0.7162	0.3866	75.20	0.7151	0.3853	75.34	0.7141
235	0.3936	76.79	0.7237	0.3921	76.91	0.7226	0.3907	77.04	0.7215
240	0.3995	78.55	0.7312	0.3979	78.65	0.7300	0.3963	78.80	0.7288
245	0.4058	80.35	0.7386	0.4040	80.45	0.7373	0.4023	80.58	0.7361
250	0.4125	82.20	0.7461	0.4105	82.30	0.7447	0.4085	82.38	0.7435
255	0.4196	84.11	0.7536	0.4173	84.16	0.7522	0.4152	84.25	0.7509
260	0.4271	86.05	0.7612	0.4246	86.08	0.7597	0.4222	86.15	0.7583
265	0.4352	88.00	0.7688	0.4323	88.01	0.7671	0.4296	88.05	0.7656
270	0.4439	90.05	0.7763	0.4406	90.00	0.7745	0.4376	90.00	0.7729
275	0.4533	92.10	0.7838	0.4496	92.02	0.7820	0.4461	92.00	0.7803
280	0.4635	94.20	0.7914	0.4591	94.10	0.7895	0.4551	94.05	0.7877
285	0.4745	96.35	0.7991	0.4694	96.23	0.7971	0.4647	96.12	0.7951

TABLE XXIV. Thermodynamic Properties of Liquid and Gaseous Xenon (*Continued*)

T, K	p = 140 bar			p = 150 bar			p = 160 bar		
	$v \cdot 10^8$	h	s	$v \cdot 10^3$	h	s	$v \cdot 10^3$	h	s
290	0.4864	98,55	0,8069	0.4806	98.39	0,8046	0.4750	98.22	0,8024
295	0.5000	100,8	0,8147	0,4929	100,6	0.8121	0,4866	100,3	0,8096
300	0,5154	103,2	0.8225	0,5069	102,9	0,8196	0.4995	102,5	0,8168
305	0,5328	105,7	0,8305	0,5225	105,2	0.8272	0,5137	104,7	0,8241
310	0,5526	108,2	0.8387	0,5400	107,5	0.8348	0,5294	106,9	0,8314
315	0,5752	110,8	0.8472	0,5595	110.0	0.8428	0.5467	109,3	0,8389
320	0,6011	113,6	0,8560	0,5815	112,6	0.8510	0,5659	111,7	0,8466
325	0,6308	116,5	0,8651	0,6064	115,3	0,8594	0,5873	114,3	0,8545
330	0,6647	119,6	0,8744	0,6343	118,1	0,8679	0,6111	116,9	0,8625
335	0,7031	122,7	0,8839	0,6655	120,9	0,8765	0,6374	119,6	0,8705
340	0,7456	125,9	0,8933	0,7001	123,9	0,8852	0,6663	122,3	0,8787
345	0,7916	129,1	0,9026	0,7378	126,9	0,8939	0,6978	125,1	0,8868
350	0,8401	132,2	0,9115	0,7782	129,8	0,9024	0,7317	127,9	0,8948
360	0,9405	138,0	0,9279	0,8642	135,5	0,9185	0,8053	133,4	0,9102
370	1,040	143,2	0,9422	0,9529	140,8	0,9328	0,8834	138,5	0,9244
380	1,136	147,8	0,9545	1,040	145,5	0,9455	0,9625	143,3	0,9372
390	1,226	151,9	0,9652	1,125	149,8	0,9566	1,041	147,7	0,9487
400	1,313	155,7	0,9747	1,206	153,7	0,9665	1,116	151,8	0,9589
420	1,473	162,3	0,9909	1,358	160,6	0,9834	1,260	158,9	0,9763
440	1,622	168,2	1,005	1,499	166,7	0,9976	1,393	165,2	0,9910
460	1,761	173,5	1,017	1,631	172,2	1,010	1,519	170,9	1,004
480	1,894	178,5	1,027	1,757	177,3	1,021	1,639	176,1	1,015
500	2,021	183,2	1,037	1,878	182,1	1,031	1,753	181,1	1,025
550	2,323	194,2	1,058	2,163	193,3	1,052	2,024	192,4	1,047
600	2,607	204,3	1,075	2,431	203,6	1,070	2,278	202,9	1,065

TABLE XXIV. Thermodynamic Properties of Liquid and Gaseous Xenon (*Continued*)

T, K	p = 140 bar			p = 150 bar			p = 160 bar		
	$v \cdot 10^3$	h	s	$v \cdot 10^3$	h	s	$v \cdot 10^3$	h	s
650	2.880	214.1	1.091	2.688	213.5	1.086	2.521	212.9	1.081
700	3.144	223.4	1.105	2.937	222.9	1.100	2.756	222.4	1.095
750	3.403	232.6	1.117	3.180	232.1	1.112	2.985	231.7	1.108
800	3.656	241.5	1.129	3.418	241.1	1.124	3.209	240.8	1.119
850	3.906	250.3	1.140	3.652	250.0	1.135	3.430	249.7	1.130
900	4.152	259.0	1.150	3.883	258.7	1.145	3.648	258.5	1.140
950	4.396	267.6	1.159	4.112	267.4	1.154	3.862	267.1	1.150
1000	4.638	276.1	1.167	4.338	275.9	1.163	4.076	275.7	1.159
1050	4.878	284.5	1.175	4.563	284.4	1.171	4.288	284.2	1.167
1100	5.117	292.9	1.183	4.786	292.8	1.179	4.497	292.7	1.175
1150	5.354	301.2	1.191	5.008	301.2	1.186	4.706	301.1	1.182
1200	5.590	309.5	1.198	5.229	309.5	1.193	4.914	309.4	1.189
1250	5.825	317.9	1.205	5.449	317.7	1.200	5.120	317.7	1.196
1300	6.059	326.0	1.211	5.668	326.0	1.207	5.326	326.0	1.202

TABLE XXIV. Thermodynamic Properties of Liquid and Gaseous Xenon (*Continued*)

T, K	p = 170 bar			p = 180 bar			p = 190 bar		
	$v \cdot 10^3$	h	s	$v \cdot 10^3$	h	s	$v \cdot 10^3$	h	s
170	0.3349	55.64	0.6130	0.3344	55.86	0.6123	0.3339	56.08	0.6117
175	0.3382	57.30	0.6227	0.3377	57.52	0.6220	0.3372	57.74	0.6213
180	0.3417	58.96	0.6320	0.3412	59.18	0.6313	0.3406	59.39	0.6306
185	0.3453	60.62	0.6411	0.3447	60.83	0.6404	0.3442	61.04	0.6397

TABLE XXIV. Thermodynamic Properties of Liquid and Gaseous Xenon (*Continued*)

T, K	p = 170 bar			p = 180 bar			p = 190 bar		
	$v \cdot 10^3$	h	s	$v \cdot 10^3$	h	s	$v \cdot 10^3$	h	s
190	0,3490	62.26	0,6499	0,3484	62.46	0,6492	0.3478	62,67	0.6484
195	0,3529	63,90	0,6585	0,3522	64,11	0,6577	0,3515	64.31	0,6568
200	0.3568	65.54	0,6667	0,3561	65,74	0,6659	0,3554	65.94	0.6651
205	0,3609	67,19	0,6749	0,3601	67,38	0,6740	0,3594	67.57	0,6732
210	0,3652	68,83	0,6828	0.3644	69.01	0,6819	0.3635	68.,20	0,6811
215	0,3696	70.47	0,6906	0,3687	70,65	0,6897	0,3678	70.82	0.6888
220	0,3742	72,13	0,6982	0,3732	72,30	0,6972	0,3723	72,47	0,6963
225	0,3791	73,80	0.7056	0.3780	73.96	0,7046	0,3769	74,12	0.7037
230	0.3841	75.48	0,7130	0,3829	75.63	0,7120	0,3817	75,79	0,7110
235	0,3893	77,18	0,7204	0,3880	77.32	0,7193	0.3867	77.46	0,7183
240	0,3948	78,90	0,7277	0,3934	79.05	0.7267	0,3920	79,18	0.7256
245	0,4006	80.69	0,7350	0,3990	80,80	0,7339	0.3975	80,90	0,7328
250	0,4067	82,50	0,7423	0,4049	82.59	0.7411	0,4032	82,69	0,7400
255	0,4131	84,34	0.7496	0,4112	84,40	0,7483	0,4093	84,49	0,7472
260	0.4199	86.20	0,7569	0,4177	86,26	0,7555	0,4156	86,36	0,7543
265	0.4271	88,10	0,7642	0.4246	88,15	0,7627	0,4223	88.21	0,7613
270	0,4347	90,05	0,7715	0.4320	90,10	0,7698	0,4294	90.10	0,7684
275	0.4428	92,04	0,7787	0,4397	92.05	0,7770	0,4369	92,02	0,7754
280	0,4514	94.05	0,7860	0,4479	94.04	0,7842	0,4448	94,06	0,7824
285	0,4605	96.07	0,7932	0.4566	96,04	0,7912	0,4531	95,90	0,7894
290	0,4702	98,10	0,8003	0.4658	98,04	0,7982	0,4619	97,85	0,7962
295	0,4810	100,1	0,8072	0,4758	100,1	0,8050	0,4714	99.85	0,8028
300	0.4930	102,2	0,8142	0,4869	102,0	0,8118	0,4818	101,8	0,8094
305	0.5062	104,4	0.8213	0,4992	104,0	0,8186	0.4931	103,8	0,8161
310	0.5203	106,5	0,8284	0,5123	106,1	0,8255	0.5053	105,8	0,8228

TABLE XXIV. Thermodynamic Properties of Liquid and Gaseous Xenon (*Continued*)

T, K	p = 170 bar			p = 180 bar			p = 190 bar		
	$v \cdot 10^3$	h	s	$v \cdot 10^3$	h	s	$v \cdot 10^3$	h	s
315	0.5359	108.7	0.8356	0.5266	108.3	0.8324	0.5184	107.9	0.8296
320	0.5530	111.1	0.8429	0.5421	110.5	0.8394	0.5326	110.1	0.8363
325	0.5719	113.5	0.8503	0.5590	112.8	0.8465	0.5480	112.3	0.8432
330	0.5926	116.0	0.8578	0.5774	115.2	0.8537	0.5645	114.6	0.8501
335	0.6153	118.5	0.8654	0.5974	117.6	0.8610	0.5825	116.9	0.8570
340	0.6401	121.1	0.8731	0.6191	120.1	0.8683	0.6018	119.2	0.8640
345	0.6670	123.7	0.8807	0.6425	122.6	0.8755	0.6225	121.6	0.8710
350	0.6960	126.3	0.8883	0.6677	125.1	0.8828	0.6447	124.0	0.8779
360	0.7590	131.6	0.9031	0.7228	130.1	0.8969	0.6933	128.8	0.8915
370	0.8278	136.6	0.9170	0.7831	135.0	0.9104	0.7466	133.6	0.9045
380	0.8988	141.4	0.9297	0.8466	139.7	0.9229	0.8023	138.2	0.9166
390	0.9701	145.9	0.9413	0.9115	144.1	0.9345	0.8623	142.6	0.9280
400	1.040	150.0	0.9517	0.9763	148.3	0.9450	0.9220	146.8	0.9388
420	1.175	157.3	0.9696	1.103	155.8	0.9633	1.040	154.3	0.9573
440	1.302	163.8	0.9846	1.223	162.4	0.9786	1.153	161.1	0.9730
460	1.422	169.6	0.9976	1.336	168.4	0.9919	1.262	167.2	0.9865
480	1.536	175.0	1.0090	1.445	173.9	1.004	1.365	172.8	0.9984
500	1.645	180.0	1.0193	1.549	179.0	1.014	1.464	178.0	1.009
550	1.902	191.6	1.0414	1.794	190.8	1.037	1.699	190.0	1.032
600	2.143	202.2	1.0599	2.024	201.6	1.055	1.918	200.9	1.051
650	2.374	212.3	1.0760	2.244	211.7	1.072	2.128	211.2	1.067
700	2.597	221.9	1.0903	2.455	221.5	1.086	2.330	221.0	1.082
750	2.814	231.3	1.1033	2.662	230.9	1.099	2.526	230.5	1.095
800	3.026	240.4	1.115	2.863	240.1	1.111	2.718	239.8	1.107
850	3.235	249.4	1.126	3.061	249.1	1.122	2.906	248.8	1.118

TABLE XXIV. Thermodynamic Properties of Liquid and Gaseous Xenon (*Continued*)

T, K	p = 170 bar			p = 180 bar			p = 190 bar		
	$v \cdot 10^3$	h	s	$v \cdot 10^3$	h	s	$v \cdot 10^3$	h	s
900	3,440	258,2	1,136	3,256	258,0	1,132	3,092	257,8	1,128
950	3,644	266,9	1,145	3,449	266,7	1,142	3,275	266,6	1,138
1000	3,845	275,6	1,154	3,640	275,4	1,151	3,456	275,2	1,147
1050	4,045	284,1	1,163	3,829	284,0	1,159	3,636	283,8	1,155
1100	4,243	292,6	1,171	4,016	292,5	1,167	3,814	292,4	1,163
1150	4,440	301,0	1,178	4,203	300,9	1,174	3,991	300,8	1,171
1200	4,635	309,3	1,185	4,388	309,3	1,181	4,167	309,2	1,178
1250	4,830	317,7	1,192	4,573	317,6	1,188	4,342	317,6	1,185
1300	5,024	326,0	1,198	4,756	325,9	1,195	4,516	325,9	1,191

TABLE XXIV. Thermodynamic Properties of Liquid and Gaseous Xenon (*Continued*)

T, K	p = 200 bar			p = 220 bar			p = 240 bar		
	$v \cdot 10^3$	h	s	$v \cdot 10^3$	h	s	$v \cdot 10^3$	h	s
170	0,3335	56,30	0,6110	0,3326	56,75	0,6097	—	—	—
175	0,3367	57,96	0,6206	0,3358	58,40	0,6193	0,3349	58,84	0,6180
180	0,3401	59,61	0,6299	0,3391	60,05	0,6286	0,3381	60,48	0,6272
185	0 3436	61,25	0,6390	0,3425	61,68	0,6376	0,3414	62,11	0,6362
190	0,3472	62,88	0,6477	0,3460	63,30	0,6463	0,3449	63,72	0,6449
195	0,3509	64,51	0,6562	0,3496	64,92	0,6547	0,3484	65,34	0,6533
200	0,3547	66,14	0,6643	0,3533	66,54	0,6628	0,3520	66,95	0,6613
205	0,3586	67,76	0,6723	0,3572	68,15	0,6708	0,3558	68,55	0,6693
210	0,3627	69,39	0,6803	0,3611	69,77	0,6786	0,3596	70,15	0,6770

TABLE XXIV. Thermodynamic Properties of Liquid and Gaseous Xenon (*Continued*)

T, K	p = 200 bar			p = 220 bar			p = 240 bar		
	$v \cdot 10^3$	h	s	$v \cdot 10^3$	h	s	$v \cdot 10^3$	h	s
215	0.3669	71.01	0.6879	0.3652	71.37	0.6862	0.3636	71.75	0.6846
220	0.3713	72.65	0.6954	0.3695	73.00	0.6937	0.3677	73.37	0.6920
225	0.3759	74.29	0.7028	0.3739	74.63	0.7010	0.3720	74.98	0.6991
230	0.3806	75.94	0.7101	0.3784	76.27	0.7082	0.3764	76.60	0.7063
235	0.3855	77.61	0.7173	0.3832	77.92	0.7153	0.3809	78.23	0.7134
240	0.3906	79.30	0.7244	0.3881	79.60	0.7223	0.3857	79.90	0.7204
245	0.3960	81.00	0.7315	0.3932	81.30	0.7294	0.3906	81.55	0.7274
250	0.4016	82.78	0.7386	0.3985	83.05	0.7364	0.3957	83.25	0.7343
255	0.4075	84.60	0.7458	0.4041	84.80	0.7434	0.4010	85.00	0.7411
260	0.4137	86.42	0.7529	0.4099	86.55	0.7503	0.4065	86.76	0.7479
265	0.4201	88.26	0.7600	0.4160	88.37	0.7572	0.4123	88.53	0.7547
270	0.4269	90.12	0.7670	0.4224	90.20	0.7641	0.4183	90.31	0.7614
275	0.4341	92.00	0.7740	0.4291	92.05	0.7709	0.4246	92.10	0.7681
280	0.4417	93.90	0.7808	0.4362	93.90	0.7776	0.4311	93.90	0.7748
285	0.4497	95.82	0.7876	0.1436	95.75	0.7842	0.4380	95.73	0.7812
290	0.4581	97.74	0.7943	0.4514	97.60	0.7908	0.4453	97.57	0.7876
295	0.4671	99.67	0.8008	0.4597	99.50	0.7972	0.4530	99.42	0.7938
300	0.4769	101.6	0.8073	0.4685	101.4	0.8034	0.4611	101.3	0.8000
305	0.4876	103.6	0.8138	0.4779	103.3	0.8096	0.4697	103.1	0.8061
310	0.4990	105.6	0.8204	0.4880	105.2	0.8159	0.4787	105.0	0.8121
315	0.5112	107.6	0.8270	0.4988	107.1	0.8222	0.4883	106.9	0.8182
320	0.5243	109.7	0.8335	0.5102	109.1	0.8285	0.4985	108.8	0.8242
325	0.5384	111.8	0.8401	0.5223	111.1	0.8347	0.5093	110.7	0.8301
330	0.5535	114.0	0.8468	0.5352	113.2	0.8410	0.5206	112.6	0.8361
335	0.5697	116.3	0.8535	0.5490	115.3	0.8473	0.5326	114.6	0.8421

TABLE XXIV. Thermodynamic Properties of Liquid and Gaseous Xenon (*Continued*)

T, K	p = 200 bar			p = 220 bar			p = 240 bar		
	$v \cdot 10^3$	h	s	$v \cdot 10^3$	h	s	$v \cdot 10^3$	h	s
340	0,5871	118.5	0,8602	0.5636	117.4	0,8536	0,5452	116.7	0,8481
345	0.6058	120,8	0.8669	0,5791	119,6	0,8599	0,5585	118.7	0,8540
350	0.6256	123.1	0.8735	0.5955	121.8	0,8661	0,5725	120.7	0.8599
360	0,6689	127,8	0.8866	0.6310	126,2	0,8784	0,6026	124.9	0.8715
370	0,7166	132.4	0,8993	0.6700	130,5	0,8903	0.6355	129,0	0.8829
380	0,7677	136,9	0.9113	0,7120	134.7	0,9017	0,6709	133.1	0,8938
390	0,8212	141.2	0,9226	0,7564	138,9	0,9126	0.7085	137.1	0,9043
400	0,8759	145.4	0,9330	0,8026	143.0	0,9229	0,7478	141.1	0,9142
420	0.9856	153.0	0,9517	0,8974	150,6	0,9415	0,8299	148,6	0,9326
440	1,093	159.8	0,9676	0.9921	157,5	0.9576	0.9137	155,5	0,9488
460	1,195	166.0	0,9813	1,085	163,9	0,9717	0,9970	162,0	0,9631
480	1,294	171,7	0,9935	1,175	169,8	0,9837	1,079	168,0	0,9758
500	1,389	177,1	1,004	1.261	175,4	0,9944	1.158	173.6	0,9872
550	1,613	189,2	1,028	1,467	187,7	1,019	1.348	186.3	1,012
600	1,823	200,3	1,047	1,661	199,0	1,039	1.527	197.9	1,032
650	2,024	210.6	1,063	1,845	209.6	1.056	1,698	208,6	1,050
700	2.217	220.6	1,078	2,022	219.7	1,071	1,860	218.9	1,065
750	2.404	230.1	1,091	2,194	229,4	1,084	2,020	228,7	1,078
800	2,587	239,5	1,103	2,362	238,8	1,096	2,176	238,3	1,090
850	2,767	248,6	1,114	2,527	248,1	1,108	2,328	247,6	1,101
900	2,944	257,5	1,125	2,689	257,1	1,118	2,477	256,7	1,112
950	3,119	266,4	1,134	2,849	266,0	1,128	2,625	265,7	1,122
1000	3,291	275,1	1,143	3,007	274,8	1,137	2,770	274,6	1,131
1050	3,462	283,7	1,152	3,163	283,5	1,145	2,914	283,3	1,139
1100	3,632	292,2	1,160	3,318	292,1	1,153	3,057	292,0	1,147

TABLE XXIV. Thermodynamic Properties of Liquid and Gaseous Xenon (*Continued*)

T, K	p = 200 bar			p = 220 bar			p = 240 bar		
	$v \cdot 10^3$	h	s	$v \cdot 10^3$	h	s	$v \cdot 10^3$	h	s
1150	3.801	300.6	1.167	3.472	300.7	1.161	3.199	300.6	1.155
1200	3.968	309.2	1.174	3.625	309.1	1.168	3.339	309.1	1.162
1250	4.135	317.6	1.181	3.777	317.6	1.175	3.479	317.5	1.169
1300	4.301	325.9	1.188	3.928	325.9	1.181	3.618	325.9	1.176

TABLE XXIV. Thermodynamic Properties of Liquid and Gaseous Xenon (*Continued*)

T, K	p = 260 bar			p = 280 bar			p = 300 bar		
	$v \cdot 10^3$	h	s	$v \cdot 10^3$	h	s	$v \cdot 10^3$	h	s
175	0.3340	59.29	0.6167	0.3331	59.74	0.6155	0.3322	60.19	0.6143
180	0.3371	60.92	0.6259	0.3362	61.37	0.6247	0.3353	61.81	0.6234
185	0.3404	62.54	0.6349	0.3394	62.98	0.6336	0.3384	63.42	0.6323
190	0.3438	64.15	0.6435	0.3427	64.58	0.6421	0.3417	65.02	0.6408
195	0.3472	65.76	0.6519	0.3461	66.18	0.6505	0.3450	66.61	0.6491
200	0.3508	67.36	0.6599	0.3496	67.77	0.6584	0.3484	68.19	0.6570
205	0.3544	68.95	0.6678	0.3531	69.36	0.6663	0.3519	69.77	0.6649
210	0.3582	70.55	0.6755	0.3568	70.95	0.6740	0.3554	71.35	0.6725
215	0.3620	72.13	0.6830	0.3605	72.52	0.6814	0.3591	72.92	0.6799
220	0.3660	73.74	0.6903	0.3644	74.12	0.6887	0.3629	74.50	0.6872
225	0.3702	75.34	0.6974	0.3684	75.71	0.6958	0.3668	76.08	0.6942
230	0.3744	76.95	0.7046	0.3725	77.30	0.7029	0.3708	77.67	0.7012
235	0.3788	78.56	0.7116	0.3768	78.91	0.7098	0.3749	79.26	0.7081

TABLE XXIV. Thermodynamic Properties of Liquid and Gaseous Xenon (*Continued*)

T, K	p = 260 bar			p = 280 bar			p = 300 bar		
	$v \cdot 10^3$	h	s	$v \cdot 10^3$	h	s	$v \cdot 10^3$	h	s
240	0.3834	80.20	0.7186	0.3812	80,50	0,7167	0.3791	80,88	0.7150
245	0,3881	81.85	0.7256	0,3857	82.15	0,7236	0.3835	82.51	0.7218
250	0,3930	83,55	0,7324	0,3904	83,85	0.7303	0,3881	84.18	0,7284
255	0,3980	85,25	0,7392	0,3953	85,55	0,7370	0,3927	85.86	0,7350
260	0,4033	87,00	0.7460	0.4003	87,25	0,7436	0,3976	87,56	0,7416
265	0,4088	88,77	0.7528	0,4056	89,00	0,7502	0,4026	89,28	0,7482
270	0,4145	90,55	0,7594	0,4110	90,75	0,7568	0,4077	91,00	0,7546
275	0,4204	92,33	0,7659	0,4166	92,50	0,7631	0.4131	92,72	0,7609
280	0,4266	94,15	0,7723	0,4224	94,25	0,7694	0,4186	94,44	0,7670
285	0,4330	95,95	0,7786	0,4285	96,00	0,7756	0,4243	96,17	0,7731
290	0,4397	97,75	0,7848	0,4348	97,75	0,7818	0,4303	97,89	0,7792
295	0,4468	99,60	0,7909	0,4414	99,50	0,7878	0,4365	99,62	0,7850
300	0,4543	101,4	0,7968	0,4484	101,2	0,7938	0,4431	101,3	0,7908
305	0,4623	103,2	0,8027	0,4558	103,0	0,7995	0.4500	103,0	0,7965
310	0,4707	105,0	0,8085	0,4636	104.7	0,8052	0,4572	104,7	0,8021
315	0,4795	106,8	0,8144	0,4717	106,5	0,8108	0,4647	106.4	0.8076
320	0,4887	108,6	0,8201	0,4801	108,3	0.8164	0,4725	108,1	0,8131
325	0,4983	110,4	0,8259	0,4889	110,1	0,8221	0,4806	109,9	0,8186
330	0,5084	112,2	0,8317	0,4981	111,9	0,8277	0,4891	111,7	0,8241
335	0,5191	114,1	0,8375	0,5077	113,8	0,8333	0.4978	113,5	0,8296
340	0,5302	116,1	0,8432	0,5177	115.7	0,8389	0,5069	115,3	0,8350
345	0.5419	118,0	0,8489	0,5282	117,5	0,8444	0,5164	117,1	0,8404
350	0,5542	120,0	0,8546	0,5391	119,4	0,8498	0,5263	119,0	0,8456
360	0.5804	123,9	0,8657	0,5623	123,2	0,8606	0,5472	122,7	0.8560
370	0,6087	127,9	0,8765	0,5873	127,0	0,8710	0,5695	126,4	0,8661

TABLE XXIV. Thermodynamic Properties of Liquid and Gaseous Xenon (*Continued*)

T, K	p = 260 bar			p = 280 bar			p = 300 bar		
	$v \cdot 10^3$	h	s	$v \cdot 10^3$	h	s	$v \cdot 10^3$	h	s
380	0.6392	131.8	0.8870	0.6140	130.8	0.8811	0.5934	130.1	0.8759
390	0.6716	135.7	0.8971	0.6424	134.6	0.8909	0.6186	133.7	0.8854
400	0.7056	139.5	0.9068	0.6723	138.3	0.9003	0.6452	137.4	0.8945
420	0.7772	146.9	0.9247	0.7354	145.5	0.9178	0.7014	144.4	0.9117
440	0.8515	153.8	0.9408	0.8015	152.3	0.9337	0.7607	151.1	0.9274
460	0.9264	160.3	0.9552	0.8690	158.8	0.9481	0.8217	157.5	0.9416
480	1.001	166.3	0.9681	0.9366	164.9	0.9610	0.8834	163.6	0.9546
500	1.074	172.0	0.9797	1.004	170.6	0.9728	0.9449	169.4	0.9664
550	1.249	185.0	1.005	1.166	183.8	0.9979	1.095	182.7	0.9918
600	1.415	196.8	1.025	1.321	195.8	1.019	1.240	194.8	1.013
650	1.574	207.7	1.043	1.469	206.9	1.037	1.379	206.0	1.031
700	1.726	218.1	1.058	1.611	217.4	1.052	1.513	216.7	1.047
750	1.874	228.1	1.072	1.750	227.5	1.066	1.642	226.9	1.061
800	2.019	237.7	1.084	1.884	237.2	1.079	1.769	236.7	1.073
850	2.160	247.1	1.096	2.016	246.7	1.090	1.893	246.3	1.085
900	2.299	256.3	1.106	2.146	256.0	1.101	2.014	255.7	1.096
950	2.435	265.4	1.116	2.274	265.1	1.111	2.134	264.9	1.106
1000	2.570	274.3	1.125	2.400	274.1	1.120	2.252	273.9	1.115
1050	2.704	283.1	1.134	2.524	282.9	1.129	2.369	282.8	1.124
1100	2.836	291.8	1.142	2.647	291.7	1.137	2.484	291.6	1.132
1150	2.967	300.5	1.150	2.770	300.4	1.145	2.599	300.3	1.140
1200	3.098	309.0	1.157	2.891	309.0	1.152	2.712	309.0	1.147
1250	3.227	317.5	1.164	3.012	317.5	1.159	2.825	317.5	1.154
1300	3.356	326.0	1.170	3.132	326.0	1.165	2.937	326.0	1.161

TABLE XXIV. Thermodynamic Properties of Liquid and Gaseous Xenon (*Continued*)

T, K	p = 320 bar			p = 340 bar			p = 360 bar		
	$v \cdot 10^3$	h	s	$v \cdot 10^3$	h	s	$v \cdot 10^3$	h	s
175	0,3314	60.64	0,6131	0,3306	61,10	0.6119	—	—	—
180	0,3344	62,26	0,6222	0,3335	62,71	0,6210	0.3327	63.16	0,6198
185	0,3375	63,86	0,6310	0,3366	64.31	0,6298	0,3357	64.76	0.6286
190	0,3407	65,45	0,6395	0,3397	65,89	0,6383	0,3387	66.33	0,6370
195	0,3439	67,04	0,6478	0,3429	67,47	0,6465	0,3419	67,91	0,6452
200	0,3472	68,62	0,6557	0,3461	69,04	0,6544	0,3451	69.47	0,6530
205	0,3506	70,19	0,6635	0,3495	70,61	0,6622	0,3483	71,04	0,6608
210	0,3541	71,76	0,6711	0,3529	72,17	0,6697	0,3517	72,59	0,6683
215	0,3577	73,32	0,6785	0,3564	73,72	0,6770	0,3551	74.13	0,6756
220	0,3614	74,89	0,6856	0,3600	75,29	0,6841	0,3586	75,69	0,6827
225	0,3652	76,46	0,6926	0,3637	76,85	0,6911	0,3622	77,24	0,6897
230	0,3691	78,04	0,6996	0,3674	78,42	0,6981	0,3659	78,80	0,6967
235	0,3731	79,62	0,7065	0,3713	79,99	0,7050	0,3696	80,36	0,7036
240	0,3772	81,22	0,7134	0,3753	81,58	0,7118	0,3735	81,95	0,7104
245	0,3814	82,85	0,7202	0,3794	83,19	0,7185	0,3775	83,55	0,7171
250	0,3858	84,51	0,7268	0,3836	84,84	0,7251	0,3816	85,19	0,7237
255	0,3903	86,19	0,7334	0,3880	86,50	0,7316	0,3858	86,84	0,7301
260	0,3949	87,97	0,7398	0,3925	88,16	0,7380	0,3901	88,49	0,7364
265	0,3997	89,66	0,7462	0,3971	89,83	0,7443	0,3945	90,15	0,7426
270	0,4047	91,36	0,7526	0,4018	91,51	0,7505	0,3991	91,81	0,7487
275	0,4098	93,05	0,7587	0,4067	93,18	0,7566	0,4038	93,46	0,7547
280	0,4150	94,74	0,7648	0,4117	94,85	0,7626	0,4086	95,11	0,7606
285	0,4204	96,43	0,7708	0,4169	96,52	0,7685	0,4136	96,76	0,7664

TABLE XXIV. Thermodynamic Properties of Liquid and Gaseous Xenon (*Continued*)

T, K	p = 320 bar			p = 340 bar			p = 360 bar		
	$v \cdot 10^3$	h	s	$v \cdot 10^3$	h	s	$v \cdot 10^3$	h	s
290	0.4260	98.12	0.7768	0.4223	98.19	0.7743	0.4188	98.40	0.7721
295	0.4319	99.80	0.7826	0.4279	99.86	0.7800	0.4241	100.1	0.7777
300	0.4382	101.5	0.7882	0.4337	101.5	0.7856	0.4296	101.7	0.7832
305	0.4447	103.1	0.7938	0.4398	103.1	0.7911	0.4354	103.3	0.7886
310	0.4515	104.8	0.7993	0.4462	104.8	0.7965	0.4414	104.9	0.7939
315	0.4585	106.5	0.8047	0.4528	106.4	0.8019	0.4476	106.5	0.7992
320	0.4658	108.2	0.8101	0.4596	108.1	0.8072	0.4541	108.2	0.8045
325	0.4733	109.9	0.8155	0.4667	109.8	0.8125	0.4607	109.8	0.8097
330	0.4811	111.6	0.8208	0.4740	111.5	0.8177	0.4675	111.5	0.8149
335	0.4892	113.4	0.8261	0.4815	113.3	0.8229	0.4746	113.2	0.8200
340	0.4976	115.1	0.8314	0.4893	115.0	0.8281	0.4819	114.9	0.8251
345	0.5063	116.9	0.8366	0.4973	116.8	0.8332	0.4894	116.7	0.8301
350	0.5153	118.7	0.8418	0.5056	118.5	0.8383	0.4971	118.4	0.8350
360	0.5343	122.3	0.8519	0.5231	122.0	0.8481	0.5132	121.8	0.8447
370	0.5545	125.9	0.8617	0.5416	125.5	0.8577	0.5303	125.2	0.8541
380	0.5757	129.5	0.8712	0.5613	129.0	0.8670	0.5484	128.6	0.8632
390	0.5980	133.0	0.8805	0.5820	132.5	0.8760	0.5674	132.0	0.8720
400	0.6227	136.5	0.8894	0.6036	135.9	0.8847	0.5873	135.4	0.8805
420	0.6733	143.4	0.9062	0.6496	142.7	0.9012	0.6293	142.0	0.8967
440	0.7269	150.0	0.9216	0.6984	149.2	0.9164	0.6740	148.5	0.9116
460	0.7823	156.4	0.9358	0.7490	155.5	0.9304	0.7206	154.7	0.9255
480	0.8387	162.5	0.9487	0.8008	161.5	0.9432	0.7684	160.7	0.9382
500	0.8953	168.3	0.9605	0.8531	167.3	0.9551	0.8168	166.5	0.9500
550	1.035	181.7	0.9861	0.9833	180.8	0.9808	0.9382	180.0	0.9758
600	1.171	193.9	1.007	1.110	193.1	1.002	1.057	192.4	0.9974

TABLE XXIV. Thermodynamic Properties of Liquid and Gaseous Xenon (*Continued*)

T, K	p = 320 bar			p = 340 bar			p = 360 bar		
	$v \cdot 10^3$	h	s	$v \cdot 10^3$	h	s	$v \cdot 10^3$	h	s
650	1,301	205.3	1,025	1,238	204.6	1.021	1,173	203,9	1,016
700	1,427	216.0	1,041	1,352	215.4	1,037	1,286	214,9	1,032
750	1,549	226.3	1,056	1.467	225,8	1,051	1,395	225,4	1,047
800	1,669	236.3	1,069	1,580	235,9	1,064	1,502	235,5	1,060
850	1,785	245.9	1,080	1,690	245,6	1,076	1,606	245,3	1,072
900	1,899	255.4	1,091	1,7980	255,1	1,087	1,709	254,8	1,083
950	2,012	264.6	1,101	1,904	264,4	1,097	1,809	264,2	1,093
1000	2,123	273,7	1,110	2,009	273,6	1,106	1,909	273,4	1,102
1050	2,233	282,7	1,119	2,113	282,6	1,115	2,007	282,5	1,111
1100	2,341	291,5	1,127	2,216	291,5	1,123	2,104	291,4	1,119
1150	2,449	300,3	1,135	2,317	300,3	1,131	2,201	300,2	1,127
1200	2,556	309,0	1,143	2,418	309,0	1,139	2,296	309,0	1,135
1250	2,662	317,6	1,150	2,519	317,6	1,146	2,391	317,7	1,142
1300	2,768	326,1	1,156	2,618	326,2	1,152	2,485	326,3	1,148

TABLE XXIV. Thermodynamic Properties of Liquid and Gaseous Xenon (*Continued*)

T, K	p = 380 bar			p = 400 bar			p = 420 bar		
	$v \cdot 10^3$	h	s	$v \cdot 10^3$	h	s	$v \cdot 10^3$	h	s
180	0.3319	63,62	0,6186	0,3311	64,07	0,6174	0,3303	64,53	0,6163
185	0,3348	65,20	0,6274	0,3340	65,66	0,6262	0,3331	66,11	0,6250

TABLE XXIV. Thermodynamic Properties of Liquid and Gaseous Xenon (*Continued*)

T, K	p = 380 bar			p = 400 bar			p = 420 bar		
	$v \cdot 10^3$	h	s	$v \cdot 10^3$	h	s	$v \cdot 10^3$	h	s
190	0.3378	66.78	0.6358	0.3369	67.22	0.6346	0.3360	67.67	0.6334
195	0.3409	68.35	0.6440	0.3399	68.79	0.6427	0.3390	69.23	0.6415
200	0.3440	69.93	0.6518	0.3430	70.34	0.6505	0.3420	70.78	0.6493
205	0.3472	71.46	0.6595	0.3461	71.89	0.6582	0.3451	72.33	0.6570
210	0.3505	73.01	0.6670	0.3493	73.43	0.6657	0.3482	73.86	0.6644
215	0.3538	74.55	0.6743	0.3526	74.97	0.6729	0.3514	75.39	0.6716
220	0.3573	76.10	0.6813	0.3560	76.50	0.6800	0.3547	76.93	0.6786
225	0.3608	77.64	0.6883	0.3594	78.04	0.6869	0.3581	78.46	0.6854
230	0.3644	79.19	0.6952	0.3629	79.57	0.6937	0.3615	79.99	0.6922
235	0.3680	80.75	0.7020	0.3665	81.12	0.7005	0.3650	81.53	0.6989
240	0.3718	82.32	0.7087	0.3702	82.68	0.7072	0.3686	83.08	0.7056
245	0.3757	83.91	0.7152	0.3739	84.26	0.7138	0.3722	84.65	0.7122
250	0.3796	85.53	0.7216	0.3778	85.86	0.7203	0.3760	86.24	0.7187
255	0.3837	87.16	0.7279	0.3817	87.48	0.7267	0.3798	87.85	0.7250
260	0.3879	88.79	0.7341	0.3858	89.10	0.7329	0.3837	89.46	0.7312
265	0.3922	90.43	0.7403	0.3899	90.73	0.7390	0.3878	91.08	0.7372
270	0.3966	92.07	0.7464	0.3941	92.36	0.7450	0.3919	92.69	0.7431
275	0.4011	93.70	0.7524	0.3985	93.97	0.7509	0.3961	94.29	0.7489
280	0.4057	95.33	0.7583	0.4030	95.59	0.7567	0.4004	95.89	0.7546
285	0.4105	96.95	0.7641	0.4076	97.20	0.7624	0.4048	97.48	0.7602
290	0.4155	98.57	0.7698	0.4122	98.80	0.7680	0.4093	99.07	0.7657
295	0.4206	100.19	0.7754	0.4170	100.4	0.7735	0.4139	100.6	0.7712
300	0.4259	101.8	0.7809	0.4220	102.0	0.7789	0.4187	102.2	0.7766
305	0.4313	103.4	0.7863	0.4272	103.6	0.7842	0.4237	103.8	0.7819
310	0.4369	105.0	0.7916	0.4326	105.2	0.7893	0.4288	105.4	0.7870

TABLE XXIV. Thermodynamic Properties of Liquid and Gaseous Xenon (*Continued*)

T, K	p = 380 bar			p = 400 bar			p = 420 bar		
	$v \cdot 10^3$	h	s	$v \cdot 10^3$	h	s	$v \cdot 10^3$	h	s
315	0.4428	106.6	0.7968	0.4383	106.8	0.7944	0.4342	107.0	0.7921
320	0.4489	108.3	0.8020	0.4442	108.4	0.7995	0.4398	108.6	0.7972
325	0.4552	109.9	0.8071	0.4502	110.0	0.8046	0.4455	110.2	0.8022
330	0.4617	111.6	0.8122	0.4563	111.6	0.8096	0.4513	111.8	0.8072
335	0.4683	113.2	0.8172	0.4626	113.3	0.8146	0.4573	113.4	0.8121
340	0.4751	114.9	0.8222	0.4690	115.0	0.8195	0.4634	115.0	0.8170
345	0.4822	116.6	0.8271	0.4756	116.6	0.8243	0.4697	116.7	0.8217
350	0.4894	118.3	0.8320	0.4824	118.3	0.8291	0.4761	118.3	0.8264
360	0.5045	121.7	0.8415	0.4966	121.6	0.8385	0.4894	121.6	0.8356
370	0.5204	125.0	0.8507	0.5114	124.9	0.8475	0.5034	124.8	0.8446
380	0.5371	128.4	0.8596	0.5270	128.2	0.8563	0.5180	128.1	0.8532
390	0.5546	131.7	0.8682	0.5434	131.5	0.8648	0.5333	131.3	0.8616
400	0.5730	135.0	0.8766	0.5604	134.7	0.8730	0.5492	134.5	0.8696
420	0.6118	141.5	0.8925	0.5964	141.1	0.8886	0.5827	140.8	0.8850
440	0.6530	147.9	0.9072	0.6346	147.4	0.9032	0.6184	146.9	0.8994
460	0.6960	154.0	0.9209	0.6745	153.4	0.9167	0.6556	152.9	0.9128
480	0.7403	160.0	0.9336	0.7157	159.4	0.9293	0.6941	158.8	0.9253
500	0.7853	165.7	0.9453	0.7577	165.1	0.9409	0.7334	164.5	0.9369
550	0.8988	179.2	0.9711	0.8641	178.6	0.9667	0.8334	178.0	0.9626
600	1.011	191.7	0.9928	0.9699	191.1	0.9885	0.9334	190.5	0.9844
650	1.122	203.4	1.011	1.074	202.8	1.007	1.032	202.3	1.003
700	1.228	214.4	1.027	1.175	213.8	1.023	1.128	213.5	1.020
750	1.331	224.9	1.042	1.273	224.5	1.038	1.222	224.2	1.035
800	1.432	235.1	1.056	1.370	234.8	1.052	1.314	234.5	1.048
850	1.531	245.0	1.068	1.464	244.7	1.064	1.404	244.5	1.060

TABLE XXIV. Thermodynamic Properties of Liquid and Gaseous Xenon (*Continued*)

T, K	p = 380 bar			p = 400 bar			p = 420 bar		
	$v \cdot 10^3$	h	s	$v \cdot 10^3$	h	s	$v \cdot 10^3$	h	s
900	1.629	254.6	1.079	1.557	254.4	1.075	1.493	254.2	1.071
950	1.725	264.0	1.089	1.648	263.9	1.085	1.580	263.7	1.082
1000	1.819	273.3	1.098	1.738	273.2	1.095	1.666	273.1	1.091
1050	1.912	282.4	1.107	1.827	282.3	1.104	1.751	282.3	1.100
1100	2.005	291.4	1.115	1.915	291.3	1.112	1.835	291.3	1.108
1150	2.096	300.2	1.123	2.002	300.2	1.120	1.918	300.3	1.116
1200	2.187	309.0	1.131	2.089	309.1	1.127	2.000	309.1	1.124
1250	2.277	317.7	1.138	2.175	317.8	1.134	2.082	317.9	1.131
1300	2.366	326.3	1.145	2.260	326.4	1.141	2.163	326.6	1.138

TABLE XXIV. Thermodynamic Properties of Liquid and Gaseous Xenon (*Continued*)

T, K	p = 440 bar			p = 460 bar			p = 480 bar		
	$v \cdot 10^3$	h	s	$v \cdot 10^3$	h	s	$v \cdot 10^3$	h	s
180	0.3295	64.99	0.6152	0.3288	65.45	0.6141	0.3280	65.91	0.6130
185	0.3323	66.56	0.6239	0.3315	67.02	0.6228	0.3307	67.47	0.6217
190	0.3352	68.12	0.6322	0.3343	68.57	0.6311	0.3335	69.03	0.6300
195	0.3381	69.68	0.6403	0.3372	70.13	0.6392	0.3363	70.58	0.6380
200	0.3410	71.22	0.6481	0.3401	71.67	0.6469	0.3392	72.11	0.6457
205	0.3441	72.76	0.6557	0.3432	73.20	0.6545	0.3421	73.64	0.6533
210	0.3472	74.29	0.6631	0.3461	74.73	0.6619	0.3451	75.16	0.6607

TABLE XXIV. Thermodynamic Properties of Liquid and Gaseous Xenon (*Continued*)

T, K	p = 440 bar			p = 460 bar			p = 480 bar		
	$v \cdot 10^3$	h	s	$v \cdot 10^3$	h	s	$v \cdot 10^3$	h	s
215	0,3503	75,81	0,6703	0,3492	76,24	0,6691	0,3481	76,67	0,6679
220	0,3535	77,35	0,6773	0,3524	77,77	0,6760	0,3512	78,20	0,6748
225	0,3568	78,87	0,6841	0,3556	79,29	0,6828	0,3544	79,71	0,6815
230	0,3602	80,40	0,6909	0,3688	80,81	0,6895	0,3576	81,23	0,6882
235	0,3636	81,93	0,6975	0,3622	82,33	0,6961	0,3608	82,74	0,6948
240	0,3671	83,48	0,7041	0,3656	83,87	0,7026	0,3642	84,28	0,7013
245	0,3706	85,04	0,7107	0,3691	85,42	0,7091	0,3676	85,83	0,7077
250	0,3743	86,62	0,7171	0,3726	86,99	0,7154	0,3710	87,39	0,7140
255	0,3780	88,22	0,7233	0,3763	88,57	0,7216	0,3746	88,96	0,7202
260	0,3818	89,82	0,7294	0,3800	90,15	0,7277	0,3782	90,54	0,7263
265	0,3857	91,42	0,7354	0,3837	91,74	0,7337	0,3819	92,12	0,7323
270	0,3897	93,02	0,7413	0,3876	93,33	0,7396	0,3856	93,70	0,7381
275	0,3937	94,61	0,7471	0,3916	94,91	0,7454	0,3894	95,27	0,7439
280	0,3979	96,19	0,7528	0,3956	96,48	0,7511	0,3933	96,83	0,7495
285	0,4022	97,76	0,7584	0,3997	98,04	0,7567	0,3973	98,38	0,7550
290	0,4065	99,33	0,7639	0,4039	99,60	0,7622	0,4014	99,93	0,7604
295	0,4109	100,9	0,7693	0,4082	101,2	0,7676	0,4055	101,5	0,7657
300	0,4155	102,5	0,7746	0,4126	102,7	0,7729	0,4097	103,0	0,7709
305	0,4202	104,0	0,7798	0,4171	104,3	0,7780	0,4140	104,5	0,7760
310	0,4251	105,6	0,7849	0,4218	105,8	0,7830	0,4185	106,0	0,7810
315	0,4303	107,1	0,7900	0,4267	107,4	0,7880	0,4232	107,5	0,7860
320	0,4357	108,7	0,7950	0,4318	109,0	0,7929	0,4281	109,1	0,7909
325	0,4411	110,3	0,8000	0,4370	110,5	0,7978	0,4332	110,7	0,7958
330	0,4467	111,9	0,8049	0,4424	112,1	0,8027	0,4383	112,2	0,8006
335	0,4524	113,5	0,8098	0,4478	113,7	0,8075	0,4436	113,8	0,8053

TABLE XXIV. Thermodynamic Properties of Liquid and Gaseous Xenon (*Continued*)

T, K	p = 440 bar			p = 460 bar			p = 480 bar		
	$v \cdot 10^3$	h	s	$v \cdot 10^3$	h	s	$v \cdot 10^3$	h	s
340	0.4582	115.1	0.8146	0.4534	115.3	0.8122	0.4489	115.4	0.8100
345	0.4642	116.7	0.8193	0.4591	116.9	0.8169	0.4543	117.0	0.8147
350	0.4703	118.4	0.8239	0.4649	118.4	0.8215	0.4597	118.6	0.8192
360	0.4829	121.6	0.8330	0.4769	121.6	0.8305	0.4713	121.7	0.8281
370	0.4961	124.8	0.8418	0.4894	124.8	0.8392	0.4833	124.8	0.8367
380	0.5099	128.0	0.8503	0.5024	128.0	0.8476	0.4957	127.9	0.8450
390	0.5242	131.2	0.8585	0.5160	131.1	0.8557	0.5085	131.0	0.8530
400	0.5391	134.3	0.8665	0.5300	134.2	0.8636	0.5217	134.1	0.8608
420	0.5706	140.5	0.8817	0.5596	140.3	0.8785	0.5497	140.2	0.8756
440	0.6039	146.6	0.8959	0.5909	146.4	0.8926	0.5792	146.2	0.8894
460	0.6388	152.6	0.9092	0.6237	152.3	0.9057	0.6102	152.0	0.9024
480	0.6749	158.4	0.9215	0.6577	158.0	0.9179	0.6422	157.7	0.9146
500	0.7118	164.0	0.9330	0.6925	163.6	0.9294	0.6751	163.3	0.9260
550	0.8060	177.5	0.9587	0.7815	177.0	0.9550	0.7594	176.7	0.9520
600	0.9006	190.1	0.9805	0.8712	189.6	0.9769	0.8447	189.2	0.9730
650	0.9941	201.9	0.9995	0.9602	201.5	0.9959	0.9295	201.1	0.9924
700	1.086	213.1	1.016	1.048	212.8	1.013	1.013	212.4	1.009
750	1.176	223.9	1.031	1.133	223.6	1.028	1.095	223.3	1.024
800	1.263	234.2	1.044	1.218	234.0	1.041	1.176	233.7	1.038
850	1.350	244.3	1.057	1.300	244.1	1.053	1.255	243.9	1.050
900	1.434	254.1	1.068	1.381	253.9	1.064	1.333	253.8	1.061
950	1.518	263.6	1.078	1.461	263.5	1.075	1.409	263.4	1.072
1000	1.600	273.0	1.088	1.540	272.9	1.085	1.485	272.9	1.081
1050	1.681	282.2	1.097	1.618	282.2	1.094	1.560	282.2	1.090
1100	1.761	291.3	1.105	1.695	291.3	1.102	1.634	291.4	1.099

TABLE XXIV. Thermodynamic Properties of Liquid and Gaseous Xenon (*Continued*)

T, K	p = 440 bar			p = 460 bar			p = 480 bar		
	$v \cdot 10^3$	h	s	$v \cdot 10^3$	h	s	$v \cdot 10^3$	h	s
1150	1.841	300.3	1.113	1.771	300.3	1,110	1,707	300,4	1.107
1200	1.920	309.2	1.121	1.846	309.3	1,118	1.779	309,3	1.115
1250	1.998	318.0	1.128	1.921	318.1	1,125	1.851	318.1	1.122
1300	2.076	326.7	1.135	1.996	326.8	1.132	1,923	326.9	1,129

TABLE XXIV. Thermodynamic Properties of Liquid and Gaseous Xenon (*Continued*)

T, K	p = 500 bar			p = 550 bar			p = 600 bar		
	$v \cdot 10^3$	h	s	$v \cdot 10^3$	h	s	$v \cdot 10^3$	h	s
185	0.3300	67.94	0,6206	0.3281	69.09	0.6179	0.3264	70,25	0.6153
190	0.3327	69.48	0,6289	0.3308	70.63	0,6262	0.3290	71.78	0.6236
195	0,3355	71.03	0,6369	0.3335	72.16	0.6342	0.3316	73.31	0.6315
200	0.3383	72.56	0.6446	0.3362	73.69	0,6418	0,3342	74.83	0.6391
205	0,3412	74,09	0.6522	0.3389	75,20	0,6493	0.3368	76.33	0.6466
210	0,3441	75.60	0,6595	0.3417	76.71	0,6566	0.3395	77.83	0.6538
215	0.3471	77.11	0,6666	0,3446	78,21	0.6637	0.3423	79.32	0.6609
220	0,3501	78,63	0,6735	0.3475	79.71	0,6705	0.3450	80.82	0.6677
225	0.3532	80.13	0.6802	0.3504	81,21	0.6772	0.3479	82.30	0,6743
230	0,3563	81,65	0.6869	0.3534	82,71	0,6838	0.3507	83.79	0,6809
235	0.3595	83,16	0.6935	0.3565	84.21	0,6903	0.3536	85.28	0.6873
240	0.3628	84.69	0,6999	0.3596	85.73	0.6967	0.3566	86.79	0,6936
245	0.3661	86,23	0.7063	0,3627	87.26	0.7030	0.3596	88.31	0.6998
250	0.3695	87.78	0.7126	0,3659	88.80	0.7092	0.3626	89.84	0.7059

TABLE XXIV. Thermodynamic Properties of Liquid and Gaseous Xenon (*Continued*)

T, K	p = 500 bar			p = 550 bar			p = 600 bar		
	$v \cdot 10^3$	h	s	$v \cdot 10^3$	h	s	$v \cdot 10^3$	h	s
255	0.3730	89.36	0.7188	0.3692	90.35	0.7153	0.3657	91.38	0.7119
260	0.3765	90.91	0.7248	0.3725	91.91	0.7213	0.3688	92.92	0.7178
265	0.3800	92.48	0.7307	0.3758	93.46	0.7271	0.3720	94.45	0.7236
270	0.3837	94.04	0.7365	0.3792	95.00	0.7329	0.3752	95.97	0.7293
275	0.3874	95.60	0.7422	0.3827	96.54	0.7385	0.3785	97.48	0.7348
280	0.3912	97.15	0.7478	0.3862	98.07	0.7440	0.3818	98.99	0.7402
285	0.3950	98.70	0.7533	0.3898	99.60	0.7494	0.3851	100.5	0.7455
290	0.3989	100.2	0.7587	0.3935	101.1	0.7547	0.3886	102.0	0.7507
295	0.4028	101.8	0.7640	0.3972	102.6	0.7599	0.3920	103.4	0.7558
300	0.4068	103.3	0.7691	0.4010	104.0	0.7650	0.3955	104.9	0.7608
305	0.4110	104.8	0.7741	0.4049	105.5	0.7700	0.3991	106.4	0.7657
310	0.4155	106.3	0.7791	0.4089	106.9	0.7749	0.4028	107.8	0.7705
315	0.4201	107.8	0.7840	0.4130	108.4	0.7797	0.4067	109.3	0.7753
320	0.4248	109.3	0.7889	0.4172	109.9	0.7844	0.4106	110.7	0.7800
325	0.4296	110.8	0.7937	0.4215	111.5	0.7891	0.4147	112.2	0.7846
330	0.4345	112.4	0.7985	0.4259	113.0	0.7938	0.4189	113.7	0.7892
335	0.4395	114.0	0.8033	0.4305	114.6	0.7984	0.4231	115.2	0.7938
340	0.4447	115.6	0.8079	0.4352	116.1	0.8030	0.4273	116.7	0.7984
345	0.4499	117.1	0.8125	0.4400	117.6	0.8075	0.4317	118.2	0.8028
350	0.4552	118.7	0.8170	0.4448	119.2	0.8119	0.4361	119.7	0.8072
360	0.4662	121.8	0.8258	0.4547	122.2	0.8205	0.4451	122.7	0.8156
370	0.4775	124.9	0.8343	0.4650	125.3	0.8288	0.4543	125.7	0.8239

TABLE XXIV. Thermodynamic Properties of Liquid and Gaseous Xenon (*Continued*)

T, K	p = 500 bar			p = 550 bar			p = 600 bar		
	$v \cdot 10^3$	h	s	$v \cdot 10^3$	h	s	$v \cdot 10^3$	h	s
380	0.4893	128,0	0,8425	0.4756	128,3	0,8369	0.4639	128,7	0,8318
390	0,5015	131,1	0,8505	0.4864	131,3	0,8446	0,4738	131,6	0,8394
400	0,5141	134,1	0,8581	0,4977	134,2	0,8521	0,4839	134,5	0,8467
420	0,5406	140,1	0,8728	0.5211	140,0	0,8664	0,5050	140,2	0,8607
440	0,5686	146,0	0,8865	0,5457	145,8	0,8798	0.5271	145,9	0,8738
460	0,5979	151,8	0.8994	0.5714	151,5	0,8924	0,5501	151,4	0,8861
480	0,6282	157,1	0.9114	0,5980	157,1	0,9044	0,5740	156,9	0,8978
500	0,6593	163,0	0.9228	0.6258	162,5	0,9154	0,5986	162,3	0,9088
550	0,7394	176,3	0,9482	0,6967	175.7	0,9405	0,6621	175,4	0.9337
600	0,8206	188,9	0,9700	0,7691	188,3	0,9624	0,7273	187,9	0,9555
650	0,9015	200,8	0,9891	0,8416	200,2	0,9815	0,7929	199,8	0,9746
700	0,9815	212,2	1,006	0,9136	211,6	0,9984	0,8583	211,3	0,9916
750	1,060	223,1	1,021	0,9847	222,6	1,014	0,9230	222,3	1,007
800	1,137	233.6	1,035	1,055	233,2	1,027	0,9870	233,0	1,021
850	1,213	243.8	1,047	1,124	243,5	1,040	1,050	243,3	1,033
900	1,288	253,7	1,058	1,192	253,5	1,051	1,112	253,4	1,045
950	1,362	263.4	1,069	1,259	263,3	1,062	1.174	263,3	1,055
1000	1,435	272.9	1,079	1,325	272,9	1,072	1.235	273,0	1,065
1050	1,506	282,2	1,088	1,391	282,3	1.081	1,295	282,4	1,075
1100	1,577	291,4	1,096	1,456	291,6	1,089	1,354	291.8	1.083
1150	1,648	300.5	1,104	1,520	300.7	1,097	1,413	301.0	1,091
1200	1,718	309,4	1,112	1,583	309,7	1,105	1,472	310,1	1,099
1250	1,787	318,3	1,119	1,646	318,6	1,113	1,530	319,0	1,107
1300	1,855	327,1	1,126	1,709	327.5	1,119	1,587	327.9	1,113

TABLE XXIV. Thermodynamic Properties of Liquid and Gaseous Xenon (*Continued*)

T, K	p = 650 bar			p = 700 bar			p = 750 bar		
	$v \cdot 10^3$	h	s	$v \cdot 10^3$	h	s	$v \cdot 10^3$	h	s
190	0.3272	72.94	0.6210	0.3256	74.11	0.6186	0.3240	75.28	0.6162
195	0.3297	74.46	0.6289	0.3280	75.63	0.6265	0.3263	76.79	0.6241
200	0.3322	75.97	0.6365	0.3304	77.13	0.6340	0.3287	78.29	0.6316
205	0.3348	77.47	0.6440	0.3329	78.62	0.6414	0.3311	79.78	0.6390
210	0.3374	78.96	0.6511	0.3354	80.11	0.6486	0.3335	81.26	0.6461
215	0.3401	80.44	0.6581	0.3380	81.58	0.6555	0.3360	82.72	0.6530
220	0.3427	81.93	0.6649	0.3405	83.06	0.6623	0.3385	84.20	0.6598
225	0.3454	83.41	0.6715	0.3432	84.53	0.6688	0.3410	85.66	0.6663
230	0.3482	84.89	0.6781	0.3458	86.00	0.6754	0.3435	87.13	0.6728
235	0.3510	86.37	0.6845	0.3485	87.47	0.6817	0.3461	88.59	0.6791
240	0.3538	87.87	0.6907	0.3512	88.96	0.6879	0.3487	90.07	0.6853
245	0.3566	89.38	0.6969	0.3539	90.46	0.6940	0.3513	91.56	0.6914
250	0.3595	90.90	0.7030	0.3567	91.97	0.7000	0.3540	93.05	0.6973
255	0.3625	92.42	0.7090	0.3595	93.47	0.7059	0.3567	94.53	0.7031
260	0.3654	93.93	0.7148	0.3623	94.96	0.7117	0.3594	96.01	0.7088
265	0.3684	95.43	0.7205	0.3652	96.44	0.7174	0.3621	97.48	0.7144
270	0.3715	96.93	0.7261	0.3681	97.92	0.7230	0.3649	98.95	0.7199
275	0.3746	98.42	0.7316	0.3710	99.39	0.7285	0.3677	100.4	0.7253
280	0.3777	99.91	0.7369	0.3740	100.8	0.7338	0.3705	101.9	0.7305
285	0.3809	101.4	0.7421	0.3770	102.3	0.7390	0.3734	103.3	0.7356
290	0.3841	102.9	0.7472	0.3800	103.8	0.7441	0.3762	104.8	0.7406
295	0.3873	104.4	0.7522	0.3831	105.2	0.7491	0.3791	106.2	0.7455
300	0.3906	105.8	0.7571	0.3862	106.7	0.7540	0.3821	107.6	0.7504
305	0.3939	107.3	0.7620	0.3893	108.1	0.7588	0.3852	109.0	0.7552
310	0.3974	108.7	0.7668	0.3927	109.6	0.7635	0.3883	110.4	0.7600

TABLE XXIV. Thermodynamic Properties of Liquid and Gaseous Xenon (*Continued*)

T, K	p = 650 bar			p = 700 bar			p = 750 bar		
	$v \cdot 10^3$	h	s	$v \cdot 10^3$	h	s	$v \cdot 10^3$	h	s
315	0,4010	110,2	0.7715	0.3961	111.0	0.7681	0.3914	111.8	0.7646
320	0,4048	111.6	0,7761	0.3996	112,4	0,7727	0,3947	113.2	0,7691
325	0,4086	113,1	0,7807	0.4031	113.9	0.7772	0,3981	114.6	0,7736
330	0,4124	114.5	0,7852	0.4066	115,3	0.7816	0.4015	116.1	0,7780
335	0,4162	116.0	0.7897	0,4102	116.8	0,7860	0.4049	117,5	0,7823
340	0,4202	117.4	0,7941	0.4139	118.2	0.7903	0.4083	119.0	0.7866
345	0.4242	118.9	0,7985	0.4176	119,7	0,7946	0,4118	120,4	0,7908
350	0,4282	120,4	0.8028	0,4214	121.1	0,7988	0,4153	121,8	0,7950
360	0,4364	123.3	0,8112	0,4291	124,0	0,8071	0,4226	124.7	0.8032
370	0,4451	126.3	0,8193	0.4370	126.9	0,8151	0.4301	127,6	0,8111
380	0,4539	129,2	0,8271	0,4452	129,8	0,8228	0,4377	130.5	0,8187
390	0,4630	132,1	0.8346	0,4537	132,7	0,8302	0,4454	133.3	0,8261
400	0,4723	134,9	0,8418	0,4622	135.5	0,8373	0,4534	136,1	0,8331
420	0.4915	140.6	0,8555	0,4799	141,0	0,8509	0,4698	141.6	0.8465
440	0,5115	146.1	0.8684	0,4982	146,5	0.8636	0.4868	147,0	0,8591
460	0,5323	151.6	0.8806	0,5172	151.9	0,8755	0.5042	152,3	0.8709
480	0.5538	157.0	0.8920	0,5368	157.2	0.8868	0.5223	157.5	0,8820
500	0.5760	162,3	0,9029	0,5570	162,4	0,8975	0,5408	162.7	0,8926
550	0,6335	175.2	0,9276	0,6094	175,3	0,9220	0.5889	175,4	0,9168
600	0,6927	187.7	0,9492	0.6636	187,6	0,9435	0.6387	187,7	0.9383
650	0,7525	199,6	0,9683	0.7185	199,6	0,9626	0,6894	199.6	0,9573
700	0,8123	211.1	0,9854	0,7734	211,1	0.9797	0,7402	211.1	0,9744
750	0,8716	222.2	1,001	0,8281	222,2	0,9953	0.7909	222,3	0,9897
800	0,9304	232.9	1.015	0,8824	232,9	1,009	0,8413	233.1	1.004
850	0,9885	243.3	1,027	0.9362	243.4	1,022	0.8913	243.6	1.016

TABLE XXIV. Thermodynamic Properties of Liquid and Gaseous Xenon (*Continued*)

T, K	p = 650 bar			p = 700 bar			p = 750 bar		
	$v \cdot 10^3$	h	s	$v \cdot 10^3$	h	s	$v \cdot 10^3$	h	s
900	1.046	253.5	1.039	0.9893	253.6	1.033	0.9408	253.8	1.028
950	1.103	263.4	1.050	1.042	263.6	1.044	0.9898	263.9	1.039
1000	1.159	273.1	1.060	1.094	273.4	1.054	1.038	273.7	1.049
1050	1.214	282.7	1.069	1.146	283.0	1.064	1.086	283.3	1.059
1100	1.269	292.1	1.078	1.197	292.4	1.072	1.134	292.8	1.068
1150	1.324	301.3	1.086	1.247	301.7	1.080	1.181	302.2	1.076
1200	1.378	310.5	1.094	1.298	310.9	1.088	1.228	311.4	1.084
1250	1.432	319.5	1.101	1.348	320.0	1.096	1.275	320.5	1.091
1300	1.485	328.4	1.108	1.397	329.0	1.103	1.321	329.6	1.098

TABLE XXIV. Thermodynamic Properties of Liquid and Gaseous Xenon (*Continued*)

T, K	p = 800 bar			p = 850 bar			p = 900 bar		
	$v \cdot 10^3$	h	s	$v \cdot 10^3$	h	s	$v \cdot 10^3$	h	s
195	0.3247	77.97	0.6217	0.3232	79.14	0.6195	0.3218	80.33	0.6173
200	0.3270	79.46	0.6292	0.3255	80.63	0.6269	0.3240	81.81	0.6247
205	0.3294	80.94	0.6366	0.3277	82.11	0.6342	0.3262	83.28	0.6320
210	0.3317	82.41	0.6437	0.3300	83.58	0.6413	0.3284	84.75	0.6391
215	0.3341	83.87	0.6506	0.3323	85.03	0.6482	0.3306	86.20	0.6459
220	0.3365	85.34	0.6573	0.3347	86.50	0.6549	0.3329	87.66	0.6526
225	0.3390	86.80	0.6638	0.3370	87.95	0.6614	0.3352	89.11	0.6591
230	0.3414	88.26	0.6703	0.3394	89.40	0.6678	0.3375	90.56	0.6655
235	0.3439	89.72	0.6766	0.3418	90.86	0.6741	0.3398	92.00	0.6718

TABLE XXIV. Thermodynamic Properties of Liquid and Gaseous Xenon (*Continued*)

T, K	p = 800 bar			p = 850 bar			p = 900 bar		
	$v \cdot 10^3$	h	s	$v \cdot 10^3$	h	s	$v \cdot 10^3$	h	s
240	0,3464	91,19	0,6827	0,3442	92,33	0,6802	0.3421	93.47	0.6779
245	0,3489	92,68	0,6888	0,3466	93,81	0,6862	0,3445	94.94	0.6839
250	0,3515	94,16	0,6947	0,3491	95,29	0,6922	0,3469	96.40	0.6898
255	0,3540	95,63	0,7005	0,3516	96,75	0,6980	0,3493	97.86	0,6956
260	0,3566	97,10	0,7062	0,3541	98,22	0,7037	0,3517	99,32	0.7012
265	0,3593	98,56	0,7119	0,3566	99,67	0,7092	0,3541	100,8	0.7067
270	0,3619	100,0	0,7174	0,3591	101,1	0,7146	0,3565	102,2	0,7121
275	0,3646	101,4	0,7227	0,3617	102,5	0,7199	0,3590	103.6	0,7174
280	0,3673	102,8	0,7279	0,3643	103,9	0,7251	0.3615	105.0	0,7226
285	0,3700	104,3	0,7330	0,3669	105,3	0,7302	0,3640	106.4	0.7277
290	0,3728	105.7	0,7380	0,3695	106,7	0,7352	0.3665	107.8	0,7327
295	0,3755	107,1	0,7429	0,3722	108,1	0,7401	0.3690	109.2	0,7376
300	0,3783	108,5	0,7477	0,3749	109,5	0,7449	0.3716	110.6	0,7424
305	0,3813	109,9	0,7524	0,3776	110,9	0,7496	0.3742	112.0	0,7471
310	0,3843	111,3	0,7570	0,3805	112,3	0,7542	0.3770	113.4	0,7517
315	0,3874	112.7	0,7615	0,3834	113,7	0,7587	0.3798	114.8	0,7561
320	0,3905	114,2	0,7660	0,3863	115,1	0,7631	0.3826	116.2	0,7604
325	0,3937	115,6	0,7704	0,3892	116,5	0,7674	0.3854	117.6	0,7647
330	0,3969	117,0	0,7748	0,3922	117,9	0,7717	0.3882	119.0	0,7689
335	0,4001	118,5	0,7791	0,3953	119.3	0,7759	0.3911	120.4	0,7731
340	0,4033	120,0	0,7833	0,3984	120,7	0,7801	0.3940	121,8	0,7772
345	0,4066	121,4	0,7875	0,4016	122,1	0,7842	0,3969	123.2	0,7813
350	0,4099	122,8	0,7916	0,4048	123,5	0,7883	0,4000	124,6	0,7853
360	0,4167	125,6	0,7996	0,4112	126,3	0,7962	0,4062	127,4	0,7931
370	0,4237	128,4	0,8074	0,4179	129,1	0,8039	0,4126	130,2	0,8007

TABLE XXIV. Thermodynamic Properties of Liquid and Gaseous Xenon *(Continued)*

T, K	p = 800 bar			p = 850 bar			p = 900 bar		
	$v \cdot 10^3$	h	s	$v \cdot 10^3$	h	s	$v \cdot 10^3$	h	s
380	0.4309	131.2	0.8149	0.4248	131.9	0.8113	0.4192	132.9	0.8080
390	0.4381	134.0	0.8222	0.4316	134.7	0.8185	0.4258	135.6	0.8151
400	0.4457	136.8	0.8292	0.4387	137.5	0.8255	0.4325	138.3	0.8220
420	0.4610	142.2	0.8425	0.4532	142.9	0.8387	0.4463	143.7	0.8352
440	0.4768	147.6	0.8549	0.4680	148.2	0.8511	0.4602	149.0	0.8474
460	0.4930	152.8	0.8666	0.4832	153.4	0.8627	0.4745	154.1	0.8589
480	0.5097	158.0	0.8777	0.4987	158.6	0.8736	0.4890	159.2	0.8698
500	0.5268	163.1	0.8881	0.5146	163.7	0.8840	0.5038	164.3	0.8801
550	0.5712	175.7	0.9121	0.5557	176.1	0.9078	0.5422	176.7	0.9037
600	0.6172	188.0	0.9334	0.5985	188.3	0.9289	0.5820	188.8	0.9247
650	0.6642	199.8	0.9524	0.6422	200.1	0.9479	0.6229	200.5	0.9436
700	0.7115	211.3	0.9695	0.6863	211.6	0.9649	0.6642	212.0	0.9606
750	0.7587	222.5	0.9849	0.7305	222.8	0.9807	0.7056	223.2	0.9760
800	0.8057	233.3	0.9988	0.7745	233.6	0.9943	0.7469	234.0	0.9901
850	0.8523	243.9	1.012	0.8182	244.2	1.007	0.7880	244.6	1.003
900	0.8986	254.2	1.023	0.8616	254.5	1.019	0.8289	254.9	1.015
950	0.9444	264.2	1.034	0.9046	264.6	1.030	0.8695	265.1	1.026
1000	0.9899	274.1	1.044	0.9473	274.5	1.040	0.9097	275.1	1.036
1050	1.035	283.8	1.054	0.9897	284.3	1.050	0.9497	284.8	1.045
1100	1.080	293.3	1.063	1.032	293.8	1.059	0.9893	294.4	1.054
1150	1.124	302.7	1.071	1.074	303.2	1.067	1.029	303.9	1.063
1200	1.168	312.0	1.079	1.115	312.6	1.075	1.068	313.2	1.071
1250	1.212	321.1	1.087	1.156	321.8	1.082	1.107	322.4	1.078
1300	1.255	330.2	1.094	1.197	330.9	1.089	1.145	331.6	1.085

TABLE XXIV. Thermodynamic Properties of Liquid and Gaseous Xenon (*Continued*)

T, K	p = 950 bar			p = 1000 bar		
	$v \cdot 10^3$	h	s	$v \cdot 10^3$	h	s
200	0.3225	82.99	0.6225	0.3211	84.18	0.6204
205	0.3246	84.46	0.6298	0.3232	85.65	0.6277
210	0.3268	85.92	0.6369	0.3253	87.10	0.6347
215	0.3290	87.37	0.6437	0.3274	88.54	0.6416
220	0.3312	88.82	0.6504	0.3296	90.00	0.6482
225	0.3334	90.27	0.6568	0.3317	91.44	0.6546
230	0.3356	91.72	0.6632	0.3339	92.88	0.6610
235	0.3379	93.16	0.6695	0.3361	94.32	0.6672
240	0.3402	94.62	0.6756	0.3383	95.78	0.6733
245	0.3424	96.09	0.6816	0.3405	97.25	0.6793
250	0.3447	97.55	0.6875	0.3427	98.71	0.6852
255	0.3470	99.01	0.6932	0.3449	100.2	0.6909
260	0.3494	100.5	0.6988	0.3472	101.6	0.6965
265	0.3517	101.9	0.7043	0.3495	103.0	0.7020
270	0.3541	103.3	0.7097	0.3517	104.4	0.7073
275	0.3564	104.7	0.7150	0.3540	105.8	0.7125
280	0.3588	106.1	0.7201	0.3563	107.2	0.7176
285	0.3612	107.5	0.7252	0.3586	108.6	0.7226
290	0.3637	108.9	0.7302	0.3610	110.0	0.7275
295	0.3661	110.3	0.7351	0.3633	111.4	0.7323
300	0.3685	111.7	0.7399	0.3657	112.8	0.7370
305	0.3711	113.1	0.7446	0.3681	114.2	0.7416
310	0.3737	114.4	0.7492	0.3705	115.6	0.7461

TABLE XXIV. Thermodynamic Properties of Liquid and Gaseous Xenon (*Continued*)

T, K	p = 950 bar			p = 1000 bar		
	$v \cdot 10^3$	h	s	$v \cdot 10^3$	h	s
315	0.3763	115.8	0.7537	0.3730	117.0	0.7505
320	0.3789	117.2	0.7581	0.3755	118.4	0.7548
325	0.3816	118.6	0.7624	0.3781	119.8	0.7591
330	0.3844	120.0	0.7666	0.3808	121.2	0.7633
335	0.3872	121.4	0.7707	0.3835	122.6	0.7675
340	0.3900	122.8	0.7748	0.3862	124.0	0.7716
345	0.3929	124.2	0.7788	0.3889	125.4	0.7757
350	0.3958	125.6	0.7827	0.3916	126.8	0.7797
360	0.4016	128.3	0.7904	0.3974	129.5	0.7876
370	0.4076	131.0	0.7979	0.4032	132.2	0.7951
380	0.4139	133.7	0.8052	0.4091	134.8	0.8024
390	0.4203	136.4	0.8122	0.4154	137.4	0.8094
400	0.4269	139.1	0.8190	0.4217	140.0	0.8162
420	0.4400	144.5	0.8319	0.4344	145.3	0.8290
440	0.4533	149.7	0.8440	0.4470	150.5	0.8409
460	0.4667	154.9	0.8554	0.4598	155.7	0.8522
480	0.4804	160.0	0.8662	0.4728	160.7	0.8629
500	0.4943	165.0	0.8765	0.4859	165.7	0.8731
550	0.5302	177.3	0.8999	0.5196	177.9	0.8964
600	0.5675	189.3	0.9208	0.5545	189.9	0.9172
650	0.6057	201.0	0.9396	0.5904	201.6	0.9359
700	0.6445	212.5	0.9566	0.6268	213.1	0.9528
750	0.6835	223.6	0.9720	0.6637	224.2	0.9682
800	0.7224	234.5	0.9861	0.7005	235.0	0.9823
850	0.7612	245.1	0.9989	0.7372	245.7	0.9952

TABLE XXIV. Thermodynamic Properties of Liquid and Gaseous Xenon (*Continued*)

T, K	p = 950 bar			p = 1000 bar		
	$v \cdot 10^3$	h	s	$v \cdot 10^3$	h	s
900	0.7998	255.5	1.011	0.7737	256.1	1.007
950	0.8381	265.7	1.022	0.8100	266.2	1.018
1000	0.8762	275.6	1.032	0.8461	276.2	1.028
1050	0.9140	285.4	1.042	0.8819	286.0	1.038
1100	0.9515	295.0	1.051	0.9176	295.7	1.047
1150	0.9888	304.5	1.059	0.9529	305.2	1.055
1200	1.026	313.9	1.067	0.9881	314.6	1.063
1250	1.063	323.1	1.075	1.023	323.9	1.070
1300	1.099	332.3	1.082	1.058	333.0	1.077

TABLE XXV. Isobaric Specific Heat c_p of Liquid and Gaseous Xenon

T, K	c_p, at p, bar								
	1	10	20	30	40	50	60	70	80
170	0.168	0.348	0.347	0.346	0.345	0.344	0.343	0.342	0.341
200	0.165	0.356	0.353	0.351	0.349	0.347	0.345	0.343	0.341
250	0.162	0.198	0.272	0.472	0.456	0.443	0.432	0.422	0.413
300	0.160	0.178	0.204	0.242	0.305	0.429	0.790	3.043	1.203
350	0.159	0.170	0.183	0.200	0.220	0.245	0.277	0.317	0.367
400	0.159	0.166	0.175	0.184	0.194	0.206	0.219	0.238	0.250
450	0.159	0.164	0.170	0.176	0.183	0.190	0.198	0.206	0.214
500	0.159	0.163	0.167	0.172	0.177	0.181	0.187	0.192	0.197
550	0.159	0.162	0.165	0.169	0.172	0.176	0.180	0.184	0.187
600	0.159	0.161	0.164	0.167	0.170	0.173	0.176	0.178	0.181
650	0.159	0.161	0.163	0.165	0.168	0.170	0.172	0.175	0.177
700	0.159	0.160	0.162	0.164	0.166	0.168	0.170	0.172	0.174
750	0.159	0.160	0.162	0.163	0.165	0.167	0.168	0.170	0.172
800	0.159	0.160	0.161	0.163	0.164	0.166	0.167	0.168	0.170
850	0.158	0.160	0.161	0.162	0.163	0.165	0.166	0.167	0.168
900	0.158	0.159	0.161	0.162	0.163	0.164	0.165	0.166	0.167
950	0.158	0.159	0.160	0.161	0.162	0.163	0.164	0.165	0.166
1000	0.158	0.159	0.160	0.161	0.162	0.163	0.163	0.164	0.165
1050	0.158	0.159	0.160	0.161	0.161	0.162	0.163	0.164	0.164
1100	0.158	0.159	0.160	0.160	0.161	0.162	0.162	0.163	0.164
1150	0.158	0.159	0.160	0.160	0.161	0.161	0.162	0.163	0.163
1200	0.158	0.159	0.159	0.160	0.161	0.161	0.162	0.162	0.163
1250	0.158	0.159	0.159	0.160	0.160	0.161	0.161	0.162	0.162
1300	0.158	0.159	0.159	0.160	0.160	0.161	0.161	0.161	0.162

TABLE XXV. Isobaric Specific Heat c_p of Liquid and Gaseous Xenon (*Continued*)

T, K	c_p, at p, bar								
	90	100	150	200	250	300	350	400	450
170	0.340	0.339	0.335	0.332	—	—	—	—	—
200	0.339	0.338	0.331	0.326	0.321	0.317	0.313	0.310	0.308
250	0.403	0.395	0.370	0.356	0.345	0.335	0.328	0.322	0.317
300	0.797	0.699	0.457	0.398	0.371	0.354	0.343	0.332	0.323
350	0.428	0.496	0.585	0.463	0.402	0.368	0.347	0.332	0.321
400	0.267	0.285	0.374	0.403	0.382	0.358	0.338	0.323	0.312
450	0.222	0.232	0.275	0.309	0.321	0.320	0.314	0.305	0.296
500	0.202	0.208	0.235	0.258	0.274	0.282	0.284	0.281	0.278
550	0.191	0.195	0.214	0.230	0.243	0.253	0.258	0.260	0.260
600	0.184	0.187	0.201	0.214	0.224	0.232	0.238	0.242	0.244

TABLE XXV. Isobaric Specific Heat c_p of Liquid and Gaseous Xenon (*Continued*)

T, K	c_p, at p, bar								
	90	100	150	200	250	300	350	400	450
650	0.179	0.182	0.193	0.202	0.210	0.218	0.224	0.228	0.231
700	0.176	0.178	0.187	0.195	0.202	0.208	0.212	0.216	0.219
750	0.173	0.175	0.182	0.189	0.195	0.200	0.204	0.208	0.211
800	0.171	0.172	0.178	0.184	0.189	0.194	0.198	0.202	0.205
850	0.169	0.170	0.176	0.181	0.185	0.189	0.193	0.196	0.199
900	0.168	0.169	0.174	0.178	0.182	0.185	0.188	0.191	0.193
950	0.167	0.168	0.172	0.176	0.179	0.182	0.184	0.187	0.189
1000	0.166	0.167	0.170	0.174	0.177	0.179	0.181	0.184	0.186
1050	0.165	0.166	0.169	0.172	0.175	0.177	0.179	0.182	0.183
1100	0.164	0.165	0.168	0.170	0.173	0.175	0.177	0.179	0.180
1150	0.164	0.164	0.167	0.169	0.172	0.174	0.175	0.177	0.178
1200	0.163	0.164	0.166	0.168	0.170	0.172	0.173	0.175	0.176
1250	0.163	0.163	0.165	0.167	0.169	0.171	0.172	0.174	0.175
1300	0.162	0.163	0.165	0.166	0.168	0.170	0.171	0.173	0.174

TABLE XXV. Isobaric Specific Heat c_p of Liquid and Gaseous Xenon (*Continued*)

T, K	c_p, at p, bar								
	500	550	600	650	700	750	800	900	1000
200	0.306	0.304	0.303	0.301	0.300	0.299	0.297	0.296	0.294
250	0.312	0.309	0.306	0.303	0.300	0.297	0.295	0.293	0.291
300	0.315	0.309	0.305	0.301	0.298	0.295	0.293	0.289	0.286
350	0.312	0.305	0.300	0.296	0.293	0.290	0.287	0.283	0.278
400	0.302	0.295	0.289	0.286	0.282	0.280	0.277	0.273	0.268
450	0.289	0.283	0.278	0.273	0.269	0.266	0.263	0.260	0.257
500	0.274	0.271	0.267	0.264	0.261	0.258	0.255	0.251	0.248
550	0.259	0.257	0.256	0.254	0.252	0.250	0.248	0.245	0.242
600	0.245	0.245	0.244	0.244	0.243	0.242	0.241	0.239	0.237
650	0.232	0.233	0.234	0.234	0.234	0.234	0.234	0.233	0.231
700	0.222	0.224	0.225	0.226	0.226	0.226	0.227	0.226	0.225
750	0.214	0.216	0.217	0.218	0.219	0.219	0.220	0.220	0.220
800	0.207	0.209	0.210	0.211	0.212	0.213	0.214	0.215	0.215
850	0.201	0.203	0.204	0.206	0.207	0.208	0.209	0.210	0.210
900	0.196	0.198	0.199	0.201	0.202	0.203	0.204	0.205	0.206
950	0.192	0.194	0.195	0.197	0.198	0.198	0.200	0.201	0.202
1000	0.188	0.190	0.191	0.193	0.194	0.194	0.196	0.197	0.198
1050	0.185	0.187	0.188	0.189	0.191	0.191	0.192	0.194	0.195
1100	0.183	0.184	0.185	0.186	0.188	0.188	0.189	0.191	0.192
1150	0.180	0.182	0.183	0.184	0.185	0.186	0.187	0.188	0.189
1200	0.178	0.180	0.181	0.182	0.183	0.183	0.184	0.186	0.187
1250	0.177	0.178	0.179	0.180	0.181	0.181	0.182	0.183	0.184
1300	0.175	0.176	0.177	0.178	0.179	0.180	0.180	0.182	0.183

TABLE XXVI. Dynamic Viscosity of Xenon

T, K	$\eta \cdot 10^8$, Pa · s at p, bar							
	1	10	20	30	40	50	60	70
170	1376	45359	45719	46073	46434	46744	47033	47407
180	1453	40187	40543	40899	41252	41566	41874	42253
190	1530	35453	35836	36166	36550	36881	37249	37591
200	1606	31169	31561	31890	32261	32634	32968	33305
210	1682	27338	27724	28108	28408	28814	29186	29517
220	1757	1830	24284	24670	25036	25394	25759	26100
230	1832	1900	21183	21579	21975	22328	22704	23000
240	1907	1972	18357	18786	19202	19604	19968	20319
250	1982	2043	2148	16224	16666	17092	17506	17880
260	2056	2115	2208	2393	14301	14782	15230	15650
270	2130	2187	2271	2420	11877	12580	13092	13576
280	2204	2259	2336	2463	2695	9716	10642	11477
290	2277	2330	2402	2513	2694	3045	7231	9131
300	2350	2402	2469	2568	2720	2962	3420	5380
310	2422	2472	2536	2626	2756	2949	3249	3780
320	2494	2543	2604	2686	2801	2962	3192	3532
330	2566	2614	2672	2748	2851	2991	3180	3437
340	2637	2684	2739	2810	2904	3028	3189	3398
350	2708	2754	2807	2874	2960	3071	3212	3389
360	2779	2822	2875	2938	3017	3118	3243	3398
370	2849	2893	2942	3003	3077	3170	3283	3420
380	2918	2961	3009	3066	3136	3222	3325	3448
390	2988	3030	3077	3131	3198	3278	3373	3484
400	3057	3098	3144	3196	3259	3338	3421	3524
420	3193	3233	3276	3325	3381	3448	3524	3612
440	3328	3367	3408	3453	3505	3565	3633	3711
460	3461	3498	3537	3580	3628	3683	3745	3814
480	3592	3628	3666	3707	3752	3802	3858	3920
500	3721	3756	3792	3831	3873	3920	3971	4028
550	4038	4071	4104	4138	4176	4216	4259	4305
600	4344	4375	4405	4437	4470	4506	4542	4582
650	4640	4669	4698	4727	4757	4788	4821	4856
700	4928	4955	4982	5009	5037	5066	5096	5127
750	5206	5232	5257	5282	5309	5335	5362	5391
800	5477	5502	5526	5550	5572	5599	5624	5650
850	5741	5764	5787	5810	5833	5856	5879	5904
900	5997	6019	6041	6062	6084	6106	6128	6150
950	6248	6269	6290	6310	6331	6352	6372	6394
1000	6492	6512	6532	6552	6571	6591	6610	6630
1050	6732	6752	6770	6789	6808	6826	6845	6864
1100	6965	6984	7002	7020	7037	7056	7073	7091
1150	7195	7213	7230	7247	7265	7282	7299	7316
1200	7420	7437	7454	7470	7487	7503	7519	7536
1250	7640	7657	7673	7689	7704	7720	7736	7752
1300	7857	7873	7888	7904	7919	7934	7950	7965

TABLE XXVI. Dynamic Viscosity of Xenon (*Continued*)

T, K	$\eta \cdot 10^8$, Pa · s at p, bar							
	80	100	120	140	160	180	200	220
170	47716	48360	48933	49597	50205	50891	51446	52083
180	42567	43282	43886	44491	45151	45772	46397	47012
190	37903	38582	39209	39873	40544	41188	41769	42376
200	33663	34356	34972	35639	36294	36930	37564	38184
210	29832	30505	31186	31837	32473	33117	33721	34360
220	26460	27146	27780	28444	29068	29713	30321	30897
230	23408	24098	24761	25425	26051	26650	27264	27890
240	20716	21389	22072	22739	23450	23979	24600	25171
250	18269	19003	19683	20361	20995	21610	22194	22775
260	16060	16830	17546	18225	18867	19474	20078	20675
270	14032	14853	15605	16310	16973	17584	18203	18777
280	12113	13045	13863	14590	15282	15921	16523	17103
290	10053	11317	12249	13050	13782	14439	15032	15606
300	6815	9529	10683	11588	12368	13061	13686	14257
310	4920	7698	9177	10216	11051	11786	12361	13038
320	4070	6007	7755	8948	9866	10638	11319	11934
330	3794	4972	6532	7807	8798	9610	10318	10953
340	36669	4473	5642	6845	7855	8698	9427	10068
350	3609	4220	5089	6099	7069	7907	8638	9291
360	3586	4078	4756	5582	6443	7243	7963	8610
370	3582	3998	4548	5224	5967	6704	7390	8019
380	3592	3952	4414	4978	5616	6272	6914	7515
390	3614	3932	4331	4813	5359	5944	6530	7095
400	3642	3926	4280	4698	5176	5692	6222	6747
420	3712	3948	4233	4566	4940	5349	5786	6229
440	3797	3998	4237	4512	4821	5156	5515	5891
460	3890	4066	4271	4506	4766	5050	5354	5670
480	3988	4144	4324	4528	4753	4998	5259	5533
500	4090	4230	4390	5470	4769	4982	5211	5453
550	4356	4466	4592	4731	4881	5045	5218	5400
600	4625	4717	4820	4932	5053	5183	5321	5464
650	4893	4972	5058	5152	5253	5360	5473	5593
700	5159	5228	5303	5384	5470	5561	5657	5757
750	5420	5481	5547	5617	5692	5771	5854	5941
800	5676	5732	5791	5853	5919	5989	6061	6136
850	5928	5979	6032	6088	6147	6209	6273	6340
900	6173	6220	6269	6320	6373	6429	6487	6547
950	6415	6458	6504	6550	6599	6650	6702	6757
1000	6650	6691	6733	6776	6822	6868	6916	6965
1050	6883	6922	6961	7001	7043	7085	7130	7176
1100	7109	7145	7182	7221	7260	7299	7340	7382
1150	7331	7368	7402	7438	7475	7512	7550	7589
1200	7552	7586	7618	7652	7686	7722	7758	7794
1250	7767	7799	7830	7862	7895	7928	7962	7996
1300	7980	8009	8040	8070	8101	8133	8164	8197

TABLE XXVI. Dynamic Viscosity of Xenon (*Continued*)

T, K	$\eta \cdot 10^8$, Pa · s at p, bar							
	240	260	280	300	320	340	360	380
180	47641	48199	48779	49303	49969	54196	51048	51523
190	42961	43593	44193	44716	45319	45848	46458	47035
200	38779	39305	39938	40566	41011	41668	42167	42750
210	34976	35507	36111	36662	37308	37806	38335	38905
220	31503	32074	32682	33209	33782	34332	34910	35377
230	28468	29015	29605	30129	30679	31241	31739	32300
240	25739	26288	26875	27409	27940	28485	29038	29501
250	23346	23909	24468	24995	25510	26047	26555	27058
260	21220	21767	22326	22850	23364	23854	24376	24859
270	19342	19879	20417	21063	21443	21939	22431	22887
280	17670	18211	18749	19248	19742	20238	20713	21179
290	16162	16727	17251	17753	18246	18722	19178	19628
300	14827	15372	15887	16388	16874	17343	17824	18236
310	13611	14139	14649	15157	15622	16103	16555	17007
320	12515	13042	13565	14050	14535	14980	15427	15877
330	11538	12082	12583	13077	13541	13989	14434	14857
340	10658	11209	11709	12208	12663	13108	13538	13958
350	9884	10429	10935	11425	11881	12326	12744	13148
360	9200	9739	10244	10727	11174	11609	12090	12431
370	8595	9132	9632	10100	10560	10983	11386	11780
380	8072	8598	9090	9546	10035	10406	10806	11192
390	7628	8135	8610	9065	9512	9898	10296	10676
400	6962	7739	8194	8633	9046	9448	9834	10205
420	6671	7104	7520	7928	8316	8692	9056	9407
440	6272	6652	7029	7395	7753	8102	8440	8769
460	6200	6334	6667	6999	7324	7643	7958	8264
480	5819	6110	6408	6705	7002	7292	7581	7865
500	5704	5960	6223	6490	6757	7025	7288	7550
550	5589	5786	5988	6196	6405	6655	6830	7041
600	5616	5773	5933	6122	6266	6440	6615	6789
650	5716	5845	5978	6114	6254	6397	6542	6677
700	5863	5970	6082	6196	6315	6335	6557	6681
750	6031	6124	6219	6318	6419	6523	6628	6736
800	6215	6296	6380	6466	6554	6644	6736	6831
850	6410	6482	6556	6631	6709	6789	6870	6954
900	6609	6673	6739	6807	6876	6947	7019	7093
950	6813	6871	6930	6991	7053	7117	7183	7248
1000	7017	7069	7123	7178	7235	7292	7352	7411
1050	7324	7373	7423	7473	7526	7580	7634	7690
1100	7425	7470	7515	7562	7609	7700	7707	7756
1150	7524	7564	7605	7648	7691	7735	7779	7825
1200	7832	7870	7909	7949	7989	8030	8073	8115
1250	8031	8067	8104	8140	8179	8217	8256	8296
1300	8230	8264	8298	8331	8368	8404	8441	8478

TABLE XXVI. Dynamic Viscosity of Xenon (*Continued*)

T, K	$\eta \cdot 10^8$, Pa · s at p, bar							
	400	420	440	460	480	500	550	600
180	52083	52568	53181	53600	54184	—	—	—
190	47562	48076	48634	49101	49658	50202	51422	52586
200	43275	43853	44401	44891	45431	45951	47199	48423
210	39468	40008	40491	41051	41544	42057	43340	44580
220	35950	36519	37043	37498	38060	38512	39810	41003
230	32813	33353	33837	34352	34852	35348	36594	37732
240	30030	30550	30995	31525	32017	32488	33691	34835
250	27556	28024	28521	29017	29488	29966	31102	31237
260	25330	25832	26311	26757	27238	27664	28796	29933
270	23398	23848	24310	24763	25217	25647	26774	27822
280	21636	22097	22565	22979	23446	23856	24929	25961
290	20098	20537	20974	21410	21832	22500	23278	24273
300	18718	19152	19586	19987	20416	20861	21814	22801
310	17452	17885	18311	18716	19142	19548	20483	21439
320	16305	16718	17134	17535	17944	18321	19276	20196
330	15270	15686	16080	16487	16870	17261	18192	19045
340	14369	14761	15145	15527	15918	16208	17185	18030
350	13553	13939	14320	14690	15060	15407	16289	17109
360	12817	13201	13577	13929	14287	14637	15484	16283
370	12167	12539	12899	13248	13597	13942	14754	15552
380	11573	11939	12288	12633	12971	13304	14095	14862
390	11043	11402	11744	12078	12411	12738	13516	14255
400	10565	10911	11256	11579	11905	12224	12978	13702
420	9828	10157	10484	10792	11100	11403	12124	12815
449	9092	9407	9713	10006	10296	10583	11271	11929
460	8565	8858	9150	9425	9703	9970	10632	11257
480	8142	8417	8686	8950	9212	9464	10098	10687
500	7809	8063	8317	8564	8806	9052	9638	10203
550	7256	7467	7679	7890	8097	8304	8825	9306
600	6966	7144	7321	7498	7676	7853	8293	8726
650	6836	6985	7137	7288	7441	7599	7973	8353
700	6812	6938	7066	7196	7328	7460	7794	8126
750	6846	6956	7067	7183	7297	7412	7703	7999
800	6926	7023	7122	7220	7321	7426	7681	7944
850	7038	7124	7210	7299	7388	7479	7707	7943
900	7168	7245	7323	7402	7481	7563	7768	7939
950	7316	7384	7454	7525	7598	7670	7856	8046
1000	7473	7534	7598	7662	7727	7792	7962	8134
1050	7639	7695	7753	7811	7870	7931	8084	8241
1100	7808	7859	7912	7965	8020	8075	8215	8362
1150	7983	8030	8078	8128	8177	8228	8358	8492
1200	8159	8204	8248	8294	8340	8386	8507	8629
1250	8336	8377	8419	8461	8504	8548	8658	8772
1300	8515	8554	8592	8632	8672	8713	8816	8922

TABLE XXVI. Dynamic Viscosity of Xenon (*Continued*)

T, K	$\eta \cdot 10^8$, Pa · s at p, bar							
	650	700	750	800	850	900	950	1000
190	53813	54955	56133	—	—	—	—	—
200	49658	50806	51930	53152	54195	55250	56322	57395
210	45777	46965	48144	49225	50361	51388	52549	53570
220	42201	43368	44442	45643	46708	47782	48918	49918
230	38915	40082	41178	42308	43380	44448	45503	46569
240	35991	37112	38211	39330	40387	41441	42419	43434
250	33389	34441	35543	36608	37591	38596	39672	40655
260	31000	32020	33081	34121	35141	36095	37088	38067
270	28855	29888	30833	31873	32883	33833	34802	35733
280	26966	27929	28961	29875	30854	31778	32699	33609
290	25250	26220	27201	28089	29009	29902	30762	31685
300	23754	24663	25577	26487	27366	28245	29117	29944
310	22360	23239	24106	24944	25806	26644	27473	28376
320	21067	21922	22776	23575	24410	25234	26077	26885
330	19916	20758	21541	22315	23166	23944	24710	25498
340	18870	19674	20455	21193	21978	22763	23507	24247
350	17936	18703	19463	20208	20934	21686	22376	23124
360	17093	17841	18567	19279	19993	20693	21386	22056
370	16310	17042	17729	18477	19121	19784	20476	21121
380	15611	16312	16996	17658	18293	18947	19629	20225
390	14974	15648	16330	16981	17599	18207	18816	19412
400	14394	15065	15710	16338	16942	17533	18101	18665
420	13476	14110	14724	15318	15889	16445	16975	17500
440	12558	13155	13738	14298	14837	15358	15850	16336
460	11854	12429	12988	13521	14031	14527	15000	15468
480	11258	11809	12342	12855	13345	13820	14280	14709
500	10750	11280	11880	12279	12751	13214	13657	14077
550	9791	10260	10720	11160	11605	12016	12431	12812
600	9152	9569	9980	10386	10800	11164	11537	11960
650	8730	9100	9468	9833	10189	10543	10888	11224
700	8457	8791	9123	9449	9774	10094	10407	10723
750	8293	8592	8889	9183	9476	9770	10060	10345
800	8208	8475	8741	9009	9277	9545	9811	10074
850	8180	8421	8663	8905	9148	9394	9636	9883
900	8192	8410	8628	8850	9074	9297	9521	9746
950	8239	8437	8638	8841	9045	9250	9458	9665
1000	8311	8491	8677	8860	9049	9234	9427	9620
1050	8405	8568	8735	8908	9080	9255	9432	9610
1100	8510	8660	8816	8971	9130	9294	9457	9622
1150	8628	8769	8910	9056	9202	9354	9507	9660
1200	8756	8885	9018	9153	9289	9429	9570	9714
1250	8890	9010	9134	9258	9386	9516	9648	9782
1300	9032	9143	9259	9376	9494	9617	9740	9864

TABLE XXVII. Thermal Conductivity of Xenon

T, K	λ · 10^8, kW/(m · K) at p, bar							
	1	10	20	30	40	50	60	70
170	327	7091	7127	7163	7199	7229	7267	7303
180	345	6667	6705	6742	6780	6819	6852	6891
190	363	6249	6292	6330	6373	6412	6451	6491
200	382	5844	5889	5932	5974	6017	6061	6100
210	400	5442	5489	5537	5585	5630	5676	5719
220	418	465	5095	5148	5200	5249	5299	5346
230	435	479	4698	4760	4817	4873	4927	4980
240	453	494	4297	4365	4431	4495	4555	4615
250	470	509	567	3961	4041	4114	4185	4251
260	489	526	578	664	3635	3725	3808	3887
270	506	541	589	660	3168	3313	3417	3513
280	524	557	601	663	761	2709	2915	3095
290	541	573	614	669	748	884	2122	2580
300	558	588	627	677	744	842	1009	
310	576	605	642	688	747	826	940	1127
320	593	621	656	698	752	820	909	1034
330	610	637	670	710	758	818	894	990
340	627	653	685	722	767	821	886	966
350	644	669	699	735	776	826	884	953
360	660	685	713	747	786	831	884	945
370	677	701	728	760	797	839	888	943
380	693	716	743	773	808	847	892	943
390	710	732	758	787	820	858	899	946
400	726	748	772	800	832	867	906	950
420	759	780	803	829	857	889	924	962
440	790	810	831	856	882	911	943	977
460	822	841	861	884	909	935	964	995
480	853	871	890	912	935	960	986	1015
500	884	901	920	940	962	985	1010	1036
550	959	974	991	1009	1028	1048	1069	1091
600	1032	1046	1061	1077	1094	1112	1130	1150
650	1103	1116	1129	1144	1159	1175	1192	1209
700	1171	1183	1196	1209	1222	1237	1252	1267
750	1237	1248	1260	1272	1285	1298	1311	1326
800	1302	1312	1323	1335	1346	1358	1371	1384
850	1364	1374	1384	1394	1405	1417	1428	1440
900	1425	1434	1444	1454	1464	1474	1485	1496
950	1485	1494	1502	1512	1522	1531	1542	1552
1000	1543	1551	1560	1569	1578	1587	1596	1606
1050	1600	1608	1616	1624	1633	1642	1651	1660
1100	1655	1662	1670	1678	1686	1695	1703	1712
1150	1710	1717	1725	1732	1740	1748	1756	1764
1200	1763	1770	1777	1784	1791	1799	1807	1815
1250	1815	1822	1828	1835	1842	1850	1857	1864
1300	1867	1873	1880	1886	1893	1900	1907	1914

TABLE XXVII. Thermal Conductivity of Xenon (*Continued*)

T, K	$\lambda \cdot 10^8$, kW/(m · K) at p, bar							
	80	100	120	140	160	180	200	220
170	7336	7406	7469	7536	7601	7661	7722	7785
180	6924	6999	7069	7133	7198	7266	7334	7397
190	6526	6603	6676	6745	6815	6887	6954	7023
200	6140	6223	6297	6373	6446	6521	6592	6665
210	5762	5847	5930	6010	6089	6164	6242	6316
220	5395	5487	5575	5661	5742	5826	5903	5979
230	5030	5129	5225	5318	5407	5491	5574	5657
240	4673	4782	4887	4984	5081	5171	5261	5344
250	4316	4438	4552	4658	4763	4860	4953	5045
260	3962	4099	4226	4345	4454	4561	4659	4757
270	3602	3761	3904	4035	4154	4267	4376	4477
280	3229	3424	3587	3732	3866	3989	4103	4209
290	2792	3071	3271	3439	3589	3721	3841	3953
300		2676	2938	3139	3307	3456	3587	3706
310	1491	2240	2598	2840	3030	3193	3336	3466
320	1217	1797	2254	2546	2764	2943	3098	3237
330	1116	1496	1936	2267	2511	2707	2874	3022
340	1065	1336	1689	2018	2279	2489	2667	2821
350	1035	1247	1522	1816	2076	2293	2477	2637
360	1016	1192	1414	1663	1903	2120	2306	2469
370	1006	1156	1342	1554	1771	1974	2157	2320
380	999	1132	1292	1473	1666	1854	2030	2189
390	997	1116	1257	1417	1586	1759	1923	2077
400	997	1105	1232	1373	1525	1681	1835	1980
420	1004	1096	1201	1317	1441	1571	1703	1832
440	1014	1094	1185	1284	1390	1500	1612	1726
460	1028	1101	1180	1267	1359	1455	1554	1653
480	1045	1111	1182	1259	1340	1425	1513	1602
500	1064	1124	1189	1258	1332	1408	1486	1567
550	1115	1164	1218	1274	1333	1394	1457	1521
600	1170	1212	1258	1305	1354	1406	1458	1512
650	1227	1264	1303	1345	1388	1432	1477	1523
700	1283	1317	1352	1388	1426	1465	1505	1545
750	1340	1370	1402	1434	1468	1503	1539	1575
800	1397	1425	1453	1483	1514	1545	1578	1611
850	1452	1478	1504	1532	1560	1588	1618	1648
900	1508	1531	1556	1581	1607	1633	1660	1688
950	1563	1584	1607	1631	1655	1679	1705	1730
1000	1616	1637	1658	1680	1703	1725	1749	1773
1050	1669	1689	1709	1729	1750	1772	1794	1816
1100	1721	1739	1758	1777	1797	1817	1838	1859
1150	1773	1790	1807	1826	1845	1864	1883	1903
1200	1823	1839	1856	1873	1891	1909	1927	1946
1250	1872	1888	1904	1920	1937	1954	1972	1990
1300	1922	1936	1952	1967	1983	2000	2017	2033

TABLE XXVII. Thermal Conductivity of Xenon (*Continued*)

T, K	$\lambda \cdot 10^8$, kW/(m · K) at p, bar							
	240	260	280	300	320	340	360	380
180	7461	7526	7584	7645	7705	7766	7822	7879
190	7086	7152	7217	7278	7340	7403	7466	7524
200	6732	6797	6862	6929	6997	7059	7118	7183
210	6388	6455	6525	6595	6663	6723	6789	6854
220	6056	6130	6201	6270	6340	6407	6475	6539
230	5734	5812	5889	5959	6031	6104	6170	6238
240	5426	5507	5586	5664	5737	5811	5884	5953
250	5130	5215	5299	5376	5454	5532	5604	5679
260	4848	4937	5023	5104	5187	5264	5343	5416
270	4573	4667	4757	4844	4927	5010	5089	5165
280	4314	4413	4506	4596	4684	4768	4850	4930
290	4062	4167	4264	4358	4452	4536	4619	4700
300	3818	3929	4031	4128	4222	4313	4400	4481
310	3587	3699	3805	3906	4003	4097	4187	4276
320	3364	3480	3591	3696	3796	3892	3983	4074
330	3154	3278	3391	3498	3601	3698	3793	3883
340	2960	3087	3205	3316	3419	3519	3614	3707
350	2780	2910	3031	3144	3250	3352	3448	3540
360	2615	2747	2870	2984	3092	3194	3292	3385
370	2467	2601	2724	2839	2950	3051	3149	3242
380	2334	2468	2591	2706	2817	2917	3015	3109
390	2218	2350	2472	2587	2699	2796	2894	2988
400	2118	2246	2365	2477	2584	2685	2782	2874
420	1957	2075	2188	2295	2397	2495	2589	2678
440	1838	1946	2050	2151	2247	2339	2429	2515
460	1753	1851	1946	2039	2129	2217	2301	2383
480	1691	1779	1867	1953	2037	2118	2197	2275
500	1647	1727	1807	1887	1964	2041	2115	2188
550	1586	1651	1716	1782	1847	1910	1974	2036
600	1566	1621	1676	1739	1786	1841	1896	1950
650	1570	1617	1665	1712	1760	1808	1856	1900
700	1586	1628	1670	1712	1754	1797	1839	1881
750	1612	1649	1686	1724	1762	1800	1838	1876
800	1644	1677	1711	1745	1780	1814	1849	1884
850	1678	1709	1740	1771	1803	1834	1866	1898
900	1716	1744	1773	1802	1831	1860	1889	1918
950	1756	1782	1809	1835	1862	1889	1917	1943
1000	1797	1821	1846	1871	1896	1921	1946	1972
1050	1839	1861	1885	1908	1931	1955	1979	2003
1100	1880	1902	1923	1945	1967	1989	2012	2034
1150	1923	1943	1963	1985	2005	2026	2047	2069
1200	1965	1984	2004	2023	2043	2063	2083	2103
1250	2008	2026	2044	2063	2082	2100	2119	2138
1300	2051	2068	2085	2103	2121	2139	2157	2175

TABLE XXVII. Thermal Conductivity of Xenon (*Continued*)

T, K	$\lambda \cdot 10^8$, kW/(m · K) at p, bar							
	400	420	440	460	480	500	550	600
180	7935	7993	8051	8107	8162			
190	7582	7643	7701	7757	7812	7868	8003	8134
200	7243	7304	7366	7422	7479	7537	7674	7810
210	6920	6981	7038	7101	7160	7219	7365	7503
220	6604	6671	6733	6792	6856	6916	7062	7207
230	6307	6372	6434	6503	6562	6627	6777	6922
240	6020	6089	6154	6221	6284	6349	6502	6651
250	5748	5818	5885	5955	6021	6085	6243	6395
260	5489	5563	5632	5700	5768	5834	5996	6152
270	5243	5315	5387	5459	5528	5596	5763	5920
280	5006	5082	5157	5228	5301	5371	5540	5699
290	4784	4861	4937	5010	5083	5157	5326	5490
300	4571	4649	4728	4803	4879	4953	5125	5293
310	4364	4445	4528	4604	4683	4757	4931	5103
320	4160	4244	4326	4407	4488	4562	4743	4914
330	3971	4057	4139	4220	4300	4377	4564	4729
340	3795	3881	3964	4045	4125	4202	4389	4560
350	3631	3716	3800	3881	3962	4039	4226	4397
360	3475	3562	3646	3728	3808	3885	4071	4244
370	3333	3420	3504	3585	3664	3743	3927	4103
380	3200	3286	3370	3452	3530	3608	3792	3968
390	3077	3164	3247	3328	3407	3485	3669	3842
400	2963	3049	3133	3214	3292	3368	3550	3723
420	2765	2849	2930	3009	3086	3161	3340	3510
440	2598	2679	2758	2834	2909	2982	3157	3322
460	2462	2539	2616	2688	2760	2830	3000	3160
480	2350	2423	2495	2565	2634	2701	2865	3019
500	2259	2328	2396	2463	2529	2593	2749	2898
550	2098	2158	2218	2277	2334	2391	2530	2665
600	2003	2056	2109	2161	2212	2262	2386	2507
650	1950	1997	2044	2090	2135	2181	2292	2401
700	1925	1965	2007	2048	2090	2130	2231	2330
750	1915	1952	1989	2028	2065	2102	2194	2284
800	1918	1953	1987	2021	2055	2090	2174	2258
850	1930	1961	1992	2024	2056	2088	2165	2243
900	1948	1977	2006	2036	2064	2094	2166	2225
950	1971	1998	2025	2053	2080	2107	2175	2243
1000	1997	2023	2048	2074	2099	2125	2189	2252
1050	2027	2050	2074	2098	2122	2147	2206	2266
1100	2057	2079	2102	2124	2147	2170	2226	2284
1150	2090	2111	2133	2154	2175	2197	2251	2305
1200	2123	2144	2164	2184	2205	2225	2276	2327
1250	2158	2177	2196	2216	2235	2254	2303	2352
1300	2194	2212	2230	2249	2267	2286	2332	2379

TABLE XXVII. Thermal Conductivity of Xenon (*Continued*)

T, K	$\lambda \cdot 10^8$, kW/(m · K) at p, bar							
	650	700	750	800	850	900	950	1000
190	8269	8393	8519					
200	7949	8079	8204	8333	8450	8569	8691	8807
210	7639	7773	7903	8032	8155	8274	8396	8514
220	7346	7483	7611	7743	7866	7991	8113	8231
230	7062	7201	7338	7468	7596	7720	7847	7965
240	6796	6936	7075	7207	7338	7467	7588	7711
250	6544	6684	6824	6958	7092	7219	7349	7472
260	6303	6448	6588	6729	6859	6988	7117	7243
270	6072	6218	6362	6503	6639	6772	6897	7027
280	5855	6003	6150	6291	6427	6561	6693	6821
290	5648	5800	5949	6088	6229	6362	6492	6622
300	5453	5605	5754	5894	6036	6173	6308	6435
310	5265	5416	5564	5706	5848	5985	6120	6257
320	5074	5228	5377	5520	5667	5802	5943	6080
330	4894	5052	5200	5342	5494	5631	5767	5903
340	4726	4884	5035	5177	5326	5466	5601	5735
350	4567	4724	4877	5019	5163	5306	5438	5577
360	4416	4573	4722	4868	5012	5151	5287	5417
370	4270	4430	4577	4724	4865	5003	5141	5268
380	4134	4293	4441	4585	4724	4860	5001	5127
390	4007	4163	4314	4459	4597	4728	4861	4985
400	3886	4043	4192	4334	4472	4602	4729	4853
420	3670	3824	3971	4111	4245	4372	4496	4613
440	3479	3630	3773	3911	4043	4170	4290	4406
460	3314	3461	3602	3737	3866	3990	4110	4224
480	3169	3312	3450	3582	3709	3832	3949	4060
500	3043	3182	3316	3445	3570	3690	3806	3916
550	2796	2923	3047	3167	3285	3398	3509	3616
600	2625	2740	2854	2965	3073	3180	3283	3385
650	2508	2613	2716	2817	2918	3016	3113	3209
700	2428	2523	2618	2711	2803	2894	2984	3074
750	2374	2462	2549	2635	2720	2804	2888	2971
800	2340	2422	2502	2582	2661	2740	2817	2895
850	2319	2395	2470	2540	2618	2690	2765	2837
900	2310	2381	2451	2521	2590	2659	2727	2796
950	2309	2376	2442	2508	2573	2638	2702	2766
1000	2315	2378	2440	2502	2563	2625	2685	2746
1050	2326	2385	2443	2502	2561	2619	2677	2734
1100	2340	2396	2452	2507	2562	2618	2673	2728
1150	2358	2412	2464	2518	2570	2623	2675	2728
1200	2378	2429	2480	2530	2581	2631	2681	2731
1250	2400	2449	2498	2546	2594	2642	2690	2738
1300	2425	2472	2519	2565	2611	2657	2703	2749

REFERENCES

Part 1

Chapter 1

1. Maradudin, A. A., Montroll, E. W. and Weiss, G. H. Theory of lattice dynamics in the harmonic approximation. Solid state physics: advances in research and applications, Suppl. 3. Academic Press, 1963.

1.2. Leibfried, G. and Ludwig, W. Theory of anharmonic effects in crystals. Solid state physics, 12, pp. 276–456. Academic Press, 1961.

1.3. Klein M. L., Horton G. K., Feldman J. L.—Phys. Rev., 1969, *184,* 968.

1.4. Horton G. K., Leech J. W.—Proc. Phys. Soc., 1963, *82,* 816.

1.5. Abovskiy, V. A. and Levin, Yu. L. Teplofiz., vys. temp., 1970, *9,* 731.

1.6. Abovskiy, V. A. Investigation of thermodynamic properties of liquids of the zero valent group at high densities. Author's abstract of candidate thesis. Moscow Energetics Institute, Moscow, 1971.

1.7. Abovskiy. Abstracts of reports at the 17th all-union conference on low-temperature physics. Donetsk Institute of Engineering Physics, USSR Academy of Sciences. Donetsk, 1972.

1.8. Barker J. A. Lattice theories of the liquid state. Oxford, 1963.

1.9. Levin, Yu. L. In: Issledovaniya v oblasti vysokikh davleniy (Studies at high pressures). Transactions of the all-union institute of physical- and radio-engineering measurements, No. 5(35), 1971.

1.10. Blackman M. Handbuch der physik. vol. 7, T. I, Berlin, 1955, p. 383.

1.11. Bhatia A. B., Horton G. K.—Phys. Rev., 1955, *98,* 1715.

1.12. Peresada, V. N. Zh. eksp. teor, fiz., 1967, *53,* 605.

1.13. Vashchenko, V. Ya. and Zubarev, V. N. Fiz. tv. tela, 1963, *5,* 886.

1.14. Zharkov, V. N. and Kalinin, V. A. Uravneniye sostoyaniya tverdykh tel pri vysokikh davleniyakh i temperaturakh (Equation of state of solids at high pressures and temperatures). Nauka Publishing House, Moscow, 1968.

1.15. Leibfried, G. Gittertheorie der mechanischen und thermischen Eigenschafter der Kristallen. Encyclopedia of Physics, S. Flugge, ed. vol. VII/1, pp. 104–324. Springer Verlag, 1955.

1.16. Horton G. K., Schiff H.—Can. J. Phys., 1958, *36*, 1127.

1.17. Losee D. L., Simmons R. O.—Phys. Rev., 1968, *172*, 944.

1.18. Manzheliye, V. G. Thermal properties of solidified gases. Author's abstract of doctoral thesis. Khar'kov, 1969.

1.19. Landau, L. D. and Lifshits, E. M. Statistical Physics. Pergamon Press, 1969.

1.20. Belogolovskiy, M. A., Tolpygo, K. B. and Troitskaya, Ye. P. Fiz. tv. tela, 1971, *13*, 2109.

1.21. DeLaunau J.—Solid State Phys., 1956, *2*, 219.

1.22. Lennard-Jones J. E., Ingham A. E.—Proc. Roy. Soc., 1925, *A107*, 636.

1.23. Bewilogua L., Gladun C.—Contemp. Phys., 1968, *9*, 277.

1.24. Rowlinson J. S.—Mol. Phys., 1965, *9*, 197.

1.25. Lidiard, A. B. Ionic conductivity of crystals. Encyclopedia of physics, S. Flugge, ed. vol. XX, pp. 246–349. Springer Verlag, 1957.

1.26. Foreman A. J., Lidiard A. B.—Phil. Mag., 1963, *8*, 97.

1.27. Peterson O. G., Batchelder D. N., Simmons R. O.—Phys. Rev., 1966, *150*, 703.

1.28. Batchelder D. N., Losee D. L., Simmons R. O.—Phys. Rev., 1967, *162*, 767.

1.29. Stewart J. W.—Phys. Rev., 1955, *97*, 578.

1.30. Stewart J. W.—J. Phys. Chem. Solids, 1958, *29*, 641.

1.31. Fagerstroem C. H., Hallet A. C. H.—Low Temp. Phys., 1965, *9*, 1092.

1.32. Fenichel C. H., Serin B.—Phys. Rev., 1966, *142*, 490.

1.33. Bolz L. H., Mauer F. A. Advances in X—Ray Analysis. 1963, *6*, 242.

1.34. Clusius K., Flubacher P., Piesbergen U., Schleich K., Sperandio A.—Z. Naturforsch., 1960, *15a*, 1.

1.35. Flubacher P., Leadbetter A. J., Morrison J. A.—Proc. Phys. Soc., 1961, *78*, 1449.

1.36. Van Witzenburg W., Stryland J. C.—Can. J. Phys., 1968, *46*, 811.

1.37. Crawford R. K., Daniels W. B.—J. Chem. Phys., 1969, *50*, 3171.

1.38. Crawford R. K., Daniels W. B.—Phys. Rev. Letters, 1968, 21, *6*, 367.

1.39. Stishov S. M., Fedosimov, V. I. and Makarenko, I. N. Equation of state and the crystallization of argon. Preprint of the Institute of Crystallography, USSR Academy of Sciences, Moscow, 1972.

1.40. Beaumont R. H., Chihara H., Morrison J. A.—Proc. Phys. Soc., 1961, *78*, 1462.

1.41. Clusius K., Riccoboni L.—Z. Phys. Chem., 1937, *B38*, 81.

1.42. Packard J. R., Swenson C. A.—J. Phys. Chem. Solids, 1963, *24*, 1405.

1.43. Eatwell A. L., Smith B. L.—Phil. Mag., 1961, *6*, 461.

1.44. Sears D. K., Klug H. P.—J. Chem. Phys., 1962, *37*, 3002.

1.45. Travers M., Jaquerod A.—Z. Phys. Chem. (Leipzig), 1903, *45*, 435.

1.46. Crommelin C.—Comm. Phys. Lab. Univ. Leiden, 1923, 162 c.

1.47. Crommelin C., Gibson R.—Comm. Phys. Lab. Univ. Leiden, 1927, 185 b.

1.48. Keesom W., Haantjes J.—Physica, 1935, *2*, 460; 1935, *2*, 986.

1.49. Henning F., Otto J.—Z. Phys., 1936, *37*, 633.

1.50. Bigeleisen J., Roth E.—J. Chem. Phys., 1961, *35*, 68.

1.51. Grilly E.—Cryogenics, 1962, *2(4)*, 226.

1.52. Crommelin C.—Comm. Phys. Lab. Univ. Leiden, 1913, 138 c; 1914, 140 a.

1.53. Born F.—Ann. Phys., 1922, *69*, 473.

1.54. Clark A. a. o.—Physika, 1951, *17*, 876.

1.55. Freeman M., Halsey G.—J. Phys. Chem., 1956, *60*, 1119.

1.56. Heastie R., Lefebvre C.—Proc. Phys. Soc. (London), 1960, *76*, 180.

1.57. Boato G., Scol'es G., Vallaury M.—Nuovo Cimento, 1962, *23*, 1041.

1.58. Ramsey W., Travers M.—Z. Phys. Chem., 1901, *38*, 641.

1.59. Petters K., Weil K.—Z. Phys. Chem., 1930, *A148*, 27.

1.60. Allen F., Moore R.—J. Am. Chem. Soc., 1931, *53*, 2522.

1.61. Keesom W., Mazur J., Meihuizen J.—Physica, 1935, *2*, 669.

1.62. Fisher B., MacMillan W.—J. Phys. Chem., 1958, *62*, 494.

1.63. Podgurscki H., Davis F.—J. Phys. Chem., 1961, *65,* 1343.
1.64. Salter L.—Trans. Farad. Soc., 1963, *59,* 657.
1.65. Grütter A., Shorrock J.—Nature, 1964, *204,* 1084.
1.66. Ebert L.—Österr. Chem. Ztg., 1954, *112,* 1.
1.67. Grilly E., Mills R.—Phys. Rev., 1957, *105(4),* 1140.
1.68. Bridgman P.—Rev. Modern Phys., 1946, *18,* 1.
1.69. Simon F.—Proc. Amer. Acad. Arts Sci., 1954, *83,* 319.
1.70. Gilvarry J.—Phys. Rev., 1956, *102,* 317.
1.71. Simon F., Ruhemann M., Edwards W.—Z. Phys. Chem., 1929, *B6,* 62.
1.72. Simon F., Glatzel G.—Z. Anorg. Allgem. Chem., 1929, *178,* 309.
1.73. Simon F., Ruhemann M., Edwards W.—Z. Phys. Chem. (Leipzig), 1930, *B6,* 331.
1.74. Keesom W., Lisman J.—Comm. Phys. Lab. Univ. Leiden, 1933, No. 224b.
1.75. Mills R., Grilly E.—Phys. Rev., 1955, *99,* 480.
1.76. Crawford R. K., Daniels W. B.—J. Chem. Phys., 1971, *55(12),* 5651.
1.77. Bridgman P.—Proc. Amer. Acad. Arts Sci., 1935, *70,* 1.
1.78. Swenson C.—Phys. Rev., 1953, *89,* 538.
1.79. Clusius K., Weigand K.—Z. Phys. Chem., 1939, *B 42,* 111; 1939, *B 43,* 79; 1940, *B 46,* 1.
1.80. Robinson D.—Proc. Roy. Soc. (London), 1954, *A225,* 393.
1.81. Michels A., Prins C.—Physica, 1962, *28,* 101.
1.82. Lahr P., Eversole W.—J. Chem. Eng. Data, 1962, *7,* 42.
1.83. Hardy W. H., Crawford R. K., Daniels W. B.—J. Chem. Phys., 1971, *54(3),* 1005.
1.84. Stryland J., Crawford R., Mastoor M.—Can. J. Phys., 1960, *38,* 1546.
1.85. Garside D., Smith B.—Physica, 1967, *37,* 369.
1.86. Michels A., Wassenaar T.—Physica, 1950, *16,* 253.
1.87. Verbeke O.—Cryogenics, 1970, *10,* 6.
1.88. Hoge H.—J. Res. NBS, 1950, *44,* 381.
1.89. Rabinovich, V. A. and Veksler, L. S. In: Teplofizicheskiye svoystva veshchestv i materialov (GSSSD. Ser. Fizicheskiye konstanty i svoystva veshchestv) (Thermophysical properties of substances and materials) (State service for standard reference data. series: physical constants and properties of substances), No. 2, p. 83. Moscow, 1970.
1.90. Kamerlingh Onnes H., Crommelin C.—Comm. Phys. Lab. Univ. Leiden, 1915, 147 d.
1.91. Cath P., Kamerlingh Onnes H.—Comm. Phys. Lab. Univ. Leiden, 1917, 152 b.
1.92. Crommelin C.—Comm. Phys. Lab. Univ. Leiden, 1924, Suppl., 60.
1.93. Verschaffelt J.—Comm. Phys. Lab. Univ. Leiden, 1928. Suppl., 64 d.
1.94. Gibbons R. M.—Cryogenics, 1969, *9(4),* 251.
1.95. Furukawa G.—Metrologia, 1971, *8,* 11.
1.96. Furukawa G., Saba W., Sweger D., Plumb H.—Metrologia, 1970, *6,* 35.
1.97. Compton J.—Metrologia, 1970, *6,* 103.
1.98. Olszewsky K.—Trans. Roy. Soc. (London), 1895, *A186,* 253.
1.99. Ramsay W., Travers M.—Trans. Roy. Soc. (London), 1901, *A197,* 47.
1.100. Clusius K., Frank A.—Z. Elektrochem., 1943, *49,* 308.
1.101. Narinskiy, G. B. Koslorod, No. 3, p. 9, 1957.
1.102. Michels A., Levelt J., de Graaff W.—Physica, 1958, *24,* 659.
1.103. Luriy, L. I. and Rabinovich, V. A. Zhurn. fiz. khim., 1966, *60(3),* 709.
1.104. Vasserman, A. A., Kazavchinskiy, Ya. Z. and Rabinovich, V. A. Thermophysical properties of air and air components. Halsted Press, New York, 1971; also available in English from NTIS as TT70-50095, A18.†
1.105. Van Itterbeek A., De Boelpaep J., Verbeke O., Theeuwes F., Staes K.—Physica, 1964, *30,* 2119.
1.106. McGain W., Ziegler W.—J. Chem. Eng. Data, 1967, *12,* 199.

† Changed in proof—T.B.S.

1.107. Verbeke O., Jansoone V., Gielen R., De Boelpaep J.—J. Phys. Chem., 1969, *73(12)*, 4076.

1.108. Van Itterbeek A., Verbeke O., Staes K.—Physica, 1963, *29,* 742.

1.109. Bowman D., Aziz R., Lim C.—Can. J. Phys., 1969, *47,* 267.

1.110. Pool R., Shields B., Staveley L.—Nature, 1958, *181,* 831.

1.111. Thomaes G., Steenwinkel R.—Mol. Phys., 1962, *5,* 301.

1.112. Radovskiy, I. S. In: Teplo- i massoperenos (Heat and mass transfer), Energiya Press, Moscow-Leningrad, 1966, *7,* 152.

1.113. Mathias E., Kamerlingh Onnes H., Crommelin C.—Comm. Phys. Lab. Univ. Leiden, 1912, 131a.

1.114. Chashkin, Yu. R. Experimental study of factors affecting the form of the singularity of specific heat C_v at the liquid-vapor critical point. Author's abstract of candidate thesis. Moscow State University, Moscow, 1968.

1.115. Justi E.—Z. Physik, 1935, *36(4)*, 571.

1.116. Meihuizen J., Crommelin C.—Physica, 1937, *4,* 1.

1.117. Michels A., Wassenaar T., Zwietering Th.—Physica, 1952, *18,* 1.

1.118. Theeuwes F., Bearman R.—J. Chem. Thermodyn., 1970, *2,* 179.

1.119. Thomaes G., Steenwinkel R.—Nature, 1962, *193,* 160.

1.120. Heuse W., Otto J.—Z. Tech. Physik, 1932, *13,* 277.

1.121. Theeuwes F., Bearman R.—J. Chem. Thermodyn. 1970, *2, 4,* 507.

1.122. Weinberger M., Schneider W.—Can. J. Chem., 1952, *30,* 422.

1.123. Whiteway S., Mason S.—Can. J. Chem., 1953, *31,* 569.

1.124. Habgood H., Schneider W.—Can. J. Chem, 1954, *32,* 98.

Chapter 2

2.1. Beattie J., Bridgeman O.—Proc. Amer. Acad. Arts Sci., 1928, *63,* 229.

2.2. Benedict M., Webb G., Rubin L.—J. Chem. Phys., 1940, *8,* 334.

2.3. Martin G., Hou Y.—AICREJ., 1955, *1,* 142.

2.4. Vukalovich, M. P. and Novikov, I. I. Uravneniye sostoyaniya real'nykh gazov (Equation of state of real gases). Gosenergoizdat Publishing House, Moscow-Leningrad, 1948.

2.5. Mayer J.—J. Chem. Phys., 1937, *5,* 67.

2.6. Bogolyubov, N. N. Problemy dinamicheskoy teorii v statisticheskoy fizike (Problems of dynamic theory in statistical physics). Gostekhizdat Publishing House, Moscow-Leningrad, 1946.

2.7. Kazavchinskiy, Ya. Z. Dokl. AN SSSR, 1954, *95,* 1005.

2.8. Kazavchinskiy, Ya. Z. Teploenergetika, 1958, No. 7, 44; 1960, No. 11, 59.

2.9. Vukalovich, M. P., Altunin, V. V. and Spiridonov, G. A. Teplofiz. vys. temp., 1967, *5(2)*, 265; 1967, *5(3)*, 528. In: Reports of a scientific-technical conference on scientific research achievements during 1966–1967. Heat power section. Thermophysics subsection. Moscow Energetics Institute, Moscow, 1967.

2.10. Altunin, V. V. and Gadetskiy, O. G. In: Teplofizicheskiye svoystva veshchestv (Thermophysical properties of substances). Transactions of the Moscow Energetics Institute, No. 75. Moscow, 1970.

2.11. Kazavchinskiy, Ya. Z. Inzh. fiz. zhurn., 1964, *7,* 129.

2.12. Vasserman, A. A. Zhurn. fiz. khim., 1964, *38(10)*, 2386.

2.13. Vasserman, A. A. and Slyn'ko, A. G. In: Teplofizicheskiye sovystva gazov (Thermophysical properties of gases), p. 99. Nauka Publishing House, Moscow, 1970.

2.14. Kazavchinskiy, Ya. Z. and Serdyuk, L. S. In: Teplofizicheskiye svoystva zhidkostey i gazov pri vysokikh temperaturakh i plazmy (Thermophysical properties of liquids and gases at high temperatures and of plasma), p. 161. Standards Publishing House, Moscow, 1969.

2.15. Kazavchinskiy, Ya. Z. and Taran, V. N. Ibid, p. 171.

2.16. Strobridge T.—NBS Tech. Note, 129, Jan., 1962; also available as PB 161630 from NTIS.†

† Changed in proof—T.B.S.

2.17. Weber L.—NBS Report, NBS Report 9710, 1968; also available is *J Res NBS*, 1970, *74A*(*1*), 96–129.†
2.18. Stewart R. The thermodynamic properties of oxygen. PhD Thesis University of Iowa, 1966.†
2.19. Goodwin R.—J. Res. NBS, 1967, *71A*, 203.
2.20. Argon, helium and the rare gases. Ed. G. A. Cook. N.Y. Interscience Publishers, 1961.
2.21. Crommelin C., Martinez J., Onnes H.—Comm. Phys. Lab. Univ. Leiden, 1920, 154a.
2.22. Holborn L., Otto J.—Z. Physik, 1924, *23*, 77.
2.23. Holborn L., Otto J.—Z. Physik, 1925, *33*, 1.
2.24. Holborn L., Otto J.—Z. Physik, 1926, *38*, 359.
2.25. Michels A., Gibson R.—Ann. Phys., 1928, *87*, 850.
2.26. Nicholson G., Schneider W.—Can. J. Chem., 1955, *33*, 589.
2.27. Michels A., Wassenaar T., Louwerse P.—Physica, 1960, *26*, 539.
2.28. Gladun A.—Cryogenics, 1966, *6*(*1*), 31.
2.29. Sullivan J., Sonntag R.—Adv. Cryogenic Eng., 1967, *12*, 706.
2.30. Runge R.—VDI Forchungsheft, 1968, 529.
2.31. Onosovskiy, Ye. V. and Moroz, A. I. In: Teplofizicheskiye svoystva veshchestv i materialov (GSSSD, Ser. Fizicheskiye konstanty i svoystva veshchestv), No. 2, p. 95. Moscow, 1970.
2.32. Onosovskiy, Ye. V. Ibid, No. 3, p. 31. Moscow, 1971.
2.33. Streett W.—J. Chem. Eng. Data, 1971, *16*, 289.†
2.34. Dedikov, Yu. A., Medvedev, V. A. and Orlova, M. P. In: Transactions of the all-union research institute of physical and radio engineering, No. 34(4). Moscow, 1972; Zhurn. fiz. khim., 1972, *46*, 2669.
2.35. Otto J. Handbuch der Experimental Physik. vol. 8. Leipzig, 1929.
2.36. Yendall E.—Adv. Cryogenic Eng., 1960, *4*, 47.
2.37. McCarty R., Stewart R., Timmerhaus K.—Adv. Cryogenic Eng., 1963, *8*, 135.
2.38. Crain R., Sonntag R.—Adv. Cryogenic Eng., 1966, *11*, 379.
2.39. Roebuck J.—Proc. Amer. Acad. Arts Sci., 1925, *60*, 537.
2.40. McCarty R., Stewart R. Advances in thermophysical properties extreme temperatures and pressures. 3rd Symp. ASME, N.Y., 1965, p. 84.†
2.41. Kamerlingh Onnes H., Crommelin C.—Comm. Phys. Lab. Univ. Leiden, 1910, 118b.
2.42. Holborn L., Schultze H.—Ann. Phys., 1915, *47*, 1089.
2.43. Holborn L., Otto J.—Z. Phys., 1922, *10*, 367; 1924, *30*, 320.
2.44. Masson I., Dolley L.—Proc. Roy, Soc. (London), 1923, *A103*, 524.
2.45. Bridgman P.—Proc. Amer. Acad. Arts Sci., 1924, *59*, 171.
2.46. Bridgman P.—Proc. Amer. Acad. Arts Sci., 1935, *70*, 1; Phys. Rev., 1934, *45*, 930.
2.47. Tanner C., Masson I.—Proc. Roy. Soc., 1930, *A126*, 268.
2.48. Roebuck J., Osterberg H.—Phys. Rev., 1934, *46*, 785.
2.49. Clark A., Katz L.—Canad. J. Res., 1943, *21A*(*1*), 1.
2.50. Michels A., Wijker Hub., Wijker Hk.—Physica, 1949, *15*, 627.
2.51. Whalley E., Lupien Y., Schneider W.—Can. J. Chem., 1953, *31*, 722.
2.52. Ishkin, I. P. and Rogovaya, I. A. Zhurn. fiz. khim., 1957, *31*, 410.
2.53. Michels A., Levelt J., DeGraaff W.—Physica, 1958, *24*, 659.
2.54. Lecocq A.—J. Rech. Centre Nat. Rech. Sci., 1960, *55*, 50.
2.55. Rogovaya, I. A. and Kaganer, M. G. Zhurn. fiz. khim., 1961, *35*, 2135.
2.56. Bagatskiy, M. I., Voronel', A. V. and Gusak, V. G. Zhurn. eksp. i teor. fiz., 1962, *43*, 728.
2.57. Lestz S.—J. Chem. Phys., 1963, *38*, 2830.
2.58. Radovskiy, I. S. Zhurn. prikl. mekh. i teor. fiz., 1963, *2*, 159; 1964, *3*, 172.
2.59. Voronel', A. V., Snigirev, V. G. and Chashkin, Yu. R. Zhurn. eksp. i teor. fiz., 1965, *48*, 981.
2.60. Van Dael W., Van Itterbeek A., Cops A., Thoen J.—Physica, 1966, *32*, 611.

† Changed in proof—T.B.S.

2.61. Polyakov, Ye. V. and Tsiklis, D. S. Zhurn., fiz. khim., 1967, *41*, 2370. In: Teplofizicheskiye svoystva veshchestv i materialov (GSSSD. Ser. Fizicheskiye konstanty i svoystva veshchestv), No. 2, p. 102. Moscow, 1970.

2.62. Carome E., Cykowski C., Havlicek J., Swyt D.—Physica, 1968, *38*, 307.

2.63. Thoen J., Van Dael W.—Physica, 1969, *45*, 339.

2.64. Robertson S., Babb S., Scott G.—J. Chem. Phys., 1969, *50*, 2160.

2.65. Cragoe C.—J. Res. NBS, 1941, *26*, 495.

2.66. Benedict M.—J. Am. Chem. Soc., 1937, *59*, 2224.

2.67. Roebuck J., Murrel K. Temperature—its measurement and control in science and industry. N.Y., Reinhold, 1941, p. 60.

2.68. Yepifanova, V. I. et al. Razdeleniye vozdukha metodom glubokogo okhlazhdeniya (Separation of air by deep freezing), *1*. Mashinostroyeniye Publishing House, Moscow, 1964.

2.69. Beattie J., Brierley J., Barriault R.—J. Chem. Phys., 1952, *20*, 1613.

2.70. Whalley E., Schneider W.—Trans. ASME, 1954, *76*, 1001.

2.71. Trappeniers N., Wassenaar T., Wolkers G.—Physica, 1966, *32*, 1503.

2.72. Theeuwes F., Bearman R.—J. Chem. Thermodyn., 1970, *2*, 171.

2.73. Streett W., Staveley L.—J. Chem. Phys., 1971, *55*(*5*), 2495.

2.74. Jepson W., Rowlinson J.—J. Chem. Phys., 1955, *23*, 1599.

2.75. Beattie J., Barriault R., Brierley J.—J. Chem. Phys., 1951, *19*, 1219.

2.76. Chynaweth A., Schneider W.—J. Chem. Phys., 1952, *20*, 1777.

2.77. Michels A., Wassenaar T., Louwerse P.—Physica, 1954, *20*, 99.

2.78. Whalley E., Lupien Y., Schneider W.—Can. J. Chem., 1955, *33*, 633.

2.79. Voronel', A. V. Zhurn. fiz. khim., 1961, *35*, 958.

2.80. Altunin, V. V. and Gvozdkov, A. V. Inzh.-fiz. zhurn., 1969, *16*(*2*), 320.†

2.81. Vukalovich, M. P. and Altunin, V. V. Teploenergetika, 1963, No. 7, 5.

2.82. Tsymarnyy, V. A. and Zagoruchenko, V. A. Teplofiz. vys. temp., 1965, *3*(*3*), 473.†

2.83. Rabinovich, V. A., Tokina, L. A. and Berezin, V. M. Ibid, 1970, *8*(*4*), 789; 1973, *11*(*1*), 64.

2.84. Vukalovich, M. P. and Altunin, V. V. Teplofizicheskiye svoystva dvuokisi ugleroda (Thermophysical properties of carbon dioxide) Atomizdat Publishing House, Moscow, 1965; also available in English from Collets, London, 1968.†

2.85. Mikhaylov-Mikheyev, P. B. Metall gazovykh turbin (Metal for gas turbines). Mashgiz Publishing House, Moscow, 1958.

2.86. Veksler, L. S., Perel'shteyn, I. I. and Rabinovich, V. A. In: Teplofizicheskiye svoystva veshchestv i materialov (GSSSD. Ser. Fizicheskiye konstanty i svoystva veshchestv), No. 3, p. 35. Moscow, 1971.

2.87. Shaber A. Zum thermischen Verhalten fluider Stoffe. Diss. Karlsruhe, 1965.

2.88. Nedostup, V. I. Investigation of the properties of gas mixtures by the method of thermodynamic similarity. Author's abstract of candidate thesis. Odessa, 1966.

2.89. Morsy T. Zum termischen und kalorischen Verhalten realer fluider Stoffe. Diss. Karlsruhe, 1963.

2.90. Nedostup, V. I. In: Proceedings of the all-union scientific and technical conference on thermodynamics. Reports of sections ''thermodynamics of phase transitions, flow and irreversible processes'' and ''thermophysical properties of substances''. Leningrad Institute of Refrigeration Technology. Leningrad, 1970.

2.91. Holleran E.—J. Chem. Phys., 1967, *47*, 5318.

2.92. Koeppe W.—Kältetechnik, 1962, *14*(*12*), 399.

2.93. Kazavchinskiy, Ya. Z. Dokl. AN SSSR, 1965, *161*, 1127.

2.94. Gyorg D., Obert E.—AICHEJ, 1964, *10*, 625.

2.95. Kazavchinskiy, Ya. Z., Kudashev, V. I. and Nedostup, V. I. Teplofiz. vys. temp., 1966, *4*(*4*), 513.

2.96. Kozlov, A. D. and Spiridonov, G. A. In: Abstracts of reports at the all-union thermo-

† Changed in proof—T.B.S.

physics conference on high-temperature properties of substances. Odessa Institute of Refrigeration Technology. Odessa, 1970.

2.97. London, F. Trans. Farad. Soc., 1937, *33,* 8.

2.98. Lippold H.—Cryogenics, 1969, *9,* 112.

2.99. Streett W., Staveley L.—J. Chem. Phys., 1969, *50,* 2302.

2.100. Blagoy, Yu. P. and Sorokin, V. A. Zhurn. fiz. khim, 1970, *44,* 2745; In: Termodinamicheskiye i termokhimicheskiye konstanty (Thermodynamic and thermophysical constants), p. 97. Nauka Publishing House, Moscow, 1970.

2.101. Gavrilko, V. G. Thermal expansion of solidified inert gases and methane. Author's abstract of candidate thesis. Khar'kov, 1969.

2.102. Fowler, R. H. and Guggenheim, E. A. Statistical thermodynamics. Cambridge University Press, 1949.

2.103. Rossi J., Danon F.—Disc. Farad. Soc., 1965, *40,* 97.

2.104. Axilrod B.—J. Chem. Phys., 1951, *19,* 724.

2.105. Brown I.—Proc. Phys. Soc., 1966, *89,* 987.

2.106. Pollack G.—Rev. Modern Phys., 1964, *36,* 748.

2.107. Shakhporonov, M. I. Molekulyarnaya teoriya rastvorov (Molecular theory of solutions). Gostekhizdat Publishing House, Moscow, 1956.

2.108. Holleran E.—J. Chem. Phys., 1968, *49(1)*, 39.

2.109. Hirschfelder, J. O. Curtiss, C. F. and Bird, R. B. Molecular theory of gases and liquids. John Wiley & Sons, New York, 1954.

2.110. Barker J., Leonard P., Pompe A.—J. Chem. Phys., 1966, *44(11)*, 4206.

2.111. Nedostup, V. I. Zhurn. fiz. khim., 1970, *44(9)*.

2.112. Holleran E., Sinka I.—J. Chem. Phys., 1971, *55(9)*, 4260.

2.113. Mason E., Spurling T. The virial equation of state. Oxford Press, 1969.

2.114. Oishi J.—J. Sci. Res. Inst. Tokio, 1949, *43,* 220.

2.115. Cottrell T., Hamilton R., Taubinger R.—Trans. Farad. Soc., 1956, *52,* 1310.

2.116. Dymond J., Smith E. The virial coefficients of gases. Oxford Press, 1969.

2.117. Pool R., Saville G., Herington T., Schields B., Staveley L.—Trans. Farad. Soc., 1962, *58,* 1692.

2.118. Fender B., Halsey G.—J. Chem. Phys., 1962, *36,* 1881.

2.119. Weir R., Wynn Jones I., Rowlinson J., Saville G.—Trans. Farad. Soc., 1967, *63,* 1320.

2.120. Birne M., Jones M., Staveley L.—Trans. Farad. Soc., 1968, *64,* 1747.

2.121. Reeves G., Whytlaw-Gray R.—Proc. Roy. Soc., 1955, *A232,* 173.

2.122. Greenlief C., Constabaris G.—J. Chem. Phys., 1966, *44,* 4649.

2.123. Holleran E.—J. Chem. Thermodyn., 1970, *2,* 779.

2.124. Michels A., Abels J., Ten Seldam C., De Graaff W.—Physica, 1960, *26,* 381.

2.125. Hall K., Canfield F.—Physica, 1967, *33,* 481.

2.126. Vasserman, A. A. Kreyzerova, A. Ya. and Nedostup, V. I. Teplofiz. vys. temp., 1971, *9,* 915.

2.127. Mason E., Munn R., Smith F.—Disc. Farad. Soc., 1965, *40,* 27.

2.128. Dymond J.—J. Chem. Phys., 1971, *54(9)*, 3675.

2.129. Rabinovich, V. A., Veksler, L. S. and Slobodyanik, P. T. In: Teplofizicheskiye svoystva gazov (Thermophysical properties of gases), p. 153. Nauka Publishing House, Moscow, 1970.

2.130. Pitzer K.—J. Chem. Phys., 1939, *7,* 583.

2.131. Guggenheim J.—J. Chem. Phys., 1945, *13,* 253.

2.132. Placzek F.—Z. Physik, 1931, *70,* 84.

2.133. Kazavchinskiy, Ya. Z., Kessel'man, P. M. and Rabinovich, V. A. Inzh.-fiz. zhurn., 1961, *4(12)*, 16.†

2.134. Kessel'man, P. M. Teplofiz. vys. temp., 1964, *2(6)*, 879.†

† Changed in proof—T.B.S.

2.135. Tsykalo, A. L. In: Teplofizicheskiye svoystva gazov. (Thermophysical properties of gases) Nauka Publishing House, Moscow, 1970.

2.136. Donth E.—Physica, 1966, *32*(*5*), 913.

2.137. Temperley, H. N., Rowlinson, J. and Rushbrooke, G. S. Physics of Simple Liquids. Wiley, New York, 1968.

2.138. Copeland D., Kestner D.—J. Chem. Phys., 1968, *49*(*12*), 5214.

2.139. Levelt M.—Physica, 1960, *26*, 5.

2.140. Rowlinson J.—Disc. Farad. Soc., 1965, *40*, 19xx.†

2.141. Nedostup, V. I. Inzh.-fiz. zhurn., 1966, *10*(*6*), 783.†

2.142. Kessel'man, P. M. and Kamenetskiy, V. R. In: Proceedings of the 3rd all-union conference on heat and mass transfer, *7*, p. 111. Nauka i Teknika Publishing House, Minsk, 1968.

2.143. Nedostup, V. I. and Slyn'kov, A. G. Teplifiz. vys. temp., 1970, *8*(*4*), 917.†

2.144. Perel'shteyn, I. I. In: Teplofizicheskiye svoystva veshchestv i materialov (GSSSD. Ser. Fizicheskiye konstanty i svoystva veshchestv), No. 4, p. 65. Moscow, 1971.

2.145. Mathias E., Crommelin C. A., Kamerlingh Onnes H.—Comm. Phys. Lab. Univ. Leiden, 1923, 162b.

2.146. Veksler, L. S. and Rabinovich, V. A. In: Teplofizicheskiye svoystva veshchestv i materialov (GSSSD. Ser Fizicheskiye svoystva veshchestv i materialov), No. 6, p. 94. Moscow, 1974.

2.147. Shavandrin, A. M., Potapova, N. M. and Chashkin, Yu. R. Ibid, No. 9. Moscow, 1975.

2.148. Mathias E., Crommelin C. A., Meihuizen J.—Physica, 1937, *4*(*11*), 1200.

2.149. Patterson H. S., Cripps R. S., Whitlaw-Gray R.—Proc. Roy. Soc., 1912, *A86*, 591.

Chapter 3

3.1. De Boer, J. Proc. Roy. Soc., 1952, *A215*, 4–29.

3.2. Hill, T. L. Statistical mechanics. McGraw-Hill, New York, 1956.

3.3. Fisher, I. Z. Statistical theory of liquids. University of Chicago Press, 1964.

3.4. Frenkel, J. Kinetic theory of liquids. Dover, New York, 1955.

3.5. Lennard-Jones J. E., Devonshire A. E.—Proc. Roy. Soc. 1937, *A163*, 53.

3.6. Kirkwood J. G.—J. Chem. Phys., 1950, *18*, 380.

3.7. Taylor W. J.—J. Chem. Phys., 1956, *24*, 454.

3.8. Vasserman, A. A. and Rabinovich, V. A. Thermophysical properties of liquid air and its components. Halsted Press, New York, 1970; also available in English from NTIS as TT69-55092, A12.

3.9. Dahler J. S., Hirshfelder J. O.—J. Chem. Phys., 1960, *32*, 330.

3.10. Barker J. A.—Proc. Roy. Soc., 1955, *A230*, 390; 1956; *A237*, 63.

3.11. Mayer, J. E. and Mayer, M. G. Statistical mechanics. Wiley, New York, 1940.

3.12. DeBoer J.—Physica, 1954, *20*, 655.

3.13. Cohen E., DeBoer J., Salsburg Z. W.—Physica, 1955, *21*, 137.

3.14. Rabinovich, V. A. and Abovskiy, V. A. Teplofiz. vys. temp., 1970, *8*(*1*), 46; Teplofizichiskiye svoystva veshchestv i materialov (GSSSD. Ser. Fizicheskiye konstanty i svoystva veshchestv), No. 3, p. 44. Moscow, 1971.

3.15. Rowlinson J. S.—Rev. Progr. Phys., 1965, *28*, 169.

3.16. Barker J. A.—Proc. Roy. Soc., 1961, *A259*, 442.

3.17. Chung H. S., Dahler J. S.—J. Chem. Phys., 1964, *40*, 2868; 1965, *43*, 2606.

3.18. Ree T. S., Ree T., Eyring H.—Proc. Nat. Acad. Sci. USA, 1962, *48*, 501.

3.19. Ree T. S., Ree T., Eyring H.—J. Phys. Chem., 1964, *68*, 1163.

3.20. Kirkwood J. G.—J. Chem. Phys., 1935, *3*, 300.

3.21. Levelt J. M. H., Cohen E. G. D. Studies in statistical mechanics. Vol. 2, part B. Amsterdam, 1964.

3.22. Verlet L.—Nuovo Cimento, 1960, *18*, 77.

† Changed in proof—T.B.S.

3.23. Verlet L., Levesque D.—Physica, 1962, *28,* 1124.
3.24. Percus J. K., Yevick G. J.—Phys. Rev., 1958, *110(1)*, 1.
3.25. Stell G.—Physica, 1963, *29,* 517.
3.26. Ornstein L. S., Zernike F.—Proc. Acad. Sci. Amst., 1914, *17,* 793.
3.27. Levesque D.—Physica, 1966, *32,* 1985.
3.28. Tsykalo, A. L. and Selevanyuk, V. I. Teplofiz. vys. temp., 1971, *9(5)*, 1068.
3.29. Zwanzig R. W.—J. Chem. Phys., 1954, *22,* 1420.
3.30. McQuarrie D. A., Katz.—J. Chem. Phys., 1966, *44,* 2393.
3.31. Barker J. A., Henderson D.—J. Chem. Phys., 1967, *47,* 2856.
3.32. Barker J. A., Henderson D.—J. Chem. Phys., 1967, *47,* 4714.
3.33. Barker J. A., Henderson D., Smith W. K.—Phys. Rev. Letters, 1968, *21,* 134.
3.34. Mansoori G. A., Canfield F. B.—J. Chem. Phys., 1969, *51,* 4958.
3.35. Ross M., Alder B. J.—J. Chem. Phys., 1967, *46,* 4203.
3.36. Oppenheim J., Mazur P.—Physica, 1957, *23,* 197.
3.37. McDonald J. R., Singer K.—Disc. Farad. Soc., 1967, *43,* 40.
3.38. Verlet L.—Phys. Rev., 1967, *159,* 98.
3.39. Yevseyev, A. M. and Chervin, V. G. Zhurn. fiz. khim., 1969, *43,* 400.
3.40. Toxvaerd S., Praestgaard E.—J. Chem. Phys., 1971, *54(12)*, 5440.
3.41. Michels A., Wassenaar T., Zwietering Th.—Physica, 1952, *18,* 67.
3.42. Rogovaya, I. A. and Kaganer, M. G. Zhurn. khim. fiz., 1960, *34,* 1933.
3.43. Van Itterbeek A., Verbeke O. Physica, 1960, *26,* 931.
3.44. Streett W. B.—J. Chem. Phys., 1967, *46(9)*, 3282.
3.45. Blagoy, Yu. P. and Sorokin, V. A. Zhurn. fiz. khim., 1968, *42,* 547.
3.46. Sorokin, V. A. and Blagoy, Yu. P. In: Termodinamicheskiye i terkhmokhimicheskiye konstanty (Thermodynamic and thermochemical constants). K. V. Astakhov, ed. Nauka Publishing House, Moscow, 1970, p. 97.
3.47. Michels A., Prins C.—Physica, 1962, *28,* 101.
3.48. Streett W., Staveley L.—Adv. Cryogenic Eng., 1968, *13,* 363.
3.49. Terry M., Lynch J., Bunclark M., Mansell K., Staveley L.—J. Chem. Thermodyn., 1969, *1,* 413.
3.50. Criffiths R. B.—Phys. Rev. Letters, 1965, *14,* 623.
3.51. Vincentini-Missoni M., Levelt-Sengers J. M. H. Green M. S.—Phys. Rev. Letters. 1969, *22,* 389.
3.52. Dobrovol'skiy, O. A. and Golubev, I. F. In: Khimiya i tekhnologiya produktov organicheskogo sinteza. Fiziko-khimicheskiye issledovaniya (Chemistry and technology of products of organic synthesis. Physico-chemical studies). Transactions of the state research and planning institute of the nitrogen industry and products of organic synthesis, No. 8, p. 14, 1971.
3.53. Golubev, I. F. and Dobrovol'skiy, O. A. Gazovaya promyshlennost', 1964, No. 5, 43.
3.54. Gladun C.—Cryogenics, 1966, *6(1)*, 27.
3.55. Gladun C.—Cryogenics, 1967, *7(7)*, 78.
3.56. Blagoy, Yu. P. and Sorokin, V. A. Zhurn. fiz. khim., 1970, *44(11)*, 2745.
3.57. Theeuwes F., Bearman R.—J. Chem. Thermodyn., 1970, *2(4)*, 501.
3.58. Leadbetter A., Thomaes H.—Trans. Faraday Soc., 1965, *61(1)*, 10.
3.59. Clusius K.—Z. Phys. Chem., 1929, *B4,* 1.
3.60. Clusius K.—Z. Phys. Chem., 1936, *B31,* 459.
3.61. Gladun C.—Cryogenics, 1971, *11(3)*, 205.
3.62. Jones G., Walker P.—Proc. Phys. Soc., 1956, *B69,* 1348.
3.63. Eucken A., Hauck F.—Z. Phys. Chem., 1928, *134,* 161.
3.64. Gladun C., Menzel F.—Cryogenics, 1970, *10(3)*, 210.
3.65. Garside D., Smith B.—Physica, 1967, *37,* 369.
3.66. Lim C., Aziz R.—Can. J. Phys., 1967, *45,* 1284.
3.67. Shenball, R. Pis'ma v ZhETF, 1967, *6(7)*, 719.
3.68. Güsewell D., Schmeissner F., Schmidt J.—Cryogenics, 1970, *10(2)*, 150.

3.69. Larson E., Naugle D., Adair T.—J. Chem. Phys., 1971, *54(6)*, 2429.

3.70. Naugle D.—J. Chem. Phys., 1972, *56(11)*, 5730.

3.71. Van Itterbeek A., Van Dael W., Grevendonk W.—Physica, 1959, *25*, 640.

3.72. Van Itterbeek A., Grevendonk W., Van Dael W., Forrez G.—Physica, 1959, *25*, 1255.

3.73. Dobbs E., Finegold L.—J. Acoust. Soc. Amer., 1960, *32*, 1250.

3.74. Van Itterbeek O., Van Dael W.—Cryogenics, 1961, *1*, 226.

3.75. Blagoy, Yu. P., Butko, A. Ye., Mikhaylenko, S. A. and Yakuba, V. V. Akusticheskiy zhurn., 1966, *12*, 405.

3.76. Aziz R., Bowman D., Lim C.—Can. J. Chem., 1967, *45*, 2079.

3.77. Bowman D., Lim C., Aziz R.—Can. J. Chem., 1968, *46*, 1175.

3.78. Blagoy, Yu. P. Butko, A. Ye., Mikhaylenko, S. A. and Yakuba, V. V. Zhurn. fiz. khim., 1967, *41*(*7*), 1699.

3.79. Vasserman, A. A. and Rabinovich, V. A. Inzh.-fiz. zhurn., 1967, *12*(*3*), 342.

3.80. Vasserman, A. A. and Rabinovich, V. A. Ibid, 1967, *13*(*2*), 190.

3.81. Vasserman, A. A. and Rabinovich, V. A. Zhurn. fiz. khim., 1968, *42*(*7*), 1616.

3.82. Zagoruchenko, V. A. and Zhuravlev, A. M. Thermophysical properties of gaseous and liquid methane. Halsted Press, New York, 1971; also available in English from NTIS as TT70-50097, A12.†

3.83. Golovskiy, Ye. A. and Tsymarnyy, V. A. Teploenergetika, 1969, *16*(*7*), 52.†

3.84. Vasserman, A. A., Golovskiy, Ye. A. and Tsymarnyy, V. A. Inzh.-fiz. zhurn., 1971, *20*(*4*), 734.

3.85. Vasserman, A. A. and Rabinovich, V. A. In: Teplofizicheskiye khrakteristiki veshchestv (GSSSD. Ser. Fizicheskiye konstanty i svoystva veshchestv) (Thermophysical parameters of substances (State service for standard reference data)), No. 1, p. 38. Standards Publishing House, Moscow, 1968.

3.86. Vasserman, A. A. and Kreyzerova, A. Ya. Izv. AN BSSR, seriya fiziko-energeticheskikh nauk, 1971, No. 1, 58.

3.87. Vasserman, A. A. and Kreyzerova, A. Ya. Zhurn., prikl. mekh. i teor. fiz., 1972, No. 2, 119.

3.88. Lanezos, C., Practical methods of applied analysis. Applied analysis. Prentice-Hall Englewood Cliffs, NJ, 1956.†

3.89. Rabinovich, A. A. Comprehensive study of thermophysical properties of helium analogs (Ne, Ar, Kr, Xe). Author's abstract of PhD thesis Azerbaijan Institute of Petroleum and Chemistry. Baku, 1970.

3.90. Goldman K., Scrase N. G.—Physica, 1969, *45*(*1*), 1.

3.91. Abovskiy, V. A. and Gopanovich, L. N. Zhurn. fiz. khim., 1971, *45*, 932.

Part 2

Chapter 4

4.1. Rankine A.—Proc. Roy. Soc. (London), 1910, *A83*, 516; Phys. Z., 1910, *11*, 497; 1910, *11*, 745.

4.2. Edwards R.—Proc. Roy. Soc. (London), 1928, *A119*, 578.

4.3. Trautz M., Baumann.—Ann. Phys., 1929, *2*, 733; Trautz M., Kipphan K.—Ann. Phys., 1929, *2*, 743; Trautz M., Ludewigs.—Ann. Phys., 1929, *3*, 409; Trautz M., Binkele H.—Ann. Phys., 1930, *5*, 561; Trautz M., Melster.—Ann. Phys., 1930, *7*, 409; Trautz M., Zink R.—Ann. Phys., 1930, *7*, 427; Trautz M., Heberling R.—Ann. Phys., 1931, *10*, 155; 1934, *20*, 118; Trautz M., Zimmerman H.—Ann. Phys., 1935, *22*, 189.

4.4 Van Itterbeek A., van Paemel O.—Physica, 1938, *5*, 1009; 1940, *7*, 273.

4.5. Wobser R., Müller F.—Kolloid-Beih., 1941, *52*, 165.

4.6. Johnston H., Grilly E.—J. Phys. Chem., 1942, *46*, 948.

† Changed in proof—T.B.S.

4.7. Kestin J., Leidenfrost W.—Trans. ASME, 1959, *81*, 321; Physica, 1959, *25*, 1033.
4.8. Thornton E.—Proc. Phys. Soc., 1960, *76*, 104; 1961, *77*, 1166.
4.9. Kestin J., Whitelaw J.—Physica, 1963, *29*, 335.
4.10. Dawe R., Smith E.—J. Chem. Phys., 1970, *52(2)*, 693.
4.11. Kestin J., Wakeham W., Watanabe K.—J. Chem. Phys., 1970, *53(10)*, 3773.
4.12. Guevara F., Stensland G.—Phys. Fluids, 1971, *14(3)*, 746.
4.13. Schultze H. Ann. Phys., 1901, *5*, 140.
4.14. Vasilesco V.—Ann. Phys. (Paris), 1945, *20*, 137–176; 292–334.
4.15. Bonilla C., Wang S., Weiner H.—Trans. ASME, 1956, *78*, 1285.
4.16. Rocco A., Halford J.—J. Chem. Phys., 1958, *28(6)*, 1152.†
4.17. Cronin D.—Amer. J. Phys., 1965, *33*, 835.
4.18. Rigby M., Smith E.—Trans. Farad. Soc., 1966, *62*, 54.
4.19. Clarke A., Smith E.—J. Chem. Phys., 1968, *49(9)*, 3988.
4.20. Timrot, D. L., Serednitskiya, M. A. and Trakuyeva, S. A. Teploenergetika, 1969, No. 1, 83.
4.21. Guevara F., McInteer B., Wageman W.—Phys. Fluids, 1969, *12(12)*, 2493.
4.22. Kalelkar A., Kestin J.—J. Chem. Phys., 1970, *52(8)*, 4248.
4.23. Nasini A., Rossi C.—Gazz. Chem. Ital., 1928, *58*, 433.
4.24. Clifton D.—J. Chem. Phys., 1963, *38*, 1123.
4.25. Goldblatt M., Guevara F., McInteer B.—Phys. Fluids, 1970, *13(11)*, 2873.
4.26. Rabinovich, V. A. and Kryuchkov, V. A. In: Teplofizicheskiye svoystva veshchestv i materialov (GSSSD. Ser. Fizicheskiye konstanty i svoystva veshchestv), No. 2, p. 118. Moscow, 1970.
4.27. Hanley H., Childs G.—Science, 1968, *159(3819)*, 1114.†
4.28. Albertoni S., Cercignani C., Gotusso L.—Phys. Fluids, 1963, *6*, 993.
4.29. Raw C., Ellis C.—J. Chem. Phys., 1958, *28*, 1198.
4.30. Macwood G.—Physica, 1938, *5*, 374, 763.
4.31. Michels A., Gibson R.—Proc. Roy. Soc., 1931, *A134*, 288.
4.32. Newell G.—Z. Angew. Math. Phys., 1959, *10*, 160.
4.33. Weber S.—Ann. Phys., 1927, *82*, 479.
4.34. Curie M., Lepape A.—J. Phys. Radium, 1931, *2*, 392.
4.35. Kannuluik W., Carman E.—Proc. Phys. Soc. (London), 1952, *B65*, 701.
4.36. Keyes F.—Trans. ASME, 1954, *76*, 809; 1955, *77*, 1395.
4.37. Zaytseva, L. S. Zhurn. tekhn. fiz., 1959, *29(4)*, 497.
4.38. Sengers J., Bolk W., Stigter C.—Physica, 1964, *30*, 1018.
4.39. Saxena V., Saksena M., Saxena S.—Indian J. Phys., 1966, *40*, 597.
4.40. Gupt G., Saxena S.—Can. J. Phys., 1967, *45*, 1418.
4.41. Gandhi J., Saxena S.—Brit. J. Appl. Phys., 1967, *18(6)*, 807; Mol. Phys., 1967, *12(1)*, 57.†
4.42. Srivastava, B. N., Das Gupta, A., Brit. J. Appl. Phys., 1967, *18(7)*, 945.†
4.43. Saxena V., Saxena S.—J. Chem., Phys., *48(12)*, 5662.†
4.44. Vargaftik, N. B. and Yakush, L. V. In: Teplofizicheskiye svoystva gazov. Nauka Publishing House, Moscow, 1970.
4.45. Schwarze W.—Ann. Phys., 1903, *11*, 303; Z. Phys., 1903, *4*, 229.
4.46. Eucken A.—Z. Phys., 1911, *12*, 1101; 1913, *14*, 324.
4.47. Weber S.—Ann. Phys., 1917, *54*, 437.
4.48. Dickins B.—Proc. Roy. Soc. (London), 1934, *A143*, 517.
4.49. Kannuluik W., Martin L.—Proc. Roy. Soc. (London). 1934, *A144*, 496.
4.50. Kannuluik W., Donald H.—Austral. J. Sci. Res., 1950, *3A*, 417.
4.51. Schottky W.—Z. Elektrochem., 1952, *56(9)*, 889.
4.52. Rothman A., Bromley L.—Ind. Eng. Chem., 1955, *47(5)*, 899.

† Changed in proof—T.B.S.

4.53. Schäfer K., Reiter F.—Naturwissenshaften, 1956, *43,* 296; Z. Elektrochem., 1957, *61,* 1230.
4.54. Vines R.—Trans. ASME, Ser. C, 1960, *2,* 48.
4.55. Vargaftik, N. B. and Zimina, N. Kh. Teplofiz. vys. temp., 1964, *2(5)*, 716.
4.56. Timrot, D. L. and Umanskiy, A. S. Ibid, 1965, *3(3)*, 381; 1966, *4(2)*, 000.†
4.57. Collins D., Greif R.—Int. J. Heat Mass Transfer, 1965, *8,* 9.
4.58. Collins D., Menard W.—Int. J. Heat Mass Transfer, 1966, *9,* 56.
4.59. Saxena V., Saxena S.—Chem. Phys. Letters, 1968, *2,* 44.
4.60. Bailey B., Kellner K.—Physica, 1968, *39,* 444.
4.61. Grambhir R., Saxena S.—Mol. Phys., 1966, *11,* 3.
4.62. Gupta G.—Int. J. Heat Mass Transfer, 1967, *10,* 7.
4.63. Saxena V., Saxena S.—J. Chem. Phys., 1969, *51(8)*, 3361.
4.64. Umanskiy, A. S., Timrot, D. L. and Koroleva, V. V. In: Teplofizicheskiye svoystva gazov. Nauka Publishing House, Moscow, 1970.
4.65. Hanley H.—NBS Tech. Note, 1966, No. 333.
4.66. Hanley H., Childs G.—NBS Tech. Note, 1967, No. 352.
4.67. Hanley H., Klein M.—NBS Tech. Note, 1967, No. 360.
4.68. Dymond J., Rigby M., Smith E.—Phys. Fluids, 1966, *9(6)*, 1222.
4.69. Dymond J., Alder B.—J. Chem. Phys., 1969, *51,* 309.
4.70. Guggenheim E., McGlashan M.—Proc. Roy. Soc., 1960, *A225,* 456.
4.71. Dymond J., Rigby M., Smith E.—J. Chem. Phys., 1965, *42,* 2801.
4.72. Transport Phenomena in Fluids. Ed. H. J. M. Hanley. Marcel Dekker, NY, 1969.†
4.73. Barua A., Chakraborti P.—Physica, 1961, *27,* 733.
4.74. Chakraborti P.—Physica, 1963, *29,* 227.
4.75. Rabinovich, V. A., Vasserman, A. A., Nedostup, V. I. and Tsesis, M. A. In: Teplofizicheskiye svoystva veshchestv i materialov (GSSSD. Ser. Fizicheskiye konstanty i svoystva veshchestv), No. 3. Moscow, 1971.
4.76. Vasil'kovskaya, T. N. and Golubev, I. F. In: Khimiya i tekhnologiya produktov organicheskogo sinteza. Fizikokhimicheskiye issledovaniya. Transactions of the state research and planning institute of the nitrogen industry and products of organic synthesis, No. 8, p. 99. Moscow, 1971.
4.77. Golubev, I. F. and Shpagina, I. B. Ibid, p. 91.
4.78. Tufeu R. Etude experimentale en fonction de la temperature et de la pression de la conductivite thermique de I'ensemble de gas rares et des melanges helium-argon. These Le Grade de Docteur es Sciences. Paris, 1971.
4.79. LeNeindre B. Contribution à l'étude expérimentale de la conductivité thermique de quelques fluides à haute température et à haute pression. These de doctorat d'etat es Sciences. Paris, 1969.
4.80. Tarzimanov, A. A. and Arslanov, V. A. In: Proceedings of the 3rd all-union conference on heat and mass transfer, *7,* p. 81. Institute of heat and mass transfer, belorussian academy of sciences. Minsk, 1972.
4.81. Golubev, I. F. and Shpagina, I. B. In: Khimiya i tekhnologiya azotnykh udobreniy i produktov organicheskogo sinteza. Fiziko-khimicheskiye issledovaniya. (Chemistry and technology of nitrogen fertilizers and products of organic synthesis. Physico-chemical studies), p. 62. State research and planning institute of the nitrogen industry and products of organic synthesis. Moscow, 1969.
4.82. Faubert F. M., Springer G. S.—J. Chem. Phys., 1972, *57(6)*, 2333.
4.83. Sprinqer G. S., Wingeier E. W.—J. Chem. Phys., 1973, *59(5)*, 2747.
4.84. Hanley H. J. M.—J. Phys. Chem. Ref. Data, 1973, *2(3)*, 619.

† Changed in proof—T.B.S.

Chapter 5

5.1. Flynn G., Hanks R., Lemaire N., Ross J.—J. Chem. Phys., 1963, *38,* 154.

5.2. Förster S.—Cryogenics, 1963, *3*(*3*), 176.

5.3. Kestin J., Nagashima A.—J. Chem. Phys., 1964, *40,* 3648.

5.4. Trappeniers N., Botzen A., Van den Berg H., Van Oosten J.—Physica, 1964, *30,* 985.

5.5 Reynes E., Thodos G.—J. Chem. Eng. Data, 1966, *11,* 137.

5.6. Rabinovich, V. A., Kryuchkov, V. A. and Zikeyev, Ye. A. In: Teplofizicheskiye svoystva veshchestv i materialov (GSSSD. Ser. Fizicheskiye konstanty i svoystva veshchestv), No. 7, Moscow, 1970.

5.7. Slyusar', V. P., Rudenko, N. S. and Tret'yekov, V. M. Ibid, No. 7, 1973.

5.8. Rudenko N. S., Schubnikow L. W.—Phys. Z. Sovjetunion, 1934, *6*(*5*), 470.

5.9. Van Itterbeek A., Van Paemel O.—Physica, 1941, *8,* 133.

5.10. Golubev, I. F. and Petrov, V. A. In: Trudy GIAP [Transactions of the state research and planning institute of the nitrogen industry and products of organic synthesis), No. 2, p. 5. Moscow, 1953.

5.11. Michels A., Botzen A., Schuurman W.—Physica, 1954, *20,* 1141.

5.12. Ivasaki H.—Sci. Repts. Tohoku Univ., Ser. 1954, *A6*(*3*), 296.

5.13. Kestin J., Pilarczyk K.—Trans. ASME, 1954, *76,* 987.

5.14. Zhdanova, N. F. Zhur. eksp. i teor. fiz., 1956, *31, 4*(*10*), 724.

5.15. Makita T.—Rev. Phys. Chem. Japan, 1957, *27,* 16.

5.16. Filippova, G. P. and Ishkin, I. P. Kislorod, 1958, *2,* 38. Inzh.-fiz. zhurn. 1961, *4*(*3*), 105.

5.17. Kestin J., Wang H.—Trans. ASME. 1958, *80,* 11.

5.18. Van Itterbeek A., Zink H., Van Paemel O.—Cryogenics, 1962, *2,* 210.

5.19. Boon J., Thomaes G.—Physica, 1963, *29,* 208.

5.20. Saji Y., Kobayashi S.—Cryogenics, 1964, *4,* 136.

5.21. Reynes E., Thodos G.—Physica, 1964, *30,* 1529.

5.22. Lowry B., Rice S., Gray P.—J. Chem. Phys., 1964, *40,* 3673.

5.23. Andreyev, I. I., Tsederberg, N. V. and Popov, V. N. Teploenergetika, 1966, No. 8, 78.

5.24. Van Itterbeek A., Hellemans J., Zink H., Van Cauteren M.—Physica, 1966, *32,* 2171.

5.25. De Bock A., Grevendonk W., Awouters H.—Physica, 1967, *34,* 49.

5.26. De Bock A., Grevendonk W., Herreman W.—Physica, 1967, *37,* 227.

5.27. Boon J., Legros J., Thomaes G.—Physica, 1967, *33,* 547.

5.28. Hellemans J., Zink H., Van Paemel O.—Physica 1970, *46*(*3*), 395.

5.29. Trappeniers N., Botzen A., Van Oosten J., Van den Berg H.—Physica, 1965, *31,* 945.

5.30. Trappeniers N., Botzen A., Ten Seldam C., Van den Berg H., Van Oosten J.—Physica, 1965, *31,* 1681.

5.31. Slyusar', V. P., Rudenko, N. S. and Tret'yakov, V. M. In: Teplofizicheskiye svoystva veshchestv pri nizkikh temperaturakh (Thermophysical properties of substance at low temperatures). All-union research institute of physical- and radio-engineering measurements, Moscow, 1972.

5.32. Slyusar', V. P., Rudenko, N. S. and Tret'yakov, V. M. Ukr. fiz. zhurn., 1972, *17*(*8*).

5.33. Golubev, I. F. Viscosity of gases and gas mixtures. Halsted Press, New York, 1971; also available in English from NTIS as TT70-50022 (245 pp).†

5.34. Golubev, I. F. and Agayev, N. A. Vyazkost' predel'nykh uglevodorodov (Viscosity of saturated hydrocarbons). Azerbaijan State Publishing House, Baku, 1964.

5.35. Vukalovich, M. P., Ivanov, A. I., Fokin, L. R. and Yakovlev, A. T. Teplofizicheskiye svoystva rtuti (Thermophysical properties of mercury). Standards Publishing House, Moscow, 1971.

† Changed in proof—T.B.S.

5.36. Born M., Green H. S.—Proc. Roy. Soc., 1946, *A188,* 10.

5.37. Kirkwood J. C.—J. Chem. Phys., 1946, *14,* 180; 1947, *15,* 72.

5.38. Yvon J. Le theorie statist. des fluides et l'equation d'Etat. Paris, 1935.

5.39. Irving I. H., Kirkwood J. C.—J. Chem. Phys., 1950, *18,* 817; 1951, *19,* 1173.

5.40. Gurikov, Yu. V. Zhurn. fiz. khim., 1963, *37,* 1455, 2420.

5.41. Choh, S. T. and Uhlenbeck G. E. The kinetic theory of dense gases. In: lectures in statistical mechanics. G. E. Uhlenbeck and G. W. Ford, eds. American Mathematical Society, 1963.

5.42. Dorfman J., Cohen E.—Phys. Letters, 1965, *16,* 124.

5.43. Dorfman J., Cohen E.—J. Math. Phys., 1967, *8*(*2*), 282.

5.44. Sengers J. V.—Phys. Fluids, 1966, *9,* 1685.

5.45. Brokaw R. S. Statistical mechanics of transport properties. NASA TM X-52478. Washington, 1968.

5.46. Hanley H. J. M., McCarty R. D., Sengers J. V.—J. Chem. Phys., 1969, *50*(*2*), 857.

5.47. Chapman, S. and Cowling, T. G. The mathematical theory of nonuniform gases, 2nd ed. Cambridge University Press, 1953.

5.48. Stogryn D. E., Hirshfelder J. C.—J. Chem. Phys., 1959, *31,* 1534, 1545; 1960, *33,* 942.

5.49. Kim S. K., Ross J.—J. Chem. Phys., 1965, *42,* 263.

5.50. Hoffman D. K., Curtiss S. F.—Phys. Fluids, 1965, *8,* 890.

5.51. Kestin J., Ro S. T., Wakeham W.—Physica, 1972, *58*(*2*), 165.

5.52. Snider R. F., Curtiss C. F.—Phys. Fluids, 1958, *1,* 122; 1960, *3,* 903.

5.53. Vargaftik, N. B. Izv. VTI (All-union heat engineering institute), No. 1, p. 13, 1952.

5.54. Rivkin, S. L. Investigation of thermophysical properties of water and steam under critical conditions. Author's abstract of PhD thesis. All-Union Heat Engineering Institute, Moscow, 1965.

5.55. Rivkin, S. L., Levin, A. Ya., Izrailevskiy, L. B. and Kharitonov, K. T. Inzh. fiz. zhurn., 1971, *21*(*3*), 405.†

5.56. Kessel'man, P. M. and Kamenetskiy, V. R. Teploenergetika, 1967, No. 9, 73.

5.57. Vasserman, A. A. and Nedostup, V. I. Zhurn. prikl. mekh. i teor. fiz., 1971, No. 3, 118.

5.58. Kamenetskiy, V. R. Compressibility factor of compressed gases (Methods of calculation and generalized data). Author's abstract of candidate thesis. Lomonosov Technological Institute, Odessa, 1969.

5.59. Glasstone, S., Laidler, K. J. and Eyring, H. Theory of rate processes. McGraw-Hill, 1941.

5.60. Frenkel', Ya. I. Teoriya tverdykh i zhidkikh tel (Theory of solids and liquids). Gostekhizdat Publishing House, Moscow, 1934.

5.61. Alder B. J., Einwohner T.—J. Chem. Phys., 1965, *43,* 3399.

5.62. Alder B. J.—Ber. Bunsen Ges. Phys. Chem., 1966, *70*(*9*/*10*), 968.

5.63. Doolittle A. K.—J. Appl. Phys., 1951, *22,* 12.

5.64. Doolittle A. K., Doolittle D. D.—J. Appl. Phys., 1957, *28,* 8.

5.65. Cohen M., Turnbull D.—J. Chem. Phys., 1959, *31*(*5*), 1164.

5.66. Macedo P. B., Litovitz T. A.—J. Chem. Phys., 1965, *42*(*1*), 245.

5.67. Matheson A. J.—J. Chem. Phys., 1966, *44*(*2*), 695.

5.68. Lyusternik, V. Ye. Teplofiz. vys. temp., 1969, *7,* 365.

5.69. Ree F. H., Hoover W. G.—J. Chem. Phys., 1967, *46,* 4181.

5.70. Hanley H. J. M., McCarty R. D., Cohen E. G. D.—Physica, 1972, *60*(*2*), 322.

5.71. Dymond J. H., Alder B. J.—J. Chem. Phys., 1966, *45*(*6*), 2061.

5.72. Rowlinson J.—Mol. Phys., 1964, *7*(*4*), 349; 1964, *8*(*2*), 107.

† Changed in proof—T.B.S.

5.73. Chen R. Y., Henderson D., Davision S. G.—Proc. Nat. Acad. Sci. USA, 1965, *54(6)*, 1514.

5.74. Wentorf R., Buehler R., Hirschfelder J., Curtiss C.—J. Chem. Phys., 1950, *18*, 1484.

Chapter 6

6.1. Sengers J. V., Bolk W. T., Stigter C. J.—Physica, 1964, *30*, 1018.

6.2. Lenoir J., Comings E.—Chem. Eng. Prog., 1951, *47*, 223.

6.3. Uhlir A.—J. Chem. Phys., 1952, *20*, 463.

6.4. Lenoir J., Junk W., Comings E.—Chem. Eng. Prog., 1953, *49*, 539.

6.5. Keyes F.—Trans. ASME, 1954, *76*, 809.

6.6 Keyes F.—Trans. ASME, 1955, *77*, 1395.

6.7. Michels A., Botzen A., Friedman A., Sengers J.—Physica, 1956, *22*, 121.

6.8. Ziebland H., Burton J.—Brit. J. Appl. Phys., 1958, *9*, 52.

6.9. Tsederberg, N. V., Popov, V. N. and Morozova, N. A. Teploenergetika, 1960, No. 6, 82.

6.10. Michels A., Sengers J., Van de Klundert I.—Physica, 1963, *29*, 149.

6.11. Ikenberry L., Rice S.—J. Chem. Phys., 1963, *39*, 1561.

6.12. Keyes F., Vines R.—Trans. ASME, Ser. C., 1965, *87(2)*, 177.

6.13. Rosenbaum B., Oshen S., Thodos G.—J. Chem. Phys. 1966, *44*, 2831.

6.14. Freud P. S., Rothberg G. M.—Rev. Sci. Instr., 1967, *38(2)*, 238.

6.15. Amirkhanov, Kh. I., Adamov, A. P. and Gasanov, G. G. Teploenergetika, 1970, No. 12, 61.

6.16. Amirkhanov, Kh. I., Adamov, A. P. and Gasanov, G. D. Inzh.-fiz. zhurn., 1972, *22(5)*, 835.

6.17. Tarzimanov, A. A. and Arslanov, V. A. In: Trudy KKhTI [Transactions of the kazan' institute of chemical technology], No. 47. Ser. Teplo- i massoobmen. Kazan', 1971.

6.18. Tufeu R., LeNeindre B., Bury P.—Compt. Rend. Acad. Sci. Paris, 1971, *B273(3)*, 113.

6.19. Ziebland H., Burton J.—Brit. J. Appl. Phys., 1955, *6*, 416.

6.20. Bailey B., Kellner K.—Brit. J. Appl. Phys., 1967, *18*, 1645.

6.21. Guildner L. A.—J. Res. NBS, 1962, *66A*, 341.

6.22. Michels A., Sengers J. V., Van der Gulik P. S.—Physica, 1962, *28*, 1216.

6.23. Golubev, I. F. Teploenergetika, 1963, No. 12, 78.

6.24. Johannin P.—J. Rech. Centre Nat. Rech. Sci., 1958. *43*, 116.

6.25. Tarzimanov, A. A. Investigation of thermal conductivity of gases over a wide range of parameters of state. Author's abstract of PhD thesis. Kazan' Institute of Chemical Technology. Kazan', 1972.

6.26. Kim S. K., Flynn G. P., Ross J.—J. Chem. Phys., 1965, *43(11)*, 4166.

6.27. Sengers J. V.—Phys. Rev. Letters, 1965, *15*, 515.

6.28. Kawasaki K., Oppenheim I.—Phys. Rev., 1965, *139(6A)*, 1763.

6.29. Weinstock J.—Phys. Rev., 1965, *140(2A)*, 460.

6.30. Gracki J. A., Flynn G. P., Ross J.—J. Chem. Phys., 1969, *51(9)*, 3856.

6.31. Kestin J., Paykoc E., Sengers J. V.—Physica, 1971, *54(1)*, 1.

6.32. Tsederberg, N. V. Teploprovodnost' gazov i zhidkostey (Thermal conductivity of gases and liquids). Gosenergoizdat Publishing House, Moscow, 1963; English translation from the M.I.T. Press, 1965.†

6.33. Diller D. E., Hanley H. J. H., Roder H. M.—Cryogenics, 1970, *10*, 286.

6.34. Vargaftik, N. B. Izv. VTI, 1951, No. 7, 13.

6.35. Vargaftik, N. B., Filippov, L. P., Tarzimanov, A. A. and Yurchak, R. P. Teplopro-

† Changed in proof—T.B.S.

vodnost' gazov i zhidkostey (Thermal conductivity of gases and liquids). Standards Publishing House, Moscow, 1970.

6.36. Kessel'man, P. M. and Kamenetskiy, V. R. In: Teplofizicheskiye svoystva veshchestv i materialov (GSSSD. Ser. Fizicheskiye konstanty i svoystva veshchestv), No. 2, p. 220. Moscow, 1970.

6.37. Vasserman, A. A. and Nedostup, V. I. Inzh.-fiz. zhurn., 1971, *20(1)*, 119.

6.38. Vasserman, A. A. Nedostup, V. I. and Tsesis, M. A. In: Teplofizicheskiye svoystva veshchestv i materialov (GSSSD. Ser. Fizicheskiye konstanty i svoystva veshchestv), No. 3, p. 85. Moscow, 1971.

6.39. Vasserman, A. A. and Nedostup, V. I. Ibid, No. 5, p. 21. Moscow, 1972.

6.40. Bach J., Grigull U.—BWK, 1966, *18,* 3.

6.41. Rivkin, S. L. Teploenergetika, 1968, No. 7, 52.

6.42. Vasserman, A. A. Teplofiz. vys. temp., 1972, *10(5)*, 1116.

Part III

III.1. Glushko, V. P. ed. Termodinamicheskiye svoystva individua'lnykh veshchestv (Thermodynamic properties of individual substances), *2*. USSR Academy of Sciences, Moscow, 1962.

III.2. Michels A., Wassenaar T., Wolkers G.—Physica, 1965, *31,* 237.

III.3. Dedikov, Yu. A. An experimental study of thermodynamic properties of neon at low temperatures and pressures to 120 MPa. Author's abstract of candidate thesis. Moscow, 1972.

III.4. Hilsenrath J. et al., Tables of thermal properties of gases. NBS Circ. 564. Washington, 1955.

III.5. Din F. Thermodynamic functions of gases. Vol. 2. L., Butterworths, London, 1956.

III.6. Gosman A. L., McCarty R. D., Hust J. G. Thermodynamic properties of argon from the triple point to 300 K at pressures to 1000 atmospheres. NSRDS—NBS 27. Washington, 1969.

III.7. International thermodynamic tables of the fluid state. Argon, 1971. Ed. Angus S., Armstrong B., from tables, prepared by Gosman A. L., McCarty R. D., Hust J. G., Vasserman A. A., Rabinovich V. A. Butterworths, London, 1972.

III.8 Vukalovich, M. P., Altunin, V. V. and Spiridonov, G. A. Inzh.-fiz. zhur., 1969, *16(3)*, 000.

III.9. Vargaftik, N. B. Handbook of Physical Properties of Liquids and Gases. Hemisphere Publishing Corporation, 1975.

III.10. Michels A., Wassenaar T., Wolkers G., Dawson J.—Physica, 1956, *22,* 17.

III.11. Vukalovich, M. P., Altunin, V. V. and Spiridonov, G. A. In: Proceedings of the all-union scientific-technological conference on thermodynamics. Reports of the section on thermodynamic properties of substances. Leningrad Institute of Refrigeration Technology, p. 167, Leningrad, 1969.